GEOGRAPHY

Realms, Regions, and Concepts

FOURTEENTH EDITION

GEOGRAPHY

Realms, Regions, and Concepts

H. J. de Blij

John A. Hannah Professor of Geography
Michigan State University

Peter O. Muller

Professor, Department of Geography and Regional Studies
University of Miami

John Wiley & Sons, Inc.

VICE PRESIDENT & EXECUTIVE PUBLISHER	Jay O'Callaghan
EXECUTIVE EDITOR	Ryan Flahive
ASSISTANT EDITOR	Veronica Armour
EDITORIAL ASSISTANT	Meredith Leo
MARKETING MANAGER	Margaret Barrett
MARKETING ASSISTANT	Susan Matulewicz
PRODUCTION MANAGER	Dorothy Sinclair
SENIOR PRODUCTION EDITOR	Sandra Dumas
SENIOR ILLUSTRATION EDITOR	Sigmund Malinowski
ELECTRONIC ILLUSTRATIONS	Mapping Specialists, Ltd.
SENIOR PHOTO EDITOR	Jennifer MacMillan
PHOTO RESEARCHER	Mary Ann Price
MEDIA EDITOR	Bridget O'Lavin
DIRECTOR CREATIVE SERVICES	Harry Nolan
SENIOR DESIGNER	Kevin Murphy
PRODUCTION SERVICES	Furino Production
COVER DESIGN	David Levy
FRONT COVER PHOTO	H. J. de Blij
BACK COVER PHOTO	H. J. de Blij

This book was set in 10.5/12.6 Times Roman by Prepare and printed and bound by RR Donnelley/Jefferson City. The cover was printed by RR Donnelley/Jefferson City.

The paper in this book was manufactured by a mill whose forest management programs include sustained yield harvesting of its timberlands. Sustained yield harvesting principles ensure that the number of trees cut each year does not exceed the amount of new growth.

This book is printed on acid free paper. ∞

ISBN 13   978-0470-46242-3
ISBN 13   978-0470-55631-3

Printed in the United States of America

10   9   8   7   6   5   4   3   2

**To the Memory of
Jack Reilly**

PREFACE

For the past four decades, *Geography: Realms, Regions, and Concepts* has reported (and sometimes anticipated) trends in the discipline of Geography and developments in the world at large. In thirteen preceding editions, *Regions*, as the book has come to be called, has explained the modern world's geographic realms (the greatest regional entities on the face of the Earth) and their natural environments and human dimensions. In the process, this book has become an introduction to Geography itself, the discipline that links the study of human societies and natural environments. We look at the ways people have organized their living space, have adapted to changing social as well as environmental circumstances, and continue to confront forces beyond their control ranging from climate change to globalization. From old and still relevant concepts to new and untested ideas, *Regions* provides geographic perspective on our transforming world.

The book before you, therefore, is an information highway to geographic literacy. The first edition appeared in 1971, at a time when school geography in the United States (though not in Canada) was a subject in decline. It was a precursor of a dangerous isolationism in America, and geographers foresaw the looming cost of geographic illiteracy. Sure enough, the media during the 1980s began to report that polls, public surveys, tests, and other instruments were recording a lack of geographic knowledge at a time when our world was changing ever faster and becoming more competitive by the day. Various institutions, including the National Geographic Society, banks, airline companies, and a consortium of scholarly organizations, mobilized to confront an educational dilemma that had resulted substantially from a neglect of the very topics this book is about.

Before we can usefully discuss such commonplace topics as our "shrinking world," our "global village," and our "distant linkages," we should know what the parts are, the components that do the shrinking and linking. This is not just an academic exercise. You will find that much of what you encounter in this book is of immediate, practical value to you—as a citizen, a consumer, a traveler, a voter, a job-seeker. North America is a geographic realm with intensifying global interests and involvements. Those interests and involvements require countless, often instantaneous decisions. Such decisions must be based on the best possible knowledge of the world beyond our continent. That knowledge can be gained by studying the layout of our world, its environments, societies, resources, policies, traditions, and other properties—in short, its regional geography.

REALMS AND REGIONS . . .

This book is organized into 13 chapters. The Introduction discusses the world as a whole, outlining the physical stage on which the human drama is being played out, providing environmental frameworks, demographic data, political background, and economic-geographical context. Each of the remaining 12 chapters focuses on one of the world's major geographic realms.

The first part of each chapter addresses the dominant geographic aspects of the realm under discussion, including considerable emphasis on its environmental geography. The second part focuses on the realm's component regions and countries, highlighting the often-fascinating diversity that marks the physical and human geography of every area of the planet. Here we get a chance to report on the hopes and the problems of peoples facing forces of change over which all too many have little or no control.

. . . AND CONCEPTS

To achieve these aims, we introduce as many as 150 geographic concepts placed in their regional settings. Most of these concepts are indeed primarily geographical, but others are ideas about which, we believe, students of geography should have some knowledge. Although such concepts are listed on the opening page of every chapter, we have not, of course, enumerated every geographic notion used in that chapter. Many colleagues, we suspect, will want to make their own realm-concept associations, and as readers will readily perceive, the book's organization is quite flexible. It is possible, for example, to focus almost exclusively on substantive regional material or, alternatively, to concentrate mainly on conceptual issues.

NEWS ABOUT THE FOURTEENTH EDITION

New editions of *Regions* have a reputation for comprehensive and detailed revision, and what you are about to read is not a mere update of the previous version. The world changes too fast to allow us to fall back on earlier interpretations. Just have a look at the very first regional chapter, focusing on Europe, and compare the Thirteenth Edition to the Fourteenth. The great story of Europe, as we have reported in the past, was its massive effort to form an economic union with increasing political coordination. Today the countries at Europe's helm, such as Germany and France, face growing objections from countries that once benefited from unification but now find themselves in debt and penury. So Europe is risking failure, and the countries at its heart, at the *core*, must find ways to temper the rising fears and doubts that prevail in the countries of the outlying *periphery*. At the root of these fears is the global economic downturn that began in 2008, which is having different impacts in different places.

As the writing continued, we were confronted by circumstances and prospects never encountered before, and the question became: to what extent is what is happening a permanent reversal, ultimately affecting the geography of entire realms, and to what degree is this a temporary setback on a well-established and durable course? Take the case of China, where everything we say about the transformation of the coastal provinces remains true: the factories are still there; the cultural landscape has been transformed; the cities still look modern, even futuristic. But behind that landscape, in the first half of 2009, lay something unaccustomed: thousands of factories are closed, workers by the millions are being laid off, and others, arriving back from the countryside assuming they will restart their old jobs, find that their jobs no longer exist. Efforts to revive the U.S. economy are crucial to Chinese producers; but at press time it was far from certain that such a revival was assured.

As a result, this edition of *Regions* is in some ways unique: we must understand how globalization created burgeoning cities and mushrooming economies, but we

do not yet know the full extent of the impact of the global recession. On the cover of the Thirteenth Edition was a dramatic photograph of the skyscrapers of Dubai, site of the world's tallest building and scene of the ultimate "service" economy. But by mid-2009 the foreign workers had left, high-rise projects lay idle, playgrounds of the rich saw their customers dwindle, and the bottom had fallen out of the real estate market. But the skyline still impresses, as if awaiting the return of the boom.

If it is too early to tell whether those Chinese factories will come back to life and Dubai's hotels will fill up again, it is useful to learn how the world economy got here. We refer to economic geography throughout this book because location and culture have much to do with success and failure. During the era of unbridled globalization, India was often criticized for its rejection of globalizing influences and its resistance to modernization. But in 2009, India found itself far more protected against the onslaught of the recession than countries more open to globalization. We revised our chapter on India exhaustively for the Thirteenth Edition, but much of what we wrote then still applies today.

No one could forecast the course of the economic downturn in progress as we wrote, but understanding the fundamentals helps us gauge the direction in which we are going. To begin with, the situation formed a forceful reminder that, when the chips are down (and were they ever down!), the state continues to play the key role in international affairs. For all the multinational efforts and organizations we discuss in this book, governments want to be seen as protectors of the interests of their "nations," and this quickly leads to protectionism that is the very antithesis of globalization. By mid-2009, it was obvious that European and American leaders tended to disagree over ways to confront the problem: America's leaders continued to favor intervention and "stimulus packages," while their European counterparts preferred less intrusion and more reliance on existing safeguards. It was also clear that China's leadership was becoming increasingly concerned over its financial entanglement with the United States, to the point that some began calling for a new global currency to replace the dollar. Meanwhile, patterns and trends of long duration suddenly shifted and changed; laborers by the hundreds of thousands were leaving employment magnets such as China's coastal cities and America's construction sites; all over the world, from Las Vegas to Dubai, work halted on megaprojects because the money to complete them ran out. The unanswered question in mid-2009 was how long and how deep the economic downturn would be, and to what extent its impact would be long-lasting. That question will undoubtedly come up time and again in your classes.

What to Look For

With a revision manuscript approaching 100,000 words and countless map modifications, no summary can adequately encapsulate all this new edition contains. In the **Introduction**, your attention will be drawn to a new map and accompanying text that focuses on the growing inequalities in our world, a phenomenon that can be recorded at all levels of scale from the global to the local. At the global level, a generally favored and fortunate "core" containing just 15 percent of the world's population annually earns about 75 percent of all the income (this in a world some prominent writers refer to as "flat"!). Even greater core-periphery disparities mark individual realms, regions, and states on a small planet with a still-growing population, creating risks to the very order and stability the world will need to make progress.

In Chapter 1, modern **Europe** is discussed within a core-periphery framework, and this could not be more relevant than it is in today's economic circumstances. Europe's grand experiment in unification and coordination faces its mightiest challenge as core-periphery contrasts and disagreements threaten six decades of progress toward integration.

Chapter 2, in which the main topic is **Russia**, looks at the implications of climate change on that country's relative location and raises the question of Russia's governance. Russian leaders proclaim a desire to revive their country's stature as a global power, but representative government, human rights, and legal institutions are among the casualties of the process. Meanwhile, Russia's 2008 armed intervention in neighboring Georgia has implications for all the countries that emerged from the 1991 breakup of the former Soviet Union.

Chapter 3, **North America**, was completely restructured and updated. New material on indigenous Americans, multiculturalism, and migration is accompanied by several new and modified maps; the momentous transformation of the North American city, still in progress, anchors much of the discussion. Structurally, the chapter is now more sharply divided into the two segments marking all regional chapters: the first part deals with the geography of the realm as a whole, the second defines and surveys its regional components.

In Chapter 4, **Middle America**, the emphasis continues on the giant of the realm, Mexico, where a long-term challenge to national cohesion, arising from the country's cultural transition from northwest to southeast, is now superseded by what some are calling its "Colombianization" as the illegal drug trade spawns a new crisis. This is not merely a matter of smuggling and profiteering: drug cartels contest their territories with armed force and combat law enforcement with ruthless efficiency,

causing thousands of casualties and seriously damaging the economy.

Less dramatic but no less significant are developments in **South America,** the subject of Chapter 5 (where, as in Chapter 4, we studiously avoid the inappropriate term "Latin" America as we focus on the indigenous state, on Amerindian ascendancy, and on the realm's multiculturalism). So much is happening in South America—from Brazil's steady emergence toward superpower status and a "crescent of populism" from Venezuela through Bolivia to Ecuador to China's growing economic presence in pursuit of commodities and influence—that this realm is now one of the world's most fascinating geographic studies.

When it comes to **Subsaharan Africa** it is well to remember that Chapter 6 deals with the very source of humanity—at our ultimate roots, we are all Africans. The world's first great globalization was its Africanization. But today Africa suffers more than any other realm from the disadvantages that prevail throughout the global periphery; thus, through significant new text we focus on the causes and consequences of Subsaharan Africa's difficulties. Major new text addresses the hopes and troubles of post-Mandela (and post-Mbeki) South Africa, the tragedy of Zimbabwe, the prospects of Kenya, and the continuing crisis in The Congo. We also take note of the remarkable success of Ghana and of the challenges in other parts of West Africa in an expanded narrative.

Chapter 7, **North Africa/Southwest Asia**, is reconfigured and rebalanced, with new material on Egypt and Sudan, greater detail on the desert fringe to complement added text on this area in Chapter 6, updated discussion of the Middle East, a total rewrite of the section on the Arabian Peninsula in light of new economic as well as cultural developments there, and important new information on Turkey as well as Iran.

Chapter 8, **South Asia**, was the subject of major revision in the Thirteenth Edition, but there are new perspectives on demography (including newly available data on rates of natural increase) and cultural geography, the important issue of India's Muslim minority, the tale of Tata Motors as an indicator of Indian economic prospects, and, in the discussion of Sri Lanka, the success (in mid-2009) of the government's counteroffensive against the Tamil Eelam Insurgent State. Moreover, our coverage of Pakistan was expanded to provide more background for that country's intensifying struggle against militant Islamic revivalists.

In Chapter 9, **East Asia**, we respond to reviewer and adopter comments by clarifying and tightening the discussion on the four major river basins and their historic as well as modern role in the evolution of the Chinese state. We also focus on China in global and local contexts, linking the incredible pace of infrastructure expansion to growing disparities of wealth: China's core-periphery contrasts are becoming ever sharper and pose a danger, in the absence of representative government, in this time of economic contraction. Tens of thousands of Pacific Rim factories lay idle in 2009, and millions of workers were forced to return to their countryside villages, with social and political consequences that are just beginning to emerge. Mongolia, South Korea, and Taiwan also are subject to significant revision, particularly Taiwan following a first-hand encounter during the spring of 2009.

Chapter 10, **Southeast Asia**, underwent less general but far more specific revision, focused mainly on its regional components in Laos, Thailand, Myanmar, Malaysia, and Singapore. The stability and economic growth of Indonesia, Southeast Asia's giant, is crucial to this realm afflicted elsewhere by ruthless dictatorship and political fragility.

In Chapter 11, the **Austral Realm**, we have used the available space to focus on three ongoing concerns: the environment (Australia has been suffering through dreadful droughts and death-dealing fires); the first inhabitants and their circumstances in this modern, globalizing society; and the country's cultural diversification as immigration continues to alter its social geography.

Chapter 12, as readers familiar with this book will notice, has a new title, taking note of the increasing importance of the world's polar regions. In **Pacific Realm and Polar Futures**, we note the effects of a growing human presence in (and designs on) both polar zones within which environmental change is facilitating entry for tourists, tens of thousands of whom are sailing Antarctic waters and landing on the Earth's southernmost shores. Climate change is also opening once-icebound waterways in the Arctic, where Russians, Canadians, Americans, and others dream of northern navigation routes and potential resources in the form of oil and natural gas. Our final map in this chapter reveals these prospects in the Arctic, but what it cannot show is the threat this invasion of the planet's last frontier poses to the fragile ecosystems prevailing there. And in the enormous oceanic realm between the poles, Pacific Islanders face other consequences of climate change as the ocean, inextricably bound to the Arctic and Antarctic, becomes an existential threat to those living on its low-lying shores.

This latest edition of *Regions* reflects our continuing commitment to bring to our readers our geographic perspectives on a fast-changing world in general as well as specific terms. Since the previous edition was written, the senior author spent useful time in India, China, Japan, Russia, Europe, Thailand, Myanmar, and Greenland, maintaining research links and strengthening impressions worldwide to build on 40 years of the most fascinating project imaginable. We trust that you will find this Fourteenth Edition as challenging and informative as you have our earlier ones.

Maps

As is implied in the overview above, keeping our maps current (to press time in mid-2009) is a vital part of the revision process. Our first task is to decide which new maps are needed. As readers of *Regions* will note, for the first time each chapter now contains a population distribution map of the part of the world it covers. Beyond that systematic cartographic enhancement, additional new maps include the global core (Fig. G-12); several maps in Chapter 1 that display the reconfiguration of Europe into a new, core-periphery regional framework; North America's settlement consolidation (Fig. 3-5), administrative frameworks (Fig. 3-6), religions (Fig. 3-11), and westward expansion (Fig. 3-13); and the changing physical, economic, and political geography of the Arctic Ocean Basin (Fig. 12-5).

The map revision process also involves modifying and updating existing maps. Every map is carefully scrutinized and few escape change. Even global maps showing the matrix of the world's countries undergo such treatment: in this new edition, we have recognized Kosovo as a full-fledged state, and the map of southeastern Europe has been modified accordingly. Many of the other map alterations we have made reflect the changing state of world affairs through the end of our new century's first decade. And the dynamism of world regional geography also requires that we anticipate developments (such as the recent redrawing of the map of Sudan's provinces [Fig. 7-11] that will figure prominently in the expected referendum on the independence of the country's southern sector). Among the many other geographic hotspots we constantly monitor are the growing European Union; the restive edges of the Russian realm; the turmoil affecting parts of Mexico, Colombia, and Bolivia; the hostilities that adhere to the Islamic Front that lies across the northern fringe of Subsaharan Africa; the many conflicts, ongoing as well as potential, that plague North Africa/Southwest Asia from Algeria in the west to Israel and Iraq in the center to Iran and Afghanistan in the east; the struggles of Pakistan against the forces unleashed by al-Qaeda and the Taliban; and the scramble that is just gearing up to claim the geographic assets of the Arctic. So, overall, the book before you records the state of the world in maps and forecasts the future.

Photography

The map undoubtedly is geography's closest ally, but there are times when photography is not far behind. Whether from space, from an aircraft, from the tallest building in town, or on the ground, a photograph can indeed be worth a thousand words. When geographers perform field research in some area of the world, they are likely to maintain a written record that correlates with the photographic one. This convergence is most visibly on display in the "Field Notes" captions that accompany photos taken by the authors, many of which are new to this Fourteenth Edition. Photographs from other sources, dozens of which are new to this edition, were very carefully selected, and are always accompanied by detailed captions that emphasize their connections to the places and concepts being discussed in the surrounding text.

Data Sources

For all matters geographical, of course, we consult *The Annals of the Association of American Geographers, The Professional Geographer, The Geographical Review, The Journal of Geography*, and many other academic journals published regularly in North America—plus an array of similar periodicals published in English-speaking countries from Scotland to New Zealand.

As with every new edition of this book, all quantitative information was updated to the year of publication and checked rigorously. In addition to the major revisions described above, hundreds of other modifications were made, many in response to readers' and reviewers' comments. Some readers found our habit of reporting urban population data within the text disruptive, so we continue to tabulate these at the beginning of each chapter's Regions of the Realm section. The stream of new spellings of geographic names continues, and we pride ourselves on being a reliable source for current and correct usage.

The statistical data that constitute the Data Table at the very end of the book beginning inside the back cover ("Area and Demographic Data for the World's States") are derived from numerous sources, and each data category is explained in the box at the end of the table. As users of such data are aware, considerable inconsistency marks the reportage by various agencies, and it is often necessary to make informed decisions on contradictory information. For example, some sources still do not reflect the declining rates of population increase or life expectancies in AIDS-stricken African countries. Others list demographic averages without accounting for differences between males and females in this regard. In formulating the table, we have used among our sources the United Nations, the Population Reference Bureau, the World Bank, the Encyclopaedia Britannica *Books of the Year*, the *Economist* Intelligence Unit, the *Statesman's Year-Book*, and *The New York Times Almanac*. Increasingly, these are supplemented by referring to literally hundreds of the most appropriate websites, both to double-check the data and obtain the very latest numbers.

The urban population figures—which also entail major problems of reliability and comparability—are mainly drawn from the most recent database published by the United Nations' Population Division. For cities of less than 750,000, we developed our own estimates from a variety of other sources. All the urban population figures used in this book are estimates for 2010, and they represent *metropolitan-area totals* unless otherwise specified.

Pedagogy

One of our aims is to help students learn important geographic concepts and ideas, and to make sense of our complex and rapidly changing world. A list of the special features in each chapter follows.

Opener Maps. A comprehensive map of the realm opens each chapter. These maps were specially created according to our specifications, and they have been updated for this Fourteenth Edition. As in the previous edition, these maps are assigned the first figure number in each chapter.

Concepts, Ideas, and Terms. Each chapter begins with a boxed sequential listing of the key geographic concepts, ideas, and terms that appear in the pages that follow. These are noted by numbers in the margins that correspond to the introduction of each item in the text.

Two-Part Chapter Organization. To help the reader to logically organize the material within chapters, we have divided the regional chapters into two distinct parts. First, "Defining the Realm" includes the general physiographic, historical, and human-geographic background common to the realm. The second section, "Regions of the Realm," presents each of the distinctive regions within the realm (denoted by the symbol ●).

List of Regions. Also on the chapter-opening page, a list of the regions within the particular realm provides a preview and helps to organize the chapter. For ease of identification, the ● symbol that denotes the regions list here also appears beside each region heading in the chapter.

Major Geographic Qualities (Boxes). Near the beginning of each realm chapter, we list in boxed format the major geographic qualities that best summarize that portion of the Earth's surface.

Sidebar Boxes. Special topical issues are highlighted in boxed sections. These boxes allow us to include interesting and current topics without interrupting the flow of material within chapters. Occasionally, these boxes will include a place that merits the reader's closer attention.

Points to Ponder (Boxes). In each chapter, near the end of the "Defining the Realm" section, we offer in boxed format a few brief statements about things of geographic import happening (or likely to happen) in a realm or region that may give one pause.

Regional Issues (Boxes). In each regional chapter, a controversial issue has been singled out for debate by two opposing voices. These commentaries are drawn from the senior author's field notes and correspondence and are intended to stimulate classroom discussion. Individual readers may "vote" their opinions by e-mail, as indicated at the bottom of each box.

Among the Realm's Great Cities (Boxes). This feature reflects the ever-expanding process and influence of urbanization worldwide. More than 30 profiles of the world's leading cities are presented, each accompanied by a specially drawn map.

Major Cities of the Realm Population Tables. Near the beginning of every "Regions of the Realm" section of each regional chapter, we have included a table reporting the most up-to-date urban population data (based on 2010 estimates drawn from the sources listed above). Readers should find this format less disruptive than citing the population when the city is mentioned in the text.

What You Can Do (Boxes). At the end of each chapter, we have provided a box that focuses on practical suggestions, ideas to get involved with, and ways to contribute. The first element is a *Recommendation* to help you get involved in and contribute to projects and/or organizations related to world regional geography. The second element, entitled *Geographic Connections*, offers a set of challenges of a practical nature that asks you to visualize how you would use what you have learned in a business, government, or educational setting. The third element, labeled *Geographic Literature*, supplies a brief guide to the key works on that chapter's topic and points the reader toward the comprehensive *References and Further Readings* component of the book's website.

From the Field Notes. In the Eighth Edition we introduced a new feature that has proven effective in some of our other books. Many of the photographs in this book were taken by the senior author while doing fieldwork. By linking them to expanded captions entitled "From the Field Notes," valuable insights are provided into how a geographer observes and interprets information in the field.

Appendices, Glossary, References, and Indexes. In previous editions, we included a number of features at the end of the book that enriched and/or supplemented the main text. These have all been preserved and updated. The increasing pressures of space, however, allow only three sections to remain in the book itself: (1) Appendix A, a *Conversions Table* linking metric and British units of measurement; (2) an extensive *Glossary* of terms; and (3) a comprehensive *general index*. Four additional sections, which should not be overlooked, now appear exclusively on the book's website ***www.wiley.com/college/deblij***: (1) Appendix B, a guide to *Using the Maps*; (2) Appendix C, an overview of *Opportunities in Geography*; (3) Appendix D, a *Pronunciation Guide*; and (4) a detailed bibliography—under the title *References and Further Readings*, and organized by chapter—that introduces the wide-ranging literature of the discipline and world regional geography.

Ancillaries

A broad spectrum of print and electronic ancillaries are available to accompany the Fourteenth Edition of *Regions*. (Additional information, including prices and ISBNs for ordering, can be obtained by contacting John Wiley & Sons.) These ancillaries are described next.

THE TEACHING AND LEARNING PACKAGE

This Fourteenth Edition of *Geography: Realms, Regions, and Concepts* is supported by a comprehensive supplements package that includes an extensive selection of print, visual, and electronic materials.

Wiley/National Geographic *College Atlas of the World*

Edited by H. J. de Blij and Roger Downs

Wiley is proud to offer the *College Atlas of the World* to your students through our exclusive partnership with the National Geographic Society. State-of-the-art cartographic technology plus award-winning design and content make this affordable, compact, yet comprehensive atlas the ultimate resource for every geography student.

The past decade has seen an explosion of innovative and sophisticated digital mapping technologies—and the Society has been at the forefront of developing these powerful new tools to create the finest, most functional, and most informative atlases available anywhere. This powerful resource can be purchased individually or at a significant discount when packaged with any Wiley textbook.

Resources That Help Teachers Teach

MEDIA-ENHANCED PowerPoint Presentations. This fresh collection of PowerPoint presentations will help you bring the striking photographs, precise maps, and vivid illustrations from *Regions* into your classroom. We understand that each professor has unique needs for their classroom presentations, so we offer three different types of presentation for each chapter:

- PowerPoints with just the text art.
- PowerPoints with text art and presentation notes.
- Media-Integrated PowerPoints with links to videos and animations.

Videos and Podcasts. We have created a series of streaming video resources and podcasts to support *Geography: Realms, Regions, and Concepts 14e*. Our streaming videos provide short lecture-launching video clips that can be used to introduce new topics, enhance your presentations, and stimulate classroom discussion. To help you integrate these resources into your syllabus, we have developed a comprehensive library of teaching resources, study questions, and assignments available in WileyPLUS.

Image Gallery. We provide online electronic files for the line illustrations and maps in the text, which the instructor can customize for presenting in class (for example, in handouts, overhead transparencies, or Power-Points).

In addition to the text images, you also have access to ***ConceptCaching***, an online database of photographs that explores what a region looks like and what it feels like to live in that region. Photographs and GPS coordinates are "cached" and categorized along core concepts of geography. Professors can access the images or submit their own by visiting ***www.ConceptCaching.com***.

Instructor's Manual and Test Bank. This manual includes a test bank, chapter overviews and outlines, and lecture suggestions. The *Test Bank* for *Regions* contains over 3000 test items, including multiple-choice, fill-in, matching, and essay questions. It is distributed via the secure Instructor's website as electronic files, which can be saved into all major word processing programs. The *Instructor's Manual* is also available on the password-protected Instructor's website.

Overhead Transparency Set. All of the book's maps and diagrams are available as print-on-demand transparencies in beautifully rendered, four-color format, and have been resized and edited for maximum effectiveness in large lecture halls.

WileyPLUS. Online course management of the educational learning package is available to accompany *Regions*. WileyPLUS combines videos, interactivities, automated grading of many questions, and grades that flow automatically into a grade book—including an electronic version of the textbook—in an all-in-one easy-to-use system. It has been used successfully in both small and large classes, and is particularly well suited as a stand-alone platform for online classes.

WileyPLUS is flexible. Its features include:

1. Map testing: develop students' spatial knowledge with interactive map tests of countries, cities, and physical features.
2. Ask the questions you've always wanted to ask by creating questions to add to any assignment. For instance, you could create your own series of questions about migration to your State or employment shares in your Province.
3. Set your own policies: give students from one to five (or even unlimited) attempts at each question, give hints after a missed attempt, set due dates, assign point values, and give individual students extensions. You can set up the assignments so that students lose some percentage of the possible points if they don't give the correct answer by the *n*th attempt.
4. Make announcements or post files for the class.
5. Draw from the instructor resources for discussion questions or PowerPoint files of graphics.
6. Organize and manage class rosters and grades. You can create rosters or allow students to pay and self-enroll online.
7. The test banks are included among the instructor resources for each chapter, and you can use these to create quizzes and tests for your students to take online. These can also be automatically graded and will flow into the grade book.

Your investment of time to develop a good course using WileyPLUS can be carried over to subsequent semesters and other instructors in your university.

Wiley Faculty Network. This peer-to-peer network of faculty is ready to support your use of online course management tools and discipline-specific software/

learning systems in the classroom. The WFN will help you apply innovative classroom techniques, implement software packages, tailor the technology experience to the needs of each individual class, and provide you with virtual training sessions led by faculty for faculty.

The Regions 14e Instructors' Site. This comprehensive website includes numerous resources to help you enhance your current presentations, create new presentations, and employ our pre-made PowerPoint presentations. These resources include:

- A complete collection of PowerPoint presentations. Three for each chapter.
- All of the line art from the text.
- Access to photographs.
- A comprehensive collection of animations.
- A comprehensive test bank.

Resources That Help Students Learn

GeoDiscoveries Website www.wiley.com/college/deblij. This easy-to-use and student-focused website helps reinforce and illustrate key concepts from the text. It also provides interactive media content that helps students prepare for tests and improve their grades. This website provides additional resources that complement the textbook and enhance your students' understanding of geography and improve their grade by using the following resources:

- **Videos** provide a first-hand look at life in other parts of the world.
- **Interactive Animations and Exercises** allow you to explore key concepts from the text.
- **Virtual Field Trips** invite you to discover different locations through photos and help you better understand how geographers view the world.
- **Podcasts, Web Cams, and Live Radio** let you see and hear what is happening all over the world in real time.
- **Flashcards** offer an excellent way to drill and practice key concepts, ideas, and terms from the text.
- **Map Quizzes** help students master the place names that are crucial to their success in this course. Three game-formatted place-name activities are provided for each chapter.
- **GeoDiscoveries Modules** allow students to explore key concepts in greater depth using videos, animations, and interactive exercises.
- **Chapter Review Quizzes** provide immediate feedback to true/false, multiple-choice, and other short-answer questions.

- **Concepts, Ideas, and Terms Interactive Flashcards** help students review and quiz themselves on the concepts, ideas, and terms discussed in each chapter.
- **Interactive Drag-and-Drop Exercises** challenge students to correctly label important illustrations from the textbook.
- **Audio Pronunciation** is provided for over 2000 key words and place names from the text.
- **Annotated Web Links** put useful electronic resources into context.
- **Area and Demographic Data** are provided for every country and world realm (see inside back cover).

Student Study Guide. This popular Study Guide supports the book content and is packed with useful study and review tools. For each chapter in the textbook, the Study Guide gives students and faculty access to chapter objectives, content questions-and-answers, outline maps of each realm, sample tests, and more. To arrange for your students to receive this free supplement with their copy of *Regions*, please contact your local Wiley sales representative.

WileyPLUS. For the student, WileyPLUS offers a number of enhanced features—including an affordable price. There are a number of bonus features like map testing, flash cards, and area and demographic data for countries around the world. The links to related web pages are live and instantaneous. Students can answer questions online and submit electronically to receive immediate feedback. When answering a question, students can click on links to refer to the text or media resources.

ACKNOWLEDGMENTS

Throughout the four decades since the publication of the First Edition of *Geography: Realms, Regions, and Concepts*, we have been fortunate to receive advice and assistance from well over a thousand people. One of the rewards associated with the publication of a book of this kind is the steady stream of correspondence and other feedback it generates. Geographers, economists, political scientists, education specialists, and others have written or e-mailed us, often with fascinating enclosures or attachments. We make it a point to respond personally to every such contact, and our editors have communicated with many of our correspondents as well. Moreover, we have considered every suggestion made, and many who wrote or transmitted their reactions through other channels will see their recommendations in print in this edition.

STUDENT RESPONSE

A major part of the correspondence we receive comes from student readers. We would like to take this opportunity to extend our deep appreciation to the several million students around the world who have studied from the first 17 editions of our World Regional Geography texts. In particular, we thank the students from nearly 150 different colleges across the United States who took the time to send us their opinions.

Students told us they found the maps and graphics attractive and functional. We have not only enhanced the map program with exhaustive updating but have added a number of new and updated maps to this Fourteenth Edition as well as making significant changes in many others. Generally, students have told us that they found the pedagogical devices quite useful. We have kept the study aids the students cited as effective: a boxed list of each chapter's key concepts, ideas, and terms (numbered for quick reference in both the box and text margins); a box summarizing each realm's major geographic qualities; and an extensive Glossary.

FACULTY FEEDBACK

In assembling the Fourteenth Edition, we are indebted to the following people for advising us on a number of matters:

THOMAS L. BELL, University of Tennessee
RICHARD A. BERRYHILL, East Central Community College (Mississippi)
KATHLEEN BRADEN, Seattle Pacific University
DEBORAH CORCORAN, Southwest Missouri State University
EDMAR BERNADES DASILVA, Broward Community College (Florida)
WILLIAM V. DAVIDSON, Louisiana State University
GARY A. FULLER, University of Hawai'i
RICHARD J. GRANT, University of Miami
MARGARET M. GRIPSHOVER, University of Tennessee
SHIRLENA HUANG, National University of Singapore
RICHARD LISICHENKO, Fort Hays State University (Kansas)
IAN MACLACHLAN, University of Lethbridge (Alberta)
DALTON W. MILLER, JR., Mississippi State University
JAN NIJMAN, University of Miami
VALIANT C. NORMAN, Lexington Community College (Kentucky)
PAI YUNG-FENG, New York City
IWONA PETRUCZYNIK, Mercyhurst College (Pennsylvania)
SHOURASENI SEN ROY, University of Miami
IRA SHESKIN, University of Miami
RICHARD SLEASE, Oakland, North Carolina
CATHY WEIDMAN, Austin, Texas
ANTOINETTE WINKLERPRINS, Michigan State University

In addition, several faculty colleagues assisted us with earlier editions, and their contributions continue to grace the pages of this book. Among them are:

JAMES P. ALLEN, California State University, Northridge
STEPHEN S. BIRDSALL, University of North Carolina at Chapel Hill
J. DOUGLAS EYRE, University of North Carolina at Chapel Hill
FANGYONG-MING, Shanghai (China)
EDWARD J. FERNALD, Florida State University
RAY HENKEL, Arizona State University
RICHARD C. JONES, University of Texas at San Antonio
GIL LATZ, Portland State University
IAN MACLACHLAN, University of Lethbridge (Alberta)
MELINDA S. MEADE, University of North Carolina at Chapel Hill
HENRY N. MICHAEL, late of Temple University
CLIFTON W. PANNELL, University of Georgia
J. R. VICTOR PRESCOTT, University of Melbourne (Australia)
JOHN D. STEPHENS, University of Washington
CANUTE VANDERMEER, University of Vermont

Faculty members from a large number of North American colleges and universities continue to supply us with vital feedback and much-appreciated advice. Our publishers arranged several feedback sessions, and we are most grateful to the following professors for showing us where the text could be strengthened and made more precise:

MARTIN ARFORD, Saginaw Valley State University (Michigan)
DONNA ARKOWSKI, Pikes Peak Community College (Colorado)
GREG ATKINSON, Tarleton State University (Texas)
DENIS BEKAERT, MIDDLE TENNESSEE STATE UNIVERSITY
DONALD J. BERG, South Dakota State University
DAVID COCHRAN, University of Southern Mississippi
MARCELO CRUZ, University of Wisconsin at Green Bay
LARRY SCOTT DEANER, Kansas State University
JASON DITTMER, Georgia Southern University
JAMES DOERNER, University of Northern Colorado
STEVEN DRIEVER, University of Missouri-Kansas City
ELIZABETH DUDLEY-MURPHY, University of Utah
DENNIS EHRHARDT, University of Louisiana-Lafayette
WILLIAM FORBES, Stephen F. Austin State University (Texas)
BILL FOREMAN, Oklahoma City Community College
ERIC FOURNIER, Samford University (Alabama)
RANDY GABRYS ALEXSON, University of Wisconsin-Superior
WILLIAM GARBARINO, Community College of Allegheny County (Pennsylvania)
JON GOSS, University of Hawai'i
SARA HARRIS, Neosho County Community College (Kansas)
INGRID JOHNSON, Towson State University (Maryland)
KRIS JONES, Saddleback College (California)
THOMAS KARWOSKI, Anne Arundel Community College (Maryland)
ROBERT KERR, University of Central Oklahoma
ERIC KEYS, University of Florida
JACK KINWORTHY, Concordia University-Nebraska
CHRISTOPHER LAINGEN, Kansas State University

UNNA LASSITER, Stephen F. Austin State University (Texas)
CATHERINE LOCKWOOD, Chadron State College (Nebraska)
GEORGE LONBERGER, Georgia Perimeter College
DALTON W. MILLER, JR., Mississippi State University
FERNANDO F. MINGHINE, Wayne State University (Michigan)
VERONICA MORMINO, Harper College (Illinois)
TOM MUELLER, California University (Pennsylvania)
IRENE NAESSE, Orange Coast College (California)
RICHARD OLMO, University of Guam
J. L. PASZTOR, Delta College (Michigan)
PAUL E. PHILLIPS, Fort Hays State University (Kansas)
ROSANN POLTRONE, Arapahoe Community College (Colorado)
JEFF POPKE, East Carolina University
DAVID PRIVETTE, Central Piedmont Community College (North Carolina)
RHONDA REAGAN, Blinn College (Texas)
A. L. RYDANT, Keene State College (New Hampshire)
NANCY SHIRLEY, Southern Connecticut State University
DEAN SINCLAIR, Northwestern State University (Louisiana)
SUSAN SLOWEY, Blinn College (Texas)
DEAN B. STONE, Scott Community College (Iowa)
JAMIE STRICKLAND, University of North Carolina at Charlotte
MORRIS O. THOMAS, Michigan State University
RUTHINE TIDWELL, Florida Community College at Jacksonville
IRINA VAKULENKO, Collin County Community College (Texas)
KIRK WHITE, York College of Pennsylvania
THOMAS WHITMORE, University of North Carolina at Chapel Hill
KEITH YEARMAN, College of DuPage (Illinois)
LAURA ZEEMAN, Red Rocks Community College (Colorado)

We also received input from a much wider circle of academic geographers. The list that follows is merely representative of a group of colleagues across North America to whom we are grateful for taking the time to share their thoughts and opinions with us:

MEL AAMODT, California State University-Stanislaus
WILLIAM V. ACKERMAN, Ohio State University
PAUL S. ADAMS, University of Pittsburgh at Greensburg
W. FRANK AINSLEY, University of North Carolina, Wilmington
CHRISTOPHER A. AIRRIESS, Ball State University (Indiana)
SIAW AKWAWUA, University of Northern Colorado
DONALD P. ALBERT, Sam Houston State University
TONI ALEXANDER, Louisiana State University
R. GABRY SALEXSON, University of Wisconsin-Superior
KHALED MD. ALI, Geologist, Dhaka, Bangladesh
NIGEL ALLAN, University of California-Davis
JOHN L. ALLEN, University of Wyoming
KRISTIN J. ALVAREZ, Keene State College (New Hampshire)
DAVID L. ANDERSON, Louisiana State University, Shreveport
KEAN ANDERSON, Freed-Hardeman University
JEFF ARNOLD, Southwestern Illinois College
JERRY R. ASCHERMANN, Missouri Western State College
JOSEPH M. ASHLEY, Montana State University
PATRICK ASHWOOD, Hawkeye Community College
THEODORE P. AUFDEMBERGE, Concordia College (Michigan)
JAIME M. AVILA, Sacramento City College
EDWARD BABIN, University of South Carolina-Spartanburg
ROBERT BAERENT, Randolph-Macon College (Virginia)
MARVIN W. BAKER, University of Oklahoma

GOURI BANERJEE, Boston University (Massachusetts)
MICHELE BARNABY, Pittsburg State University (Kansas)
J. HENRY BARTON, Thiel College (Pennsylvania)
CATHY BARTSCH, Temple University (Pennsylvania)
STEVEN BASS, Paradise Valley Community College (Arizona)
THOMAS F. BAUCOM, Jacksonville State University (Alabama)
KLAUS J. BAYR, Keene State College (New Hampshire)
DENIS A. BEKAERT, Middle Tennessee State University
JAMES BELL, Linn Benton Community College (Oregon)
KEITH M. BELL, Volunteer State Community College (Tennessee)
WILLIAM H. BERENTSEN, University of Connecticut
DONALD J. BERG, South Dakota State University
ROYAL BERGLEE, Morehead State University (Kentucky)
RIVA BERLEANT-SCHILLER, University of Connecticut
LEE LUCAS BERMAN, Southern Connecticut State University
RANDY BERTOLAS, Wayne State College (Nebraska)
THOMAS BITNER, University of Wisconsin-Marshfield/Wood County
WARREN BLAND, California State University, Northridge
DAVIS BLEVINS, Huntington College (Alabama)
HUBERTUS BLOEMER, Ohio University
S. BO JUNG, Bellevue College (Nebraska)
JOHN BOYER, Virginia Tech
R. DENISE BLANCHARD-BOEHM, Texas State University
MARTHA BONTE, Clinton Community College (Idaho)
GEORGE R. BOTJER, University of Tampa (Florida)
R. LYNN BRADLEY, Belleville Area College (Illinois)
KEN BREHOB, Elmhurst College (Illinois)
JAMES A. BREY, University of Wisconsin-Fox Valley
ROBERT BRINSON, Santa Fe Community College (Florida)
REUBEN H. BROOKS, Tennessee State University
PHILLIP BROUGHTON, Arizona Western College
LARRY BROWN, Ohio State University
LAWRENCE A. BROWN, Troy State-Dothan (Alabama)
ROBERT N. BROWN, Delta State University (Mississippi)
STANLEY D. BRUNN, University of Kentucky
RANDALL L. BUCHMAN, Defiance College (Ohio)
MICHAELEANN BUELL, Northwest Arkansas Community College
DANIEL BUYNE, South Plains College
MICHAEL BUSBY, Murray State University (Kentucky)
MARY CAMERON, Slippery Rock University (Pennsylvania)
MICHAEL CAMILLE, University of Louisiana at Monroe
MARY CARAVELIS, Barry University (Florida)
DIANA CASEY, Muskegon Community College (Michigan)
DIANN CASTEEL, Tusculum College (Tennessee)
BILL CHAPPELL, Keystone College
MICHAEL SEAN CHENOWETH, University of Wisconsin-Milwaukee
DAVE CLABORN, Olivet Nazarene University
STANLEY CLARK, California State University, Bakersfield
JOHN E. COFFMAN, University of Houston (Texas)
DWAYNE COLE, Cornerstone University (Michigan)
JERRY COLEMAN, Mississippi Gulf Coast Community College
JONATHAN C. COMER, Oklahoma State University
BARBARA CONNELLY, Westchester Community College (New York)
FRED CONNINGTON, Southern Wesleyan University
WILLIS M. CONOVER, University of Scranton (Pennsylvania)
OMAR CONRAD, Maple Woods Community College (Missouri)
ALLAN D. COOPER, Otterbein College
WILLIAM COUCH, University of Alabama, Huntsville
BARBARA CRAGG, Aquinas College (Michigan)

GEORGES G. CRAVINS, University of Wisconsin-La Crosse
ELLEN K. CROMLEY, University of Connecticut
JOHN A. CROSS, University of Wisconsin-Oshkosh
SHANNON CRUM, University of Texas at San Antonio
WILLIAM CURRAN, South Suburban College (Illinois)
KEVIN M. CURTIN, University of Texas at San Antonio
JOSE A. DA CRUZ, Ozarks Technical Community College (Arkansas)
ARMANDO DA SILVA, Towson State University (Maryland)
MARK DAMICO, Green Mountain College (Vermont)
DAVID D. DANIELS, Central Missouri State University
RUDOLPH L. DANIELS, Morningside College (Iowa)
SATISH K. DAVGUN, Bemidji State University (Minnesota)
WILLIAM V. DAVIDSON, Louisiana State University
CHARLES DAVIS, Mississippi Gulf Coast Community College
JAMES DAVIS, Illinois College
JAMES L. DAVIS, Western Kentucky University
PEGGY E. DAVIS, Pikeville College
ANN DEAKIN, State University of New York, College at Fredonia
KEITH DEBBAGE, University of North Carolina at Greensboro
MOLLY DEBYSINGH, California State University, Long Beach
DENNIS K. DEDRICK, Georgetown College (Kentucky)
JOEL DEICHMANN, Bentley College
STANFORD DEMARS, Rhode Island College
TOM DESULIS, Spoon River College
RAMESH DHUSSA, Drake University (Iowa)
THOMAS DIMICELLI, William Paterson College (New Jersey)
MARY DOBBS, Highland Community College, Wamego
SCOTT DOBLER, Western Kentucky University
DANIEL F. DOEPPERS, University of Wisconsin-Madison
JAMES DOERNER, University of Northern Colorado
ANN DOOLEN, Lincoln College (Illinois)
STEVEN DRIEVER, University of Missouri-Kansas City
WILLIAM DRUEN, Western Kentucky University
ALASDAIR DRYSDALE, University of New Hampshire
KEITH A. DUCOTE, Cabrillo Community College (California)
ELIZABETH DUDLEY-MURPHY, University of Utah
WALTER N. DUFFET, University of Arizona
MIKE DUNNING, University of Alaska Southeast
CHRISTINA DUNPHY, Champlain College (Vermont)
ANTHONY DZIK, Shawnee State University (Kansas)
DENNIS EDGELL, Bowling Green State University, Firelands (Ohio)
RUTH M. EDIGER, Seattle Pacific University
JAMES H. EDMONSON, Union University (Tennessee)
M. H. EDNEY, State University of New York-Binghamton
HAROLD M. ELLIOTT, Weber State University (Utah)
JAMES ELSNES, Western State College
SALVATORE ENGL-DI MAURO, State University of New York at New Paltz
ROBERT W. EVANS, Fresno City College (California)
EVERSON DICKEN, California State University, San Bernardino
DINO FIABANE, Community College of Philadelphia
G. A. FINCHUM, Milligan College (Tennessee)
CAITLIN FINLAYSON, Florida State University
IRA FOGEL, Foothill College (California)
RICHARD FOLEY, Cumberland College
ROBERT G. FOOTE, Wayne State College (Nebraska)
RONALD FORESTA, University of Tennessee
ELLEN J. FOSTER, Texas State University
G. S. FREEDOM, McNeese State University (Louisiana)

EDWARD T. FREELS, Carson-Newman College
PHILIP FRIEND, Inver Hills Community College (Minnesota)
JAMES FRYMAN, University of Northern Iowa
OWEN FURUSETH, University of North Carolina at Charlotte
RICHARD FUSCH, Ohio Wesleyan University
GARY GAILE, University of Colorado-Boulder
EVELYN GALLEGOS, Eastern Michigan University & Schoolcraft College
GAIL GARBRANDT, Mount Union College, University of Akron
CHAD GARICK, Jones County Junior College
RICHARD GARRETT, Marymount Manhattan College (New York)
JERRY GERLACH, Winona State University (Minnesota)
MARK GISMONDI, Northwest Nazarene University
LORNE E. GLAIM, Pacific Union College (California)
SHARLEEN GONZALEZ, Baker College (Michigan)
DANIEL B. GOOD, Georgia Southern University
GARY C. GOODWIN, Suffolk Community College (New York)
S. GOPAL, Boston University (Massachusetts)
MARVIN GORDON, Lake Forest Graduate School of Business (Illinois)
ROBERT GOULD, Morehead State University (Kentucky)
MARY GRAHAM, York College of Pennsylvania
GORDON GRANT, Texas A&M University
PAUL GRAY, Arkansas Tech University
DONALD GREEN, Baylor University (Texas)
GARY M. GREEN, University of North Alabama
STANLEY C. GREEN, Laredo State University (Texas)
RAYMOND GREENE, Western Illinois University
MARK GREER, Laramie County Community College (Wyoming)
WALT GUTTINGER, Flagler College (Florida)
JOHN E. GYGAX, Wilkes Community College (Pennsylvania)
LEEANN HAGAN, College of Southern Idaho
RON HAGELMAN, University of New Orleans
W. GREGORY HAGER, Northwestern Connecticut Community College
RUTH F. HALE, University of Wisconsin-River Falls
JOHN W. HALL, Louisiana State University-Shreveport
PETER L. HALVORSON, University of Connecticut
DAVID HANSEN, Pennsylvania State University, Harrisburg & Schuylkill
ELLEN HANSEN, Emporia State University (Kansas)
MERVIN HANSON, Willmar Community College (Minnesota)
ROBERT C. HARDING, Lynchburg College (Virginia)
MICHELLE D. HARRIS, Bob Jones University (South Carolina)
SCOTT HARRIS, CGCS Drury University
ROBERT J. HARTIG, Fort Valley State College (Georgia)
SUZANNA HARTLEY, Shelton State Community College
TRUMAN A. HARTSHORN, Georgia State University
CAROL HAZARD, Meredith College
HARLOW Z. HEAD, BARTON COLLEGE
ISAAC HEARD, University of North Carolina at Charlotte
DOUG HEFFINGTON, Middle Tennessee State University
JAMES G. HEIDT, University of Wisconsin Center-Sheboygan
EARL HEIDTKE, Martin Luther College
CATHERINE HELGELAND, University of Wisconsin-Manitowoc
NORMA HENDRIX, East Arkansas Community College
JAMES E. HERRELL, Otero Junior College (Colorado)
JAMES HERTZLER, Goshen College (Indiana)
JOHN HICKEY, Inver Hills Community College (Minnesota)
THOMAS HIGGINS, San Jacinto College (Texas)
EUGENE HILL, Westminster College (Missouri)

MIRIAM HELEN HILL, Indiana University Southeast

SUZY HILL, University of South Carolina-Spartanburg

ROBERT HILT, Pittsburg State University (Kansas)

SOPHIA HINSHALWOOD, Montclair State University (New Jersey)

PRISCILLA HOLLAND, University of North Alabama

MARK R. HOOPER, Freed-Hardeman University

R. HOSTETLER, Fresno City College (California)

ERICK HOWENSTINE, Northeastern Illinois University

LLOYD E. HUDMAN, Brigham Young University (Utah)

JANIS W. HUMBLE, University of Kentucky

JUANA IBANEZ, University of New Orleans

TONY IJOMAH, Harrisburg Area Community College (Pennsylvania)

WILLIAM IMPERATORE, Appalachian State University (North Carolina)

RICHARD JACKSON, Brigham Young University (Utah)

MARY JACOB, Mount Holyoke College (Massachusetts)

GREGORY JEANE, Samford University (Alabama)

SCOTT JEFFREY, Catonsville Community College (Maryland)

JERZY JEMIOLO, Ball State University (Indiana)

NILS I. JOHANSEN, University of Southern Indiana

BRIAN E. JOHNSON, Rhode Island College

DAVID JOHNSON, University of Southwestern Louisiana

INGRID JOHNSON, Towson State University (Maryland)

RICHARD JOHNSON, Oklahoma City University

SHARON JOHNSON, Marymount College (New York)

JEFFREY JONES, University of Kentucky

KRIS JONES, Saddleback College (California)

MARCUS E. JONES, Claflin College (South Carolina)

MOHAMMAD S. KAMIAR, Florida Community College, Jacksonville

MATTI E. KAUPS, University of Minnesota-Duluth

JOANNE W. KAY, Brigham Young University, Idaho

PHILIP L. KEATING, Indiana University

DAVID KEELING, Western Kentucky University

COLLEEN KEEN, Gustavus Adolphus College (Minnesota)

ARTIMUS KEIFFER, Wittenberg University (Ohio)

GORDON F. KELLS, Mott Community College (Michigan)

KAREN J. KELLY, Palm Beach Community College, Boca Raton (Florida)

RYAN KELLY, Lexington Community College (Kentucky)

VIRGINIA KERKHEIDE, Cuyahoga Community College and Cleveland State University (Ohio)

TOM KESSENGER, Xavier University (Ohio)

MASOUD KHEIRABADI, Maryhurst University

SUSANNE KIBLER-HACKER, Unity College (Maine)

CHANGJOO KIM, Minnesota State University

JAMES W. KING, University of Utah

THOMAS KING, Doane College

JOHN C. KINWORTHY, Concordia College (Nebraska)

ALBERT KITCHEN, Paine College (Georgia)

CLEM KLAPHAKE, Bellevue University

MARTI KLEIN, Saddleback College

TED KLIMASEWSKI, Jacksonville State University (Alabama)

ROBERT D. KLINGENSMITH, Ohio State University-Newark

LAWRENCE M. KNOPP, JR., University of Minnesota-Duluth

LYNN KOEHNEMANN, Gulf Coast Community College

TERRILL J. KRAMER, University of Nevada

JOE KRAUSE, Community College of Indiana at Lafayette

BRENDER KREKELER, Northern Kentucky University; Miami University (Ohio)

ARTHUR J. KRIM, Cambridge, Massachusetts

MICHAEL A. KUKRAL, Rose-Hulman Institute of Technology

CHRIS LANEY, Berkshire Community College (Massachusetts)

ELROY LANG, El Camino Community College (California)

RICHARD L. LANGILL, Saint Martin's University (Washington)

CHRISTOPHER LANT, Southern Illinois University

A. J. LARSON, University of Illinois-Chicago

PAUL R. LARSON, Southern Utah University

LARRY LEAGUE, Dickinson State University (North Dakota)

DAVID R. LEE, Florida Atlantic University

WOOK LEE, Texas State University

JOE LEEPER, Humboldt State University (California)

YECHIEL M. LEHAVY, Atlantic Community College (New Jersey)

SCOTT LEITH, Central Connecticut State University

JAMES LEONARD, Marshall University (West Virginia)

ELIZABETH J. LEPPMAN, St. Cloud State University (Minnesota)

JOHN C. LEWIS, Northeast Louisiana University

DAN LEWMAN JR., Southwest Mississippi Community College

CAEDMON S. LIBURD, University of Alaska-Anchorage

T. LIGIBEL, Eastern Michigan University

Z. L. LIPCHINSKY, Berea College (Kentucky)

ALLAN L. LIPPERT, Manatee Community College (Florida)

RICHARD LISICHENKO, Fort Hays State University (Kansas)

JOHN L. LITCHER, Wake Forest University (North Carolina)

LEE LIU, Central Montana State University

LI LIU, Stephen F. Austin State University (Texas)

WILLIAM R. LIVINGSTON, Baker College (Michigan)

CATHERINE M. LOCKWOOD, Chadron State College (Nebraska)

GEORGE E. LONGENECKER, Vermont Technical College

CYNTHIA LONGSTREET, Ohio State University

JACK LOONEY, University of Massachusetts, Boston

TOM LOVE, Linfield College (Oregon)

K. J. LOWREY, Miami University (Ohio)

JAMES LOWRY, Stephen F. Austin State University (Texas)

MAX LU, Kansas State University

ROBIN R. LYONS, University of Hawai'i-Leeward Community College

SUSAN M. MACEY, Texas State University

MICHAEL MADSEN, Brigham Young University, Idaho

RONALD MAGDEN, Tacoma Community College (Washington)

CHRISTIAN EMAINZER, Oxnard College (California)

LAURA MAKEY, California State University, San Bernardino

HARLEY I. MANNER, University of Guam

ANTHONY PAUL MANNION, Kansas State University

GARY MANSON, Michigan State University

CHARLES MANYARA, Radford University (Virginia)

W. ANDREW MARCUS, University of Oregon

JAMES T. MARKLEY, Lord Fairfax Community College (Virginia)

SISTER MAY LENORE MARTIN, Saint Mary College (Kansas)

KENT MATHEWSON, Louisiana State University

PATRICK MAY, Plymouth State College (New Hampshire)

DICK MAYER, Maui Community College (Hawai'i)

SARA MAYFIELD, San Jacinto College, Central (California)

DEAN R. MAYHEW, Marine Maritime Academy

J. P. MCFADDEN, Orange Coast College (California)

BERNARD MCGONIGLE, Community College of Philadelphia

PAUL D. MEARTZ, Mayville State University (North Dakota)

TRINA MEDLEY, Oklahoma City Community College

DIANNE MEREDITH, California State University-Sacramento

GARY C. MEYER, University of Wisconsin-Stevens Point

JUDITH L. MEYER, Southwest Missouri State University

MARK MICOZZI, East Central University (Oklahoma)

JOHN MILBAUER, Northeastern State University

DALTON W. MILLER, JR., Mississippi State University

RAOUL MILLER, University of Minnesota-Duluth

ROGER MILLER, Black Hills State University (South Dakota)

JAMES MILLS, State University of New York, College at Oneonta

INES MIYARES, Hunter College, CUNY (New York)

BOB MONAHAN, Western Carolina University

KEITH MONTGOMERY, University of Wisconsin-Madison

DAN MORGAN, University of South Carolina, Beaufort

DEBBI MORIMOTO, Merced College (California)

JOHN MORTRON, Benedict College (South Carolina)

ANNE MOSHER, Syracuse University

BARRY MOWELL, Broward Community College (Florida)

TOM MUELLER, California University (Pennsylvania)

DONALD MYERS, Central Connecticut State University

ROBERT R. MYERS, West Georgia College

GARY NACHTIGALL, Fresno Pacific University (California)

YASER M. NAJJAR, Framingham State College (Massachusetts)

KATHERINE NASHLEANAS, University of Nebraska-Lincoln

JEFFREY W. NEFF, Western Carolina University

DAVID NEMETH, University of Toledo (Ohio)

ROBERT NEWCOMER, East Central University (Oklahoma)

WILLIAM NIETER, St. John's University (New York)

WILLIAM N. NOLL, Highland Community College

VALIANT C. NORMAN, Lexington Community College (Kentucky)

RAYMOND O'BRIEN, Bucks County Community College (Pennsylvania)

PATRICK O'SULLIVAN, Florida State University

DIANE O'CONNELL, Schoolcraft College

JOHN ODLAND, Indiana University

DOUG OETTER, Georgia College and State University

ANNE O'HARA, Marygrove College (Michigan)

PATRICK OLSEN, University of Idaho

JOSEPH R. OPPONG, University of North Texas

LYNN ORLANDO, Holy Family University (Pennsylvania)

MARK A. OUMETTE, Hardin-Simmons University (Texas)

RICHARD OUTWATER, California State University, Long Beach

CISSIE OWEN, Lamar University (Texas)

MARYANN OWOC, Mercyhurst College (Pennsylvania)

THOMAS OWUSU, William Paterson University (New Jersey)

DAVID A. PADGETT, Tennessee State University

STEVE PALLADINO, Ventura College (California)

BIMAL K. PAUL, Kansas State University

SELINA PEARSON, Northwest-Shoals Community College (Alabama)

MAURI PELTO, Nichols College (Massachusetts)

JAMES PENN, Southeastern Louisiana University

LINDA PETT-CONKLIN, University of St. Thomas (Minnesota)

DIANE PHILEN, Lower Brule Community College (South Dakota)

PAUL PHILLIPS, Fort Hays State University (Kansas)

MICHAEL PHOENIX, ESRI, Redlands, California

JERRY PITZL, Macalester College (Minnesota)

BRIAN PLASTER, Texas State University

ARMAND POLICICCHIO, Slippery Rock University (Pennsylvania)

ROSANN POLTRONE, Arapahoe Community College (Colorado)

BILLIE E. POOL, Holmes Community College (Mississippi)

GREGORY POPE, Montclair State University (New Jersey)

JEFF POPKE, East Carolina University

WILLIAM PRICE, North Country Community College

VINTON M. PRINCE, Wilmington College (North Carolina)

GEORGE PUHRMANN, Drury University (Missouri)

DONALD N. RALLIS, Mary Washington College (Virginia)

MADHU RA, Bridgewater State College (Massachusetts)

RHONDA REAGAN, Blinn College (Texas)

DANNY I. REAMS, Southeast Community College (Nebraska)

JIM RECK, Golden West College (California)

ROGER REEDE, Southwest State University (Minnesota)

JOHN RESSLER, Central Washington University

JOHN B. RICHARDS, Southern Oregon State College

DAVID C. RICHARDSON, Evangel University (Missouri)

GRAY RINGLEY, Virginia Highlands Community College

ROBERT F. RITCHIE, IV, Liberty University

SUSAN ROBERTS, University of Kentucky

CURT ROBINSON, California State University, Sacramento

WOLF RODER, University of Cincinnati

JAMES ROGERS, University of Central Oklahoma

PAUL A. ROLLINSON, AICP[, Southwest Missouri State University

JAMES C. ROSE, Tompkins/Cortland Community College (New York)

THOMAS E. ROSS, Pembroke State University (North Carolina)

THOMAS A. RUMNEY, State University of New York, College at Plattsburgh

RUSHBROOK, DEREKA, University of Arizona

GEORGE H. RUSSELL, University of Connecticut

ANNE RUSZKIEWICZ, Sullivan County Community College (New York)

BILL RUTHERFORD, Martin Methodist College

RAJAGOPAL RYALI, Auburn University at Montgomery (Alabama)

PERRY RYAN, Mott Community College (Michigan)

JAMES SAKU, Frostburg State University (Maryland)

DAVID R. SALLEE, University of North Texas

RICHARD A. SAMBROOK, Eastern Kentucky University

EDUARDO SANCHEZ, Grand Valley State University (Michigan)

JOHN SANTOSUOSSO, Florida Southern College

JUSTIN SCHEIDT, Ferris State University

GINGER SCHMID, Texas State University

BRENDA THOMPSON SCHOOLFIELD, Bob Jones University (South Carolina)

ADENA SCHUTZBERG, Middlesex Community College (Massachusetts)

WILLIAM SCHWEITZER, Pennsylvania State University

ROGER M. SELYA, University of Cincinnati

RENEE SHAFFER, University of South Carolina

WENDY SHAW, Southern Illinois University, Edwardsville

SIDNEY R. SHERTER, Long Island University (New York)

HAROLD SHILK, Wharton County Jr. College (Texas)

NANDA SHRESTHA, Florida A&M University

WILLIAM R. SIDDALL, Kansas State University

DAVID SILVA, Bee County College (Texas)

JOSEANTONIO SIMENTAL, Marshall University (West Virginia)

MORRIS SIMON, Stillman College (Alabama)

ROBERT MARK SIMPSON, University of Tennessee at Martin

KENN E. SINCLAIR, Holyoke Community College (Massachusetts)

ROBERT SINCLAIR, Wayne State University (Michigan)
JIM SKINNER, Southwest Missouri State University
BRUCE SMITH, Bowling Green State University (Ohio)
EVERETT G. SMITH, JR., University of Oregon
PEGGY SMITH, California State University, Fullerton
RICHARD V. SMITH, Miami University (Ohio)
JAMES SNADEN, Charter Oak State College (Connecticut)
DAVID SORENSON, Augustana College (Illinois)
SISTER CONSUELO SPARKS, Immaculata University (Pennsylvania)
CAROLYN D. SPATTA, California State University, Hayward
M. R. SPONBERG, Laredo Junior College (Texas)
DONALD L. STAHL, Towson State University (Maryland)
DAVID STEA, Texas State University
ELAINE STEINBERG, Central Florida Community College
D. J. STEPHENSON, Ohio University Eastern
HERSCHEL STERN, Mira Costa College (California)
REED F. STEWART, Bridgewater State College (Massachusetts)
NOEL L. STIRRAT, College of Lake County (Illinois)
JOSEPH P. STOLTMAN, Western Michigan University
DEAN B. STONE, Scott Community College
WILLIAM M. STONE, Saint Xavier University, Chicago
GEORGE STOOPS, Minnesota State University
DEBRA STRAUSSFOGEL, University of New Hampshire
JAMIE STRICKLAND, University of North Carolina at Charlotte
WAYNE STRICKLAND, Roanoke College (Virginia)
PHILIP STURM, Ohio Valley College, Vienna
PHILIP SUCKLING, Texas State University
ALLEN SULLIVAN, Central Washington University
SELIMA SULTANA, University of North Carolina at Greensboro
RAY SUMNER, Long Beach City College (California)
CHRISTOPHER SUTTON, Western Illinois University
MARY B. TACY, James Madison University (Virginia)
T. L. TARLOS, Orange Coast College (California)
WESLEY TERAOKA, Leeward Community College
MICHAEL THEDE, North Iowa Area Community College
DERRICK J. THOM, Utah State University
CURTIS THOMSON, University of Idaho
BEN TILLMAN, Texas Christian University
CLIFF TODD, University of Nebraska, Omaha
STANLEY TOOPS, Miami University (Ohio)
RICHARD J. TORZ, St. Joseph's College (New York)
HARRY TRENDELL, Kennesaw State University (Georgia)
ROGER T. TRINDELL, Mansfield University (Pennsylvania)
WEI TU, Georgia Southern University
DAN TURBEVILLE, East Oregon State College
NORMAN TYLER, Eastern Michigan University
IRINA VAKULENKO, University of Texas at Dallas
GEORGE VAN OTTEN, Northern Arizona University
GREGORY VEECK, Western Michigan University
C. S. VERMA, Weber State College (Utah)
KELLY ANN VICTOR, Eastern Michigan University
SEAN WAGNER, Tri-State University (Indiana)
GRAHAM T. WALKER, Metropolitan State College of Denver
MONTGOMERY WALKER, Yakima Valley Community College (Washington)
DEBORAH WALLIN, Skagit Valley College (Washington)
RICHARD WALSEK, University of Wisconsin, Parkside
MIKE WALTERS, Henderson Community College (Kentucky)
LINDA WANG, University of South Carolina, Aiken
J. L. WATKINS, Midwestern State University (Texas)

DAVID WELK, Reedley College (California)
KIT W. WESLER, Murray State University (Kentucky)
PETER W. WHALEY, Murray State University (Kentucky)
MACEL WHEELER, Northern Kentucky University
P. GARY WHITE, Western Carolina University (North Carolina)
W. R. WHITE, Western Oregon University
GARY WHITTON, Fairbanks, Alaska
WYNDHAM WHYNOT, Livingston College
REBECCA WIECHEL, Wilmington College (North Carolina)
MARK WILJANEN, Eastern Kentucky University
GENE C. WILKEN, Colorado State University
FORREST WILKERSON, Texas State University
P. WILLIAMS, Baldwin-Wallace College (Ohio)
STEPHEN A. WILLIAMS, Methodist College (North Carolina)
DEBORAH WILSON, Sandhills Community College (Nebraska)
MORTON D. WINSBERG, Florida State University
ROGER WINSOR, Appalachian State University (North Carolina)
WILLIAM A. WITHINGTON, University of Kentucky
A. WOLF, Appalachian State University (North Carolina)
JOSEPH WOOD, University of Southern Maine
RICHARD WOOD, Seminole Junior College (Florida)
GEORGE I. WOODALL, Winthrop College (North Carolina)
STEPHEN E. WRIGHT, James Madison University (Virginia)
LEON YACHER, Southern Connecticut State University
KYONGYUP CHU, Bergen Community College (New Jersey)
FIROOZ E. ZADEH, Colorado Mountain College
DONALD J. ZEIGLER, Old Dominion University (Virginia)
YU ZHOU, Bowling Green State University (Ohio)
ROBERT C. ZIEGENFUS, Kutztown University (Pennsylvania)
JACKSON ZIMMERMAN, University of Wisconsin, Platteville

PERSONAL APPRECIATION

For assistance with the map of North American indigenous people (Fig. 3-4), we are greatly indebted to Jack Weatherford, Professor of Anthropology at Macalester College (Minnesota); Henry T. Wright, Professor and Curator of Anthropology at the University of Michigan; and George E. Stuart, President of the Center for Maya Research (North Carolina). The map of Russia's federal regions (Fig. 2-10) could not have been compiled without the invaluable help of David B. Miller, Senior Edit Cartographer at the National Geographic Society, and Leo Dillon of the Russia Desk of the U.S. Department of State. The map of Russian physiography (Fig. 2-4) was updated thanks to the suggestions of Mika Roinila of the State University of New York, College at New Paltz. And special thanks also go to Charles Pirtle, Professor of Geography at Georgetown University's School of Foreign Service for his advice on Chapter 4; to Antoinette WinklerPrins of the Department of Geography at Michigan State University for her recommendations on Chapter 5; and to Charles Fahrer of Georgia College and State University for his suggestions on Chapter 6.

We also record our appreciation to those who ensured the quality of this book's ancillary products: Dmitrii Sidorov (California State University, Long Beach) prepared the PowerPoint presentation for this edition; Blake Harrison (Southern Connecticut State University) updated the test bank; Andrew Shears (Kent State University) updated the Clicker (PRS) questions; Randy Gabrys-Alexson (University of Wisconsin, Superior) prepared the Instructor's Manual; and Justin Scheidt (Ferris State University) prepared the Study Guide.

At the University of Miami's Department of Geography and Regional Studies, Peter Muller is most grateful for the advice and support of faculty colleagues Richard Grant, Larry Kalkstein, Miguel Kanai, Mazen Labban, Jan Nijman, Shouraseni Sen Roy, and Ira Sheskin, and particularly GIS Lab Manager Chris Hanson.

We are privileged to work with a team of professionals at John Wiley & Sons that is unsurpassed in the college textbook publishing industry. As authors we are acutely aware of these talents on a daily basis during the crucial production stage, especially the coordination and leadership skills of Senior Production Editor Sandra Dumas, Senior Illustration Editor Sigmund Malinowski, and Senior Photo Editor Jennifer MacMillan. Others who played a leading role in this process were Senior Designer Kevin Murphy and Interior Designer Nancy Field, copyeditor Betty Pessagno, and Don Larson and Terry Bush at Mapping Specialists, Ltd., in Madison, Wisconsin.

On the eve of the 40th anniversary of this book, we would also like to acknowledge the outstanding contributions of a number of former (and many still-current) Wileyans and their associates to the production of earlier editions of this book. In alphabetical order, we salute: Karen Ayoub, Johnna Barto, Martin Bentz, Barbara Bredenko, Bethany Brooks, Ellen Brown, Andrea Bryant, Olivia Buehl, Lucille Buonocore, Edward Burke, Ken Burke, Martha Cooley, Christine Cordek, Malcolm Easterlin, Joyce Franzen, Micheline Frederick, Jeanine Furino, Gigi Ghriskey, Robert Goff, Marge Graham, Karen Grant, Maura Grant, Erin Grattan, Barbara Heaney, Lisa Heard, Carmen Hernandez, Jill Hilycord, Elizabeth Hovinen, Dennis Hudson, Laura Ierardi, Suzanne Ingrao, Alex Juliani, Vivian Kahane, Joan Kalkut, Pam Kennedy, Kim Khatchatourian, Kenneth Kimmel, Karin Kincheloe, Tom Kulesa, Stella Kupferberg, Audrey Labaton, Don Larson, Terri Leigh, Rick Leyh, Madelyn Lesure, Lindsay Lovier, Jerry McCarthy, Tammy McGinnis, Pam McGreevy, Beth Meder, Joan Meisel, Alissa Mend, Elaine Miller, Ishaya Monokoff, Linda Muriello, Courtney Nelson, Hilary Newman, Harry Nolan, Ellen O'Neill, Safra Nimrod, Lisa Passmore, Eugene Patti, Betty Pessagno, Denise Powell, Mary Ann Price, Felicia Reid, Connie Rende, Ann Marie Renzi, Joan Rosenberg, Rebecca Rothaug, Barbara Russiello, Rosie Hirsch Scarano, Rachel Schneider, Debra Schwartz, Josh Spieler, Gilda Stahl, Edward Starr, Fadia Stermasi, Elizabeth Swain, Kelly Tavares, Alexandra Truitt, Aliyah Vinikoor, Andrew Yockers, and Patricia Young.

Getting back to the Fourteenth Edition, we much appreciated the leadership of Executive Geosciences Editor Ryan Flahive, who was the prime mover in launching and guiding this latest edition, and was effectively assisted throughout the revision process by Meredith Leo. Our College Marketing Manager, Danielle Torio, gave us considerable attention throughout the book's preparation, and her advice and enthusiasm were most welcome. Beyond this immediate circle, we acknowledge the support and encouragement we continue to receive over the years from many others at Wiley, including Vice-President for Production Ann Berlin and Publisher Jay O'-Callaghan.

As is stated on an earlier page, this edition of *Regions* is dedicated to Jack Reilly, Executive Producer of ABC Television's *Good Morning America* during the 1980s and 1990s, when the program earned an Emmy for its innovative contents. Reilly, who passed away in May 2009, established the position of Geography Editor on "GMA" that continued for seven years, until the end of his tenure at ABC. In 1990 he received a Special Award from the Association of American Geographers, during its annual meeting in Toronto, in recognition of his support for the discipline in the national and international broadcast media.

Finally, and most of all, we express our gratitude to our wives, Bonnie and Nancy, for seeing us yet again through the challenging schedule of our eighteenth collaboration in the past 27 years.

H. J. de Blij
Boca Grande, Florida

Peter O. Muller
Coral Gables, Florida

August 7, 2009

BRIEF TABLE OF CONTENTS

CONTENTS

CHAPTER 10

SOUTHEAST ASIA   528

CHAPTER 11

THE AUSTRAL REALM   578

LIST OF MAPS AND FIGURES

LIST OF MAPS AND FIGURES *(continued)*

LIST OF MAPS AND FIGURES *(continued)*

GEOGRAPHY

Realms, Regions, and Concepts

Introduction

WORLD REGIONAL GEOGRAPHY: GLOBAL PERSPECTIVES

0 km 1000 2000 3000 4000 5000 6000

0 miles 1000 2000 3000 4000

Goode Homolosine Projection
Scale 1:103,750,000

Elevation (m)

Ice cap

6000
3000
1500
600
0
-150
-1500
-3000
-6000

FIGURE G-1

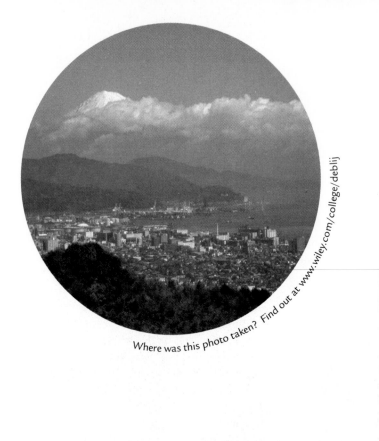

Where was this photo taken? Find out at www.wiley.com/college/deblij

WHAT ARE YOUR expectations as you open this book? Many students come to college or university planning to concentrate on some other field but graduate as geography majors—because they discover how interesting and unexpectedly challenging the discipline of geography is. Maybe you will be one of them. Whatever you decide, we hope that this course, and this book, will open new vistas and bring new perspectives, and help you navigate this increasingly complex and often daunting world.

THE UNITED STATES AND THE WORLD

You could not have chosen a better time to be studying geography. The world is changing on many fronts, and so is the United States. Still the most formidable of all countries, the United States remains a superpower capable of influencing nations and peoples, lives and livelihoods from pole to pole. That power confers on Americans the responsibility to learn as much as they can about those nations and livelihoods, so that the decisions of their government representatives are well-informed. But in this respect, the United States is no superpower. Geographic literacy is a measure of international comprehension and awareness, and Americans' geographic literacy ranks low among countries of consequence. For the world, that is not a good thing, because such geographic fogginess tends to afflict not only voters but also the representatives they elect, from the school board to the White House.

In This Chapter

- Maps in our minds
- Why *tsunamis* happen and where they get started
- The world's most dangerous places to live
- Global core and periphery: Power and place
- How countries become mired in debt
- Is globalization good or bad?

Photos: © H. J. de Blij

So you are about to strike a blow against geographic illiteracy, and at the end of this term you will find yourself among the still-small minority of Americans who have spent even just one semester studying what you will learn in the weeks ahead. When you get into discussions about world affairs, you will be able to bring to the table a dimension not usually weighed during such conversations. Enjoy!

The World and the State

Take a look at a map of the countries of the world today, such as the one just inside this book's front cover, and you are reminded of a jigsaw puzzle of about 200 pieces. Getting to know these players on the global stage is not as challenging as it may at first appear, and by the end of the term you will even be familiar with the most important subdivisions of some of them. Officially, a sovereign country is referred to as a *state*, for example, the state of Mexico or the state of Japan. But as all Americans know, that same term is used to designate constituent parts of some states, for example, the State of California or the State of Texas. Some other countries, such as Nigeria and India, also call their largest subdivisions States. Note the difference: when we are writing about a country, we call it a state. When we refer to a constituent part, we call it a *State*, capitalized to emphasize the difference.

Some states use other designations for their largest administrative components, and it is consistent to capitalize them as well, such as the Province of Ontario in Canada or the Autonomous Community of Catalonia, a part of Spain. The Italians call them Regions; in Myanmar (Burma) they are called Divisions. We'll have to get used to the different terms. But the important point is this: the state will continue to play a key role in the fortunes of our planet. In the early 1990s, when the Soviet Union collapsed and the Cold War ended, there was much theorizing about the end of the state as we have known it and the emergence of a "New World Order" that would signal a new era of amalgamation and cooperation. Two decades later, states such as Iran and Venezuela, North Korea and Somalia remind us that power still centers in states and that a new world order remains a distant objective.

As we will see, a truly new world order would require global adoption of human-rights standards, the universal acceptance of representative government, the worldwide alleviation of poverty, the raising of public-health standards, and other goals yet unmet. We have a long way to go, and our task is to understand how the world is changing and where we are headed. Geography, as you will discover, is a powerful ally in this mission.

A WORLD ON MAPS

Just a casual glance at the pages that follow reveals a difference between this and other textbooks: there are almost as many maps as there are pages. Geography is more closely identified with maps than any other discipline, and we urge you to give as much (or more!) attention to the maps in this book as you do to the text. It is often said that a picture is worth a thousand words, but a map can be worth a million. When we write "see Figure XX," we really mean it . . . and we hope that you will get into the habit. We humans are territorial creatures, and the boundaries that fence off our 200 or so countries reflect our divisive ways. But other, less visible borders—between religions, languages, rich and poor—partition our planet as well. When political and cultural boundaries are at odds, there is nothing like a map to summarize the circumstances. Just look at the one on page 374. How clear were the implications of this map to those who made the decision to send American troops into war?

Maps in Our Minds

All of us carry in our minds maps of what psychologists call our "activity space": the apartment building or the house we live in, the streets nearby, the way to school or business, the general layout of our hometown or city. You will know what lane to use when you turn into a shopping mall, or where to park at the movie theater. You can probably draw from memory a pretty good map of your home State. These **mental maps** allow you to navigate your activity space with efficiency, predictability, and safety. When you arrive as a first-year student on a college or university campus, a new mental map will start forming. At first you may need a hard-copy map to find your way around, but soon you will dispense with that because your mental map will be sufficient. And it will continue to improve as your activity space expands.

If a well-formed mental map is useful for decisions in daily life, then an adequate mental map is surely indispensable when it comes to decision-making in the wider world. You can give yourself an interesting test. Choose some part of the world, beyond North America, in which you have an interest or about which you have a strong opinion—for example, Israel, Taiwan, Afghanistan, North Korea, or Venezuela. On a blank piece of paper, draw a map that reflects your impression of the regional layout there: the country, its internal divisions, major cities, neighbors, seas (if any), and so forth. That is your mental map of the place. Put it away for future reference, and try it again at the end of this course. You will have proof of your improved mental-map inventory.

The Map Revolution

The maps in this book show larger and smaller parts of the world in various contexts, some depicting political frameworks, others displaying ethnic, cultural, economic, environmental, and other features unevenly distributed across our world. But cartography (the making of maps) has undergone a dramatic technological revolution—a revolution that continues. Earth-orbiting satellites with special on-board scanners and television cameras transmit information to computers on the surface, recording the expansion of deserts, the shrinking of glaciers, the depletion of forests, the growth of cities, and myriad other geographic phenomena. The Earthbound computers possess ever-expanding capabilities not only to sort this information but also to display it graphically. This allows geographers to develop *geographic information systems (GIS)*, presenting on-screen information within seconds that would have taken months to assemble just a few decades ago.

Nonetheless, satellites—even spy satellites—cannot record everything that occurs on the Earth's surface. Sometimes the transition zones between ethnic groups or cultural sectors can be discerned by satellites, for example, in changing types of houses or religious shrines, but this kind of information tends to require on-the-ground verification through field research and reporting. No satellite view of Iraq could show you the distribution of Sunni and Shi'ite Muslim adherents. Many of the boundaries you see on the maps in this book cannot be seen from space because long stretches are not even marked on the ground. So the maps you are about to "read" have their continued uses. They summarize complex situations and allow us to begin forming lasting mental maps of the areas they represent.

GEOGRAPHY'S PERSPECTIVE

Geography has been described as the most interdisciplinary of disciplines. That is a testimonial to geography's historic linkages to many other fields, ranging from geology to economics and from sociology to political science. And, as has been the case so often in the past, geography is in the lead on this point. Today, interdisciplinary studies and research are more prevalent than ever. The old barriers between disciplines are breaking down.

This should not suggest that college and university departments are no longer relevant; they are just not as exclusive as they used to be. These days, you can learn some good geography in economics departments and some useful economics in geography departments. But

FROM THE FIELD NOTES

"From the observation platform atop the Seoul Tower one would be able to see into North Korea except for the range of hills in the background: the capital lies in the shadow of the DMZ (demilitarized zone), relic of one of the hot conflicts of the Cold War. The vulnerable Seoul-Incheon metropolitan area ranks among the world's largest, its population approaching 20 million. I asked my colleague, a professor of geography at Seoul National University, why the central business district, the cluster of buildings in the center of this photo, does not seem to reflect the economic power of this city (note the sector of traditional buildings in the foreground, some with blue roofs, the culture's favorite color). 'Turn around and look across the [Han] River,' he said. 'That's the new Seoul, and there the skyline matches that of Singapore, Beijing, or Tokyo.' Height restrictions, disputes over land ownership, and congestion are among the factors that slowed growth in this part of Seoul and caused many companies to build in the Samsung area. Looking in that direction, you cannot miss the large U.S. military facility, right in the heart of the urban area, on some of the most valuable real estate and right next to an upscale shopping district. In the local press, the debate over the presence of American forces dominated the letters page day after day."   © H. J. de Blij

Concept Caching

www.conceptcaching.com

each discipline still has its own particular way of looking at the world.

In a very general way, we can visualize three key perspectives when we try to figure out how the world works. One is the historic or chronological (you have heard the expression "if you do not learn the lessons of history, you will be doomed to repeat them"). History's key question is when. A second perspective centers on the systems people have invented to stabilize their interactions, from the economic to the political. Here the question is how. The third perspective is the geographic— the spatial—and here the key question is *where*.

2 This **spatial perspective** has pervaded geography from its beginnings, and while it is less dominant today it still forms a bond for the discipline. It is a good place to start our global journey.

Environment and Society

There is another "glue" that binds geography and has done so for a very long time: an interest in the relationships between human societies and ever-changing natural (physical) environments. You can see how the spatial perspective comes into play here: environmental change is in the news on a daily basis in the form of global warming, but this current surge of global warming is only the latest phase of endless climatic and ecological fluctuation. When the first migrating humans arrived in Europe, the climate was much colder than it is today—and it was to get colder still. Just 18,000 years ago, an eyeblink in the planet's history, icesheets covered most of present-day Canada and reached southward to the Ohio River, and a large icecap covered the Alps of Switzerland. A mere 5000 years ago, the Sahara had woods and rivers and wildlife and human settlements, but then, in a very short time, the climate changed and a desert formed, splitting Africa's human population into "Mediterranean" (North) and "Subsaharan."

Much of the epochal story of humanity's migration from its African sources into Eurasia, Australia, and ultimately the Americas still needs to be unraveled, and for this knowledge of the physical geography of the planet is crucial. As we will discover, such knowledge is useful even in the interpretation of current events. A volcanic eruption in the Philippines in 1991 changed the geopolitical landscape in the region in ways that still endure. But let us wait with the details until Chapter 10.

Location and Distribution

Geographers, therefore, need to be conversant with the location and distribution of salient features on the Earth's surface. This includes the natural (physical) world simplified in Figure G-1 as well as the human world, and our inquiry will view these in temporal (historical) as well as spatial perspective. We take a penetrating look at the overall geographic framework of the contemporary world, the still-changing outcome of thousands of years of human achievement and failure, movement and stagnation, stability and revolution, interaction and isolation. The spatial structure of cities, the layout of farms and fields, the networks of transportation, the configurations of rivers, the patterns of climate—all these form part of

our investigation. As you will find, geography employs a comprehensive spatial vocabulary with meaningful terms such as area, distance, direction, clustering, proximity, accessibility, and many others we will encounter in the pages that follow. For geographers, some of these terms have more specific definitions than is generally assumed. There is a difference, for example, between *area* and *region*, and between *boundary* and *frontier*. Other terms, such as *location* and *pattern*, can have multiple meanings. The vocabulary of geography holds some surprises.

GEOGRAPHIC REALMS

Ours may be a globalizing, interconnected world, a world of international trade and travel, migration and movement, tourism and television, financial flows and telephone traffic, a world that, in some contexts, is taking on the properties of a "global village"—but that village still has neighborhoods. Their names are Europe, South America, Southeast Asia, and others familiar to us all. Like the neighborhoods of a city or town, these global neighborhoods may not have sharply defined borders, but their persistence, after tens of thousands of years of human dispersal, is beyond doubt. Geographers call such global neighborhoods **geographic realms**. **3** Each of these realms possesses a particular combination of environmental, cultural, and organizational properties. These characteristic qualities are imprinted on the landscape, giving each realm its own traditional attributes and social settings. As we come to understand the human and environmental makeup of these geographic realms, we learn not only *where* they are located (as we noted, a key question in geography), but also *why they are located where they are*, how they are constituted, and what their future is likely to be in our fast-changing world. Accordingly, Figure G-2 forms the framework for our investigation.

REALMS AND REGIONS

Geographers, like other scholars, seek to establish order from the countless data that confront them. Biologists have established a system of classification to categorize the many millions of plants and animals into a hierarchical system of seven ranks. In descending order, we humans belong to the animal *kingdom*, the *phylum* (division) named chordata, the *class* of mammals, the *order* of primates, the *family* of hominids, the *genus* designated *Homo*, and the *species* known as *Homo sapiens*. Geologists classify the Earth's rocks into three major (and many subsidiary) categories, and then fit these categories into a

FIGURE G-2

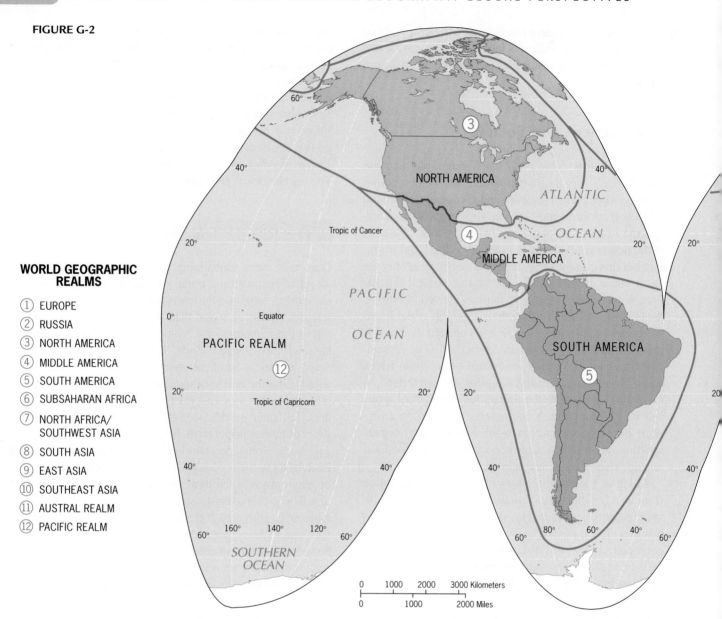

WORLD GEOGRAPHIC REALMS

1. EUROPE
2. RUSSIA
3. NORTH AMERICA
4. MIDDLE AMERICA
5. SOUTH AMERICA
6. SUBSAHARAN AFRICA
7. NORTH AFRICA/ SOUTHWEST ASIA
8. SOUTH ASIA
9. EAST ASIA
10. SOUTHEAST ASIA
11. AUSTRAL REALM
12. PACIFIC REALM

complicated geologic time scale that spans hundreds of millions of years. Historians define eras, ages, and periods to conceptualize the sequence of the events they study.

Geography, too, employs systems of classification. When geographers deal with urban problems, for instance, they use a classification scheme based on the sizes and functions of the places involved. Some of the terms in this classification are part of our everyday language: megalopolis, metropolis, city, town, village, hamlet.

Regional Classification

In regional geography, the focus of this book, our challenge is different. We, too, need a hierarchical framework for the areas of the world we study, from the largest to the smallest. But our classification system is horizontal, not vertical. It is spatial. There are four levels:

1. **Landmasses and Oceans** (Fig. G-1).This is geography's equivalent to the biologists' overarching kingdoms (of plants and animals), although the issue arises as to whether the icy surface of the Antarctic "continent" constitutes a landmass. Since no permanent human population has become established on these glaciers (some permanent stations exist in the ice-free parts of the Antarctic Peninsula), no regional geography has as yet evolved.

2. **Geographic Realms** (Fig. G-2). Based on a combination of physical and human factors, these are the most comprehensive divisions of the inhabited world.

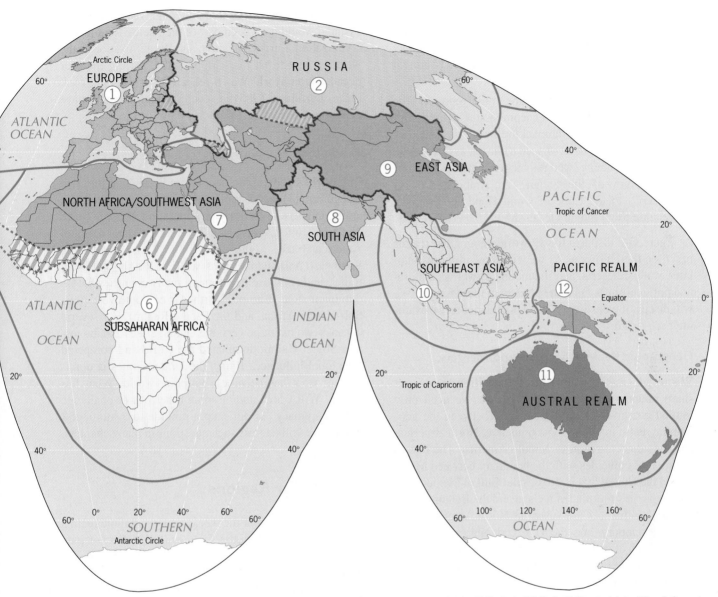

© H. J. de Blij, P. O. Muller, and John Wiley & Sons, Inc.

3. **Geographic Regions.** More specific criteria divide the great geographic realms into smaller regions, as will be demonstrated at the conclusion of this chapter.

4. **Subregions, Domains, Districts.** Subdivisions of regions, sometimes based on single factors, are mapped for specific purposes.

Criteria for Geographic Realms

In any classification system, criteria are the key. Not all animals are mammals; the criteria for inclusion in that biological class are more specific and restrictive. A dolphin may look and act like a fish, but both anatomically and functionally dolphins belong to the class of mammals.

- **Physical and Human** Geographic realms are based on sets of spatial criteria. They are the largest units into which the inhabited world can be divided. The criteria on which such a broad regionalization is based include both physical (that is, natural) and human (or social) yardsticks. On the one hand, South America is a geographic realm because physically it is a continent and culturally it is dominated by a set of social norms. The realm called South Asia, on the other hand, lies on a Eurasian landmass shared by several other geographic realms; high mountains, wide deserts, and dense forests combine with a distinctive social fabric to create this well-defined realm centered on India.

- **Functional** Geographic realms are the result of the interaction of human societies and natural environments, a *functional* interaction revealed by farms, mines, fishing ports, transport routes, dams, bridges, villages, and countless other features that mark the landscape. According to this criterion, Antarctica is a continent but not a geographic realm.

- **Historical** Geographic realms must represent the most comprehensive and encompassing definition of the great clusters of humankind in the world today. China lies at the heart of such a cluster, as does India. Africa constitutes a geographic realm from the southern margin of the Sahara (an Arabic word for "desert") to the Cape of Good Hope and from its Atlantic to its Indian Ocean shores.

Figure G-2 displays the 12 world geographic realms based on these criteria. As we will show in more detail later, waters, deserts, and mountains as well as cultural and political shifts mark the borders of these realms. We will discuss the position of these boundaries as we examine each realm.

Geographic Realms: Margins and Mergers

4 Where geographic realms meet, **transition zones**, not sharp boundaries, mark their contacts. We need only remind ourselves of the border zone between the geographic realm in which most of us live, North America, and the adjacent realm of Middle America. The line in Figure G-2 coincides with the boundary between Mexico and the United States, crosses the Gulf of Mexico, and then separates Florida from Cuba and the Bahamas. But Hispanic influences are strong in North America north of this boundary, and the U.S. economic influence is strong south of it. The line, therefore, represents an ever-changing zone of regional interaction. Again, there are many ties between South Florida and the Bahamas, but the Bahamas resemble a Caribbean more than a North American society.

In Africa, the transition zone from Subsaharan to North Africa is so wide and well defined that we have put it on the world map; elsewhere, transition zones tend to be narrower and less easily represented. In these early years of the twenty-first century, such countries as Belarus (between Europe and Russia) and Kazakhstan (between Russia and Muslim Southwest Asia) lie in inter-realm transition zones. Remember that over much (though not all) of their length, borders between realms are zones of regional change.

Geographic Realms: Changing Times

Had we drawn Figure G-2 before Columbus made his voyages (1492–1504), the map would have looked different: Amerindian states and peoples would have determined the boundaries in the Americas; Australia and

New Guinea would have constituted one realm, and New Zealand would have been part of the Pacific Realm. The colonization, Europeanization, and Westernization of the world changed that map dramatically. During the four decades after World War II relatively little change took place, but since 1985 far-reaching realignments have again been occurring.

Geographic Realms: Two Categories

The world's geographic realms can be divided into two categories: (1) those dominated by one major political entity, in terms of territory or population or both (North America/United States, Middle America/Mexico, South America/Brazil, South Asia/India, East Asia/China, Southeast Asia/Indonesia as well as Russia and Australia), and (2) those that contain many countries but no dominant state (Europe, North Africa/Southwest Asia, Subsaharan Africa, and the Pacific Realm). For several decades two major powers, the United States and the former Soviet Union, dominated the world and competed for global influence. Today, the United States is dominant in what has become a unipolar world. What lies ahead? Will China and/or some other power challenge U.S. supremacy? Is our map a prelude to a multipolar world? We will address such questions in many chapters.

Criteria for Regions

The spatial division of the world into geographic realms establishes a broad global framework, but for our purposes a more refined level of spatial classification is needed. This brings us to an important organizing concept in geography: the **regional concept**. To continue the **5** analogy with biological taxonomy, we now go from phylum to order. To establish regions within geographic realms, we need more specific criteria.

Let us use the North American realm to demonstrate the regional idea. When we refer to a part of the United States or Canada (e.g., the South, the Midwest, or the Prairie Provinces), we employ a regional concept—not scientifically but as part of everyday communication. We reveal our perception of local or distant space as well as our mental image of the region we are describing.

But what exactly is the Midwest? How would you draw this region on the North American map? Regions are easy to imagine and describe, but they can be difficult to outline on a map. One way to define the Midwest is to use the borders of States: certain States are part of this region, others are not. You could also use agriculture as the chief criterion: the Midwest is where corn and/or soybeans occupy a certain percentage of the farmland. Each method results in a different delimitation; a Midwest based on States is different from a Midwest based on farm production. Therein lies an important principle: regions are scientific devices that allow us to make spatial generalizations, and they are based on artificial criteria we establish to help us construct them. If you were studying the geography behind politics, then a Midwest region defined by State boundaries would make sense. If you were studying agricultural distributions, you would need a different definition.

Area

Given these different dimensions of the same region, we can identify properties that all regions have in common. To begin with, all regions have *area*. This observation would seem obvious, but there is more to this idea than meets the eye. Regions may be intellectual constructs, but they are not abstractions: they exist in the real world, and they occupy space on the Earth's surface.

Boundaries

It follows that regions have *boundaries*. Occasionally, nature itself draws sharp dividing lines, for example, along the crest of a mountain range or the margin of a forest. More often, regional boundaries are not self-evident, and we must determine them using criteria that we establish for that purpose. For example, to define a citrus-growing agricultural region, we may decide that only areas where more than 50 percent of all farmland stands under citrus trees qualify to be part of that region.

Location

All regions also have *location*. Often the name of a region contains a locational clue, as in the Amazon Basin or Indochina (a region of Southeast Asia lying between **6** India and China). Geographers refer to the **absolute location** of a place or region by providing the latitudinal and longitudinal extent of the region with respect to the Earth's grid coordinates. A far more practical mea- **7** sure is a region's **relative location**, that is, its location with reference to other regions. Again, the names of some regions reveal aspects of their relative location, as in *Eastern* Europe and *Equatorial* Africa.

Homogeneity

Many regions are marked by a certain *homogeneity* or sameness. Homogeneity may lie in a region's human (cultural) properties, its physical (natural) characteristics, or both. Siberia, a vast region of northeastern Russia, is marked by a sparse human population that resides in widely scattered, small settlements of similar form, frigid climates, extensive areas of permafrost (permanently frozen subsoil), and cold-adapted vegetation. This dominant uniformity makes it one of Russia's natural and cultural regions, extending from the Ural Mountains in the west to the Pacific Ocean in the east. When regions display a measurable and often visible internal homogeneity, they are called **formal** **8** **regions**. But not all formal regions are visibly uniform. For example, a region may be delimited by the area in which, say, 90 percent or more of the people speak a particular language. This cannot be seen in the landscape, but the region is a reality, and we can use this criterion to draw its boundaries accurately. It, too, is a formal region.

Regions as Systems

Other regions are marked not by their internal sameness, but by their functional integration—that is, the way they work. These regions are defined as **spatial systems** and **9** are formed by the areal extent of the activities that define them. Take the case of a large city with its surrounding zone of suburbs, urban-fringe countryside, satellite towns, and farms. The city supplies goods and services to this encircling zone, and it buys farm products and other commodities from it. The city is the heart, the core of this region, and we call the surrounding zone of interaction the city's **hinterland**. But the city's influence **10** wanes on the outer periphery of that hinterland, and there lies the boundary of the functional region of which the city is the focus. A **functional region**, therefore, is forged **11** by a structured, urban-centered system of interaction. It has a *core* and a *periphery*. As we shall see, core-periphery contrasts in some parts of the world are becoming strong enough to endanger the stability of countries.

Interconnections

All human-geographic regions are *interconnected*, being linked to other regions. We know that the borders of geographic realms sometimes take on the character of transition zones, and so do neighboring regions. Trade, migration, education, television, computer linkages, and other interactions blur regional boundaries. These are just some of the links in the fast-growing interdependence among the world's peoples, and they reduce the differences that still divide us. Understanding these differences will lessen them further.

THE PHYSICAL SETTING

This book focuses on the geographic realms and regions produced by human activity over thousands of years. But we should never forget the natural environments in which all this activity took place because we can still recognize the role of these environments in how people make their living. Certain areas of the world, for example, presented opportunities for plant and animal domestication that other areas did not. The people who happened to live in those favored areas learned to grow wheat, rice, or root crops, and to domesticate oxen, goats, or llamas. We can still discern those early "patterns of opportunity" on the map in the twenty-first century. From such opportunities came adaptation and invention, and thus arose villages, towns, cities, and states. But people living in more difficult environments found it much harder to achieve this organization. Take tropical Africa. There, the many species of wildlife, from gazelles and zebras to giraffes and buffalo, could not be domesticated; that is, they would not breed in captivity. Wild animals were a threat, not an opportunity. Eventually, humans domesticated only one African animal, the guinea fowl. Early African peoples faced environmental disadvantages that persist today. The modern map carries many imprints of the past.

Natural (Physical) Landscapes

The landmasses of Planet Earth present a jumble of **natural landscapes** ranging from rugged mountain chains to smooth coastal plains (Fig. G-1). The map of natural landscapes reminds us that the names of the con-

tinents and their mountain backbones are closely linked: North America and its Rocky Mountains, South America and the Andes, Eurasia with its Alps (in the west) and Himalayas (toward the east), Australia and its Great Dividing Range. There is, however, a prominent exception: Africa, the Earth's second-largest landmass, has peripheral mountains in the north (the Atlas) and south (the Cape Ranges), but no linear chain from coast to coast. We will soon discover why.

In his book *Guns, Germs, and Steel*, UCLA geographer Jared Diamond argues that the orientation of these continental "axes" has had "enormous, sometimes tragic, consequences [that] affected the rate of spread of crops and livestock, and possibly also of writing, wheels, and other inventions. That basic feature of geography thereby contributed heavily to the very different experiences of Native Americans, Africans, and Eurasians in the last 500 years." Inevitably, it also affected the evolution of the regional map we are set to investigate.

And it affected not only the evolution of the regional map, but also its future. Mountain ranges formed barriers to movement but, as Diamond stresses, also channeled the spread of agricultural and technical innovations. Rugged mountains obstructed contact but also protected peoples and cultures against their enemies (today Taliban, al-Qaeda, and Chechnyan fugitives are using remote and rugged mountain refuges for such purposes). As we

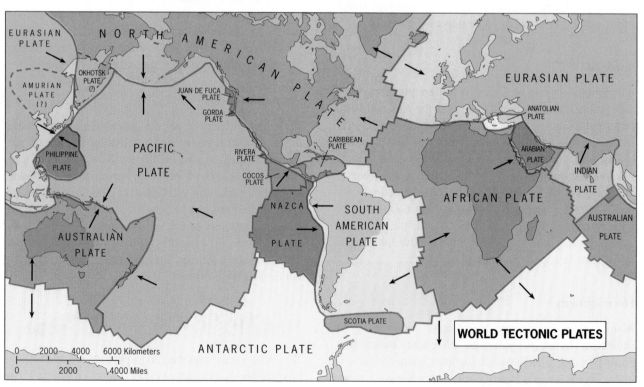

FIGURE G-3   © H. J. de Blij, P. O. Muller, and John Wiley & Sons, Inc.

study each of the world's geographic realms, we will find that physical landscapes continue to play significant roles in this modern world. River basins in Asia still contain several of the planet's largest population concentrations: the advantages of fertile soils and ample water that first enabled clustered human settlement now sustain hundreds of millions in crowded South and East Asia. But river basins in Africa and South America support no such numbers because the combination of natural and historical factors is different. That is one reason why the study of world regional geography is important: it puts the human map in environmental as well as regional perspective.

Natural Hazards

Our planet may be 4.6 billion years old, but it is far from placid. As you read this, Earth tremors are shaking the still-thin crust on which we live, volcanoes are erupting, storms are raging. Even the very continents are moving measurably, pulling apart in some areas, colliding in others. Hundreds of thousands of human lives are lost to natural calamities in almost every decade, and such calamities have at times altered the course of history.

About a century ago a geographer named Alfred Wegener, a German scientist whose specialty actually was climatology rather than the study of surface landforms, used spatial analysis to explain something that is obvious even from a small-scale map like Figure G-1: the

apparent, jigsaw-like "fit" of the landmasses, especially across the South Atlantic Ocean. He concluded that the landmasses on the map are actually pieces of a supercontinent that existed hundreds of millions of years ago (he called it *Pangaea*) and that "drifted" away when, for some reason, that supercontinent broke up. His hypothesis of **continental drift** set the stage for scientists in **13** other disciplines to search for a mechanism that might make this possible, and much of the answer to that search proved to lie in the crust beneath the ocean surface. Today we know that the continents are "rafts" of relatively light rock that rest on slabs of heavier rock called **tectonic plates** (Fig. G-3) whose movement is propelled **14** by giant circulation cells in the red-hot magma below (when this molten magma reaches the surface through volcanic vents, we call it *lava*).

Inevitably, moving tectonic plates collide. When they do, earthquakes and volcanic eruptions result, and the physical landscape is thrown into spectacular relief. Compare Figures G-3 and G-4, and you can see the outlines of the tectonic plates in the distribution of these hazards to human life. The December 26, 2004 *tsunami* (earthquake-generated ocean wave) that claimed some 300,000 lives around the Indian Ocean Basin resulted from an undersea quake off the Indonesian island of Sumatera. One of the Earth's oceans is almost completely encircled by active volcanoes and earthquake epicenters. Appropriately, this is called the **Pacific Ring of Fire**. **15**

It is useful to compare Figure G-4 to Figure G-2 to see which of the world's geographic realms are most susceptible to the hazards inherent in crustal instability.

FIGURE G-4   © H. J. de Blij, P. O. Muller, and John Wiley & Sons, Inc.

Russia, Europe, Africa, and Australia are relatively safe; in other realms the risks are far greater in one sector than in others (western as opposed to eastern North and South America, for example). As we will find, for certain parts of the world the details on Figure G-4 present a clear and present danger. Some of the world's largest cities lie in areas most vulnerable to sudden disaster.

Climate

The prevailing climate constitutes a key factor in the geography of realms and regions (as we will discover, some regions are essentially *defined* by climate). But we should always remember that climates change and that the climate predominating in a certain region today may not be the climate prevailing there several thousand years ago. Any map of climate, including the ones in this chapter, is but a still-picture of our always-changing world.

Ice Ages and Climate Change

16 Climatic conditions have swung back and forth for as long as the Earth has had an atmosphere. Periodically, an **ice age** lasting tens of millions of years chills the planet and caus-

es massive ecological change. One such ice age occurred while Pangaea was still in one piece, between 250 and 300 million years ago. Another started about 35 million years ago, and we are still experiencing it. The current epoch of this ice age, on average the coldest yet, is called the *Pleistocene* and has been going on for nearly two million years.

In our time of global warming this may come as a surprise, but we should remember that an ice age is not a period of unbroken, bitter cold. Rather, an ice age consists of surges of cold, during which glaciers expand and living space shrinks, separated by warmer phases when the ice recedes and life spreads poleward again. The cold phases are called **glaciations**, and they tend to last **17** longest, although milder spells create some temporary relief. The truly warm phases, when the ice recedes poleward and mountain glaciers melt away, are known as **interglacials**. We are living in one of these interglacials **18** today. It even has a geologic name: the *Holocene*.

Imagine this: just 18,000 years ago, great icesheets had spread all the way to the Ohio River Valley, covering most of the Midwest, the zenith of a glaciation that had lasted about 100,000 years, the *Wisconsinan Glaciation* (Fig. G-5). The Antarctic Icesheet was bigger than ever, and even in the tropics, great mountain

EXTENT OF GLACIATION DURING THE PLEISTOCENE

- Land areas covered by glaciers
- Present-day glaciers

FIGURE G-5

© H. J. de Blij, P. O. Muller, and John Wiley & Sons, Inc.

FIGURE G-6

© H. J. de Blij, P. O. Muller, and John Wiley & Sons, Inc.

glaciers pushed down valleys and onto plateaus. But then Holocene warming began, the continental and mountain glaciers receded, and ecological zones that had been squeezed between the advancing glaciers now spread north and south. In Europe, particularly, where humans had arrived from Africa via Southwest Asia during one of the milder spells of the Wisconsinan Glaciation, living space expanded and human numbers grew.

Global Warming

Consider the transformation that our planet—and our ancestors—experienced as the ice receded. Huge slabs of ice thousands of miles across slid into the oceans from Canada and Antarctica. Rivers raged with silt-clogged meltwater and formed giant deltas. Sea level rose rapidly, submerging vast coastal plains. Animals and plants migrated into newly opened lands at higher elevations and latitudes. As climatic zones shifted, moist areas fell dry, and arid areas turned wet. In Southwest Asia, where humans had begun to cultivate crops and towns signaled the rise of early states, such sudden changes often destroyed what had been achieved. In northern tropical Africa, where rivers flowed and grasses fed wildlife, a region extending from Atlantic shores to the Red Sea fell dry in a very short time around 5000 years ago. We know it today as the Sahara. We call the **19** process **desertification**.

Temperatures reached their present-day levels about 7000 years ago, but the effects of this Holocene inter-

glacial warmth took more time to take hold, as the Sahara's later desiccation shows. Soils formed on newly exposed rock, but generally faster in warm and moist areas than in dry and still-cooler zones. Pine forests that had migrated southward during the Wisconsinan Glaciation now moved north, and equatorial rainforests expanded in all directions. The world's first empires formed in areas that had been inhospitable during the Wisconsinan: the Roman Empire at the western end of Eurasia, the Han Empire at the eastern end. Europeans and Chinese traded with each other along the Silk Route, now open for business.

A great deal of this environmental history remains imprinted in Figure G-2 today. The great river basins of East Asia, where humans exploited agricultural opportunities thousands of years ago, still anchor populous China today. The great Ganges Basin, where one of early humankind's great population explosions may have taken place, remains the core of a huge modern state, India. The Roman Empire's legacies infuse much of European culture.

The past 1000 years have witnessed some troubling environmental developments. Around 1300, the first warning signals of a return to cooler conditions caused crop failures and social dislocation, first in Europe and later in China. This episode, which worsened during the 1600s and has come to be known as the *Little Ice Age*, had major impacts on the human geography of Eurasia and on peoples elsewhere in the world as well, accompanied as it was by long-term changes in climate. A return to pre-Little Ice Age

"Walking the hilly countryside anywhere in Indonesia leaves you in no doubt about the properties of *Af* climate. The sweltering sun, the hot, humid air, the daily afternoon rains, and the lack of relief even when the sun goes down—the still atmosphere lies like a heavy blanket on the countryside to make this a challenging field trip. Deep, fertile soils form rapidly here on the volcanic rocks, and the entire landscape is green, most of it now draped by rice fields and terraced paddies. The farmer whose paddies I photographed here told me that his land produces three crops per year. Consider this: the island of Jawa, about the size of Louisiana, has a population of well over 130 million—its growth made possible by this combination of equatorial circumstances... Alaska, almost a dozen times as large as Jawa, has a population under three-quarters of a million. Here climates range from *Cfc* to *E*, soils are thin and take thousands of years to develop, and the air is arctic. We sailed slowly into Glacier Bay, in awe of the spectacular, unspoiled scenery, and turned into a bay filled with ice floes, some of them serving as rafts for sleeping seals. Calving (breaking up) into the bay was the Grand Pacific Glacier, with evidence of its recent recession all around. Less than 300 years ago, all of Glacier Bay was filled with ice; today you can sail miles to the Johns Hopkins (shown here) and Margerie tidewater glaciers' current outer edges. Global warming in action—and a reminder that any map of climate is a still picture of a changing world." © H. J. de Blij

www.conceptcaching.com

conditions commenced in the early nineteenth century, but then the exponential growth of the Earth's human population began to affect the composition of the atmosphere, an impact that grew as the Industrial Revolution proceeded. Now humanity itself became a factor in global climate change, and most scientists conclude that the accelerated warming over the past several decades is largely due to human activity.

Climate on the Map

We have just learned how variable climate can be, but in a human lifetime we see little evidence of this variability. We talk about the *weather* (the immediate state of the atmosphere) in a certain place at a given time, but as a technical term **climate** defines the aggregate, total record of weather conditions at a place, or in a region, over the entire period during which records have been kept.

The key records are those of temperature and precipitation, their seasonal variations, and extremes, but other data (humidity, wind direction, and the like) also come into play at the larger scale. Precipitation across the planet, for example, is quite variable (Fig. G-6), but this map cannot show when, seasonally, this precipitation arrives. Average temperatures, too, mean very little: What is needed at minimum is information on daily and monthly ranges.

For our discussion of global realms and regions, a comprehensive map of climates is enormously important and useful. Efforts to create such a map have gone on for about a century, and Figure G-7, based on a system devised by Wladimir Köppen and modified by Rudolf Geiger, is one of the most useful because of its comparative simplicity.

The Köppen-Geiger map represents climatic regions through a system of letter symbols. The first (capital) letter is the critical one: the *A* climates are humid and tropical, the *B* climates are arid, the *C* climates are mild and humid, the *D* climates show increasing extremes of seasonal heat and cold, and the *E* climates reflect the frigid conditions at and near the poles.

Figure G-7 merits your attention because familiarity with it will help you understand much of what follows in this book. The map has practical utility as well. Although it depicts climatic regions, daily weather in each color-coded region is relatively standard. If, for example, you are familiar with the weather in the large area mapped as *Cfa* in the southeastern United States, you will feel at home in Uruguay (South America), Kwazulu-Natal (South Africa), New South Wales (Australia), and Fujian Province (China). Let us look at the world's climatic regions in some detail.

Humid Equatorial (A) Climates   The humid equatorial, or tropical, climates are characterized by high temperatures all year and by heavy precipitation. In the *Af* subtype, the rainfall arrives in substantial amounts every month; but in the *Am* areas, the arrival of the annual wet ***monsoon*** (the Arabic word for "season" [see p. 407]) marks a sudden enormous increase in precipitation. The *Af* subtype is named after the vegetation that develops there—the tropical rainforest. The *Am* subtype, prevailing in part of peninsular India, in a coastal area of West Africa, and in sections of Southeast Asia, is appropriately referred to as the monsoon climate. A third tropical climate, the savanna (*Aw*), has a wider daily and annual temperature range and a more strongly seasonal distribution of rainfall. As Figure G-6 indicates, savanna rainfall totals tend to be lower than those in the rainforest zone, and savanna seasonality is often expressed in a "double maximum." Each year produces two periods of increased rainfall separated by pronounced dry spells. In many savanna zones, inhabitants refer to the "long rains" and the "short rains" to identify those seasons; a persistent problem is the unpredictability of the rain's arrival. Savanna soils are not among the most fertile, and when the rains fail hunger looms. Savanna regions are far more densely peopled than rainforest areas, and millions of residents of the savanna subsist on what they cultivate. Rainfall variability is their principal environmental problem.

Dry (B) Climates   Dry climates occur in both lower and higher latitudes. The difference between the *BW* (true ***desert***) and the moister *BS* (semiarid ***steppe***) varies but may be taken to lie at about 25 centimeters (10 in) of annual precipitation. Parts of the central Sahara in North Africa receive less than 10 centimeters (4 in) of rainfall. Most of the world's arid areas have an enormous daily temperature range, especially in subtropical deserts. In the Sahara, there are recorded instances of a maximum daytime shade temperature of more than 50°C (122°F) followed by a nighttime low of less than 10°C (50°F).

Soils in these arid areas tend to be thin and poorly developed; soil scientists have an appropriate name for them—aridisols.

Humid Temperate (C) Climates   As the map shows, almost all these mid-latitude climate areas lie just beyond the Tropics of Cancer and Capricorn (23.5° North and South latitude, respectively). This is the prevailing climate in the southeastern United States from Kentucky to central Florida, on North America's west coast, in most of Europe and the Mediterranean, in southern Brazil and northern Argentina, in coastal South Africa, in eastern Australia, and in eastern China and southern Japan. None of these areas suffers climatic extremes or severity, but the winters can be cold, especially away from water bodies that moderate temperatures. These areas lie midway between the winterless equatorial climates and the summerless polar zones. Fertile and productive soils have developed under this regime, as we will note in our discussion of the North American and European realms.

The humid temperate climates range from moist, as along the densely forested coasts of Oregon, Washington, and British Columbia, to relatively dry, as in the so-called Mediterranean (dry-summer) areas that include not only coastal southern Europe and northwestern Africa but also the southwestern tips of Australia and Africa, central Chile, and Southern California. In these Mediterranean environments, the scrubby, moisture-preserving vegetation creates a natural landscape different from that of richly green Western Europe.

Humid Cold (D) Climates   The humid cold (or "snow") climates may be called the continental climates, for they seem to develop in the interior of large landmasses, as in the heart of Eurasia or North America. No equivalent land areas at similar latitudes exist in the Southern Hemisphere; consequently, no *D* climates occur there.

Great annual temperature ranges mark these humid continental climates, and cold winters and relatively cool summers are the rule. In a *Dfa* climate, for instance, the warmest summer month (July) may average as high as 21°C (70°F), but the coldest month (January) might average only –11°C (12°F). Total precipitation, much of it snow, is not high, ranging from about 75 centimeters (30 in) to a steppe-like 25 centimeters (10 in). Compensating for this paucity of precipitation are cool temperatures that inhibit the loss of moisture from evaporation and evapotranspiration (moisture loss to the atmosphere from soils and plants).

Some of the world's most productive soils lie in areas under humid cold climates, including the U.S. Midwest,

WORLD CLIMATES
After Köppen–Geiger

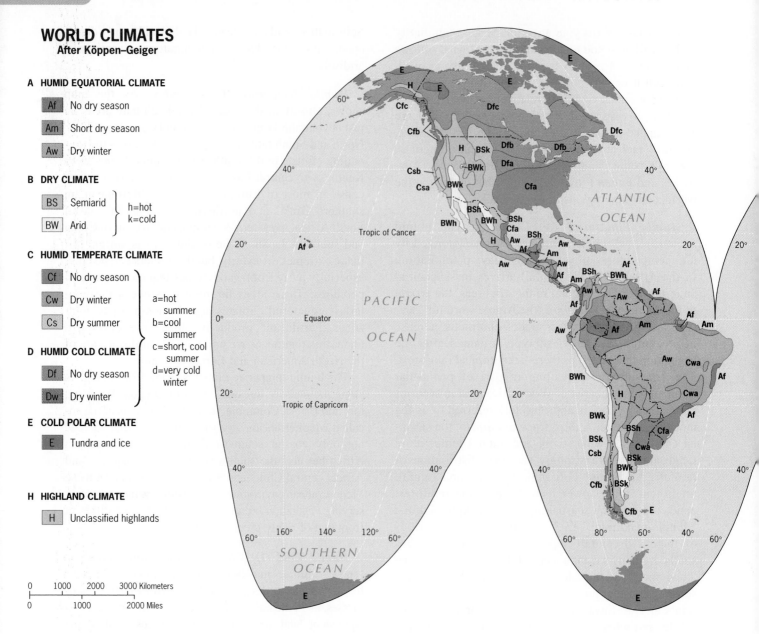

A   HUMID EQUATORIAL CLIMATE

Af	No dry season
Am	Short dry season
Aw	Dry winter

B   DRY CLIMATE

| BS | Semiarid | | h=hot |
| BW | Arid | | k=cold |

C   HUMID TEMPERATE CLIMATE

Cf	No dry season
Cw	Dry winter
Cs	Dry summer

a=hot summer
b=cool summer
c=short, cool summer
d=very cold winter

D   HUMID COLD CLIMATE

| Df | No dry season |
| Dw | Dry winter |

E   COLD POLAR CLIMATE

| E | Tundra and ice |

H   HIGHLAND CLIMATE

| H | Unclassified highlands |

```
0    1000    2000    3000 Kilometers
0    1000    2000 Miles
```

FIGURE G-7

parts of southern Russia and Ukraine, and Northeast China. The winter dormancy (when all water is frozen) and the accumulation of plant debris during the fall balance the soil-forming and enriching processes. The soil differentiates into well-defined, nutrient-rich layers, and substantial organic humus accumulates. Even where the annual precipitation is light, this environment sustains extensive coniferous forests.

Cold Polar (E) and Highland (H) Climates   Cold polar (***E***) climates are differentiated into true icecap conditions, where permanent ice and snow keep vegetation from gaining a foothold, and the tundra, which may have average temperatures above freezing up to four months of the year. Like

rainforest, savanna, and steppe, the term ***tundra*** is vegetative as well as climatic, and the boundary between the ***D*** and ***E*** climates in Figure G-7 corresponds closely to that between the northern coniferous forests and the tundra.

Finally, the ***H*** climates—unclassified highlands mapped in gray (Fig. G-7)—resemble the ***E*** climates. High elevations and the complex topography of major mountain systems often produce near-Arctic climates above the tree line, even in the lowest latitudes such as the equatorial section of the high Andes of South America.

Let us not forget an important qualification concerning Figure G-7: this is a still-picture of a changing scene, a single frame from an ongoing film. Climate is still

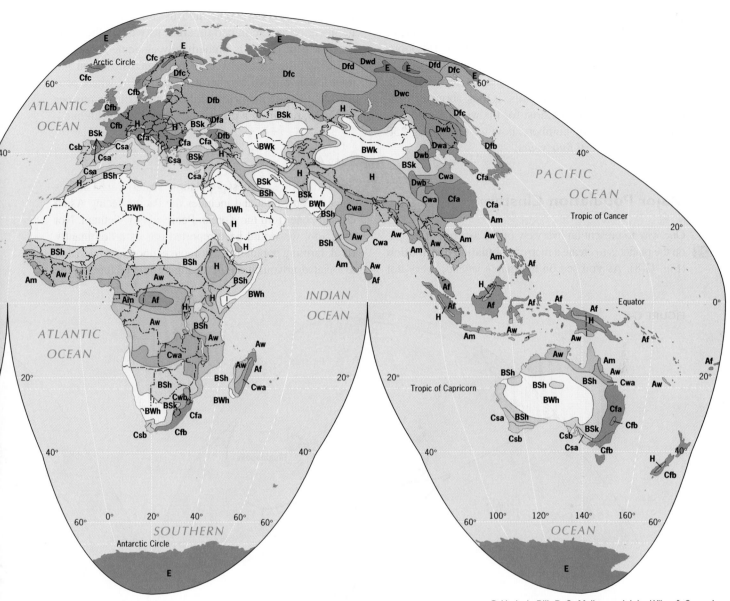

© H. J. de Blij, P. O. Muller, and John Wiley & Sons, Inc.

changing, and less than a century from now climatologists are likely to be modifying the climate maps to reflect new data. Who knows: we may have to redraw even those familiar coastlines. Environmental change is a never-ending challenge.

REALMS OF POPULATION

Earlier we noted that population numbers by themselves do not define geographic realms or regions. Population distributions, and the functioning society that gives them common ground, are more significant criteria. That is why we can identify one geographic realm (the Austral) with less than 30 million people and another (East Asia) with more than 1.5 billion inhabitants. Neither population numbers nor territorial size alone can delimit a geographic realm. Nevertheless, the map of world population distribution (Fig. G-8) suggests the relative location of several of the world's geographic realms, based on the strong clustering of population in certain areas. Before we examine these clusters in some detail, remember that the Earth's human population now totals just under 6.9 billion—six thousand nine hundred million people confined to the landmasses that constitute less than 30 percent of our planet's surface, much of which is arid desert, rugged mountain terrain, or frigid tundra. (Remember that Figure G-8 is another still-picture of an ever-changing scene: the rapid growth of humankind continues.) After thousands of years of slow growth, world population during the nineteenth and twentieth centuries grew at an increasing rate. That rate has recently been slowing down, but consider this: it took about 17 centuries after the birth of Christ for the world to add 250 million people to its numbers; now we are adding 250 million about every *four years*.

Major Population Clusters

One way to present an overview of the location of people on the planet is to create a map of **population distribution** (Fig. G-8). As you see on the map's legend, every dot

represents 100,000 people, and the clustering of large numbers of people in certain areas as well as the near-emptiness of others is immediately evident. There is a technical difference between population distribution and *population density*, another way of showing where people are. Density maps reveal the number of persons per unit area, requiring a different cartographic technique.

East Asia

Still the world's greatest population cluster, *East Asia* lies centered on China and includes the Pacific-facing Asian coastal zone from the Korean Peninsula to Vietnam. Not long ago, we would have reported this as a dominantly rural, farming population, but rapid economic growth and associated urbanization are changing this picture. In the

FIGURE G-8

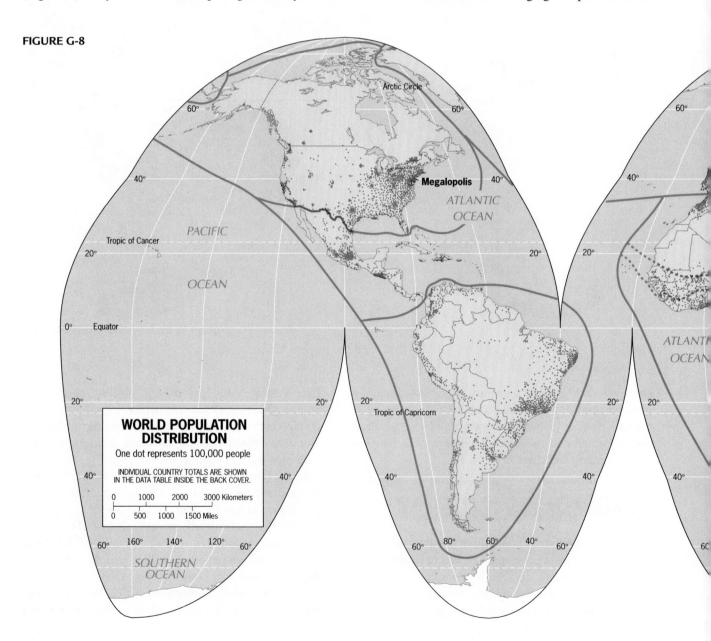

WORLD POPULATION DISTRIBUTION
One dot represents 100,000 people

INDIVIDUAL COUNTRY TOTALS ARE SHOWN IN THE DATA TABLE INSIDE THE BACK COVER.

0   1000   2000   3000 Kilometers
0   500   1000   1500 Miles

interior river basins of the Huang (Yellow) and Chang/Yangzi (**A** and **B** on the map), and in the Sichuan Basin between these two letters, most of the people remain farmers, and, as the Data Table inside the back cover shows, farmers still outnumber city-dwellers in China as a whole. But the great cities of coastal and near-coastal China are attracting millions of new inhabitants, and interior cities are growing rapidly as well. Before 2020, the East Asia cluster will become more urban than rural.

South Asia

The *South Asia* population cluster lies centered on India and includes its populous neighbors, Pakistan and Bangladesh. This huge agglomeration of humanity focuses on the wide plain of the Ganges River (**C** in Fig.

G-8). It is nearly as large as that of East Asia and at present growth rates is poised to overtake East Asia in 2011. A larger percentage of the people remain farmers, although pressure on the land is greater because farming is less efficient than in East Asia.

Europe

The third-ranking population cluster, *Europe*, also lies on the Eurasian landmass but at the opposite end from China. The European cluster, including western Russia, counts over 700 million inhabitants, which puts it in a class with the two larger Eurasian concentrations—but there the similarity ends. In Europe, the key to the linear, east-west orientation of the axis of population (**D** in Fig. G-8) is not a fertile river basin but a zone of raw

© H. J. de Blij, P. O. Muller, and John Wiley & Sons, Inc.

materials for industry. Europe is among the world's most highly urbanized and industrialized realms, its human agglomeration sustained by factories and offices rather than paddies and pastures.

The three world population concentrations just discussed (East Asia, South Asia, and Europe) account for more than 3.7 billion (54 percent) of the world's 6.9 billion people. No other cluster comes close to these numbers. The next-ranking cluster, *Eastern North America*, is only about one-quarter the size of the smallest of the Eurasian concentrations. As in Europe, the population in this area is concentrated in major metropolitan complexes; the rural areas are now relatively sparsely settled. Geographic realms and regions, therefore, display varying **22** levels of **urbanization**, the percentage of the total population living in cities and towns. And some regions are urbanizing much more rapidly than others, a phenomenon we will explain as we examine the world's realms.

The quantitative dominance of Eurasia's population clusters is revealed dramatically by a special map trans- **23** formation called a **cartogram** (Fig. G-9). Here the map of the world is redrawn, so that the area of each country reflects *not* territory but population numbers. China and India stand out because, unlike the European population cluster, their populations are not fragmented among many national political entities. But neither Figure G-8 nor Figure G-9 can tell us how the populations they exhibit are changing. As we shall see, varying rates of population growth mark the world's realms and regions: some countries' populations are growing rapidly, while populations in other countries are actually shrinking.

REALMS OF CULTURE

Imagine yourself in a boat on the Nile River, headed upstream (south) from Khartoum, Sudan. The desert sky is blue, the heat is searing. You pass villages on the shore that look much the same: low, square or rectangular dwellings, some recently whitewashed, others gray, with flat roofs, wooden doors, and small windows. The minaret of a modest mosque may rise above the houses, and you get a glimpse of a small central square. There is very little vegetation; here and there a hardy palm tree stands in a courtyard. People on the paths wear long white robes and headgear, also white, that looks like a baseball cap without the visor. A few goats lie in the shade. Along the river's edge lie dusty farm fields that yield to the desert in the distance. At the foot of the river's bluff lie some canoes.

FROM THE FIELD NOTES

"The Atlantic-coast city of Bergen, Norway displayed the Norse cultural landscape more comprehensively, it seemed, than any other Norwegian city, even Oslo. The high-relief site of Bergen creates great vistas, but also long shadows; windows are large to let in maximum light. Red-tiled roofs are pitched steeply to enhance runoff and inhibit snow accumulation; streets are narrow and houses clustered, conserving warmth... The coastal village of Mengkabong on the Borneo coast of the South China Sea represents a cultural landscape seen all along the island's shores, a stilt village of the Bajau, a fishing people. Houses and canoes are built of wood as they have been for centuries. But we could see some evidence of modernization: windows filling wall openings, water piped in from a nearby well." © H. J. de Blij

CARTOGRAM OF THE WORLD'S NATIONAL POPULATIONS, 2010

1	Guatemala	16	Paraguay	31	Slovakia	46	Jordan	61	Guinea-Bissau	76	Malawi	91	Maldives
2	El Salvador	17	Uruguay	32	Slovenia	47	United Arab Emirates	62	Sierra Leone	77	Angola	92	Laos
3	Honduras	18	Chile	33	Croatia	48	Yemen	63	Liberia	78	Namibia	93	Cambodia
4	Nicaragua	19	Guadeloupe	34	Bosnia	49	Saudia Arabia	64	Ivory Coast	79	Botswana	94	Malaysia
5	Costa Rica	20	Barbados	35	Serbia	50	Oman	65	Burkina Faso	80	Mocambique	95	Singapore
6	Panama	21	Netherlands Antilles	36	Montenegro	51	Libya	66	Togo	81	Lesotho	96	Brunei
7	Jamaica	22	Bahamas	37	Kosovo	52	Niger	67	Benin	82	Swaziland	97	Papua New Guinea
8	Haiti	23	Martinique	38	Hungary	53	Chad	68	Equatorial Guinea	83	Comoros	98	New Zealand
9	Dominican Republic	24	Norway	39	Macedonia	54	Tunisia	69	Cameroon	84	Madagascar	99	Belarus
10	Puerto Rico	25	Sweden	40	Albania	55	Central African Republic	70	Gabon	85	Mauritius	100	Estonia
11	Trinidad and Tobago	26	Finland	41	Bulgaria	56	Senegal	71	Congo	86	Djibouti	101	Latvia
12	Guyana	27	Austria	42	Greece	57	Mauritania	72	Rwanda	87	Eritrea	102	Lithuania
13	Suriname	28	Switzerland	43	Cyprus	58	Mali	73	Burundi	88	Mongolia	103	Moldova
14	Ecuador	29	Portugal	44	Lebanon	59	Guinea	74	Zambia	89	Afghanistan	104	Georgia
15	Bolivia	30	Czech Republic	45	Israel	60	Gambia	75	Zimbabwe	90	Bhutan	105	Armenia
												106	Azerbaijan
												107	Turkmenistan
												108	Tajikistan
												109	Kyrgyzstan

FIGURE G-9

© H. J. de Blij, P. O. Muller, and John Wiley & Sons, Inc.

21

FIGURE G-10

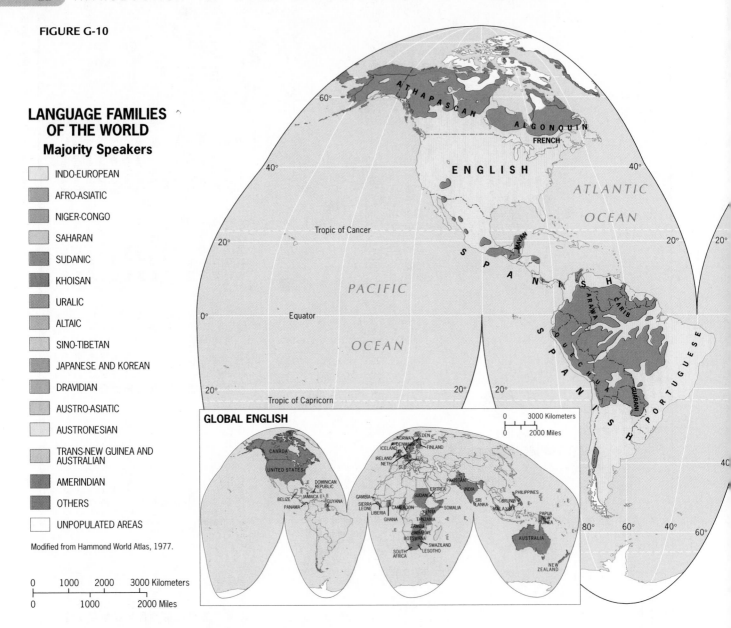

LANGUAGE FAMILIES OF THE WORLD
Majority Speakers

- INDO-EUROPEAN
- AFRO-ASIATIC
- NIGER-CONGO
- SAHARAN
- SUDANIC
- KHOISAN
- URALIC
- ALTAIC
- SINO-TIBETAN
- JAPANESE AND KOREAN
- DRAVIDIAN
- AUSTRO-ASIATIC
- AUSTRONESIAN
- TRANS-NEW GUINEA AND AUSTRALIAN
- AMERINDIAN
- OTHERS
- UNPOPULATED AREAS

Modified from Hammond World Atlas, 1977.

```
0    1000   2000   3000 Kilometers
0         1000      2000 Miles
```

GLOBAL ENGLISH

```
0         3000 Kilometers
0         2000 Miles
```

24 All this is part of central Sudan's rural **cultural landscape**, the distinctive attributes of a society imprinted on its portion of the world's physical stage. The cultural landscape concept was first articulated in the 1920s by a University of California geographer named Carl Ortwin Sauer, who defined it as "the forms superimposed on the physical landscape by the activities of man." In those days "man" was shorthand for "humanity," but there are other reasons to revise Sauer's definition. The cultural landscape consists of more than "forms" such as buildings, gardens, and roads; also part of it are modes of dress, aromas of food, and sounds of music.

Continue your journey southward on the Nile, and you will witness a remarkable transition. Quite suddenly, the square, solid-walled, flat-roofed houses of central Sudan give way to round, wattle-and-thatch, conical-roofed

dwellings of the south. You may note that clouds have appeared in the sky: it rains more here, and flat roofs will not do. The desert has given way to green. Vegetation, natural as well as planted, grows between houses, flanking even the narrow paths. The villages seem less orderly, more varied. People ashore wear a variety of clothes, the women often in colorful dresses, the adult men in shirts and slacks, but shorts when they work the fields, although you see more women wielding hoes than men. You have traveled from one cultural landscape into another, from Arabized, Islamic North Africa to animist or Christian Subsaharan Africa. You have crossed the boundary between two geographic realms.

No geographic realm, not even the Austral Realm, has just one single cultural landscape, but cultural landscapes help define realms as well as regions. The cultural land-

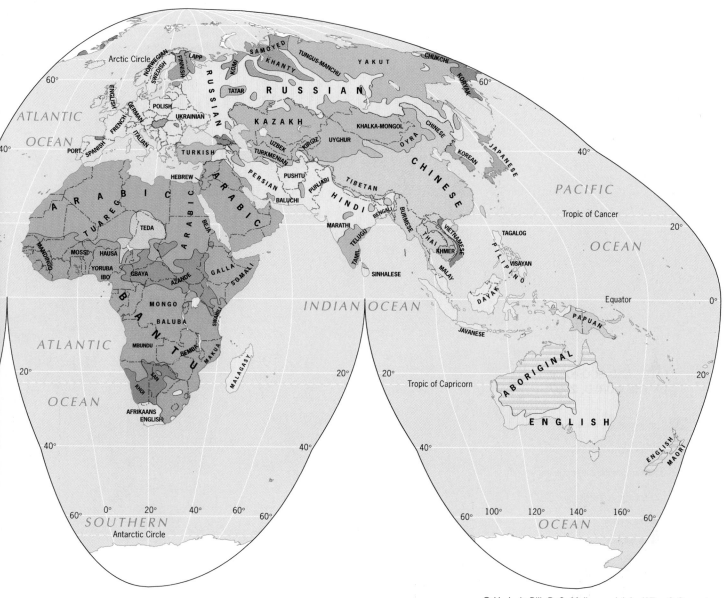

© H. J. de Blij, P. O. Muller, and John Wiley & Sons, Inc.

scape of the high-rise North American city with its sprawling suburbs differs from that of Brazil or East Asia; the meticulously terraced paddies of Southeast Asia are unlike anything to be found in the rural cultural landscape of neighboring Australia. Variations of cultural landscapes *within* geographic realms, such as between highly urbanized and dominantly rural (and more traditional) areas, help us define the world's regions.

The Geography of Language

Language is the essence of culture. People tend to feel passionately about their mother tongue, especially when they believe it is threatened in some way. In the United States today, movements such as "English Only" and "English First" reflect many people's fears that the primacy of English as the national language is under threat as a result of immigration. As we will see in later chapters, some governments try to suppress the languages (and thus the cultures) of minorities in mistaken attempts to enforce national unity, provoking violent reactions.

In fact, languages emerge, thrive, and die out over time, and linguists estimate that the number of lost languages is in the tens of thousands—a process that continues. One year from the day you read this, about 25 more languages will have become extinct, leaving no trace. Some major ones of the past, such as Sumerian and Etruscan, have left fragments in later languages. Others, like Sanskrit and Latin, live on in their modern successors. At present, about 7000 languages remain, half of them classified by linguists as endangered. By

FIGURE G-11

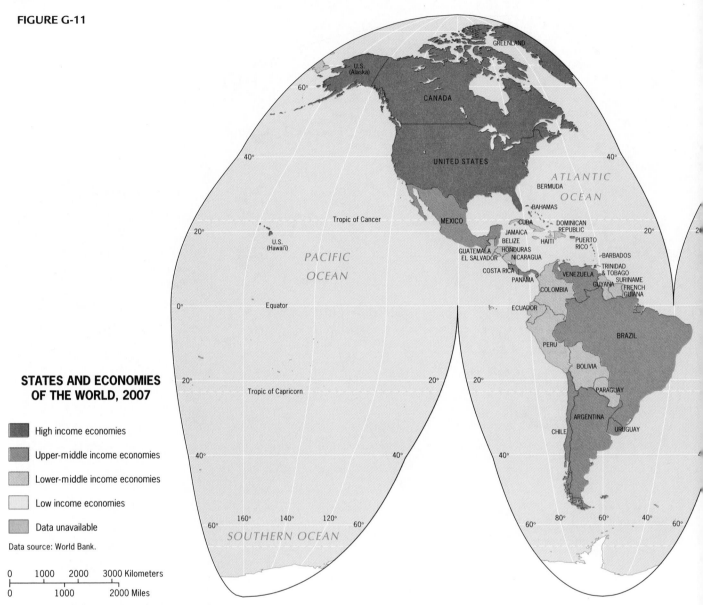

**STATES AND ECONOMIES
OF THE WORLD, 2007**

■ High income economies

■ Upper-middle income economies

☐ Lower-middle income economies

☐ Low income economies

▨ Data unavailable

Data source: World Bank.

0   1000   2000   3000 Kilometers

0   1000   2000 Miles

© H.J. de Blij, P.O. Muller, and John Wiley & Sons, Inc.

the end of this century, the world is likely to be left with just a few hundred languages, so billions of people will no longer be speaking their ancestral mother tongues.

Scholars have tried for many years to unravel the historic roots and branches of the "language tree," and their debates continue. Geographers trying to map the outcome of this research keep having to modify the pattern, so you should take Figure G-10 as a work in progress, not the final word. At minimum, there are some 15 so-called *language families*, groups of languages with a shared but usually distant origin. The most widely distributed language family, the Indo-European (shown in yellow on the map), includes English, French, Spanish, Russian, Persian, and Hindi. This encompasses the languages of European colonizers that were carried and implanted worldwide, English most of all. Today, English serves as

the national or official language of many countries and outposts and remains the *lingua franca* of government, commerce, and higher education in many multicultural societies (Fig. G-10 inset map). In the postcolonial era, English became the chief medium of still another wave of ascendancy now in progress: globalization.

But even English is going the way of Latin, morphing into versions you will hear (and learn to use) as you travel, forms of English that may, generations from now, be the successors that Italian and Spanish are to Latin. In Hong Kong, Chinese and English are producing a local "Chinglish" you may hear in the first taxi you enter. In Lagos, Nigeria, where most of the people are culturally and ethnically Yoruba, a language called "Yorlish" is emerging. No map can keep up with the continuing evolution of language.

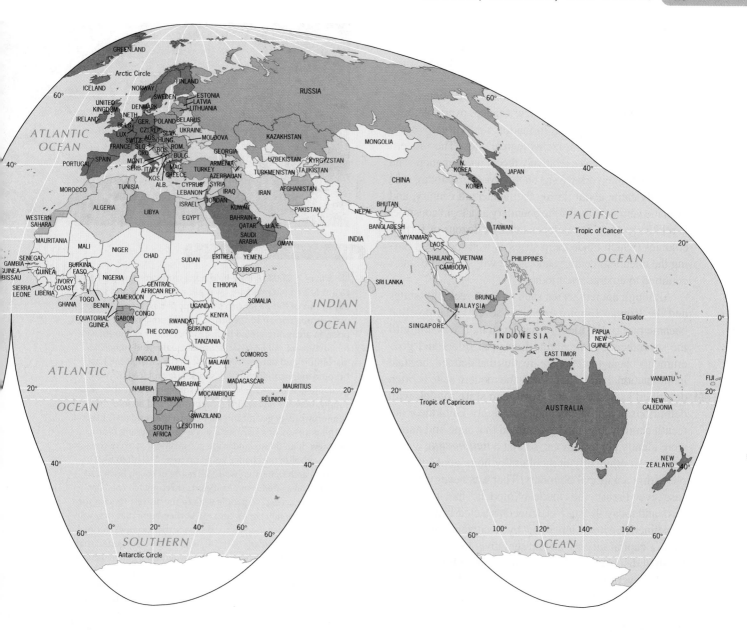

REALMS, REGIONS, AND STATES

Our analysis of the world's regional geography requires data, and it is crucial to know the origin of these data. Unfortunately, we do not have a uniformly-sized grid to superimpose over the globe: we must depend on the world's nearly 200 countries to report vital information. Irregular as the boundary framework shown on the world map (see the one inside the book's front cover or Fig. G-11) may be, it is all we have. Fortunately, all large and populous countries are subdivided into States, provinces, or other major internal entities, and their governments provide information on each of these subdivisions when they periodically conduct their census.

The State

As we noted at the beginning of this introductory chapter, although we often refer to the political entities on the world map as *countries*, the appropriate geographic term for them is *states*. To avoid confusion with subdivisions under the same rubric, we capitalize internal States such as the U.S. State of Pennsylvania or Nigeria's Cross River State.

As a political, social, and economic institution, the **state** has been developing for thousands of years, ever since farm surpluses made possible the growth of large and prosperous towns that could command hinterlands and control peoples far beyond their protective walls. Eventually, clusters of towns banded together and expanded their power, one among them grew larger than the others and became the capital and seat of authority, and if conditions of productivity and security permitted, a state was in the making. The process occurred almost simultaneously in several areas of the world, and some ancient states proved remarkably durable.

Environmental changes, invasions, migrations, epidemics, conflicts, and other circumstances overtook those early experiments in state formation, and virtually nothing survives, except in archeological sites. The modern state as we know it is a relatively recent phenomenon. The boundary framework we see on the world map today substantially came about during the nineteenth century, and even the very concept of the boundary as a legal construct is relatively new. The independence of dozens of former colonies (which made them states as well) occurred during the twentieth century, and as we will see, some states continue to emerge even today.

Although ancient states such as the Greek city-states and the Roman Empire exhibited several qualities that became part of the modern state system, it was not until the seventeenth century that European rulers and governments began to negotiate treaties that defined the state in international law. That is why the modern state is often described as based on the **European state model**, with definitions of nationality and sovereignty. Until the nineteenth century, this model assumed that state and nation were ideally conterminous, so that a *nation-state* would enclose an ethnically and culturally homogeneous people within a national boundary. That was never truly the case (even France, the "model of models," had its minorities), and today the state is defined as a clearly and legally defined territory inhabited by a citizenry governed from a capital city by a representative government. As we shall discover in Chapter 1, not even in Europe itself are all governments truly representative, but the European state model has become the aspiration for the world.

Power and the State

As we will observe in the pages that follow, states vary not only in terms of their dimensions, domestic resource base, productive capacity, and other physical and cultural properties, but also in terms of their influence in world affairs. That influence may not seem proportional to these properties: some relatively small countries carry a lot of weight in international affairs, while other larger and more populous ones are much less influential. The United Kingdom or UK (Great Britain plus Northern Ireland), for instance, is about the size of the U.S. State of Oregon and has a population smaller than a half-dozen of the States of India, but on the global stage it is a major player and its capital, London, is one of the world's leading cities. Russia, on the other hand, has more than twice the population of the UK and is nearly double the territorial size of the United States—but its role in the world has far less impact than Britain's.

FROM THE FIELD NOTES

"February 1, 2003. A long-held hope came true today: thanks to a Brazilian intermediary I was allowed to enter and spend a day in two of Rio de Janeiro's hillslope *favelas*, an eight-hour walk through one into the other. Here live millions of the city's poor, in areas often ruled by drug lords and their gangs, with minimal or no public services, amid squalor and stench, in discomfort and danger. And yet life in the older *favelas* has become more comfortable as shacks are replaced by more permanent structures, electricity is sometimes available, water supply, however haphazard, is improved, and an informal economy brings goods and services to the residents. I stood in the doorway of a resident's single-room dwelling for this overview of an urban landscape in transition: satellite-television disks symbolize the change going on here. The often blue cisterns catch rainwater; walls are made of rough brick and roofs of corrugated iron or asbestos sheeting. No roads or automobile access, so people walk to the nearest road at the bottom of the hill. Locals told me of their hope that they will some day have legal rights to the space they occupy. During his campaign for president of Brazil, Lula da Silva suggested that long-term inhabitants should be awarded title, and in 2003 his government approved the notion. It will be complicated: as the photo shows, people live quite literally on top of one another, and mapping the chaos will not be simple (but will be made possible with geographic information systems). This would allow the government to tax residents, but it would also allow residents to obtain loans based on the value of their *favela* properties, and bring millions of Brazilians into the formal economy. The hardships I saw on this excursion were often dreadful, but you could sense the hope for and anticipation of a better future."   © H. J. de Blij

www.conceptcaching.com

Concept Caching

Geography has much to do with such circumstances, but the key lesson is this: in the world of today, for all our efforts to cooperate diplomatically (the United Nations), economically (the European Union), strategically (the North Atlantic Treaty Organization), and in other ways, it is the state and its government—not regions or realms—that holds the *power* and makes decisions in the global arena.

Defining power is a difficult matter because power takes various forms and is exercised in different ways. A state's power may be defined as its capacity to use its human and natural resources to influence the behavior of other states, but there are always constraints: despite its superpower status, the United States could not influence a large number of state governments to join its war on Saddam Hussein's Iraq. And notwithstanding its unrepresentative regime, its moribund economy, and its physical and resource limitations, the small state of North Korea managed to mount a military threat complete with missiles and nuclear arms that alarmed the world and mobilized six powers in response. There was a time when soldiers, armaments, and reserves measured power. Today, weapons of mass destruction, oil, and rogue regimes create a very different measure of the power potential of states.

States and Realms

Realms are assemblages of regions, and regions are clusters of states. As Figures G-2 and G-11 indicate, the borders between geographic realms frequently coincide with boundaries between individual countries—for example, between North America and Middle America along the U.S.-Mexico boundary.

But a realm boundary can also cut across states, as does the one between Subsaharan Africa and the Muslim-dominated realm we call North Africa/Southwest Asia. Here the border between realms takes on the properties of a transition zone, fragmenting states such as Chad and Sudan. Along the border of the Russian realm, similar transitions mark Kazakhstan in Asia and Belarus in Europe. Countries that lie in such transition zones experience strongly divisive forces.

States and Regions

The global boundary framework is even more useful in delimiting regions *within* geographic realms. We shall dis-

cuss regional divisions every time we introduce a geographic realm, but an example is appropriate here. In the Middle American realm, we recognize four regions. Two of these lie on the mainland: Mexico, the giant of the realm, and Central America, which consists of the seven comparatively small countries located between Mexico and the Panama-Colombia border (which is the boundary with the South American realm). Central America often is misdefined in news reports; the correct regional definition is based on the politico-geographical framework.

Political Geography

To our earlier criteria of physical geography, population distribution, and cultural geography, therefore, we now add political geography as a determinant of world-scale geographic regions. In doing so, we should be aware that the global boundary framework continues to change and that boundaries are created (as in 1993 between the Czech Republic and Slovakia) as well as eliminated (e.g., between former West and East Germany in 1990). But the overall system, much of it resulting from colonial and imperial expansionism, has endured, despite the predictions of some geographers that the "boundaries of imperialism" would be replaced by newly negotiated ones in the postcolonial period.

PATTERNS OF ECONOMIC DEVELOPMENT

Finally, as we prepare for our study of world regional geography, it is all too clear that realms, regions, and states do not enjoy the same level of prosperity. The field of **economic geography** focuses on spatial aspects of the   27

The Gini Coefficient

ECONOMIC GEOGRAPHERS study regional disparities and their causes, and one of these causes lies in variations of income. But it can be difficult to determine who is earning how much across sizeable populations. Enter Corado Gini, an Italian statistician, who made pioneering contributions—including a mathematical formula to measure the degree of dispersion of a phenomenon through a population, including economic gains. His name is forever linked to an index that reveals what proportion of a population is sharing in the wealth, and who is not.

This index ranges from 0.0 (no differences at all; everyone earns the same amount) to 1.0 (one earner takes all). A country in which a few tycoons control all the wealth and everyone else labors for a pittance will have a "GC" of close to 1; but a country with a more equitable spread of income will be much closer to 0.

As important as the actual number is the way the GC is changing. When China was under strict communist rule and before its modern economic boom began, its GC was low (nobody, of course, actually measured it). By 1993, however, it was reported to be 0.41, and today it is approaching 0.5. China's incomes are increasingly concentrated in the country's wealthier Pacific Rim. India's GC, probably already underestimated at 0.38, may even be rising faster than China's. But Brazil, long exhibiting one of the world's highest (if not the highest) GC approaching 0.6, is showing signs of a decrease partly as a result of social programs we discuss in Chapter 5. The GC for some states is unavailable, even as an estimate. Certain governments prefer not to let the Gini out of the bottle.

A Core-Periphery World

IF REGIONS AND STATES today are marked by comparatively prosperous cores and poorer peripheries, what does this portend for the future? Are the old rules about exploiting cores and exploited peripheries giving way to a "flatter" and fairer world? Some analysts of world affairs think so: Thomas Friedman's popular book, *The World is Flat*, argues that the world is becoming so mobile, so interconnected, and so integrated that historic barriers are falling, including those separating cores from peripheries. Interaction, say these observers, is now global, ever-freer trade is the new rule, and the flow of people, ideas, money, and jobs accelerates by the day. Core-periphery notions are becoming irrelevant as contrasts between them fade. "Geography is history," some of them like to say.

But is it? If the world is to become "flatter," and if regional disparities and income inequalities are to decline, it is going to take more than flowing money and freer trade (some of the richer states that advocate free trade protect their own inefficient industries, farm as well as factory, against foreign competition). It will require changes in attitude as well. Core-area residents are used to their advantages and comforts, and in general they guard against losing them. Not only do they take a dim view of cheap imports that threaten local industries; they want to control immigration, and they tend to resist cultural infusions in the form of unfamiliar languages and unconventional religions.

Looking at it this way, even the world as a whole has a core-periphery dichotomy that counters notions of a global flattening (Fig. G-12). Note that we have shifted the perspective from that used on our other world maps, and the best way to see what Figure G-12 reveals is to outline it on a globe. The *global core*, anchored by North America and flanked by Europe to the east and Japan and Australia to the west, not only constitutes an assemblage of the most affluent states and the most prosperous cities, but also hosts the bulk of the financial and corporate empires that drive economic globalization. As mapped in Figure G-12, the core contains about 15 percent of the world's population, but that population earns some 75 percent of total annual income. The few core-like outliers in the periphery—for example, Singapore, Dubai (United Arab Emirates)—barely affect this disproportionate picture.

Naturally, the global core is a magnet for migrants. Millions of workers hope to leave the periphery, which contains the world's poorest regions, to seek a better life somewhere in the core. Many of them die every year in the waters of the Mediterranean, the Caribbean, and the Atlantic. Others risk their lives at barriers and barricades that encircle the global core, from the "security fence" between Mexico and the United States to the walls that guard Israel's safety to the razor wire that encircles Spain's outposts on North Africa's shores. Perhaps the most vivid expression of core-periphery contrast straddles the DMZ

ways people make their living and deals with patterns of production, distribution, and consumption of goods and services. As with all else in this world, these patterns reveal much variation. Individual states report the nature and value of their imports and exports, farm and factory output, and many other economic data to the United Nations and other international agencies. From such information, economic geographers can measure the comparative well-being of the world's countries (see box titled "The Gini Coefficient" on p. 27). The concept of **development** is used to gauge a state's economic, social, and institutional growth.

A Caution

The concept of development as measured by data that reflect totals and averages for entire national populations entails some risks of which we should be aware from the start. When a state's economy is growing as a whole, and even when it is "booming" by comparison to other states, this does not automatically mean that every citizen is better off and the income of every worker is rising. Averages have a way of concealing region-

al variability and local stagnation. In very large states such as India and China, it is useful to assess regional, provincial, and even local economic data to discover to what extent the whole country is sharing in "development." In the case of India, we should all know that the State of Maharashtra (containing the largest city, Mumbai) is far in the lead when it comes to its share of the national economy. In China, the coastal provinces of the Pacific Rim far outstrip those of the interior. In Spain, the people of the Autonomous Community of Catalonia (focused on Barcelona) never tire of telling you that theirs is the most productive entity in the whole country. The question is: to what extent do these economic successes better the lives of all of the state's inhabitants? That information is not as easily obtained as are those "national" averages.

Realms of Development

Various schemes to group the world's states into economic-geographic categories have come and gone, and others will probably arise in the future. For our pur-

FIGURE G-12   © H.J. de Blij, P.O. Muller, and John Wiley & Sons, Inc.

on the Korean Peninsula (see Chapter 9). Look at the map above carefully and you can probably find a dozen places where, in the recent past or the present, efforts have been made to inhibit migration into the core, including the waters between Cuba and Florida, the "green line" across Cyprus, and the Arafura Sea north of Australia.

And it has worked pretty well. The immigration issue roils U.S. politics today, but globally, according to UN estimates, only about 3 percent of the world's population live in a country other than the one in which they were born. The global core can be seen as the ultimate geographic expression of a phenomenon as old has human avarice.

poses, the classification scheme used by the World Bank (one of the agencies that monitor economic conditions across the globe) is the most effective. It sorts countries into four categories based on the success of their economies: (1) high-income, (2) upper-middle-income, (3) lower-middle-income, and (4) low-income. These categories, when mapped, display interesting regional clustering (Fig. G-11). Compare this map to our global framework (Fig. G-2), and you can see the role of economic geography in the layout of the world's geographic realms. Also evident are regional boundaries within realms—for instance, between Brazil and its western neighbors, between South Africa and most of the rest of Subsaharan Africa, and between west and east in Europe.

Economic geography is not the whole story, but along with factors of physical geography (such as climate), cultural geography (including resistance or receptivity to change and innovation), and political geography (history of colonialism, growth of democracy), it plays a powerful role in shaping our variable world.

Cores and Peripheries

It has been obvious for a very long time that human success on the Earth's surface has focused on certain areas and bypassed others. The earliest cities and states of the Fertile Crescent, the empires of the Incas and the Aztecs, the dominance of ancient Rome, and many other hubs of activity tell the story of development and decay, of growth and collapse. In their heyday, such centers of authority, innovation, production, and expansion were the earliest **core areas**, places of dominance whose **29** inhabitants exerted their power over their surroundings near and far.

Such core areas grew rich and, in many cases, endured for long periods because their occupants skillfully exploited these surroundings—controlling and taxing the local population, forcing workers to farm the land and mine the resources at their command. This created a **periphery** that sustained the core for as long as the system lasted, so that core-periphery interactions, one-sided **30** though they were, created wealth for the former and enforced stability in the latter.

The sociologist Immanuel Wallerstein explored core-periphery relationships in his book *World Systems Theory*, pointing out that core areas need peripheries in order to stay in business (as it were), so that exploitation—of the periphery by the core—was and remains the norm. The nature of this exploitation has changed over time and with changing technologies: the peripheries of the earliest core areas were local, but that of Rome was regional, and today a case can be made that the entire world forms a core-periphery framework (see box titled "A Core-Periphery World").

Core and Scale

The core-periphery phenomenon, therefore, exists at all levels of generalization, from the local to the global scale. Except for a few special cases, all states contain core areas. These national core areas are often anchored by the capital and/or largest city in the country, for example Paris (France), Tokyo (Japan), Buenos Aires (Argentina), Bangkok (Thailand). Larger countries may have more than one core area, such as Australia with its East and West Coast core areas and intervening periphery. Even individual States such as Michigan or Mississippi have core-periphery properties. In Michigan, the State core area lies in the southeast, centered on Detroit; its periphery lies in the northwest, especially in its distant Upper Peninsula, exploited for its natural resources.

The colonial era created a new kind of core area, where the colonizing power took environmentally favored land, started farms and plantations, and built colonial headquarters that grew into major cities. At the time, few imagined that those colonial cities would some day become the capitals of independent countries, but by the time that occurred, core-periphery relationships had developed that outlasted the colonial era. Cities such as Nairobi (Kenya) and Harare (Zimbabwe) house privileged elites whose relationships with the peoples in their peripheries are sometimes even more exploitive than they were during colonial times.

So uneven is the landscape of economic geography, at every scale, that those data on national income averages are anything but reliable when it comes to the average worker. A much more reliable index would be one
31 that reports **regional disparities**, because even high-income countries still have areas of poverty, while low-income countries have given rise to bustling, skyscrapered cities whose streets are jammed with luxury automobiles and shops selling opulent goods.

Globalization

If a geographic concept can arouse strong passions,
32 **globalization** is it. To leading economists, politicians, and businesspeople, this is the best of all possible worlds: the march of international capitalism, open markets, and free trade. To millions of poor farmers and powerless citizens in debt-mired African and Asian countries, it represents a system that will keep them in poverty and subservience forever. Economic geographers can prove that global economic integration allows the overall economies of poorer countries to grow faster: compare their international trade to their national income, and you will find that the *gross national income (GNI)** of those that engage in more international trade (and thus are more "globalized") rises, while the GNI of those with less international trade actually declines. The problem is that those poorest countries contain more than 2 billion people, and their prospects in this globalizing world are worsening, not rising. In countries like the Philippines, Kenya, and Nicaragua, income per person has shrunk, and there globalization is seen as a culprit, not a cure.

One-Way Street?

Globalization in the economic sphere is proceeding under the auspices of the World Trade Organization (WTO), of which the United States is the leading architect. To join, countries must agree to open their economies to foreign trade and investment and to adhere to a set of economic rules discussed annually at ministerial meetings. As of mid-2009, the WTO had 153 member-states, all expecting benefits from their participation. But the driving forces of globalization, including European countries and Japan, did not always do their part when it came to giving the poorer countries the chances the WTO seemed to promise. The case of the Philippines is often cited: joining the WTO, its government assumed, would open world markets to Filipino farm products, which can be brought to those markets cheaply because farm workers' wages in the Philippines are very low. Economic planners forecast a huge increase in farm jobs, a slow rise in wages, and a major expansion in produce exports. Instead, they found themselves competing against North American and European farmers who receive subsidies toward the production as well as the export of their products—and losing out. Meanwhile, low-priced, subsidized U.S. corn appeared on Filipino markets. As a result, the Philippine economy lost several hundred thousand farm jobs, wages went down, and WTO membership had the effect of severely damaging its agricultural

*Gross national income (GNI) is the total income earned from all goods and services produced by the citizens of a country, within or outside of its borders, during a calendar year. *Per capita GNI* is a widely used indicator of the variation of spendable income around the globe, and is reported for each country in the world in the farthest-right column of the Data Table at the very end of the book.

sector. Not surprisingly, the notion of globalization is not popular among rural Filipinos.

This story has parallels among banana and sugar growers in the Caribbean, cotton and sugar producers in Africa, and fruit and vegetable farmers in Mexico. Economic globalization has not been the two-way street most WTO members, especially the poorer countries, had hoped. Nor is rich-country protectionism confined to agriculture. When cheap foreign steel began to threaten what remained of the American steel industry in 2001, the U.S. government erected tariffs to guard domestic producers against unwanted competition.

An Open World

When meetings of economic ministers take place in major cities, they attract thousands of vocal and sometimes violent opponents who demonstrate and occasionally riot to express their opposition (see photo p. 32). This happens because globalization is seen as more than an economic strategy to perpetuate the advantages of rich countries: it is also viewed as a cultural threat. Globalization constitutes Americanization, these protesters say, eroding local traditions, endangering moral standards, and menacing the social fabric. Such views are not confined to the disadvantaged, poorer countries still losing out in our globalizing world: in France the leader of a so-called Peasant Federation became a national hero after his sympathizers destroyed a McDonald's restaurant in what he described as an act of cultural self-defense. McDonald's restaurants, he argued, promote the consumption of junk food over better, more traditional fare. Elsewhere the target is violent American movies or the lyrics of certain American popular music.

Nevertheless, the process of globalization—in cultural as well as economic spheres—continues to change our world, for better and for worse. (In the United States, opponents of globalization bemoan the loss of American jobs to foreign countries where labor is cheaper.) In theory, globalization breaks down barriers to international trade, stimulates commerce, brings jobs to remote places, and promotes social, cultural, political, and other kinds of exchanges. High-tech workers in India are employed by computer firms based in California. Japanese cars are assembled in Thailand. American shoes are made in China. Fast-food restaurant chains spread standards of service and hygiene as well as familiar (and standard) menus from Tokyo to Tel Aviv. If wages and standards of employment are lower in newly involved countries than in the global core, the gap is shrinking, employment conditions are more open to scrutiny, and global openness is a byproduct of globalization.

FROM THE FIELD NOTES

"It was an equatorial day here today, in hot and humid Singapore. Walked the three miles to the Sultan Mosque this morning and observed the activity arising from the Friday prayers, then sat in the shade of the large ficus tree on the corner of Arab Street and watched the busy pedestrian and vehicular traffic. An Indian man saw me taking notes and sat down beside me. For some time he said nothing, then asked whether I was writing a novel. 'If only I could!' I answered, and explained that I was noting locations and impressions for photos and text in a geography book. He said that he was Hindu but liked living in this area because some of what he called 'Old Singapore' still survived. 'No problems between Hindus and Muslims here,' he said. 'The police maintain strict security here around the mosques, especially after September 2001. The government tolerates all religions, but it doesn't tolerate religious conflict.' He offered to show me the Hindu shrine where he and his family joined others from Singapore's Hindu community. As we walked down Arab Street, he pointed skyward. 'That's the future,' he said. 'The symbol of globalization. Do you mention it in your book? It's in the papers every day, supposedly a good thing for Singapore. Well, maybe. But a lot of history has been lost to make space for what you see here.' I took the photo, and promised to mention his point."   © H. J. de Blij

POINTS TO PONDER

- Around the middle of 2012, the global human population will reach and exceed 7 billion.

- As the second decade of our new century opened, nearly one billion people continued to survive on less than one American dollar a day.

- The planet appears to be entering a period of increasing environmental fluctuation, posing growing threats to future economic and political stability.

- Globalization is glorified as well as denigrated, but its march is inexorable. Those who support it say that its "rising tide lifts all boats."

A New Revolution

We should be aware that the current globalization process is a revolution—but also that it is not the first of its kind. The first "globalization revolution" occurred during the nineteenth and early twentieth centuries, when Europe's colonial expansion spread ideas, inventions, products, and habits around the world. Colonialism transformed the world as the European powers built cities, transport networks, dams, irrigation systems, power plants, and other facilities, often with devastating impact on local traditions, cultures, and economies. From goods to games (soap to soccer) people in much of the world started doing similar things. The largest of all colonial empires, that of Britain,

made English a worldwide language (see Fig. G-10 inset map), a key element in the current, second globalization revolution.

The current globalization is even more revolutionary than the colonial phase because it is driven by more modern, higher-speed communications. When the British colonists planned the construction of their ornate Victorian government and public buildings in Bombay (now called Mumbai), the architectural drawings had to be prepared in London and sent by boat to India. When the Chinese government in the 1990s decided to create a Manhattan-like commercial district on the riverfront in Shanghai, the plans were drawn in the United States, Japan, and Western Europe and transmitted to Shanghai via the Internet. Today, one container ship carrying products from China to the U.S. market hauls more cargo than a hundred colonial-era boats. And, as the chapters that follow will show frequently, the world's national political boundaries are becoming increasingly porous. Economic alliances enable manufacturers to send raw materials and finished products across borders that once inhibited such exchanges. Groups of countries forge economic unions whose acronyms (NAFTA, Mercosur) stand for freer trade. The ultimate goal of the World Trade Organization is to lower remaining barriers the world over, thereby boosting not just regional commerce but also global trade.

The Future

As with all revolutions, the overall consequences of the ongoing globalization process are uncertain. Critics underscore that one of its outcomes is a growing gap

This edition of *Regions* appears at a time when the global economy is in severe recession and debate rages over the merits (and demerits) of globalization, American-model capitalism, European-style socialism, and other systems of economics and governance. As the crisis deepened, leaders of the so-called Group of 20 Countries, comprising the world's largest and most influential economies, met in London—one of the "capitals of globalization"—in April 2009 to coordinate their responses. But thousands of protestors, including anti-capitalist, anti-war, and environmental activists, took to the streets of London, blocking traffic, and kindling fierce violence in the city's financial district. Barricades around bank buildings and angry slogans on banners made London the hub of global discord.
© Zhang Yuwei/Xinhua/Landov.

between rich and poor, a polarization of wealth that will destabilize the world. Core-periphery contrasts are intensified, not lessened, by globalization as the poor in peripheral societies are exploited by core-based corporations. Proponents argue that, as with the Industrial Revolution, it will take time for the benefits to spread— but that globalization's ultimate effects will be advantageous to all. Indeed, the world is functionally shrinking, and we will find evidence for this throughout the book. But the "global village" still retains its distinctive neighborhoods, and two revolutionary globalizations have failed to erase their particular properties. In the chapters that follow we use the vehicle of geography to visit and investigate them.

THE REGIONAL FRAMEWORK

At the beginning of this Introduction, we outlined a map of the great geographic realms of the world (Fig. G-2). We then addressed the task of dividing these realms into regions, and we used criteria ranging from physical geography to economic geography. The result is Figure G-13. Before we begin our survey, here is a summary of the 12 geographic realms and their regional components.

Europe (1)

Territorially small and politically fragmented, Europe remains disproportionately influential in global affairs. A key component of the global core, Europe today consists of two regions: a mainland and offshore Core in the west, and a wide Periphery curving across the realm's southern, eastern, and far northern domains.

Russia (2)

Territorially enormous and politically unified, Russia was the dominant force in the former Soviet Union that disbanded in 1991. Undergoing a difficult transition from dictatorship to democracy and from communism to capitalism, Russia is geographically complex and changing. We define four regions: the Russian Core and Peripheries, the Eastern Frontier, Siberia, and the Far East.

North America (3)

Another realm in the global core, North America consists of the United States and Canada. We identify nine regions: the North American Core, the Maritime Northeast, French Canada, the South, the Southwest, the Pacific Hinge, the Western Frontier, the Continental Interior, and the Northern Frontier. Five of these regions extend across the U.S.-Canada border.

Middle America (4)

Nowhere in the world is the contrast between the global core and periphery as sharply demarcated as it is between North and Middle America. This small, fragmented realm clearly divides into four regions: Mexico, Central America, and the Greater Antilles and Lesser Antilles of the Caribbean Basin.

South America (5)

The continent of South America also defines a geographic realm in which Iberian (Spanish and Portuguese) influences dominate the cultural geography but Amerindian imprints survive. We recognize four regions: the North, composed of Caribbean-facing states; the Andean West, with its strong Amerindian influences; the Southern Cone; and Brazil, the realm's giant.

Subsaharan Africa (6)

Between the African Transition Zone in the north and the southernmost Cape of South Africa lies Subsaharan Africa. The realm consists of five regions: Southern Africa, East Africa, Equatorial Africa, West Africa, and the African Transition Zone.

North Africa/Southwest Asia (7)

This vast geographic realm has several names, extending as it does from North Africa into Southwest and, indeed, Central Asia. Some geographers call it *Naswasia* or Afrasia. There are six regions: Egypt and the Lower Nile Basin plus the Maghreb in North Africa, the Middle East, the Arabian Peninsula, the Empire States, and Turkestan in Southwest Asia.

South Asia (8)

Physically, South Asia is one of the most clearly defined geographic realms, but has a complex cultural geography. It consists of five regions: India at the center, Pakistan to the west, Bangladesh to the east, the mountainous North, and the southern region that includes the islands of Sri Lanka and the Maldives.

East Asia (9)

The vast East Asian geographic realm extends from the deserts of Central Asia to the tropical coasts of the South

FIGURE G-13

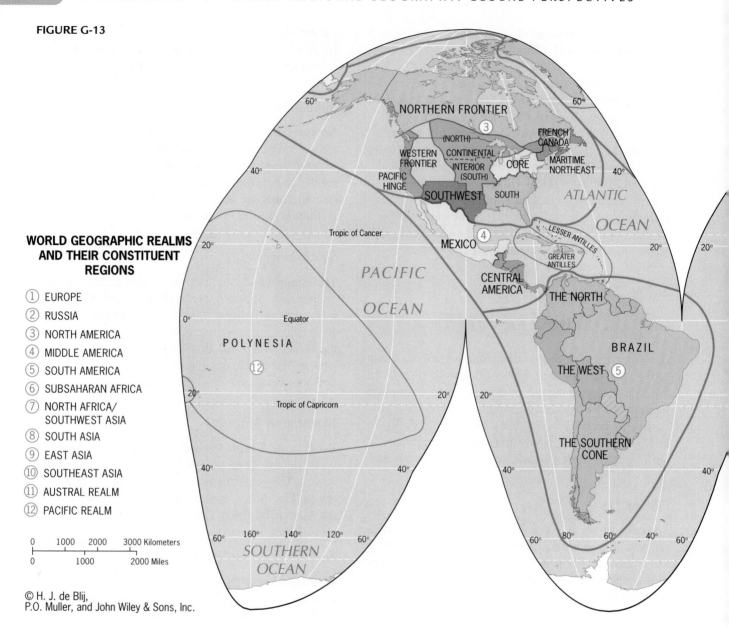

WORLD GEOGRAPHIC REALMS AND THEIR CONSTITUENT REGIONS

① EUROPE
② RUSSIA
③ NORTH AMERICA
④ MIDDLE AMERICA
⑤ SOUTH AMERICA
⑥ SUBSAHARAN AFRICA
⑦ NORTH AFRICA/ SOUTHWEST ASIA
⑧ SOUTH ASIA
⑨ EAST ASIA
⑩ SOUTHEAST ASIA
⑪ AUSTRAL REALM
⑫ PACIFIC REALM

© H. J. de Blij,
P.O. Muller, and John Wiley & Sons, Inc.

China Sea and from Japan to the Himalayan border with India. We identify five regions: China Proper, including North Korea; Xizang (Tibet) in the southwest; desert Xinjiang in the west; Mongolia in the north; and the *Jakota Triangle* (*Ja*pan, South *Ko*rea, and *Ta*iwan) in the east.

Southeast Asia (10)

Southeast Asia is a varied mosaic of natural landscapes, cultures, and economies. Influenced by India, China, Europe, and the United States, it includes dozens of religions and hundreds of languages plus economies representing both the global core and periphery. Physically, Southeast Asia consists of a broad peninsular mainland

and an offshore arc consisting of thousands of islands. The two regions (Mainland and Insular) are based on this distinction.

Austral Realm (11)

Australia and its neighbor New Zealand form the Austral geographic realm by virtue of continental dimensions, insular separation, and predominantly Western cultural heritage. The regions of this realm are defined by physical as well as cultural geography: in Australia, a highly urbanized, two-part core and a vast, desert-dominated interior; and in New Zealand, two main islands that exhibit considerable geographic contrast.

Pacific Realm (12)

The vast Pacific Ocean, larger than all the landmasses combined, contains tens of thousands of islands large and small. Dominant cultural criteria warrant three regions: Melanesia, Micronesia, and Polynesia.

THE PERSPECTIVE OF GEOGRAPHY

As this introductory chapter demonstrates, our world regional survey is no mere description of places and areas. We have combined the study of realms and regions with a look at geography's ideas and concepts—the notions, generalizations, and basic theories that make the discipline what it is. We continue this method in the chapters ahead so that we will become better acquainted with the world and with geography. By now you are aware that geography is a wide-ranging, multifaceted discipline. It is often described as a social science, but that is only half the story: in fact, geography straddles the divide between the social and the physical (natural) sciences. Many of the ideas and concepts you will encounter have to do with the multiple interactions between human societies and natural environments.

33 **Regional geography** allows us to view the world in an all-encompassing way. As we have seen, regional geography borrows information from many sources to create an overall image of our divided world. Those sources are not random but constitute topical or 34 **systematic geography**. Research in the systematic fields of geography makes our world-scale generalizations possible. As Figure G-14 shows, these systematic fields relate closely to those of other disciplines. Cultural geography, for example, is allied with anthropology and sociology; it is the spatial perspective that distinguishes cultural geography. Economic geography focuses on the spatial dimensions of economic activity; political geography concentrates on the spatial imprints of political behavior. Other systematic fields include historical, medical, and behavioral geography (as well as environmental, agricultural, coastal, and additional branches of geography not shown in Fig. G-14). Moreover, we will also draw on information from biogeography, marine geography, population geography, and climatology (as we did earlier in this chapter).

These systematic fields of geography are so named because their approach is global, not regional. Take the geographic study of cities, urban geography. Urbanization is a worldwide process, and urban geographers can identify certain human activities that all cities in the world exhibit in one form or another. But cities also display regional properties. The typical Japanese city is quite distinct from, say, the African city. Regional

THE RELATIONSHIP BETWEEN REGIONAL AND SYSTEMATIC GEOGRAPHY

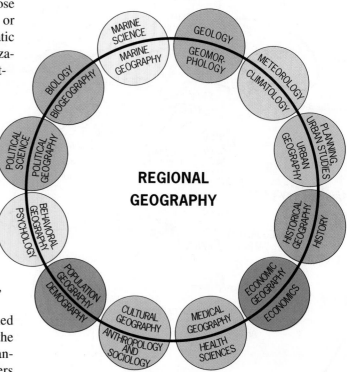

© H. J. de Blij, P.O. Muller, and John Wiley & Sons, Inc.

FIGURE G-14

What Do Geographers Do?

A SYSTEMATIC SPATIAL perspective and an interest in regional study are the unifying themes and enthusiasms of geography. Geography's practitioners include physical geographers, whose principal interests are the study of geomorphology (land surfaces), research on climate and weather, vegetation and soils, and the management of water and other natural resources. There also are geographers whose research and teaching concentrate on the ecological interrelationships between the physical and human worlds. They study the impact of humankind on our globe's natural environments and the influences of the environment (including such artificial contents as air and water pollution) on human individuals and societies.

Other geographers are regional specialists, who often concentrate their work for governments, planning agencies, and multinational corporations on a particular region of the world. Still other geographers—who now

constitute the largest group of practitioners—are devoted to topical or systematic subfields such as urban geography, economic geography, and cultural geography (see Fig. G-14 above). They perform numerous tasks associated with the identification and resolution (through policy-making and planning) of spatial problems in their specialized areas. And, as in the past, there remain many geographers who combine their fascination for spatial questions with technical expertise.

Computerized cartography, geographic information systems (GIS), remote sensing, and even environmental engineering are among the specializations listed by the 10,000-plus professional geographers of North America. In Appendix C, you will find considerable information on the discipline, how one trains to become a geographer, and the many exciting career options that are open to the young professional.

geography, therefore, borrows from the systematic field of urban geography, but it injects this regional perspective.

In the following chapters we call upon these systematic fields to provide us with a better understanding of the world's realms and regions. As a result, you will gain insights into the discipline of geography as well as the regions we investigate. This will prove that geography is a relevant and practical discipline when it comes to comprehending, and coping with, our fast-changing world (see box titled "What Do Geographers Do?" and Appendix C for a discussion of career opportunities).

What You Can Do

RECOMMENDATION: If you are not already doing so, please allow us to urge you to begin keeping a diary. Several decades from now, the most interesting and perhaps even useful parts of your diary may not be time and place, but your feelings and opinions about events and experiences you are having and will have over the years to come. What is important enough to warrant recording your own views? The intervention in Afghanistan? The worldwide economic recession that began in 2008? The ever-widening impact of globalization? The effect on you of an unusually stimulating course? The comments of a particularly opinionated visiting speaker? Many years from now you may be surprised about your concerns and perceptions in these early years of our new century.

GEOGRAPHIC CONNECTIONS

1 If you were asked by someone without a geographic background to define the geographic characteristics of your home region, how would you respond? Where do you perceive its boundaries to lie, on what criteria do you base these, and are they sharp or transitional? (You will be tempted to draw a sketch map to support this part of your answer.) Where is your home with respect to the urban framework in your region, and do you consider yourself a resident of a city, suburbia, exurbia, or a rural hinterland? What elements of the formal and the functional can you discern in your regional experience? How would you summarize your region's natural environment(s), and, in your lifetime, have you noticed changes that might be considered trends? If so, how have these affected you?

2 Imagine that you are working in your university's study-abroad office. Students will be planning to study for a semester in numerous countries, and they know that you have some geographic experience. How would you answer their questions about the climate and weather they are likely to experience in those countries, using Figure G-7? What should a student from Vancouver in western Canada's British Columbia expect in Auckland, New Zealand? Would a student from Charleston, South Carolina feel at home (weatherwise) in Shanghai, China? What can you tell a student from San Francisco about the weather in Cape Town, South Africa? How would you prepare a student from Phoenix, Arizona for the weather in Paris, France?

3 A well-known economist once wrote the following: "Virtually all of the rich countries of the world are outside the tropics, and virtually all the poor countries are in them . . . climate, then, accounts for quite a significant proportion of the cross-national and cross-regional disparities of world income." Comment on this assertion, using Figures G-7 and G-11 as well as the text and box on pp. 28–29. How might a geographer's perspective on such an issue differ from an economist's?

Geographic Literature: A comprehensive listing of noteworthy books and recent articles on global geography, as well as some introductory works on the discipline itself, can be found in the *References and Further Readings* section of this book's website at **www.wiley.com/college/deblij.**

EUROPE

GREENLAND
(KALAALLIT NUNAAT)
(DENMARK)

Denmark
Strait

Jan Mayen
(NORWAY)

Barents Sea

Hammerfest

KOLA
PENINSULA

Murmansk

Arctic Circle

ICELAND

Reykjavik
Akureyri

Vík

Narvik

Bodø

Kiruna

Rovaniemi

Kemi

Petrozavodsk

Lake
Ladoga

Onega Ba

Whi

ATLANTIC
OCEAN

Faeroe Islands
(DENMARK)

Trondheim

Östersund

Luleå

Oulu

NORWAY

SWEDEN

FINLAND

ST. PETERS

Shetland Islands
(U.K.)

Bergen

Sundsvall

Gävle

Åland
(FINLAND)

Tampere

Turku

Lahti

Helsinki

Lake
Peipus

Norwegian
Sea

Gulf of Bothnia

Vaasa

Umeå

Orkney Islands

Stavanger

Oslo

Skien

Uppsala

Stockholm

Hiiumaa

Tallinn

ESTONIA

Tartu

Pskov

Outer Hebrides

Inverness

SCOTTISH
HIGHLANDS

Glasgow

Edinburgh

North
Sea

Kristiansand

Göteborg

Jönköping

Visby

Gotland
(SWEDEN)

Saaremaa

Gulf of
Riga

Riga

LATVIA

Daugavpils

Vitsyebsk

Inner Hebrides

Newcastle

Aalborg

Växjö

Öland

Klaipeda

LITHUANIA

Kaunas

Mahily

British Isles

Londonderry

Belfast

Isle
of Man
(U.K.)

Leeds

Manchester

DENMARK
COPENHAGEN

Aarhus

Kattegat

Malmö

Bornholm
(DENMARK)

Baltic
Sea

Kaliningrad

RUSSIA

Vilnius

MENSK
(MINSK)

Ireland

IRELAND

Limerick

Dublin

Liverpool

Irish Sea

Odense

Gdansk

BELARU

Cork

St. George's
Channel

Cardiff

Cambrian
Plateau

Birmingham

UNITED
KINGDOM

Frisian Islands

HAMBURG

Szczecin

Bydgoszcz

WARSAW

Brest

Bremen

NORTHERN

POLAND

EUROPE

Rivne

Plymouth

LONDON

Bristol

Amsterdam

NETHERLANDS

Hannover

BERLIN

Poznan

Lodz

Celtic
Sea

Land's End

Isle of Wight

English Channel

Strait of Dover

The Hague

Rotterdam

Antwerp

Wista

Wroclaw

Katowice

Krakow

Lviv

Channel Islands
(U.K.)

Le Havre

Rouen

BRUSSELS

BELGIUM

Lille

Essen

Düsseldorf

Cologne

Bonn

GERMANY

Leipzig

Dresden

Oder

Notec

Vinnyts

Brest

Seine

LUXEMBOURG

Luxembourg

Frankfurt

PRAGUE

CZECH
REPUBLIC

Brno

Dniester

U

Brest

Nantes

PARIS

Strasbourg

Nuremberg

Stuttgart

Danube

SLOVAKIA

Kosice

Chernivtsi

Tours

Loire

Munich

Salzburg

VIENNA

Bratislava

Oradea

Cluj

FRANCE

Zürich

Bern

Innsbruck

AUSTRIA

Graz

BUDAPEST

Debrecen

TRANSYLVANIAN ALPS

ROMANIA

Brasc

Limoges

Lyon

Geneva

LIECH.

SWITZERLAND

Mt. Blanc
(4,807 m, 15,771 ft.)

MILAN

Trieste

SLOVENIA

Ljubljana

Pécs

HUNGARY

Mures

Timisoara

Bay of
Biscay

Bordeaux

MASSIF
CENTRAL

Turin

Verona

Venice

Zagreb

CROATIA

BELGRADE

BUCHAREST

La Coruña

Gijón

CANTABRIAN MOUNTAINS

Bilbao

Pico de Aneto
(3,404 m
11,168 ft.)

Montpellier

Nice

Cannes

MONACO

Genoa

Bologna

Rijeka

DINARIC ALPS

BOSNIA

Sarajevo

SERBIA

Danube

Craiova

Vigo

PYRENEES

Toulouse

Marseille

Florence

SAN
MARINO

Split

Nis

BALKAN

BULGA

Porto

ANDORRA

Andorra la Vella

Elba

Adriatic
Sea

KOSOVO

SOFIA

Plovdiv

PORTUGAL

IBERIAN

Valladolid

Zaragoza

BARCELONA

Corsica
(FRANCE)

Bastia

Podgorica

Pristina

PENINSULA

Douro

MADRID

Ligurian Sea

ROME

Bari

MONTENEGRO

ALBANIA

MACEDONIA

Skopje

GREECE

Lisbon

Tagus

SPAIN

Balearic Sea

Sassari

Olbia

ITALY

Naples

Tirane

Thessaloniki

Mount Olympus
(2,917 m, 9,570

PENINSULA

Quadiana

Valencia

Majorca

Minorca

Sardinia
(ITALY)

Gulf of
Taranto

PINDUS MTS.

Larisa

Córdoba

Seville

Guadalquivir

SIERRA
NEVADA

Alicante

Murcia

Palma de
Majorca

Ibiza

Balearic Islands
(SPAIN)

Tyrrhenian
Sea

Cagliari

Patrai

PELOPONNESUS

Cycla

Athen

Cádiz

Algeciras

Málaga

Cartagena

Mediterranean

Palermo

Messina

Sicily

Mt. Etna (3,350 m, 10,991 ft.)

Ionian
Sea

Kálamai

Strait of Gibraltar

GIBRALTAR (U.K.)

Tangier

Ceuta (SPAIN)

Alboran Sea

ALGIERS

Bejaia

Constantine

Tunis

Catania

Strait of Sicily

MOROCCO

Melilla (SPAIN)

Tétouan

ALGERIA

Oran

TUNISIA

Pantelleria
(ITALY)

MALTA

Valletta

CONCEPTS, IDEAS, AND TERMS

1. Land hemisphere
2. Physiography
3. Infrastructure
4. Local functional specialization
5. *The Isolated State*
6. Model
7. Industrial Revolution
8. Nation-state
9. Nation
10. Centrifugal forces
11. Centripetal forces
12. Indo-European language family
13. Complementarity
14. Transferability
15. Intervening opportunity
16. Primate city
17. Metropolis
18. Central business district (CBD)
19. Supranationalism
20. Devolution
21. Four Motors of Europe
22. Regional state
23. Site
24. Situation
25. Conurbation
26. Landlocked location
27. Break-of-bulk
28. *Entrepôt*
29. Shatter belt
30. Balkanization
31. Exclave
32. Irredentism

REGIONS

- STATES OF THE MAINLAND CORE
- THE CORE OFFSHORE: THE BRITISH ISLES
- THE CONTIGUOUS CORE IN THE SOUTH
- THE DISCONTINUOUS CORE IN THE NORTH
- THE EASTERN PERIPHERY

FIGURE 1-1 © H. J. de Blij, P. O. Muller, and John Wiley & Sons, Inc.

Where was this photo taken? Find out at www.wiley.com/college/deblij

In This Chapter

- Why Europe is becoming the world's home for the aged
- Europe's unifying express hits several roadblocks
- Provinces behaving like countries rile national governments
- How Islam is changing Europe's cultural geography
- Why the British are different

I**T IS APPROPRIATE** to begin our investigation of the world's geographic realms in Europe because over the past five centuries Europe and Europeans have influenced and changed the rest of the world more than any other realm or people has done. European empires spanned the globe and transformed societies far and near. European colonialism propelled an early wave of globalization. Millions of Europeans migrated from their homelands to the Old World as well as the New, changing (and sometimes nearly obliterating) traditional communities and creating new societies from Australia to North America. Colonial power and economic incentive combined to impel the movement of millions of imperial subjects from their ancestral homes to distant lands: Africans to the Americas, Indians to Africa, Chinese to Southeast Asia, Malays to South Africa's Cape, Native Americans from east to west. In agriculture, industry, politics, and other spheres, Europe generated revolutions—and then exported those revolutions across the world, thereby consolidating the European advantage.

But throughout much of that 500-year period of European hegemony, Europe also was a cauldron of conflict. Religious, territorial, and political disputes precipitated bitter wars that even spilled over into the colonies. And during the twentieth century, Europe twice plunged the world into war. The terrible, unprecedented toll of World War I (1914–1918) was not enough to stave off World War II (1939–1945), which drew in the United States and ended with the first-ever use of nuclear weapons in Japan. In the aftermath of that war, Europe's weakened powers lost most of their colonial possessions and a new rivalry emerged: an ideological Cold War between the communist Soviet Union and the capitalist United States. This Cold War lowered an Iron Curtain across the heart of Europe, leaving most of the east under Soviet control and most of the west in the American camp. Western Europe proved resilient, overcoming the destruction of war and the loss of colonial power to

Photos: © H. J. de Blij

1. The European geographic realm lies on the western extremity of the Eurasian landmass.
2. Though territorially small, Europe is heavily populated and is fragmented into 40 states.
3. Europe's enduring world influence results mainly from advantages accrued over centuries of colonial and imperial domination.
4. European natural environments are highly varied, and Europe's resource base is rich and diverse.
5. Europe's geographic diversity, cultural as well as physical, created strong local identities, specializations, and opportunities for trade and commerce.
6. European nation-states, based in durable and powerful core areas, survived the loss of colonies and evolved into modern democratic states.
7. Europe's states are engaged in a historic effort to achieve multinational economic integration and, to a lesser degree, political coordination.
8. Europe's relatively prosperous population is highly urbanized, rapidly aging, and in demographic decline partly offset by significant immigration.
9. Local demands for greater autonomy, and cultural challenges posed by immigration, are straining the European social fabric.
10. Despite Europe's momentous unification efforts, east-west contrasts still mark the realm's regional geography.
11. Relations between Europe and neighboring Russia are increasingly problematic.

regain economic strength. Meanwhile, the Soviet communist experiment failed at home and abroad, and in 1990 the last vestiges of the Iron Curtain were lifted. Since then, a massive effort has been underway to reintegrate and reunify Europe from the Atlantic coast to the Russian border, the key geographic story of this chapter.

Defining the Realm

GEOGRAPHICAL FEATURES

As Figure 1-1 shows, Europe is a realm of peninsulas and islands on the western margin of the world's largest landmass, Eurasia. It is a realm of just under 600 million people and 40 countries, but it is territorially quite small. Yet despite its modest proportions it has had—and continues to have—a major impact on world affairs. For many centuries Europe has been a hearth of achievement, innovation, and invention.

Europe's Eastern Border

The European realm is bounded on the west, north, and south by Atlantic, Arctic, and Mediterranean waters, respectively. But where is Europe's eastern limit? Some scholars place it at the Ural Mountains, deep inside Russia, thereby recognizing a "European" Russia and, presumably, an "Asian" one as well. Our regional definition places Europe's eastern boundary between Russia and its numerous European neighbors to the west. This definition is based on several geographic factors including European-Russian contrasts in territorial dimensions, population size, cultural properties, and historic aspects, all discernible on the maps in Chapters 1 and 2.

Resources

Europe's peoples have benefited from a large and varied store of raw materials. Whenever the opportunity or need arose, the realm proved to contain what was required. Early on, these requirements included cultivable soils, rich fishing waters, and wild animals that could be domesticated; in addition, extensive forests provided wood for houses and boats. Later, coal and mineral ores propelled industrialization. More recently, Europe proved to contain substantial deposits of oil and natural gas.

Climates

From the balmy shores of the Mediterranean Sea to the icy peaks of the Alps, and from the moist woodlands and moors of the Atlantic fringe to the semiarid prairies north of the Black Sea, Europe presents an almost infinite range of natural environments (Fig. 1-2). Compare western Europe and eastern North America in Figure G-7 (pp. 16–17) and you will see the moderating influence of the warm ocean current known as the North Atlantic Drift and its onshore windflow.

Human Diversity

The European realm is home to peoples of numerous cultural-linguistic stocks, including not only Latins, Germanics, and Slavs but also minorities such as Finns, Magyars (Hungarians), Basques, and Celts. This diversity of ancestries continues to be an asset as well as a liability. It has generated not only interaction and exchange, but also conflict and war.

Locational Advantages

Europe also has outstanding locational advantages. Its *relative location*, at the heart of the **land hemisphere**, creates maximum efficiency for contact with much of the rest of the world (Fig. 1-3). A "peninsula of peninsulas," Europe is nowhere far from the ocean and its avenues of seaborne trade and conquest. Hundreds of kilometers of navigable rivers, augmented by an unmatched system of canals, open the interior of Europe to its neighboring seas and to the shipping lanes of the world.

Also consider the scale of the maps of Europe in this chapter. Europe is a realm of moderate distances and close proximities. Short distances and large cultural differences make for intense interaction, the constant circulation of goods and ideas. That has been the hallmark of Europe's geography for more than a millennium.

FIGURE 1-2

© H. J. de Blij, P. O. Muller, and John Wiley & Sons, Inc.

© H. J. de Blij, P. O. Muller, and John Wiley & Sons, Inc. 3000 kilometers=ca. 1900 miles

FIGURE 1-3

LANDSCAPES AND OPPORTUNITIES

In the Introduction we noted the importance of physical geography in the definition of geographic realms. The natural landscape with its array of landforms (such as mountains and plateaus) is a key element in the total physical geography— or **physiography**—of any part of the terrestrial world. Other physiographic components include climate and the physical features that mark the natural landscape, such as vegetation, soils, and water bodies.

Europe's area may be small, but its landscapes are varied and complex. Regionally, we identify four broad units: the Central Uplands, the southern Alpine Moun-

tains, the Western Uplands, and the North European Lowland (Fig. 1-4.)

The *Central Uplands* form the heart of Europe. It is a region of hills and low plateaus loaded with raw materials whose farm villages grew into towns and cities when the Industrial Revolution transformed this realm.

The *Alpine Mountains*, a highland region named after the Alps, extend from the Pyrenees on the French-Spanish border to the Balkan Mountains near the Black Sea, and include Italy's Appennines and the Carpathians of eastern Europe.

The *Western Uplands*, geologically older, lower, and more stable than the Alpine Mountains, extend from Scandinavia through western Britain and Ireland to the heart of the Iberian Peninsula in Spain.

The *North European Lowland* extends in a lengthy arc from southeastern Britain and central France across Germany and Denmark into Poland and Ukraine, from where it continues well into Russia. Also known as the Great European Plain, this has been an avenue for human migration time after time, so that complex cultural and economic mosaics developed here together with a jigsaw-like political map. As Figure 1-4 shows, many of Europe's major rivers and connecting waterways serve this populous region, where a number of Europe's leading cities (London, Paris, Amsterdam, Copenhagen, Berlin, Warsaw) are located.

Many picturesque southern European towns and villages are not only many centuries old but are also located in the realm's most dangerous tectonic zone (see Figs. G-3 and G-4 on pages 10 and 11). With hilly terrain, many older buildings, and narrow streets, this can be a lethal combination when earthquakes strike. In April 2009, an earthquake in central Italy devastated many historic mountain towns including L'Aquila (a 90-minute drive northeast of Rome), shown here during the ensuing search and rescue operation. Nearly 300 people lost their lives and more than 1500 sustained serious injuries. From Portugal to Greece, Mediterranean Europe's scenic beauty comes with ever-present risk.
© Marco DiLauro/Getty Images, Inc.

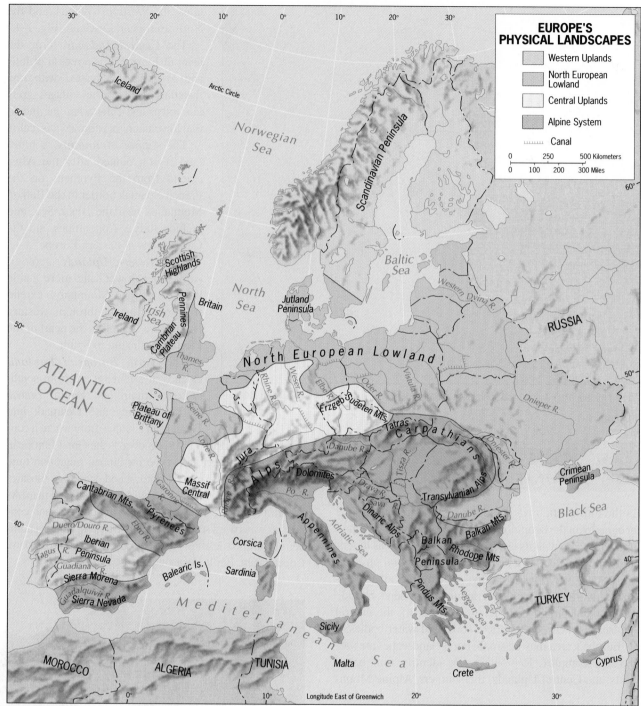

FIGURE 1-4

© H. J. de Blij, P. O. Muller, and John Wiley & Sons, Inc.

HISTORICAL GEOGRAPHY

Modern Europe was peopled in the wake of the Pleistocene's most recent glacial retreat and global warming—a gradual warming that caused tundra to give way to deciduous forest and ice-filled valleys to turn into grassy vales. On Mediterranean shores, Europe witnessed the rise of its first great civilizations, on the islands and peninsulas of Greece and later in what is today Italy.

Ancient Greece and Imperial Rome

Ancient Greece lay exposed to influences radiating from the advanced civilizations of Mesopotamia and the Nile Valley, and in their fragmented habitat the Greeks laid the foundations of European civilization. Their achievements in political science, philosophy, the arts, and other spheres have endured for 25 centuries. But the ancient Greeks never managed to unify their domain, and their persistent conflicts proved fatal when the Romans challenged them from the west. By 147 BC, the last of the sovereign Greek intercity leagues (alliances) had fallen to the Roman conquerors.

The center of civilization and power now shifted to Rome in present-day Italy. Borrowing from Greek culture, the Romans created an empire that stretched from Britain to the Persian Gulf and from the Black Sea to Egypt; they made the Mediterranean Sea a Roman lake carrying armies to distant shores and goods to imperial Rome. With an urban population that probably exceeded 1 million, Rome was the first metropolitan-scale urban center in Europe.

The Romans founded numerous other cities throughout their empire and linked them to the capital through a vast system of overland and water routes, facilitating political control and enabling economic growth in their provinces. It was an unparalleled **infrastructure**, much of which long outlasted the empire itself.

Triumph and Collapse

Roman rule brought disparate, isolated peoples into the imperial political and economic sphere. By guiding (and often forcing) these groups to produce particular goods or materials, the Romans launched Europe down a road for which it would become famous: **local functional specialization**. The workers on Elba, a Mediterranean island, mined iron ore. Those near Cartagena in Spain produced silver and lead. Certain farmers were taught irrigation to produce specialty crops. Others raised livestock for meat or wool. The *production of particular goods by particular people in particular places* became and remained a hallmark of the realm.

The Romans also spread their language across the empire, setting the stage for the emergence of the *Romance languages*; they disseminated Christianity; and they established durable systems of education, administration, and commerce. But when their empire collapsed in the fifth century, disorder ensued, and massive migrations soon brought Germanic and Slavic peoples to their present positions on the European stage. Capitalizing on Europe's weakness, the Arab-Berber Moors from North Africa, energized by Islam, conquered most of Iberia and penetrated France. Later the Ottoman Turks invaded eastern Europe and reached the outskirts of Vienna.

Rebirth and Royalty

Europe's revival—its *Renaissance*—did not begin until the fifteenth century. After a thousand years of feudal turmoil marking the "Dark" and "Middle" Ages, powerful monarchies began to lay the foundations of modern states. The discovery of continents and riches across the oceans opened a new era of *mercantilism*, the competitive accumulation of wealth chiefly in the form of gold and silver. Best placed for this competition were the kingdoms of western Europe. Europe was on its way to colonial expansion and world domination.

THE REVOLUTIONS OF MODERNIZING EUROPE

Even as Europe's rising powers reached for world domination overseas, they fought with each other in Europe itself. Powerful monarchies and land-owning ("landed") aristocracies had their status and privilege challenged by ever-wealthier merchants and businesspeople. Demands for political recognition grew; cities mushroomed with the development of industries; the markets for farm products burgeoned; and Europe's population, more or less stable at about 100 million since the sixteenth century, began to increase.

The Agrarian Revolution

We know Europe as the focus of the Industrial Revolution, but before this momentous development occurred another revolution was already in progress: the *agrarian revolution*. Port cities and capital cities thrived and expanded, and their growing markets created economic opportunities for farmers. This led to revolutionary changes in land ownership and agricultural methods. Improved farm practices, better equipment, superior storage facilities, and more efficient transport to the urban markets marked a revolution in the countryside. The colonial merchants brought back new crops (the American potato soon became a European staple), and market prices rose, drawing more and more farmers into the economy.

Agricultural Market Model

The transformation of Europe's farmlands reshaped its economic geography, producing new patterns of land use and market links. The economic geographer Johann Heinrich von Thünen (1783–1850), himself an estate farmer who had studied these changes for several decades, published his observations in 1826 in a pioneering work entitled *The Isolated State*, chronicling the geography of Europe's agricultural transformation.

Von Thünen used information from his own farmstead to build what today we call a **model** (an idealized representation of reality that demonstrates its most important properties) of the location of productive activities in Europe's farmlands. Since a model is an abstraction that must always involve assumptions, von Thünen postulated a self-contained area (hence the "isolation") with a single market center, surrounded by flat and uninterrupted land without impediments to cultivation or transportation. In such a situation, transport costs would be directly proportional to distance.

Location Theory

Von Thünen's model revealed four zones or rings of land use encircling the market center (Fig. 1-5). Innermost and directly adjacent to the market would lie a zone of intensive farming and dairying, yielding the most perishable and highest-priced products. Immediately beyond lay a zone of forest used for timber and firewood (still a priority in von Thünen's time). Next there would be a ring of field crops, for example, grains or potatoes. A fourth zone would contain pastures and livestock. Beyond lay wilderness, from where the costs of transport to market would become prohibitive.)

In many ways, von Thünen's model was the first analysis in a field that would eventually become known as *location theory*. Von Thünen knew, of course, that the real Europe did not present the conditions postulated in his model. But it did demonstrate the economic-geographic forces that shaped the new Europe, which is why it is still being discussed today. More than a century after the publication of *The Isolated State*, geographers Samuel van Valkenburg and Colbert Held produced a map of twentieth-century European agricultural intensity, revealing a striking, ring-like concentricity focused on the vast urbanized area lining the North Sea—now the dominant market for a realmwide, macroscale "Thünian" agricultural system (Fig. 1-5, inset map).

The Industrial Revolution

7 The term **Industrial Revolution** suggests that an agrarian Europe was suddenly swept up in wholesale industrialization that changed the realm in a few decades. In

FIGURE 1-5

© H. J. de Blij, P. O. Muller, and John Wiley & Sons, Inc.

reality, seventeenth- and eighteenth-century Europe had been industrializing in many spheres, long before the chain of events known as the Industrial Revolution began. From the textiles of England and Flanders to the iron farm implements of Saxony (in present-day Germany), from Scandinavian furniture to French linens, Europe had already entered a new era of local functional specialization. It would therefore be more appropriate to call what happened next the period of Europe's *industrial intensification*.

British Primacy

In the 1780s, the Scotsman James Watt and others devised a steam-driven engine, which was soon adopted for numerous industrial uses. At about the same time, coal (converted into carbon-rich coke) was recognized as a vastly superior substitute for charcoal in smelting iron. These momentous innovations had a rapid effect. The power loom revolutionized the weaving industry. Iron smelters, long dependent on Europe's dwindling forests for fuel, could now be concentrated near coalfields.

Engines could move locomotives as well as power looms. Ocean shipping entered a new age.

Britain had an enormous advantage, for the Industrial Revolution occurred when British influence reigned worldwide and the significant innovations were achieved in Britain itself. The British controlled the flow of raw materials, they held a monopoly over products that were in global demand, and they alone possessed the skills necessary to make the machines that manufactured the products. Soon the fruits of the Industrial Revolution were being exported, and the modern industrial spatial organization of Europe began to take shape. In Britain, manufacturing regions developed near coalfields in the English Midlands, at Newcastle to the northeast, in southern Wales, and along Scotland's Clyde River around Glasgow.

Resources and Industrial Development

In mainland Europe, a belt of major coalfields extends from west to east, roughly along the southern margins of the North European Lowland, due eastward from southern England across northern France and Belgium, Germany (the Ruhr), western Bohemia in the Czech Republic, Silesia in southern Poland, and the Donets Basin (Donbas) in eastern Ukraine. Iron ore is found in a broadly similar belt and together with coal provides the key raw material for the manufacturing of steel. As in Britain, this cornerstone industry now spawned new concentrations of economic activity, which grew steadily as millions migrated in from the countryside to fill expanding employment opportunities. Densely populated and heavily urbanized, these emerging agglomerations became the backbone of Europe's world-scale population cluster (as shown in Fig. G-8).

Two centuries later, this east-west axis along the coalfield belt remains a major feature of Europe's population distribution map (Fig. 1-6). It should also be noted that while industrialization produced new cities, another set of manufacturing zones arose in and near many existing urban centers. London—already Europe's leading urban focus and Britain's richest domestic market—was typical of these developments. Many local industries were established here, taking advantage of the large supply of labor, the ready availability of capital, and the proximity of so great a number of potential buyers. Although the Industrial Revolution thrust other places into prominence, London did not lose its primacy: industries in and around the British capital multiplied.)

Consequences of Clustering

The industrial transformation of Europe, like the agrarian revolution, became the focus of geographic research. One of the leaders in this area was the economic geographer Alfred Weber (1868–1958), who published a spatial analysis of the process titled *Concerning the Location of Industries* (1909). Unlike von Thünen, Weber focused on activities that take place at particular points rather than over large areas. His model, therefore, represented the factors of industrial location, the clustering or dispersal of places of intense manufacturing activity.

One of Weber's most interesting conclusions has to do with what he called *agglomerative* (concentrating) and *deglomerative* (dispersive) forces. It is often advantageous for certain industries to cluster together, sharing equipment, transport facilities, labor skills, and other assets of urban areas. This is what made London (as well as Paris and other cities that were not situated on rich deposits of industrial raw materials) attractive to many manufacturing plants that could benefit from agglomeration and from the large markets that these cities anchored. As Weber found, however, excessive agglomeration may lead to disadvantages such as competition for increasingly expensive space, congestion, overburdening of infrastructure, and environmental pollution. Manufacturers may then move away, and deglomerative forces will increase.

The Industrial Revolution spread eastward from Britain onto the European mainland throughout the middle and late nineteenth century (Fig. 1-7). Population skyrocketed, emigration mushroomed, and industrializing cities burst at the seams. European states already had acquired colonial empires before this revolution started; now colonialism gave Europe an unprecedented advantage in its dominance over the rest of the world.

Political Revolution and Evolution

Revolution in a third sphere—the political—had been going on in Europe even longer than the agrarian or industrial revolutions. Europe's *political revolution* took many different forms and affected diverse peoples and countries, but in general it headed toward parliamentary representation and democracy.

Historical geographers point to the Peace (Treaty) of Westphalia in 1648 as a key step in the evolution of Europe's state system, ending decades of war and recognizing territories, boundaries, and the sovereignty of

FIGURE 1-6

© H. J. de Blij, P. O. Muller, and John Wiley & Sons, Inc.

countries. This treaty's stabilizing effect lasted until 1806, by which time revolutionary changes were again sweeping across Europe.

Most dramatic was the French Revolution (1789–1795), but political transformation had come much earlier to the Netherlands, Britain, and the Scandinavian countries. Other parts of Europe remained under the control of authoritarian (dictatorial) regimes headed by monarchs or despots. Europe's patchy political revolution lasted into the twentieth century, and by then *nationalism* (national spirit, pride, patriotism) had become a powerful force in European politics.

FIGURE 1-7

© H. J. de Blij, P. O. Muller, and John Wiley & Sons, Inc.

A Fractured Map

As a geographic realm, Europe occupies only about 5 percent of the Earth's land area; but that tiny area is fragmented into some 40 countries—more than one-fifth of all the states in the world today. Therefore, when you look at Europe's political map, the question that arises is how did so small a geographic realm come to be divided into so many political entities? Europe's map is a legacy of its feudal and royal periods, when powerful kings, barons, dukes, and other rulers, rich enough to fund armies and powerful enough to exact taxes and tribute from their domains, created bounded territories in which they reigned supreme. Royal marriages, alliances, and conquests actually simplified Europe's political map. In the early nineteenth century there still were 39 German states; a unified Germany as we know it today did not emerge until the 1870s.

State and Nation

Europe's political revolution produced a form of political-territorial organization known as the nation-state, a state embodied by its culturally distinctive popula-

tion. But what is a **nation-state** and what is not? The term **nation** has multiple meanings. In one sense it refers to a people with a single language, a common history, a similar ethnic background. In the sense of *nationality* it relates to legal membership in the state, that is, citizenship. Very few states today are so homogeneous culturally that the culture is conterminous with the state. Europe's prominent nation-states of a century ago—France, Spain, the United Kingdom, Italy—have become multicultural societies, their nations defined more by an intangible "national spirit" and emotional commitment than by cultural or ethnic homogeneity. [8] [9]

Division and Unity

Mercantilism and colonialism empowered the states of western Europe; the United Kingdom (Britain) was the superpower of its day. But all countries, even Europe's nation-states in their heyday, are subject to divisive stresses. Political geographers use the term **centrifugal forces** [10] to identify and measure the strength of such division, which may result from religious, racial, linguistic, political, economic, or regional factors.

11 Centrifugal forces are measured against **centripetal forces**, the binding, unifying glue of the state. General satisfaction with the system of government and administration, legal institutions, and other functions of the state (notably including its treatment of minorities) can ensure stability and continuity when centrifugal forces threaten. In the recent case of Yugoslavia, the centrifugal forces unleashed after the end of the Cold War exceeded the weak centripetal forces in that relatively young state, and it disintegrated with dreadful consequences.

Europe's political evolution continues, but overall it is going in a different direction. Today a growing majority of European states and their leaders recognize that closer association and regional coordination form the key to a more stable and secure future. A realmwide union is in the making, and it is not inconceivable that a European superstate may at some future time emerge from this process.

CONTEMPORARY EUROPE

If you were to arrive by air or ship in Europe for a visit, you would probably be impressed by the modernity of its infrastructure. Airports tend to be modern and spacious, and security operates smoothly. Public transport, from high-speed intercity trains to local buses, is efficient and seems to reach even the most remote places. Superhighways connect major cities, although Europe's relatively compact, old, and densely-built-up urban areas tend to delay incoming automobiles and cause traffic jams that will look familiar. Ultramodern high-rise architecture and historic preservation give Europe a cultural landscape welding the future to the past. From nuclear power plants (France) and offshore oil platforms (Norway) to flood-control technology (Netherlands) and mega-ship construction (Finland), as well as in countless other enterprises, Europe is in the vanguard of the world. In 2008, eight of Europe's national economies ranked among the top 20 in the world; collectively, Europe's economies outrank even the United States.

Collectively—but that's the problem. Europe does not have a long-term history of collective action. Europe has historically been a crucible of political revolution and evolution. Its nation-states, enriched and empowered by their colonial conquests, fought with each other at home and abroad, dragging distant peoples into their wars and plunging the world into global conflict twice during the twentieth century. After the end of World War II in the mid-1940s, Europeans saw their ravaged realm divided in the Cold War between Western and Soviet spheres. The countries on the western side of the "Iron Curtain" embarked on a massive effort not only to recover, but also to put old divisions permanently behind them. Assisted by the United States, they moved far along the road to unification—especially in the economic arena. But, as we will see later in this chapter, Europe's old fractiousness has not been overcome.

Language and Religion

It is worth remembering that Europe's territory is just over 60 percent the size of the United States, but that the population of Europe's 40 countries is about twice as large as America's. This population of 594 million speaks numerous languages, almost all of which belong to the **Indo-European language family** (Figs. 1-8, **12** G-10). But most of those languages are not mutually understandable; some, such as Finnish and Hungarian, are not even members of this Indo-European family. When the unification effort began, one major problem was to determine which languages to recognize as "official." That problem still prevails, although English has become the realm's unofficial *lingua franca*. During your visit to Europe, though, you would find that English is more commonly usable in western Europe than farther east. Europe's multilingualism remains a major barrier to integration.

Another centrifugal force confronting Europeans involves religion. Europe's cultural heritage is steeped in Christian traditions, but sectarian strife between Catholics and Protestants that plunged parts of the realm into bitter and widespread conflict still divides communities and, as until recently in Northern Ireland, can still arouse violence. Some political parties still carry the name "Christian," for example, Germany's Christian Democrats. But today, a new factor roils the religious landscape: the rise of Islam. In the east, this takes the form of new Islamic assertiveness in an old Muslim bastion: the (Turkish) Ottoman Empire left behind millions of converts from Bosnia to Bulgaria among whom many are demanding greater political power. In the west, this Islamic resurgence results from the relatively recent immigration of millions of Muslims from North Africa and other parts of the Islamic world. Here, as mosques overflow with the faithful, churches stand nearly empty

LANGUAGES OF EUROPE

MAJOR INDO-EUROPEAN BRANCHES	OTHER INDO-EUROPEAN BRANCHES	URALIC LANGUAGE FAMILY	

GERMANIC GROUP

WESTERN GERMANIC
1 Dutch
2 German
3 Frysian
4 English

NORTHERN GERMANIC
5 Danish
6 Swedish
7 Norwegian
8 Icelandic
9 Faeroese

ROMANCE GROUP

10 Portuguese
11 Spanish
12 Catalan
13 Provençal
14 French
15 Italian
16 Rhaeto-Romansch
17 Romanian
18 Corsican-Italian
19 Sardinian-Italian
20 Walloon

SLAVIC GROUP

WEST SLAVONIC
21 Polish
22 Slovak
23 Czech
24 Lusatian

EAST SLAVONIC
25 Russian
26 Ukrainian
27 Belarussian

SOUTH SLAVONIC
28 Slovene
29 Serbo-Croatian
30 Macedonian
31 Bulgarian

CELTIC GROUP

BRITANNIC
32 Breton
33 Welsh

GAELISH
34 Irish Gaelic
35 Scots Gaelic

BALTIC GROUP

36 Latvian
37 Lithuanian

HELLENIC

38 Greek

THRACIAN/ILLYRIAN GROUP

39 Albanian

INDO-IRANIAN GROUP

40 Romani (dispersed)

FINNO-UGRIC GROUP

41 Finnish
42 Karelian
43 Saami
44 Estonian
45 Hungarian
46 Komi

SAMOYEDIC GROUP

47 Samoyedic

ALTAIC LANGUAGE FAMILY

TURKIC GROUP

48 Turkish

OTHER LANGUAGES

BASQUE

49 Basque

Areas with significant concentrations of other languages (usually adjacent national languages)

Boundary between languages

After Murphy, 1998.

FIGURE 1-8

© H. J. de Blij, P. O. Muller, and John Wiley & Sons, Inc.

as secularism among Europeans is on the rise. Some scholars have gone so far as to describe Europe as having entered a "post-Christian" era.

For so small a realm, Europe's cultural geography is sharply varied. The popular image of Europe tends to be formed by British pageantry, French countrysides, German cities—but go beyond this core, and you will find isolated Slavic communities in the mountains facing the Adriatic, Muslim towns in poverty-mired Albania, Roma (Gypsy) villages in the interior of Romania, farmers using traditional methods unchanged for centuries in rural Poland. That map of 40 countries does not begin to reflect the diversity of European cultures.

Spatial Interaction

If not culture, what does unify Europe? The answer lies in this realm's outstanding opportunities for productive interaction. The ancient Romans realized it, but they

The French high-speed train TGV (*Trés Grande Vitesse*, meaning Very High Speed) symbolizes France's modernization as does its global leadership in nuclear-power generation. The expanding TGV rail network, requiring the construction of elaborate viaducts and dedicated tracks, provides intercity passengers with an alternative to air travel. This train, seen as it speeds over the Ventabren Viaduct in Provence, left Calais on France's north coast at 4:30 pm, stopped in Paris and Lyon, and reached Marseille on the Mediterranean, 1067 km (662 mi) away, three and a half hours later. The TGV is a project of the France's state-run rail company, not a private initiative. © AP/Wide World Photos.

developed a system of production that focused on Rome, not primarily on regional exchange. Modern Europeans seized on the same opportunities to create a realmwide network of spatial interaction whose structure intensified as time went on, linking places, communities, and countries in countless ways. The geographer Edward Ullman envisaged this process as operating on three principles:

- **Complementarity** When one area produces a surplus of a commodity required by another area, regional complementarity prevails. The mere existence of a particular resource or product is no guarantee of trade: it must be needed elsewhere. When two areas each require the other's products, double complementarity occurs. Europe exhibits countless examples of this complementarity, from local communities to entire countries. Industrial Italy needs coal from western Europe; western Europe needs Italy's farm products. **13**

- **Transferability** The ease with which a commodity can be transported by producer to consumer defines its transferability. Distance and physical obstacles can raise the cost of a product to the point of unprofitability. But Europe is small, distances are short, and Europeans have built the world's most efficient transport system of roads, railroads, and canals linking navigable rivers. **14**

- **Intervening Opportunity** The third of Ullman's spatial interaction principles holds that potential trade between two places, even if they are in a position of strong complementarity and high transferability, will develop only in the absence of a closer, intervening source of supply. To use our earlier example, if Switzerland proved to contain large coal reserves close to the Italian border, Italy would avail itself of that opportunity, reducing or eliminating its imports from more distant western Europe. **15**

A Highly Urbanized Realm

Such opportunities for spatial interaction and exchange contribute to an ever-increasing level of urbanization in Europe. Overall, about three of every four Europeans live in towns and cities, an average that is far exceeded in the west (see Data Table inside back cover) but not yet attained in much of the east. Large cities are production centers as well as marketplaces, and they also form the crucibles of their nations' cultures. By several measures, European cities are more diverse than their American counterparts.

The geographer Mark Jefferson during the 1930s studied the pivotal role of great cities in the development of

16 national cultures. He postulated a "law" of the **primate city**, which stated that "a country's leading city is always disproportionately large and exceptionally expressive of national culture." This obviously debatable "law" still has relevance: Paris personifies France in countless ways; nothing in the United Kingdom rivals London; Athens anchors and dominates Greece. Europe's primate cities have also grown disproportionately since the end of World War II, gaining rather than losing primacy in this still-urbanizing realm.

Cityscapes

Primate cities tend to be old, and in general the European cityscape looks quite different from the American. Seemingly haphazard inner-city street systems impede traffic; central cities may be picturesque, but they are also cramped. The urban layout of the London region (Fig. 1-9) reveals much about the internal spatial struc-

METROPOLITAN LONDON

Built-up area	Road
Greenbelt	Railroad
Central Business District	

0 5 10 15 20 Kilometers
0 5 10 Miles

FIGURE 1-9 © H. J. de Blij, P. O. Muller, and John Wiley & Sons, Inc.

17 ture of the European **metropolis** (the central city plus its suburban ring). Such a *metropolitan area* remains focused on the large city at its center, especially the **18** downtown **central business district (CBD)**, which is the oldest part of the urban agglomeration and contains the region's largest concentration of business, government, and shopping facilities as well as its wealthiest and most prestigious residences. Wide residential sectors radiate outward from the CBD across the rest of the central city,

each one home to a particular income group. Beyond the central city lies a sizeable suburban ring, but residential densities are much higher here than in the United States because the European tradition is one of setting aside recreational spaces (in "greenbelts") and living in apartments rather than in detached single-family houses. There also is a greater reliance on public transportation, which further concentrates the suburban development pattern. That has allowed many nonresidential activities

If you were to be asked what city might be shown here, would Paris spring to mind? Most images of France's capital show venerable landmarks such as the Eiffel Tower or the Arc de Triomphe or other landmarks of the Old City. But this is another Paris, the district called *La Défense*, where Paris shows its ultramodern face. This is where Paris escapes the height restrictions and architectural limitations of the historic center, where glass-boxed highrises reflect the vibrant global city this is, and where the landmark is the "Cube," a huge open structure admired as well as reviled (as was the Eiffel Tower in its time). Set just slightly off the axis created by the Champs Elysées beyond, the Cube is the focus for the high-tech, financial, and service hub *La Défense* has become. Look in the distance and you will see the Arc de Triomphe appropriately (as some would see it) diminished, and the Eiffel Tower to its right, just a needle rising from the vast urban tapestry that is Paris today.
© Pascal Crapet/Stone/Getty Images, Inc.

to suburbanize as well, and today ultramodern outlying business centers increasingly compete with the CBD in many parts of urban Europe (see photo on page 53).

A Changing Population

When a population urbanizes, average family size declines, and so does the overall rate of natural increase. There was a time when Europe's population was (in the terminology of population geographers) exploding, sending millions to the New World and the colonies and still growing at home. But today Europe's indigenous population, unlike most of the rest of the world's, is actually shrinking. To keep a population from declining, the (statistically) average woman must bear 2.1 children. For Europe as a whole, that figure was 1.5 in 2008. But several large countries recorded 1.3, including Germany, Italy, and Poland. And two eastern European countries (Slovakia and Bosnia) even recorded 1.2—the lowest ever seen in any human population.

Such *negative population growth* poses serious challenges for any nation. When the population pyramid becomes top-heavy, the number of workers whose taxes pay for the social services of the aged goes down, leading to reduced pensions and dwindling funds for health care. Governments that impose tax increases endanger the business climate; their options are limited. Europe, especially western Europe, is experiencing a *population implosion* that will be a formidable challenge in decades to come.

Immigrant Infusions

Meanwhile, *immigration* is partially offsetting the losses European countries face. Millions of Turkish Kurds (mainly to Germany), Algerians (France), Moroccans (Spain), West Africans (Britain), and Indonesians (Netherlands) are changing the social fabric of what once were nearly unicultural nation-states. As noted above, one key dimension of this change is the spread of Islam in Europe (mapped in Fig. 1-10). Muslim populations in eastern Europe (such as Albania's, Kosovo's, and Bosnia's) are local, Slavic communities converted during the period of Ottoman rule. The Muslim sectors of western European countries, on the other hand, represent more recent immigrations.

The vast majority of these immigrants are intensely devout, politically aware, and culturally insular. They continue to arrive in a Europe where native populations are stagnant or declining, where religious institutions are weakening, where secularism is rapidly rising, where political positions often appear to be anti-Islamic, and where cultural norms are incompatible with Muslim traditions. Many young men, unskilled and uncompetitive

in a European Union (EU) where unemployment is already high, get involved in petty crime or the drug trade, are harassed by the police, and turn to their faith for solace and reassurance. Unlike most other immigrant groups, Muslim communities also tend to resist assimilation, making Islam the essence of their identity. In Britain alone there are more than 1500 mosques; and wherever they appear, local cultural landscapes are transformed (see photo below).

The Growing Multicultural Challenge

In truth, European governments have not done enough to foster the very integration they see Muslims rejecting. The French assumed that their North African

FROM THE FIELD NOTES

"After taking the train from Schiphol Airport to Rotterdam and checking into my downtown hotel, I took a train from the 'Central Station' (which was in the middle of one huge construction site) to what the Dutch call 'Hoek van Holland' ('Corner of Holland') on the North Sea coast, where huge ferries disembark thousands of travelers from England and elsewhere daily. A colleague had advised me to get off at Maassluis and walk the streets for a sense of the history as well as the modern changes going on here. Certainly the growing Islamic presence is transforming cultural landscapes in Europe, but not just in the cities. Even in small towns like this, the minarets of mosques rise between the spires of churches and, as is the case here, the local mosque is the centerpiece for a multipurpose Muslim cultural center that serves as a social (and political) center in ways that churches do not. A couple of generations ago, the people living in those apartments might have expected to overlook a church, but today their vista is quite different." © H. J. de Blij.

www.conceptcaching.com

EUROPE: MUSLIMS AS PERCENTAGE OF NATIONAL POPULATION, 2010

- Less than 1%
- 1–4%
- 4–7%
- 7–12%
- 12–25%
- More than 25%

0   250   500 Kilometers
0   100   200   300 Miles

FIGURE 1-10

© H. J. de Blij, P. O. Muller, and John Wiley & Sons, Inc.

immigrants would aspire to assimilation as Muslim children went to public schools from their urban public-housing projects; instead, they got into disputes over the dress codes of Muslim girls. The Germans for decades would not award German citizenship to children of immigrant parents born on German soil. The British found their modest efforts at integration stymied by the Muslim practice of importing brides from Islamic countries, which had the effect of perpetuating segregation. In many European cities, the housing projects that form the living space of Muslim residents (by now many of them born and retired there) are dreadful, barren, impoverished environments unseen by Europeans whose images of Paris, Manchester, or Frankfurt are far different.

The social and political implications of Europe's demographic transformation are numerous and far-reaching. Long known for tolerance and openness, European societies are attempting to restrict immigration in various ways; political parties with anti-immigrant platforms are gaining ground. Multiculturalism poses a growing challenge in a changing Europe.

EUROPE'S MODERN TRANSFORMATION

At the end of World War II, much of Europe lay shattered, its cities and towns devastated, its infrastructure wrecked, its economies ravaged. The Soviet Union had taken control over the bulk of eastern Europe, and communist parties seemed poised to dominate the political life of major western European countries.

In 1947, U.S. Secretary of State George C. Marshall proposed a European Recovery Program designed to help counter all this dislocation and to create stable political conditions in which democracy would survive. Over the next four years, the United States provided about $13 billion in assistance to Europe (about $100 billion in today's money). Because the Soviet Union refused U.S. aid and forced its eastern European satellites to do the same, the Marshall Plan applied solely to 16 European countries, including defeated (West) Germany and Turkey.

European Unification

The Marshall Plan did far more than stimulate European economies. It confirmed European leaders' conclusion that their countries needed a joint economic-administrative structure not only to coordinate the financial assistance, but also to ease the flow of resources and products across Europe's mosaic of boundaries, to lower restrictive trade tariffs, and to seek ways to improve political cooperation.

Benelux Precedent

For all these needs Europe's governments had some guidelines. While in exile in Britain, the leaders of three small countries—Belgium, the Netherlands, and Luxembourg—had been discussing an association of this kind even before the end of the war. There, in 1944, they formulated and signed the Benelux Agreement, intended to achieve total economic integration. When the Marshall Plan was launched, the Benelux precedent helped speed the creation of the Organization for European Economic Cooperation (OEEC), which was established to coordinate the investment of America's aid (see the box entitled "Supranationalism in Europe").

Supranationalism

Soon the economic steps led to greater political cooperation as well. In 1949, the participating governments created the Council of Europe, the beginnings of what was to become a European Parliament meeting in Strasbourg, France. Europe was embarked on still another political revolution, the formation of a multinational union involving a growing number of European states. Geographers define **supranationalism** as the voluntary association in economic, political, or cultural spheres of three or more independent states willing to yield some measure of sovereignty for their mutual benefit. In later chapters we will encounter other supranational organizations, including the North American Free Trade Agreement (NAFTA), but none has reached the plateau achieved by the European Union (EU).

The EU's administrative, economic, and even political framework has become so advanced that the organization now has a headquarters with many of the trappings of a capital city. Early on, EU planners chose Brussels (already the national capital of Belgium, a member of Benelux) as the organization's center of governance. But then, in order to avoid giving Brussels too much prominence, they chose Strasbourg, in the northeast corner of France, as the seat of the European Parliament, whose elected membership represents all EU countries.

Policies and Priorities

In Europe, the key initiatives arose from the Marshall Plan and lay in the economic arena; political integration came more haltingly. Under the Treaty of Rome, six countries joined to become the European Economic Community (EEC) in 1957, also called the "Common Market." In 1973 the United Kingdom, Ireland, and Denmark joined, and the renamed European Community (EC) had nine members. As Figure 1-11 shows, membership reached 15 countries in 1995, after the organization had been renamed yet one more time to become the European Union (EU). Since then, the number of member-states has climbed to 27.

The European Union is not just a paper organization for bankers and manufacturers. It has a major impact on the daily lives of its member-countries' citizens in countless ways. (You will see one of these ways when you arrive at an EU airport and find that EU-passport holders move through inspection at a fast pace, while non-EU citizens wait in long immigration lines.) Taxes tend to be high in Europe, and those collected in the richer member-states are used to subsidize growth and development in the less prosperous ones. This is one of the

burdens of membership that is not universally popular in the EU, to say the least. But it has strengthened the economies of Portugal, Greece, and other national and regional economies to the betterment of the entire organization. Some countries also object to the terms and rules of the Common Agricultural Policy (CAP), which, according to critics, supports farmers far too much and, according to others, far too little. (France in particular obstructs efforts to move the CAP closer to consensus, subsidizing its agricultural industry relentlessly while arguing that this protects its rural cultural heritage as well as its farmers.)

New Money

Ever since the first steps toward European unification were taken, EU planners dreamed of a time when the EU would have a single currency not only to symbolize

Supranationalism In Europe

1944	Benelux Agreement signed.
1947	Marshall Plan created (effective 1948–1952).
1948	Organization for European Economic Cooperation (OEEC) established.
1949	Council of Europe created.
1951	European Coal and Steel Community (ECSC) Agreement signed (effective 1952).
1957	Treaty of Rome signed, establishing European Economic Community (EEC) (effective 1958), also known as the Common Market and "The Six." European Atomic Energy Community (EURATOM) Treaty signed (effective 1958).
1959	European Free Trade Association (EFTA) Treaty signed (effective 1960).
1961	United Kingdom, Ireland, Denmark, and Norway apply for EEC membership.
1963	France vetoes United Kingdom EEC membership; Ireland, Denmark, and Norway withdraw applications.
1965	EEC–ECSC–EURATOM Merger Treaty signed (effective 1967).
1967	European Community (EC) inaugurated.
1968	All customs duties removed for intra-EC trade; common external tariff established.
1973	United Kingdom, Denmark, and Ireland admitted as members of EC, creating "The Nine." Norway rejects membership in the EC by referendum.
1979	First general elections for a European Parliament held; new 410-member legislature meets in Strasbourg. European Monetary System established.
1981	Greece admitted as member of EC, creating "The Ten."
1985	Greenland, acting independently of Denmark, withdraws from EC.
1986	Spain and Portugal admitted as members of EC, creating "The Twelve." Single European Act ratified, targeting a functioning European Union in the 1990s.

1987	Turkey and Morocco make first application to join EC. Morocco is rejected; Turkey is told that discussions will continue.
1990	Charter of Paris signed by 34 members of the Conference on Security and Cooperation in Europe (CSCE). Former East Germany, as part of newly reunified Germany, incorporated into EC.
1991	Maastricht meeting charts European Union (EU) course for the 1990s.
1993	Single European Market goes into effect. Modified European Union Treaty ratified, transforming EC into EU.
1995	Austria, Finland, and Sweden admitted into EU, creating "The Fifteen."
1999	European Monetary Union (EMU) goes into effect.
2001	Denmark's voters reject EMU participation by 53 to 47 percent.
2002	The euro is introduced as historic national currencies disappear in 12 countries.
2003	First draft of a European constitution is published to mixed reviews from member-states.
2004	Historic expansion of EU from 15 to 25 countries with the admission of Cyprus, the Czech Republic, Estonia, Hungary, Latvia, Lithuania, Malta, Poland, Slovakia, and Slovenia.
2005	Proposed EU Constitution is disapproved by voters in France and the Netherlands.
2007	Romania and Bulgaria are admitted, bringing total EU membership to 27 countries. Slovenia adopts the euro.
2008	Cyprus and Malta adopt the euro.
2009	Slovakia becomes the sixteenth adopter of the euro.

FIGURE 1-11

© H. J. de Blij, P. O. Muller, and John Wiley & Sons, Inc.

its strengthening unity but also to establish a joint counterweight to the mighty American dollar. A European Monetary Union (EMU) became a powerful objective of EU planners, but many observers thought it unlikely that member-states would in the end be willing to give up their historic marks, francs, guilders, liras, and escudos. Yet it happened. In 2002, twelve of the (then) 15 EU countries withdrew their currencies and began using the *euro*, with only the United Kingdom (Britain), Denmark, and Sweden staying out (Fig. 1-11). Add nonmember Norway to this group, and not

a single Scandinavian country converted to the euro. On the other hand, Slovenia adopted the euro in 2007, Cyprus and Malta in 2008, and Slovakia became the sixteenth adopter in 2009.

Momentous Expansion

Expansion has always been an EU objective, and the subject has always aroused passionate debate (see the Issue Box on p. 60 titled "Europe: How Desirable Is Economic and Political Union?"). Will the incorporation of weaker economies undermine the strength of the whole? Despite such misgivings, negotiations to expand the EU have long been in progress and still continue. In 2004 a momentous milestone was reached: ten new members were added, creating a greater European Union with 25 member-states. Geographically, these ten came in three groups: three Baltic states (Estonia, Latvia, and Lithuania); five contiguous states in eastern Europe, extending from Poland and the Czech Republic through Slovakia, Hungary, and Slovenia; and two Mediterranean island-states, the ministate of Malta and the still-divided state of Cyprus. And in 2007, both Romania and Bulgaria were incorporated, raising the number of members to 27 and extending the EU to the shores of the Black Sea (Fig. 1–11).

Numerous structural implications arise from this, affecting all EU countries. A common agricultural policy is now even more difficult to achieve, given the poor condition of farming in most of the new members. Also, some of the former EU's poorer states, which were on the receiving end of the subsidy program that aided their development, now will have to pay up to support the much poorer new eastern members. And disputes over representation at EU's Brussels headquarters arose quickly. Even before the newest members' accession, Poland was demanding that the representative system favor medium-sized members (such as Poland and Spain) over larger ones such as Germany and France.

The Remaining Outsiders

This momentous expansion is having major geographic consequences for all of Europe, and not just the EU. As Figure 1–11 shows, following the admission of Romania and Bulgaria in 2007, two groups of countries and peoples are still left outside the Union:

1. *States emerging from the former Yugoslavia plus Albania.* In this cluster of western Balkan states, most of them troubled politically and economically, only Slovenia has achieved full membership. The remainder rank among Europe's poorest and most ethnically fractured states, but they contain nearly 25 million people whose circumstances could worsen outside "the club." Better, EU leaders reason, to move them toward membership by demanding political, social, and economic reforms, although the process will take many years. Discussions with Croatia began in 2005 and with newly independent Montenegro in 2006, but the aftermath of war and other problems are holding up Serbia's invitation. Elsewhere, ethnically and politically divided Bosnia seemed far from ready for negotiations, Albania was even less prepared, and un-

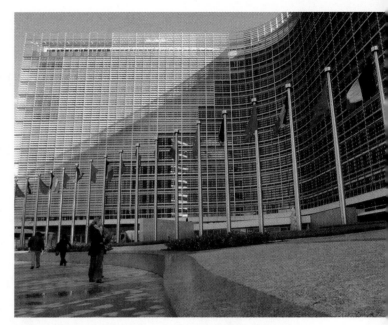

The Berlaymont building in Brussels, renovated for use by the European Commission and now known as the Berlaymont Commission Building, serves as the EU's executive headquarters. The renovation itself (some would say true to EU form) was attended by years of chaos and scandal, cost overruns, and lengthy delays, and locals refer to it as the "Berlaymonster." You might think that the EU flag and the national flags of the member–states would fly from the poles, but look carefully and you will see another pattern. Strange, *n'est ce pas?*
© Associated Press.

resolved issues continue to roil newly-independent Kosovo (see p. 95). Note that this area of least-prepared countries also is the most Islamic corner of Europe.

2. *Ukraine and its neighbors.* Four former Soviet republics in Europe's "far east" could some day join the EU, (even Belarus, until recently the most disinterested, may now be leaning in that direction). The government of Ukraine, however, has shown interest, although its electorate is strongly (and regionally) divided between a pro-EU west and a pro-Russian east. As we shall see later, Moldova views the EU as a potential supporter in a struggle against devolution along its eastern flank, and in 2006 a country not even on the map—Georgia, across the Black Sea from Ukraine—proclaimed its interest in EU membership for similar reasons: to find an ally against Russian intervention in its internal affairs.

An Islamic EU Member?

We should take note of still another potential candidate for EU membership: Turkey. EU leaders would like to include a mainly Muslim country in what Islamic states

Europe: How Desirable Is Economic and Political Union?

IN SUPPORT OF UNION

"As a public servant in the French government I have been personally affected by the events of the past half-century. I am in favor of European unification because Europe is a realm of comparatively small countries, all of which will benefit from economic and political union. Europe's history is bedeviled by its fragmentation—its borders and barriers and favored cores and disadvantaged peripheries, its different economies and diverse currencies, its numerous languages and varied cultures. Now we have the opportunity and determination to overcome these divisions and to create an entity whose sum is greater than its parts, a supranational Europe that will benefit all of its members.

"In the process we will not only establish a common currency, eliminate trade tariffs and customs barriers among our members, and enable citizens of one member-state to live and work in any of the others. We will also make European laws that will supersede the old "national" laws and create regulations that will be adhered to by all members on matters ranging from subsidies to help small farms survive to rules regarding environmental protection. In order to assist the poorer members of this European Union, the richer members will send billions of euros to the needy ones, paid for from the taxes levied on their citizens. This has already lifted Spain and Ireland from poverty to prosperity. But to ensure that "national" economies do not get out of line, there will be rules to govern their fiscal policies, including a debt limit of 3 percent of GDP.

"The EU will require sacrifices from people and governments, but the rewards will make these worth it. Our European Parliament, with representatives from all member-countries, is the harbinger of a truly united Europe. The European Commission deals with the practical problems as we move toward unification, including the momentous EU enlargement of 2004. A half-century after the first significant steps were taken, the EU has 27 member-states including former wartime enemies, reformed communist economies, and newly independent entities. Rather than struggling alone, these member-states will form part of an increasingly prosperous and economically powerful whole.

"I hope that administrative coordination and economic union will be followed by political unification. The ultimate goal of this great movement should be the creation of a federal United States of Europe, a worthy competitor and countervailing force in a world dominated by another federation, the United States of America."

IN OPPOSITION TO THE EUROPEAN UNION

"I am a taxi driver in Vienna, Austria, and here's something to consider: the bureaucrats, not the workers, are the great proponents of this European Union plan. Let me ask you this: how many member-countries' voters have had the chance to tell their governments whether they want to join the EU and submit to all those confounded regulations? It all happened before we realized the implications. At least the Norwegians had common sense: they stayed out of it, because they didn't want some international body to tell them what to do with their oil and gas and their fishing industry. But I can tell you this: if the workers of Europe had had the chance to vote on each step in this bureaucratic plot, we'd have ended it long ago.

"So what are we getting for it? Higher prices for everything, higher taxes to pay for those poor people in southern Italy and needy Portugal, costly environmental regulations, and—here's the worst of it—a flood of cheap labor and an uncontrolled influx of immigrants. Notice that this recent 2004 expansion wasn't voted on by the Germans (they're too small to matter much, but Germany is the largest EU member), and do you know why? Because the German government knew well and good that the people wouldn't support this ridiculous scheme. This whole European Union program is concocted by the elite, and we have to accept it whether we like it or not.

"Does make me wonder, though. Do these EU planners have any notion of the real Europe they're dealing with? They talk about how those 'young democracies' joining in 2004 will have their democratic systems and their market reforms locked into place by being part of the Union, but I have friends who try to do business in Poland and Slovakia, and they tell me that there may be democracy, but all they experience is corruption, from top to bottom. I'll bet that those subsidies we'll be paying for to help these 'democracies' get on their feet will wind up in Swiss bank accounts. By the way, the Swiss, you'll notice, aren't EU members either. Wonder why?

"A United States of Europe? Not as far as I am concerned. I am Austrian, and I am and will always be a nationalist as far as my country is concerned. I speak German, but already English is becoming the major EU language, though the Brits were smart enough not to join the euro. I'm afraid that these European Commission bureaucrats have their heads in the clouds, which is why they can't see the real Europe they're putting in a straitjacket."

Regional **ISSUE**

Vote your opinion at www.wiley.com/college/deblij

sometimes call the "Christian Club," but Turkey needs progress on social standards, human rights, and economic policies before its accession can be contemplated. Still, the fact that the EU has now reached deeply into eastern Europe, encompasses 27 members, has a common currency and a parliament, is developing a constitution, and is even considering expansion beyond the realm's borders constitutes a tremendous and historic achievement in this fragmented, fractious part of the world. The EU now has a combined population of just under 500 million constituting one of the world's richest markets; its member-states account for more than 40 percent of the world's exports.

It is remarkable that all this has been accomplished in little more than half a century. Some of the EU's leaders want more than economic union: they envisage a United States of Europe, a political as well as an economic competitor for the United States. To others, such a "federalist" notion is an abomination not even to be mentioned (the British in general are especially wary of such an idea). Whatever happens, Europe is going through still another of its revolutionary changes, and when you study its evolving map you are looking at history in the making.

Centrifugal Forces

For all its dramatic progress toward unification, Europe remains a realm of geographic contradictions. Europeans are well aware of their history of conflict, division, and repeated self-destruction. Will supranationalism finally overcome the centrifugal forces that have so long and so frequently afflicted this part of the world?

Devolutionary Pressures

Even as Europe's states have been working to join forces in the EU, many of those same states are confronting **20** severe centrifugal stresses. The term **devolution** has come into use to describe the powerful centrifugal forces whereby regions or peoples within a state, through negotiation or active rebellion, demand and gain political strength and sometimes autonomy at the expense of the center. Most states exhibit some level of internal regionalism, but the process of devolution is set into motion when a key centripetal binding force—the nationally accepted idea of what a country stands for—erodes to the point that a regional drive for autonomy, or for outright secession, is launched. As Figure 1-12 shows, numerous European countries are affected by devolution.

States respond to devolutionary pressures in various ways, ranging from accommodation to suppression. One way to deal with these centrifugal forces is to give his-toric regions (such as Scotland or Catalonia) certain rights and privileges formerly held exclusively by the national government. Another answer is for EU member-states to create new administrative divisions that will allow the state to meet some (or all) regional demands. The EU's central administration in Brussels has its own ideas as to what such regions or provinces should look like; indeed, many of the new State maps you will see later in this chapter (Figs. 1-15, 1-17, 1-20, 1-21) often result from negotiations between the EU and the national government. But not all EU members are satisfied with the wisdom of the administrators in Brussels.

In some instances, the devolutionary forces were just too strong for any accommodation. Even before they became members of the EU, the Czechs and Slovaks, who had both been part of a country called Czechoslovakia, peacefully decided in 1993 that it was better to separate in what became known as the "velvet divorce" (today you see them on the EU map as the Czech Republic and Slovakia). Elsewhere, as we will see, devolutionary forces produced violent fragmentation, and the potential exists for more of this. So even as the EU is consolidating, the threat of disintegration still haunts parts of the realm.

The EU's New Economic Geography

Up to this point in the chapter, we have on occasion made casual references to longstanding distinctions between western and eastern as well as northern and southern Europe. Certainly those old regional designations still have some validity: "Western Europe" remains the core of the EU; "Eastern Europe" still is, on average, the poorer, lagging part of the realm; and the cultural landscapes of northern (Nordic) and southern (Mediterranean) Europe, though converging, still differ markedly. But the establishment and continuing growth of the European Union have generated a new economic landscape that today not only transcends the old but is fundamentally reshaping the realm's regional geography (as we will see in the next section).

Engines of Growth

By investing heavily in new infrastructure and by smoothing the flows of money, labor, and products, European planners have dramatically reduced the divisive effects of their national boundaries. And by acknowledging demands for greater freedom of action by their provinces, States, departments, and other administrative

FIGURE 1-12

© H. J. de Blij, P. O. Muller, and John Wiley & Sons, Inc.

units of their countries, European leaders unleashed a wave of economic energy that transformed some of these units into powerful engines of growth. When we look at regions and states in the second part of this chapter, we will have to pay attention to these emerging powerhouses, which tend to be focused on major cities.

Four of these growth centers are especially noteworthy, **21** to the point that geographers refer to them as the **Four Motors of Europe**: (1) France's southeastern *Rhône-Alpes Region*, centered on the country's second-largest city, Lyon; (2) *Lombardy* in north-central Italy, focused on the industrial city of Milan; (3) *Catalonia* in northeastern Spain, anchored by the cultural and manufacturing center of Barcelona; and (4) *Baden-Württemberg* in southwestern Germany, headquartered by the high-tech city of Stuttgart. Often, the local governments in these subregions simply bypass the governments in their national capitals, dealing not only with each other but even with foreign governments as their business networks span the globe. In this they are imitated by other provinces, all seeking to foster their local economies and, in the process, strengthen their political position relative to the state.

Regional States

As Europe's internal political barriers to trade disappear, and regional transferability grows accordingly, the realm's economic geography changes in important ways. Provinces, States, and other subnational political units on opposite sides of international boundaries can now cooperate in pursuit of shared economic goals. Such cross-border cooperation creates a new economic map that seems to ignore the older political one, creating what the Japanese economist Kenichi Ohmae calls **22** **regional states**, meaning regional *economic* states. Moreover, this can occur at every geographic scale—from small, start-up, localized cross-border investments to massive growth engines such as the Four Motors—and it is further changing the realm's economic landscape.

All these developments underscore how far-reaching and irreversible Europe's current transformation is. European leaders understood the need for it; the Marshall Plan jump-started it; good economic times sustained it; and the end of the Cold War galvanized it. But Europe's ultimate supranational goals have not yet been attained. The problems of the old Eastern Europe remain evident in the Data

Table at the end of this book. The sense of have and have-not, core and periphery, still pervades EU operations and negotiations. Many Europeans, perhaps a majority, feel that the EU project is a bureaucratic edifice constructed by leaders who are out of touch with grass-roots Europe. Thus Europe's quest for enduring unity continues.

POINTS TO PONDER

- New states are still forming in Europe, where the modern state originated; Kosovo is the latest, but cultural passions may yet endanger that project.

- A resurgence of Scotland's independence movement may imperil the unity of the "United" Kingdom.

- Public support in Islamic Turkey in favor of joining the (mostly Christian) European Union has dropped from more than 70 to less than 30 percent.

- Belarus, quarreling with Russia, shows signs of tilting toward the EU.

Regions of the Realm: Core and Periphery

Our objective in this and later regional discussions is to become familiar with the regional and national frameworks of the realms under investigation, and to discover how individual states fit into and function within the larger mosaic. Even (perhaps especially) in this era of globalization, and even in integrating Europe, the state still plays a key role in the process. Where these states are located, how they interact, and what their prospects may be in this changing world are among the key questions we will address.

Europe presents us with a particular challenge because of its numerous countries and territories, and because of the continuing changes affecting its geography. Readers of this book who reside in North America tend to be accustomed to a familiar and relatively stable map of two countries, with the United States divided into 50 States and Canada into 10 provinces and 3 territories; the biggest change in decades came in 1999 when Canada created the Territory of Nunavut. In Europe, on the other hand, 15 new national names have appeared on the political map since 1990, some of them revivals of old entities such as Estonia and Lithuania, others new and less familiar (Montenegro, Moldova). And the process appears far from over: Kosovo became nominally independent in 2008, and Belgium may yet fracture the way Czechoslovakia did.

And Europe displays something else North America does not: *microstates* that do not have the attributes of "complete" states but are on the map as tiny yet separate entities

MAJOR CITIES OF THE REALM

City	Population* (in millions)
Amsterdam, Netherlands	1.0
Athens, Greece	3.2
Barcelona, Spain	5.0
Berlin, Germany	3.4
Brussels, Belgium	1.7
Frankfurt, Germany	3.7
London, UK	8.6
Lyon, France	1.4
Madrid, Spain	5.7
Milan, Italy	4.0
Paris, France	9.9
Prague, Czech Republic	1.2
Rome, Italy	3.3
Stuttgart, Germany	2.6
Vienna, Austria	2.3
Warsaw, Poland	1.7

*Based on 2010 estimates.

FIGURE 1-13

© H. J. de Blij, P. O. Muller, and John Wiley & Sons, Inc.

nonetheless, such as Monaco, San Marino, Andorra, and Liechtenstein. Add to these the Russian exclave (outlier) of Kaliningrad (on the Baltic Sea) and the British dependency of Gibraltar (at the entrance to the Mediterranean Sea), and Europe seems to have a bewildering political mosaic indeed. The good news is that no comparable part of any other world geographic realm is as complicated as this.

How can we make it easier to refer to these countries and territories locationally? In the old days, Europe divided rather easily into Western, Northern (Nordic), Medi-

terranean (Southern), and Eastern, and its countries could conveniently be grouped into these regions. Some of this historical geography lingers in cultural landscapes and still has residual relevance. But in the era of the European Union and globalization, that regionalization scheme has become outdated—except as a very general designation (it helps to know that Moldova lies in the east, Gibraltar in the south, Luxembourg in the west). So the "historic regions" shown in Fig. 1-13 are not regions in the functional sense.

Core and Periphery

In the Introduction, we encountered the core-periphery phenomenon (pp. 28–30) that presents itself at all levels of spatial generalization. Europe is no exception, and the original Common Market still anchors what has become a *core area* for the entire European realm (Fig. 1-13). The British Isles form part of this Core, but the British—whose EU membership was delayed by a French veto and who never adopted the euro—still are not the full-fledged members the Germans, the French, and the Irish are. It is therefore appropriate to consider the United Kingdom as a geographically distinct component of the European Core.

In the discussion that follows, we will focus first on countries such as Germany and France that lie entirely within the Core region we have delimited. Next, we discuss countries with significant regions inside as well as outside the Core, for example Sweden, Spain, and Italy. Finally, our attention focuses on Europe's Periphery.

● STATES OF THE MAINLAND CORE

Not counting the microstate of Liechtenstein, eight states form the Mainland Core of Europe: dominant Germany and France, the three Benelux countries, the two landlocked mountain states of Switzerland and Austria, and the Czech Republic (Fig. 1-14). This is the European region sometimes still referred to as Western Europe, and its most populous country (Germany) also has Europe's largest economy.

Reunited Germany

Twice during the twentieth century Germany plunged Europe and the world into war, until, in 1945, the defeated and devastated German state was divided into two parts, West and East (see the delimitation in red on Fig. 1-15). Its eastern boundaries were also changed, leaving the industrial district of Silesia in newly defined Poland, that of Saxony in communist-ruled and Soviet-controlled East Germany, and the Ruhr in West Germany (see Fig. 1-14). Aware that these were the industrial centers that had enabled Nazi Germany to seek world domination through war, the victorious allies laid out this new boundary framework to make sure this would never happen again.

In the aftermath of World War II, Soviet and Allied administration of East and West Germany differed. Soviet rule in East Germany was established on the Russian-communist model and, given the extreme hardships the USSR had suffered at German hands during the war, harshly punitive. The American-led authority in West Germany was less strict and aimed more at rehabilitation. When the Marshall Plan was instituted, West Germany was included, and its economy recovered rapidly. Meanwhile, West Germany was reorganized politically into a modern federal state on democratic foundations.

Revival

West Germany's economy thrived. Between 1949 and 1964 its gross national income (GNI) tripled while industrial output rose 60 percent. It absorbed millions of German-speaking refugees from eastern Europe (and many escapees from communist East Germany as well). Since unemployment was virtually nonexistent, hundreds of thousands of Turkish Kurds and other foreign workers arrived to take jobs Germans could not fill or did not want.

Simultaneously, West Germany's political leaders participated enthusiastically in the OEEC and in the negotiations that led to the six-member Common Market. Geography worked in West Germany's favor: it had common borders with all but one of the EEC member-states. Its transport infrastructure, rapidly rebuilt, was second to none in the realm. More than compensating for its loss of Saxony and Silesia were the expanding Ruhr (in the hinterland of the Dutch port of Rotterdam) and the newly emerging industrial complexes centered on Hamburg in the north, Frankfurt (the leading financial hub as well) in the center, and Stuttgart in the south. West Germany exported huge quantities of iron, steel, motor vehicles, machinery, textiles, and farm products. To this day, Germany has the largest export economy by value in the world.

Setback and Shock

No economy grows without setbacks and slowdowns, however, and Germany experienced such problems in the 1970s and 1980s, when energy shortages, declining competitiveness on world markets, lagging modernization (notably in the aging Ruhr), and social dilemmas involving rising unemployment, an aging population, high taxation, and a backlash against foreign resident workers plagued West German society. And then, quite suddenly, the collapse of the communist Soviet Union opened the door to reunification with East Germany.

Germany Restored

In 1990, West Germany had a population of about 62 million and East Germany 17 million. Communist misrule in the East had yielded outdated factories, crumbling infrastructures, polluted environments, drab cities, inefficient

FIGURE 1-14

© H. J. de Blij, P. O. Muller, and John Wiley & Sons, Inc.

farming, and inadequate legal and other institutions. Reunification was more a rescue than a merger, and the cost to West Germany was enormous. When the West German government imposed sales-tax increases and an income-tax surcharge on its citizens, many Westerners doubted the wisdom of reunification. It was projected that it would take decades to reconstruct Virginia-sized East Germany: ten years later, exports from the former East still contributed only about 7 percent of the national total. Regional disparity would afflict Germany for a very long time to come.

The Federal Republic

Before reunification, West Germany functioned as a federal state consisting of ten States or *Länder* (Fig. 1-15). East Germany had been divided under communist rule

into 15 districts including East Berlin. Upon reunification, East Germany was reorganized into six new States based on traditional provinces within its borders. Figure 1-15 makes a key point: regional disparity in terms of gross domestic product (GDP) per person* remains a serious problem between the former East and West. Note that five of former East Germany's six States (Berlin being the sole exception) are in the lowest income category, while most of the ten former West German States rank in the two higher income categories. In the first decade of this century, Germany's economy was stagnant, raising unem-

Gross domestic product (GDP) is the total value of all goods and services produced in a country (or subnational entity) by that political unit's economy during a given calendar year. GDP per capita is that total divided by the resident population.

FIGURE 1-15

© H. J. de Blij, P. O. Muller, and John Wiley & Sons, Inc.

ployment and slowing former East Germany's recovery. Nonetheless, the gap continues to narrow, and with 81.9 million inhabitants including over 7 million foreigners and more than 4 million ethnic Germans born outside the country, Germany is again exerting its dominance over a mainland Europe in which it has no peer.

France

German dominance in the European Union is a constant concern in the other leading Mainland Core country. The French and the Germans have been rivals in Europe for centuries. France (population: 62.6 million) is an old state, by most measures the oldest in this region. Ger-

many is a young country, created in 1871 after a loose association of German-speaking states had fought a successful war against . . . the French.

Territorially, France is much larger than Germany, and the map suggests that France has a superior relative location, with coastlines on the Mediterranean Sea, the Atlantic Ocean, and, at Calais, even a window on the North Sea. But France does not have any good natural harbors, and oceangoing ships cannot navigate its rivers and other waterways far inland.

The map of the Mainland Core region (Fig. 1-14) reveals a significant demographic contrast between France and Germany. France has one dominant city, Paris, at the heart of the Paris Basin, France's core area.

AMONG THE REALM'S GREAT CITIES . . . Paris

IF THE GREATNESS of a city were to be measured solely by its number of inhabitants, Paris (9.9 million) would not even rank in the world's top 20. But if greatness is measured by a city's historic heritage, cultural content, and international influence, Paris has no peer. Old Paris, near the Île de la Cité that housed the original village where Paris began (Fig. 1-16A) and carries the eight-century-old Notre Dame Cathedral, contains an unparalleled assemblage of architectural and artistic landmarks old and new. The Arc de Triomphe, erected by Napoleon in 1806 (though not completed until 1836), commemorates the emperor's victories and stands as a monument to French neoclassical architecture, overlooking one of the world's most famous streets, the Champs Elysées, which leads to the grandest of city squares, the Place de la Concorde, and on to the magnificent palace-turned-museum, the Louvre.

Even the Eiffel Tower, built for the 1889 International Exposition over the objections of Parisians who regarded it as ugly and unsafe, became a treasure. From its beautiful Seine River bridges to its palaces and parks, Paris embodies French culture and tradition. It is perhaps the ultimate primate city in the world.

As the capital of a globe-girdling empire, Paris was the hearth from which radiated the cultural forces of Francophone assimilation, transforming much of North, West, and Equatorial Africa, Madagascar, Indochina, and many smaller colonies into societies on the French model. Distant cities such as Dakar, Abidjan, Brazzaville, and Saigon acquired a Parisian atmosphere. France, meanwhile, spent heavily to keep Paris, especially Old Paris, well maintained—not just as a relic of history, but as a functioning, vibrant center, an example to which other cities can aspire.

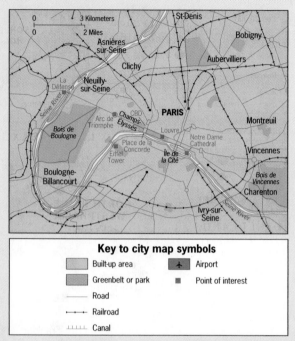

© H. J. de Blij, P. O. Muller, and John Wiley & Sons, Inc.

Today, Old Paris is ringed by a new and different Paris. Stand on top of the Arc de Triomphe and turn around from the Champs Elysées, and the tree-lined avenue gives way to La Défense, an ultramodern high-rise complex that is one of Europe's leading business districts (see photo, p. 53). But from atop the Eiffel Tower you can see as far as 80 kilometers (50 mi) and discern a Paris visitors rarely experience: grimy, aging industrial quarters, and poor, crowded neighborhoods where discontent and unemployment fester—and where Muslim immigrants cluster in a world apart from the splendor of the old city.

No other city in France comes close to Paris in terms of population or centrality: Paris has 9.9 million residents, whereas its closest rival, Lyon, has only 1.4 million. Germany has no city to match Paris, but it does have a number of cities with populations between 1 and 5 million, and is much more highly urbanized overall (89 percent) than France (77 percent).

Paris: Site and Situation

Why should Paris, without major raw materials nearby, have grown so large? Whenever geographers investigate the evolution of a city, they focus on two important loca-

23 tional qualities: its **site** (the physical attributes of the place

it occupies) and its **situation** (its location relative to surrounding areas of productive capacity, other cities and towns, barriers to access and movement, and other aspects of the greater regional framework in which it lies). **24**

The site of the original settlement at Paris lay on an island in the Seine River, a defensible place where the river was often crossed. This island, the *Île de la Cité*, was a Roman outpost 2000 years ago; for centuries its security ensured continuity. Eventually the island became overcrowded, and the city expanded along the banks of the river (Fig. 1-16A).

Soon the settlement's advantageous situation stimulated its growth and prosperity. Its fertile agricultural hinterland thrived, and, as an enlarging market, Paris's focality increased steadily. The Seine River is joined near

A　　　　　　　　　　　SITE　|　SITUATION　　　　　　　　　　**B**

SITE AND SITUATION OF PARIS, FRANCE

▪▪▪▪ 12th Century Wall	—— Roads	▮ Forests, Parks
—— Wall of 1840s	—— Railroads	▯ Built-up area

FIGURE 1-16　　　　　　　　　　　　　　© H. J. de Blij, P. O. Muller, and John Wiley & Sons, Inc.

Paris by several navigable tributaries (the Oise, Marne, and Yonne). When canals extended these waterways even farther, Paris was linked to the Loire Valley, the Rhône-Saône Basin, Lorraine (an industrial area in the northeast), and the northern border with Belgium. When Napoleon reorganized France and built a radial system of roads—followed later by railroads—that focused on Paris from all parts of the country, the city's primacy was assured (Fig. 1-16B). The only disadvantage in Paris's situation lies in its seaward access: oceangoing ships can sail up the Seine River only as far as Rouen.

Modern France

Paris, in accordance with Weber's agglomeration principle, grew into one of Europe's greatest cities. French industrial development was less spectacular, but northern French agriculture remained Europe's most productive and varied, exploiting the country's wide range of soils and climates and enjoying state subsidies and protections. Today France's economic geography is marked by new high-tech industries. It is a leading producer of high-speed trains, aircraft, fiber-optic Communications systems, and space-related technologies. It also is the world leader in nuclear power, which currently supplies more than 80 percent of its electricity and thereby reduces its dependence on foreign oil imports.

Napoleon's Legacy

When Napoleon reorganized France in the early 1800s, he broke up the country's large traditional subregions and established more than 80 small *départements* (additions and subdivisions later increased this number to 96). Each *département* had representation in Paris, but the power was concentrated in the capital, not in the individual *départements*. France became a highly centralized state and remained so for nearly two centuries (see inset map, Fig. 1-17). Only the island *département* of Corsica produced a rebel movement, whose violent opposition to French rule continued for decades and even touched the mainland. In 2003 the voters in Corsica rejected an offer of special status for their island, including limited autonomy. They want more, and trouble lies ahead.

Decentralizing the State

Today, France is decentralizing. A new subnational framework of 22 historically significant provinces, groupings of *départements* called **regions** (Fig. 1-17), has been established to accommodate the devolutionary forces felt throughout Europe and, indeed, throughout the world. These regions, though still represented in the Paris government, have substantial autonomy in

REGIONS OF FRANCE

GDP PER CAPITA, IN EUROS, 2005

- Over 26,000
- 22,000–26,000
- Under 22,000

—— Railroad

Data Source: Eurostat, 2008

City population

- Under 50,000
- 50,000–250,000
- 250,000–1,000,000
- 1,000,000–5,000,000
- Over 5,000,000

National capitals are underlined

0   100   200   300 Kilometers
0   50   100   150 Miles

DÉPARTEMENTS

FIGURE 1-17

© H. J. de Blij, P. O. Muller, and John Wiley & Sons, Inc.

such spheres as taxation, borrowing, and development spending. The cities that anchor them benefit because they are the seats of governing regional councils that can attract investment, not only within France but also from abroad.

Lyon, France's second city and headquarters of the region named Rhône-Alpes, has become a focus for growth industries and multinational firms. This region is evolving into a self-standing economic powerhouse that is becoming a driving force in the European economy; indeed, it is one of the Four Motors of Europe

(see p. 62) with its own international business connections to countries as far away as China and Chile.

France has one of the world's most productive and diversified economies, based in one of humanity's richest cultures and vigorously promoted and protected (notably its heavily subsidized agricultural sector). And although France and Germany agree on many aspects of EU integration, they tend to differ on important issues. Old, historically centralized France is less eager than young, federal Germany to push political integration in supranational Europe.

Benelux

Three small countries are crowded into the northwest corner of the Mainland Core—the Netherlands, Belgium, and Luxembourg—and are collectively referred to by their first syllables (*Be-Ne-Lux*). Their total population, 27.7 million, reminds us that this is one of the most densely peopled corners of our planet. They are also a highly productive trio: both the Netherlands and Belgium rank among the top 20 economies of the world, and tiny Luxembourg has the world's highest per-capita gross national income. The seafaring Dutch had a thriving agricultural economy and a rich colonial empire; the Belgians forged ahead during the Industrial Revolution; and Luxembourg came into its own after the Second World War as a financial center for the evolving European Union and, indeed, the world.

The Netherlands, one of Europe's oldest democracies and a constitutional monarchy today, has for centuries been expanding its living space—not by warring with its neighbors but by wresting land from the sea. Its greatest project so far, the draining and reclaiming of almost the entire Zuider Zee (Southern Sea), began in 1932 and continues. In the southwestern province of Zeeland, islands are being connected by dikes and the water is being pumped out, adding still more *polders* (as the Dutch call inhabited land claimed from the sea lying behind dikes and below sea level) to the national territory. Among the technologies in which the Dutch lead the world, not surprisingly, is the engineering of coastal systems to control the sea and protect against storms.

The regional geography of this highly urbanized country (16.5 million) is noted for the *Randstad*, a roughly triangular urban core area anchored by Amsterdam, the constitutional capital, Rotterdam, Europe's largest port, and The Hague, the seat of government. This **conurbation**, as geographers call large urban areas when two or more cities merge spatially, now forms a ring-shaped complex that surrounds a still-rural center (in Dutch, *rand* means edge or margin; *stad* means city).

The economic geography of the Netherlands, like that of the other states in this region, is heavily dominated by services, finance, and trade; manufacturing contributes about 15 percent of the value of the GDP annually, and farming, once a mainstay, less than 3 percent. Amsterdam's airport, Schiphol, is regularly recognized as Europe's best; Rotterdam, one of the world's busiest ports, is also rated as the most efficient.

Belgium also has a thriving economy and a major port, Antwerp. But Belgium is a much younger state than the Netherlands, becoming independent within its present borders as recently as 1830. With 10.7 million people, Belgium's regional geography is dominated by a cultural fault line that cuts diagonally across the country, sep-

No country has a more appropriate name than the Netherlands: most of its land, and the majority of its population, exists below sea level. Over centuries of increasingly sophisticated engineering, the Dutch created a system of artificial levees, dikes, pumps (originally windmill-driven), drainage canals, and locks that kept the water out. But that system failed disastrously in February 1953 when a storm with hurricane-force winds breached the dikes of numerous polders, flooding 162,000 hectares (400,000 acres) of southwestern Holland, killing more than 1800 people, drowning tens of thousands of livestock, and causing incalculable property damage. In response, the government imposed a "disaster tax" and, drawing on Marshall Plan funds and national reserves, began a comprehensive project to prevent future calamities of this kind. What you see here is part of this gigantic Delta Plan, consisting of a system of storm-surge barriers containing 61 enormous sluices normally open but closed when a threat arises. The Delta Plan shortens the Dutch North Sea coastline by 720 kilometers (450 mi), mainly by closing off the channels between the islands of the province of Zeeland (see Fig. 1-14). © AP/Wide World Photos.

arating a Flemish-speaking majority (58 percent) centered on Flanders in the northwest from a French-speaking minority in southeastern Wallonia (31 percent). Brussels, the mainly French-speaking capital, lies like a cultural island in the Flemish-speaking sector; but the city also is one of Belgium's greatest assets because it serves as the headquarters, and in many ways as the functional capital, of the European Union. Still, devolution is a looming problem for Belgium, with political parties espousing Flemish separatism roiling the social landscape.

Luxembourg, one of Europe's many ministates, lies landlocked between Germany, Belgium, and France, with a Grand Duke as head of state (its official name is the Grand Duchy of Luxembourg), a territory of only 2600 square kilometers (1000 sq mi), and a population of just half a million. Luxembourg has translated sovereignty, relative location, and stability into a haven for financial, service, and information-technology industries. In 2007 there were more than 160 banks in this tiny country and nearly 14,000 holding companies (corporations that hold controlling stock in other businesses in Europe and worldwide). With its unmatched per-capita income (U.S. $64,400 in 2007) Luxembourg is in some ways the greatest beneficiary of the advent of the European Union. And in no country in Europe is support for the EU stronger than it is here.

The Alpine States

Switzerland, Austria, and the *microstate* of Liechtenstein on their border share an absence of coasts and the mountainous topography of the Alps—and little else (Fig. 1-14). Austria speaks one language; the Swiss speak German in their north, French in the west, Italian in the southeast, and even a bit of Rhaeto-Romansch in the remote central highlands (Fig. 1-8). Austria has a large primate city; multicultural Switzerland does not. Austria has a substantial range of domestic raw materials; Switzerland does not. Austria is twice the size of Switzerland and has a larger population, but far more trade crosses the Swiss Alps between western and Mediterranean Europe than crosses Austria.).

Switzerland, not Austria, is in most ways the leading state in the Alpine subregion of the Mainland Core. Mountainous terrain and **landlocked location** can constitute crucial barriers to economic development, tending to inhibit the dissemination of ideas and innovations, obstruct circulation, constrain farming, and divide cultures. That is why Switzerland is such an important lesson in human geography. Through the skillful maximization of their opportunities (including the transfer needs of their neighbors), the Swiss have transformed their seemingly restrictive environment into a prosperous state. They used the waters cascading from their mountains to generate hydroelectric power to develop highly specialized industries. Swiss farmers perfected ways to optimize the productivity of mountain pastures and valley soils. Swiss leaders converted their country's isolation into stability, security, and neutrality, making it a world banking giant, a global magnet for money. Zürich, in the German sector, is the financial center; Geneva, in the French sector, is one of the world's most internationalized cities. The Swiss feel that they do not need to join the EU—and they have not done so.

Austria, which joined the EU in 1995, is a remnant of the Austro-Hungarian Empire and has a historical geography that is far more reminiscent of unstable eastern Europe than that of Switzerland. Even Austria's physical geography seems to demand that the country look eastward: it is at its widest, lowest, and most productive in the east, where the Danube links it to Hungary, its old ally in the anti-Muslim wars of the past.

Vienna, by far the Alpine subregion's largest city, also lies on the country's eastern perimeter. One of the world's most expressive primate cities with magnificent architecture and monumental art, Vienna today is the Mainland Core's easternmost city, but Vienna's relative location changed dramatically with EU enlargement in 2004 and 2007. Peripheral to the EU and a vanguard of the Core until 2004, Vienna found itself in a far more central position when the EU border moved eastward. But many Austrians had their doubts about their neighbors to the east becoming EU members of potentially equal standing, and in Austria public support for the European Union has fallen to the lowest level of any member-state.

The Czech Republic

The Czech Republic, product of the 1993 Czech-Slovak "velvet divorce," centers on the historic province of Bohemia, the mountain-encircled national core area that focuses on the capital, Prague. This is a classic primate city, its cultural landscape faithful to Czech traditions; but it is also an important industrial center. The surrounding mountains contain many valleys with small towns that specialize, Swiss-style, in fabricating high-quality goods. In the old Eastern Europe, even during the communist period, the Czechs always were leaders in technology and engineering; their products could be found on markets in foreign countries near and far.

Bohemia always was cosmopolitan and Western in its exposure, outlook, development, and linkages; Prague lies in the basin of the Elbe River, its traditional outlet through northern Germany to the North Sea. Today, the Czech Republic (10.4 million) is reclaiming its position at the center of European action.

● THE CORE OFFSHORE: THE BRITISH ISLES

As Figure 1-13 shows, two countries form the maritime portion of the European core area: the United Kingdom and Ireland (Fig. 1-18). These countries lie on two major islands, surrounded by a constellation of tiny ones. The larger island, a mere 34 kilometers (21 mi) off the mainland at the closest point, is called *Britain*; its smaller neighbor to the west is *Ireland*.

States and Peoples

The names attached to these islands and the countries they encompass are the source of some confusion. They still are called the British Isles, even though British dominance over

FIGURE 1-18

© H. J. de Blij, P. O. Muller, and John Wiley & Sons, Inc.

most of Ireland ended in 1921. The state that occupies Britain and the northeastern corner of Ireland is officially called the United Kingdom of Great Britain and Northern Ireland—United Kingdom for short and UK by abbreviation. But this country often is referred to simply as Britain, and its people are known as the British. The state of Ireland officially is known as the Republic of Ireland (*Eire* in Irish Gaelic), but it does not include the entire island of Ireland.

How convenient it would be if physical and political geography coincided! Unfortunately, the two do not. During the long British occupation of Ireland, which is overwhelmingly Catholic, many Protestants from northern Britain settled in northeastern Ireland. In 1921, when British domination ended, the Irish were set free—except in that corner in the north, where London kept control to protect the area's Protestant settlers. That is why the

AMONG THE REALM'S GREAT CITIES . . . London

Sail **WESTWARD UP** the meandering Thames River toward the heart of London, and be prepared to be disappointed. London does not overpower or overwhelm with spectacular skylines or beckoning beauty. It is, rather, an amalgam of towns—Chelsea, Chiswick, Dulwich, Hampstead, Islington—each with its own social character and urban landscape. Some of these towns come into view from the same river the Romans sailed 2000 years ago: Silvertown and its waterfront urban renewal; Greenwich with its famed Observatory; Thamesmead, the model modern commuter community. Others somehow retain their identity in the vast metropolis that remains (Fig. 1-9), in many ways Europe's most civilized and cosmopolitan city. And each contributes to the whole in its own way: every part of London, it seems, has its memories of empire, its memorials to heroes, its monuments to wartime courage.

Along the banks of the Thames, London displays the heritage of state and empire: the Tower and the Tower Bridge, the Houses of Parliament (officially known as the Palace of Westminster), the Royal Festival Hall. Step ashore, and you find London to be a memorable mix of the historic and (often architecturally ugly) modern, of the obsolete and the efficient, of the poor and the prosperous. Public transportation, by world standards, is excellent, even though facilities are aging rapidly; traffic, however, often is chaotic, gridlocked by narrow streets. Recreational and cultural amenities are second to none. London seems to stand with one foot in the twenty-first century and the other in the nineteenth. Its airports are ultramodern. But the Eurostar TGV train from Paris had to wait for decades until upgraded tracks were finally opened in 2007 to allow it to operate at its normal high-

© H. J. de Blij, P. O. Muller, and John Wiley & Sons, Inc.

speed between the Channel Tunnel and London's St. Pancras Station; the extension of such a rail network into other parts of Britain remained only a dream in 2010.

London remains one of the world's most livable metropolises for its size (8.6 million) largely because of farsighted urban planning that created and maintained, around the central city, a so-called Greenbelt set aside for recreation, farming, and other nonresidential, noncommercial uses (Fig. 1-9). Although London's growth eroded this Greenbelt, in places leaving only "green wedges," the design preserved crucial open space in and around the city, channeling suburbanization toward a zone at least 40 kilometers (25 mi) from the center. The map reveals the design's continuing impact.

country to this day is officially known as the United Kingdom of Great Britain and Northern Ireland.

Roots of Devolution

Northern Ireland (Fig. 1-18) was home not only to Protestants from Britain, but also to a substantial population of Irish Catholics who found themselves on the wrong side of the border when Ireland was liberated. Ever since, conflict has intermittently engulfed Northern Ireland and spilled over into Britain and even, in the form of terrorism, into nearby mainland Europe.

Although all of Britain lies in the United Kingdom, political divisions exist here as well. England is the largest of these units, the center of power from which the rest of the region was originally brought under unified control. The English conquered Wales in the Mid-

dle Ages, and Scotland's link to England, cemented when a Scottish king ascended the English throne in 1603, was ratified by the Act of Union of 1707. Thus England, Wales, Scotland, and Northern Ireland became the United Kingdom.

A Discrete Region

The British Isles form a distinct entity in Europe for several reasons. Britain's insularity provided centuries of security from turbulent mainland Europe, protecting the evolving British nation as it achieved a system of parliamentary government that had no peer in the Western world. Having united the Welsh, Scots, and Irish, the British set out to forge what would become the world's largest colonial empire. An era of mercantilism and domestic manufacturing (the latter based on water

THE CORE OFFSHORE: THE BRITISH ISLES

power from streams flowing off the Pennines, Britain's mountain backbone) foreshadowed the momentous Industrial Revolution, which transformed Britain—and much of the world. British cities became synonyms for specialized products as the smokestacks of factories rose like forests over the urban scene. London on the Thames River anchored an English core area that mushroomed into the headquarters of a global political, financial, and cultural empire. As recently as World War II, the narrow English Channel ensured the United Kingdom's impregnability against German invasion, giving the British time to organize their war machine. When the United Kingdom emerged from that conflict as a leading power among the victorious allies, it seemed that its superpower role in the postwar era was assured.

End of Empire

Two unanticipated developments changed that prospect: the worldwide collapse of colonial empires and the rapid resurgence of the European mainland. Always ambivalent about the EC and EU, and with its first membership application vetoed by the French in 1963, Britain (admitted in 1973) has worked to restrain moves toward tighter integration. When most member-states adopted the new euro in favor of their national currencies, the British kept their pound sterling and delayed their participation in the EMU. As for a federalized Europe, to Britain this is out of the question. In this as in other respects, Britain's historic, insular standoffishness continues.

The United Kingdom

As we have noted, the British Isles as a region consists of two political entities: the United Kingdom and Ireland. The UK, with an area about the size of Oregon and a population of 61.8 million, is by European standards quite a large country. Based on a combination of physiographic, historical, cultural, economic, and political criteria, the United Kingdom can be divided into four subregions (numbered in Fig. 1-18):

1. *England* So dominant is this subregion of the United Kingdom that the entire country is sometimes referred to by this name. Small wonder: England is anchored by the huge London metropolitan area, which by itself contains more than one-seventh of the UK's total population. Indeed, along with New York and Tokyo, London is regarded as one of the three leading "world cities" of the current era of globalization, with financial, high-technology, communications, engineering, and related industries reflecting the momentum of its long-term growth and agglomera-

tion. To the north and west, England was the hearth of the Industrial Revolution, its cities synonymous with manufactures of matchless variety. Today obsolescence, economic stagnation, and social problems trouble this part of the subregion.

2. *Wales* This nearly rectangular, rugged territory was a refuge for ancient Celtic peoples, and in its western counties more than half the inhabitants still speak Welsh. Because of the high-quality coal reserves in its southern tier, Wales too was engulfed by the Industrial Revolution, and Cardiff, the capital, was once the world's leading coal exporter. But the fortunes of Wales also declined, and many Welsh emigrated. Among the 3 million who remained, however, the flame of Welsh nationalism survived, and in 1997 the voters approved the establishment of a Welsh Assembly to administer public services in Wales, a first devolutionary step.

3. *Scotland* Nearly twice as large as the Netherlands and with a population about the size of Denmark's, Scotland is a major component of the United Kingdom. As Figure 1-18 suggests, most of Scotland's more than 5 million people live in the Scottish Lowlands anchored by Edinburgh, the capital, in the east and Glasgow in the west. Attracted there by the labor demands of the Industrial Revolution (coal and iron reserves lay in the area), the Scots developed a world-class shipbuilding industry. Decline and obsolescence were followed by high-tech development, notably in the hinterland of Glasgow, and Scottish participation in the exploitation of oil and gas reserves under the North Sea (Fig. 1-18), which transformed the eastern ports of Aberdeen and Leith (Edinburgh). But many Scots feel that they are disadvantaged within the UK and should play a major role in the EU. Therefore, when the British government put the option of a Scottish parliament before the voters in 1997, 74 percent approved. Many Scots still hope that sovereignty lies in their future, and in local elections in 2007 the independence-minded Scottish National Party won more seats in the parliament than any of the other parties.

4. *Northern Ireland* Prospects of devolution in Scotland pale before the devastation caused by political and sectarian conflict in Northern Ireland. With a population of 1.8 million occupying the northeastern one-sixth of the island of Ireland, this area represents the troubled legacy of British colonial rule. A declining majority, now about 53 percent of the people in Northern Ireland, trace their ancestry to Scotland or England and are Protestants; a growing minority, currently around 46 percent, are Roman Catholics, who share their Catholicism with virtually the entire population

"The Troubles" between Protestants and Catholics, pro- and anti-British factions that have torn Northern Ireland apart for decades at a cost of more 3000 lives, are etched in the cultural landscape. The so-called "Peace Wall" across West Belfast, shown here separating Catholic and Protestant neighborhoods, is a tragic monument to the failure of accommodation and compromise, a physical manifestation of the emotional divide that still runs deep—despite the political compromise engineered by former Prime Minister Tony Blair and his negotiators. In May 2007, a Northern Ireland Assembly began functioning again, after hardline Catholic and Protestant political parties were persuaded to join in the effort. But what an analyst said to the author back in 2005 still is the view across the cityscape you see here: "Good fences make good neighbors . . . the Peace Wall will have to stay up for a few more decades yet." © Barry Chin/The Boston Globe/Redux.

of the Irish Republic on the other side of the border. Although Figure 1-18 suggests that there are majority areas of Protestants and Catholics in Northern Ireland, no clear separation exists; mostly they live in clusters throughout the territory, including walled-off neighborhoods in the major cities of Belfast and Londonderry (see photo above). Partition is no solution to a conflict that has raged for four decades at a cost of thousands of lives; Catholics accuse London as well as the local Protestant-dominated administration of discrimination, whereas Protestants accuse Catholics of seeking union with the Republic of Ireland. The persistent mediation efforts of former Prime Minister Tony Blair were crucial in the creation of a Northern Ireland Assembly to which powers will be devolved from London. After its first launch failed in 2003, forcing the British government to resume direct rule, a second try succeeded in 2007 when all parties agreed to cooperate, and the Assembly began functioning again.

Republic of Ireland

What Northern Ireland has been missing through its conflicts is evident to the south, in the Irish Republic itself. Here participation in the EU, adoption of the euro, business-friendly tax policies, comparatively low wages, an English-speaking workforce, and an advantageous relative location combined to produce, around the turn of this century, the highest rate of economic growth in the entire European Union. This booming, service-based economy, accompanied by burgeoning cities and towns, fast-rising real estate prices, mushrooming industrial parks, and bustling traffic, transformed a country long known for emigration into a magnet for industrial workers, producing new social challenges for a closely knit, long-isolated society. Among these immigrants were thousands of people of Irish descent returning from foreign places to take jobs at home, workers from elsewhere in the European Union (including large numbers of Poles), and job-seekers from African and Caribbean countries.

The Republic of Ireland fought itself free from British colonial rule just three generations ago. Its cool, moist climate had earlier led to the adoption of the American potato as the staple crop, but excessive rain and a blight in the late 1840s, coupled with colonial mismanagement, caused famine and cost over a million lives. Another 2 million Irish emigrants left for North American and other shores.

Hard-won independence in 1921 did not bring economic prosperity, and Ireland stagnated into the 1990s, when its newfound advantages boosted the economy to the point that Ireland became known as the *Celtic Tiger*. But toward the end of the first decade of the twenty-first century Ireland's overheated economy began to struggle as the real estate market declined, service industries found more favorable conditions in eastern Europe, and unemployment rose. Coupled with the global recession that began in 2008, this triggered another of Ireland's many waves of emigration: not only did Poles return to their homeland, but some Irish workers themselves looked eastward for jobs lacking at home. By 2009, Ireland's first economic boom had faded.

● THE CONTIGUOUS CORE IN THE SOUTH

As Figures 1-13 and 1-19 show, the northern sectors of two major southern states form parts of the European Core: northern Italy and northern Spain. Italy and Spain are two of the four countries that constitute southernmost Europe, and in both important urbanized and industrialized subregions have become integral parts of Europe's core area.

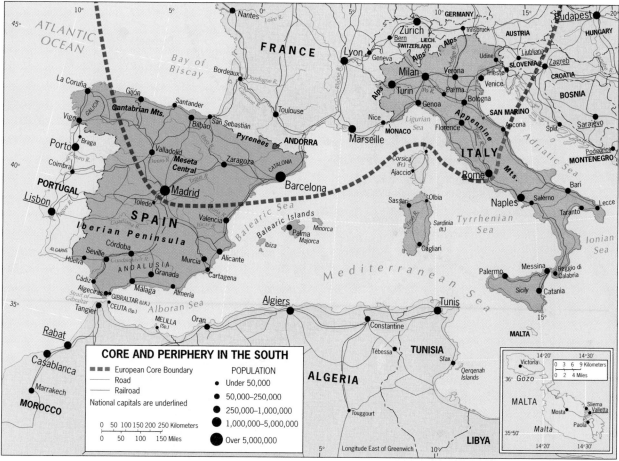

FIGURE 1-19

© H. J. de Blij, P. O. Muller, and John Wiley & Sons, Inc.

Portugal, on the western flank of the Iberian Peninsula, remains outside the European core, far less urbanized, much more agrarian, and not strongly integrated into it. The fourth national entity in this southern domain is the island ministate of Malta, south of Sicily, a historically important crossroads with a population of about 400,000 and a booming tourist industry.

Italy

Centrally located in Mediterranean Europe, most populous of the realm's southern states, best connected to the European Core, and economically most advanced is Italy (59.8 million), a charter member of Europe's Common Market.

Administratively, Italy is organized into 20 internal regions, many with historic roots dating back centuries (Fig. 1-20). Several of these regions have become powerful economic entities centered on major cities, such as Lombardy (Milan) and Piedmont (Turin); others are historic hearths of Italian culture, including Tuscany

(Florence) and Veneto (Venice). These regions in the northern half of Italy stand in strong social, economic, and political contrast to such southern regions as Calabria (the "toe" of the Italian "boot") and Italy's two major Mediterranean islands, Sicily and Sardinia. Not surprisingly, Italy is often described as two countries—a progressive north and a stagnant south or *Mezzogiorno*. The urbanized, industrialized north is part of Europe's Core; the low-income south typifies the Periphery.

North of the Ancona Line

North and south are bound by the ancient headquarters, Rome, which lies astride the narrow transition zone between Italy's contrasting halves. This clear manifestation of Europe's Core-Periphery contrast is referred to in Italy as the *Ancona Line*, named after the city on the Adriatic coast where it reaches the other side of the peninsula (Fig. 1-20, blue line). Whereas Rome remains Italy's capital and cultural focus, the functional core area of Italy has shifted northward into Lombardy in the basin of the

FIGURE 1-20   © H. J. de Blij, P. O. Muller, and John Wiley & Sons, Inc.

Po River. Here lies southern Europe's leading manufacturing complex, in which a large skilled labor force and ample hydroelectric power from Alpine and Appennine slopes combine with a host of imported raw materials to produce a wide range of machinery and precision equipment. The Milan–Turin–Genoa triangle exports appliances, instruments, automobiles, ships, and many specialized products. Meanwhile, the Po Basin, lying on the margins of southern Europe's dominant Mediterranean climatic regime (with its hot, dry summers), enjoys a more even pattern of rainfall distribution throughout the year, making it a productive agricultural zone as well.

Metropolitan Milan embodies the new, modern Italy. Not only is Milan (at 4.0 million) Italy's largest city and leading manufacturing center—making Lombardy one of Europe's Four Motors—but it also is the country's financial and service-industry headquarters. Today the Milan area, a cornerstone of the European Core, has just 7 percent of Italy's population but accounts for fully one-third of the entire country's national income.

AMONG THE REALM'S GREAT CITIES . . . Rome

FROM A HIGH VANTAGE point, Rome seems to consist of an endless sea of tiled roofs, above which rise numerous white, ochre, and gray domes of various sizes; in the distance, the urban perimeter is marked by high-rises fading in the urban haze. This historic city lives amid its past as perhaps no other as busy traffic encircles the Colosseum, the Forum, the Pantheon, and other legacies of Europe's greatest empire.

Founded about 3000 years ago at an island crossing point on the Tiber River about 25 kilometers (15 mi) from the sea, Rome had a high, defensible site. A millennium later, with a population some scholars estimate as high as 1 million, it was the capital of a Roman domain that extended from Britain to the head of the Persian Gulf and from the shores of the Black Sea to North Africa. Rome's emperors endowed the city with magnificent, marble-faced, columned public buildings, baths, stadiums, obelisks, arches, and statuary; when Rome became a Christian city, the domes of churches and chapels added to its luster.

It is almost inconceivable that such a city could collapse, but that is what happened after the center of Roman power shifted eastward to Constantinople (now Istanbul). By the end of the sixth century, Rome probably had fewer than 50,000 inhabitants, and in the thirteenth, a mere 30,000. Papal rule and a Renaissance revival lay ahead, but in 1870, when Rome became the capital of newly united Italy, it still had a population of only 200,000.

Now began a growth cycle that eclipsed all previous records. As Italy's political, religious, and cultural focus (though not an industrial center to match), Rome grew to 1 million by 1930, to 2 million by 1960, and subsequently to 3.3 million where it has leveled off today. The religious enclave of Vatican City, Roman Catholicism's headquarters, makes Rome a twin capital; the Vatican

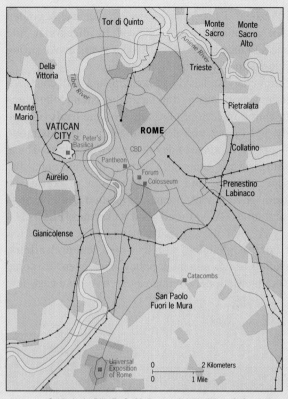

© H. J. de Blij, P. O. Muller, and John Wiley & Sons, Inc.

functions as an independent entity and has a global influence that Italy cannot equal.

Rome today remains a city whose economy is dominated by service industries: national and local government, finance and banking, insurance, retailing, and tourism employ three-quarters of the labor force. The new city sprawls far beyond the old, walled, traffic-choked center where the Roman past and the Italian future come face to face.

The Mezzogiorno

As in Germany, the lowest-income regions of Italy lie concentrated in one part of the country: the Mezzogiorno in the south. But Italy's north-south disparity continues to grow, which has led to taxpayers' revolts over the subsidies the state pays to the poorer southern regions (a separatist movement has even espoused an independent *Padania* in the north). In truth, the south receives the bulk of Italy's illegal immigrants, whether from Africa across the Mediterranean or from the states of former Yugoslavia and Albania across the Adriatic, and it is these workers, willing to work for low wages, who move north to take jobs in the factories of Milan

and Turin. Italian southerners tend to stay where they were born. Sicily and the Mezzogiorno underscore the problems of a periphery.

Spain

At the western end of southern Europe lies the Iberian Peninsula, separated from France and western Europe by the rugged Pyrenees and from North Africa by the narrow Strait of Gibraltar (Fig. 1-19). Spain (population: 46.7 million) occupies most of this compact Mediterranean landmass, and peripheral Portugal lies in its southwestern corner. Three additional tiny entities also mark Spain's political map (see the box titled "Gibraltar, Ceuta, and Melilla").

Imperial Romans, Muslim Moors, and Catholic kings left their imprints on Iberia; the boundary between Spain and Portugal dates from the twelfth century. A golden age of colonialism was followed by dictatorial rule (Spain's Franco and Portugal's Salazar still evoke strong emotions) and by economic stagnation.

Democracy replaced dictatorship, and both countries benefited enormously from their admission to the European Union in 1986. Spain followed the leads of Germany and France and decentralized its administrative structure in response to devolutionary pressures. These pressures were especially strong in Catalonia, the Basque country, and Galicia, and in response the Madrid government created so-called Autonomous Communities (ACs) for all 17 of its regions (Fig. 1-21). Every AC has its own parliament and administration that control planning, public works, cultural affairs, education, environmental policy, and even, to some extent, international commerce. Each AC can negotiate

its own degree of autonomy with the central government in the capital. Some Spanish observers feel that devolution has gone too far and that a federal system (such as Germany's) would have been preferable, but now there is no turning back.

Even so, the AC framework has not totally defused the most violent of the secessionist-minded movements, propelled by a small minority of Basques. On the other hand, relations between Madrid and Catalonia have improved. Centered on the prosperous, productive coastal city of Barcelona, the AC named Catalonia is Spain's leading industrial area and has become one of Europe's Four Motors. Catalonia is endowed with its own distinctive language and culture that find vivid expression in Barcelona's urban landscape.

As Catalans like to remind visitors to their corner of Spain, most of the country's industrial raw materials are found in the northwest, but most of its major industrial development has occurred in the northeast, where innovations and skills propel a high-technology-driven regional economy. In recent years, Catalonia—with

FIGURE 1-21

© H. J. de Blij, P. O. Muller, and John Wiley & Sons, Inc.

6 percent of Spain's territory and 16 percent of its population—has annually produced 25 percent of all Spanish exports and nearly 40 percent of its industrial exports. Such economic strength translates into political power, and in Spain the issue of Catalonian separatism is never far from the surface.

As Figure 1-21 shows, Spain's capital and largest city, Madrid, lies near the geographic center of the state. It also lies within an economic-geographic transition zone. In terms of people's annual income, Spain's north is far more affluent than its south, a direct result of the country's distribution of resources, climate (the south suffers from drought and poorer soils), and overall development opportunities. The most prosperous ACs are Barcelona-centered Catalonia and Madrid; between them, the contiguous group of ACs extending from the Basque Country to Aragon rank next. In the ACs of northern Spain, economies tend to do better in the east, where tourism and winegrowing (especially in La Rioja, a famous name in Spanish wine) are among the mainstays, than in the west, where industrial obsolescence, dwindling raw-material sources, the decline of the fishing industry, and emigration plague local economies. Spain's southernmost ACs have long been the least developed economically and are symptomatic of Europe's Periphery.

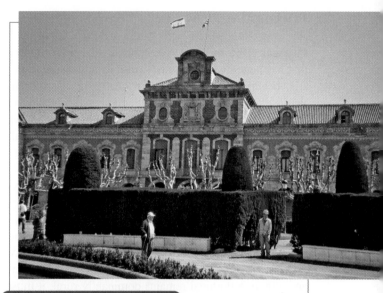

FROM THE FIELD NOTES

"Catalonian nationalism is visible both obviously and subtly in Barcelona's urban landscape. Walking toward the Catalonian Parliament, I noticed that the flags of Spain (left) and Catalonia (right) flew from slightly diverging flagpoles above the entrance to the historic building. Is there a message here?" © H. J. de Blij.

www.conceptcaching.com

Concept Caching

Gibraltar, Ceuta, and Melilla

ALTHOUGH SPAIN AND the United Kingdom are both EU members committed to cooperative resolution of territorial problems, the two countries are embroiled in a dispute over a sliver of land at the southern tip of Iberia: legendary *Gibraltar* (see lower inset map, Fig. 1-21). "The Rock" was ceded by Spain to the British in perpetuity in 1713 (though not the neck of the peninsula linking it to the mainland, which the British later occupied) and has been a British colony ever since. Its 30,000-odd residents are used to British institutions, legal rules, and schools. Successive Spanish governments have demanded that Gibraltar be returned to Madrid's rule, but the colony's residents are against it. And under their 1969 constitution, Gibraltarians have the right to vote on any transfer of sovereignty.

Now the British and Spanish governments are trying to reach an agreement under which they would share the administration of Gibraltar for an indefinite time; since both are EU members, little would have to change. The advantages would be many: the economic slowdown caused by the ongoing dispute would ease, border checks would be lifted, EU benefits would flow. But Gibraltarians have their doubts and want to put the issue to a referendum. Spain will not accept the idea of a referendum, and the matter is far from resolved.

Spain's refusal to allow, and abide by, a referendum in Gibraltar stands in marked contrast to its demand for a referendum in its own outposts, two small African *exclaves* on the coast of Morocco, *Ceuta* and *Melilla* (Fig. 1-21). Morocco has been demanding the return of these two small cities, but Spain has refused on the grounds that the local residents do not want this. The matter has long simmered quietly but came to international attention in 2002 when a small detachment of Moroccan soldiers seized the island of Perejil, an uninhabited Spanish possession off the Moroccan coast also wanted by Morocco. In the diplomatic tensions that followed, the entire question of Spain's holdings in North Africa (and its anti-Moroccan stance in the larger matter of Western Sahara [an issue discussed in Chapter 7]) exposed some contradictions Madrid had preferred to conceal.

Portugal and Malta

Portugal (10.6 million), a comparatively poor country that has benefited enormously from its admission to the EU, occupies the southwestern corner of the Iberian Peninsula. Since one rule of EU membership is that the richer members assist the poorer ones, Portugal shows the results in a massive renovation project in its capital, Lisbon, as well as in the modernization of surface transport routes.

Unlike Spain, which has major population clusters on its interior plateau as well as its coastal lowlands, the Portuguese are concentrated along and near the Atlantic coast. Lisbon and the second city, Porto, are coastal cities. The best farmlands lie in the moister western and northern zones of the country; but the farms are small and inefficient, and although Portugal remains dominantly rural, it must import as much as half of its foodstuffs. Exporting textiles, wines, corks, and fish, and running up an annual deficit, the indebted Portuguese economy remains a far cry from those of other European countries of similar dimensions—typical of the European Periphery rather than the Core.

Southernmost Europe also contains the ministate of Malta, located in the central Mediterranean Sea just south of Sicily. Malta is a small archipelago of three inhabited and two uninhabited islands with a population of just over 400,000 (Fig. 1-19, inset map). An ancient crossroads and culturally rich with Arab, Phoenician, Italian, and British infusions, Malta became a British dependency and served British shipping and its military. It suffered terribly during World War II bombings, but despite limited natural resources recovered strongly during the postwar era. Today Malta has a booming tourist industry and a relatively high standard of living, and was one of the ten new member-states to join the European Union as part of the historic expansion of 2004.

● THE DISCONTINUOUS CORE IN THE NORTH

Take another look at Figure G-11, and it is clear that the countries of northern Europe have national incomes and standards of living that are representative of the European Core. But consider the implications of Figures 1-13 and 1-22, and it is obvious that most of northern Norway, Sweden, and Finland form part of the European Periphery, not the Core.

The six countries of this northern domain of Europe (sometimes called *Nordic* Europe or *Norden*) have a combined population of only 26.4 million, which is less than half of Italy's. Nothing here compares to Italy's north or Spain's Catalonia, but the core areas of the three Scandinavian coun-

tries (Sweden, Norway, and Denmark) make up in prosperity and external linkages what they lack in dimensions. It is not even unreasonable to include the North Sea in the European Core because much of Europe's domestic energy resources have come from its oil- and gasfields.

As Figure 1-22 shows, national core areas here are coastal, low-latitude (insofar as possible), and centered on the capital cities. The language map (Fig. 1-8) reminds us why Estonia is part of this northern cluster of countries even though it is not a component of the European Core. The north's remoteness, isolation, and environmental severity also have had positive, binding effects for this domain. The countries of the Scandinavian Peninsula lay removed from the wars of mainland Europe (although Norway was overrun by Nazi Germany during World War II). The three major languages—Swedish, Norwegian, and Danish—are mutually intelligible, and in terms of religion there is overwhelming adherence to the same Lutheran church in the three Scandinavian countries as well as Iceland and Finland. Furthermore, democratic and representative governments emerged early, and individual rights and social welfare have long been carefully protected. Women participate more fully in government and politics here than in any other part of the world.

Sweden is the largest Nordic country in terms of both population (9.2 million) and territory. Most Swedes live south of 60° North latitude (which passes through Uppsala just north of the capital, Stockholm), in what is climatically the most moderate part of the country (Fig. 1-22). Here lie the primate city, core area, and the main industrial districts; here, too, are the main agricultural areas that benefit from the lower relief, better soils, and milder climate.

Sweden long exported raw or semifinished materials to industrial countries, but today the Swedes are making finished products themselves, including automobiles, electronics, stainless steel, furniture, and glassware. Much of this production is based on local resources, including a major iron ore reserve at Kiruna in the far north (there is a steel mill at Luleå). Swedish manufacturing, in contrast to that of several western European countries, is based in dozens of small and medium-sized towns specializing in particular products. Energy-poor Sweden was a pioneer in the development of nuclear power, but a national debate over the risks involved has reversed that course.

Norway does not need a nuclear power industry to supply its energy needs. It has found its economic opportunities on, in, and beneath the sea. Norway's fishing industry, now augmented by highly efficient fish farms, long has been a cornerstone of the economy, and its merchant marine spans the world. But since the 1970s, Norway's economic life has been transformed by the bounty of oil and natural gas discovered in its sector of the North Sea.

FIGURE 1-22

© H. J. de Blij, P. O. Muller, and John Wiley & Sons, Inc.

With its limited patches of cultivable soil, high relief, extensive forests, frigid north, and spectacularly fjorded coastline, Norway has nothing to compare to Sweden's agricultural or industrial development. Its cities, from the capital Oslo and the North Sea port of Bergen to the historic national focus of Trondheim as well as Arctic Hammerfest, lie on the coast and have difficult overland connections. The isolated northern province of Finnmark has even become the scene of an autonomy movement among the reindeer-herding indigenous Saami (Fig. 1-12). The distribution of Norway's population of 4.8 million has been described as a necklace, its beads linked by the thinnest of strands. But this has not constrained national development. Norway in the mid-2000s had the second-lowest unemployment rate in Europe (after tiny Luxembourg). In terms of income per capita, Norway is one of the richest countries in the world.

Norwegians have a strong national consciousness and a spirit of independence. In the mid-1990s, when Sweden and Finland voted to join the European Union, the Norwegians again said no. They did not want to trade their economic independence for the regulations of a larger, even possibly safer, Europe.

Denmark, territorially small by Scandinavian standards, has a population of 5.5 million, second largest in Norden after Sweden. It consists of the Jutland Peninsula and several islands to the east at the gateway to the Baltic Sea; it is on one of these islands, Sjaelland, that the capital of Copenhagen is located. Copenhagen, the "Singapore of the Baltic," has long been a port that collects, stores, and transships large quantities of goods. This **break-of-bulk** function exists because many oceangoing vessels cannot enter the shallow Baltic Sea, making the city an *entrepôt* where transfer facilities and activities prevail. The completion of the Øresund bridge-tunnel link to southern Sweden in 2000 enhanced Copenhagen's situation (Fig. 1-22).

Denmark remains a kingdom, and in centuries past Danish influence spread far beyond its present confines. Remnants of that period now challenge Denmark's governance. Greenland came under Danish rule after union with Norway (1380) and remained a Danish possession when that union ended (1814). In 1953, Greenland's status changed from colony to province, and in 1979 the 60,000 inhabitants were given home rule with an Inuit name: *Kalaallit Nunaat*. They promptly exercised their rights by withdrawing from the European Union, of which they had become a part when Denmark joined. Another restive dependency is the Faroe Islands, located between Scotland and Iceland. These 17 small islands and their 45,000 inhabitants were awarded self-government in 1948, complete with their own flag and currency, but even this was not enough to defuse demands for total independence. A referendum in 2001 confirmed that even Denmark is not immune from Europe's devolutionary forces (Fig. 1-12).

Finland, territorially almost as large as Germany, has only 5.3 million residents, most of them concentrated in the triangle formed by the capital, Helsinki, the textile-producing center, Tampere, and the shipbuilding center, Turku (Fig. 1-22). A land of evergreen forests and glacial lakes, Finland has an economy that has long been sustained by wood and wood product exports. But the Finns, being a skillful and productive people, have developed a diversified economy in which the manufacture of precision machinery and telecommunications equipment (prominently including cell phones) as well as the growing of staple crops are key.

As in Norway and Sweden, environmental challenges and relative location have created Nordic cultural landscapes in Finland, but the Finns are not a Scandinavian people; their linguistic and historic links are instead with the Estonians to the south across the Gulf of Finland. As we will see in Chapter 2, ethnic groups speaking Finno-Ugric languages are widely dispersed across what is today western Russia.

Estonia, northernmost of the three "Baltic states," is part of Norden by virtue of its ethnic and linguistic ties to Finland. But during the period of Soviet control from 1940 to 1991, Estonia's demographic structure changed drastically: today about 25 percent of its 1.3 million inhabitants are Russians, most of whom came there as colonizers.

After a difficult period of adjustment, Estonia today is forging ahead of its Baltic neighbors (see Data Table inside back cover) and catching up with its Nordic counterparts. Busy traffic links Tallinn, the capital, with Helsinki, and a new free-trade zone at Muuga Harbor facilitates commerce with Russia. But more important for Estonia's future was its entry into the European Union in 2004.

Iceland, the volcanic, glacier-studded island in the frigid waters of the North Atlantic just south of the Arctic Circle, is the sixth Nordic country. Inhabited by people with Scandinavian ancestries (population: 340,000), Iceland and its small neighboring archipelago, the Westermann Islands, are of special scientific interest because they lie on the Mid-Atlantic Ridge, where the Eurasian and North American tectonic plates of the Earth's crust are diverging and new land can be seen forming (see Fig. G-3).

Iceland's population is almost totally urban, and the capital, Reykjavik, contains about half the country's inhabitants. The nation's economic geography is almost entirely oriented to the surrounding waters, whose seafood harvests give Iceland one of the world's highest standards of living—but at the risk of overfishing. Disputes over fishing grounds and fish quotas have intensified in recent decades; the Icelanders argue that, unlike the Norwegians or the British, they have little or no alternative economic opportunity.

● THE EASTERN PERIPHERY

In the past, "Eastern Europe" was one of those regional designations that had various interpretations. Figure 1-13 suggests that it incorporated all of Europe east of Germany, Austria, and Italy, north of Greece, and south of Finland. The Soviet communist domination of Eastern Europe (1945–1990) behind an ideological and strategic "Iron Curtain" served to reinforce the division between "West" and "East." That fateful division entrenched the peripheral condition of the eastern states, and not even the subsequent eastward expansion of the EU could, in the short term, erase the disadvantages with which they were burdened.

Today, with the Soviet occupation mostly a distant memory and the European Union extending from the

British Isles to the Black Sea, it might seem that the geographic notion of an Eastern Europe is defunct. But not so fast. Take a look at the statistics in the Data Table at the end of the book, and you will see that, with some notable but still minimal exceptions, there remain significant contrasts between Europe's west and east. True, Estonia has taken its place as one of Nordic (no longer Eastern) Europe's economic success stories. And in the south, Slovenia is another winner. But these small countries in favorable locations are less representative of Europe's eastern Periphery than poorly governed Poland, deeply indebted Hungary, or corrupt Slovakia—not to mention barely post-communist Romania and economically troubled Bulgaria (with a per-capita GNI less than one-third of that of the Core).

Eastern Europe's tumultuous history played itself out on a physical stage of immense diversity, its landscapes ranging from open plains and wide river basins to strategic mountains and crucial corridors. Epic battles fought centuries ago remain fresh in the minds of many people living here today; pivotal past migrations are celebrated as though they happened yesterday. Nowhere in Europe is the cultural geography as complex. Illyrians, Slavs, Turks, Magyars, and other peoples converged on this region from near and far. Ethnic and cultural differences kept them in chronic conflict.

Geographers call such a region a **shatter belt**, a zone [29] of persistent splintering and fracturing. Geographic terminology uses several expressions to describe the breakup of established order, and these tend to have their roots in this part of the world. One of those expressions is **balkanization**. The southern half of eastern Europe is [30] referred to as the Balkans or Balkan Peninsula, after the name of a mountain range in Bulgaria. Balkanization denotes the recurrent division and fragmentation of this part of the region, and it is now applied to any place where such processes take place. A more recent term is *ethnic cleansing*—the forcible ouster of entire populations from their homelands by a stronger power bent on taking their territories. The term may be new, but the process is as old as eastern Europe itself.

Cultural Legacies

Each episode in the historical geography of eastern Europe has left its legacy in the cultural landscape. Twenty centuries ago the Roman Empire ruled much of it

(Romania is a cartographic reminder of this period); for most of the second half of the twentieth century, the Soviet Empire controlled nearly all of it. In the intervening two millennia, Christian Orthodox church doctrines spread from the southeast, and Roman Catholicism advanced from the northwest. Turkish (Ottoman) Muslims invaded and created an empire that reached the environs of Vienna. By the time the Austro-Hungarian Empire ousted the Turks, millions of eastern Europeans had been converted to Islam. Albania and Kosovo today remain predominantly Muslim countries.

The collapse of the Soviet Empire in the early 1990s freed the European countries that had been under Soviet rule, and with only one exception—Belarus—those countries turned their gaze from Moscow to the west. Meanwhile, the European Union had been expanding eastward, and membership in the EU became an overriding goal for the majority of the liberated eastern states. This generated a new map of the region (Fig. 1-23), integrating most of embattled old Eastern Europe into the EU and creating, as we saw earlier, a new layout of insiders and outsiders.

The Geographic Framework

As Figure 1-13 reminds us, the Eastern Europe of old, like the realm's southern and northern peripheries, also has countries that are contiguous to the European Core. Especially since the momentous EU expansions of 2004 and 2007, several of these countries are likely to become part of the European Core as their governments improve, economies grow, and incomes rise. As we saw, the Czech Republic has already become integrated into the Core region, and Slovenia is likely to be the first post-Yugoslavia state to merit similar inclusion. Things are changing fast in the eastern Periphery.

On the basis of this combination of dynamics and location, the countries of Europe's eastern Periphery can be grouped as follows (Fig. 1-23). As you will see, each of these groups has a *key state* whose geographic importance outweighs all others:

1. EU countries contiguous to the European Core (4)
   Key state: Poland
2. EU countries disconnected from the European Core (6)
   Key state: Greece
3. Non-EU countries in the South (7)
   Key state: Serbia
4. Non-EU countries in the East (3)
   Key state: Ukraine

When studying Figure 1-23, it is important to take note of still another one of those anomalous territories that mark the map of Europe: *Kaliningrad*, located on the Baltic Sea between Poland and Lithuania. This is an exclave of Russia, acquired following the end of World War II, and potentially important as an outpost in Europe of Russian influence at a time when the EU as well as NATO* are advancing toward Russia's borders in what was once the Soviet Union's front yard.

EU Countries Contiguous to the European Core

As Figures 1-13 and 1-23 show, the four states in this group all share borders with EU Core country Germany or Austria. Of these four, the most important state is Poland, which was also the largest and most populous country among the ten that joined the EU in 2004. With 38.1 million people and lying between two historic enemies, Poland has borders that have shifted time and again, but its current status may be more durable than in the past. As the map shows, the historic and once-central capital of Warsaw now lies closer to Russia than to Germany, but the country looks west, not east. During the Soviet-communist period, Silesia became its industrial heartland, and Katowice, Wroclaw, and Krakow grew into major industrial cities amid some of the world's worst environmental degradation. The Soviets invested far less in agriculture, collectivizing farms without modernizing technologies and leaving post-Soviet farming in abysmal condition. All this made governing Poland difficult, but the prospect of EU membership (with the promise of EU subsidies) motivated the government to get its house in order. After entering the EU, Poland saw hundreds of thousands of workers leave for jobs in the European Core, but many have returned and the economy is growing.

Government also was the problem in neighboring, landlocked Slovakia, where during the communist period the people were far more pro-Soviet than the Czechs next door. Many observers of the 2004 EU expansion wondered whether misgoverned Slovakia should be admitted because the capital, Bratislava, had become synonymous with corruption and inefficiency. Slovakia's Hungarian minority, comprising about 10 percent of the population of 5.4 million and concentrated in the south along the Danube River, was at odds with the Slovak regime. Moreover, additional concerns were raised over reports of mistreatment of the smaller Roma (Gypsy) minority (see the box titled "Europe's Stateless Nation"). But again the promise of

*NATO (North Atlantic Treaty Organization) is a military alliance that was established at the height of the Cold War in 1949. This U.S.-led supranational defense pact shielded postwar Europe against the Soviet military threat. Today, in the post-Soviet era, NATO is modifying its objectives as well as expanding its membership (up from the original 12 countries in 1949 to 28 in 2009).

FIGURE 1-23

© H. J. de Blij, P. O. Muller, and John Wiley & Sons, Inc.

EU membership led to some reforms, and after 2004 the economy perked up. Following another downturn in 2007, a more enlightened administration took over, and in 2008 Slovakia surprised many by meeting the terms of admission to the European Monetary Union (EMU), adopting the euro in 2009 before the Poles or the Hungarians could.

Many economic geographers anticipated a bright future for also-landlocked Hungary following the collapse of the Soviet Empire and the opening of the door

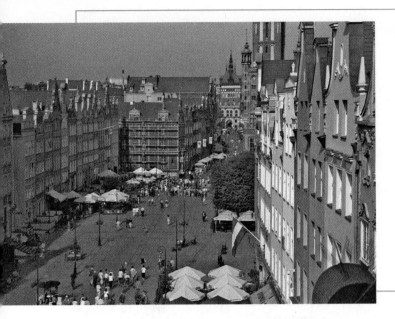

"Turbulent history and prosperous past are etched on the cultural landscape of the old Hanseatic city of Gdansk, Poland. Despite major wartime destruction, much of the old architecture survives, and restoration has revived not only Gdansk but also its twin (port) city, Gdynia. Gdansk was the initial stage for the rebellion in the 1980s of the Solidarity labor union against Poland's communist regime, and after that regime was toppled, the country went through difficult economic times. Today Poland, a full member of the European Union, is making rapid progress; its historic cities, from Gdansk in the north to Krakow in the south, attract throngs of foreign visitors and the tourist industry is booming. Note: don't let that Dutch tricolor mislead you. Attracting foreign visitors is one way to raise revenues, and the historic old city on the Baltic is becoming a tourist draw."   © Sigmund Malinowski.

www.conceptcaching.com

Concept Caching

Europe's Stateless Nation

EVERY ETHNIC GROUP in the European population, it seems, has its homeland, whether it is a state or a sub-unit. From the Frisians to the Corsicans and from the Norwegians to the Greeks, everyone has a historic base, whether it is a State or a province or some other entity, no matter how small.

But there is one significant exception: the *Roma*. About 8 million Roma, formerly (and derisively) called *gypsies*, constitute Europe's troubled, stateless nation. Although their origins are uncertain, the Roma are believed to have originated in India and migrated westward along several routes, one of which carried them via present-day Iran and Turkey into Europe. They never established a domain; they have retained their mobile, nomadic lifestyle to the present day; and they live, mostly in poverty, in a discontinuous arc across Bulgaria, Romania, Hungary, Slovakia, and the Czech Republic. Wherever they have gone, they have found themselves facing discrimination, resentment, unemployment, and poor health conditions. Their mobile lifestyle contributes to low educational levels and to the perception and reality of associated crime.

The Roma were an issue during the debate preceding the EU's 2004 enlargement and again prior to the accession of Bulgaria and Romania in 2007, when many Europeans talked of a "Roma deluge" as a reason to oppose expansion. Even before 2004, some Roma had been entering the European Core as asylum-seekers, arriving in English towns and setting up their wagons and encampments in public parks and village squares. When several Core countries strength-

Europe's largest minority, the stateless Roma, also are the realm's poorest. Slovakia is one of several European countries with substantial Roma populations, and its government has been criticized by the EU for its treatment of Roma citizens. This depressing photograph of a Roma settlement in Hermanovce shows a cluster of makeshift dwellings virtually encircled by a moat bridged only by a walkway. The village of Hermanovce may itself not be very prosperous, but that moat represents a social chasm between comparative comfort and inescapable deprivation.   © Tomasz Tomaszewski/ngs/Getty Images, Inc.

ened their asylum rules, their objective was not only to stem the flow from Islamic countries but also within the EU itself, with the Roma the primary target. EU efforts to help Roma in their source countries through subsidies and other assistance have been hampered by the high levels of corruption in those countries where Roma minorities are largest.

FIGURE 1-24

© H. J. de Blij, P. O. Muller, and John Wiley & Sons, Inc.

to Europe. The Hungarians (Magyars) moved into the middle Danube River Basin more than a thousand years ago from an Asian source; they have neither Slavic nor Germanic roots. They converted their fertile lowland into a thriving nation-state and created an imperial power that held sway over an area far larger than present-day Hungary. Ethnic Magyar remnants of this Greater Hungary can still be found in parts of Romania, Serbia, and Slovakia (Fig. 1-24), and the government in the twin-cities capital astride the Danube River, Budapest, has a history of irredentism toward these external minorities.

The concept of **irredentism**—a government's support for ethnic or cultural cohorts in neighboring or distant countries, derives from a nineteenth-century campaign by Italy to incorporate the territory inhabited by an Italian-speaking minority of Austria, calling it *Italia Irredenta* or "Unredeemed Italy." One advantage of joining the EU, obviously, was that much of the reason for Hungarian irredentism disappeared. When and if Serbia joins the EU, it may disappear altogether.

With a population of 9.9 million, a distinctive culture, and a considerable and varied resource base, Hungary should have good prospects, and its economic potential was a strong factor in its 2004 admission into the EU. But mismanagement, political corruption, and growing indebtedness have set Hungary back. Behind the elegance of Budapest, a primate city nearly ten times the size of Hungary's next-largest urban center, lies an economy still mired in its rural past.

The success story among the four countries in this grouping remains modest, with most of it accounted for by progressive Slovenia, which lies wedged against Austria and Italy in the hilly terrain near the head of the Adriatic Sea (Fig. 1-23). With 2.0 million people, a nearly homogeneous ethnic complexion, and a productive economy, Slovenia was the first "republic" of the seven that emerged from the collapse of Yugoslavia to be invited to join the EU. Shortly thereafter, Slovenia became part of the euro zone.

EU Countries Disconnected from the European Core

In our Introduction we emphasized that core areas are a matter of scale, and that smaller core areas and peripheries exist within larger ones. Here is a good example of this phenomenon. By definition, we are dealing here with the European Periphery. The key state in this group, Greece, forms a subsidiary core among the countries not directly connected to the European Core itself.

Greece's land boundaries are with Turkey, Bulgaria, Macedonia, and Albania; as Figure 1-25 reveals, it also owns islands just offshore from mainland Turkey. Altogether, the Greek archipelago numbers some 2000 islands, ranging in size from Crete (8260 square kilometers [3190 sq mi]) to small specks of land in the Cyclades. In addition, Greeks represent the great majority on the now-divided island of Cyprus.

Ancient Greece was a cradle of Western civilization, and later it was absorbed by the expanding Roman Empire. For some 350 years beginning in the mid-fifteenth century, Greece was under the sway of the Ottoman Turks. Greece regained independence in 1827, but not until nearly a century later, through a series of Balkan wars, did it acquire its present boundaries. During World War II, Nazi Germany occupied and ravaged the country, and in the postwar era the Greeks quarreled with the Turks, the Albanians, and the newly independent Macedonians. Today, Greece finds itself between the Muslim world of Southwest Asia and the Muslim communities of the eastern European Periphery.

Volatile as Greece's surroundings are, Greece itself is a country on the move, another EU success story. Political upheavals in the 1970s and economic stagnation in the 1980s are all but forgotten in the new century: now Greece, its economy thriving, is described as the loco-

motive for the Balkans, a beacon for the EU in a crucial part of the world. Recent infrastructure improvements, in conjunction with hosting the 2004 Olympics, have focused on metropolitan Athens, where nearly one-third of the population is concentrated, and include new subways, a new beltway, and a new airport.

Greece is strongly affected by the recent enlargement of the EU, particularly the admission of Romania and its neighbor, Bulgaria, in 2007. Until 2007, Greece lay separated from the contiguous EU by nonmember-states, but when Romania and Bulgaria joined the EU, Greece became part of the conterminous Union. The effect of what will eventually become a far more porous border with Bulgaria may still be distant, but more immediately, Greece will lose much of its EU subsidy as the poorer new members start receiving EU financial support. Meanwhile, Greece's democratic institutions still need strengthening, its educational institutions need modernizing (street riots greeted the introduction of private university education, an EU regulation), and corruption remains a serious problem. And although Greece is on better terms today with its fractious neighbors and invests in development projects in Macedonia and Bulgaria (for example, an oil pipeline from Thessaloniki to a Greek-owned refinery in Skopje, Macedonia), it will take skillful diplomacy to navigate the shoals of historic discord.

Modern Greece is a nation of 11.2 million centered on historic Athens, one of the realm's great cities. With its

The Mediterranean Sea, a maritime link between global core and global periphery, has claimed an unknown number of victims trying to reach European coasts. Overloaded fishing boats such as the one shown here, carrying more than 230 hopeful (but illegal) migrants, often fail to make the crossing in a sea known for its sudden storms. If they do reach European shores, their occupants risk arrest and deportation; this boat was apprehended and escorted into the Sicilian port of Licata. EU countries, less in need of unskilled labor than they once were and more inclined to close their borders, are taking a harder line against undocumented immigrants and asylum-seekers.
© Antonio Parrinello/Reuters/Landov

FIGURE 1-25

© H. J. de Blij, P. O. Muller, and John Wiley & Sons, Inc.

port of Piraeus, metropolitan Athens contains about 30 percent of the Greek population, making it one of Europe's most congested and polluted urban areas. Athens is the quintessential primate city; the monumental architecture of ancient Greece still dominates its cultural landscape. The Acropolis and other prominent landmarks attract a steady stream of visitors; tourism is one of Greece's leading sources of foreign revenues, and Athens is only the beginning of what the country has to offer.

Deforestation, soil erosion, and variable rainfall make farming difficult in much of Greece, but the country remains strongly agrarian, reflective of Europe's Periphery. It is self-sufficient in staple foods, and farm products continue to figure prominently among exports. But other sectors of the economy, including manufacturing (textiles) and the service industries, are growing rapidly. Volatile politically, Greece will be challenged to maintain its growth after the momentous recent EU

AMONG THE REALM'S GREAT CITIES . . . Athens

TAKE A MAP of Greece. Draw a line to encompass its land area as well as all the islands. Now find the geographic center of this terrestrial and maritime territory, and you will find a major city nearby. That city is Athens, capital of modern Greece, culture hearth of ancient Greece, source of Western civilization.

Athens today is only part of a greater metropolis that sprawls across, and extends beyond, a mountain-encircled, arid basin that opens southward to the Bay of Phaleron, an inlet of the Aegean Sea. Here lies Piraeus, Greece's largest port, linked by rail and road to the adjacent capital. Other towns, from Keratsinion in the west to Agios Dimitrios in the east, form part of an urban area that not only lies at, but is, the heart of modern Greece. With a population of 3.2 million, most of the country's major industries, its densest transport network, and a multitude of service functions ranging from government to tourism, Greater Athens is a well-defined national core area.

Arrive at the new Eleftherios Venizelos Airport east of Athens and drive into the city, and you can navigate on the famed Acropolis, the 150-meter (500-ft) high hill that was ancient Athens' sanctuary and citadel, crowned by one of humanity's greatest historic treasures, the Parthenon, temple to the goddess Athena. Only one among many relics of the grandeur of Athens 25 centuries ago, it is now engulfed by mainly low-rise urbanization and all too often obscured by smog generated by factories and vehicles and trapped in the basin. Conservationists decry the damage air pollution has inflicted on the city's remaining historic structures, but in truth the greatest destruction was wrought by the Ottoman Turks, who

© H. J. de Blij, P. O. Muller, and John Wiley & Sons, Inc.

plundered the country during their occupation of it, and the British and Germans, who carted away priceless treasures to museums and markets in western Europe.

Travel westward from the Middle East or Turkey, and Athens will present itself as the first European city on your route. Travel eastward from the heart of Europe, however, and your impression of its cultural landscape may differ: behind the modern avenues and shopping streets are bazaar-like alleys and markets where the atmosphere is unlike that of any other in this geographic realm. And beyond the margins of Greater Athens lies a Greece that lags far behind this bustling, productive coreland.

expansions and against the backdrop of a still-unstable Balkan neighborhood.

Also in this southeastern corner of the European Periphery lies the island country of Cyprus, whose political geography merits special attention because of the complications it created, and continues to create, for the EU. As Figure 1-25 shows, Cyprus lies closer to Turkey, Syria, and Lebanon than it does to any part of Europe, but it is peopled dominantly by Greeks. In 1571, the Turks conquered Cyprus, then ruled by Venice, and controlled it until 1878 when the British took over. Most of the island's Turks arrived during the Ottoman period; the Greeks have been there longest.

When the British were ready to give Cyprus independence after World War II, the 80-percent-Greek majority mostly preferred union with Greece. Ethnic conflict followed, but in 1960 the British granted Cyprus

independence under a constitution that prescribed majority rule but guaranteed minority rights.

This fragile order broke down in 1974, and civil war engulfed the island. Turkey sent in troops and massive dislocation followed, resulting in the partition of Cyprus into northern Turkish and southern Greek sectors (Fig. 1-25, inset map). In 1983, the 40 percent of Cyprus under Turkish control, with about 100,000 inhabitants (plus some 30,000 Turkish soldiers), declared itself the independent *Turkish Republic of Northern Cyprus*. Only Turkey recognizes this ministate (which now contains a population of about 200,000); the international community recognizes the government on the Greek side as legitimate. This legitimacy, however, is questionable in light of the Greek side's rejection of the United Nations plan to allow both the Greek and the Turkish side of the island to join the EU in 2004. The Turkish-Cypriot voters accepted it, yet they were left out when the Greek side of the island was admitted to the EU in 2004. Resentment was high on the Turkish side and in Turkey itself as well—just as discussions on Turkey's own admission to the EU were getting under way. It was—and remains—a reminder that Cyprus's "Green Line" separating the Greek and Turkish communities constitutes not just a regional border but a boundary between geographic realms.

Just how Romania managed to persuade EU leaders to endorse its 2007 accession remains a question for many Europeans. As the Data Table inside the back cover indicates, Romania has some of Europe's worst social indicators; its economy is weak, its incomes are low, its political system has not been sufficiently upgraded from communist times (many "apparatchiks" have acquired state assets under the guise of "privatization" and are controlling the political process), and political infighting and corruption are endemic.

But Romania is an important country, located in the lower basin of the Danube River and occupying much of the heart of eastern Europe. With 21.4 million inhabitants and a pivotal situation on the Black Sea, Romania is a bridge between central Europe and the realm's southeastern corner, where EU member Greece and would-be member Turkey face each other across land and water.

Romania's drab and decaying capital of Bucharest (once known as the Paris of the Balkans) and its surrounding core area lie in the interior, linked by rail to the Black Sea port of Constanta. The country's once-productive oilfields have been depleted, about a third of the labor force works in agriculture, poverty is widespread in the countryside as well as the towns, and unemployment is high. Many talented Romanians continue to leave the country in search of opportunities elsewhere. Romania has been described as the basket case of the Balkans, but that did not derail its EU membership drive.

Across the Danube lies Romania's southern neighbor, Bulgaria. The rugged Balkan Mountains form Bulgaria's physiographic backbone, separating the Danube and Maritsa basins. As the map shows, Bulgaria has five neighbors, several of which are in political turmoil.

The Bulgarian state appeared in 1878, when the Russian czar's armies drove the Turks out of this area. The Slavic Bulgars, who form 85 percent of the population of 7.5 million, were (unlike the Romanians) loyal allies of Moscow during the Soviet period. But they did not treat their Turkish minority, about 10 percent of the population, very kindly, closing mosques, prohibiting use of the Turkish language, and forcing Turkish families to adopt Slavic names. Conditions for the remaining Turks improved somewhat after the end of the Soviet period. They improved further as Bulgaria became a candidate and then in 2007 a member of the European Union, which required judicial and other social reforms.

Bulgaria has a Black Sea coast and an outlet, the port of Varna, but the main advantage it derives from its coastal location is the tourist trade its beaches generate. The capital, Sofia, lies not on the coast but near the opposite border with Serbia, and in its core-area hinterland some foreign investment is changing the economic landscape—but slowly. Bulgaria's GNI is on a par with Romania's, although the telltale agricultural sector is much smaller here.

Bulgarians, too, are emigrating in droves—and this worries the countries of western Europe, where an uncontrolled influx of immigrants with newly-won access to their job markets would create serious problems. In 2006 the United Kingdom announced that it would place severe restrictions on workers from Romania and Bulgaria—a surprising reversal for a country that has long championed openness in the EU job market. It was a signal that, by admitting these two states, the EU's leaders had crossed a line.

As Figure 1-13 reminds us, the boundary of the European Core that traverses the Baltic Sea includes southern Norway and Sweden but excludes Latvia and Lithuania. The economies of these two Baltic states are improving but are not yet on a par with those of Scandinavia. Latvia, centered on the port of Riga, was tightly integrated into the Soviet system during Moscow's long domination. Still today, only 59 percent of the population of 2.3 million is Latvian, and nearly 30 percent is Russian. After independence ethnic tension arose between these Baltic and Slav sectors, but the prospect of EU membership required the end of discriminatory practices. Latvians concentrated instead on the economy, which had been left in dreadful shape, with the result that it qualified for 2004 admission. Consider this: 20 years ago, virtually all of Latvia's trade was with the Soviet Union. Today its principal trading partners are Germany, the United Kingdom, and Sweden. Russia figures in only one category: Latvia's import of oil and gas.

Lithuania (3.4 million) has a residual Russian minority of only about 7 percent, but relations with its giant neighbor are worse than Latvia's—this despite Lithuania's greater dependence on Russia as a trading partner. One reason for this bad relationship has to do with neighboring Kaliningrad, Russia's Baltic Sea exclave (Fig. 1-23). When Kaliningrad became a Russian territory after World War II, Lithuania was left with only about 80 kilometers (50 mi) of Baltic coastline and a small port that was not even connected by rail to the interior capital of Vilnius. Lithuania in 2005 called for the demilitarization of Kaliningrad as a matter of national security. Despite these problems, Lithuania's economy during 2003 had the highest growth rate in Europe, spurred by foreign investment and by profits from its oil refinery at Mazeikiai, facilitating EU admission in 2004.

Non-EU Countries in the South

We now turn to a part of the European Periphery that has undergone wrenching changes since 1990, a process that started with the violent dismantling of communist Yugoslavia during the last decade of the twentieth century and continues, fortunately less violently, today. Here the European Union has made only slight progress, and just one of the countries that emerged from Yugoslavia's disintegration, Slovenia, has joined the EU. Discussions with Croatia, Slovenia's neighbor, have repeatedly been put on hold. No other state in this area is a serious candidate for admission at present.

The key state on the new map is Serbia, the name of what is left of a much larger domain once ruled by the Serbs—who were dominant in the former Yugoslavia. Centered on the historic capital of Belgrade on the Danube River, Serbia is the largest and potentially the most important country of this subregion of the old Eastern Europe. But the Serbs are having to accommodate some major changes. First, more than 1 million of them live in neighboring Bosnia, where they have an uneasy relationship with the local Muslims and Croats (Fig. 1-26). Second, the coastal province named *Montenegro* broke away from Serbia in 2006, when voters there opted to form an independent state. Third, its Muslim-majority province of *Kosovo* declared independence in 2008, its sovereignty immediately recognized by the United States and a majority of (but not all) European governments. And fourth, Serbia still incorporates a Hungarian minority of some 400,000 in its northern province of *Vojvodina* on the northern side of the Danube at a time when Hungary has already joined the EU.

Serbs have a history of prominence and dominance in the former Yugoslavia and even earlier, but the Serbian sphere of influence continues to shrink. Still, with a population of 7.3 million today, this remains former Yugoslavia's key country. The Serbs objected vigorously to Kosovo's secession, in which they were supported by their fellow Slavs, the Russians. EU planners worry that the Russians may encourage the Serbs to defer membership in the EU even if it were offered, and there are also concerns that Serbs may face new difficulties in Bosnia.

FIGURE 1-26

© H. J. de Blij, P. O. Muller, and John Wiley & Sons, Inc.

If EU expansion occurs here in the south, it may involve Croatia, the crescent-shaped country with prongs along the Hungarian border as well as the Adriatic coast. But here, too, there are problems. About 90 percent of Croatia's 4.4 million citizens are Croats, but the country's Serb minority has faced discrimination and has declined from 12 percent of the population to under five. The EU took a dim view of human rights issues in Croatia, but these improved when EU membership beckoned. The next stumbling block came in the form of a wave of organized crime and unpunished corruption, which has stalled progress since 2007. Meanwhile, about 800,000 Croats live outside Croatia, in Bosnia, where their relationships with Muslims and Serbs are not always cooperative.

When former Yugoslavia collapsed, Bosnia was the cauldron of calamity. No ethnic group was overwhelmingly dominant here, and this multicultural, effectively landlocked triangle of territory, lying between the Serbian stronghold to the east and the Croatian republic to the west and north, fell victim to disastrous conflict among Serbs, Croats, and *Bosniaks* (now the official name for Bosnia's Muslims, who constitute about 50 percent of the population of 3.8 million). As many as 250,000 people perished in concentration camps associated with ethnic cleansing practices; in 1995, a U.S. diplomatic effort resulted in a truce that partitioned the country as shown in Figure 1-26. Serbian leaders were indicted for war crimes but not captured or extradited, which delayed any EU-admission initiatives. One leader, former Bosnian Serb president Radovan Karadzic, was arrested in 2008, but another, General Ratko Mladic, remained at large in mid-2009. Meanwhile, Serbia's former president, Slobodan Milosevic, was tried in The Hague for war crimes and died there in captivity. This is one rough corner of Europe.

The southernmost "republic" of former Yugoslavia was Macedonia, which emerged from the collapse as a state with a mere 2 million inhabitants, of whom two-thirds are Macedonian Slavs. As the map shows, Macedonia adjoins Muslim Albania and Kosovo, and its northwestern corner is home to the 30 percent of Macedonians who are nominally Muslims. The remainder of this culturally diverse population are Turks, Serbs, and Roma. Macedonia is one of Europe's poorest countries, landlocked and powerless. Even its very name caused it problems; Macedonia's Greek neighbors argued that this name was Greek property and so would not recognize it. Next, Macedonia faced an autonomy movement among its Albanian citizens, requiring allocation of scarce resources in the effort to hold the fledgling state together. Macedonians cling to the hope that eventual EU admission will bring it subsidies and better economic times.

The ministate of Montenegro has a mere 620,000 inhabitants, one-third of them Serbs, a small capital (Podgorica), some scenic mountains, and a short but spectacular Adriatic coastline—and very little else to justify its position as one of Europe's 40 countries. Tourism, a black market, some Russian investment, and a scattering of farms sum up the assets of this country.

In 1999, amid the chaos of disintegrating Yugoslavia, NATO took charge in Kosovo, then still a formal Serbian province, after a brief but damaging military campaign. With an overwhelmingly Albanian-Muslim population of about 2 million and a small Serb minority in its northern corner, landlocked Kosovo has few of the attributes of a full-fledged state, but balkanization continues in this part of Europe. NATO eventually turned over Kosovo's administration to the United Nations, and in 2008 the capital of Pristina witnessed independence celebrations as the UN administration yielded to a newly elected national government.

The only other dominantly Muslim country in Europe is Albania, where some 70 percent of the population of 3.2 million adhere to Islam. Albania also shares with one other country—Moldova—the status of being Europe's poorest state. Albania has one of Europe's highest rates of natural population growth, and many Albanians try to emigrate to the EU by crossing the Adriatic to Italy. Most Albanians subsist on livestock herding and farming, but the poverty-stricken Gegs in the north lag behind the somewhat better-off Tosks in the south, with the capital of Tirane lying close to this cultural divide. For all Europe's globalization, Albania would represent the symptoms of the global periphery anywhere in the world.

Non-EU Countries in the East

West of Russia and east of EU states Poland and Romania lie three countries of which one, the key state of Ukraine, is territorially the largest in all of Europe. Demographically, with a population of 45.6 million, Ukraine also firmly ranks in the second tier of European states, along with Poland and Spain. As Figure 1-23 shows, Ukraine's relative location is crucial. Not only does it link the core of Russia, to the periphery of the European Union: it forms the northern shore of the Black Sea from Russia to Romania, including the strategically important Crimea Peninsula. And most importantly, oil and gas pipelines connect Russian fields to European markets across Ukrainian territory.

Ukraine's capital, Kiev (Kyyiv), is a major historic, cultural, and political focus. Briefly independent before the communist takeover in Russia, Ukraine regained its sovereignty as a much-changed country in 1991. Once a land of farmers tilling its famously fertile soils, Ukraine emerged from the Soviet period with a huge industrial complex in its east—and with a large (17 percent) Russian minority. Ukraine's boundaries also changed during the Soviet era. In 1954, a Soviet dictator capriciously transferred the entire Crimea Peninsula, including its Russian inhabitants, to Ukraine as a reward for its productivity.

The Dnieper River forms a useful geographic reference to comprehend Ukraine's spatial division (Fig. 1-23). To its west lies agrarian, rural, mainly Roman Catholic Ukraine; in its great southern bend and eastward lies industrial, urban, Russified (and Russian Orthodox) Ukraine. Soviet planners made eastern Ukraine a communist Ruhr based on abundant local coal and iron ore, making the Donets Basin (*Donbas* for short) a key industrial complex. Meanwhile, the Russian Soviet Republic supplied Ukraine with oil and gas.

As Figure 1-24 shows, Ukraine, with the exception of its eastern and urban-concentrated Russian minority (still 17 percent today), has an ethnically homogeneous population by eastern European standards. Ukraine is a critically important country for Europe's future, but it suffers from numerous problems ranging from political mismanagement and corruption to a faltering economy and rising crime. Yet this country has access to international shipping lanes, a large resource base, massive farm production, educated and skilled labor, and a large domestic market.

A presidential election in 2004 first drew international attention to Ukraine's political geography. Figure 1-27 indicates voter preferences in the west and east; in effect, Ukraine's electorate is divided diagonally between a pro-Western (and pro-EU) west and a pro-Russian east, resulting in a situation reminiscent of Czechoslovakia's before its "velvet divorce." In Ukraine's case, however, the transit of energy supplies creates a complication. Ukraine's own dependence on Russian supplies makes it vulnerable to Moscow's decisions on prices as well as supplies. Some of Ukraine's leaders urge speedy integration into the European sphere; others prefer a middle road between Europe and Russia.

Moldova, Ukraine's small and impoverished neighbor (which is by many measures Europe's poorest country) was a Romanian province seized by the Soviets in 1940 and made into a landlocked "Soviet Socialist Republic." A half-century later, along with other such "republics," Moldova gained independence when the Soviet Union disintegrated. Romanians remain in the majority among its 4.1 million people, but most of the Russians and Ukrainians (each about 13 percent) have moved across the Dniester River to an industrialized strip of land

UKRAINE'S DIVIDED ELECTORATE

Majority supporting Western-oriented candidate

Majority supporting candidates favored by Russia

© H. J. de Blij, P. O. Muller, and John Wiley & Sons, Inc.

FIGURE 1-27

between that river and the Ukrainian border, proclaiming there a "Republic of Transdniestria" (see Fig. 1-12). But such separatist efforts form only one of Moldova's many problems. Its economy, dominated by farming, is in decline; an estimated 40 percent of the population works outside its borders because unemployment in Moldova is as high as 30 percent; smuggling and illegal arms trafficking are rife. These misfortunes translate into weakness, and Russia's malign influence in the form of support for Transdniestria's separatists keeps the country in turmoil. In fact, in 2007 the elected president of Moldova was persuaded by Russian pressure to recognize the regime of breakaway Transdniestria as legitimate.

The third state in this part of the European Periphery in many ways is the most peripheral of all: autocratic, backward Belarus. Landlocked and misgoverned, sustained in part by the transit of energy supplies from Russia to Europe's Core countries, Belarus has few functional links to Europe and little prospect of progress. More than 80 percent of Belarus's 9.6 million citizens are Belarussians ("White" Russians), a West Slavic people; only some 11 percent are (East Slavic) Russians. A small Polish minority occasionally complains of mistreatment and discrimination. Devastated during World War II, Belarus became one of Moscow's most loyal satellites, and the Soviets built Mensk (Minsk), the country's capital, into a large industrial center. But in the post-Soviet era, Belarus has lagged badly, and its government functioned in ways reminiscent of the communist era.

Unlike Ukrainians, Belarussians so far express little interest in joining the EU, and Belarus cannot even be viewed as a functioning part of the European Periphery. On the contrary, its long-time authoritarian leader has made overtures toward Moscow, seeking to join the Russian Federation in some formal way. The Russian leadership, however, has not responded with enthusiasm.

With nearly 600 million inhabitants in 40 countries, including some of the world's highest-income economies, a politically stable and economically integrated Europe would be a superpower in any new world order. However, Europe's political geography is anything but stable as devolutionary forces and cultural stresses continue to trouble the realm even as EU expansion proceeds. Europeans have not yet found a way to give voice to collective viewpoints in the world, or to generate collective action in times of crisis. Importantly, even after the expansion of 2007, the EU incorporates only two-thirds of the realm's national economies, and further enlargement will become increasingly difficult for economic, political, and cultural reasons. Europe has always been a realm of revolutionary change, and it remains so today.

What You Can Do

RECOMMENDATION: Make your point in the press! Most newspapers publish letters to the editor, and often that page is the most interesting of all. Writing a letter to the editor, however, is not as easy as it might appear. When you have to make your argument in just 150 to 200 words (sometimes a few dozen more, depending on editorial policy), this has a way of focusing the mind. But if you have a geographically informed opinion on a matter of current concern, or if you have spotted an error on a map or in a story, the editor would like to hear from you (submission instructions are found on the newspaper's editorial page and website). It takes practice, and it is best to start with a local or regional newspaper, but the effort may pay off in several unexpected ways because even a local newspaper today attracts a global audience through its website and other electronic links. And you will be amazed how practice in accurate, concise writing under space limitations and for a wide audience will improve your exam-writing skills.

GEOGRAPHIC CONNECTIONS

1. You are a member of the staff of a multinational corporation doing business in several European countries. Your management has been used to dealing with governments in the capitals of a number of countries including the Netherlands (The Hague), Poland (Warsaw), and Hungary (Budapest). Now the company has announced plans to expand operations in Spain at Barcelona, in Belgium at Antwerp, in Ukraine at Donetsk, and in the United Kingdom at Edinburgh. You have been asked to prepare a "hurdles and pitfalls" report as part of the preparation for this initiative. What advice will you provide based on your knowledge of Europe's cultural geography and devolutionary pressures?

2. The European Union expanded from 15 to 27 member-states during the middle years of the last decade, and voters in every one of the joining countries endorsed their governments' desire for membership. But even as other countries hope to enter the EU in the foreseeable future, certain states have actually declined membership. What countries are these, and why would they stay out of the EU? Under what circumstances might a member-state decide to leave the EU? Can you relate these considerations to the decision by 3 of the 15 pre-2004 members not to join the EMU and to reject the use of the euro?

3. The largest and most populous country in Europe not to be a member of the EU is largely Christian Ukraine, and in 2009 Ukraine was not even in discussions aimed at EU membership. Yet Turkey, not in Europe and almost entirely Muslim, is in the early stages of negotiations toward eventual membership. What cultural-geographic and geopolitical factors explain this remarkable situation, and in what ways does this illustrate the ever-greater difficulties confronted by the EU as it makes an ever-greater impact in and on its periphery?

Geographic Literature: The key introductory works on this realm were authored by Berentsen; Gottmann; Murphy, Jordan-Bychkov, & Bychkova Jordan; McDonald; Ostergren & Rice; and Shaw. These works, together with a comprehensive listing of noteworthy books and recent articles on the geography of Europe, can be found in the *References and Further Readings* section of this book's website at www.wiley.com/college/deblij.

RUSSIA

Elevation (m)

3000
1500
600
300
150
0
below sea level
0
-150
-1500
-3000

—— Roads
—— Railroads

0 km   200   400   600   800   1000   1200
0 miles   200   400   600   800

Lambert Azimuthal Equal-Area Projection
Scale 1:20,000,000

NORWAY
SWEDEN
SCANDINAVIA
FINLAND
Gulf of Bothnia

DENMARK
COPENHAGEN
North Sea
Baltic Sea
Kaliningrad
POLAND
LITHUANIA
VILNIUS
RUSSIA
LATVIA
Riga
Gulf of Riga
ESTONIA
Tallinn
Gulf of Finland
Helsinki
Stockholm

ARCTIC
Svalbard (NORWAY)
George Land
Alexandra Land
Franz Josef Wilczek Land
Graham Bell I.
Land
Ushakov I.
Nor
(Seve
Kosc
October Revolution I.
Bolshe

Barents Sea
Murmansk
Lovozero
KOLA PENINSULA
Arkhangelsk
White Sea
Onega
Severodvinsk
Kolguyev I.
Naryan Mar
Gornyatskiy
Vorkuta
Novaya Zemlya
Belyy I.
Vaygach I.
Kara Sea
Dickson
TAYMYR PE
YAMAL PENINSULA
GYDA PENINSULA
Novyy Port
Salekhard
Dudinka
Norilsk
PUTORAN PLATEAU

BELARUS
MINSK
KYYIV (KIEV)
UKRAINE
DNIEPER LOWLAND
Dnieper
Smolensk
Pskov
Lake Peipus
Novgorod
ST. PETERSBURG
Lake Ladoga
Lake Onega
Petrozavodsk
Cherepovets
Rybinsk
Vologda
Northern Dvina
Syktyvkar
PECHORA BASIN
Mt. Narodnaya
(1,895 m, 6,214 ft.)
Urengoy
Sergino
Khanty-Mansiysk
Ob
WEST SIBERIAN
RUS
Nizhnevartovsk
Surgut

Tver
Zelenograd
Volga
Yaroslavl
Kostroma
Ivanovo
Dzerzhinsk
Vladimir
MOSCOW
Kaluga
Tula
Bryansk
Orel
Kursk
Staryy Oskol
Belgorod
KHARKIV
DNIPROPETROVSK
DONETSK
Taganrog

Lipetsk
Voronezh
Tambov
Ryazan
NIZHNIY NOVGOROD
Cheboksary
Yoshkar-Ola
Vyatka
Kudymkar
Izhevsk
KAZAN
Ulyanovsk
Saransk
Penza
Naberezhnye Chelny
Kama
PERM
YEKATERINBURG
Nizhny-Tagil
Tyumen
Irtysh

Sea of Azov
ROSTOV
Shakhty
Krasnodar
Armavir
Maykop
Cherkessk
Stavropol
Elista
Nalchik
Nazran
Groznyy
Vladikavkaz
Makhachkala
CASPIAN DEPRESSION
Astrakhan
VOLGOGRAD
Volzhsky
Engels
Balakovo
SAMARA
Tolyatti
Saratov
Don
Volga
Syzran
Sterlitamak
Ufa
Magnitogorsk
Orenburg
Orsk
CHELYABINSK
Kurgan
URAL MOUNTAINS
Petropavl
OMSK
Kolpashevo
Lesosibirsk

GEORGIA
TBILISI
CAUCASUS MOUNTAINS
ARMENIA
YEREVAN
AZERBAIJAN
BAKI
TABRIZ
Caspian Sea
USTYURT PLATEAU
Aral Sea
Garabogaz Bay
KAZAKHSTAN
Astana
KAZAKH UPLANDS
Semey
Rubtsovsk
Gorno-Altaysk
Barnaul
Biysk
NOVOSIBIRSK
Prokopyevsk
Novokuznetsk
Tomsk
Kemerovo
Krasnoyarsk
Abakan
Yenisey
Kyzyl
SAYAN MOU
(3,491
Mt. Belukha
(4,506 m, 14,783 ft.)
ALTAY MOUNT

IRAN
ELBURZ MOUNTAINS
TEHRAN
IRANIAN PLATEAU
MASHHAD
SALT DESERT
ISFAHAN
ARAL
KARAJ
QOM
TURKMENISTAN
Ashgabat
GARAGUM
Amu Darya
QIZILQUM
UZBEKISTAN
TASHKENT
MOYYNQUM
SARYESIK-ATYRAU DESERT
Lake Balqash
Bishkek
KYRGYZSTAN
ALMATY
ÜRÜMQI
JUNGGAR BASIN
SHAN
Ysyk-Köl
NORTH POLE

OCEAN

Begichev I.

Nordvik

NORTH SIBERIAN LOWLAND

Laptev Sea

Gulf of Yana

Tiksi

New Siberian Islands

Kotelnyy I.

Novaya Sibir I.

Malyy Lyakhovskiy I.

Bolshoy Lyakhovskiy I.

SULA

Zemlya]

170°

180°

170°

160°

150°

140°

130°

120°

East Siberian Sea

Wrangel I.

Chukchi Sea

Ambarchik

Pevek

Chukchi Peninsula

70°

Bering Strait

UNITED STATES

St. Lawrence I.

Gulf of Anadyr

Anadyr

St. Matthew I.

60°

Bering Sea

Aleutian Islands

CHUKCHI RANGE

Arctic Circle

KORYAK RANGE

Mt. Ledyanaya (2,562 m, 8,406 ft.)

Karaginskiy I.

Palana

Commander Islands (RUSSIA)

170°

50°

180°

KOLYMA RANGE

VERKHOYANSK RANGE

CHERSKIY RANGE

Ust Nera

Mt. Mus Khaya (2,959 m, 9,708 ft.)

S I B E R I A

Kolyma

Omsukchan

Okhotsk

Magadan

Shelikhov Gulf

KAMCHATKA PENINSULA

Mt. Klyuchevskaya (4,750 m, 15,584 ft.)

Petropavlovsk-Kamchatskiy

Bolsheretsk

160°

CENTRAL

YAKUTSK BASIN

Lena

Namtsy

Yakutsk

Ust Maya

Aldan

Sea of Okhotsk

Paramushir I.

Cape Lopatka

B

SIBERIAN

PLATEAU

Vilyuy

Mirnyy

Suntar

Lensk

Lena

DZHUGDZHUR RANGE

Uda Bay

Shantar Islands

Nogliki

Sakhalin Island

Kurile Islands

Angara

Ust Ilimsk

Ust Kut

het

Bratsk

Lake Baykal

STANOVOY RANGE

Mt. Skalistyy (2,467 m, 8,094 ft.)

Skovorodino

Amur

Blagoveshchensk

Heihe

Komsomolsk

Birobidzhan

Khabarovsk

Mt. Tardoki-Yani (2,090 m, 6,857 ft.)

SIKHOTE-ALIN RANGE

Vanino

Tatar Strait

LaZarevo

Yamino

Uyuzhno Sakhalinsk

Novikovo

Urup

La Perouse Strait

Wakkanai

Etorofu

Kunashiri

Shikotan

Habomai

Kushiro

HOKKAIDO

SAPPORO

150°

Sardyk

453 ft.)

Angarsk

Irkutsk

Ulan-Ude

Aginskoye

Chita

YABLONOVYY RANGE

GREATER KHINGAN RANGE

Songhua

HARBIN

Lake Khanka

Dalnegorsk

Vladivostok

Nakhodka

J A P A N

Aomori

PACIFIC OCEAN

INS

Ulaanbaatar

NORTHEAST CHINA PLAIN

CHANGCHUN

JILIN

Xiliao

East Sea (Sea of Japan)

SENDAI

140°

MONGOLIA

MONGOLIAN PLATEAU

Liao

SHENYANG

TANGSHAN

Korea Bay

NORTH KOREA

PYONGYANG

SEOUL

SOUTH KOREA

TAEGU

Mt. Fuji (3,776 m, 12,388 ft.)

NAGOYA

KOBE

OSAKA

KYOTO

KAWASAKI

YOKOHAMA

TOKYO

NS

GOBI DESERT

CHINA

BEIJING

TANGSHAN

HISHIMA

2

REGIONS

- **RUSSIAN CORE AND PERIPHERIES**
- **EASTERN FRONTIER**
- **SIBERIA**
- **RUSSIAN FAR EAST**

Where were these photos taken? Find out at www.wiley.com/college/deblij

CONCEPTS, IDEAS, AND TERMS

1	Near Abroad
2	Oligarch
3	Climatology
4	Continentality
5	Weather
6	Tundra
7	Taiga
8	Permafrost
9	Forward capital
10	Colonialism
11	Imperialism
12	Russification
13	Federation
14	Collectivization
15	Command economy
16	Unitary state system
17	Distance decay
18	Population decline
19	Heartland theory
20	Core area

In This Chapter

- The geography of Russian nationalism
- Russia's colonies and the Soviet Empire
- How Russia copes with distance and isolation
- Moscow and the challenge of terrorism
- Where Russia meets China
- Trouble in Transcaucasia

Photos: © H. J. de Blij

FIGURE 2-1 *Map:* © H. J. de Blij,
P. O. Muller, and John Wiley & Sons, Inc.

TAKE ANOTHER LOOK at Figure G-2 (p. 7) and you might wonder whether Russia should be counted as one of the world's geographic realms. Does this high-latitude country qualify as one of the dozen great building blocks of the early-twenty-first-century world? Or is it merely another region of Europe, just an exceptionally large one? Well, here we have a geographic problem. If Russia is part of Europe, then Europe extends all the way across Eurasia to the Pacific Ocean. If it isn't, then Russia is either part of some other realm or it stands alone. Have a look at this issue on a globe (two-dimensional map projections have a way of distorting reality, and Russia in Figure G-2 looks a lot smaller than it really is).

In fact, Russia is territorially so large that the area of this single country exceeds that of East Asia (including China) and South Asia (including India) combined. Russia is almost three times the size of Europe, nearly as large as all of South America, and more than twice the size of continental Australia. From this perspective, there is no doubt that Russia is a world realm.

But Russia does not have a population to match its enormous territorial dimensions. East Asia and South Asia combined have a still-growing population of more than 3 billion; Russia has a shrinking population of just 141 million. That is less than the total of Brazil or Indonesia or Pakistan or Nigeria! How can Russia be regarded as a major political, economic, and social force in the contemporary world?

That is not just a geographic question. It vexes Russians as well. During the twentieth century Russia rose to global power, victorious in World War II, in control of a colonial empire that extended from the Baltic countries to Kazakhstan, empowered by massive armies and nuclear weapons, possessor of vast energy resources, a legitimate contender for world domination. Russia was the cornerstone of the Soviet Union, its communist ideology the key issue of the Cold War that followed World War II.

When the Soviet Union collapsed in 1991 and Russia lost its empire, chaos ensued. As millions of Russians streamed back to Russia from the former empire, Russia sought to keep some form of control over events in its **Near Abroad**, the so-called republics that previously fell under communist rule (and where millions of Russians today remain under the jurisdiction of non-Russian governments [Table 2-1]). At home, Russians struggled to reform their government toward some semblance of democracy even as minorities demanded rights ranging from autonomy to outright independence. A war broke out in the south, where Muslim rebels in Chechnya, one of Russia's internal republics, resisted continued Russian rule. The Chechens responded to Russian intervention by mounting a terrorist campaign that took hundreds of lives in the capital, Moscow, alone. Meanwhile, rapacious opportunists took advantage of Russia's weak economic controls to acquire huge components of the energy industry under the guise of "privatization." These newly rich **oligarchs** then tried to use their wealth to gain political power as well. Russia's first elected president, the ineffectual Boris Yeltsin, presided

1

2

TABLE 2-1

Social Characteristics of the Other 14 Ex-Soviet Republics

Name	Population (Millions)	Percent Russians	Religion	Official Language
Armenia	3.1	2	Armenian Orthodox (Christian) 73%	Armenian
Azerbaijan	8.9	3	Shi'ite Muslim 53%; Sunni Muslim 34%	Azeri
Belarus	9.6	11	Belarussian Orthodox 49%; Roman Catholic 13%	Belarussian
Estonia	1.3	26	Estonian Orthodox 17%; Estonian Lutheran 17%	Estonian
Georgia	4.6	6	Georgian Orthodox, 55%; Sunni Muslim 15%	Georgian
Kazakhstan	16.0	27	Sunni Muslim 43%; Russian Orthodox 9%	Kazakh; Russian
Kyrgyzstan	5.4	10	Sunni Muslim 61%; Russian Orthodox 8%	Kyrgyz; Russian
Latvia	2.3	28	Christian Churches 33%; Russian Orthodox 16%	Latvian
Lithuania	3.4	7	Roman Catholic 79%	Lithuanian
Moldova	4.1	13	Moldovan Orthodox, 32%; Other Orthodox, 31%	Moldovan
Tajikistan	7.6	1	Sunni Muslim 78%	Tajik
Turkmenistan	5.4	6	Sunni Muslim 87%	Turkmen
Ukraine	45.6	17	Ukrainian Orthodox 46%	Ukrainian
Uzbekistan	28.1	6	Sunni Muslim 88%	Uzbek

MAJOR GEOGRAPHIC QUALITIES OF

Russia

1. Russia is the largest territorial state in the world. Its area is nearly twice as large as that of the next-ranking country (Canada).

2. Russia is the northernmost large and populous country in the world; much of it is cold and/or dry. Extensive rugged mountain zones separate Russia from warmer subtropical air, and the country lies open to Arctic air masses.

3. Russia was one of the world's major colonial powers. Under the czars, the Russians forged the world's largest contiguous empire; the Soviet rulers who succeeded the czars took over and expanded that empire.

4. For so large an area, Russia's shrinking population of just over 140 million is comparatively small. The population remains heavily concentrated in the westernmost one-fifth of the country.

5. Development in Russia is concentrated west of the Ural Mountains; here lie the major cities, leading industrial regions, densest transport networks, and most productive farming areas. National integration and economic development east of the Urals extend mainly along a narrow corridor that stretches from the southern Urals region to the southern Far East around Vladivostok.

6. Russia is a multicultural state with a complex domestic political geography. Twenty-one internal Republics, originally based on ethnic clusters, function as politico-geographical entities.

7. Its large territorial size notwithstanding, Russia suffers from land encirclement within Eurasia; it has few good and suitably located ports.

8. Regions long part of the Russian and Soviet empires are realigning themselves in the post-communist era. Eastern Europe and the heavily Muslim Southwest Asia realm are encroaching on Russia's imperial borders.

9. The failure of the Soviet communist system left Russia in economic disarray. Many of the long-term components described in this chapter (food-producing areas, railroad links, pipeline connections) broke down in the transition to the post-Soviet order.

10. Russia long has been a source of raw materials but not a manufacturer of export products, except weaponry. Few Russian automobiles, televisions, cameras, or other consumer goods reach world markets.

over a state in dangerous turmoil. In the world at large, Russia's status was at a low ebb.

Russian citizens, surveys indicated, yearned for a stronger government, better security, greater stability, and more effective delivery of social services ranging from education to pensions. Stronger government is what they got when Vladimir Putin succeeded Yeltsin as Russia's second elected president in 2000. Putin arrested oligarchs (some fled to other countries, but others wound up in jail), clamped down on the governments of Russia's regions, pursued Chechnya's rebels into the mountains, muzzled the once-freewheeling media, and vowed to the nation that he would restore Russia's powerful influence in world affairs. He managed to get Russia invited to join the so-called G-7, a group of the world's leading economies, making this the G-8 and raising more than a few eyebrows, because Russia's economy hardly qualified. Russia began to intervene forcefully in the Near Abroad, including efforts to influence national elections (notably Ukraine's, see p. 96), as well as domestic affairs in such places as neighboring Georgia and non-neighbor Moldova. And Russia obstructed U.S. and European designs, ranging from politics in the Balkans to nuclear development in Iran.

As President Putin's second term drew to a close in 2008, many Russians, including appointed and elected political leaders, were reluctant to see him leave center stage. The Russian constitution, however, limits the president to two consecutive terms. But there was another way to keep Putin near the levers of power: by "electing" as president his former chief of staff (and later deputy prime minister), Dmitri Medvedev, and then having Medvedev appoint former President Putin the new prime minister. On paper, this maneuver was an effective way to ensure continuity of the machinery of state; in practice, it was still another setback for democracy in Russia.

As you study the geography of this gigantic country, it is well to remember that Russia also remains one of the world's nuclear powers with the capacity to back up its renewed quest for global influence with military might. As we will see, post-Soviet Russia is a country troubled in many ways even as its leaders seek to resurrect its global geopolitical stature.

Defining the Realm

Although the name we give to this world realm is Russia, the geographic situation actually is not so simple (Fig. 2-1). The name is appropriate because Russia is the overwhelming component of this realm. But, as we noted in the Introduction, geographic realms usually are not bounded by sharply defined political or physical features: they tend to display zones of transition on their margins. And so it is with Russia. In Chapter 1 we noted that Russia's western neighbor, Belarus, satisfies neither European nor Russian criteria. It was part of the Soviet Union and currently remains a throwback to the days of authoritarian politics and stagnant economics. In its western border areas, however, there is a European minority (mostly Polish) and some evidence of European influences. As the century's second decade opened, Belarus's convergence with Russia seemed to be in reverse, but for now we continue to map Belarus as one of the transition zones marking the "Russian" realm.

A second transition zone lies in so-called *Transcaucasia,* the mountainous area south of Russia between the Black Sea and the Caspian Sea. Historic conflicts between local peoples and outside invaders (including Russians), as well as among locals themselves, have created a complicated geopolitical mosaic plagued by instability. This lack of stability affects both Russia itself and its neighbors. In the 1990s, Russia waged a terrible war to suppress independence-minded Chechnya. After the turn of the millennium, Chechnya's neighbor, Ingushetiya, became a scene of strife. Then in 2008 Russia undertook military action beyond its border in Georgia, where pro-Russian minorities claimed discrimination by Georgian authorities. It was just the latest eruption of violence in what is truly a zone of transition—where Russian power does not end at its border but peters out in the mountains and valleys beyond.

In Chapter 7, we will encounter a third transition zone on Russia's perimeter, another relict of this realm's Soviet past. During communist rule, millions of Russians moved to the Soviet "republics" outside Russia. But in one such republic, Kazakhstan, many of them moved into a zone right across the border from Russia and proceeded to "Russify" that area so completely that its cultural landscape became Russian. Many of those Russians and their descendants have now moved back to Russia, but northern Kazakhstan still carries its Russian imprint and constitutes another one of the realm's peripheral transition zones.

And still another transition zone may be developing, this one internal. As we will discover, Russia's population is shrinking, and this decline is especially rapid in Russia's Far East, in its Pacific margin where Russia borders China and North Korea. With North Korea in disarray and refugees crossing the border, and with Chinese workers and tradespeople filtering into Russia from China's economically lagging Northeast, Russia's distant frontier is taking on new demographic and social characteristics. The immigration question is fiercely debated in Moscow, but the forces of change are already altering the cultural landscape of Russia's remote eastern corner.

An Internal Empire

And within its borders, Russia still is an empire. It was the Soviet Empire that disintegrated, not the Russian one: about 18 percent of Russia's more than 140 million inhabitants are non-Russians, including numerous minorities large and small whose presence is reflected by the administrative map. The Chechens are but one of those minorities, many of which look back upon a history of mistreatment but most of which have accommodated themselves to the reality of Russian dominance. Russia's failure in Chechnya should not obscure what was in fact a remarkable success of the Yeltsin administration: avoiding the collapse of administrative order and the secession of minority-based "republics" throughout the country. For a time, it seemed that Tatarstan, the historic home of Muslim Tatars, would declare independence along with other minorities long unhappy under Soviet-Russian rule. But Yeltsin's government gave these peoples ways to express their demands and to negotiate their terms, and in the end the Russian map stayed intact—except in the Transcaucasus. Still, one of the risks inherent in a return to "strong" government is a reversal of this good fortune. Chechnya's rebellion and Muslim Chechens' terrorist actions have already made life much more difficult for other minorities on the streets of Moscow and other cities, where many ethnic Russians equate minority appearance and dress with rebellion and risk. Alienating Russia's other minorities risks the future of the state itself.

ROOTS OF THE RUSSIAN REALM

The name *Russia* evokes cultural-geographic images of a stormy past: terrifying czars, conquering Cossacks, Byzantine bishops, rousing revolutionaries, clashing cultures. Russians repulsed the Tatar (Mongol) hordes,

In the 1960s visitors to the (then) Soviet Union, one of the authors of this book among them, were shocked by the fate of virtually all of the country's Christian churches, whose condition ranged from bad repair to outright ruin. Architecturally ornate and historic church buildings with superb murals and exquisite glasswork stood empty, subject to the elements and vandalism. Doors and furniture had been taken for firewood; some churches served as barns, others as storage. In his atheist zeal, Josef Stalin ordered the demolition of one of Moscow's religious treasures, the Christ-the-Savior Cathedral (other Moscow cathedrals fared better and were turned into museums). This photograph symbolizes the reversal of fortunes following the collapse of the Soviet Union: religion is again a part of Russian life, the Russian Orthodox Church again dominates the religious map, and a reconstructed Cathedral of the Savior again stands where it did before the communist era. © Harf Zimmermann.

forged a powerful state, colonized a vast contiguous empire, defeated Napoleon, adopted communism, and, when the communist system failed, lost most of their imperial domain. Today, Russia seeks a new place in the globalizing world.

A Vibrant Culture

Pre-communist Russia may be described as a culture of extremes. It was a culture of strong nationalism, resistance to change, and despotic rule. Enormous wealth was concentrated in a small elite. Powerful rulers and bejeweled aristocrats perpetuated their privileges at the expense of millions of peasants and serfs who lived in dreadful poverty. The Industrial Revolution arrived late in Russia, and a middle class was slow to develop. Yet the Russian nation gained the loyalty of many of its citizens, and its writers and artists were among the world's greatest. Authors such as Tolstoy and Dostoyevsky chronicled the plight of the poor; composers celebrated the indomitable Russian people and, as Tchaikovsky did in his *1812 Overture*, commemorated their victories over foreign foes.

Empire of the Czars

Under the czars, Russia grew from nation into empire. The czars' insatiable demands for wealth, territory, and power sent Russian armies across the plains of Siberia, through the deserts of interior Asia, and into the mountains along Russia's rim. Russian pioneers ventured even farther, entering Alaska, traveling down the Pacific coast of North America, and planting the Russian flag near San Francisco in 1812. But as Russia's empire expanded, its internal weaknesses gnawed at the power of the czars. Peasants rebelled. Unpaid (and poorly fed) armies mutinied. When the czars tried to initiate reforms, the aristocracy objected. The empire at the beginning of the twentieth century was ripe for revolution, which began in 1905 (see the box titled "The Soviet Union, 1924–1991").

Communist Victory

The last czar, Nicholas II, was overthrown in 1917, and civil war followed. The victorious communists led by V.I. Lenin soon swept away much of the Russia of the past. The Russian flag disappeared. The czar and his family were executed. The old capital of Russia, St. Petersburg, was renamed Leningrad in honor of the revolutionary leader. Moscow, in the interior of the country, was chosen as the new capital for a country with a new name, the ***Soviet Union***. Eventually, this Union consisted of 15 political entities, each a Soviet Socialist Republic (SSR). Russia was just one of these republics, and the name *Russia* disappeared from the international map.

But on the Soviet map, Russia was the giant, the dominant republic, the Slavic center. Not for nothing was the communist revolution known as the Russian Revolution. The other republics of the Soviet Union were for minori-

The Soviet Union, 1924–1991

FOR 67 YEARS RUSSIA was the cornerstone of the Soviet Union, the so-called Union of Soviet Socialist Republics (USSR). The Soviet Union was the product of the Revolution of 1917, when more than a decade of rebellion against the rule of Czar Nicholas II led to his abdication. Russian revolutionary groups were called *soviets* ("councils"), and they had been active since the first workers' uprising in 1905. In that year, thousands of Russian workers marched on the czar's palace in St. Petersburg in protest, and soldiers fired on them. Hundreds were killed and wounded. Russia descended into chaos.

A coalition of military and professional men forced the czar's abdication in 1917. Then Russia was ruled briefly by a provisional government. In November 1917, the country held its first democratic election—and, as it turned out, its last for more than 70 years.

The provisional government allowed the exiled activists in the Bolshevik camp to return to Russia (there were divisions among the revolutionaries): Lenin from Switzerland, Trotsky from New York, and Stalin from internal exile in Siberia. In the ensuing political struggle, Lenin's Bolsheviks gained control over the revolutionary soviets, and this ushered in the era of communism. In 1924, the new communist empire was formally renamed the Union of Soviet Socialist Republics, or Soviet Union for short.

Lenin the organizer died almost immediately in 1924 and was succeeded by Stalin the tyrant, and many of the peoples under Moscow's control suffered unimaginably. In pursuit of communist reconstruction, Stalin and his adherents starved millions of Ukrainian peasants to death, forcibly relocated entire ethnic groups (including a people known as the Chechens), and exterminated "uncooperative" or "disloyal" peoples. The full extent of these horrors may never be known.

On December 25, 1991, the inevitable occurred: the Soviet Union ceased to exist, its economy a shambles, its political system shattered, the communist experiment a failure. The last Soviet president, Mikhail Gorbachev, resigned, and the Soviet hammer-and-sickle flag flying atop the Kremlin was lowered for the last time and immediately replaced by the white, red, and blue Russian tricolor. A new and turbulent era had begun, but Soviet structures and systems will long cast their shadows over transforming Russia.

---

ties the czars had colonized or for countries that fell under Soviet sway later, but none could begin to match Russia. The Soviet Empire was the legacy of czarist expansionism, and the new communist rulers were Russian first and foremost (with the exception of Josef Stalin, who was a Georgian).

Seven Fateful Decades

Officially, the *Union of Soviet Socialist Republics* endured from 1924 to 1991. When the USSR collapsed, the republics were set free and world maps had to be redrawn. Now Russia and its former colonies had to adjust to the politico-geographical realities of a changing world. It proved to be a difficult, sometimes desperate journey.

RUSSIA'S PHYSICAL GEOGRAPHY

Russia's physiography is dominated by vast plains and plateaus rimmed by rugged mountains (Fig. 2-1). Only the Ural Mountains break an expanse of low relief that extends from the Polish border to central Siberia. In the western part of this huge plain, where the northern pine forests meet the midlatitude grasslands, the Slavs established their domain.

Harsh Environments

The historical geography of Russia is the story of Slavic expansion from its populous western heartland across interior Eurasia to the east, and into the mountains and deserts of the south. This eastward march was hampered not only by vast distances but also by harsh natural conditions. As the northernmost populous country on Earth, Russia has virtually no natural barriers against the onslaught of Arctic air. Moscow lies farther north than Edmonton, Canada, and St. Petersburg lies at latitude 60° North—the latitude of the southern tip of Greenland. Winters are long, dark, and bitterly cold in most of Russia; summers are short and growing seasons limited. Many a Siberian frontier outpost was doomed by cold, snow, and hunger.

It is therefore useful to view Russia's past, present, and future in the context of its **climatology**. This field **3** of geography investigates not only the distribution of climatic conditions over the Earth's surface but also the processes that generate this spatial arrangement. The Earth's atmosphere traps heat received as radiation from the Sun, but this greenhouse effect varies over planetary space—and over time. As we noted in the introductory chapter, much of what is today Russia was in the grip of a glaciation until the onset of the warmer Holocene. But

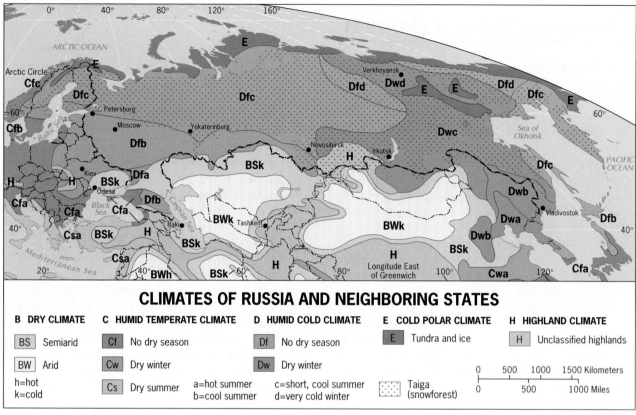

FIGURE 2-2

© H. J. de Blij, P. O. Muller, and John Wiley & Sons, Inc.

even today, with a natural warming cycle in progress augmented by human activity, Russia still suffers from severe cold and associated drought.

Currently, precipitation totals, even in western Russia, range from modest to minimal because the warm, moist air carried across Europe from the North Atlantic Ocean loses much of its warmth and moisture by the time it reaches Russia. Figures G-6, G-7, and 2-2 reveal **4** the consequences. Russia's climatic **continentality** (inland climatic environment remote from moderating and moistening maritime influences) is expressed by its prevailing *Dfb* and *Dfc* conditions. Compare the Russian map to that of North America (Fig. G-7), and you note that, except for a small corner off the Black Sea, Russia's climatic conditions resemble those of the Upper Midwest of the United States and interior Canada. Along its entire northern edge, Russia has a zone of *E* climates, the most frigid on the planet. In these Arctic latitudes originate the polar air masses that dominate its environments.

Looking North

For as long as Russia has been inhabited, the north has been the country's forbidding frontier, its harsh conditions legendary, its resources extracted with great dif-

ficulty, and its remote "gulag" prisons feared as the ultimate in punishment. Nothing blocks those frigid air masses of the Arctic from pushing southward, freezing rivers and soils and propelling blizzards that can last for weeks. Many research teams exploring Russia's Arctic shores and the islands beyond were lost, never to return. Thick ice on the Arctic Ocean defied even the most modern icebreakers, sometimes crushing those that ventured out too far.

But the Arctic environment is changing, and the changes have economists as well as politicians thinking hard. Take a look at a globe from the top (north), and you can see how many countries would be affected by a sustained warming trend in the Arctic that would melt ice permanently, open up sea routes still seasonally closed, and bring oil and gas resources below the ocean floor within the reach of modern extractive technology (these issues are discussed in the final section of Chapter 12). No state has a longer coastline on the Arctic Ocean than Russia, and the Russians are busy staking claims and drawing seaward borders in anticipation of the opportunities that may soon open and the competition that lies ahead (see Fig. 12-5 on p. 617). Already, Canada and the United States, among others, are contesting what the Russians are doing in the Arctic arena. Some experts claim that as much as 20 percent of the remaining

RUSSIAN REALM POPULATION DISTRIBUTION: 2010
One dot represents 50,000 persons

FIGURE 2-3

© H. J. de Blij, P. O. Muller, and John Wiley & Sons, Inc.

"potential" oil reserves on the planet may be locked inside the Arctic Circle. The stakes are high indeed.

We return to the topic of maritime claims and seaward boundaries in the final chapter, because the Arctic is not the only polar zone where actual and potential competition for resources are growing concerns. The Antarctic, too, has its claims and counterclaims, although international agreements there have forestalled much of what might have happened. Those Antarctic agreements may not be of much help in the far north, and the globe tells you why: Antarctica is a continent, a landmass still covered by ice that spreads out over the adjoining ocean. The Arctic, however, is an ice-covered ocean, where different international rules and laws apply.

Climates and Peoples

Much of what will happen in the Russian Arctic still lies in the future, perhaps the distant future. When we study Russia's natural environments and population clusters, we still focus on those areas where conditions are most favorable for habitation and where people have settled over a long span of postglacial time. By studying the map of Russia's climates (Fig. 2-2) we can begin to understand what the map of population distribution (Fig. 2-3)

shows. The overwhelming majority of Russia's 140-plus million people remain concentrated in the west and southwest, where climatic conditions always were least difficult when farming was the mainstay of most of the people. The map in Figure 2-3 is a legacy of millennia that will continue to mark Russia's living space for generations to come.

Environment and Politics

Climate and weather (there is a distinction: *climate* is a long-term average, whereas **weather** refers to existing atmospheric conditions at a given place and time) have always challenged Russia's farmers. Conditions are most favorable in the west, but even there temperature extremes, variable and undependable rainfall, and short growing seasons make farming difficult. During the Soviet period, fertile and productive Ukraine supplied much of Russia's food needs, but even then Russia often had to import grain. Soviet rulers wanted to reduce their country's dependence on imported food, and their communist planners built major irrigation projects to increase crop yields in the colonized republics of Central Asia. As we will see in Chapter 7, some of these attempts to overcome nature's limitations spelled disaster for the local people.

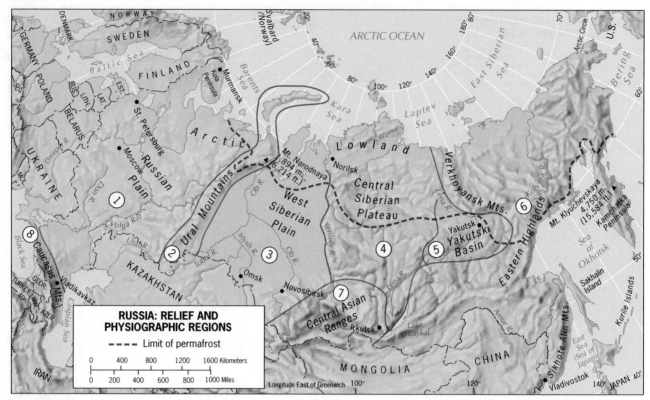

FIGURE 2-4

© H. J. de Blij, P. O. Muller, and John Wiley & Sons, Inc.

Physiographic Regions

To assess the physiography of this vast country, please refer to Figure 2-4. Note how mountains and deserts encircle Russia: the Caucasus in the southwest ⑧; the Central Asian Ranges in the center ⑦; the Eastern Highlands facing the Pacific from the Bering Sea to the East Sea (Sea of Japan) ⑥. The Kamchatka Peninsula has a string of active volcanoes in one of the world's most earthquake-prone zones (Fig. G-4). Warm subtropical air thus has little opportunity to penetrate Russia, whereas cold Arctic air sweeps southward without impediment. Russia's Arctic north is a gently sloping lowland broken only by the Urals and Eastern Highlands.

Russia's vast and complex physical stage can be divided into eight physiographic regions, each of which, at a larger scale, can be subdivided into smaller units. In the Siberian region, one criterion for such subdivision would be the vegetation. The Russian language has given us two terms to describe this vegetation: **tundra**, the treeless plain along the Arctic shore where mosses, lichens, and some grasses survive; and **taiga**, the mostly coniferous forests that begin to the south of where the tundra ends, and extend over vast reaches of Siberia, which means the "sleeping land" in Russian.

The Russian Plain

The Russian Plain ① is the eastward continuation of the North European Lowland, and here the Russian state formed its first core area. Travel north from Moscow at its heart, and the countryside soon is covered by needle-leaf forests like those of Canada; to the south lie the grain fields of southern Russia and, beyond, those of Ukraine. Note the Kola Peninsula and Barents Sea in the far north: warm water from the North Atlantic comes around northern Norway and keeps the port of Murmansk ice-free most of the year. The Russian Plain is bounded on the east by the Ural Mountains ②; though not a high range, it is topographically prominent because it separates two extensive plains. The main range of the Urals is more than 3200 kilometers (2000 mi) long and reaches from the shores of the Kara Sea to the border with Kazakhstan. It is not a barrier to east-west transportation, and its southern end is densely populated. Here the Urals yield minerals and fossil fuels.

Siberia

East of the Urals lies Siberia. The West Siberian Plain ③ has been described as the world's largest unbroken lowland; this is the vast basin of the Ob and Irtysh rivers.

Over the last 1600 kilometers (1000 mi) of its course to the Arctic Ocean, the Ob falls less than 100 meters (330 ft). In Figure 2-4, note the dashed line that extends from west of the northern Urals to the Bering Sea and includes the Arctic Lowland. North of the line, water in the ground is permanently frozen; this **permafrost** creates another obstacle to permanent settlement. Looking again at the West Siberian Plain, we see that the north is permafrost-ridden and the central zone is marshy. The far south, however, has such major cities as Omsk and Novosibirsk, beyond the reach of permafrost conditions.

East of the West Siberian Plain the country begins to rise, first into the Central Siberian Plateau ④, another sparsely settled, remote, permafrost-affected region. Here winters are long and cold, and summers are short; the area remains barely touched by human activity. Beyond the Yakutsk Basin ⑤, the terrain becomes mountainous and the relief high. The Eastern Highlands ⑥ are a jumbled mass of ranges and ridges, precipitous valleys, and volcanic mountains. Lake Baykal lies in a trough that is over 1500 meters (5000 ft) deep—the deepest rift lake in the world (see Chapter 6). On the far eastern Kamchatka Peninsula, volcanic Mount Klyuchevskaya reaches 4750 meters (15,584 ft).

Encircling Mountains

The northern part of region ⑥ is Russia's most inhospitable zone, but southward along the Pacific coast the climate is less severe. Nonetheless, this is a true frontier region. The forests provide opportunities for lumbering, a fur trade exists, and there are gold and diamond deposits.

Mountains also mark the southern margins of much of Russia. The Central Asian Ranges ⑦, from the Kazakh border in the west to Lake Baykal in the east, contain many glaciers whose annual meltwaters send alluvium (silt)-laden streams to enrich farmlands at lower elevations. The Caucasus ⑧, in the land corridor between the Black and Caspian seas, form an extension of Europe's Alpine Mountains and exhibit a similarly high *relief* (range of elevations). Here Russia's southern border is sharply defined by *topography* (surface configuration).

As our physiographic map suggests, the more habitable terrain in Russia becomes latitudinally narrower from west to east. Beyond the southern Urals, the zone of settlement in Russia becomes discontinuous (in Soviet times it extended into northern Kazakhstan, where a large Russian population still lives). Isolated towns did develop in Russia's vast eastern reaches, even in Siberia, but the tenuous ribbon of settlement does not widen again until it reaches the country's Far Eastern Pacific Rim (Fig. 2-3).

EVOLUTION OF THE RUSSIAN STATE

Four centuries ago, when European kingdoms were sending fleets to distant shores to search for riches and capture colonies, there was little indication that the largest of all empires would one day center on a city in the forests halfway between Sweden and the Black Sea. The plains of Eurasia south of the taiga had seen waves of migrants sweep from east to west: Scythians, Sarmatians, Goths, Huns, and others came, fought, settled, and were absorbed or driven off. Eventually, the Slavs, heirs to these Eurasian infusions, emerged as the dominant culture in what is today Ukraine, south of Russia and north of the Black Sea.

From this base on fertile, productive soils, the Slavs expanded their domain, making Kiev (Kyyiv) their southern headquarters and establishing Novgorod on Lake Ilmen in the north. Each was the center of a state known as a *Rus*, and both formed key links on a trade route between the German-speaking Hanseatic ports of the Baltic Sea and the trading centers of the Mediterranean. During the eleventh and twelfth centuries, the Kievan Rus and the Novgorod Rus united to form a large and comparatively prosperous state astride forest and *steppe* (short-grass prairie).

The Mongol Invasion

Prosperity attracts attention, and far to the east, north of China, the Mongol Empire had been building under Genghis Khan. In the thirteenth century, the Mongol (Tatar) "hordes" rode their horses into the Kievan Rus, and the state fell. By mid-century, Slavs were fleeing into the forests, where the Mongol horsemen had less success than on the open plains of Ukraine. Here the Slavs reorganized, setting up several new Russes. Still threatened by the Tatars, the ruling princes paid tribute to the Mongols in exchange for peace.

Moscow, deep in the forest on the Moscow River, was one of these Russes. Its site was remote and defensible. Over time, Moscow established trade and political links with Novgorod and became the focus of a growing area of Slavic settlement. When the Mongols, worried about Moscow's expanding influence, attacked near the end of the fourteenth century and were repulsed, Moscow emerged as the unchallenged center of the Slavic Russes. Soon its ruler, now called a Grand Duke, took control of Novgorod, and the stage was set for further growth.

FIGURE 2-5

© H. J. de Blij, P. O. Muller, and John Wiley & Sons, Inc.

Grand Duchy of Muscovy

Moscow's geographic advantages again influenced events. Its centrality, defensibility, links to Slavic settlements including Novgorod, and open frontiers to the north and west, where no enemies threatened, gave Moscow the potential to expand at the cost of its old enemies, the Mongols. Many of these invaders had settled in the basin of the Volga River and elsewhere, where the city of Kazan had become a major center of Islam, the Tatars' dominant religion. During the rule of Ivan the Terrible (1547–1584), the Grand Duchy of Muscovy became a major military power and imperial state. Its rulers called themselves czars and claimed to be the heirs of the Byzantine emperors. Now began the expansion of the Russian domain (Fig. 2-5), marked by the defeat of the Tatars at Kazan and the destruction of hundreds of mosques.

The Cossacks

This eastward expansion of Russia was spearheaded by a relatively small group of semi-nomadic peoples who came to be known as Cossacks and whose original home

was in present-day Ukraine. Opportunists and pioneers, they sought the riches of the eastern frontier, chiefly fur-bearing animals, as early as the sixteenth century. By the mid-seventeenth century they reached the Pacific Ocean, defeating Tatars in their path and consolidating their gains by constructing *ostrogs* (strategic fortified waystations) along river courses. Before the eastward expansion halted in 1812 (Fig. 2-5), the Russians had moved across the Bering Strait to Alaska and down the western coast of North America into what is now northern California (see the box titled "Russians in North America").

Czar Peter the Great

When Peter the Great became czar (he ruled from 1682 to 1725), Moscow already lay at the center of a great empire—great, at least, in terms of the territories it controlled. The Islamic threat had been ended with the defeat of the Tatars. The influence of the Russian Orthodox Church was represented by its distinctive religious architecture and powerful bishops. Peter consolidated Russia's gains and hoped to make a modern, European-style state out of his loosely knit country. He built St. Petersburg

Russians in North America

THE FIRST WHITE settlers in Alaska were Russians, not Western Europeans, and they came across Siberia and the Bering Strait, not across the Atlantic and North America. Russian hunters of the sea otter, valued for its high-priced pelt, established their first Alaskan settlement at Kodiak Island in 1784. Moving south along the North American coast, the Russians founded additional villages and forts to protect their tenuous holdings until they reached as far as the area just north of San Francisco Bay, where they built Fort Ross in 1812.

But the Russian settlements were isolated and vulnerable. European fur traders began to put pressure on their Russian competitors, and St. Petersburg found the distant settlements a burden and a risk. In any case, American, British, and Canadian hunters were decimating the sea otter population, and profits declined. When U.S. Secretary of State William Seward offered to purchase Russia's North American holdings in 1867, St. Petersburg quickly agreed—for $7.2 million. Thus Alaska, including its lengthy southward coastal extension, became U.S. territory and, in 1959, the forty-ninth State. Although Seward was ridiculed for his decision—Alaska was called "Seward's Folly" and "Seward's Icebox"—he was vindicated when gold was discovered there in the 1890s. The twentieth century further underscored the wisdom of Seward's action, strategically as well as economically. At Prudhoe Bay off Alaska's northern Arctic slope, large oil reserves were tapped in the 1970s and are still being exploited. And like Siberia, Alaska probably contains yet-unknown riches.

9 as a **forward capital** on the doorstep of Swedish-held Finland, fortified it with major military installations, and made it Russia's leading port.

Czar Peter the Great, an extraordinary leader, was in many ways the founder of modern Russia. In his desire to remake Russia—to pull it from the forests of the interior to the waters of the west, to open it to outside influences, and to relocate its population—he left no stone unturned. Prominent merchant families were forced to move from other cities to St. Petersburg. Ships and wagons entering the city had to bring building stones as an entry toll. The czar himself, aware that to become a major power Russia had to be strong at sea as well as on land, went to the Netherlands to work as a laborer in the famed Dutch shipyards to learn the most efficient method for building ships. Meanwhile, the czar's forces continued to conquer people and territory: Estonia was incorporated in 1721, widening Russia's window to the west, and major expansion soon occurred south of the city of Tomsk (Fig. 2-5).

Czarina Catherine the Great

Under Czarina Catherine the Great, who ruled from 1760 to 1796, Russia's empire in the Black Sea area grew at the expense of the Ottoman Turks. The Crimea Peninsula, the port city of Odesa (Odessa), and the whole northern coastal zone of the Black Sea fell under Russian control. Also during this period, the Russians made a fateful move: they penetrated the area between the Black and Caspian seas, the mountainous Caucasus with dozens of ethnic and cultural groups, many of which were Islamized. The cities of Tbilisi (now in Georgia), Baki (Baku) in Azerbaijan, and Yerevan (Armenia) were captured. Eventually, the Russian push toward an Indian Ocean outlet was halted by the British, who held sway in Persia (modern Iran), and also by the Turks.

Meanwhile, Russian colonists entered Alaska, founding their first American settlement at Kodiak Island in 1784. As they moved southward, numerous forts were built to protect their tenuous holdings against indigenous peoples. Eventually, they reached nearly as far south as San Francisco Bay where, in the fateful year 1812, they erected Fort Ross.

Catherine the Great had made Russia a colonial power, but the Russians eventually gave up on their North American outposts. The sea-otter pelts that had attracted the early pioneers were running out, European and white American hunters were cutting into the profits, and Native American resistance was growing. When U.S. Secretary of State William Seward offered to purchase Russia's Alaskan holdings in 1867, the Russian government quickly agreed—for $7.2 million. Thus Alaska and its Russian-held panhandle became U.S. territory and, ultimately, the forty-ninth State.

A Russian Empire

Although Russia withdrew from North America, Russian expansionism during the nineteenth century continued in Eurasia. While extending their empire southward, the Russians also took on the Poles, old enemies to the west, and succeeded in taking most of what is today the Polish state, including the capital of Warsaw. To the northwest, Russia took over Finland from the Swedes in 1809.

Penetration of Muslim Asia

During most of the nineteenth century, however, the Russians were preoccupied with Central Asia—the region between the Caspian Sea and western China—where Tashkent and Samarqand (Samarkand) came under St. Petersburg's control (Fig. 2-5). The Russians here were still bothered by raids of nomadic horsemen, and they sought to establish their authority over the Central Asian steppe country as far as the edges of the high mountains that lay to the south. Thus Russia gained many Muslim subjects, for this was Islamic Asia they were penetrating. Under czarist rule, these people retained some autonomy.

Confronting China and Japan

Much farther to the east, a combination of Japanese expansionism and a decline of Chinese influence led Russia to annex from China several provinces east of the Amur River. Soon thereafter, in 1860, the port of Vladivostok on the Pacific was founded.

Now began the events that were to lead to the first involuntary halt to the Russian drive for territory. As Figure 2-5 shows, the most direct route from western Rus-

sia to the port of Vladivostok lay across northeastern China, the territory then still called Manchuria. The Russians had begun construction of the Trans-Siberian Railroad in 1892, and they wanted China to permit the track to cross Manchuria. But the Chinese resisted. Then, taking advantage of the Boxer Rebellion in China in 1900 (see Chapter 9), Russian forces occupied Manchuria so that railroad construction might proceed.

This move, however, threatened Japanese interests in this area, and the Japanese confronted the Russians in the Russo-Japanese War of 1904–1905. Not only was Russia defeated and forced out of Manchuria: Japan even took possession of the southern part of Sakhalin Island, which they named Karafuto and retained until the end of World War II.

THE COLONIAL LEGACY

Thus Russia, like Britain, France, and other European powers, expanded through **colonialism**. Yet whereas the other European powers expanded overseas, Russian influence traveled overland into Central Asia, Siberia, China, and the Pacific coastlands of the Far East. What

10

emerged was not the greatest empire but the largest territorially contiguous empire in the world. At the time of the Japanese war, the Russian czar controlled more than 22 million square kilometers (8.5 million sq mi), just a tiny fraction less than the area of the Soviet Union after the 1917 Revolution. Thus the communist empire, to a large extent, was the legacy of St. Petersburg and European Russia, not the product of Moscow and the socialist revolution.

The czars embarked on their imperial conquests in part because of Russia's relative location: Russia always lacked warm-water ports. Had the Revolution not intervened, their southward push might have reached the Persian Gulf or even the Mediterranean Sea. Czar Peter the Great envisaged a Russia open to trading with the entire world; he developed St. Petersburg on the Baltic Sea into Russia's leading port. But in truth, Russia's historical geography is one of remoteness from the mainstreams of change and progress, as well as one of self-imposed isolation.

An Imperial, Multinational State

Centuries of Russian expansionism did not confine itself to empty land or unclaimed frontiers. The Russian state became an imperial power that annexed and incorporated many nationalities and cultures. This was done by employing force of arms, by overthrowing uncooperative rulers, by annexing territory, and by stoking the fires of ethnic conflict. By the time the ruthless Russian regime began to face revolution among its own people, czarist
[11] Russia was a hearth of **imperialism**, and its empire contained peoples representing more than 100 nationalities. The winners in the ensuing revolutionary struggle—the communists who forged the Soviet Union—did not liberate these subjugated peoples. Rather, they changed the empire's framework, binding the peoples colonized by the czars into a new system that would in theory give them autonomy and identity. In practice, it doomed those peoples to bondage and, in some cases, extinction.

When the Soviet system failed and the Soviet Socialist Republics became independent states, Russia was left without the empire that had taken centuries to build and consolidate—and that contained crucial agricultural and mineral resources. No longer did Moscow control the farms of Ukraine and the oil and natural gas reserves of Central Asia. But look again at Figure 2-5 and you will see that, even without its European and Central Asian colonies, Russia remains an empire. Russia lost the "republics" on its periphery, but Moscow still rules over a domain that extends from the borders of Finland to North Korea. Inside that domain Russians are in the overwhelming majority, but many subjugated nationalities,

from Tatars to Yakuts, still inhabit ancestral homelands. Accommodating these many indigenous peoples is one of the challenges facing the Russian Federation today.

In the 1990s, Russia began to reorganize in the aftermath of the collapse of the Soviet Union. This reorganization cannot be understood without reference to the seven decades of Soviet communist rule that went before. We turn next to this crucial topic.

THE SOVIET LEGACY

The era of communism may have ended in the Soviet Empire, but its effects on Russia's political and economic geography will long remain. Seventy years of centralized planning and implementation cannot be erased overnight; regional reorganization toward a market economy cannot be accomplished in a day.

While the world of capitalism celebrates the failure of the communist system in the former Soviet realm, we should remember why communism found such fertile ground in the Russia of the 1910s and 1920s. In those days Russia was infamous for the wretched serfdom of its peasants, the cruel exploitation of its workers, the excesses of its nobility, and the ostentatious palaces and riches of the czars. Ripples from the western European Industrial Revolution introduced a new age of misery for those laboring in factories. There were workers' strikes and ugly retributions, but when the czars finally tried to better the lot of the poor, it was too little too late. There was no democracy, and the people had no way to express or channel their grievances. Europe's democratic revolution passed Russia by, and its economic revolution touched the czars' domain only slightly. Most Russians, and tens of millions of non-Russians under the czars' control, faced exploitation, corruption, starvation, and harsh subjugation. When the people began to rebel in 1905, there was no hint of what lay in store; even after the full-scale Revolution of 1917, Russia's political future hung in the balance.

The Russian Revolution was no unified uprising. There were factions and cliques; the Bolsheviks ("Majority") took their ideological lead from Lenin, while the Mensheviks ("Minority") saw a different, more liberal future for their country. The so-called Red army factions fought against the Whites, while both battled the forces of the czar. The country stopped functioning; the people suffered terrible deprivations in the countryside as well as the cities. Most Russians (and other nationalities within the empire as well) were ready for radical change.

That change came when the Revolution succeeded and the Bolsheviks bested the Mensheviks, most of whom were exiled. In 1918, the capital was moved from

"June 25, 1964. We left Moscow this morning after our first week in the Soviet Union, on the road to Kiev. Much has been seen and learned after we arrived in Leningrad from Helsinki (also by bus) last Sunday. Some interesting contradictions: in a country that launched *Sputnik* nearly a decade ago, nothing mechanical seems to work, from elevators to vehicles. The arts are vibrant, from the ballet in the Kirov Theatre to the Moscow Symphony. But much of what is historic here seems to be in disrepair, and in every city and town you pass, rows and rows of drab, gray apartment buildings stand amid uncleared rubble, without green spaces, playgrounds, trees. My colleagues here talk about the good fortune of the residents of these buildings: as part of the planning of the 'socialist city,' their apartments are always close to their place of work, so that they can walk or ride the bus a short distance rather than having to commute for hours. Our bus broke down as we entered the old part of Tula, and soon we learned that it would be several hours before we could expect to reboard. We had stopped right next to a row of what must have been attractive shops and apartments a half-century ago; everything decorative seems to have been removed, and nothing has been done to paint, replace rotting wood, or fix roofs. I walked up the street a mile or so, and there was the 'new' Tula—the Soviet imprint on the architecture of the place. This scene could have been photographed on the outskirts of Leningrad, Moscow, or any other urban center we had seen. What will life be like in the Soviet city of the distant future?"   © H. J. de Blij.

Concept Caching

Petrograd (as St. Petersburg had been renamed in 1914, to remove its German appellation) to Moscow. This was a symbolic move, the opposite of the forward-capital principle: Moscow lay deep in the Russian interior, not even on a major navigable waterway (let alone a coast), amid the same forests that much earlier had protected the Russians from their enemies. The new Soviet Union would look inward, and the communist system would achieve with Soviet resources and labor the goals that had for so long eluded the country. The chief political and economic architect of this effort was the revolutionary leader who had prevailed in the power struggle: V. I. Lenin (born Vladimir Ilyich Ulyanov).

The Political Framework

Russia's great expansion had brought many nationalities under czarist control; now the revolutionary government sought to organize this heterogeneous ethnic mosaic into a smoothly functioning state. The czars had conquered, but they had done little to bring Russian culture to the peoples they ruled. The Georgians, Armenians, Tatars, and residents of the Muslim states of Central Asia were among dozens of individual cultural, linguistic, and religious groups that had not been "Russified." In 1917, however, the Russians themselves constituted only about one-half of the population of the entire realm. Thus it was impossible to establish a Russian state instantly over this vast political region, and these diverse national groups had to be accommodated.

The question of the nationalities became a major issue in the young Soviet state after 1917. Lenin, who brought the philosophy of Karl Marx to Russia, talked from the beginning about the "right of self-determination for the nationalities." The first response by many of Russia's subject peoples was to proclaim independent republics, as they did in Ukraine, Georgia, Armenia, Azerbaijan, and even in Central Asia. But Lenin had no intention of permitting the Russian state to break up. In 1923, when his blueprint for the new Soviet Union went into effect,

Here is another view of Tula, four decades after the photographs on the left were taken, and it provides some clues to the question posed at the end of the caption. Tula lies about 160 kilometers (100 mi) south of Moscow and was Russia's second city during the Middle Ages, endowed with an impressive *kremlin* (fortress) and thriving as a commercial center. Tula's twenty-first-century townscape is a mixture of historic buildings (built between 1514 and 1521), drab Soviet-era tenements (unchanged since 1964), and a few more modern apartment buildings. Old habits persist: where there is some fertile soil, you are likely to see vegetable plots, not flower gardens. © Mauro Galligani/Contrasto/Redux Pictures.

the last of these briefly independent units was fully absorbed into the sphere of the Moscow regime. Ukraine, for example, declared itself independent in 1917 and managed to sustain this initiative until 1919. But in that year the Bolsheviks set up a provisional government in Kiev, the Ukrainian capital, thereby ensuring the incorporation of the country into Lenin's Soviet framework.

The Communist System

The Bolsheviks' political framework for the Soviet Union was based on the ethnic identities of its many incorporated peoples. Given the size and cultural complexity of the empire, it was impossible to allocate territory of equal political standing to all the nationalities; the communists controlled the destinies of well over 100 peoples, both large nations and small isolated groups. It was decided to divide the vast realm into Soviet Socialist Republics (SSRs), each of which was delimited to correspond broadly to one of the major nationalities. At the time, Russians constituted about half of the developing Soviet Union's population, and, as Figure 2-6 shows, they also were (and still are) the most widely dispersed ethnic group in the realm. The Russian Republic, therefore, was by far the largest designated SSR, comprising just under 77 percent of total Soviet territory.

Within the SSRs, smaller minorities were assigned political units of lesser rank. These were called Autonomous Soviet Socialist Republics (ASSRs), which in effect were republics within republics; other areas were designated Autonomous Regions or other nationality-based units. It was a complicated, cumbersome, often poorly designed framework, but in 1924 it was launched

officially under the banner of the *Union of Soviet Socialist Republics (USSR).*

A Soviet Empire

Eventually, the Soviet Union came to consist of 15 SSRs (shown in Fig. 2-7), including not only the original republics of 1924 but also such later acquisitions as Moldova (formerly Moldavia in Romania), Estonia, Latvia, and Lithuania. The internal political layout often was changed, sometimes at the whim of the communist empire's dictators. But no communist apartheid-like system of segregation could accommodate the shifting multinational mosaic of the Soviet realm. The republics quarreled among themselves over boundaries and territory. Demographic changes, migrations, war, and economic factors soon made much of the layout of the 1920s obsolete. Moreover, the communist planners made it Soviet policy to relocate entire peoples from their homelands in order to better fit the grand design, and to reward or punish—sometimes capriciously. The overall effect, however, was to move minority peoples eastward and to replace them with Russians. This **Russification** of the [12] Soviet Empire produced substantial ethnic Russian minorities in all the non-Russian republics.

Phantom Federation

The Soviet planners called their system a **federation**. We [13] focus in more detail on this geographic concept below, but let us note here that *federalism* involves the sharing of power between a country's central government and its political subdivisions (provinces, States, or, in the Soviet case, "Socialist Republics"). Study the map of the former

FIGURE 2-6

© H. J. de Blij, P. O. Muller, and John Wiley & Sons, Inc.

Soviet Union (Fig. 2-7), and an interesting geographic corollary emerges: every one of the 15 Soviet Republics had a boundary with a non-Soviet neighbor. Not one was spatially locked within the others. This seemed to give geographic substance to the notion that any Republic was free to leave the USSR if it so desired. Reality, of course, was different. Moscow's control over the republics made the Soviet Union a federation in theory only.

The centerpiece of the tightly controlled Soviet "federation" was the Russian Republic. With half the vast state's population, the capital city, the realm's core area, and over three-quarters of the Soviet Union's territory, Russia was the empire's nucleus. In other republics, "Soviet" often was simply equated with "Russian"—it was the reality with which the lesser republics lived. Russians came to the other republics to teach (Russian was taught in the colonial schools), to organize (and often dominate) the local Communist Party, and to implement Moscow's economic decisions. This was colonialism, but somehow the communist disguise—how could socialists, as the communists called themselves, be colonialists?—and the contiguous spatial nature of the empire made it appear to the rest of the world as something else. Indeed, on the world stage the Soviet Union became a champion of oppressed peoples, a force in the decolonization process. It was an astonishing contradiction that would, in time, be exposed.

The Soviet Economic Framework

The geopolitical changes that resulted from the founding of the Soviet Union were accompanied by a gigantic economic experiment: the conversion of the empire from a czarist autocracy with a capitalist veneer to communism. From the early 1920s onward, the country's economy would be centrally planned—the communist leadership in Moscow would make all decisions regarding economic planning and development. Soviet planners had two principal objectives: (1) to accelerate industrialization and (2) to **collectivize** agriculture. For the first time ever on such a scale, and for the first time in accordance with Marxist-Leninist principles, an entire country was organized to work toward national goals prescribed by a central government.

Land and the State

The Soviet planners believed that agriculture could be made more productive by organizing it into huge state-run enterprises. The holdings of large landowners were expropriated, private farms were taken away from the farmers, and the land was consolidated into collective farms. Initially, all such land was meant to be part of a *sovkhoz*, literally a grain-and-meat factory in which agricultural efficiency, through maximum mechaniza-

FIGURE 2-7

© H. J. de Blij, P. O. Muller, and John Wiley & Sons, Inc.

tion and minimum labor requirements, would be at its peak. But many farmers opposed the Soviets and tried to sabotage the program in various ways, hoping to retain their land.

The farmers and peasants who obstructed the communists' grand design suffered a dreadful fate. In the 1930s, for instance, Stalin confiscated Ukraine's agricultural output and then ordered part of the border between the Russian and Ukrainian republics sealed—thereby leading to a famine that killed millions of farmers and their families. In the Soviet Union under communist totalitarianism, the ends justified the means, and untold hardship came to millions who had already suffered under the czars. In his book *Lenin's Tomb*, David Remnick estimates that between 30 and 60 million people lost their lives from imposed starvation, political purges, Siberian exile, and forced relocation. It was an incalculable human tragedy, but the secretive character of Soviet officialdom made it possible to hide it from the world.

The Soviet planners hoped that collectivized and mechanized farming would free hundreds of thousands of workers to labor in factories. Industrialization was the prime objective of the regime, and here the results were superior. Productivity rose rapidly, and when World War II engulfed the empire in 1941, the Soviet manufacturing sector was able to produce the equipment and weapons needed to repel the German invaders.

A Command Economy

Yet even in this context, the Soviet grand design entailed liabilities for the future. The USSR practiced a **command** **15** **economy**, in which state planners assigned the production of particular manufactures to particular places, often disregarding the rules of economic geography. For example, the manufacture of railroad cars might be assigned (as indeed it was) to a factory in Latvia. No other factory anywhere else would be permitted to produce this equipment—even if supplies of raw materials would make it cheaper to build them near, say, Volgograd 1900 kilometers (1200 mi) away. Despite an expanded and improved transport network (Fig. 2-8), such practices made manufacturing in the USSR extremely expensive, and the absence of competition made managers complacent and workers less productive.

THE RUSSIAN REALM

POPULATION

- Under 50,000
- 50,000–250,000
- 250,000–1,000,000
- 1,000,000–5,000,000
- Over 5,000,000

Railroad built before 1917
Railroad built 1917–1991
Road

National capitals are underlined

0   200   400   600   800   1000 Kilometers
0   100   200   300   400   500   600 Miles

FIGURE 2-8

Of course, the Soviet planners (political or economic) never imagined that their experiment would fail. By the time it did, Russia and its co-Soviet Socialist Republics were in an economic straitjacket that made any transition to democratic government and an open, diversified economy very difficult. When the USSR collapsed and the state lost control over resources and production, some well-placed opportunists enriched themselves by buying mines and factories. Others seized political control in some of the former Soviet regions and used their power to enrich themselves and their cohorts. Few of these oligarchs and opportunists seemed to have Russia's national interests at heart. For some time, it seemed that gigantic Russia would fracture and disintegrate. That did not happen, but Russia today still is far from the democratic society and balanced economy that would make it an esteemed member of the international community.

RUSSIA'S CHANGING POLITICAL GEOGRAPHY

When the USSR dissolved in 1991, Russia's former empire devolved into 14 independent countries, and Russia itself was a changed nation. Russians now made up about 83 percent of the population of less than 150 million, a far higher proportion than in the days of the Soviet Union. But numerous minority peoples remained under Moscow's new flag, and millions of Russians found themselves under new national governments in the 14 former Republics.

Soviet planners had created a complicated administrative structure for their "Russian Soviet Federative Socialist Republic," and Russia's post-communist leaders had to use this framework to make their country function. In 1992, most of Russia's internal "republics," autonomous regions, oblasts, and krays (all of them components of the administrative hierarchy) signed a document known as the Russian Federation Treaty, committing them to cooperate in the new federal system. At first, a few units refused to sign, including Tatarstan, scene of Ivan the Terrible's brutal conquest more than four centuries ago, and a republic in the Caucasus periphery, then known as Chechenya-Ingushetia, where Muslim rebels waged a campaign for independence. As the map shows, the latter split into two separate republics whose names soon came to be respelled Chechnya and Ingushetiya (Fig. 2-9). Eventually, only Chechnya refused to sign the Russian Federation Treaty, and subsequent Russian military intervention led to a prolonged and violent conflict, with disastrous consequences for Chechnya's people and infrastructure (the capital, Groznyy, was completely destroyed).

After President Putin took office in 2000, he managed to bring the Chechnya conflict to a conclusion through a combination of massive military action and shrewd political manipulation. This involved the "Chechenization" of the conflict, which had been driven by Muslim separatists using the high mountains of southern Chechnya as their base. By giving free rein to the pro-Russian north, and by supporting the election of a strongman president from this sector of the population, Moscow could stand back as its surrogates prevailed. But the cost was high: by most estimates, more than 100,000 people had lost their lives during the two decades of strife.

Today Chechnya is relatively calm and its capital is being rebuilt, but now there is growing violence in neighboring Ingushetiya and Dagestan, two other "republics" in this contentious Transcaucasian border zone. We will take a closer look at this drama in the regional section of the chapter.

Russia's Federal Framework—Or Is It?

Great size and vast distances do not necessarily spell advantage and opportunity when it comes to a country's national integration. The Soviet communists had created a hierarchy of republics, regions, oblasts, and other administrative entities as a way to facilitate the transmittal of decisions made by the Politburo in Moscow to places near and far. During the communist period, Russia was officially called the "Russian Soviet Federative Socialist Republic," but it was "federative" in name only.

As we noted earlier, in a *federal state system* power is shared between the central government and the subnational political units that make up the administrative framework. Usually the national government is responsible for matters such as defense, foreign policy, and foreign trade. The lesser administrative components may have authority over affairs ranging from education to transportation (in the United States, your driver's license has the name of a State on it, not the name of the country). Through negotiation and accommodation, this balance of power can shift without producing crises. Some countries owe their very survival to their federal frameworks; two examples are India and Australia (in our coverage of Australia in Chapter 11 we will further explore the concept of federalism). As is pointed out in Chapter 1, some European states have actually gone *beyond* federation in the degree of autonomy they have awarded to their regions and provinces, some of which can now control their own foreign trade.

Russia never was such a federation, its communist-era name notwithstanding. Rather, the Soviet Union func-

16 tioned as a **unitary state system**, in which government is centralized and authority over the state's component parts is imposed, not negotiated. Kingdoms of the past were unitary states, with authority vested in royalty. Today, many unitary states are democracies, including Japan and Uruguay. Their provinces or regions may not have federal rights, but government is nevertheless representative of the people. In some unitary states, however, power is so concentrated in the ruler or rulers that they become dictatorships. North Korea, Myanmar (Burma), Belarus, and Equatorial Guinea prove that this can happen anywhere in the world.

When Russia emerged from the collapse of the Soviet Union in 1991, its new leaders had to govern a country that had no history of federal give-and-take. But it would have been impractical to immediately restructure the communist-era administrative framework they had to work with. So President Yeltsin and his government tried to give the Republics and Regions a measure of the federal-style autonomy some were demanding, and introduced democracy by encouraging all subnational entities within Russia to elect representatives to the Parliament in Moscow.

Problems of Transition

None of this went easily or well. When President Putin succeeded President Yeltsin in 2000, Putin said that he found a country in chaos and in need of total reorganization.

Part of the problem is Russia's enormous size. Russia's core area lies in the far west of the country, centered on Moscow. Beyond the Ural Mountains, the periphery stretches thousands of kilometers to the shores of the Pacific Ocean. Effective government requires efficient 17 interaction, but the geographic principle of **distance decay** confirms that increasing distances between places tend to reduce the interactions among them. Far from the capital, the "elected" leaders in Russia's Far East simply ignored the laws made in Moscow, running their Regions as though they were their private domains. In the early years there was little the central government could do about it.

Another problem lay in Russia's mismanaged economy. Other than commodities such as oil or natural gas, you would not find Russian products on world markets; the communist economy had been locked in a cocoon. During the communist period, other governments purchased Soviet weapons and communist allies might order railroad cars, but consumer products were nowhere to be found. The new Russian administration wanted to sell off (privatize) former Soviet factories, farms, and energy providers, but that process soon was

rife with corruption, making billionaires out of well-placed opportunists but denying the government badly needed revenues. And very little progress was made toward diversification of the economy. Russia lived off the sale of oil and gas.

This combination of circumstances created severe strains within the fabric of Russian society. In the periphery, the capital was seen as the privileged playground for those who benefited most from the post-Soviet transition: bureaucrats and oligarchs. In the core area, there was frustration about the government's inability to run the country, wage the war in Chechnya, pay wages, and issue pension benefits. Many Russians began to yearn for the "good old days" of communism, when you could depend on your job, your paycheck, and your pension—if not your freedom to complain or to object to the system.

Back to the Future

When President Vladimir Putin was elected in 2000—the first open, orderly, and relatively democratic leadership transfer in Russia's history—he moved almost immediately to bring order to the country's chaotic economy and turbulent politics. Prominent oligarchs, some of whom were trying to use their money to buy influence in the political system, were imprisoned; others went into exile. To gain control over the unruly regions and their leaders, the Putin administration created a new "federal" framework that combined Russia's 89 Republics, Regions, and other entities into seven new *Federal Districts*. Compare Figures 2-9 and 2-10, and you will see what Putin had in mind. The new Districts were "Federal" in name only: they did not enhance regional influence in Moscow, but rather increased Moscow's authority over the Districts. As the maps show, each Federal District consists of a cluster of the old communist-era Regions, and each has a capital that symbolizes Moscow's presence in the provinces.

Putin also persuaded the compliant Parliament to end the local election of regional governors, who would henceforth be appointed by the central government. These and other moves had the effect of concentrating ever-greater power in the presidency, so that Russia seemed to be moving away from federalism and toward a more unitary form of government. Such concentration of authority was nothing new in the Russian experience, and Putin had strong support among the Russian majority when he was reelected in 2004. Even when such authoritarianism began to diminish the freedom of the media, inhibit the right of public protest, and erode the newfound independence of the courts, public approval of the country's overall direction remained strong. Among ethnic, cultural, and other minorities, the picture looked

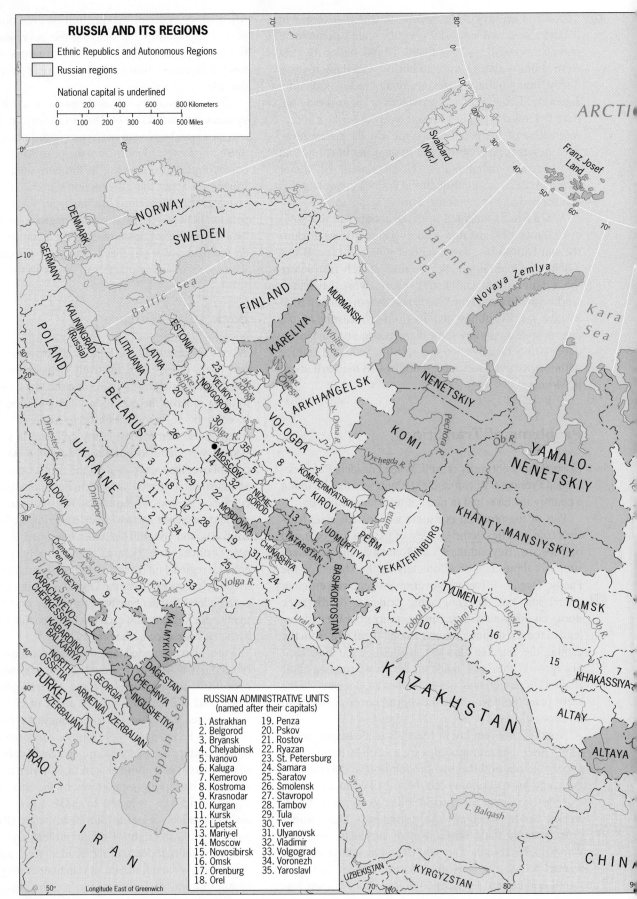

RUSSIA AND ITS REGIONS

Ethnic Republics and Autonomous Regions

Russian regions

National capital is underlined

```
0    200   400   600   800 Kilometers
0   100   200   300   400   500 Miles
```

RUSSIAN ADMINISTRATIVE UNITS
(named after their capitals)

1. Astrakhan	19. Penza
2. Belgorod	20. Pskov
3. Bryansk	21. Rostov
4. Chelyabinsk	22. Ryazan
5. Ivanovo	23. St. Petersburg
6. Kaluga	24. Samara
7. Kemerovo	25. Saratov
8. Kostroma	26. Smolensk
9. Krasnodar	27. Stavropol
10. Kurgan	28. Tambov
11. Kursk	29. Tula
12. Lipetsk	30. Tver
13. Mariy-el	31. Ulyanovsk
14. Moscow	32. Vladimir
15. Novosibirsk	33. Volgograd
16. Omsk	34. Voronezh
17. Orenburg	35. Yaroslavl
18. Orel	

Longitude East of Greenwich

FIGURE 2-9

RUSSIA: THE NEW FEDERAL DISTRICTS PROCLAIMED IN 2000 AND THEIR CAPITALS

| 37.2 | Population estimate for 2007, in millions |
| -9% | Percent decline in population, since 1991 |

FIGURE 2-10

© H. J. de Blij, P. O. Muller, and John Wiley & Sons, Inc.

bleaker, but still not as bad as it had been under communist rule. As we will see, Putin's Russia now was set to reassert its role on the global stage.

CHANGING GEOGRAPHIES

If you were living in Russia today, you would be riveted—and buffeted—by the changes the country is undergoing. Freedoms unimaginable under communist rule are routine even as other liberties are curtailed. Supermarkets, luxury hotels, traffic jams, and exclusive residences reflect the new era of globalization. Inefficiencies reminiscent of the communist era still intrude: arriving at or departing from a Russian "international" airport is not a pleasant experience. Bureaucratic hurdles and endless waits still slow the pace of life. But social, cultural, economic, and political geographies are changing fast, and the changes are etched in the landscape:

1. *The Revival of Religion*. The communist regime was avowedly atheist, but the end of Soviet rule revived the long-suppressed Russian Orthodox Church. Urban as well as rural cultural landscapes are being transformed by the reconstruction of gold-domed cathedrals and

churches (see photo p. 104); elaborately robed religious figures are participating in public ceremonies and other events; the Russian-Orthodox bureaucracy is influencing public-school curricula; prominent church leaders are promoting a revived Russian nationalism based on historic Slavic-Orthodox linkages.

2. *Organized Crime.* Organized crime exploded after the fall of the Soviet system, in part because of the end of communism's iron-fisted rule and also because law enforcement in the new, chaotic Russia could not cope with this new epidemic. Crime syndicates thrived in Russia and extended their networks abroad, forming alliances with criminal networks elsewhere. Increased street crime eroded Russians' quality of life.

3. *Endemic Corruption.* Corruption and crime are related in many ways, and during communist rule corruption was rife throughout the Soviet Union. Despite official efforts to attack this problem, corruption at all levels constitutes a major detriment to the quality of life of ordinary Russians. During the turbulent 1990s, freewheeling newspapers and other media uncovered corruption in the sale of state enterprises, the distribution of state pensions, the awarding of college degrees, and in other spheres. This

eroded public confidence and contributed to the popularity of Putin's policies, but even his administration could not stem the tide.

4. *Freedoms Threatened.* Russian citizens began to experience freedoms unimaginable under communist rule, and surveys confirm that a large majority preferred the new circumstances, despite the problems arising from the transition. Greater freedom of the press and other media; unobstructed street protests against governmental regulations; political campaigns and rallies and other forms of public participation signaled a new era in what some called Russia's "liberal revolution." But during the later years of the Putin administration, such liberties came under attack and others were misused. The media came under government pressure and individual reporters were intimidated, even murdered. Democratic reforms were reversed. Ethnic relations, especially in the large cities, deteriorated as minorities found themselves threatened and inadequately protected by law enforcement agencies. Russia's new democracy was still in difficult transition nearly two decades after the collapse of the communist system.

5. *Race Relations.* Fear or anger toward foreigners (*xenophobia*) grew in Russia as a result of a combination of circumstances: the arrival in Russia's core-area cities of traders, workers, and others from non-Slavic areas of the vast country; the war in Chechnya and associated terrorism in Moscow and other Russian cities; and perceptions that a disproportionate amount of rising street crime was attributable to "immigrants" from non-Slavic Republics and Regions. Assaults by locals on outsiders identifiable through their work, garb, or ethnicity, and the failure of law enforcement to limit this violence, increased urban tensions throughout the new Russia

RUSSIA'S DEMOGRAPHIC DILEMMA

Take another look at Figure 2-10 and note a set of figures in red for each of the Federal Districts on the map. These numbers strike fear in the heart of every Russian planner: they represent the percentage change in the population of the seven Districts, and every one of them is negative. You are seeing evidence of what some geographers refer to as Russia's demographic disaster: its rapid decline in population since the Soviet Union's collapse in 1991. In that year, Russia's population totaled about 149 million. Less than 20 years later, the number was down to 141 million—despite the in-migration of

several million ethnic Russians from the 14 former Soviet Republics beyond Russia's borders. Since the end of communist rule, Russia has experienced over ten million more deaths than births. Severe social dislocation drives this trend.

Undoubtedly, the transition from Soviet rule is one cause: uncertainty tends to cause families to have fewer children, and abortion is widespread in Russia. But the birth rate has stabilized at around 12 per thousand; it is Russia's death rate that has skyrocketed, now exceeding more than 15 per thousand. This produces an annual loss of population of over 0.3 percent, or nearly half a million per year. Russian males are the ones most affected. Average male life expectancy dropped from 71 in 1991 to 60 in 2008; it was at its lowest in the Irkutsk Region, where it was a mere 53. (Female life expectancy also declined, but markedly less, to 73.) Males are more likely to be affected by alcoholism, heavy smoking, suicide (Russia's rate is 15 times Europe's), accidents, and murder. On average, a Russian male is nine times more likely to die a violent or an accidental death than his European counterpart. Fewer than half of today's male Russian teenagers will survive to age 60.

Another cause of the high death rate among males, and the declining life expectancy even among females, is AIDS, whose impact is difficult to gauge because of Russia's high levels of tuberculosis and the questionable quality of official reporting. Estimates of the total number of Russians infected with HIV/AIDS range from 940,000 to several million, although the official number in 2008 was given as 416,000 and only a fraction of those infected were receiving treatment.

When we encounter the realm worst affected by HIV/AIDS in the world today, Subsaharan Africa, we will see that Russia's figures—even at the high end of these estimates—are still far below those of Africa. In Africa, too, life expectancies are declining markedly. Yet Russia's population is dropping while Africa's continues to rise (see the Data Table inside the back cover). The key difference lies in the birth rates, which in Subsaharan Africa are still high enough (40 per thousand) to replace those who die. That is not the case in Russia (12 per thousand).

If this **population decline** continues, Russia will **18** have fewer than 130 million citizens by 2025 and barely 110 million by 2050, raising doubts about the future of the state itself. Look again at that map of the Federal Districts (Fig. 2-10) and consider this: since 1990, the Far East District lost 17 percent of its population, Siberia 5 percent, the Northwest 9 percent, and the South 12 percent. Only the Central District, the one around Moscow, had a loss on a par with European countries (0.2 percent).

What is the answer? Russia's leaders hope that improvements in the social circumstances of the average Russian will reduce the rate of population loss. Campaigns against alcoholism and careless lifestyles are under way. In 2006, the Russian government started spending some of its plentiful oil revenues on a program that awards well over U.S. $100 per month to mothers who decide to have a second child, plus a one-time award of almost $10,000 to be put toward family expenses such as children's education or a mortgage. Such schemes, however, have a history of causing a quick "baby rush" but little long-term impact on the problem. Immigration is another option, but it is alarming that the decline just discussed has taken place even as several million Russians returned home from the Near Abroad. As for immigration by other peoples, such as Chinese and North Koreans, the barriers against it remain strong, and the reception of those who do manage to enter Russia is generally not hospitable. Russians in the western core area are not eager to see their Far Eastern frontier transformed into an extension of East Asia. And in our discussion of race relations above, we noted that xenophobic Russians in the cities of the west do not welcome non-Slavic migrants, including those from within Russia itself. Indeed, many Russian citizens were not especially welcoming to ethnic Russians arriving from the former Soviet Republics either. These are huge obstacles in the search for a demographic revival.

When the Soviet Union collapsed in 1991, Russia inherited only about half of the empire's population but more than 75 percent of its territory. Among the things Russia lacked was a true network of large, region-anchoring cities served by an efficient interconnecting transport system stimulating growth and development. Russia today has just two cities with populations exceeding 2 million, the same number as Australia, which contains less than one-sixth of Russia's population. To once again become the world power Russia's leaders envisage, Russia must reverse its demographic decline, but at present no solution is in sight.

RUSSIA'S PROSPECTS

When Peter the Great began to reorient his country toward Europe and the outside world, he envisaged a Russia with warm-water ports, a nation no longer encircled by Swedes, Lithuanians, Poles, and Turks, and a force in European affairs. Core-periphery relationships always have been crucial to Russia, and they remain so today.

Heartland and Rimland

Russia's relative location has long been the subject of study and conjecture by geographers. More than a century ago, the British geographer Sir Halford Mackinder (1861–1947) argued in a still-discussed article entitled "The Geographical Pivot of History" that western Russia and eastern Europe enjoyed a combination of natural protection and resource wealth that would someday propel its occupants to world power. This protected core, he reasoned, overshadowed the exposed periphery. Eventually, this pivotal interior region of Eurasia, which he later called the *heartland*, would become a stage for world domination.

Mackinder's **heartland theory** was published in **19** 1904 when Russia was a weak, economically backward society ripe for revolution, but Mackinder stuck to his guns. When Russia, in control of eastern Europe and with a vast colonial empire, emerged from World War II as a superpower, Mackinder's conjectures seemed prophetic.

But not all geographers agreed. Probably the first scholar to use the term *rimland* (today *Pacific Rim* is part of everyday language) was Nicholas Spykman, who in 1944 countered Mackinder by calculating that Eurasia's periphery, not its core, held the key to global power. Spykman foresaw the rise of rimland states as superpowers and viewed Japan's emergence to wealth and power as just the beginning of that process. In that perspective, Russia's twentieth-century superpower status was just a temporary phenomenon: China and India would be the forces to reckon with.

The New Russia on the World Stage

When former President Putin arrived in office, he made it clear that he wished to revive Russia's role as one of the major powers on the global stage. He worked to repair and restructure Russia's discredited military, accused internally of disciplinary failures and internationally of brutality in the Chechnya war. Not only did he manage to get Russia invited to attend the annual meetings of the G-7 group of leading economic powers; he also raised Russia's profile in other international arenas ranging from the nuclear program of Iran to the independence of Kosovo. In 2008, Russian forces intervened outside Russia on behalf of pro-Russian minorities in neighboring Georgia, a crucial moment in Russia's reassertive role. By that time, Putin had been succeeded by President Dmitri Medvedev, although Putin's hand remained on Russia's rudder.

Meanwhile, significant developments in Europe were stoking Russian nationalism. Russia has two main strategic concerns: (1) the expansion of the European

Union (EU) and (2) the growth of the North Atlantic Treaty Organization (NATO), the military organization that, after the collapse of the Soviet Union, incorporated several former Soviet satellites as full members. You can draw a series of sequential maps and, from the Russian perspective, watch Europe "encroach" on Russia over the years since 1991. When the 2004 expansion of the EU added ten countries, two of them bordered Russia and included significant Russian minorities; and four others adjoined Belarus and/or Ukraine, countries that Russia regards as parts of its sphere of interest. Already, nine former Soviet satellites (including the Czech Republic, Poland, and Hungary) are part of NATO. Russia is loath to see a U.S.-led military alliance infringe upon what it regards as its security perimeter, even though Russia since 2002 has been a "participant" in the organization—that is, Russia has been attending NATO meetings without having a veto over its military decisions.

As we noted in Chapter 1, Europe is one arena where strategic and other interests of Russia diverge from those of the United States. While Russia acquiesced in EU expansion and NATO encroachment, Russian leaders saw little reciprocation from "the West." Harsh criticism of Russia's conduct of the war in Chechnya and of the treatment of minorities by the Russian state fired up Russian nationalism. Russia, in turn, was unhelpful during the collapse of Yugoslavia and not especially cooperative in the multinational effort to constrain Iran's drive toward nuclear capability. In 2008, Russia supported Serbia in its objections to the independence of Kosovo. When the United States proposed placing missile-defense systems in NATO member-states Poland and the Czech Republic, President Medvedev threatened to put missile launchers in Kaliningrad. It was an ominous escalation of discord.

Economic Geography and Energy Issues

Russian frustration with its European neighbors and with the United States had much to do with Russia's economic problems, Europe's energy needs, and America's strategic objectives. We noted earlier that, during communist times, you would not have been likely to find Soviet consumer goods in the stores of London or New York. Well, you will still find very few Russian manufactures on world markets today. Russia's endemic corruption, stifling bureaucracy, and weak institutions (notably its legal system) make it difficult, to say the least, to get a private company started there, and few investors are willing to take the risk. As a result, Russian high-tech products contribute less than 1 percent of Russian exports by value. And try naming a single medical drug that Russia's research laboratories have brought to the world market.

But Russia does have what Europe and the world need: plentiful oil and natural gas. Following the period during the 1990s when rich entrepreneurs bought oil and gas companies and oligarchs became a force in Russian politics as well as economics, such privatization was forcefully reversed and the state again took control of most of the vast country's huge energy reserves.

That was both good and bad for Russia—good in the sense that national revenues burgeoned when energy prices on world markets rose, but bad because Russia came to depend excessively on the sale of these commodities. The Russian economy failed to diversify partly because easy money flowed from those sales, allowing Moscow to pay its pensioners (this had been a problem after 1991), satisfy social needs, and rebuild its military. Russia's economy was comparable to that of oil-rich Arab states and running the same risks: when prices fall, as they did following unprecedented highs in mid-2008, the lack of alternative income sources strains budgets.

Economic and political geographies are always intertwined, and Russia's oil exports embroiled it in difficulties with its neighbors nearby as well as its customers farther away. The member-states of the European Union form a major market for Russia's energy exports, but, as Figure 1-23 reminds us, pipelines first have to go through Belarus and Ukraine to reach those consuming countries. Both Ukraine and Belarus siphon off some of those supplies for themselves, and in the past Russia charged those Near Abroad neighbors much less for these imports than they charged "regular" consumers. But when Ukraine's democratic elections began to show a pro-European tilt, the Russians upped the price—and then turned off the supply when the Ukrainians balked. A similar quarrel (but for different reasons) with Belarus also led to supply interruptions. All this led the Germans, who heavily depend on Russian energy provisions, to start negotiating with the Russians for the construction of a pipeline that would run along the seafloor of the Baltic—thereby avoiding problems of transit overland.

Issues in the Near Abroad

Russia has significant oil and gas reserves, but it also has competitors with their own energy resources—some of them right across Russia's borders. As if the physiographic, ethnic, and cultural mosaic of Transcaucasia were not complicated enough, energy resources and exports form still another bone of contention in this much-contested transition zone. We focus later on the details, but take note here of the overall issue: once again, the interests of Russia and its international-stage competitors diverge.

The particular matter of contention involves the oil reserves near and beneath the Caspian Sea, especially those controlled by Azerbaijan, a former Soviet Republic. The most efficient export route for this oil would be a pipeline across southern Russia to the Black Sea, from where it would be sent to foreign markets in tankers. That, indeed, was the route during the Soviet period, and one of the cities that benefited along the way was Groznyy, the capital of Chechnya. But this route also gave the Russians the advantage of control. They could shut it down if they saw a reason for doing so.

This led American and other consumers to invest in an alternate pipeline route across the country named Georgia as well as neighboring Turkey to the coast of the Mediterranean Sea. This move was greatly resented in Moscow but benefited Georgia especially, because it enhanced the importance of this country in Russia's Near Abroad in U.S. and European eyes. Georgians began talking about their country joining the European Union, and Georgia was even mentioned as a potential NATO member.

But Putin's newly assertive Russia had its own Transcaucasian cards to play. Like other countries in the Near Abroad, Georgia has pro-Russian minorities whose continuing loyalty to Moscow is rewarded with favors ranging from diplomatic pressure to economic advantage—and in 2008, for the first time in post-Soviet history, Russia's armed forces crossed the country's border and intervened in Georgia on behalf of such a minority. Russian forces pushed southward to within a few kilometers of the Caspian-Mediterranean oil pipeline, sending a message to the world that Moscow's interests do not stop at the Russian border.

Beyond the Near Abroad: Challenges and Choices

Many geographers describe contemporary Russia as a combination of contradictions, a state in demographic decline and democratic decay, a people willing to accept a return to authoritarianism, an economy dangerously one-sided—and yet a force to be reckoned with, a nuclear power, a revitalized military, an aggressive counterweight to the strategic designs of competitors. In the Americas, Russia is reanimating its alliance with Cuba and finding common cause with Venezuela. In Southwest Asia, Russia's position on Iran has at times troubled the international community. In Europe, Russia takes positions that oppose EU and NATO designs. In East Asia, Russia continues to hold several islands in the Kurile chain that were taken from Japan during World War II (see Chapter 9), and resists Chinese penetration of its depopulating Pacific periphery.

Russia certainly has reclaimed a place on the global stage. The question is whether Russia will find a niche in the international community.

POINTS TO PONDER

● Organized crime with international links, local gangsterism and street crime, and corruption in government afflict Russia, but polls show that a majority of Russians still feel that life is better than it was under communism.

● The fate of hundreds of thousands of ethnic Russians stranded outside contiguous Russia's borders in Europe, including in Kaliningrad, is arousing passions in Russia.

● Embattled Chechnya's parliament has voted to rename the capital, Groznyy (which means "dreadful"), to Akhmad-Kala after an assassinated president. But the Kremlin would have to approve.

● Russia has encouraged separatism in neighboring Georgia's Abkhazia region, going so far as to issue Russian passports to Abkhazian residents.

Regions of the Realm

So vast is Russia, so varied its physiography, and so diverse its cultural landscape that regionalization requires a small-scale perspective and a high level of generalization. Figure 2-11 outlines a four-region framework: the Russian Core west of the Urals, the Eastern Frontier, Siberia, and the Far East. As we will see, each of these massive regions contains major subregions.

● RUSSIAN CORE AND PERIPHERIES

The heartland of a state is its **core area**. Here much of [20] the population is concentrated, and here lie the leading cities, major industries, densest transport networks, most intensively cultivated lands, and other key components of the country. Core areas of long standing

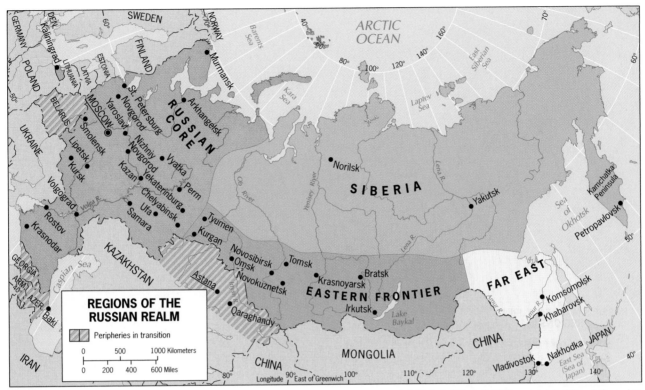

FIGURE 2-11

strongly reflect the imprints of culture and history. The Russian Core, broadly defined, extends from the western border of the Russian realm to the Ural Mountains in the east (Fig. 2-11). This is the Russia of Moscow and St. Petersburg, of the Volga River and its industrial cities.

MAJOR CITIES OF THE REALM

City	Population (in millions)
Baki (Baku), Azerbaijan	2.0
Chelyabinsk, Russia	1.1
Kazan, Russia	1.1
Moscow, Russia	10.5
Nizhniy Novgorod, Russia	1.3
Novosibirsk, Russia	1.4
St. Petersburg, Russia	4.6
Tbilisi, Georgia	1.1
Vladivostok, Russia	0.6
Volgograd, Russia	1.0
Yekaterinburg, Russia	1.3
Yerevan, Armenia	1.1

*Based on 2010 estimates.

Central Industrial Region

At the heart of the Russian Core lies the Central Industrial Region (Fig. 2-12). The precise definition of this subregion varies, for all regional definitions are subject to debate. Some geographers prefer to call this the Moscow Region, thereby emphasizing that for over 400 kilometers (250 mi) in all directions from the capital, everything is oriented toward this historic focus of the state. As both Figures 2-1 and 2-8 show, Moscow has maintained its decisive centrality: roads and railways converge in all directions from Ukraine in the south; from Mensk (Belarus) and the rest of eastern Europe in the west; from St. Petersburg and the Baltic coast in the northwest; from Nizhniy Novgorod (formerly Gorkiy) and the Urals in the east; from the cities and waterways of the Volga Basin in the southeast (a canal links Moscow to the Volga, Russia's most important navigable river); and even from the subarctic northern periphery that faces the Barents Sea.

Two Great Cities

Moscow (population: 10.5 million) is the focus of an area that includes some 50 million inhabitants (more than one-third of the country's total population), many of them concentrated in such major cities as Nizhniy Novgorod, the automobile-producing "Soviet Detroit";

FIGURE 2-12

© H. J. de Blij, P. O. Muller, and John Wiley & Sons, Inc.

Yaroslavl, the tire-producing center; Ivanovo, the heart of the textile industry; and Tula, the mining and metallurgical center where lignite (brown coal) deposits are worked (Fig. 2-12).

St. Petersburg (the former Leningrad) remains Russia's second city, with a population of 4.6 million. Under the czars, St. Petersburg was the focus of Russian political and cultural life, and Moscow was a distant second city. Today,

however, St. Petersburg has none of Moscow's locational advantages, at least not with respect to the domestic market. It lies well outside the Central Industrial Region near the northwestern corner of the country, 650 kilometers (400 mi) from Moscow. Neither is it better off than Moscow in terms of resources: fuels, metals, and foodstuffs must all be brought in, mostly from far away. The former Soviet emphasis on self-sufficiency even reduced

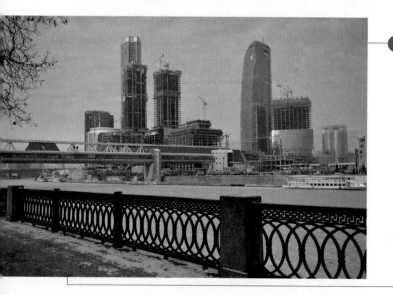

FROM THE FIELD NOTES

"The first time I stood on a bank of the Moscow River, in June 1964, I described in my notes the buildings across the water as 'historic czarist-era ornate, Soviet-image grandiose, and current-socialist-city bleak.' Now, two decades after the collapse of the USSR, central Moscow is embarked on modernization to symbolize Russia's new era, new aspirations, and new capacities. Elsewhere, architects are recreating and restoring religious structures, but here, in what will be known as the Moscow City Business and Administrative Center, the Federation Tower (seen under construction at right) is only one of a cluster of skyscrapers rising to symbolize this new era. The tallest building in this district tops out at 340 meters (1,115 ft), taller than New York's Chrysler Building. Moscow is acquiring a skyline to match its role as the capital of a muscle-flexing power."
© H. J. de Blij.

www.conceptcaching.com

Concept Caching

AMONG THE REALM'S GREAT CITIES . . . Moscow

IN THE VASTNESS of Russia's territory, Moscow, capital of the Federation, seems to lie far from the center, close to its western margin. But in terms of Russia's population distribution (Fig. 2-3), Moscow's centrality is second to none among the country's cities. Moscow lies at the heart of Russia's primary core area and at the focus of its circulation systems (Fig. 2-8).

On the banks of the meandering Moscow River, Moscow's skyline of onion-shaped church domes and modern buildings rises from the forested flat Russian Plain like a giant oasis in a verdant setting. Archeological evidence points to ancient settlement of the site, but Moscow enters recorded history only in the middle of the twelfth century. Forest and river provided defenses against Tatar raids, and when a Muscovy force defeated a Tatar army in the late fourteenth century, Moscow's primacy was assured. A huge brick Kremlin (citadel; fortress), with walls 2 kilometers (1.2 mi) in length and with 18 towers, was built to ensure the city's security. From this base Ivan the Terrible expanded Muscovy's domain and laid the foundations for Russia's vast empire.

The triangular Kremlin and the enormous square in front of it (Red Square of Soviet times), flanked by St. Basil's Cathedral and overlooking the Moscow River, is the center of old Moscow and is still the heart of this metropolitan complex of 10.5 million. From here, avenues and streets radiate in all directions to the Garden Ring and beyond. Until the 1970s, the Moscow Circular Motorway encircled most of the built-up area, but today the metropolis sprawls far beyond this beltway. For all its size, Moscow never developed a world-class downtown skyline. Communist policy was to create a "socialist city"

in which neighborhoods—consisting of apartment buildings, workplaces, schools, hospitals, shops, and other amenities—would be clustered so as to obviate long commutes into a high-rise city center.

Moscow may not be known for its architectural appeal, but the city contains noteworthy historic as well as modern structures, including the Cathedral of the Archangel and the towers of Moscow State University.

© H. J. de Blij, P. O. Muller, and John Wiley & Sons, Inc.

St. Petersburg's asset of being on the Baltic coast because some raw materials could have been imported much more cheaply across the Baltic Sea from foreign sources than from domestic sites in distant Central Asia (only bauxite deposits lie nearby, at Tikhvin).

Yet St. Petersburg was at the vanguard of the Industrial Revolution in Russia, and its specialization and skills have remained important. Today, the city and its immediate environs contribute about 10 percent of the country's manufacturing, much of it through fabricating high-quality machinery.

Facing the Barents Sea

North of the latitude of St. Petersburg, the region we have named the Russian Core takes on Siberian properties.

However, the Russian presence is much stronger in this remote northern periphery than it is in Siberia proper. Two substantial cities, Murmansk and Arkhangelsk, are road- and rail-connected outposts in the shadow of the Arctic Circle.

Murmansk lies on the Kola Peninsula not far from the border with Finland (Fig. 2-8). In its hinterland lie a variety of mineral deposits, but Murmansk is particularly important as a naval base. During World War II, Allied ships brought supplies to Murmansk; the city's remoteness shielded it from German occupation. After the war, it became a base for nuclear submarines. This city is also an important fishing port and a container facility for cargo ships.

Arkhangelsk is located near the mouth of the Northern Dvina River where it reaches an arm of the White Sea (Fig. 2-8). Ivan the Terrible chose its site during Muscovy's

CZAR PETER THE GREAT and his architects transformed the islands of the Neva Delta, at the head of the Baltic Sea's Gulf of Finland, into the Venice of the North, its palaces, churches, waterfront façades, bridges, and monuments giving St. Petersburg a European look unlike that of any other city in Russia. Having driven out the Swedes, Peter laid the foundations of the Peter and Paul Fortress on the bank of the wide Neva River in 1703, and the city was proclaimed the capital of Russia in 1712.

Peter's "window on Europe," St. Petersburg (named after the saint, not the czar) was to become a capital to match Paris, Rome, and London. During the eighteenth century, St. Petersburg acquired a magnificent skyline dominated by tall, thin spires and ornate cupolas, and graced by baroque and classical architecture. The Imperial Winter Palace and the adjoining Hermitage Museum at the heart of the city are among a host of surviving architectural treasures.

Revolution and war lay in St. Petersburg's future. In 1917, the Russian Revolution began in the city (then named Petrograd), and following the communist victory it lost its functions as a capital to Moscow and its name to Lenin. Renamed Leningrad, it suffered through the 872-day Nazi siege during World War II, holding out heroically through endless bombardment and starvation that took nearly a million lives and severely damaged many of its buildings.

The communist period witnessed the neglect and destruction of some of Leningrad's most beautiful churches, the intrusion and addition of crude monuments and bleak apartment complexes, and the rapid industrialization and growth of the city (which now totals 4.6 million). Immediately after the collapse of the Soviet Union, it

regained its original name and a new era opened. The Russian Orthodox Church revived, building restoration went forward, and tourism boomed. But the transformation from communist to capitalist ways has been accompanied by social problems that include crime and poverty.

© H. J. de Blij, P. O. Muller, and John Wiley & Sons, Inc.

early expansion; the czar wanted to make this the key port on a route to maritime Europe. Yet Arkhangelsk, mainly a port for lumber shipments, has a shorter ice-free season than Murmansk, whose port can be kept open with the help of the warm North Atlantic Drift ocean current (and by icebreakers if necessary).

Nothing in Siberia east of the Urals rivals either of these cities—yet. But their existence and growth prove that Siberian barriers to settlement can be overcome.

Povolzhye: The Volga Region

A second subregion lying within the Russian Core is the *Povolzhye*, the Russian name for an area that extends along the middle and lower valley of the Volga River. It

would be appropriate to call this the Volga Region, for that greatest of Russia's rivers is its lifeline and most of the cities that lie in the Povolzhye are on its banks (Fig. 2-12). In the 1950s, a canal was completed to link the lower Volga with the lower Don River (and thereby the Black Sea).

The Volga River was an important historic route in old Russia, but for a long time neighboring regions overshadowed it. The Moscow area and Ukraine were far ahead in industry and agriculture. The Industrial Revolution came late in the nineteenth century to the Moscow Region and did not have much effect in the Povolzhye. Its dominant function remained the transit of foodstuffs and raw materials to and from other regions.

Changing Times

This transport function is still important, but the Povolzhye has changed. First, World War II brought furious development because the Volga River, located east of Ukraine, was far from the German armies that invaded from the west. Second, in the postwar era the Volga-Urals Region for some time was the largest known source of petroleum and natural gas in the entire Soviet Union. From near Volgograd (formerly Stalingrad) in the southwest to Perm on the Urals' flank in the northeast lies a belt of major oilfields (Fig. 2-13).

Third, the transport system has been greatly expanded. The Volga-Don Canal directly connects the Volga waterway to the Black Sea; the Moscow Canal extends the northern navigability of this river system into the heart of the Central Industrial Region; and the Mariinsk canals provide a link to the Baltic Sea. Today, the Volga Region's population exceeds 20 million, and the cities of Samara (formerly Kuybyshev), Volgograd, Kazan, and Saratov all have populations between 1.0 and 1.3 million. Manufacturing has also expanded into the middle Volga Basin, emphasizing more specialized engineering industries. The huge Fiat-built auto assembly plant in Tolyatti, for example, is one of the world's largest of its kind.

The Internal Southern Periphery

We turn now to a subregion that has generated some of the most difficult politico-geographical problems the new Russia has faced since the collapse of the Soviet Union. Between the Black Sea to the west and the Caspian Sea to the east, and with the Caucasus Mountains to the south, Moscow seeks to stabilize and assert its authority over a tier of minority "republics," holdovers from the Soviet era (Fig. 2-14).

Islam and Oil

A combination of geographic factors causes the problems Moscow confronts in this peripheral zone. Here the Russians (and later the Soviets) met and stalled the advance of Islam. Here the cultural mosaic is as

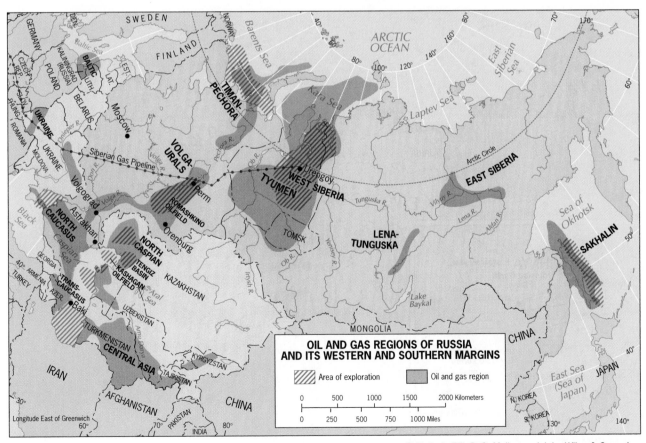

FIGURE 2-13   © H. J. de Blij, P. O. Muller, and John Wiley & Sons, Inc.

SOUTHERN RUSSIA:
INTERIOR AND EXTERIOR PERIPHERIES

POPULATION

● Under 50,000
● 50,000–250,000
● 250,000–1,000,000
● 1,000,000–5,000,000
● Over 5,000,000

National capitals are underlined

—— Railroad
—— Road
┬┬┬┬ Canal
●–●–● Oil pipeline
○–○–○ Proposed oil pipeline

FIGURE 2-14

© H. J. de Blij, P. O. Muller, and John Wiley & Sons, Inc.

intricate as anywhere in the Federation: Dagestan, the southernmost Republic fronting the Caspian Sea, has 2.8 million inhabitants comprising some 30 ethnic groups speaking about 80 languages. And beneath the land as well as under the waters of the Caspian Sea lie oil and gas reserves that make this a crucial corner of the vast country. Indeed, Russia has a stake even in the oilfields belonging to neighboring Azerbaijan: as Figure 2-14 shows, the Russians control the pipelines leading from those reserves across Russian territory to the Black Sea oil terminal at Novorossiysk. But, as we noted earlier, the Moscow government watches with great concern as new pipelines from Azerbaijan now avoid Russian territory and traverse Georgia and Turkey to reach alternate outlets.

Disaster in Chechnya

The presence of oil and oil installations endow this Southern Periphery with far greater significance in the Russian Federation than would otherwise be the case. Note again the route of the pipeline to Novorossiysk: it crosses the Republic called Chechnya whose war-devastated capital, Groznyy, was a major oil-industry center and service hub during the Soviet era (Fig. 2-14, including inset).

But Chechnya also contains a sizeable Muslim population, and fiercely independent Muslim Chechens used Caucasus mountain hideouts to resist Russian colonization during the nineteenth century. Accused of collaboration with the Nazis during World War II, on Stalin's orders the Soviets exiled the entire Chechen population to Central Asia, with much loss of life. Rehabilitated and allowed to return by Premier Nikita Khrushchev in the 1950s, the Chechens seized their opportunity in 1991 after the collapse of the Soviet Union and fought the Russian army to a stalemate. But Moscow never granted Chechnya independence (one-quarter of the population of 1.2 million was Russian), and an uneasy standoff continued. Meanwhile, Chechen hit-and-run attacks on targets in neighboring Republics continued.

In 1999, following comparatively minor terrorist incidents in Moscow, the conflict took its first major toll as three apartment buildings were bombed, resulting in 230 deaths. In 2002, Chechen terrorists took more than 700 patrons and performers hostage in a Moscow theater, and a bungled rescue effort led to the deaths of nearly 130 of the captives and several dozen terrorists. Meanwhile, the conflict showed signs of spreading beyond Chechnya's borders in other ways: in 2004 an attack by Islamic militants on a school in Beslan, a small town in the nearby Republic of North Ossetia (see Fig. 2-14), killed as many as 350 people, half of them schoolchildren.

Thus a minority "republic" the size of Connecticut with a population of little more than 1 million threatened the security and stability of the largest territorial state in the world with a population of more than 140 million. When President Putin took Russia's helm in 2000, he committed himself to ending the Chechen conflict, and he did so by giving the Russian army free rein and by co-opting Chechnya's pro-Russian elite. The results were devastating but politically effective; however, in 2009 Chechnya's neighborhood still was not pacified. Deathly incidents in Ingushetiya, Dagestan, and more recently in Kabardino-Balkariya continued to plague this fractious tier of peripheral southern Republics.

A Complex Mosaic

It is important to familiarize yourself with the layout of Russia's Internal Southern Periphery because the problems here not only afflict Russia but spill over into its neighbors. The southern Republics, colored pink in Figure 2-14, all present their particular (sometimes intertwined) problems. About half the population in Kalmykia, north of Dagestan, is Buddhist, and local leaders presented Moscow with a dilemma when they decided in 2004 to invite Tibet's Dalai Lama to visit (China, which controls Tibet and calls it Xizang, does not recognize the Dalai Lama's position and denigrates his role). Also bordering the Caspian Sea is Dagestan which, as we noted, contains more than two dozen ethnic groups. Ingushetiya, which was once part of the previous Chechenya republic, has a Muslim majority. North Ossetia is called *North* Ossetia because its kinspeople live outside Russia on the other side of the Caucasus range in *South* Ossetia—which is part of neighboring Georgia. Russian leaders argue that terrorists pass through South Ossetia into North Ossetia and on into Chechnya, and that the government in Georgia is not doing enough to prevent this movement. The Muslim Balkar people of Kabardino-Balkariya, like the Chechens, were accused by Stalin of collaboration with the Nazis and were forcibly exiled with great loss of life; today they constitute only about 12 percent of the Republic's population. The same thing happened to the Muslim Karachay of Karachayevo-Cherkessiya: many were deported, and today they constitute a minority in their Republic, where Russians and Christian Cherkess form the majority. Only in the separate Republic of Adygeya is there no significant Muslim element: Cherkess form the non-Russian population sector here.

This summation of the microregional geography of the Internal Southern Periphery helps explain why Russia faces such severe difficulties here. Even after a solution to the Chechnya problem is found, the challenge will continue: stability in this physiographically and culturally fractured subregion has always been elusive.

The Urals Region

The Ural Mountains form the eastern limit of the Russian Core. They are not particularly high; in the north they consist of a single range, but southward they broaden into a hilly zone. Nowhere are they an obstacle to east-west transportation. An enormous storehouse of metallic mineral resources located in and near the Urals has made this area a natural place for industrial development. Today, the Urals Region, well connected to the Volga and Central Industrial Region, extends from Serov in the north to Orsk in the south (Fig. 2-12).

The Central Industrial, Volga, and Urals subregions form the anchors of the Russian Core. For decades they have been spatially expanding toward one another, their interactions ever more intensive. These subregions of the Russian Core stand in sharp contrast to the comparatively less developed, forested, Arctic north and the remote upland of the Southern Periphery lying to the southwest between the Black and Caspian seas. Thus even within this Russian coreland, frontiers still await growth and development.

As noted in Chapter 1, because the Ural Mountains form a prominent north-south divide between Russia's western heartland and its eastern expanses, geographers sometimes use this to separate a "European" Russia from its presumably non-European, "Asian" east. (The National Geographic Society on its maps and in its atlases even draws a green line along the crest of the Urals marking this transition.) But geographically this makes no sense. Take the train or drive eastward across the southern end of the Urals, and you will see no changes in cultural landscapes, no geographic evidence on which to base such a partition. Indeed, as we will see, Russia remains Russia all the way to the end of its transcontinental railroad at Vladivostok on the Pacific coast.

● THE EASTERN FRONTIER

From the eastern flanks of the Ural Mountains to the headwaters of the Amur River, and from the latitude of Tyumen to the northern zone of neighboring Kazakhstan, lies Russia's vast Eastern Frontier Region, product of a gigantic experiment in the eastward extension

of the Russian Core (Fig. 2-11). As the map of cities and surface communications (Fig. 2-8) suggests, this Eastern Frontier is more densely peopled and more fully developed in the west than in the east. At the longitude of Lake Baykal, settlement has become linear, marked by ribbons and clusters along the east-west railroads. Two subregions dominate the geography: the Kuznetsk Basin in the west and the Lake Baykal area in the east.

The Kuznetsk Basin (Kuzbas)

Some 1500 kilometers (950 mi) east of the Urals lies another of Russia's primary regions of heavy manufacturing resulting from the communist period's national planning: the Kuznetsk Basin, or *Kuzbas* (Fig. 2-12). In the 1930s, it was opened up as a supplier of raw materials (especially coal) to the Urals, but that function became less important as local industrialization accelerated. The original plan was to move coal from the Kuzbas west to the Urals and allow the returning trains to carry iron ore east to the coalfields. However, good-quality iron ore deposits were subsequently discovered near the Kuznetsk Basin itself. As the new resource-based Kuzbas industries grew, so did its urban centers. The leading city, located just outside the region, is Novosibirsk, which stands at the intersection of the Trans-Siberian Railroad and the Ob River as the symbol of Russian enterprise in the vast eastern interior. To the northeast lies Tomsk, one of the oldest Russian towns in all of Siberia, founded in the seventeenth century and now caught up in the modern development of the Kuzbas Region. Southeast of Novosibirsk lies Novokuznetsk, a city that produces steel for the region's machine and metal-working plants and aluminum products from Urals bauxite.

The Lake Baykal Area (Baykaliya)

East of the Kuzbas, development becomes more insular, and distance becomes a stronger adversary. North of the Tyva Republic and eastward around Lake Baykal, larger and smaller settlements cluster along the two railroads to the Pacific coast (Fig. 2-12). West of the lake, these rail corridors lie in the headwater zone of the Yenisey River and its tributaries. A number of dams and hydroelectric projects serve the valley of the Angara River, particularly the city of Bratsk. Mining, lumbering, and some farming sustain life here, but isolation dominates it. The city of Irkutsk, near the southern end of Lake Baykal, is

As Figure 2-2 shows, Siberia is cold and, in Russia's far northeast, dry as well. In the north lies the *tundra*, treeless and windswept, and beyond is the ice of the Arctic. But where somewhat more moderate conditions prevail (moderate being a relative concept in northwestern Russia and Siberia), coniferous forests known as *taiga* cover the countryside. Also called *boreal* (cold-temperate) forests, these evergreen, needleleaf pines and firs create a dense and vast high-latitude girdle of vegetation across northern Eurasia as well as northern North America. This view from the air shows how tightly packed the trees are; they are slow-growing, but long-lived. Most of the world's taiga forest, among the largest surviving stands of primary forest on the planet, remains protected by distance from the threat of exploitation—but lumbering is nevertheless taking its toll. Recent studies, however, indicate that climate change in high latitudes is enabling the forest to expand northward at a faster rate than it diminishes due to lumbering activities elsewhere, a rare case of good news relating to global warming.
© John Cancalosi/DRK Photo

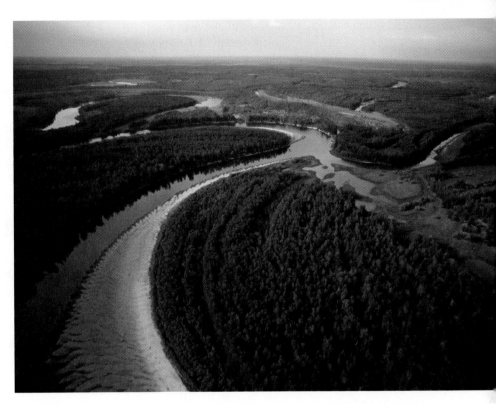

the principal service center for a vast Siberian region to the north and for a lengthy east-west stretch of southeastern Russia.

Beyond Lake Baykal, the Eastern Frontier really lives up to its name: this is southern Russia's most rugged, remote, forbidding country. Settlements are rare, many being mere camps. The Republic of Buryatiya (Fig. 2-9) is part of this zone; the territory bordering it to the east was taken from China by the czars and may become an issue in the future. Where the Russian-Chinese boundary turns southward, along the Amur River, the region called the Eastern Frontier ends and Russia's Far East begins.

● SIBERIA

Before we assess the potential of Russia's Pacific Rim, we should remember that the ribbons of settlement just discussed hug the southern perimeter of this giant country, avoiding the vast Siberian region to the north (Fig. 2-11). Siberia extends from the Ural Mountains to the Kamchatka Peninsula—a vast, bleak, frigid, forbidding land. Larger than the conterminous United States but inhabited by a dwindling population now estimated to be less than 15 million, Siberia quintessentially symbolizes the Russian environmental plight:

vast distances, cold temperatures worsened by strong Arctic winds, difficult terrain, poor soils, and limited options for survival.

But Siberia also has resources. From the days of the first Russian explorers and Cossack adventurers, Siberia's riches have beckoned. Gold, diamonds, and other precious minerals were found there, and later metallic ores including iron and bauxite were discovered. Still more recently, the Siberian interior proved to contain sizeable quantities of oil and natural gas (Fig. 2-13) and began to contribute significantly to Russia's energy supply. As the physiographic map (Fig. 2-4) shows, major rivers—the Ob, Yenisey, and Lena—flow gently northward across Siberia and the Arctic Lowland into the Arctic Ocean. Hydroelectric power development in the basins of these rivers has generated electricity used to extract and refine local ores, and run the lumber mills that have been set up to exploit the vast Siberian forests (see photo above).

The Future

The human geography of Siberia is fragmented and scattered; most of the region remains uninhabited (Fig. 2-3). Ribbons and clusters of Russian settlement did develop during the Soviet period, for example, along the Yenisey

River, which can be traced on the ethnic map (Fig. 2-6) as a ribbon of small settlements north of Krasnoyarsk, and along the Lena River, likewise flanked by patches of habitation. These villages, like the penal colonies that were part of the infamous Gulag Archipelago, lay separated by hundreds of kilometers of empty territory.

During the communist era, the Yakut Republic (later named Sakha) was the main administrative entity here, as Figure 2-9 shows, but today this Republic is part of the Far East District and its identity has been diminished by political developments. But the same global warming that may open the Russian Arctic to resource exploration and development is likely to enable new finds in Siberia itself, so that the current outflow of Siberia's inhabitants may yet be reversed. Conceivably, eastern Siberia's still-rudimentary economy may eventually be reoriented toward the Pacific coast through the export of metals and fuels yet to be discovered.

The Soviets were well aware of this potential, and they prepared for it by building the BAM (Baykal-Amur Mainline) Railroad in the 1980s. This route, lying north of and parallel to the old Trans-Siberian Railroad, extends 3540 kilometers (2200 mi) eastward from Tayshet (near the important center of Krasnoyarsk) directly to the Far East city of Komsomolsk (Fig. 2-8). In the post-Soviet era, the BAM Railroad has been bedeviled by equipment breakdowns and workers' strikes. Nonetheless, it is a key element of the infrastructure that will serve the Eastern Frontier's economic growth in the twenty-first century.

● THE RUSSIAN FAR EAST

Imagine this: a country with 8000 kilometers (5000 mi) of Pacific coastline, two major ports, interior cities nearby, huge reserves of resources ranging from minerals to fuels to timber, directly across from one of the world's largest economies—all this at a time when the Asian Pacific Rim was the world's fastest-growing economic

FROM THE FIELD NOTES

"Standing in an elevated doorway in the center of Vladivostok, you can see some of the vestiges of the Soviet period: the omnipresent, once-dominant, GUM department store behind the blue seal on the left, and the communist hammer-and-sickle on top of the defunct hotel on the right. But in the much more colorful garb of the people, and the private cars in the street, you see reflections of the new era. Once a city closed to foreigners, Vladivostok is now open to visitors—and has become a major point of entry for contraband goods."   © George W. Moore.

Concept Caching

region. Would not that country have burgeoning cities, busy harbors, growing industries, and expanding trade?

In the Russian Far East (Fig. 2-11), the answer is— no. Activity in the port of Vladivostok is a shadow of what it was during the Soviet era, when it was the communists' key naval base. The nearby container terminal at Nakhodka suffers from breakdowns and inefficiencies. The railroad to western Russia carries just a fraction of the trade it did during the 1970s and 1980s. Cross-border trade with China is minimal (although Chinese traders are finding their way to the cities by the thousands). Trade with Japan is inconsequential. The region's cities are grimy, drab, moribund. Utilities are shut off for hours at a time because of fuel shortages and system breakdowns. Outdated factories are closed down, their workers dismissed. Russians are leaving this region in large numbers.

Mainland and Island

As a region, the Russian Far East consists of two parts: the mainland area extending from Vladivostok to the Stanovoy Range and the large island of Sakhalin (Figs. 2-1, 2-15). This is cold country: icebreakers have to keep the ports of Vladivostok and Nakhodka open throughout the winter. Winters here are long and bitterly cold; summers are brief and cool. Although the population is small (less than 7 million), food must be imported because not much can be grown. Most of the region is rugged, forested, and remote. Vladivostok, Khabarovsk, and Komsomolsk are the only cities of consequential size. Nakhodka and the newer railroad terminal at Vanino are merely towns; the entire population of Sakhalin is only about 550,000 (on an island the size of Caribbean Hispaniola [19.7 million]). Offshore lie

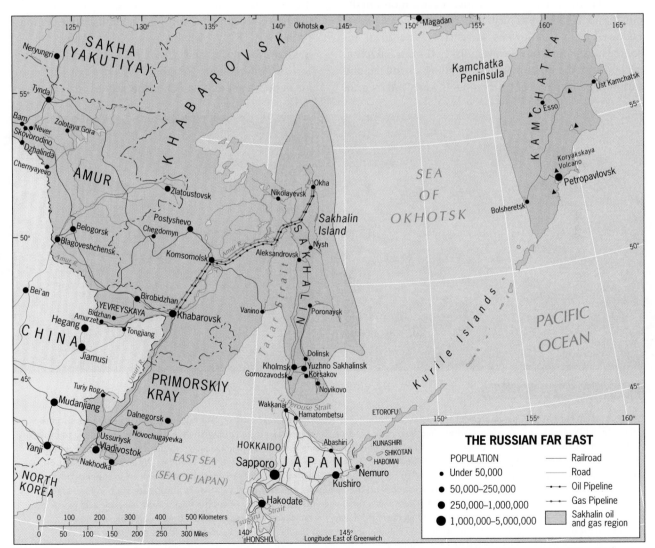

FIGURE 2-15

© H. J. de Blij, P. O. Muller, and John Wiley & Sons, Inc.

productive fishing grounds. Most important of all are the huge oil and gas reserves recently discovered on and around Sakhalin: a new era may be dawning as Russian-partnered foreign companies move in to exploit them.

Post-Soviet Malaise

The Soviet regime rewarded people willing to move to this region with housing and subsidies. The communists, like the czars before them, realized the importance of this frontier (Vladivostok means "We Own the East"), and they used every possible incentive to develop it and link it ever closer to Russia's distant western core. Freight rates on the Trans-Siberian Railroad, for example, were about 10 percent of their real costs; the trains were always loaded in both directions. Vladivostok was a military base and a city closed to foreigners, and Moscow invested heavily in its infrastructure. Komsomolsk in the north and Khabarovsk near the region's center were endowed with state-owned industries using local resources: iron ore from Komsomolsk, oil from Sakhalin, timber from the ubiquitous forests. Steel, chemical, and furniture industries sent their products westward by train

and received food and other consumer goods from the Russian heartland.

For several reasons, the post-Soviet transition has been especially difficult here in the Far East. The new economic order has canceled the region's communist-era advantages: the Trans-Siberian Railroad now must charge the real cost of hauling products from this Pacific-fronting outpost to the Russian Core. State-subsidized industries must compete on market principles, their subsidies having ended. The decline of Russia's armed forces has hit Vladivostok hard. The fleet lies rusting in port; service industries have lost their military markets; the shipbuilding industry has no government contracts. Coal miners in the Bureya River Valley (a tributary of the Amur) go unpaid for months and go on strike; coal-fired power plants do not receive fuel shipments, and cities and towns go dark.

Locals put much of the blame for their region's failure on Moscow, and with reason. As Figure 2-9 shows, the Far East contains only five administrative regions: Primorskiy Kray; Khabarovsk Kray; Amur Oblast; Sakhalin Oblast; and Yevreyskaya, originally the Jewish Autonomous Region. This does not add up to much political clout, which is the way Moscow appears

 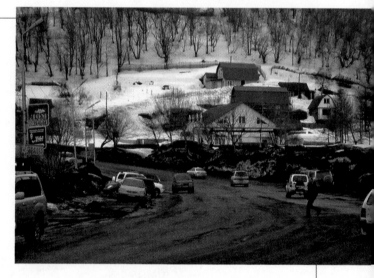

Concept Caching

to want it. Compare Figures 2-10 and 2-15 and you will note that the city of Khabarovsk, not the Primorskiy Kray capital of Vladivostok, has been made the headquarters of the Far Eastern Region as defined by Moscow.

For all its stagnation, Russia's Far East will figure prominently in Russian (and probably world) affairs. Here Russia meets China on land and Japan at sea (an unresolved issue between Russia and Japan involves several of the Kurile islands that separate the Sea of Okhotsk from the Pacific Ocean [Fig. 2-15]). Here lie vast resources ranging from Sakhalin's fuels to Siberia's lumber, Sakha's gold to Khabarovsk's metals. Here Russia has a foothold on the Pacific Rim, a window on the ocean along whose shores the world is being transformed.

TRANSCAUCASIA: THE EXTERNAL SOUTHERN PERIPHERY

The Russian geographic realm, as Figures G-2 and 2-11 show, extends beyond the confines of the Russian state. The Near Abroad is not just a figment of Russian imagination: Moscow has real and crucial interests in its external peripheries. These interests include: (1) the fate of Russian minorities remaining in former Soviet Socialist Republics (SSRs) like Latvia and Kazakhstan; (2) the intentions of governments in former Soviet-controlled states to join organizations such as the EU and NATO, for example, Ukraine and Georgia; (3) the routing of commodity exports from neighboring countries via pathways that avoid Russian territory; (4) the territorial and boundary issues remaining from old conflicts and arising from new, post-Soviet relationships; and (5) the prospect that the demographic vacuum resulting from population declines in remote peripheries will result in unwanted immigration from neighboring states.

Although Russian influence extends beyond the Russian border in several places along Russia's rim, there are particular reasons for assigning special status to the three countries of Transcaucasia between the Black Sea and the Caspian Sea. All three—Armenia, Georgia, and Azerbaijan—were former SSRs (Fig. 2-14). Their historical geographies are inextricably bound up with the colossus to the north. Following the collapse of the Soviet Union in 1991, all three were left with assets as well as liabilities resulting from the Soviet geopolitical design. Today, all three must consider Moscow in their strategies.

Armenia

Russia and Russians are popular in Armenia, and for good reason. The Soviet planners conferred SSR status on the Armenians, enabling it to become the independent Republic of Armenia in 1991, centered on Yerevan, the culture's primate city. Many Armenians feel that without Soviet intervention, they would have shared the fate of the Kurds (see Chapter 7) and be stateless today.

As Figure 2-1 indicates, landlocked Armenia (population: 3.1 million) occupies some of the most rugged and mountainous terrain in the earthquake-prone Transcaucasus. The Armenians are an embattled people who adopted Christianity 17 centuries ago and for more than a millennium sought to secure their ancient homeland here on the margins of the Islamic world. During World War I, the Ottoman Turks massacred much of the Christian Armenian minority and drove the survivors from eastern Anatolia (Turkey) and what is now Iraq into the Transcaucasus. At the end of that war in 1918, an independent Armenia arose, but its autonomy lasted only two years. In 1920, Armenia was taken over by the Soviets; in 1936, it became one of the 15 constituent SSRs of the Soviet Union. The collapse of the Soviet Empire gave Armenia what it had lost three generations earlier: independence.

Embattled Exclave

Or so it seemed. Soon afterward, the Armenians found themselves at war with neighboring Azerbaijan over the fate of some 150,000 Armenians living in Nagorno-Karabakh, a pocket of territory surrounded by Azerbaijan. Such a separated territory is called an *exclave*, and this one had been created by Soviet sociopolitical planners who, while acknowledging the cultural (Christian) distinctiveness of this cluster of Armenians, nevertheless gave (Muslim) Azerbaijan jurisdiction over it.

That was a recipe for trouble. In the ensuing conflict, Armenian troops entered Azerbaijan and gained control over the exclave, even ousting Azerbaijanis from the zone between the main body of Armenia and Nagorno-Karabakh (Fig. 2-14, vertical-striped zone). The international community, however, has not recognized Armenia's occupation, and officially the territory remains a part of Azerbaijan. In 2010, this issue is still unresolved.

Georgia

Of the three former SSRs in Transcaucasia, only Georgia has a Black Sea coast and thus an outlet to the wider world via the Turkish Straits and the Mediterranean Sea. Smaller than South Carolina, Georgia is a country of high mountains and fertile valleys. Its social and political geographies are complicated. The population of 4.6

million is more than 70 percent Georgian but also includes Armenians (8 percent), Russians (6 percent), Ossetians (3 percent), and Abkhazians (2 percent). The Georgian Orthodox Church dominates the religious community, but more than 10 percent of the people are Muslims, most of them concentrated in Ajaria in the southwest.

Unlike Armenia and Azerbaijan, Georgia has no exclaves, but its political geography is problematic nonetheless (Fig. 2-14). Within Georgia's borders lie three minority-based autonomous entities: the Abkhazian and Ajarian Autonomous Republics, and the South Ossetian Autonomous Region.

Sakartvelos, as the Georgians call their country, has a long and turbulent history. Tbilisi, the capital for 15 centuries, lay at the core of an empire around the turn of the thirteenth century, but the Mongol invasion ended that era. Next, the Christian Georgians found themselves in the path of wars between Islamic Turks and Persians. Turning northward for protection, the Georgians were annexed in 1800 by the Russians, who were looking for warm-water ports. Like other peoples overpowered by the czars, the Georgians took advantage of the Russian Revolution to reassert their independence; but the Soviets reincorporated Georgia in 1921 and proclaimed a Georgian Soviet Socialist Republic in 1936. Josef Stalin, the communist dictator who succeeded Lenin, was a Georgian.

Georgia is renowned for its scenic beauty, warm and favorable climates, agriculture (especially tea), timber, manganese, and other products. Georgian wines, tobacco, and citrus fruits are much in demand. The diversified economy could support a viable state but has yet to prosper.

Georgia and Russia

To see why, take a careful look at Figure 2-14. Georgia's political geography reveals that two of its three minority-based autonomous entities lie on its border with Russia and that one of these, in the Caucasus Mountains, cuts across the homeland of the Ossetians, dividing them into a North Ossetia in Russia and a South Ossetia in Georgia. Since the Ossetians are pro-Russian, Moscow took a strong interest in the fate of the restive minority on the Georgian side of the border, to the point of sending in army units as "peacekeepers." Meanwhile, the Russians sided with the also-defiant Abkhaz minority on the Black Sea coast in Georgia's northwest corner, going so far as to issue Russian passports to locals who were nominally Georgian citizens. Economically, the Russians played hardball by closing their border to all Georgian products.

In 2008, this situation reached the boiling point. Georgia's president, Mikheil Saakashvili, upon defeating his pro-Russian predecessor, vowed to take effective control of the minority entities and sent the army to take charge of Ajaria. Next he tackled the two pro-Russian territories in the north, thereby enraging Russian President Putin. In early 2008, NATO declared that Georgia should become a member at some future time, raising the ante still further. A few months later a brief but

August 2008 was a fateful month for Georgia. Following a still-disputed series of incidents in South Ossetia, the enclave whose population is overwhelmingly pro-Russian and anti-Georgian, Russian forces intervened, drove the Georgian military out of the territory, and entered Georgia itself. Also in control of the pro-Russian territory of Abkhazia, Moscow followed its intervention by declaring both South Ossetia and Abkhazia—internationally recognized components of Georgia no matter what their political preferences—to be independent "states." In late 2009 this key issue remained unresolved, with troubling implications for several other states bordering Russia. The sight of Russian armored vehicles in a village a mere 50 kilometers (30 mi) from Georgia's capital of Tbilisi suggests a new phase in Russia's role in the world.
© Associated Press

How Far Do Russia's Rights in the Near Abroad Really Go?

IN FAVOR OF STRONG RUSSIAN INFLUENCE

I am delighted that President Medvedev showed the world that we Russians will protect ourselves and our allies in the Near Abroad wherever it's necessary. Those Georgians who think they can ride roughshod over our friends the Ossetians have learned a lesson, and I hope they heed it. As a former officer in the Soviet Army I have no doubt that nations have to show their strength or they'll get trampled on. I'm still sorry we left Afghanistan in 1989 the way we did. And even if the USSR is no more, we Russians earned the right not to be bullied by upstart leaders like that dreadful Saakashvili in Georgia or that supposedly democratically elected Yushchenko in Ukraine. Have those people forgotten what we did for them and their countries? We created their national identities, established their republics, built their economies, put them on the map. We built their roads, railroads, bridges, and airports; we taught them Russian and nourished their cultures. We saved the Armenians from the Turks, the Estonians from the Germans, the Mongolians from the Chinese. If the Americans hadn't armed the Afghan extremists, we wouldn't have the Taliban threatening the place today.

So what thanks do we get? Saakashvili starts a war in South Ossetia and kills Russian soldiers and Ossetian villagers. Chechen terrorists kill Russians in the Caucasus, and all we hear about is that we mistreat minorities. The Latvians insult and offend us at every turn, treating its Russian-speaking residents as second-class citizens. The Ukrainians are cozying up to the European Union even though the millions of Russians in the country oppose it. Now the Americans want to put so-called missile-defense systems in Poland and the Czech Republic, and I'm pleased that President Medvedev played our Kaliningrad card. I've had it with the high-handed treatment we're getting around our perimeter, and it's time to put a stop to it.

You can look at the Near Abroad in two ways: as the ring of countries that encircles us or as the Slavic outposts that persist from Kazakhstan to Kosovo. Whichever way you define it, we Russians reserve the right to take care of our own, and we will not stand by as our kinspeople are mistreated, or our homeland threatened, by those who think we are weak and impotent. I hope the Ukrainians and the Moldovans and the Albanians and others who might miscalculate take a lesson from what happened in Georgia. If we have to, we will take appropriate action, and that includes military action. Our military forces suffered during the Yeltsin years, but those days are over. We have the money to rebuild our forces and we're doing so. We will not be pushed around, and it would be a grave mistake for the rest of the world to see us as weak or lacking resolve. The Soviet Union may have collapsed, but Russia will always be a force to reckon with.

OPPOSED TO STRONG RUSSIAN INFLUENCE

Talk about the Near Abroad is all the rage in Russia these days. Whether it's the television news, the magazines and newspapers, or talk radio, the Near Abroad is the topic that gets everyone riled up. In truth, this is nothing new—it started even before the USSR broke up nearly two decades ago. Not only were Russians marooned in the former republics, but their friends and allies among the locals had a tough time as well. Those who tried to help save the situation Gorbachev had created were seen as traitors to their people, and in places like Lithuania and Georgia it got pretty ugly.

As a history teacher, I try to put things in long-term perspective. I don't care what state or nation it is, when you've got the power you tend to abuse it, and even if you don't, you're accused of doing so by association. That certainly was true of the Western European colonial powers. Look at France and its "Near Abroad" in its former colony of Algeria. How many tens of thousands of Frenchmen were killed in the run-up to Algeria's independence, and how many afterward? We Russians feel that we weren't in a similar situation, because our Soviet republics really were not colonies. But millions of Russians did go to the republics to govern, to build infrastructure, to teach, or just to work for the fatherland, and you can be sure that many locals saw these "foreigners" as power-hungry outsiders. In the Muslim societies we never managed to persuade the people of the irrelevance of religion. In other republics, Latvia for example, there had been sympathies for Nazi Germany and we were viewed as occupiers. Get this: the Lithuanians not long ago launched a (U.S.) $34 billion claim against us for what they call Russia's "50-year occupation."

Make no mistake: I don't like it when I hear or read about the mistreatment of Russians or our allies in countries of the Near Abroad. Trying to force Russians to abandon their citizenship and compelling them to pass difficult language tests as a condition for local citizenship is no way to get over the past. But I feel that nothing is gained by the kind of violence that recently took place in South Ossetia and Abkhazia. Some reports say that Georgia's president started the whole thing by unleashing attacks on Russian protective forces; others tell a different story. But the fact is that Russian forces were within Georgia's international borders even before the conflict took its deadly turn, and that was asking for trouble. I have no doubt that the Georgian regime was guilty of discrimination against South Ossetians (and Abkhaz as well), but we're talking here about only tens of thousands of people in a tiny enclave—was it worth risking a wider war over what is essentially a minority-treatment issue?

I don't think so, but let me tell you—I'm in a small minority here. The small shooting war in Georgia has galvanized Russians. I hope it's not the beginning of a new era in the Near Abroad.

Regional ISSUE

Vote your opinion at www.wiley.com.college.deblij

costly war broke out in South Ossetia, and Russian troops invaded Georgian territory from their "peace-keeping" bases there.

This armed intervention in the Near Abroad was condemned by the international community, but Russia was unrepentant and continued to obstruct Georgian intentions in Abkhazia as well (see the Issue Box titled "How Far Do Russia's Rights in the Near Abroad Really Go?"). Russia went so far as to declare South Ossetia and Abkhazia "independent," arguing that these territories and their occupants had as much right to sovereignty as did Kosovo in southeastern Europe. And by blockading Georgia's coast with ships from Black Sea bases on the Crimean Peninsula Russia had leased from Ukraine, it was not difficult to see the implications for other states in the Near Abroad. This was an ominous confirmation of Putin's pledge to raise Russia's profile on the global stage.

Azerbaijan

Azerbaijan is the name of an independent state and of a province in neighboring Iran. The *Azeris* (short for Azerbaijanis) on both sides of that border have the same ancestry: they are a Turkish people divided between the (then) Russian and Persian empires under a treaty signed in 1828. By that time, the Azeris had become Shi'ite Muslims, and when the Soviet communists laid out their grand design for the USSR, they awarded the Azeris full SSR status. On the Persian side, the Azeris were assimilated into the Persian Empire, and their domain became a province. Today, the former Soviet Socialist Republic is the independent state of Azerbaijan (population: 8.9 million), and the more than 12 million Azeris to the south largely reside in the Iranian province.

During the brief transition to independence and at the height of their war with the Armenians, the predominantly Muslim Azeris tended to look southward, toward Iran. But geographic realities dictate a more practical orientation. Azerbaijan possesses huge reserves of oil and natural gas; under the Soviets it was one of Moscow's chief regional sources of fuels. The center of the oil industry is Baki (Baku), the capital on the shore of the Caspian Sea—but the Caspian Sea is a lake. To export its oil, Azerbaijan needs pipelines, but those of Soviet vintage link Baki to Russia's Black Sea terminal of Novorossiysk.

Playing the Energy Card

In the post-Soviet period, Azerbaijan has played its energy card to national advantage. Its authoritarian government is sometimes compared to that of Belarus, but unlike Belarus it has tried to establish some independence from Moscow. Oil and gas have made this easier: Washington, which supposedly promotes democracy, does not press the issue when it negotiates for Azerbaijan's oil, and America has become its most important customer. During Soviet and the first 15 years of post-Soviet times, oil from Baki continued to flow exclusively through Novorossiysk (Fig. 2-14). But the United States negotiated the construction of an alternate pipeline route across Georgia and Turkey to the Turkish oil terminal at Ceyhan, thus avoiding Russian transit altogether. In 2006, oil began flowing through this new outlet located at the northeastern corner of the Mediterranean Sea. Soviet state companies once exclusively extracted Azerbaijan's oil, but now American, French, British, and Japanese oil companies are developing the country's vast Caspian reserves.

Azerbaijan's nearly 9 million inhabitants have not yet benefited greatly from their country's energy riches. Given its Muslim roots and relative location, the time may come when this country turns toward the Islamic world. For the present, it is caught in the global web of energy demand and supply.

What You Can Do

RECOMMENDATION: Do you have a good-sized globe and/or a large wall map of the world's countries in your home or room? If not, we recommend that you do so (we are aware of how crowded desktop and table space can become in a room, especially a shared one). There even are self-standing globes you can place on the floor. We predict that you will be referring to this useful addition to your scholarly arsenal more than you can imagine, and you will find the map equally useful. National Geographic Maps, the Society's cartographic division, produces world maps on a sensible projection (see Appendix B) at several scales, so you can select one from their website that will fit on your wall. In general, the bigger the better, and you will soon be challenging guests to find places you have discovered in this course.

GEOGRAPHIC CONNECTIONS

1 The American Geographical Society in New York, the country's oldest geographical society, presents travelers the opportunity not only to visit remote parts of the world, but to do so accompanied by a professional geographer who specializes in the region or area being visited. Other organizations also appoint geographers as lecturers and guides on ships, planes, and even trains and buses. Imagine that you have been asked to accompany a group of your fellow students taking a riverboat down the Volga and Don rivers in July from Moscow to Rostov near the Sea of Azov, an arm of the Black Sea. In your first briefing, what will you tell them about the weather they should expect, about the minority "republics" of which they will get a glimpse, the crops and land uses they will encounter, the cities and towns they will see, and how they will get from the Volga to the Don? What might be the most interesting side trips along this itinerary?

2 Former President (and now Prime Minister) Vladimir Putin of Russia soon after his reelection proclaimed his intention to restore Russia to superpower status, to recapture much of the global might the Soviet Union once possessed. How are Russia's leaders now using their vast country's natural resources to advance this plan? Is any Soviet-era legacy likely to be helpful in achieving it? In what ways will this plan affect future relations between the European Union and Russia? Might the boundaries of the Russian realm as mapped in this chapter change as a result, and if so, where and how?

3 Russia has experienced the impact of Islamic terrorism principally in two locales: metropolitan Moscow and the Interior Southern Periphery. But Muslim minorities are far more widely distributed throughout this vast country beyond those two areas. Why is Islamic terrorism in Russia so prevalent in these locales and virtually (though not totally) absent elsewhere? What role does physiography play in the persistence of terrorism in the Interior Southern Periphery?

Geographic Literature on Russia: The key introductory works on this realm were authored by Bater, Remnick (both titles), Shaw (both titles), and Symons. These works, together with a comprehensive listing of noteworthy books and recent articles on Russia's geography, can be found in the *References and Further Readings* section of this book's website at **www.wiley.com/college/deblij**.

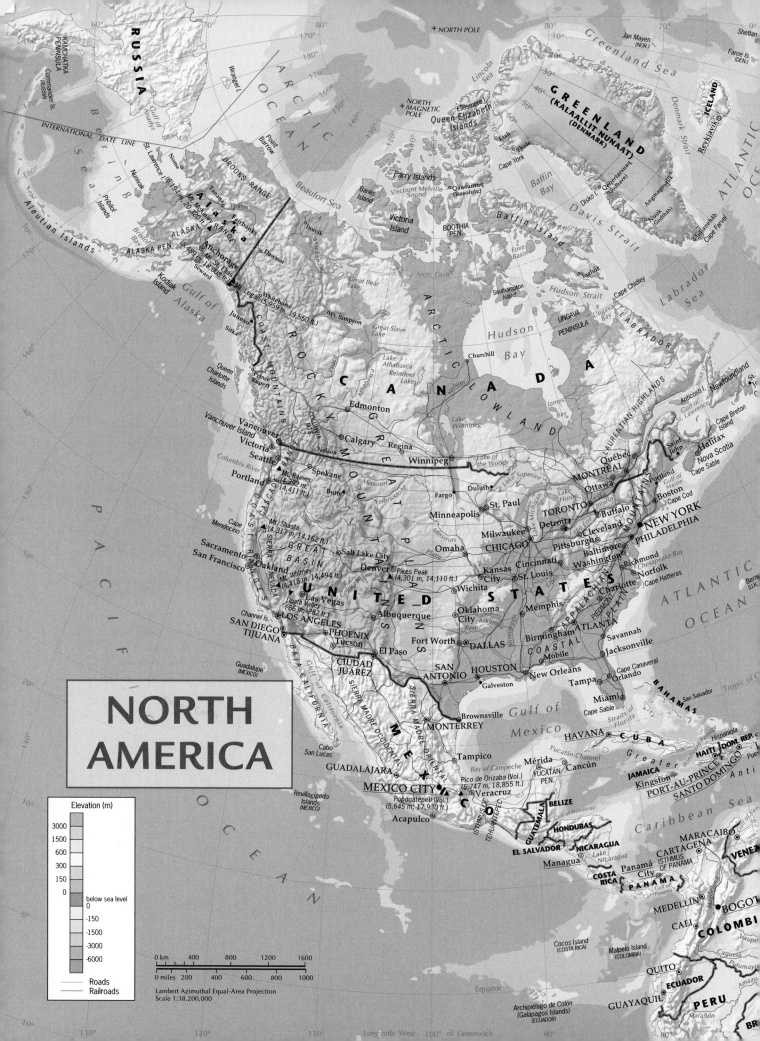

NORTH AMERICA

Elevation (m)

3000
1500
600
300
150
0
below sea level
0
-150
-1500
-3000
-6000

Roads
Railroads

0 km   400   800   1200   1600
0 miles   200   400   600   800   1000

Lambert Azimuthal Equal-Area Projection
Scale 1:38,200,000

RUSSIA

KAMCHATKA
PENINSULA
(RUSSIA)

Commander Is.
(RUSSIA)

INTERNATIONAL DATE LINE

Bering
Sea

Gulf of Anadyr

Wrangel I.

Aleutian Islands

Pribilof
Islands

Nunivak I.

St. Lawrence I.

Nome

ARCTIC OCEAN

Point Barrow

BROOKS RANGE

Alaska

Mt. McKinley
(6,194 m, 20,320 ft.)

Tanana

Fairbanks

Yukon

ALASKA RANGE

Anchorage

ALASKA

Kodiak
Island

ALASKA PEN.

Seward

Mt. St. Elias
(5,489 m, 18,008 ft.)

Inuvik

Dawson

Whitehorse

Mt. Logan
(5,959 m, 19,550 ft.)

Juneau

Sitka

Fort Simpson

COAST MOUNTAINS

Queen
Charlotte
Islands

Prince
Rupert

NORTH
MAGNETIC
POLE

+ NORTH POLE

Banks
Island

Parry Islands

Viscount Melville
Sound

Victoria
Island

BOOTHIA
PEN.

Great Bear
Lake

Arctic Circle

Beaufort Sea

Mackenzie

Queen Elizabeth
Islands

Ellesmere
Island

Etah

Thule

Qausuittuq
(Resolute)

Baffin
Bay

Cape York

GREENLAND
(KALAALLIT NUNAAT)
(DENMARK)

Jan Mayen
(NOR.)

Greenland Sea

Shetlan

Faroe Is.
(DEN.)

ICELAND

Reykjavík

Denmark Strait

Iqaluit

Disko I.

Qeqertarsuaq
(Godhavn)

Angmagssalik

Nuuk
(Godthåb)

Cape Farvel

Julianehåb

Davis Strait

Cape Chidley

Labrador
Sea

Southampton
Island

Foxe
Basin

Hudson Strait

UNGAVA
PENINSULA

Ungava
Bay

LABRADOR

Great Slave
Lake

Lake Athabasca

Reindeer
Lake

Churchill

CANADA

Hudson
Bay

HUDSON BAY
LOWLAND

LAURENTIAN HIGHLANDS

Anticosti I.

Gulf of
St. Lawrence

Newfoundland

Cape Breton
Island

ROCKY MOUNTAINS

Edmonton

Calgary

Nelson

Peace

Athabasca

Nelson

Lake
Winnipeg

James
Bay

GREAT PLAINS

Saint
John

Nova Scotia

Halifax

Cape Sable

St.

Vancouver

Vancouver Island

Victoria

Seattle

Portland

CASCADE RANGE

Mt. Rainier
(4,392 m, 14,411 ft.)

Spokane

Butte

Columbia River

COLUMBIA MTS.

Regina

Winnipeg

Lake of
the Woods

Fargo

Duluth

MONTREAL

Québec

Ottawa

Portland
Gulf of
Maine

Boston

Cape Cod

Mt. Shasta
(4,317 m, 14,162 ft.)

Cape
Mendocino

SIERRA NEVADA

COAST RANGES

Sacramento

San Francisco

Oakland

GREAT BASIN

Salt Lake City

Mt. Whitney
(4,418 m, 14,494 ft.)

Death Valley
(-86 m, -282 ft.)

Las Vegas

Yellowstone

Missouri

Snake

Platte

Denver

Pikes Peak
(4,301 m, 14,110 ft.)

UNITED STATES

Minneapolis

St. Paul

Omaha

Milwaukee

CHICAGO

Lake Superior

Lake Michigan

Lake Huron

Detroit

TORONTO

Lake Ontario

Lake Erie

Buffalo

Cleveland

Pittsburgh

NEW YORK

PHILADELPHIA

Baltimore

Washington

APPALACHIAN MOUNTAINS

Richmond

Norfolk

Chesapeake Bay

Cape Hatteras

ATLANTIC OCEAN

Bermuda
(U.K.)

Channel Is.

LOS ANGELES

SAN DIEGO

TIJUANA

Phoenix

Tucson

El Paso

CIUDAD
JUÁREZ

Albuquerque

Kansas
City

St. Louis

Wichita

Oklahoma
City

Red

Arkansas

Memphis

Cincinnati

Charlotte

PIEDMONT

ATLANTA

Birmingham

Fort Worth

DALLAS

SAN
ANTONIO

HOUSTON

Galveston

Brownsville

New Orleans

Mobile

COASTAL PLAIN

Savannah

Jacksonville

Tampa

Orlando

Cape Canaveral

Miami

Cape Sable

Mississippi

BAJA CALIFORNIA

Guadalupe
(MEXICO)

Gulf of California

SIERRA MADRE OCCIDENTAL

SIERRA MADRE ORIENTAL

Cabo
San Lucas

MONTERREY

Tampico

GUADALAJARA

MEXICO CITY

MEXICO

Popocatépetl (Vol.)
(5,645 m, 17,930 ft.)

Pico de Orizaba (Vol.)
(5,747 m, 18,855 ft.)

Veracruz

Bay of Campeche

Mérida

YUCATÁN PEN.

Cancún

Acapulco

Revillagigedo
Islands
(MEXICO)

ISTHMUS OF
TEHUANTEPEC

Gulf of
Mexico

Straits of
Florida

HAVANA

CUBA

BAHAMAS

San Salvador

Tropic of Cancer

Greater Antilles

JAMAICA

Kingston

HAITI

PORT-AU-PRINCE

DOM. REP.

SANTO DOMINGO

Hispaniola

Windward Passage

Yucatán Channel

GUATEMALA

BELIZE

Gulf of Honduras

HONDURAS

EL SALVADOR

NICARAGUA

Lake
Nicaragua

Managua

COSTA
RICA

PANAMA

Panamá
City

ISTHMUS
OF PANAMA

Gulf of
Panama

Caribbean Sea

MARACAIBO

CARTAGENA

VENE

BOGOT

MEDELLÍN

CALI

COLOMBI

Cocos Island
(COSTA RICA)

Malpelo Island
(COLOMBIA)

Equator

Archipiélago de Colón
(Galápagos Islands)
(ECUADOR)

QUITO

ECUADOR

GUAYAQUIL

PERU

BR

Marañón

Putumayo

Caquetá

PACIFIC OCEAN

Longitude West of Greenwich

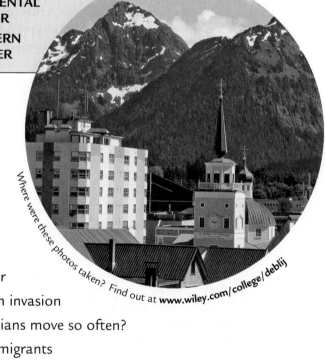

CONCEPTS, IDEAS, AND TERMS

1. Cultural pluralism
2. Physiographic province
3. Rain shadow effect
4. Fossil fuels
5. Sunbelt
6. Migration
7. Push/pull factors
8. American Manufacturing Belt
9. Ghetto
10. Outer city
11. Suburban downtown
12. Postindustrial
13. Primary activities
14. Secondary activities
15. Tertiary activities
16. Quaternary activties
17. Mosaic culture
18. Technopole
19. Pacific Rim

REGIONS

- NORTH AMERICAN CORE
- MARITIME NORTHEAST
- FRENCH CANADA
- SOUTH
- SOUTHWEST
- PACIFIC HINGE
- WESTERN FRONTIER
- CONTINENTAL INTERIOR
- NORTHERN FRONTIER

Where were these photos taken? Find out at **www.wiley.com/college/deblij**

In This Chapter

- Spectacular scenery, violent weather
- North America before the European invasion
- What makes Americans and Canadians move so often?
- The ruckus over undocumented immigrants
- Canada and Quebec: Still together, but forever?
- China's impact on the Pacific Hinge

Photos: © H. J. de Blij

FIGURE 3-1   © H. J. de Blij, P. O. Muller, and John Wiley & Sons, Inc.

WE TURN NEXT to the Western Hemisphere, the two great triangular continents that separate the Atlantic and Pacific oceans and extend, very nearly, from pole to pole, flanked by numerous islands large and small, indented by gulfs and bays of historic and economic import, and endowed with an enormous range of natural resources.

Collectively, these lands are called the Americas, one of those accidents of nomenclature that mark the world map. *Amerigo* was the first name of the Italian-born navigator and explorer Vespucci who conferred with Columbus after his second voyage, later led his own explorations, and concluded that South America was a separate landmass, not (as many, including Columbus, assumed) part of Asia. Cartographers in the early 1500s first put Amerigo's name on the southern landmass only, but before long the whole New World acquired a designation that, as we will see, entails some controversy: *the Americas*.

Two continents—North and South—form the Americas, but three geographic realms occupy them. There is, as we have already seen, a difference between continents and realms: Europe is a geographic realm but not, as is sometimes written, a continent. Europe's continent is Eurasia. So it is in North America. In the context of physical (natural) geography, North America from Canada's Ellesmere Island in the far north to Panama in the south is a continent (Fig. 3-1). In terms of modern human geography, this continent is divided into North American and Middle American realms along a transition zone marked by a political as well as physical boundary between the United States and Mexico. From the Gulf of Mexico to El Paso/Ciudad Juarez, the Rio Grande forms this bor-

MAJOR GEOGRAPHIC QUALITIES OF

North America

1. North America encompasses two of the world's biggest states territorially (Canada is the second largest in size; the United States is third).

2. Both Canada and the United States are federal states, but their systems differ. Canada's is adapted from the British parliamentary system and is divided into ten provinces and three territories. The United States separates its executive and legislative branches of government, and it consists of 50 States, the Commonwealth of Puerto Rico, and a number of island territories under U.S. jurisdiction in the Caribbean Sea and the Pacific Ocean.

3. Both Canada and the United States are plural societies. Although ethnicity is increasingly important, Canada's pluralism is most strongly expressed in regional bilingualism. In the United States, divisions occur largely along racial and income lines.

4. A substantial number of Quebec's French-speaking citizens supports a movement that seeks independence for the province. The movement's high-water mark was reached in the 1995 referendum in which (minority) non-French-speakers were the difference in the narrow defeat of separation. Prospects for a breakup of the Canadian state have diminished since 2000 but have not disappeared.

5. North America's population, not large by international standards, is one of the world's most highly urbanized and mobile. Largely propelled by a continuing wave of immigration, the realm's population total is expected to grow by more than 40 percent over the next half-century.

6. By world standards, this is a rich realm where high incomes and high rates of consumption prevail. North America possesses a highly diversified resource base, but nonrenewable fuel and mineral deposits are consumed prodigiously.

7. North America is home to one of the world's great manufacturing complexes. The realm's industrialization generated its unparalleled urban growth, but today both of its countries are experiencing the emergence of a new postindustrial society and economy.

8. The two countries heavily depend on each other for supplies of critical raw materials (e.g., Canada is a leading source of U.S. energy imports) and have long been each other's chief trading partners. Today, the North American Free Trade Agreement (NAFTA), which also includes Mexico, is linking all three economies ever more tightly as the remaining barriers to international trade and investment flow are dismantled.

9. North Americans are the world's most mobile people. Although plagued by recurrent congestion problems, the realm's networks of highways, commercial air routes, and cutting-edge telecommunications are still the most efficient on Earth.

der. From El Paso westward to the Pacific Ocean, straight-line boundaries, reinforced by fences and walls, separate the North from the Middle. Here, as we noted in the Introduction, global core and global periphery meet, contentiously and sometimes violently.

If European explorers, conquerors, and settlers had not invaded the Americas and overpowered their native inhabitants, it is just possible that we would still recognize three geographic realms. American "Indians" (another European fallacy) had made major cultural and technological achievements, created states, built cities, and developed sophistication in farming, medicine, astronomy, and other spheres. But there was nothing in either Middle or North America comparable to the Inca Empire of the South, and

nothing in North or South America like the Aztec and Maya states of the Middle. There were some similarities, to be sure, but South, Middle, and North already had their own indigenous core areas.

Today, each of the three realms of the Americas is dominated by a giant on the regional as well as the global stage. North America, the topic of this chapter, is overshadowed by the United States, still the world's leading power by most criteria. In Middle America, Mexico is territorially and demographically larger than all of the rest of the realm combined. In South America, Brazil, a state on the way to great-power status, occupies almost exactly half the realm and contains just over half its population. We begin in North America.

Defining the Realm

Defined in the context of human geography, North America is constituted by two of the world's most highly advanced countries by virtually every measure of social and economic development. Blessed by an almost unlimited range of natural resources and bonded by trade as well as culture, Canada and the United States are locked in a mutually productive embrace that is reflected by the statistics: in an average year of the recent past, more than 80 percent of Canadian exports went to the United States and about two-thirds of Canada's imports came from its southern neighbor. For the United States, Canada is the leading export market and its number-one source of imports, China notwithstanding.

Both countries rank among the world's most highly urbanized: nothing symbolizes the North American city as strongly as the skyscrapered panoramas of New York, Chicago, and Toronto—or the vast, beltway-connected suburban expanses of Los Angeles, Washington, and Houston.

North Americans also are the most mobile people in the world. Commuters stream into and out of central-city downtowns and suburban business centers by the millions each working day; most of them still drive cars, whose numbers have multiplied more than six times faster than the population since 1970. Moreover, each year about one out of every eight individuals changes his or her residence, a proportion that has recently declined but still leads the world.

The central business districts (CBDs) of North America's cities have distinctive skylines, often featuring landmark buildings. From Montreal to Baltimore, cities of the realm's original core area see their histories etched in still-surviving architectural icons like the Empire State Building and the Chrysler Building. Go west, however, and you will see less history and more modernity in CBDs from Vancouver to San Diego and, perhaps, rather less iconography. You might have trouble identifying this city from its skyline: this is an unusual perspective. But look on the left, and you see the spire of the Transamerica Building's pyramid rising above San Francisco's scenically situated but architecturally unremarkable CBD.
© AP/Wide World Photos

POPULATION GROWTH AND CLUSTERING

Although Canada and the United States share many historical, cultural, and economic qualities, they also differ in significant ways, diversifying this realm. The United States, somewhat smaller territorially than Canada, occupies the heart of the North American continent and, as a result, encompasses a greater environmental range. The U.S. population is dispersed across most of the country, forming major concentrations along both the (north-south trending) Atlantic and Pacific coasts (Fig. 3-2). The overwhelming majority of Canadians, on the other hand, live in an interrupted east-west corridor that extends across southern Canada, mostly within 300 kilometers (200 mi) of the U.S. border (Fig. 3-2). The United States, again unlike Canada, is a fragmented country in that the peninsula of Alaska is part of it (offshore Hawai'i, however, belongs in the Pacific Realm).

NORTH AMERICA
POPULATION DISTRIBUTION: 2010

One dot represents 10,000 persons

FIGURE 3-2

© H. J. de Blij, P. O. Muller, and John Wiley & Sons, Inc.

Figure 3-2 reveals that the great majority of both the United States and Canadian population resides to the east of a line drawn down the middle of the realm, reflecting the historic core-area development in the east and the later and still-continuing shift to the west, and, in the United States, to the south. Certainly this map shows the urban concentration of North America's population: you can easily identify cities such as Toronto, Chicago, Denver, Dallas-Fort Worth, San Francisco, and Vancouver. Just under 80 percent of the realm's population is concentrated in cities and towns, a higher proportion even than Europe's.

As the Data Table inside the back cover indicates, the population of the United States, which quite recently passed the 300-million mark, is growing at a rate 50 percent higher than Canada's, so that the 400-million mark may be reached as soon as 2040. This is an unusually high rate of growth for a high-income economy, resulting from a combination of natural increase (about 60 percent) and substantial immigration.

Although Canada's overall growth rate is significantly lower than that of the United States, immigration contributes proportionally even more to this increase than in the United States. With just under 34 million citizens, Canada, like the United States, has been relatively open to legal immigration, and as a result both **1** societies exhibit a high degree of **cultural pluralism** (a diversity of ancestral and traditional backgrounds). Indeed, Canada recognizes two official languages, English and French (the United States does not even designate English as such); and by virtue of its membership in the British Commonwealth, East and South Asians form a larger sector of Canada's population than Asians and Pacific Islanders do in the United States. On the other hand, cultural pluralism in the United States reflects large Hispanic (15 percent) and African American (13 percent) minorities and a wide range of other ethnic backgrounds.

Robust urbanization, substantial immigration, and cultural pluralism are defining properties of the human geography of the North American realm. We turn now to the physical stage on which the human drama is unfolding.

NORTH AMERICA'S PHYSICAL GEOGRAPHY

Physiographic Regions

One of the distinguishing properties of the North American landmass, all of which, as Figure G-3 reminds us, lies on the North American tectonic plate, is its remarkable variegation into regional landscapes. So well-defined are many of these landscape regions that we use their names in everyday language—for example, when we say that we flew over the Rocky Mountains or drove across the Great Plains or hiked in the Appalachians. These and other landscape regions are called **physiographic provinces**, and their diversity in **2** the Americas is nowhere greater than north of the Rio Grande (Fig. 3-3).

The map shows that the political geography that evolved in North America paid little heed to this physiography. In the far west, the Pacific Mountains extend all the way from Southern California through coastal Canada to Alaska. In the western interior, the Rocky Mountains form a continental backbone from central Alaska to New Mexico. Around the Great Lakes, the low-relief landscapes of the Interior Lowlands and the Great Plains to the west are shared by Canada and the United States, and the international boundary even divides the Great Lakes. In the east, the Appalachian Mountains (as the cross-section inset shows, no match for the Rockies) form a corridor of ridges, valleys, and plateaus that represent a familiar topography from Alabama and Georgia to Nova Scotia and Newfoundland. If there is a major physiographic province that "belongs" to one of the realm's two countries, it is the Canadian Shield, scoured bare by the glaciers that deposited the pulverized rocks as fertile soil in the U.S. Midwest, that is, in the Interior Lowlands. And even this bastion of ancient crystalline rock has corners in Michigan's Upper Peninsula as well as the upper Midwest's northernmost Wisconsin and northeastern Minnesota.

Climate

This diversity of landscape is matched by a variety of climates. Take another look at Figure G-7, and you can see the lineaments of landscape mirrored in the contrasts of climate. North America may not have it all—there are no areas of true tropical environment in the North American realm except at the southern tip of Florida—but it has a lot of variation. North America has moist coastal zones and arid interiors, well-watered plains, and even bone-dry deserts. On the world map, *Cf* and *Df* climates are especially good for commercial farming; note how large North America's share of these environments is.

The map leaves no doubt: the farther north you go, the colder it gets, and even though coastal areas derive moderation from warmer offshore waters, which is why there is a Vancouver and a Halifax, the rigors of *continentality* set in not far from the coast. Hot summers, frigid winters,

FIGURE 3-3

© H. J. de Blij, P. O. Muller, and John Wiley & Sons, Inc.

and limited precipitation make high-latitude continental interiors tough places to make it on the land. Figure G-7 has much to do with the southerly concentration of Canadian population, and with the lower densities in the interior throughout the realm.

In the west, especially in the United States, you can see what the Pacific Mountains do to areas inland. Moisture-laden air arrives from the ocean, the mountain wall

forces the air upward, cools it, condenses the moisture in it, and produces rain—the rain for which Seattle and Portland and other cities of the Northwest are (in)famous. By the time the air crosses the mountains and descends on the landward side, most of the moisture has been drawn from it, and the forests of the ocean side give way to scrub and brush. This **rain shadow effect** extends all the way across the Great Plains; North America does not

3

On August 29, 2005, Category-4 Hurricane Katrina cut a 230,000-square-kilometer (90,000-sq-mi) path of destruction along the Louisiana and Mississippi Gulf Coast centered around landfall just west of the mouth of the Mississippi River. Most devastated was the city of New Orleans, 80 percent of which lay underwater in the immediate aftermath of the storm. The pair of aerial photos above shows central New Orleans under normal conditions in the spring of 2004 (left) and two days after the tropical cyclone struck (right). Katrina was by far the most costly natural disaster in U.S. history, and its toll was truly staggering: more than 1800 lives lost, at least 200,000 homes destroyed, nearly a million people displaced, and more than $25 billion in insured property damage. Despite $110 billion in federal aid during the year following the disaster, rebuilding has been patchy and painfully slow. More than four years after the storm, the impact of Katrina continued to weigh heavily on the Crescent City. Roughly one-third of the pre-2005 population (ca. 150,000) has never returned. More than 30 percent of residential properties remain unoccupied, most of them still damaged. Less than half of all transit service has been restored. More children attend private rather than public schools, with the latter's enrollment barely 50 percent of its 2004-2005 total. Only about half of all medical facilities have reopened, propelling an intensifying health crisis that affects two-thirds of the city's residents (double the number reported before Katrina). The proportion of the population living in poverty is nearly 20 percent, and the number of homeless people exceeds 12,000. The tourist economy of New Orleans has fared better, but the Big Easy remains highly vulnerable to hurricanes as it continues its grim struggle toward an acceptable level of restoration. © Digital Globe/Eurimage/Photo Researchers, Inc.

turn moist again until the Gulf of Mexico sends humid tropical air into the eastern interior via the Mississippi-Missouri River Basin.

In a very general way, therefore, and not including the coastal areas on the Pacific, nature divides North America into an arid west and a humid east. Again the population map reveals more than a hint of this: draw a line approximately from Lake Winnipeg to the mouth of the Rio Grande, and look at the contrast between the (comparatively humid) east and drier west when it comes to population density. Water is a large part of this story.

Between the Rocky Mountains and the Appalachians, North America lies open to air masses from the frigid north and tropical south. In winter, southward-plunging Polar fronts send frosty, bone-dry air masses deep into the heart of the realm, turning places like Memphis and Atlanta into iceboxes; in summer, hot and humid tropical air surges northward from the Gulf of Mexico, giving Chicago and Toronto a taste of the tropics. Such air masses clash in low-pressure systems along weather fronts loaded with lightning, thunder, and, frequently, dangerously destructive tornadoes. And the summer heat brings an additional threat to the Gulf-Atlantic Coastal Plain (Fig. 3-3): hurricanes capable of inflicting catastrophic devastation on low-lying areas. Such hurricanes also prune natural vegetation, replenish underground water reservoirs, fill natural lakes, and flush coastal channels.

Great Lakes and Great Rivers

Two great drainage systems lie between the Rockies and the Appalachians: (1) the five Great Lakes that drain into the St. Lawrence River and the Atlantic Ocean, and (2) the mighty Mississippi-Missouri system that carries water from a vast interior watershed to the Gulf of Mexico, where the Mississippi forms one of the world's major *deltas*. Both natural systems have been enhanced by human engineering. In the case of the St. Lawrence Seaway, a series of locks and canals has created a direct shipping route from the Midwest to the Atlantic. The Mississippi and Missouri rivers have been fortified by artificial levees that, while failing to contain the worst of flooding, have enabled farmers to cultivate the most fertile of American soils.

Other North American rivers of note, including the Columbia and the Colorado in the west and shorter rivers such as the Hudson, Delaware, and Potomac flowing off the eastern Appalachians onto the Atlantic Coastal Plain (where their upstream waterfalls marked the limit of tidewater navigation and *Fall Line Cities* formed), had much to do with the way the realm's post-Columbian human geography unfolded.

INDIGENOUS NORTH AMERICA

When the first Europeans set foot on North American soil, the continent was inhabited by millions of people whose ancestors had reached the Americas from Asia via Alaska, and possibly also across the Pacific, more than 14,000 years before (and perhaps as long as 30,000 years ago). In search of Asia, the Europeans misnamed them "Indians," but the historic affinities of these earliest Americans were with the peoples of eastern and northeastern Asia, not India.

In North America, these Native Americans or First Nations—as they are now called in the United States and Canada, respectively—had organized themselves into hundreds of nations with a rich mosaic of languages and a great diversity of cultures (Fig. 3-4). Native American farmers grew crops the Europeans had never seen; other nations depended chiefly on fishing, herding, or hunting, or some combination of these. In the interior plains, the American bison was the mainstay of life for a large population; the white invaders shot the herds by the hundreds of thousands and doomed the people that had depended on them for thousands of years.

Elaborate houses, efficient watercraft, decorative and protective clothing, effective weaponry, and wide-ranging art forms distinguished the aboriginal nations. Certain nations had formulated sophisticated health and medical practices; ceremonial life was complex and highly developed; and political institutions were mature and elaborate. The Iroquois were practicing a form of federalism long before the Europeans thought of it.

The eastern nations were the first to bear the brunt of the European invasion. By the end of the eighteenth century, ruthless and land-hungry settlers had driven most of the Native American peoples living along the Atlantic and Gulf coasts from their homes and lands, initiating a west-ward push that was to devastate indigenous society. The United States Congress in 1789 proclaimed that "Indian . . . land and property shall never be taken from them without their consent," but in fact this is just what happened. One of the sorriest episodes in American history involved the removal of the eastern Cherokee, Chickasaw, Choctaw, Creek, and Seminole from their homelands in forced marches more than 1500 kilometers (1000 mi) westward to Oklahoma. One-fourth of the entire Cherokee population died along the way from exposure, starvation, and disease, and the survivors had little to comfort themselves. Again, Congress approved treaties that would at least protect the native peoples of the plains and those farther west, but after the mid-nineteenth century the white settlers ignored those "guarantees" as well. A half-century of war left what remained of North America's native nations with about 4 percent of U.S. territory in the form of mostly impoverished reservations.

In what is today Canada, the First Nations,* smaller in number and more widely dispersed, were also overwhelmed by the power of European settlers and decimated by the diseases they introduced. Efforts at restitution and recognition of the rights of First Nations have gone farther in Canada than in the United States. However, First Nation reserves in Canada, while more numerous than those in the United States, are smaller in area and less populous. More than half of all indigenous Canadians now reside in metropolitan areas.

THE COURSE OF EUROPEAN SETTLEMENT AND EXPANSION

The current geography of North American population is rooted in the colonial era of the seventeenth and eighteenth centuries that was dominated by Britain and France. The French sought mainly to organize a lucrative fur-trading network, while the British established settlements along the coast of what is today the eastern seaboard of the United States (the oldest, Jamestown, was founded in Virginia more than four centuries ago).

The British colonies soon became differentiated in their local economies, a diversity that was to endure and later shape much of North America's cultural geography. The northern colony of New England (Massachusetts Bay and environs) specialized in commerce; the southern Chesapeake Bay colony (tidewater Virginia and Maryland) emphasized the plantation farming of tobacco; the Middle Atlantic area lying in between (south-

---

*The indigenous peoples of Canada, numerically dominated by First Nations, also include Métis (of mixed native and European descent) and the Inuit (formerly Eskimos) of the far north.

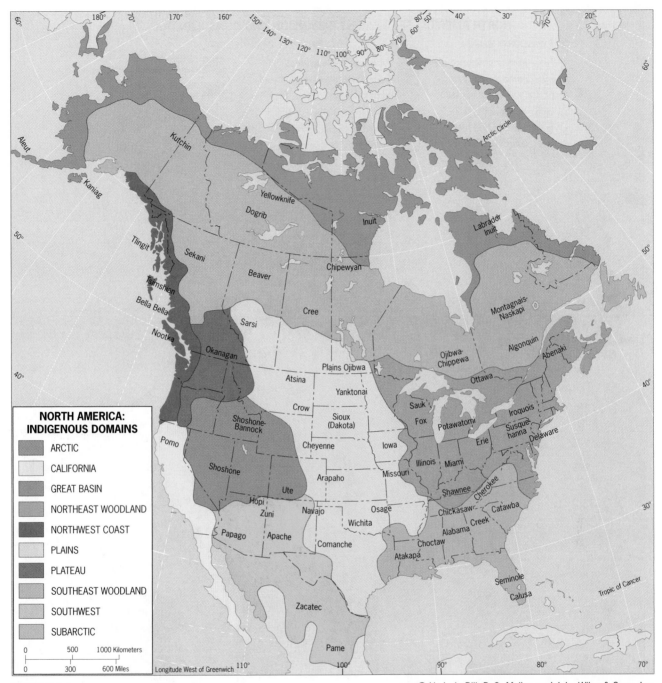

NORTH AMERICA: INDIGENOUS DOMAINS

- ARCTIC
- CALIFORNIA
- GREAT BASIN
- NORTHEAST WOODLAND
- NORTHWEST COAST
- PLAINS
- PLATEAU
- SOUTHEAST WOODLAND
- SOUTHWEST
- SUBARCTIC

FIGURE 3-4

© H. J. de Blij, P. O. Muller, and John Wiley & Sons, Inc.

eastern New York, New Jersey, eastern Pennsylvania) formed the base for a number of smaller, independent-farmer colonies.

These neighboring colonies soon thrived and yearned to expand, but the British government responded by closing the inland frontier and tightening economic controls. This policy provoked general opposition in the colonies, followed by the revolutionary challenge that was to lead to independence and the formation of the United States of America. The western frontier of the fledgling nation now swung open, and the area north of the Ohio River was occupied soon after it was discovered that soils and climate in these Interior Lowlands (Fig. 3-3) were more favorable to farming than those of the Atlantic Coastal Plain and the Appalachian foothills of the Piedmont to its west. Deforestation cleared huge tracts for farming and livestock, and new cultural landscapes spread apace. The acquisition of more distant western frontiers followed in short order (Fig. 3-5).

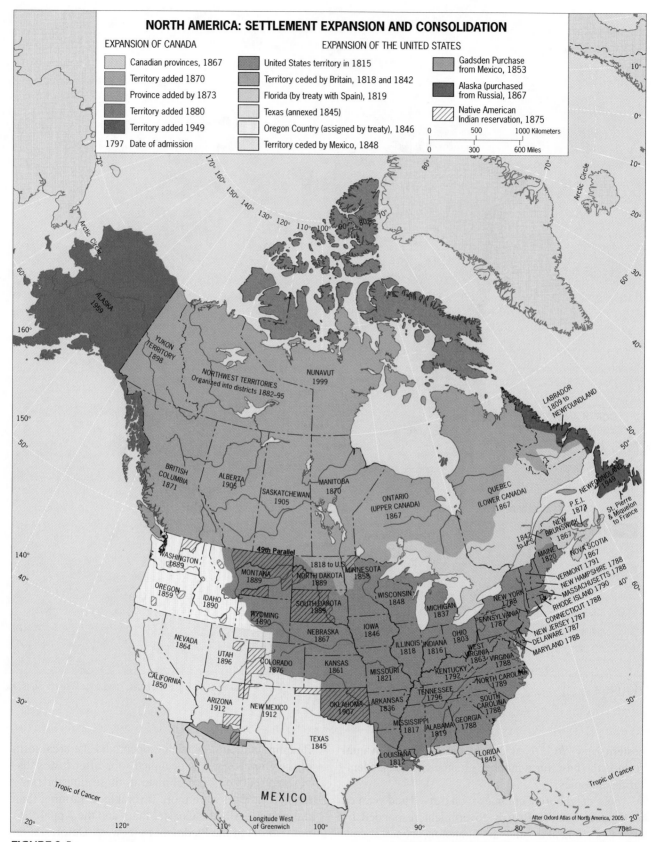

NORTH AMERICA: SETTLEMENT EXPANSION AND CONSOLIDATION

EXPANSION OF CANADA
- Canadian provinces, 1867
- Territory added 1870
- Province added by 1873
- Territory added 1880
- Territory added 1949
- 1797 Date of admission

EXPANSION OF THE UNITED STATES
- United States territory in 1815
- Territory ceded by Britain, 1818 and 1842
- Florida (by treaty with Spain), 1819
- Texas (annexed 1845)
- Oregon Country (assigned by treaty), 1846
- Territory ceded by Mexico, 1848
- Gadsden Purchase from Mexico, 1853
- Alaska (purchased from Russia), 1867
- Native American Indian reservation, 1875

0   500   1000 Kilometers
0   300   600 Miles

FIGURE 3-5

© H. J. de Blij, P. O. Muller, and John Wiley & Sons, Inc.

In present-day Canada, the course of events was rather different, but the effect—land alienation and ethnic cleansing on a vast scale and significant impact on natural environments—was similar. The French, not the British, were the first European colonizers in the north, beginning in the 1530s. *New France*, as the locals called it, grew during the seventeenth century into a vast if poorly defined sphere of influence that encompassed the St. Lawrence Valley, the Great Lakes Basin, and the Mississippi Valley all the way south to *Nouvelle* Orleans. You can see numerous French names on the "American" map to this day, their pronunciations mangled by their non-French inhabitants, from Detroit to Bourbonnais and from "Champaign" to Metairie. When the English challenged the French for primacy in this far-flung domain, a series of wars that started in the 1680s led to victory and the cession of New France to the British Crown in 1763. But by then, the French colony in the St. Lawrence Valley was no longer merely a fragile settlement. Populous, urbanized, extending from the Great Lakes to the mouth of the St. Lawrence, organized under French legal and land-tenure systems, and strongly Catholic, this Francophone cultural transplant would not be easy to control. The British decided to give Quebec enough autonomy and self-administration to stave off a war of suppression. It was the start of an uneasy relationship between expatriate cultures that continues to this day.

Even as the European settlers fought among themselves and, in the case of the Thirteen Colonies, revolted against their European rulers, and as white settlement and land acquisition marched westward, a momentous development was altering the cultural fabric of the southeastern periphery of North America. In 1619, a Dutch ship arriving at Jamestown from the coast of Guinea offered several of its Africans in bondage for sale to the local tobacco planters. That moment marks the beginning of the slave trade in North (not Middle) America. By the end of the eighteenth century, there were more than 200,000 Africans in the State of Virginia alone, and slave labor became the indispensable factor in the profitable cotton and tobacco export trade of the "American South." By the time the issue of slavery and its associated barbarism instigated the Civil War (1861–1865) that was to decide the future of the United States, the basis for an African American diaspora had been created.

As Figure 3-5 shows, the westward consolidation of organized space in North America took place more rapidly south of the latitude of the Great Lakes than to the north. In 1850, to take a benchmark year, not only the eastern United States but also much of the Pacific coast had been reached and settled. But in Canada, the settle-ment frontier had only reached Lake Huron. The Canadian push to the west was delayed in part by the comparative harshness of the rugged Canadian Shield north and west of the Great Lakes and by the barrier of the Canadian Rocky Mountains, a formidable obstacle without convenient mountain passes. Nor did there seem to be any urgency to this initiative—until concerns arose that the United States might expand in a northerly (as well as westerly) direction.

THE FEDERAL MAP OF NORTH AMERICA

The two states that constitute North America may have arrived at their administrative frameworks with different motives and at different rates of speed, but the result is unmistakably similar: their internal political geographies are dominated by straight-line boundaries of administrative convenience. In comparatively few places, physical features such as rivers or the crests of mountain ranges mark internal boundaries, but by far the greater length of boundaries is ruler-straight. Even most of the international boundary between Canada and the United States west of the Great Lakes coincides with a parallel of latitude (49°N). Certainly the resulting map (Fig. 3-6) looks nothing like the one of indigenous domains (Fig. 3-4).

The reasons are all too obvious: the framework was laid out before, in some areas long before, significant white settlement occurred. And by delimiting the internal administrative units early and clearly, governments precluded later disputes over territory and resources. In any case, neither Canada nor the United States was to become a unitary state: both countries are *federal states*. One practical difference is this: the Canadians call their primary subdivisions *provinces*, whereas in the United States they are, as the country's name implies, *States*.

It is no matter of trivial pursuit to take a careful look at the names of the provinces and States that compartmentalize this realm, because some are of greater consequence than others. We have already noted the special significance of mainly French-cultured Quebec, still in a special position in federal Canada. In many ways the key Canadian province is Ontario, containing the country's largest and most globalized city (Toronto), and with the capital (Ottawa) on its eastern river border with Quebec. In the United States, key States in the political and economic geography of the federation include Texas bordering Mexico, California facing the Pacific, New York

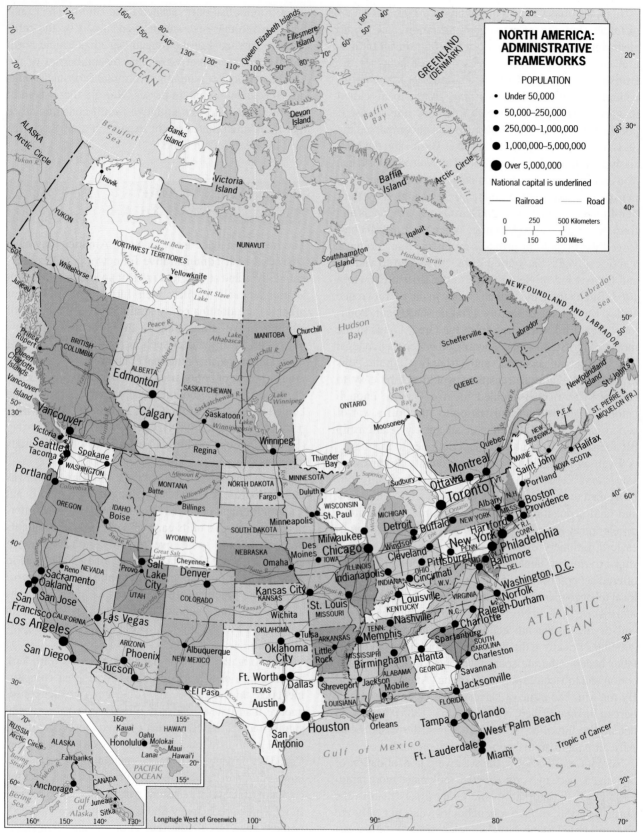

NORTH AMERICA: ADMINISTRATIVE FRAMEWORKS

POPULATION

- ● Under 50,000
- ● 50,000–250,000
- ● 250,000–1,000,000
- ● 1,000,000–5,000,000
- ● Over 5,000,000

National capital is underlined

Railroad   Road

| 0 | 250 | 500 Kilometers |
| 0 | 150 | 300 Miles |

FIGURE 3-6

© H. J. de Blij, P. O. Muller, and John Wiley & Sons, Inc.

with its urban window on the Atlantic, and Florida pointing toward the Caribbean (Miami is often referred to as the "gateway to Latin America"). The rectangular layout of North America seems to symbolize the realm's modernity and its stability. If federalism works in these high-income economies, shouldn't the rest of the world emulate it? As we shall see, the system has its assets but also some significant liabilities.

NORTH AMERICA'S NATURAL RESOURCES

One of these liabilities has to do with the variable allotment of natural resources among the States and provinces. Obviously, when the framework in Figure 3-6 was laid out, much remained unknown about mineral, fuel, and other resources contained in the North American realm. In Canada, Alberta is favored with large energy reserves (and more may be in the offing); Alberta's provincial government is not always eager to see the federal administration take part of that wealth to assist poorer provinces. In the United States, when oil prices are high, Texas and Oklahoma boom while California's budget suffers. So it is important to compare the map of natural resources (Fig. 3-7) with that of States and provinces.

Water certainly is a natural resource, and North America as a realm is comparatively well supplied with it despite concerns over long-term prospects in the American Southwest and the Great Plains, where some States depend on sources in other States. Ongoing climate change may alter this assessment, but in 2010 North America's water supplies remained adequate. The westward and southward movement of population referred to earlier, however, is creating concerns over the future as burgeoning urbanization in the arid U.S. Southwest continues to bolster demand. Another concern focuses on lowering water tables in some of North America's most crucial *aquifers* (underground water reserves) in which a combination of overuse and decreasing replenishment suggests that supply problems may arise.

In terms of other natural resources, North America is endowed with abundant reserves of minerals as well as **fossil fuel** (oil, natural gas, coal) energy sources, although the realm's voracious demand for energy generates an enormous need for fuel imports. Mineral resources are concentrated in three zones: the Canadian Shield north of the Great Lakes, the Appalachian Mountains in the east, and the mountain ranges of the west (Fig. 3-7). The Canadian Shield contains substantial iron ore, nickel, copper, gold, uranium, and diamonds. The Appalachians yield iron ore, lead, and zinc. And the western mountain zone has significant deposits of copper, lead, zinc, molybdenum, uranium, silver, and gold.

In terms of *fossil-fuel energy resources*, North America is well endowed, although, as noted, the realm's consumption, the highest in the world, cannot be met by domestic supplies. Figure 3-7 displays the distribution of oil, oil-yielding tar sands, natural gas, and coal.

The leading *oil*-production areas lie (1) along and offshore from the Gulf Coast, where the floor of the Gulf of Mexico is yielding a growing share of the output; (2) in the Midcontinent District, from western Texas to eastern Kansas; and (3) along Alaska's "North Slope" facing the Arctic Ocean. As we note in Chapter 12, new discoveries in the Arctic may in the near future alter the balance of known reserves. Another developing discovery is the Canadian crescent of oilfields curving southeastward from northern Alberta to southern Manitoba; still another recent find focuses on the seafloor off Newfoundland.

An important development is taking place in Canada's northeastern Alberta, where oil is being drawn from deposits of *tar sands* in the vicinity of the boomtown of Fort McMurray. The process is expensive and can reward investors only when the price of oil is comparatively high, but the reserves of oil estimated to be contained in the tar sands may exceed those of Saudi Arabia. In 2008, when the price of oil skyrocketed, activity in this area

Wind "farms" are arising in many more places in North America as the clean-energy campaign to generate electricity from sources other than fossil fuels accelerates. This cluster of modern wind turbines on a cattle ranch in southern Wyoming (the livestock provide a sense of scale) exploits a flat-floored windstream corridor between areas of higher relief. © Jim Harvey/Alamy

FIGURE 3-7

© H. J. de Blij, P. O. Muller, and John Wiley & Sons, Inc.

soared. When the price collapsed later that year, production slowed. It is a pattern that will probably continue for decades to come.

The distribution of *natural gas* reserves resembles that of oilfields because petroleum and natural gas tend to be found in similar geological formations (that is, the floors of ancient, shallow seas). What the map cannot reveal is the volume of production, in which this realm leads the world (Russia and Iran lead in terms of proven reserves). Natural gas has risen rapidly as the fuel of choice for elec-

tricity generation in North America, raising fears that the supply will not keep up with growing demand.

The **coal** reserves of North America may be the largest on the planet, although coal discoveries continue to be made in many other parts of the world. Even if major deposits are found elsewhere, North America's reserves—in Appalachia, under the Great Plains of the United States as well as Canada, and in the southern Midwest among other places—guarantee an adequate supply for centuries to come.

POPULATION AND MULTICULTURALISM

Given the mobility of North America's population, the immigrant streams that continue to alter it, the economic changes affecting it, and the forces acting upon it (similar to those we noted in Europe), Figure 3-2 should be viewed as the latest "still" in a motion picture, one that has been unreeling for four centuries. Slowly at first, then with accelerating speed after 1800 as one transportation breakthrough followed another and other technological innovations appeared, confident North Americans pushed their settlement frontier westward to the Pacific. Even today, such shifts continue. Not only does the "center of gravity" of population continue to move westward, but within the United States it is also moving southward—the latter a drift that gained momentum in the 1970s as the advent of universal air conditioning made the so-called **Sunbelt** States of the U.S. southern tier ever more attractive to internal migrants.

The current population map is the still-changing product of numerous forces. For centuries, North America has attracted a pulsating influx of immigrants who, in the faster-growing United States, were rapidly assimilated into the societal mainstream. Throughout the realm, people have sorted themselves out to maximize their proximity to evolving economic opportunities, and they have shown little resistance to relocating as these opportunities successively favored new locales. Today, Calgary in Alberta Province and its energy resources are attracting workers from all over Canada, and Fort McMurray (the tar sands headquarters) has become one of the newest North American boomtowns.

During the past century such transforming forces have generated a number of major migrations, of which the still-continuing shift to the west and south is only the latest. Five others are: (1) the persistent growth of metropolitan areas, first triggered by the late-nineteenth-century Industrial Revolution's impact in North America; (2) the massive movement of African Americans from the rural South to the urban North during the period of industrialization and in response to labor shortages in the North when the United States in the 1920s sharply curtailed immigration; (3) the shift of tens of millions of urban residents from central cities to suburbs and subsequently to "exurbs" even farther away from the urban core; (4) the return migration of millions of African Americans from the deindustrializing North back to the growing opportunities in the South; and (5) the strong influx of immigrants from outside North America that waxed and waned over time and brought to North America, in addition to those

in bondage from Africa, Europeans in great numbers from several sources; Middle Americans from Cuba, and more recently from Mexico and other Hispanic countries; South Asians from India and Pakistan; and East Asians from Japan, Vietnam, and Hong Kong at various times.

Several of the migration streams affecting North America rank among the largest in the history of the world (see the box entitled "The Migration Process"), creating the pair of plural societies that characterize this realm but periodically challenging their cultural fabric. In the United States, the issue of illegal immigration, principally from or via Mexico, has engendered a sometimes acrimonious debate over border security and law enforcement. In 2010, an estimated 12 million illegal immigrants from Mexico were in the country at a time when the Hispanic minority had already become the country's largest, transforming cities and neighborhoods and arousing fears among locals ranging from job losses and language dilution to rising crime and social expenditures. It was not the first time an immigrant flow caused such apprehensions, but the magnitude of the "invasion" and the widespread assertiveness of the immigrants created circumstances that put the reputed American "melting pot" to the test.

In Canada, similar antagonisms arose during the 1990s when East Asian immigrants from Hong Kong arrived during and following the 1997 takeover of the British colony by Beijing's communist government. Affluent Chinese families arriving in Vancouver proceeded to purchase and renovate (or replace) traditional homes in long-stable Vancouver neighborhoods, arousing the ire of some locals who would countenance residential integration but not disfigurement. But in general, Canadian views of immigration differ from those in the United States. Canada (before the economic downturn starting in 2008) has long faced critical labor shortages, especially in its western provinces including energy-booming Alberta and Asian-trade-burgeoning British Columbia. In need not only of professionals but also truck drivers and dock workers, Canada tries to balance its legal immigration process to keep it compatible with the country's employment as well as demographic needs.

CITIES AND INDUSTRIES

The Industrial Revolution occurred almost a century later in the United States than in Europe, but when it finally did cross the Atlantic in full force starting in the 1870s, it took hold so successfully and advanced so robustly that only 50 years later North America was surpassing Europe

The Migration Process

6

BOTH CANADA AND THE UNITED STATES are products of international **migration** (residential relocation intended to be permanent). Europeans first crossed the Atlantic with the intent to establish permanent colonies in the early 1600s, and from these colonies evolved the two countries of North America. The Europeanization of North America doomed the realm's indigenous societies, but this was only one of many areas around the world where local cultures and foreign invaders came face to face. Between 1835 and 1935, perhaps as many as 75 million Europeans departed for distant shores—most of them bound for the Americas (Fig. 3-8). Some sought religious freedom, others escaped poverty and famine, still others simply hoped for a better life. A comparative few were transported against their will to penal colonies.

Studies of the *migration decision* show that migration flows vary in size with (1) the perceived difference between home (source) and destination; (2) the effectiveness of information flow, that is, the news about the destination that emigrants send back to those left behind waiting to decide; and (3) the distance between source and destination (shorter moves attract many more migrants than longer ones).

Every migration stream produces a counter-stream of returning migrants who cannot adjust, are unsuccessful, or are otherwise persuaded or compelled to return home. Migration studies also conclude that several discrete factors are at work in the process. **Push factors** motivate people to move away from an undesirable locale that may be afflicted by famine, armed conflict, religious persecution, or some other adversity. **Pull factors** attract them to destinations perceived to hold a promise of security, opportunity, or another desired goal.

7

To the early (and later) European immigrants, North America was a new frontier, a place to escape persecution and acquire a piece of land. Opportunities were reported to be unlimited. That perception of opportunity has never changed. Immigration continues to significantly shape the human-geographic complexion of the United States as well as Canada. Today's immigrants account for at least 40 percent of the annual population growth of the United States. Never in its history, however, has the United States received so large an undocumented (illegal) immigration flow in addition to the legitimate stream, and the ongoing influx from Middle to North America is one of the largest population shifts in human history.

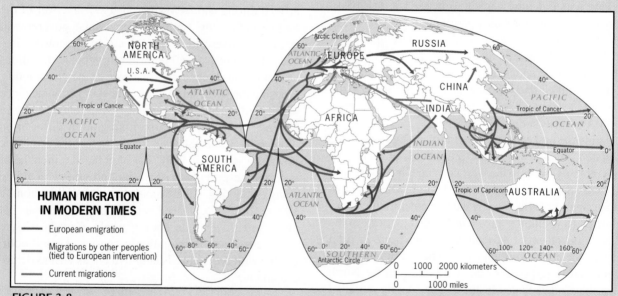

HUMAN MIGRATION IN MODERN TIMES

— European emigration

— Migrations by other peoples (tied to European intervention)

— Current migrations

FIGURE 3-8

© H. J. de Blij, P. O. Muller, and John Wiley & Sons, Inc.

as the world's mightiest industrial complex. The impact of industrial urbanization occurred simultaneously at two scales. At the *macroscale*, a system of new cities swiftly emerged, specializing in the collection and processing of raw materials and the distribution of manufactured products, for which they were linked by an increasingly efficient network of railroads. Within this realmwide urban system, at the local or *microscale*, individual cities and towns prospered in their roles as manufacturing centers and were often known (as British cities already were) for their particular products. This spatial structure still forms the geographic framework of most of the older central cities of North America's large metropolitan areas, although many associated industries have closed down as urban functions have changed in recent decades.

In fact, change is the norm for cities. Even before the full impact of the Industrial Revolution, established cities had been adjusting to the arrival of the steamship and the railroad. Later, the expansion of the rail network and the increased speed and efficiency of the trains enlarged their hinterlands. As new technologies and innovations emerged, and specializations such as Detroit's auto industry **8** strengthened, an **American Manufacturing Belt** (that extended into southern Ontario) evolved into the foundation of a North American Core (Fig. 3-9). This core area, on the way to becoming the world's most productive and important, contained the majority of the realm's industrial activity and leading cities, including New York, Chicago, Toronto, and Pittsburgh (the "steel city").

As the manufacturing cities thrived, they grew larger, vertically as well as horizontally. City centers acquired characteristic and symbolic skylines (Manhattan was a synonym for the greatest of all), and urban peripheries expanded to the point that relatively nearby cities

coalesced. The geographer Jean Gottmann combined *megalo* (very large) and *polis* (as in metropolis) to coin the term *megalopolis* for such coalescing metropolitan areas. The one extending along the Atlantic seaboard from north of Boston to south of Washington, D.C., has become known as the "Bosnywash" Megalopolis. This was the economic anchor of the North American core area: the seat of the U.S. government, the nucleus of business and finance, the hearth of culture, and the trans-Atlantic trading interface between much of the realm and Europe. In Figure 3-9, note that Canada's predominant megalopolis is its most highly urbanized zone extending from Windsor through Toronto to Montreal and Quebec City; urban geographers call this Windsor-Quebec axis *Main Street*, and it also forms part of the realm's core area.

As the map shows, other megalopolis-like conurbations are emerging elsewhere in North America, including the Pacific Northwest, the San Francisco Bay Area and Southern California, east-central Texas, and much of peninsular Florida. As we noted in the Introduction, subsidiary core areas develop within regions already dominated by an overarching core, and the North American realm presents ample evidence for this principle.

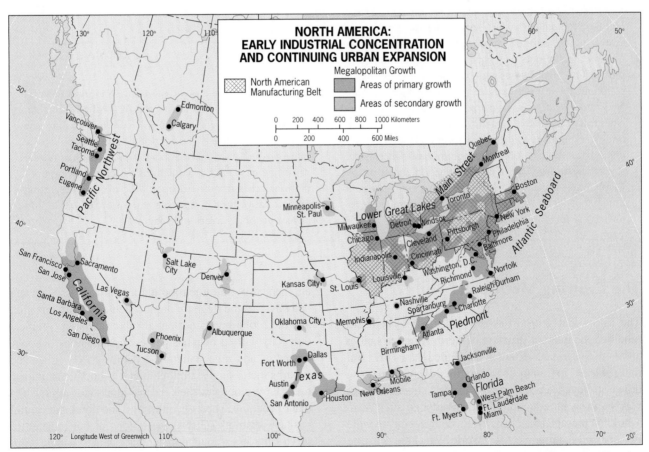

NORTH AMERICA:
EARLY INDUSTRIAL CONCENTRATION
AND CONTINUING URBAN EXPANSION

Megalopolitan Growth
North American Manufacturing Belt
Areas of primary growth
Areas of secondary growth

FIGURE 3-9

© H. J. de Blij, P. O. Muller, and John Wiley & Sons, Inc.

AMONG THE REALM'S GREAT CITIES . . . New York

NEW YORK IS much more than the largest city of the North American realm. It is one of the most famous places on Earth; it is the hemisphere's gateway to Europe and the rest of the Old World; it is a tourist mecca; it is the seat of the United Nations; it is one of the globe's most important financial centers—a true "world city" in every sense of that overused term. Unfortunately, the terrorists who planned and executed the attacks of 9/11 were only too well aware of New York's symbolic importance on the international stage.

New York City consists of five boroughs, centered by the island of Manhattan, which contains the CBD south of 59th Street (the southern border of Central Park). Here is a skyscrapered landscape unequaled anywhere, studded with more fabled landmarks, streets, squares, and commercial facilities than any other cities except London and Paris. This is also the cultural and media capital of the United States, which means that the city's influence constantly radiates across the planet thanks to New York-based television networks, newspapers and magazines, book publishers, fashion and design leaders, and artistic and new media trendsetters.

At the metropolitan scale, New York forms the center of a vast urban region, 250 kilometers (150 mi) square (population: 22.6 million), which sprawls across parts of three States in the heart of Megalopolis. That outer city has become a giant in its own right with its population of 14-plus million, massive business complexes, and flourishing suburban downtowns. Thus, despite its global connections, New York is caught in the currents that affect U.S. central cities today. Its ghettos of disadvantaged populations continue to expand; its aging port facilities and industrial base are deteriorating; its corporate community is increasingly stressed by the high cost of doing business in ultra-expensive Manhattan—and its Wall Street banking and finance community may never regain its international prominence following the economic crisis that began in 2008.

© H. J. de Blij, P. O. Muller, and John Wiley & Sons, Inc.

These challenges notwithstanding, New York today is seeking to reinforce its position as a great, sustainable, twenty-first-century city that will continue to lure the hypermobile, super-affluent global elite. Its bold new plans are geared to meet its biggest problems head-on: a major increase in affordable housing for a growing population, across-the-board upgrading of a disintegrating infrastructure, and a sizeable expansion of parks and other outdoor recreational spaces.

The Evolving Metropolis

The forces that shaped this regional urban system also modified the internal structure of North America's cities, although to a greater extent in the United States than (initially at least) in Canada. Mobility improved when horse-drawn trolleys were replaced by electric streetcars, but it was the mass introduction of the automobile after World War I that presaged the change from compact city to widely dispersed metropolis. After World War II the construction of the interstate highway system in the United States, augmented by intrametropolitan radial expressways and beltways, set the stage for unprecedented suburbanization and the evolution of residential suburbia into a complete outer city with its own businesses and industries, schools, hospitals, and other amenities. As the newly urbanized suburbs increasingly captured major economic activities, many large cities saw their comparative status diminish to that of coequal. Their once-thriving and dominant central business districts (CBDs) were now all but reduced to serving the less affluent populations that increasingly dominated the central city's close-in neighborhoods.

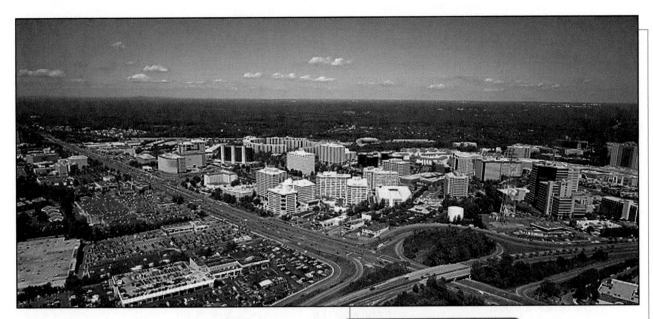

In the United States, where foreign immigration had been curtailed in the 1920s even as cotton production in the South went into decline, manufacturers in the North had recruited millions of unemployed African American workers to take jobs in their factories. This strongly impacted the social geography of industrial cities because most whites were unwilling to share their living space with the ethnically different newcomers. The result was the involuntary segregation of these latest immigrants. By the 1950s, their mostly inner-city neighborhoods had

9 become large and expanding **ghettos**—prompting millions of white central-city residents to move to the fast-developing suburbs, which further reinforced the trend toward a racially divided urban society. It was a cycle that proved difficult to reverse: as the central cities' more affluent residents moved beyond the municipal limits, the central cities' tax base shrank; this reduced the services and amenities the cities could afford; and this in turn impelled even more residents to leave. Businesses closed or followed their customers, the local job base shrank further, and severe distress afflicted the CBDs. The fate of Detroit is often cited as exemplifying this sequence of events, but to a certain degree all major U.S. cities suffered.

The suburbs, meanwhile, received not only a massive infusion of relocating urban residents but migrating outsiders as well. That happened because they became full-

10 fledged **outer cities** whose jobs (in office parks, shopping centers, high-tech industrial complexes, hotels, restaurants, and entertainment facilities) attracted workers and professionals from inside as well as outside the transforming suburban ring. Fly into a contemporary U.S. or

11 Canadian city, and you will see the high-rises of **suburban downtowns** encircling the old central-city CBD, some of

FROM THE FIELD NOTES

"Monitoring the urbanization of U.S. suburbs for the past four decades has brought us to Tyson's Corner, Virginia on many a field trip and data-gathering foray. It is now hard to recall from this 2008 view that less than 50 years ago this place was merely a near-rural crossroads. But as nearby Washington, D.C. steadily decentralized, *Tyson's* capitalized on its unparalleled regional accessibility (its Capital Beltway location at the intersection with the radial Dulles Airport Toll Road) to attract a seemingly endless parade of high-level retail facilities, office complexes, and a plethora of supporting commercial services.' [Today, this suburban downtown ranks among the largest business districts in all of North America. But it also exemplifies urban sprawl, and in 2009 its developers, tenants, and county government formed a coalition to transform Tyson's into a true city characterized by smarter, greener development. The elements of the new plan include much higher residential densities, mass transit in the form of four stations on a new Metro line connecting to Washington and Dulles Airport, and major new facilities to accommodate pedestrians and bicycles. Details can be found in a prominent article in the June 22, 2009 issue of *Time*.]
© Courtesy of Kgp Design Studio

www.conceptcaching.com

them boasting their own impressive skylines (see photo above). In Canada, where the process has had more to do with lower land values outside the city centers than with ethnic separation, the spatial effect has become similar, although most Canadian cities still remain more compact than the far-flung U.S. metropolis.

Thus the overall structure of the modern North American metropolis is multinodal, and resembles a pepperoni pizza (Fig. 3-10) in its ideal form. The traditional CBD still tends to be situated at the center, much of its former cross-traffic diverted by beltways; but the outer city's CBD-scale nodes are ultramodern

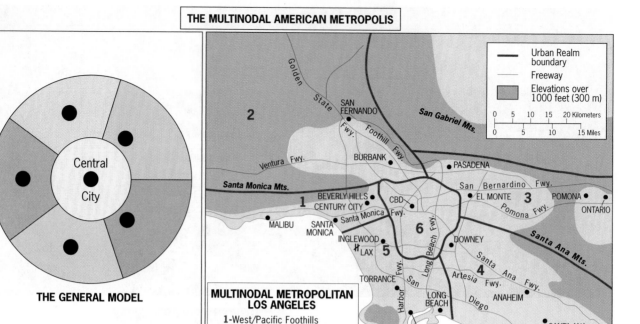

THE MULTINODAL AMERICAN METROPOLIS

THE GENERAL MODEL

MULTINODAL METROPOLITAN LOS ANGELES
1-West/Pacific Foothills
2-Northwest/San Fernando
3-East/San Gabriel Valley
4-Southeast/Orange County
5-Southwest/Pacific Lowlands
6-Central Los Angeles

A   B

FIGURE 3-10

© H. J. de Blij, P. O. Muller, and John Wiley & Sons, Inc.

and thriving. Efforts to attract businesses and more affluent residents back to the old CBD sometimes involve the construction of multiple-use high-rises that displace low-income residents, resulting in conflicts and lawsuits. The *gentrification* of nearby, crumbling neighborhoods raises real estate values as well as taxes and tends to drive lower-income locals from their homes. Reversing the tide of suburbanization and restoring the old CBDs may be unattainable objectives in the current cultural and economic climate.

CULTURAL GEOGRAPHY

North America is a realm of two countries and, to be sure, Canadians and Americans have their differences. You can even argue over names: if this entire hemisphere is called the Americas, aren't Canadians, Mexicans, and Brazilians, and others Americans too? Some geographers take this very seriously and propose that citizens of the United States call themselves "United Statesers," but that is not likely to catch on. And the United States is the only state in the Americas that actually has *America* in its official name. Furthermore, the USA is *not* the only country in the hemisphere to call itself a United States: the official name of Mexico is the United Mexican States.

So in this chapter we refer to Canadians and Americans, and when it comes to culture, their two countries have a lot in common. For all their shared multilingualism, in most areas English is both countries' *lingua franca*. For all the religions followed by minorities, the great majority of both Canadians and Americans are Christians. Both states are stable democracies; we have already noted that both have federal systems of government. Both are full partners in the North American Free Trade Agreement. Standards of living are very much the same; have a look at the social indices of the two societies in the Data Table at the end of the book, and it is clear that they have much in common. The boundary between the two states is porous, even after the events of 9/11. Americans living near the border shop for more affordable medicines in Canada; Canadians who can afford it seek medical treatment in the United States. Americans earn somewhat more annually than Canadians do, but both economies rank near the top of the high-income category in Figure G-11. Canada ranks higher (better) on the corruption index, and Canadians live longer.

As is inevitable, given demographic and economic realities, American cultural influence radiates more strongly into Canada than Canadian norms infuse America. Like America but unlike most other members of the British Commonwealth, Canada drives on the right, as

America does. Baseball and a modified version of American football are popular sports in Canada; in 1992, the Toronto Blue Jays became the first team outside the U.S. to win the World Series. Canada's national sport, ice hockey, now has more NHL teams in America than in Canada, but curling, a slower-paced Canadian pastime, has not exactly captured the American imagination. In addition, American media have a far stronger presence in Canada than Canadian media do in America.

Americans visiting Canadian cities (and vice versa) find themselves in mostly familiar settings. Canadian cities have fewer impoverished neighborhoods, no ethnic ghettos (although low-income ethnic districts do exist), lower crime rates, and, in general, better public transportation; but rush hour in Toronto very much resembles rush hour in Chicago. Central business districts have fared better in Canada than, on average, in America, although in Canada, too, new outer cities challenge older CBDs for primacy. As for rural areas, driving eastward from Michigan through southern Ontario to upstate New York does not involve a sharp contrast in cultural landscapes.

Cultural Foundations: Language and Religion

Americans and Canadians founded their societies on agriculture and rural life, then carried their values and beliefs into their evolving towns and cities. The urban geographer Brian Berry identified America's hybrid culture as exhibiting a continuing desire to be near nature, a liking for things new, an ability to move, a sense of individualism, an aggressive pursuit of goals, a need for societal acceptance, and a firm sense of destiny. These qualities are not unique to modern North American cultures, of course, but in combination they create a particular and pervasive mind-set that is reflected in many ways in this geographic realm. Berry discerned such cultural traits at various stages throughout the evolution of the urbanized life most Americans now lead—from mobility inherent in the shift to the (newer) suburbs where nature is a larger part of life to the now-eroding nuclear family's intense pursuit of educational and other goals and their aspirations for acceptance into a higher stratum of society.

Facilitating these aspirations is *language*. None of these goals would be within reach for so many were it not for the use of English throughout most of the realm. As we noted in the Introduction, North America lies at the very heart of the global core, at a time when English has become the language of globalization. In Europe, language inhibits the mobility so routinely practiced by Americans: workers moving from Poland to Ireland find

themselves at a disadvantage in competition with immigrants from Ghana or Sri Lanka. In North America, a worker from Arkansas would not even consider language to be an issue when applying for a job in Calgary.

But the near-universality of English in North America is corroding, and according to recent estimates more than one-fifth of the realm's population speaks a language other than English at home. This includes the French-speaking minority in Canada as well as the Spanish-speaking minority in the United States (the Hispanic minority is now larger than Canada's entire population) and also smaller Asian immigrant populations. Yet people whose mother tongue is a language other than English may still have ability in English (many Canadians are schooled in both official languages), so English remains the dominant medium of interaction. Interestingly, English in the United States is undergoing a change that is affecting it worldwide: in areas where it is the second language it is blending with the local tongue, producing hybrids just as is happening in Nigeria ("Yorlish"), Singapore ("Singlish"), and the Philippines ("Taglish"). In this respect, English may become the Latin of the present day: Latin, too, produced blends that eventually consolidated into Italian, French, Spanish, and the other Romance languages.

Also reflecting the cultural values cited earlier is the role of *religion*, which sets American (more so than Canadian) culture apart from much of the rest of the high-income global core. The overwhelming majority of Americans express a belief in God, and a large majority regularly attend church, in contrast to much of Europe. Religious observance is a virtual litmus test for political leaders; no other "developed" country prints "In God We Trust" on its currency. Sects and denominations are important in the life of Christian-dominated North America. Figure 3-11 shows the mosaic of Christian faiths that blankets the realm. A map at the scale of Figure 3-11 can only suggest the broadest outlines of a much more intricate pattern: Protestant denominations are estimated to number in the tens of thousands in North America, and no map could show them all. Southern (and other) Baptists form the majority across the U.S. Southeast from Texas to Virginia, Lutherans in the Upper Midwest and northern Great Plains, Methodists in a belt across the Lower Midwest, and Mormons in the interior West centered on Utah. Roman Catholicism prevails in most of Canada as well as the U.S. Northeast and Southwest, where ethnic Irish and Italian adherents form majorities in the Northeast and Hispanics in the Southwest.

Behind Figure 3-11 lie histories of proselytism, migration, conflict, and competition, but tolerance of diverse religious (even nonreligious) views and practices is a hallmark of this realm. Changing religions is not, as

FIGURE 3-11

© H. J. de Blij, P. O. Muller, and John Wiley & Sons, Inc.

it is in parts of the Islamic periphery, a matter of life and death. Intermarriage occurs frequently. Acceptance of non-Christian faiths and accommodating their needs is the norm rather than the exception. The more critical debate in this changing realm involves the separation between (Christian) church and state, a partition that, especially in the United States, is under pressure at a time when the mosaic of religions is diversifying.

THE CHANGING GEOGRAPHY OF ECONOMIC ACTIVITY

Make no mistake: economic activity is substantially a matter of culture. Among those North American cultural foundations enumerated earlier, the aggressive pursuit of goals was cited prominently—and economic goals rank at the top in this highly material realm. Virtually unrestrained pursuit of real estate ownership, even by those who would never be able to afford it and aided by remorseless companies enabled by an economic climate of deregulation, had much to do with the economic collapse that started in the United States in 2008. Energetic globalizers sometimes complain about the reluctance of some cultures in the global periphery to embrace the benefits of the corporate world. They are right: some societies are slow to accommodate change, innovation, newness, and uncertainty. Not Americans.

North America's economic geography has been in transition ever since the Industrial Revolution arrived on its shores, and the evidence is on the map. This is not a small realm: North America as here defined is more than three times the size of Europe. But it is unified by the world's most efficient four-lane highway system; it still has a major railroad network, and it is interconnected by airlines from Alaska to Florida and from Newfoundland to California. Overcoming the obstacles of distance and terrain was the great accomplishment that enabled the formation of a realmwide spatial economy that took maximum advantage of agricultural, mineral, energy, and other resources to create the world's leading economy.

Industries of all kinds were the key, but as we have noted certain industries in North America are in decline and a new economy is once again developing. Economic geographers call this the **postindustrial** era, and nobody is sure where exactly it will lead. **12**

The Spatial Economy

In economic geography, attention focuses on the unifying question—*where* is it happening? In technical language, this question involves the locational analysis of productive activities. There is a big difference between, say, farming and mining on the one hand, and teaching and banking on the other, but all are industries. Here are the four major categories of productive activities:

13 • **Primary:** the extractive sector of the economy in which workers and the environment come into direct contact, especially in *mining* and *farming*.

14 • **Secondary:** the sector in which raw materials are transformed into finished industrial products, the process of *manufacturing*.

• **Tertiary:** the sector encompassing a wide range of **15** activities such as retailing, education, finance, entertainment, and hospitality, collectively known as *services*.

• **Quaternary:** today's increasingly dominant sector, **16** involving the collection, processing, and manipulation of *information*.

As is obvious from this list, each of these activities dominates developing economies at successive stages of their growth. In industrial and postindustrial societies, primary activities employ just a tiny fraction of the labor force. But in societies still undergoing industrialization, the majority of workers are in primary industries. For example, in the United States at the beginning of this new century, only 1.9 percent of workers were engaged in farming. In India, on the other hand, 56.7 percent of all workers still labored on the land to produce food. At the same time, in the United States 82 percent of workers were employed in the services (tertiary) and information (quaternary) sectors; in India, that figure stood at 25.2 percent.

In North America, agriculture gave way to manufacturing as the dominant activity around 1900. Services surpassed manufacturing in the 1950s, but the quaternary sector overtook services only two decades later. That growth trend has continued, and today the North American realm has transformed into what many call a "new economy."

The Postindustrial Revolution

The signs of postindustrialism are unmistakable throughout the United States and Canada: this is the "computer age," the "high-tech era"—expressed not only as the "new economy" but also the new society. The term *postindustrial* tells us what the North American economies no longer are. When politicians seeking election complain about jobs leaving the country, they focus on the things North American factories no longer can produce competitively in a globalizing world. But new jobs are created in large numbers in the quaternary sector. Once again, the workforce needs retraining.

High-technology, white collar, office-based activities are the leading growth industries of the postindustrial economy. This means that the jobs they offer are not tied to resources or infrastructure, but can be done where natural and social environments are attractive. Northern California's *Silicon Valley*—the world's leading center for computer research and development and the headquarters of the U.S. microprocessor industry—is an example. Proximity to Stanford, a world-class research university, not far from cosmopolitan San Francisco, with the presence of a large pool of highly educated and skilled workers, a strong business culture, a lot of available capital, good housing, and a scenic area with good weather combined to make Silicon Valley a prototype for similar developments elsewhere, and not just in North America. Under different names (*technopolis* is used in Brazil, France, and Japan; *science park* in China, Taiwan, and South Korea) such ultramodern, campus-like complexes symbolize the postindustrial era just as the smoke-belching factory did the industrial age.

One of the hallmarks of the postindustrial revolution is the emergence of the Internet, which is being called the defining technology of the twenty-first century. The continuing growth of the Internet is being driven by the need to transmit ever greater quantities of data at ever higher speeds. Fiber-optic cables are best able to meet these needs and have given rise to a network of linkages that connect information-rich centers across the planet. North America, and more specifically the United States, is the leading component of this new global infrastructure.

The North American realm is therefore undergoing yet another momentous transformation, whose outcome will affect the entire world. As we will see, however, even modernizing North America still has its own cores and peripheries, and its own diagnostic regional and cultural landscapes.

POINTS TO PONDER

- The United States does not have an official language. Should it?

- Global warming may create an ice-free passage around northern Canada. Goodbye Panama Canal?

- Quebec's off-again, on-again independence movement is off again in the aftermath of the latest provincial election.

- Three hundred million people in the USA as of October 2006. Get ready for 400 million in 2043.

Regions of the Realm

In the discussion of the regional geography of Europe in Chapter 1, we took note of the core-periphery relationships between clusters of countries forming the more powerful, and the weaker, states of that realm. North America, consisting of only two countries, exhibits a much narrower range of economic and social indices. But as much as Canada and the United States have in common, there is ample reason to focus on each individually. In the discussion that follows, therefore, we highlight the regional geographies of two concordant neighbors—in the knowledge that Canadians tend to know more about the United States than Americans know about Canada. Next we define what we see as the emerging regions of the realm as a whole, well aware that others have tried this too, all in the effort to learn more about this remarkable corner of the world.

CANADA

Territorially the second-largest state in the world after Russia, Canada is administratively divided into only 13 entities—10 provinces and 3 territories (Russia contains nearly 90; the slightly smaller United States has 50). The

MAJOR CITIES OF THE REALM

City	Population (in millions)*
Atlanta	5.3
Boston	6.2
Chicago	10.2
Dallas—Ft. Worth	6.1
Denver	3.3
Detroit	5.4
Houston	5.6
Los Angeles	18.1
Montreal, Canada	3.7
New York	22.6
Ottawa, Canada	1.1
Philadelphia	6.4
San Diego	3.4
San Francisco	7.9
Seattle	4.1
Toronto, Canada	5.3
Vancouver, Canada	2.1
Washington, DC	5.8

*Based on 2010 estimates.

FIGURE 3-12

© H. J. de Blij, P. O. Muller, and John Wiley & Sons, Inc.

provinces—where 99.7 percent of all Canadians live—range in dimensions from tiny, Delaware-sized Prince Edward Island to Quebec, nearly twice the size of Texas (Fig. 3-12).

As in the United States, the smallest provinces lie in the northeast. Canadians call these four the Atlantic Provinces (or simply Atlantic Canada): Prince Edward Island, Nova Scotia, New Brunswick, and the mainland-and-island province named Newfoundland and Labrador. Gigantic, mainly Francophone (French-speaking) Quebec and populous, strongly urbanized Ontario, both flanking Hudson Bay, form the heart of the country. Most of western Canada (which is what Canadians call everything west of Ontario) is organized into three Prairie Provinces: Manitoba, Saskatchewan, and Alberta. In the far west, beyond the Canadian Rockies and extending to the Pacific Ocean, lies the tenth province, British Columbia.

Of the three territories in Canada's Arctic North, Yukon, adjacent to eastern Alaska, is the one you cross if you take the adventurous Alcan Highway from Dawson Creek, B.C., to Fairbanks, Alaska via Whitehorse. Yukon also is the smallest of the three, but nothing is small here in Arctic Canada. With the Northwest Territories and recently created Nunavut, the three Territories cover almost 40 percent of Canada's total area. But what *is* small is the population: only about 100,000 people inhabit this vast, frigid frontier zone.

The case of Nunavut merits special mention. Created in 1999, this newest Territory is the outcome of a major aboriginal land claim agreement between the *Inuit* people (formerly called Eskimos) and the federal government, and encompasses all of Canada's eastern Arctic as far north as Ellesmere Island (Fig. 3-12), an area far larger than any other province or territory. With about one-fifth of Canada's total area, Nunavut—which means "our land"—has

Located at the mouth of the Fraser River and alongside one of the best natural harbors on the British Columbia coast, Vancouver's regional urban prominence was assured when it became the western terminus of Canada's transcontinental railroad 125 years ago. Today it ranks as the third-largest Canadian metropolis after Toronto and Montreal, and features an impressive downtown skyline framed by the equally imposing skyline of the closest range of the (Pacific) Coast Mountains. The domed stadium at the right is BC Place, the world's largest air-supported sports facility. It is best known as the headquarters of the 2010 Winter Olympic Games, site of the opening and closing ceremonies for the myriad events taking place in the spectacular surrounding highlands that make southwesternmost Canada one of the realm's scenic glories. © Chris Cheadle/Alamy

31,000 residents (of whom 80 percent are Inuit). Not only does Nunavut as a whole have a substantial degree of native self-government, but within it the Inuit have outright ownership of an area half the size of Texas.

As we note in Chapter 12, the importance of this northern Canadian frontier is growing as climate change reduces the area of permanent as well as seasonal ice north of the Arctic Circle. The so-called Northwest Passage winds its way between islands now part of Nunavut, and in Canada's capital concerns are growing that Canadian ownership of this potentially crucial waterway may be challenged.

Canada's Spatial Structure

Earlier we noted that most of Canada's population clusters in a discontinuous ribbon, 300 km (ca. 200 mi) wide, nestled against the U.S. border, but we should not discount its more northerly outliers and the functions these perform. As Figure 3-7 reminds us, much of Canada's energy reserves, in the form of oil, tar sands, and natural gas, lies far to the north of this ribbon, which is why Edmonton and Fort McMurray are among exceptions. Major hydroelectric projects in Quebec and new energy- and mineral-related activity in Newfoundland and Labrador also lie well to the north of Canada's southern heartland.

Nevertheless, any map of Canada's human geography bears out the generalization and emphasizes the importance of environment. It seems as though nature is the core and humanity is the periphery—nature in the form of a vast, taiga-and-tundra interior and humanity on its margins. From Quebec City and Montreal, Ottawa and Toronto, Winnipeg and Calgary to Vancouver and Victoria in British Columbia's southernmost corner, Canadians cluster in the south and along ocean shores. Most of Ontario's 13.1 million residents and the great majority of Quebec's 7.8 million concentrate in these provinces' lowest latitudes. This pattern creates cross-border affinities with major American cities in several places (Toronto-Buffalo, Windsor-Detroit, Vancouver-Seattle), although none of this happens in the three Prairie Provinces of the Canadian West, which adjoin North Dakota and Montana.

Canada's capital, Ottawa, unlike the capitals of the United States and a number of other federations, does not lie in its own federal territory separated from the country's subnational administrative units. Rather than establishing a District of Columbia-type administrative entity, the Canadians were content to locate their capital astride the Ottawa River, which marks the boundary between Ontario and Quebec. Even though Parliament and other leading capital-city functions are situated on the Ontario side in Ottawa itself, many federal government operations have long been based in Gatineau and nearby communities on the Quebec side of the border.

Cultural Contrasts

Though comparatively small, Canada's population is markedly divided by culture and tradition, and this division has a pronounced regional expression. Just under 60 percent of Canada's citizens speak English as their mother tongue, 23 percent speak French, and the remaining

AMONG THE REALM'S GREAT CITIES . . . **Toronto**

TORONTO, CAPITAL OF Ontario and Canada's largest metropolis (5.3 million), is the historic heart of English-speaking Canada. The landscape of much of its center is dominated by exquisite Victorian-era architecture and surrounds a healthy downtown that is one of Canada's leading economic centers. Landmarks abound in this CBD, including the SkyDome stadium (Rogers Centre), the famous City Hall with its facing pair of curved high-rises, and mast-like CN Tower, which at 553 meters (1815 ft) is the world's tallest freestanding structure. Toronto also is a major port and industrial complex, whose facilities line the shore of Lake Ontario.

Livability is one of the first labels Torontonians apply to their city, which has retained more of its middle class than central cities of its size in the United States. *Diversity* is another leading characteristic because this is North America's richest urban ethnic mosaic. Among the largest of more than a dozen thriving ethnic communities are those dominated by Italians, Portuguese, Chinese, Greeks, and Ukrainians; overall, Toronto now includes residents from about 170 countries who speak more than 100 languages; and the immigrant inflow continues strongly, with those born outside Canada constituting nearly half of the city's population (in all the world only two other million-plus cities exhibit a higher percentage). *Vibrancy* is yet another hallmark of this remarkable city, whose resiliency was tested but not broken by the SARS-virus outbreak of 2003.

Toronto has worked well in recent decades, thanks to a metropolitan government structure that fostered cen-

© H. J. de Blij, P. O. Muller, and John Wiley & Sons, Inc.

tral city–suburban cooperation. But that relationship is increasingly stressed as the outer city gains a critical mass of population, economic activity, and political clout. Toronto has responded with plans to become a "city-state," but first it must overcome problems of municipal finance and aging infrastructure.

17 percent other languages. About 18 percent of the population is thoroughly bilingual in English and French, but about four times as many French speakers are fluent in English as English speakers have full command of French (40 percent as compared to 10 percent). The spatial clustering of more than 85 percent of the country's Francophones in Quebec accentuates this Canadian social division along linguistic and ethnic lines (Fig. 3-12). Quebec's population is 80 percent French-Canadian, and Quebec remains the historic, traditional, and emotional focus of French culture in Canada. Over the past half-century a strong nationalist movement has emerged in Quebec, and at times it has demanded outright separation from the rest of Canada. In 1995, independence was narrowly defeated in a provincial referendum, and since then support for the notion has fallen—but could rise again. This ethnolinguistic division continues to be the litmus test for Canada's federal system, and the

issue continues to pose a latent threat to the country's future national unity.

As we noted earlier, the American War of Independence had realmwide consequences of which Quebec's status is only one. British North America (the name *Canada*, derived from an indigenous word meaning "settlement," was not yet in use) received an influx of many thousands of English refugees. To forestall ever-greater difficulties between the English and French the British Parliament in 1791 divided the region affected into Upper Canada (the Ontario of the future) from a point upstream from Montreal westward, and Lower Canada (Quebec). Although this plan did defuse the threat of war, it did not work well politically. After several other tries, in 1867 London finally formulated the British North America Act establishing a four-member Canadian federation that, between 1870 and 1999, saw all the present provinces and territories join. To this day, the British monarch remains the head of state, represented by the Governor General. It is one of the issues that riles Quebec nationalists.

In recent years a number of factors have contributed to an overall decline in support for Quebec's secession. These include: (1) the implementation of provincial laws that firmly established the use of French and the primacy of Québécois culture; (2) a substantial and reassuring increase in the bilingual abilities of Quebec's Anglophones (English speakers); and (3) the arrival of a new wave of immigrants—some Francophone, some Anglophone—from Asia, Africa, the Caribbean, and Europe, who have accelerated the weakening of language barriers by settling in both French-speaking and English-speaking neighborhoods. At the same time, these immigrants diverted local preoccupation with the main issue by exposing intolerance for multiculturalism in Quebec—exemplified by local restrictions on such practices as the wearing of face veils or the carrying of ceremonial weapons during festivals.

Analyses of public opinion in Quebec suggest that what Quebecers really want is a better relationship with the rest of Canada that falls short of independence. The Québécois want to regard themselves as a *distinct nation* and want to be treated accordingly. That is precisely the recognition they received under a parliamentary resolution in 2006—but with the added caveat . . . *within a united Canada*. Thus continues a struggle whose implications, as we will see, reach far beyond the English-French divide.

The Ascendancy of Indigenous Peoples

On the wider Canadian scene, the culture-based Quebec struggle has stirred ethnic consciousness among the country's 1.4 million native (First Nations, Métis, and Inuit) peoples. They, too, have received a sympathetic hearing in Ottawa, and in recent years breakthroughs have been achieved with the creation of Nunavut and local treaties for limited self-government in northern British Columbia.

A foremost concern among these peoples is that their aboriginal rights be protected by the federal government against the provinces. This is especially true for the First Nations of Quebec's northern frontier, the Cree, whose historic domain covers more than half of the province of Quebec as it appears on current maps. Administration of the Cree was assigned to Quebec's government in 1912, a responsibility that a move toward independence might well invalidate. In any case, the Cree would probably be empowered to seek independence themselves. This would leave the French-speaking remnant of Quebec with only about 45 percent of the province's present territory.

As Figure 3-12 shows, the territory of the Cree is no unproductive wilderness: it contains the James Bay Hydroelectric project, a massive scheme of dikes and dams that has transformed much of northwestern Quebec and generates electric power for a huge market within and outside the province. In 2002, after the federal courts supported its attempts to block construction of the project, the Cree negotiated a groundbreaking treaty with the Quebec government whereby this First Nation dropped its opposition in return for a portion of the income earned from electricity sales. The Cree also secured the right to control their own economic and community development.

Centrifugal Forces

The events of the past four decades have had a profound impact on Canada's national political geography, and not only in Quebec. Leaders of the western provinces objected to federal concessions to Quebec, insisting that the equal treatment of all ten provinces is a basic principle that precludes the designation of special status for any single one. Recent federal elections and opinion surveys have revealed the emerging fault lines that surround Quebec and set the western provinces and British Columbia off from the rest of the country. Even the Ontario-led center and the eastern bloc of Atlantic Provinces voted divergently, and today the politico-geographical hypothesis of "Four Canadas" seems to be gaining credibility. Thus, with national unity always a challenge in this comparatively young federation, Canada confronts forces of *devolution* that threaten to weaken the state.

Economic Geography

Canada's spatial economy, like that of the United States, is supported by a rich, diversified resource base. Earlier we noted the mineral content of the Canadian Shield and the oil reserves of Alberta (Fig. 3-7). Canada also has long been a major agricultural producer and exporter, especially of wheat and other grains from its breadbasket in the Prairie Provinces. Postindustrial developments have caused employment in the manufacturing sector to decline substantially, with southern Ontario's industrial heartland the most severely affected area. On the other hand, Canada's robust tertiary and quaternary sectors now employ more than 70 percent of workers and create new economic opportunities. Southern Ontario is also one of the beneficiaries because it is home to the country's leading high-technology, research-and-development complex around Waterloo and Guelph, two university towns about 100 kilometers (60 mi) west of Toronto.

Trade and NAFTA

Canada's economic future will also be strongly affected by the continuing development of its trade relationships. The landmark United States-Canada Free Trade Agreement, signed in 1989, phased out most tariffs and investment restrictions between the two countries, whose annual cross-border flow of goods and services is the largest in the international trading arena. As noted earlier, in an average year more than four-fifths of Canadian exports go to the United States, from which Canada derives at least three-fifths of its imports.

In 1994, these economic linkages were further strengthened through the implementation of the ***North American Free Trade Agreement (NAFTA)***, which consolidated the gains of the 1989 pact and opened major new opportunities for both countries by adding Mexico to the trading partnership. Because these free-trade agreements have the effect of redirecting the Canadian economy, they tend to weaken domestic east-west linkages and strengthen north-south international ties. Since many local cross-border linkages built on historical and geographical commonalities are already well developed, they can be expected to intensify in the future: the Atlantic Provinces with neighboring New England; Quebec with New York State; Ontario with Michigan and adjacent Midwestern States; the Prairie Provinces with the Upper Midwest; and British Columbia with the (U.S.) Pacific Northwest. Such functional reorientations constitute yet another set of potentially powerful devolutionary challenges confronting the Ottawa government because the deepest economic fault lines coincide with those that politically delimit the "Four Canadas."

Canada must balance the economic power and cultural influence of its southern neighbor with its own interests and goals, and it has shown a strong measure of independence when it needed to do so. During the crisis surrounding 9/11, Canadians graciously hosted thousands of American airline passengers involuntarily disembarked at Canadian airports, their flights to the United States grounded on Washington's orders. But in 2003, when then-president George W. Bush initiated the invasion of Iraq, Canada refused to participate. Canada did, however, join the multinational force simultaneously operating in Afghanistan. In the international arena, Canada has an enviably high reputation for sound judgment and constructive action.

THE UNITED STATES OF AMERICA

Nothing in the United States compares with the devolutionary challenges Canada has faced throughout its evolution as a modern state. When we examined Figure 3-6, we noted that the United States has about four times as many subnational units as Canada, but none has mounted a campaign for secession of the kind Quebec has, and in only one (Hawai'i) have indigenous people demanded recognition of a part of the State as sovereign territory. Like Canada, the United States evolved incrementally, the last major acquisition involving Alaska, purchased from Russia in 1867 for a nickel a hectare (2 cents an acre). States joined the Union intermittently, the last two being Alaska and Hawai'i in 1959 (the latter was originally annexed in 1900). Voters in the Commonwealth of Puerto Rico have the option of approving Statehood, but have so far refrained from making the island the fifty-first State. In Canada, Newfoundland did not join the federation until 1949, and Nunavut's status did not become official until 1999.

On the other hand, nothing in Canada compares numerically to the immigration streams that continue to transform the United States. Migrants by the millions have arrived in surges, imprinting their traditions on the country's cultural mosaic. From Holland, Michigan and Denmark, Wisconsin to Dublin, New Hampshire and Berlin, Pennsylvania, the Dutch, Danes, Irish, Germans, and other Europeans left their marks on the ground and on the map. No country in the world has more place names starting with "New" than the United States.

Other place names, however, tell a different story. Very few African place names commemorate the arrival of hundreds of thousands of Africans in bondage, although some streets and neighborhoods now do. And in the Southwest, many Hispanic place names reflect Mexico's ouster from a vast region acquired in 1848 by the United States at the end of the Mexican War through the Treaty of Guadalupe Hidalgo, named after the town near Mexico City where the treaty was signed. This treaty established the Mexican-American border at the Rio Grande and assigned to the United States over 1.3 million square kilometers (525,000 sq mi) of Mexican territory including present-day California, Arizona, New Mexico, Nevada, western Colorado, Texas (which had already declared independence from Mexico), and Utah (Fig. 3-13). In Mexico, this disastrous event, which precipitated a civil war, is much more vivid in the national consciousness than it is in the United States. Some Mexican nationalists proclaim that much of the current migration across the U.S.-Mexican border, illegal though it may be, is a matter of "resettlement" of unjustly lost lands.

This is indeed the era of migration, and the U.S. Census Bureau tries to keep track of the process. You will often see projections charting the decline of "whites" in America's population, down to 75.1 percent in 2000 from 87.5 percent in 1980, with predictions that "whites" will cease to be a majority within a few decades. But in a population

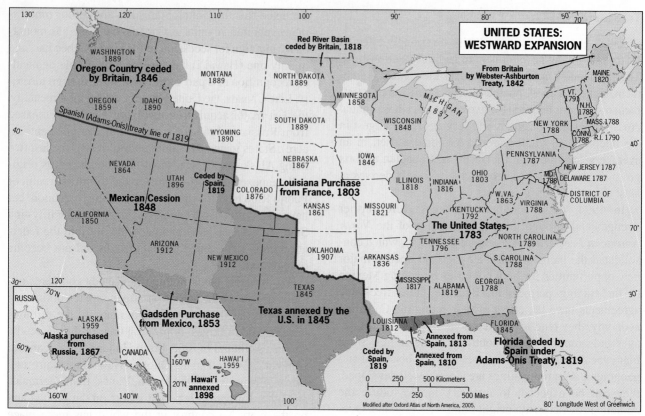

FIGURE 3-13

© H. J. de Blij, P.O. Muller, and John Wiley & Sons, Inc.

as diverse as America's, this is a rather meaningless identification, and we should try to forget about it. Far more important is the prospect of Americans' aging and its implications, how many Americans remain poor, and how many will be adequately educated by mid-century.

Ethnicity and Geography

The sheer dimensions of immigration into the United States, from virtually all corners of the world, creates an increasingly complex ethnic and cultural mosaic that tests "melting pot" assertions at every geographic scale. With more than 300 million residents, America has the third-largest national population in the world, and some States have populations larger than many countries (California alone has a population larger than Canada). Today there are more people of African descent in the United States than live in Kenya. There are just about as many Hispanic residents in America as there are in Spain. In short, there are sufficient immigrant numbers in America for them to create durable societies within the national society. The challenge for the United States is to ensure that the great majority of the immigrants become full participants in that larger society.

This certainly has not happened with the Native American peoples. They were not immigrants (or rather, they are descendants of the first immigrants who arrived at a time when there was no one else around). Of the four maps in Figure 3-14, the one showing the percentage distribution of Native Americans reflects the story of the entire realm: pushed relentlessly westward by the European settlers or overpowered when they already inhabited the West, they found themselves in one of more than 300 reservations or, mostly penuriously, in society at large. Their relatively small numbers (4.8 million, which includes Alaska and counts those who are Native American or Native American mixed with another ethnic ancestry) include a mere 200,000 in North Carolina and New York State combined. The two largest surviving nations (see Fig. 3-4) are the Navajo and the Cherokee. America's oldest residents and earliest immigrants did not find each other in the melting pot.

The next arrivals, the forced African immigrants, appeared to have even less prospect of participating in the larger society, even after the Civil War terminated slavery. Earlier we noted the major internal migration of African Americans from the declining cotton plantations of the South to the factories of the industrialized North,

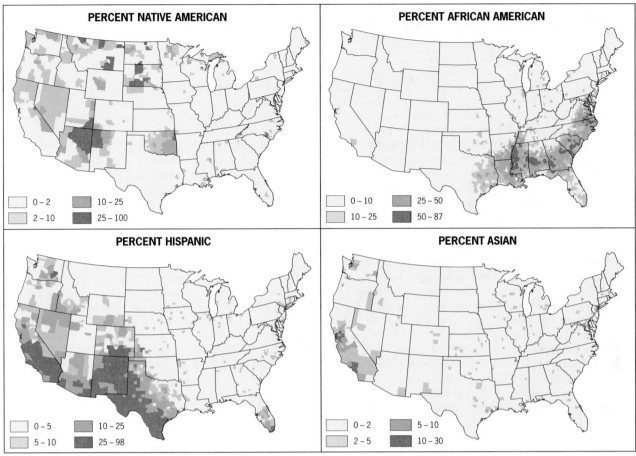

FIGURE 3-14

© H. J. de Blij, P. O. Muller, and John Wiley & Sons, Inc.

but it was not until after the civil rights movement of the 1960s that African American horizons brightened. Even so, few at the time would have believed that an African American citizen would be elected president of the United States in 2008. As the map shows, even today the African American sector of the population remains heavily concentrated in a zone from eastern Texas to Maryland. Indeed, the most recent census data report that the States of Mississippi, Louisiana, South Carolina, Georgia, and Maryland to this day have the country's largest African American minorities as a percentage of their populations. At this small scale, however, significant African American populations in metropolitan areas cannot be shown, but constitute a major presence in the cities of the Manufacturing Belt and the West Coast. In terms of the fabric of American culture, African Americans take their place in virtually all arenas, from politics to music to education.

Forerunners of what is now the largest of all minorities in America arrived in the United States before the mid-nineteenth century (New Mexico's upper Rio Grande Valley has had a significant and prosperous Hispanic population since it was annexed as a territo-ry following the 1848 Mexican War). But even as recently as 1980 this minority numbered less than 15 million. Since 1980, however, the number of Hispanics has more than tripled and now exceeds 45 million, outnumbering the African American minority and changing the ethnic-cultural map of America. And although Hispanics are now dispersing into many parts of the country—including rural areas in the Southeast and Midwest as well as urban communities from Massachusetts to Washington State—they remain strongly clustered in the Southwest, California, and Florida. In fact, along virtually the entire multi-State tier north of the Mexican border, America has become a cultural transition zone, where immigration, legal as well as illegal, has become a contentious issue (see the Regional Issue Box, "Immigrants: How Many Can North America Accommodate?"). Language, not a point of contention for the African American community, is an emotional matter for Anglophone Americans who fear erosion of the status of English in their communities and in the country as a whole.

The Asian immigrant population, whose arrival rate has also risen in recent decades, is geographically the

Immigrants: How Many Can North America Accommodate?

IMMIGRATION BRINGS BENEFITS— THE MORE THE MERRIER!

"The United States and Canada are nations of immigrants. What would have happened if our forebears had closed the door to America after they arrived and stopped the Irish, the Italians, the Eastern Europeans, and so many other nationalities from entering this country? Now we're arguing over Latinos, Asians, Russians, Muslims, you name it. Fact is, newcomers have always been viewed negatively by most of those who came before them. When Irish Catholics began arriving in the 1830s, the Protestants already here accused them of assigning their loyalty to some Italian pope rather than to their new country, but Irish Catholics soon proved to be pretty good Americans. Sound familiar? Muslims can be very good Americans too. It just takes time, longer for some immigrant groups than others. But don't you see that America's immigrants have always been the engine of growth? They become part of the world's most dynamic economy and make it more dynamic still.

"My ancestors came from Holland in the 1800s, and the head of the family was an architect from Rotterdam. I work here in western Michigan as an urban planner. People who want to limit immigration seem to think that only the least educated workers flood into the United States and Canada, depriving the less-skilled among us of jobs and causing hardship for citizens. But in fact America attracts skilled and highly educated as well as unskilled immigrants, and they all make contributions. The highly educated foreigners, including doctors and technologically skilled workers, are quickly absorbed into the workforce; you're very likely to have been treated by a physician from India or a dentist from South Africa. The unskilled workers take jobs we're not willing to perform at the wages offered. Things have changed! A few decades ago, American youngsters on summer break flooded the job market in search of temporary employment in hotels, department stores, and restaurants. Now they're vacationing in Europe or trekking in Costa Rica, and the managers of those establishments bring in temporary workers from Jamaica and Romania.

"And our own population is aging, which is why we need the infusion of younger people immigration brings with it. We don't want to become like Japan or some European countries, where they won't have the younger working people to pay the taxes needed to support the social security system. I agree with opponents of immigration on only one point: what we need is legal immigration, so that the new arrivals will get housed and schooled, and illegal immigration must be curbed. Otherwise, we need more, not fewer, immigrants."

LIMIT IMMIGRATION NOW

"The percentage of recent immigrants in the U.S. population is the highest it has been in 70 years, and in Canada in 60 years. America is adding the population of San Diego every year, over and above the natural increase, and not counting illegal immigration. This can't go on. By 2010, 15 percent of the U.S. population will consist of recent immigrants. One-third of them will not have a high-school diploma. They will need housing, education, medical treatment, and other social services that put a huge strain on the budgets of the States they enter. The jobs they're looking for often aren't there, and then they start displacing working Americans by accepting lower wages. It's easy for the elite to pontificate about how great immigration is for the American melting pot, but they're not the ones affected on a daily basis. Immigration is a problem for the working people. We see company jobs disappearing across the border to Mexico, and at the same time we have Mexicans arriving here by the hundreds of thousands, legally and illegally, and more jobs are taken away.

"And don't talk to me about how immigration now will pay social security bills later. I know a thing or two about this because I'm an accountant here in Los Angeles, and I can calculate as well as the next guy in Washington. Those fiscal planners seem to forget that immigrants grow older and will need social security too. And as for that supposed slowdown in the aging of our population because immigrants are so young and have so many children, over the past 20 years the average age in the United States has dropped by four months. So much for that nonsense. What's needed is a revamping of the tax structure, so those fat cats who rob corporations and then let them go under will at least have paid their fair share into the national kitty. There'll be plenty of money to fund social services for the aged. We don't need unskilled immigrants to pay those bills.

"And I'm against this notion of amnesty for illegal immigrants being talked about these days. All that would do is to attract more people to try to make it across our borders. I heard the president of Mexico propose opening the U.S.-Mexican border the way they're opening borders in the European Union. Can you imagine what would happen? What our two countries really need is a policy that deters illegal movement across that border, which will save lives as well as jobs, and a system that will confine immigration to legal channels. These days, that's not just a social or economic issue; it's a security matter as well."

Regional ISSUE

Vote your opinion at www.wiley.com/college/deblij

most agglomerated of the main ethnic minorities. Chinese remain the most numerous among the 15-plus million Americans who identify themselves as Asian, but this diverse minority also includes Japanese who have

been in the United States for generations, Filipinos who began arriving after World War II, Vietnamese who were resettled after the Indochina War, and Pacific Islanders including Hawaiians. Whatever their origins,

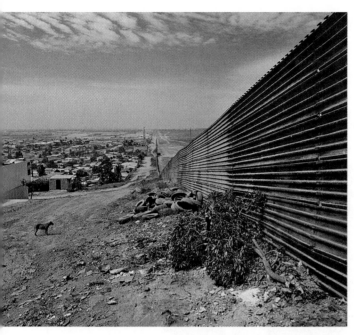

People have walled and fenced themselves off since the earliest days, as the Great Wall of China and Hadrian's Wall (built by the Romans in northern Britain) remind us. But no wall—whether the Berlin Wall during the Cold War or the Israeli "Security Barrier" of today—has ever halted all migration. Neither will migration be stopped by a fortified fence along the Mexican border from the Pacific coast to the Gulf of Mexico. However, long stretches of the U.S.–Mexico boundary already look like this, and more construction will take place. This photo shows a ground-level view of the improvised wall that slashes westward, between Tijuana, Mexico and the southernmost suburbs of San Diego, California, as it makes its way toward the Pacific shore on the horizon. The poverty-stricken landscape of this part of Tijuana extends almost to the rusty barrier itself, while development on the U.S. side has kept its distance from this well-patrolled section of the border.   © Diane Cook & Ken Jenshel.

the great majority have remained in California, where many have achieved satisfactory integration into society and economic success.

These successive immigration waves are only the largest among many inbound streams that continue to variegate America's cultural fabric. Physicians from India, nurses from South Africa, businesspeople from the Arabian Peninsula, caretakers from the Philippines, and numerous other skilled and unskilled workers arrive in numbers that make the United States one of the leading immigrant destinations on Earth.

Melting Pot or Mosaic?

Such are the numbers and so diverse are the cultural traditions that the "melting pot" that America once was is changing into something different. The United States is becoming a **mosaic culture**, an increasingly heterogeneous complex of separate, more or less uniform "tiles"

that represent more specialized groups than ever before—and that spend less time interacting and "melting." No longer are race, ethnicity, and cultural imperatives the bases for clustering. Today's communities are increasingly shaped by circumstances of occupational status, level of income, age, and especially lifestyle. Their proliferation (for example, in the form of "gated" communities now seen throughout the country) reflect what the population seems to want, but there is a serious downside to this, as Bill Bishop writes in his book *The Big Sort: Why the Clustering of Like-Minded America Is Tearing Us Apart.* As is the case in the world at large, balkanization fueled by people wanting to interact only with others closely resembling themselves leads to misunderstanding of others, miscalculations when it comes to decisions and policies, and widening fissures between communities. This threatens the very survival of the democratic values that have underpinned the evolution of American society.

Rural and Urban America

Immigrants, as every admirer of American wines knows, have always played important roles on the land. Without seasonal migrants, many farmers could not harvest their vegetables, pick the grapes, or ship their produce to market—at least not at competitive prices. With the progression from primary to quaternary activities, less than 2 percent of American workers remain on the land. But do not be deceived: U.S. farm output remains huge by world standards, and the range of farm products is enormous. Not for nothing is a large part of the Midwest known as the "Corn Belt"—today soybeans are competing for the land in this wide crescent extending from Ohio to South Dakota—and wheat exports from the Great Plains States fill the needs of customers worldwide. Mechanization and "corporate" farming have altered the agricultural landscape for livestock as well as crops, endangering the family farm, intensifying pollution problems, and unimaginably worsening conditions for factory-raised animals.

The map of "Farm Resource Regions" published by the U.S. Department of Agriculture, therefore, is of more than passing interest. As Figure 3-15 shows, the geographers who drew this map struggled to delimit formal regions on a highly diversified landscape, and they came up with some imaginative names (such as "Northern Crescent" and "Prairie Gateway"). It is worth looking at the fine print and checking it against your own experience. Note that some durable regions, such as the Eastern Uplands with its high percentage of remaining small, private farms, go all the way back to the nation's formative days of agricultural land use.

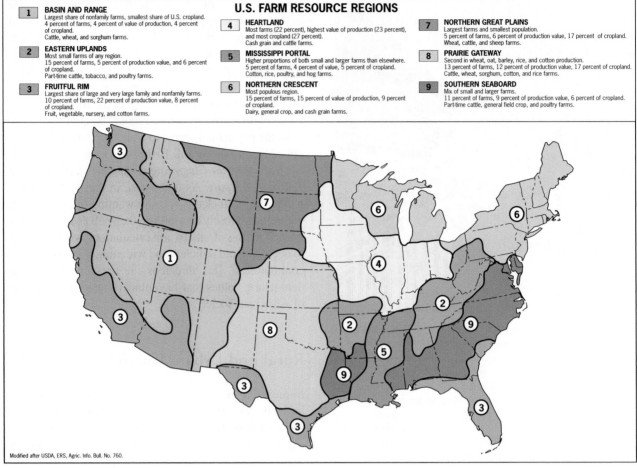

U.S. FARM RESOURCE REGIONS

1 BASIN AND RANGE
Largest share of nonfamily farms, smallest share of U.S. cropland. 4 percent of farms, 4 percent of value of production, 4 percent of cropland.
Cattle, wheat, and sorghum farms.

2 EASTERN UPLANDS
Most small farms of any region.
15 percent of farms, 5 percent of production value, and 6 percent of cropland.
Part-time cattle, tobacco, and poultry farms.

3 FRUITFUL RIM
Largest share of large and very large family and nonfamily farms.
10 percent of farms, 22 percent of production value, 8 percent of cropland.
Fruit, vegetable, nursery, and cotton farms.

4 HEARTLAND
Most farms (22 percent), highest value of production (23 percent), and most cropland (27 percent).
Cash grain and cattle farms.

5 MISSISSIPPI PORTAL
Higher proportions of both small and larger farms than elsewhere.
5 percent of farms, 4 percent of value, 5 percent of cropland.
Cotton, rice, poultry, and hog farms.

6 NORTHERN CRESCENT
Most populous region.
15 percent of farms, 15 percent of value of production, 9 percent of cropland.
Dairy, general crop, and cash grain farms.

7 NORTHERN GREAT PLAINS
Largest farms and smallest population.
5 percent of farms, 6 percent of production value, 17 percent of cropland.
Wheat, cattle, and sheep farms.

8 PRAIRIE GATEWAY
Second in wheat, oat, barley, rice, and cotton production.
13 percent of farms, 12 percent of production value, 17 percent of cropland.
Cattle, wheat, sorghum, cotton, and rice farms.

9 SOUTHERN SEABOARD
Mix of small and larger farms.
11 percent of farms, 9 percent of production value, 6 percent of cropland.
Part-time cattle, general field crop, and poultry farms.

Modified after USDA, ERS, Agric. Info. Bull. No. 760.

FIGURE 3-15

© H. J. de Blij, P. O. Muller, and John Wiley & Sons, Inc.

But the great majority of immigrants find their way to towns and cities, where they imprint their presence on the urban landscape in very different ways. From Miami's Little Havana to L.A.'s Koreatown and the older Italian neighborhoods and Chinatowns of Manufacturing Belt cities, these urban-ethnic clusters continue to form the landing strips for current arrivals, where shopkeepers speak the immigrant's language and locals know places to rent. The cities are the crucibles of changing America, and not only because of the structural transformation we described earlier.

In any case, the American city is obviously in yet another phase of reform, and what will emerge from this is still uncertain. But here is something to think about: when you plan to go to a city in some other part of the country and it is too far to drive to in a few hours—what is your first thought? Flying into an airport. But that airport may be many kilometers from the city (Denver, for example, abandoned close-in Stapleton Airport and built an ultramodern one much more distant). There was a time when cities were tied to ports, then to railways, and now it is the airport that matters. To reinvigorate the obsolete parts of those cities that are not doing well today, a "downtown airport" may be the answer. Policy scientist John Kasarda calls such a city of the future an *aerotropolis*, suggesting that local airports with large terminals and nearby amenities will attract high-tech development, business, educational and residential facilities, and perhaps generate one more stage in the transformation of metropolitan America.

For this to become reality, air travel will have to change as well, and there is some evidence that this change may be happening (though the economic downturn starting in 2008 was a major setback because many smaller cities lost air service). But harbingers of this coming change abounded as the use of private "air taxi" services at small private airports in the suburbs of larger cities was growing rapidly and prices were coming down. We may have glimpsed the future, and not too long from now that crowded airbus to a faraway terminal may become a thing of the past. The United States is constantly changing.

AMONG THE REALM'S GREAT CITIES . . . Chicago

CHICAGO lies near the southern end of Lake Michigan not far from the geographic center of the conterminous United States. Its centrality and crossroads location were evident to its earliest indigenous settlers, who developed the site as a portage where canoes could be hauled from the lake to the headwaters of a nearby stream that led to the Mississippi River. Centuries later, when the modern U.S. spatial economy emerged, Chicago became the leading hub on the continental transport network. As a freight-rail node, it still reigns supreme. And even though most long-distance passengers switched to jet planes a half-century ago, Chicago remains the quintessential North American hub city, with O'Hare International consistently ranking among the world's busiest airports.

Poet Carl Sandburg described Chicago as "the city of big shoulders," underscoring its personality and prowess as the leading U.S. manufacturing center during the century following the Industrial Revolution of the 1870s and 1880s. Among the myriad products that emanated from Chicago's industrial crucible were the first steel-frame skyscraper, elevated railway, refrigerated boxcar, cooking range, electric iron, nuclear reactor, and window envelope. Chicago also became a major commercial center, spawning a skyscraper-dominated central business district second only to New York's in size and influence.

Chicago today (metro population: 10.2 million) is trying hard to make the transition to the postindustrial age. Its diversified economy has proven to be a precious asset in this quest, and the city is determined to reinvent itself as a first-order service center. The new forces it must contend with, however, are symbolized in the fate of the emptying Sears (now Willis) Tower, the city's—and the nation's—tallest building. Sears & Roebuck itself abandoned the structure in 1992 for a new headquarters campus along Interstate-90 in the far northwestern suburb of Hoffman Estates. This shift acknowledged the rise of

Chicagoland's thriving outer city, one of the country's largest and most successful. Around O'Hare, a huge and still-expanding suburban office/industrial-park complex has truly become Chicago's "Second City," underscoring that major airports themselves will increasingly function as urban growth magnets in the ever more globalized spatial economy of the twenty-first century (see p. 180 and the top photo on p. 183).

© H. J. de Blij, P. O. Muller, and John Wiley & Sons, Inc.

NORTH AMERICAN REGIONS: CORE AND PERIPHERIES

Ample justification exists for recognizing the two national entities of the North American realm as discrete regions, based on contrasts we have enumerated. There is, however, another way to conceptualize the regional geography of North America: by combining functional and formal principles. The result, Figure 3-16, is by no means the only solution to the problem. Others have submitted alternate interpretations, including Joel Garreau

in his classic 1981 book, *The Nine Nations of North America*. As we note in the following brief narrative based on our map, there is more than ethnicity and politics to the framework of North America.

● NORTH AMERICAN CORE (1)

All of North America, like most of Europe, forms part of what we have delimited as the global core (Fig. G-12). But all core areas, including Europe, have subsidiary cores. The Core region of the North American realm combines

FIGURE 3-16

© H. J. de Blij, P. O. Muller, and John Wiley & Sons, Inc.

those of the United States and Canada, the largest cities and federal capitals of both countries, the leading financial markets, the largest number of corporate headquarters, dominant media centers, prestigious universities, cutting-edge research complexes, and the busiest airports and intercity expressways. Moreover, both the U.S. and Canadian portions of the North American Core region still contain more than one-third of their respective national populations.

Political and business decisions, investments, and other commitments made in this Core affect not just North America, but the world at large. This region is one of the world's centers of globalization, and from its Megalopolis innovations radiate outward in countless fields of endeavor. High-technology centers near Washington (Internet Alley lining the expressway to Dulles Airport) and Boston (circumferential Route 128) rival Silicon Valley.

Nonetheless, the Core's dominance over North America has been declining. As was true of the American Manufacturing Belt, it had unrivaled supremacy. But the rise of postindustrialism has diminished that regional role as competitors to the south and west continue to siphon away some of its key functions.

Flying into any major North American airport, you will see businesses clustering near the terminal complex—businesses that take advantage of proximity to national and international travel networks. High-technology firms rank high among these businesses, and this photo shows the corporate headquarters of America Online (AOL) near Dulles International Airport. AOL is located in the heart of "Internet Alley," the burgeoning corridor that stretches more than 50 kilometers (30 mi) across Northern Virginia from Washington, D.C. in the east to Dulles in the west—which has evolved into one of North America's leading technopoles (defined on p. 185). Nonetheless, like other high-tech companies, AOL has been buffeted by the economic downturn that began in 2008.   © Stephen Jaffe/AFP/Getty Images, Inc.

MARITIME NORTHEAST (2)

When you travel north from Boston into New Hampshire, Vermont, and Maine, you move across one of North America's historic culture hearths whose identity has remained strong for four centuries. Although Massachusetts, Connecticut, and Rhode Island—traditional New England States—share these qualities, they have become part of the Core, so that on a functional as well as formal basis, the Maritime Northeast extends from the northern

An unmistakably French cultural landscape in North America: the Rue Saint-Louis in Quebec City, the capital of Quebec Province. Not an English sign in sight in this center of French/Québécois culture. The green spires in the background belong to Le Chateau Frontenac, a grand hotel built on the site of the old Fort St. Louis (the Count of Frontenac was a prominent governor of what was then known as New France). French-speaking, Roman Catholic Quebec lies at the heart of French Canada. © SUPERSTOCK.

border of Massachusetts to Newfoundland, thus incorporating all four Canadian Atlantic Provinces.

Difficult environments, a maritime orientation, limited resources, and a persistent rural character have combined to slow economic development here. Primary industries—fishing, logging, some farming—are mainstays, although recreation and tourism have boosted the regional economy in recent decades. Upper New England experiences the spillover effect of the Boston area's prosperity to some degree, but not enough to ensure steady and sustained economic growth. Atlantic Canada has also endured economic hard times in the recent past. For example, overfishing has depleted fish stocks. Alternate opportunities focus on the region's spectacular scenery and the tourism it attracts. Economic prospects have also been boosted in Canada's poorest province, Newfoundland and Labrador, by the discovery of significant offshore oil reserves.

FRENCH CANADA (3)

Francophone Canada constitutes the inhabited, southern portion of Quebec from near Montreal to the mouth of the St. Lawrence River (Fig. 3-16). It includes the French-speaking Acadians who reside in neighboring New Brunswick (Fig. 3-12). The French cultural imprint on the region's cities and towns (see photo below) is matched in the rural areas by the narrow, rectangular *long lots* perpendicular to the river, also of French origin.

The economy of French Canada is no longer predominantly rural, although industrialization has not reached the level of the adjacent Core region in terms

AMONG THE REALM'S GREAT CITIES . . . Montreal

MONTREAL LIES ON a large triangular island in the St. Lawrence River, its historic core laid out along the base of what Montrealers call "the mountain"—flat-topped Mount Royal, from which the city takes its name. Culturally, Montreal (3.7 million) is the largest French-speaking city in the world after Paris, and therein lie its opportunities and challenges. Without a doubt, this is still one of the realm's most cosmopolitan cities, a strong European flavor pervading its stately CBD, its lively neighborhoods, and its bustling street life. To escape the long, harsh winters here, much of downtown Montreal has become an underground city with miles of handsome passageways lined with shops, restaurants, and cinemas, all of it served by one of the world's most modern and attractive subway systems.

Yet, despite the sparkle and continental ambience, Montreal has not been a happy place in recent times. With the escalation of tensions between Francophone and Anglophone Canada over the past four decades, Quebec's largest metropolis became a crucible of confrontation. For most of that period, the ethnolinguistic division was most visible along the Boulevard St. Laurent, the longstanding boundary between the French-speakers of Montreal's East End (who constitute about 75 percent of the city's population) and the English-speakers who clustered in the West End (the remnant of a larger Anglophone community that was reduced by 40 percent when the French-supremacy language laws were implemented). These days, however, things are improving as linguistically mixed communities proliferate throughout Montreal, driven by an influx of immigrants from outside North America who settle in both English- and French-speaking neighborhoods. At the same time, polarization continues to abate as stores, newspapers, and universities cater to increasingly bilingual clienteles,

and intermarriage between Anglophones and Franco-phones is on the rise.

Montreal's economy has also been affected. During the 1990s, hundreds of Anglophone firms (and thousands of their workers) fled to Ontario and points west, leaving in their wake a commercial real estate tailspin, rising unemployment, and a noticeable decline in the city's prosperity. Today, with the secession issue no longer on the front burner in Quebec, Montreal is steadily progressing in its comeback. Tourists have returned in droves; new foreign-trading ties, especially with the United States, are reviving the business community; and vibrant suburbs are leading the metropolis to reinvent itself as a hub for information technology, telecommunications, and biopharmaceutical companies.

© H. J. de Blij, P. O. Muller, and John Wiley & Sons, Inc.

of scale or technology (Toronto far outranks Montreal here). Tertiary and postindustrial activities are concentrated in and around Montreal, and tourism and recreation are growing in importance. But Quebec's economic prospects are shadowed by the region's political uncertainties, and bursts of nationalism (such as laws prohibiting shops from advertising in English) have eroded business confidence. When the provincial government sought to address the declining birth rate resulting from accelerating urbanization by encouraging immigration, tolerance for multiculturalism in

Quebec proved to be so low that a special commission began to hold hearings in 2007 to investigate the matter. Parochialism has cost Quebec dearly and sets this region off from the rest of the Canada that virtually encircles it.

The Acadians in the neighboring province of New Brunswick, Canada's largest cluster of French-speakers outside Quebec, take a different view. Not only do they shun the notion of independence for themselves, but they also actively promote all efforts to keep Quebec within the Canadian federation. Accommodation with the Anglophone majority and acceptance of multiculturalism in New Brunswick set an example for the rest of French Canada.

THE SOUTH (4)

In the other direction from the Core, the American South for more than a century following the Civil War remained in economic and cultural isolation from the rest of the country. In the 1970s, however, things changed so fast that a New South arose almost overnight. The Sunbelt migration drove people and enterprises into the long-stagnant cities. Core-region companies looking for headquarters or subsidiary offices found Atlanta, Charlotte, Miami-Fort Lauderdale, Tampa, and other urban areas economical and attractive, turning them into boomtowns virtually overnight. Racial segregation had been dismantled in the wake of the civil rights movement. A new social order was matched by new facilities ranging from airports (Atlanta's quickly became one of the world's busiest) to theme parks open to all. Soon the nation was watching Atlanta-based CNN on television and following space flights that originated at Cape Canaveral.

The geography of development in the South, however, is very uneven. While many cities and some agricultural areas have benefited, others have not; the South contains some of the realm's poorest rural areas, and the gap between rich and poor here is wider than in any other region in the realm. Virginia's Washington suburbs, North Carolina's Research Triangle, Tennessee's Oak Ridge complex, and Atlanta's corporate campuses are locales that represent the New South; Appalachia and rural Mississippi represent the Old South, where depressed farming areas and stagnant small industries restrict both incomes and change.

But change is inexorable, and it comes from many directions. Northerners of all backgrounds and income levels are arriving in large numbers. Asian and particularly Hispanic immigrants are transforming the ethnic mix, taking unskilled as well as skilled jobs and altering daily life in ways Southerners often find hard to accept. Foreign investors build automobile factories in Kentucky and Alabama. But the South overall ranks low among the realm's regions in terms of exports and global linkages. Meanwhile, the influx of retirees is changing the population's age structure, and not just in Florida. The new era of massive urban and suburban construction, perhaps to a greater degree than anywhere else in North America, has entrenched the communities and lifestyles of the new mosaic culture.

THE SOUTHWEST (5)

The Southwest has taken on its own regional identity relatively recently, but today it is firmly established in the North American consciousness for several reasons. One is environmental: here, steppe (**BS**) and desert (**BW**) dominate (see Fig. G-7). Another is cultural: Anglo-American, Hispanic, and Native American cultures coexist in uneasy accommodation.

As Figure 3-16 shows, three States constitute this region, although its borders do not, of course, exactly coincide with political boundaries. By some criteria, the Southwest starts in eastern Texas between Dallas (often called the westernmost Eastern city) and adjacent Fort Worth (by the same reasoning, the easternmost Western city). But the region's identity is based on the air conditioner and the *electricity* to power it; the *water* to satisfy demand that includes irrigated golf courses and ubiquitous swimming pools; and the *automobile* that has enabled the wide dispersal of population in vast low-density communities. That certainly applies to Texas as well as New Mexico and Arizona.

Texas leads in every respect. Its economy, once oil- and gas-dependent, has been restructured and diversified so that today the Dallas-Forth Worth (DFW)–Houston–San Antonio triangle has become one of the world's most productive postindustrial complexes, with several **technopoles** (state-of-the-art, high-technology centers) including Austin at the heart of it. This is also a hub of international trade and the northern anchor of a NAFTA-generated transnational growth corridor that extends into Mexico as far as Monterrey. New Mexico, in the middle of the region, is economically least developed and ranks low on many U.S. social indices, but its environmental and cultural attractions benefit Albuquerque and Santa Fe, with Los Alamos a famous name in research and development. In the west, Arizona's technologically transformed desert now harbors two large coalescing metropolises, Phoenix and Tucson. Growth continues, but climate change adds to the uncertainties of water supplies already stretched to the limit.

PACIFIC HINGE (6)

The Southwest meets the Far West near San Diego, California, and from there the region we call the Pacific Hinge extends all the way up to Vancouver, its northern anchor in southwestern British Columbia. We include all

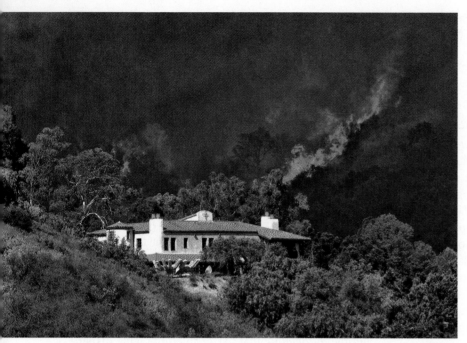

Wildfires form a constant threat in areas where dense vegetation, high summer temperatures, long dry seasons, and high winds combine to intensify this deadly natural hazard. In 2009 the worst wildfire disaster struck Australia, where the risk is always high; but in North America, California suffered casualties and property damage was well. This dramatic photo of a wind-driven wall of flames threatening a home near Santa Barbara, taken in May 2009, suggests why efforts to extinguish the flames are all too often doomed without nature's help in the form of wind reversals and rainstorms.   © AP/World Wide Photos

of California in this region but only (for environmental as well as economic reasons) the western portions of Oregon and Washington State.

In many parts of the world, this region would constitute a primary core area, with such leading cities as Los Angeles, San Francisco, Portland, and Seattle plus America's most populous State, whose economy in recent times has ranked among the world's ten largest—by country. The Pacific Hinge also includes one of the realm's most productive agricultural areas in California's Central Valley, magnificent scenery, agreeable climates, and a culturally diverse population drawn by its spectacular economic growth and pleasant living conditions.

Today this region is in a still-new phase of development because of its involvement in the economic growth of countries on opposite shores of the Pacific Ocean. Earlier, Japan's success had a salutary impact here, but what has since happened in China, South Korea, Taiwan, Vietnam, and other distant Pacific Rim economies has created unprecedented opportunities (see the box titled "The Pacific Rim"). That is why we call this region the *Pacific Hinge* of North America, now the key interface between the realm and the Rim.

The Pacific Rim

TODAY WE CONTINUE to witness the rise of a far-flung region born of a string of economic miracles on the shores of the Pacific Ocean. It is a still-discontinuous region, led by its foci on the Pacific's western margins: Japan, coastal China, South Korea, Taiwan, Thailand, Malaysia, and Singapore. The regional term **Pacific Rim** has come into use to describe this dramatic development, which over the past three decades has redrawn the map of the Pacific periphery—not only in East and Southeast Asia but also in the United States and Canada, and even in such Southern Hemisphere locales as Australia and South America's Chile.

The Pacific Rim also is a superb example of a functional region, with economic activity in the form of capital flows, raw-material movements, and trade linkages generating urbanization, industrialization, and labor migration. Within this process, human landscapes from Sydney to Santiago are being transformed inside a 32,000-kilometer (20,000-mi) corridor that girdles the globe's largest body of water. In later chapters, wherever the Pacific Rim intersects Pacific-bordering geographic realms, we will continue to discuss this important regional development and its spatial impacts.

19

AMONG THE REALM'S GREAT CITIES . . . Los Angeles

IT IS HARD to think of another city that has been in more headlines in the past several years than Los Angeles. Much of this publicity has been negative. But despite its earthquakes, mudslides, wildfires, pollution alerts, showcase trials, and riots, L.A.'s glamorous image has hardly been dented. This is compelling testimony to the city's resilience and vibrancy, and to its unchallenged position as the Western world's entertainment capital.

The plane-window view during the descent into Los Angeles International Airport, almost always across the heart of the metropolis, gives a good feel for the immensity of this urban landscape. It not only fills the huge natural amphitheater known as the Los Angeles Basin, but it also oozes into adjoining coastal strips, mountain-fringed valleys and foothills, and even the margins of the Mojave Desert more than 90 kilometers (60 mi) inland. This quintessential spread city, of course, could only have materialized in the automobile age. In fact, most of it was built rapidly over the past 80 years, propelled by the swift expansion of a high-speed freeway network unsurpassed anywhere in metropolitan America. In the process, Greater Los Angeles became so widely dispersed that today it has reorganized within a sprawling, multinodal geographic framework (see Fig. 3-10B on p. 166).

The metropolis as a whole is home to 18.1 million Southern Californians, constitutes North America's second largest urban agglomeration, and forms the southern anchor of the huge California conurbation that lines the Pacific coast from San Francisco southward to the Mexican border (Fig. 3-9). It also is the Pacific Hinge's leading trade, manufacturing, and financial center. In the global arena, Los Angeles is the eastern Pacific Rim's biggest city and the origin of the largest number of transoceanic flights and sailings to Asia.

© H. J. de Blij, P. O. Muller, and John Wiley & Sons, Inc.

Foreign trade has been a large part of this story, but other older components of this region's economy have also played their roles. The port of Los Angeles/Long Beach is the largest in the United States, but Greater Los Angeles today is also the realm's largest single metropolitan manufacturing complex, its products ranging from clothing churned out by immigrant-staffed, low-tech garment factories to high-tech medical equipment. Southern California, transformed by waves of legal and illegal immigrants from Middle America, Asia, and other parts of the world, is the center of Pacific Hinge action. Farther north, the San Francisco Bay Area has seen rather more orderly expansion, with the Silicon Valley technopole the key component of its high-technology success. In what Americans refer to as the Pacific Northwest, the Seattle-Tacoma area originally benefited from cheap hydroelectricity generated by Columbia River dam projects, attracting aluminum producers and aircraft manufacturers. Boeing remains one of the world's two leading aircraft manufacturers, the East Asian market having boosted its sales. Meanwhile, the technopole centered on suburban Redmond (Microsoft's headquarters) heralds this area's transformation into a prototype metropolis of the postindustrial economy.

Vancouver, on the northern flank of the Pacific Hinge (photo p. 172), has the locational advantage of being closer, on air and sea routes, to East Asia than any other large North American city. Ethnic Chinese constitute more than 20 percent of the metropolitan population, whose total Asian sector now approaches 40 percent. These factors gave the city a head start in forging trade and investment linkages to the western Pacific Rim, although, as we noted earlier, friction arose between Asians and Anglos concerning neighborhood standards. Over time, an east-west hybrid culture may develop, but the current focus remains on economic prospects. Beijing is proposing to build pipelines from Alberta to the British Columbia coast, where terminals would transfer oil and natural gas to tankers bound for China. The United States, Canada's biggest customer, watches with concern.

● THE WESTERN FRONTIER (7)

Where the forests of the Pacific Hinge yield to the scrub of the rain shadow on the inland slopes of the mountain wall that parallels the West Coast, and the relief turns into intermontane ("between the mountains") basins and plateaus, lies a region stretching from the Sierra Nevada and Cascades to the Rockies, encompassing parts of southern Alberta and British Columbia, eastern Washington and Oregon, all of Nevada, Utah, and Idaho plus western Montana, Wyoming, and Colorado. In Figure 3-16, you can see Edmonton, Calgary, and Denver situated at the margin of this aptly named region, and Salt Lake City, anchoring the Software Valley technopole and symbolizing its new high-tech era, at the heart of it.

Remoteness, dryness, and sparse population typified this region for a very long time. It became the redoubt of the Mormon faith and a place known for boom-and-bust mining, logging, and, where possible, livestock raising. However, in recent decades advances in communications and transportation technologies, sunny climates, open spaces, lower costs of living, and growing job opportunities have constituted effective "pull" factors here. Every time an earthquake struck in California, eastward migration got a boost. But over the past decade, Californians by the millions have emigrated from the overcrowded, overpriced Pacific coast to the high desert.

Development in this region centers on its widely dispersed urban areas, which are making this the realm's fastest-growing region (albeit from a low base)—and where thousands of high-tech manufacturing and specialized service jobs drove a two-decade-long influx, slowed but not reversed by the economic crisis that began in 2008. This growth has not come without problems as locals see land values rise, conservative ideas challenged, and lifestyles endangered.

The fastest-growing large metropolis in North America, Las Vegas, lies on the border between the Western Frontier and Southwest regions. Far more than a gambling and entertainment center that welcomes some 40 million visitors annually, Las Vegas attracts immigrants because of its history of job creation, relatively cheap land and low-cost housing, moderate taxes, and sunny if seasonally hot weather. No doubt, proximity to Southern California has had much to do with its explosive growth, but Las Vegas is also known far and wide as the ultimate frontier city—where (almost) anything goes.

For years the hottest growth area in the Western Frontier region was its southernmost corner—the Las Vegas Valley. Nevada was one of America's fastest-growing States, attracting new residents by the hundreds of thousands and visitors to its burgeoning recreation industry by the tens of millions annually. Between 1990 and 2000, the Las Vegas suburb of Henderson (left) ranked as the fastest-growing municipality in the United States. Sprawling development on a massive scale converted this patch of high desert into a low-density urban checkerboard whose expansion seemed without end. But by 2009 the growth cycle had ceased as thousands of homeowners were forced into foreclosure, unemployment was rising, and national and global economic crises were clouding the future of boomtown USA. © Mike Yamashita/Woodfin Camp & Associates.

CONTINENTAL INTERIOR (8)

The Western Frontier meets the Continental Interior region along a line marked by cities that influence both regions, from Denver in the south to Edmonton in the north. This vast North American heartland extends from interior Canada to the borders of the South and Southwest. As Figure 3-16 shows, this region has many noteworthy cities such as Kansas City, Omaha, Minneapolis, and Winnipeg, but agriculture is the dominant story here. This is North America's breadbasket. This is the land of the Corn Belt (which extends into the western half of the Core region), of the soybean (the fastest-expanding crop in the past century), of spring wheat in the Dakotas and Canada and winter wheat in Kansas, of beef and pork. Indeed, the cities and towns of this region share histories of food processing, packing, and marketing, of flour milling and soybean, sunflower, and canola oil production. It is also the scene of the struggle for survival of the family farm. The unrelenting, aggressive incursion of large-scale corporate farming threatens a way of life.

This may be farmland, but it is not immune from what might seem to be the nonagricultural realities of the modern world, not even the energy issue. Corn used to be grown for people and pigs; now it is increasingly used to make ethanol, the gasoline substitute (or additive). Farmers hope that rising corn prices will revitalize the Corn Belt. Others worry that rising prices will put a strain on consumers who eat it, mostly in the form of meat produced from corn-fed hogs and beef cattle. In any case, the amount of corn needed to make a real dent in our gasoline consumption is many times larger than the entire annual corn crop. In 2007, when around half of all corn grown in the Midwest was turned into biofuels, it replaced only about 4 percent of gasoline consumption while raising concerns about possible food shortages.

Take a look at population statistics, and it is clear that many States in the Continental Interior are losing people or gaining far fewer than the national average. The Great Plains, the western zone of this region, is especially hard hit: younger people and those better off are leaving, and older and poorer residents are staying behind. Villages and small towns are dying, and the notion of abandoning certain areas and allowing them to return to their natural state is being seriously discussed.

The contrast with the western Canadian sector of the Continental Interior could not be sharper. Alberta, despite the wild swings in the world price of oil, is experiencing one of the most spectacular economic booms in the realm's history. Recent estimates suggest that Alber-

ta's reserves, including the tar sands around Fort McMurray (see photo below), are second only to those of Saudi Arabia, and possibly even larger. Alberta has become Canada's fastest-growing province, Calgary its

As the price of a barrel of oil (and a gallon of gasoline) rises, it becomes profitable to derive oil from sources other than liquid reserves. The Canadian province of Alberta contains vast deposits of "oil sands" in which the petroleum is mixed with sand, requiring a relatively expensive and complicated process to extract it. The quantity of oil locked in these Athabasca Tar Sands is estimated to constitute one of the world's largest reserves, and this huge open-pit mining project is under way around the town of Fort McMurray to recover it. Those who assume that Canada will sell this oil to the United States should be aware that the Chinese are offering to fund construction of a pipeline across the mountains to Canada's Pacific coast; the rising cost of oil has political as well as economic ramifications.   © Jim Wark.

fastest-expanding large city. The boom makes Calgary a magnet for business and highly-skilled workers; the city's skyline is in spectacular transformation, comparable to what is happening on the Arabian Peninsula. The question is: how long will the good times last?

● THE NORTHERN FRONTIER (9)

Figure 3-16 leaves no doubt as to the dimensions of this final North American region: it is by far the largest of the realm, covering nearly 90 percent of Canada and all of the largest U.S. State, Alaska, whose population of just over 700,000 accounts for more than one-third of the region's total. Here also lies the only city of any size, Anchorage (290,000). Not only does this region include the northern parts of seven of Canada's provinces, but it also comprises the Yukon and Northwest Territories as well as recently established Nunavut.

The sparsely-peopled Northern Frontier remains a land of isolated settlement based on the exploitation of newly discovered resources. As noted earlier, the Canadian Shield, which underlies most of the eastern half of the region, is a rich storehouse of mineral resources including metallic ores such as nickel, uranium, copper, gold, silver, lead, and zinc. The Yukon and Northwest Territories have proven especially bountiful, with gold- and diamond-mining (Canada now ranks among the world's top five producers); at the opposite, eastern edge of the Shield, near Voisey's Bay on the central coast of Labrador, the largest body of high-grade nickel ore ever discovered is being opened to production.

Productive locations in the Northern Frontier comprise a far-flung network of mines, oil and gas installations, pulp mills, and hydropower stations that have spawned hundreds of small settlements and thousands of

kilometers of interconnecting transport and communications lines. Inevitably, these activities have infringed on the lands of indigenous peoples without the preparation of treaties or agreements, which has led to recent negotiations between the government and leaders of First Nations over resource development in the Northern Frontier. But essentially this vast region remains a frontier in the truest sense of the term.

Alaska's regional geography differs from the rest of the Northern Frontier in that the State contains several urban settlements and an incipient core area in the coastal Anchorage area. Its internal communications (air and surface) also are rather more developed. Another difference lies in Alaska's North Slope oil exploitation, one of the hemisphere's key energy sources since the 1300-kilometer (800-mi) Trans-Alaska Pipeline began operating in 1977. Dwindling supplies at the main reserve at Prudhoe Bay are now compelling producers to turn to huge additional reserves in Alaska's north, but opposition from preservationist groups has slowed their plans. The issue remains in contention—as is another project to construct a natural gas pipeline to parallel the oil pipeline to send gas, a by-product of the oil drilling, to the growing market in the Lower 48.

Climate change is likely to affect the Northern Frontier in as-yet unforeseeable ways. As the map shows, this region extends all the way north to Ellesmere Island and thus includes waters that, in the event of longer-term sustained global warming and associated recession of Arctic ice, will open, possibly year-round, to create natural waterways that would dramatically alter shipping routes and intercontinental distances. As we note in Chapter 12, the question of ownership of these waters—and the submerged seafloor beneath them, coming within reach of exploitation—will be a matter of significant international contention in the decades ahead.

What You Can Do

RECOMMENDATION: Get involved! Help a Geographic Alliance! Because "geographic illiteracy" still prevails in the United States, the National Geographic Society in the 1980s launched a project to help teachers prepare themselves for geography instruction. Teachers from every State in the Union were invited to participate in seminars at NGS headquarters in Washington, D.C., and these teachers subsequently presented similar seminars for colleagues in their home States. Thus a State-based network of Geographic Alliances was created and still functions today. These Geographic Alliances are always looking for assistance, resources, ideas, and other contributions. It would be an interesting research project to discover whether geography is taught in the schools in the town where your college is located, and whether the teachers teaching it were part of the Alliance network.

GEOGRAPHIC CONNECTIONS

1 Imagine that you are the chief executive of a successful young company that manufactures software for the newest high-speed computers. Where would you locate your company and why? Establish a "short list" of three possible sites, and choose one by a process of elimination in which your decision is built upon the most important locational variables for your plant and its highly-skilled workforce.

2 Immigration has long been a potent force in shaping the population mosaic of North America. Why is the "geography of demography" in the United States today so different from that of stagnant Europe? How does that difference express itself regionally? Does the immigration process affect your daily life, and how might it affect your competitive position when you are ready to enter the job market following completion of your education?

3 Enhance your experience in considering these questions by surfing the Internet and consulting appropriate websites. As an extra endeavor, ponder these same questions as they would apply to the contemporary Canadian scene, keeping an eye out for similarities and differences between the two countries of this realm.

Geographic Literature on North America: The key introductory works on this realm were authored by Birdsall et al., Boal & Royle, Bone (both titles), Getis & Getis, Hardwick et al., Hudson, McGillivray, and McKnight. These works, together with a comprehensive listing of noteworthy books and recent articles on the geography of the United States and Canada, can be found in the *References and Further Readings* section of this book's website at **www.wiley.com/college/deblij.**

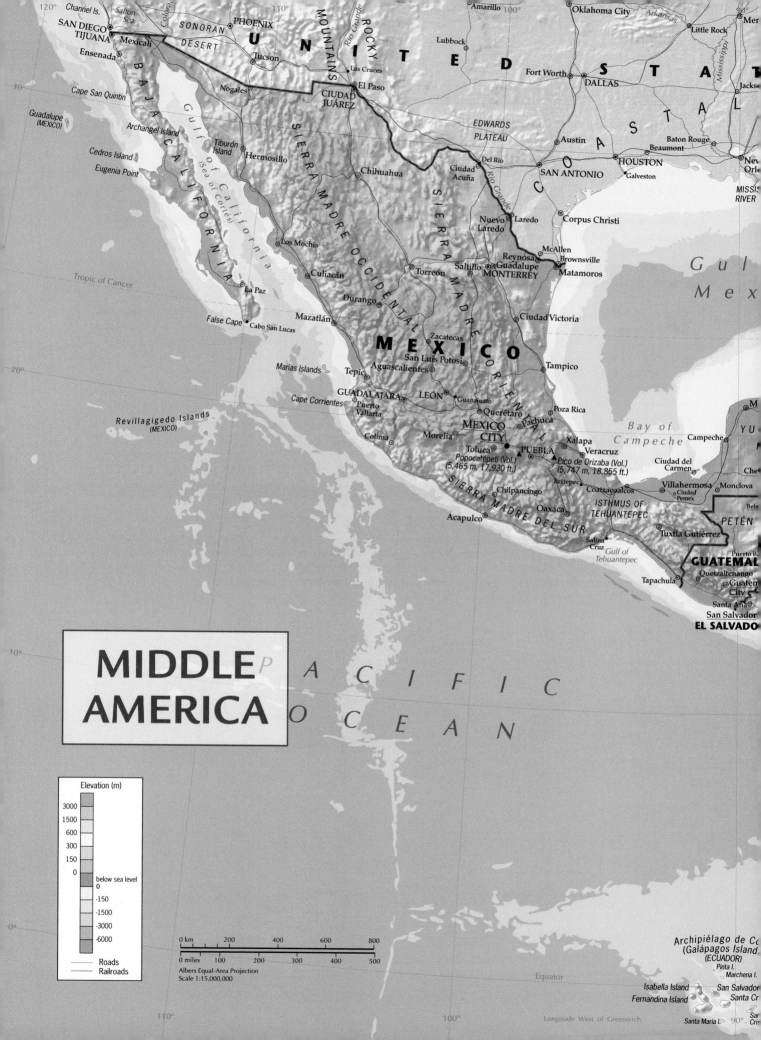

MIDDLE AMERICA

Elevation (m)

3000	
1500	
600	
300	
150	
0	
	below sea level
	0
-150	
-1500	
-3000	
-6000	

Roads
Railroads

0 km   200   400   600   800
0 miles   100   200   300   400   500
Albers Equal-Area Projection
Scale 1:15,000,000

UNITED STATES

MEXICO

Channel Is.
SAN DIEGO
TIJUANA
Ensenada
Mexicali
Cape San Quintin
Guadalupe (MEXICO)
Archangel Island
Cedros Island
Eugenia Point
Tiburón Island
Hermosillo
BAJA CALIFORNIA
Gulf of California (Sea of Cortés)
Tropic of Cancer
La Paz
False Cape • Cabo San Lucas
Marias Islands
Revillagigedo Islands (MEXICO)
Tepic
Cape Corrientes
GUADALAJARA
Puerto Vallarta
Colima
Morelia
Toluca
MEXICO CITY

SONORAN DESERT
Colo.
Rio Grande
PHOENIX
Tucson
Nogales
Las Cruces
El Paso
CIUDAD JUÁREZ
ROCKY MOUNTAINS
SIERRA MADRE OCCIDENTAL
Chihuahua
Los Mochis
Culiacán
Durango
Mazatlán
Zacatecas
San Luis Potosí
Aguascalientes
LEÓN
Guanajuato
Querétaro
Pachuca
Popocatépetl (Vol.) (5,465 m, 17,930 ft.)

Amarillo
Oklahoma City
Arkansas
Little Rock
Lubbock
Fort Worth
DALLAS
Mer
EDWARDS PLATEAU
Austin
Baton Rouge
Beaumont
Del Rio
Ciudad Acuña
Rio Grande
SAN ANTONIO
HOUSTON
Galveston
New Orl
MISSI
RIVER
COASTAL
Nuevo Laredo
Laredo
Corpus Christi
Saltillo
Guadalupe
MONTERREY
Reynosa
McAllen
Brownsville
Matamoros
Ciudad Victoria
Tampico
Poza Rica
SIERRA MADRE ORIENTAL
Gul Mex
Bay of Campeche
Campeche
YU
Ciudad del Carmen
Che
Xalapa
Veracruz
Pico de Orizaba (Vol.) (5,747 m, 18,855 ft.)
PUEBLA
Juxtepec
Chilpancingo
Oaxaca
Coatzacoalcos
Villahermosa
Ciudad Pemex
Monclova
Belm
PETÉN
GUATEMAL
Acapulco
SIERRA MADRE DEL SUR
ISTHMUS OF TEHUANTEPEC
Salina Cruz
Gulf of Tehuantepec
Tuxtla Gutiérrez
Puerto B.
Tapachula
Quetzaltenango
Guatem City
Santa Ana
San Salvador
EL SALVADO

PACIFIC OCEAN

P

Equator
Longitude West of Greenwich

Archipiélago de Co (Galápagos Island) (ECUADOR)
Pinta I.
Marchena I.
San Salvador
Santa Cr
Isabella Island
Fernandina Island
Santa María I.
Cris

120°   110°   100°   90°

30°   20°   10°

4

REGIONS

- **MEXICO**
- **CENTRAL AMERICA**
- **CARIBBEAN BASIN**
    - **GREATER ANTILLES**
    - **LESSER ANTILLES**

CONCEPTS, IDEAS, AND TERMS

1. Land bridge
2. Archipelago
3. Culture hearth
4. Mainland-Rimland framework
5. Mestizo
6. Hacienda
7. Plantation
8. Acculturation
9. Transculturation
10. Maquiladora
11. Dry canal
12. Altitudinal zonation
13. *Tierra caliente*
14. *Tierra templada*
15. *Tierra fría*
16. *Tierra helada*
17. *Tierra nevada*
18. Tropical deforestation
19. Mulatto

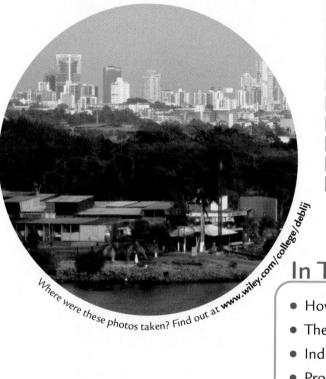

Where were these photos taken? Find out at www.wiley.com/college/deblij

Photos: © H. J. de Blij

FIGURE 4-1 *Map:* © H. J. de Blij, P. O. Muller, and John Wiley & Sons, Inc.

In This Chapter

- How NAFTA changed the economic geography of Mexico
- The many problems of gigantic Mexico City
- Indigenous peoples demand recognition and rights
- Projected Panama Canal expansion fuels boom in Panama City
- Cuba: Future star of the Caribbean?
- The debate over Puerto Rico's status continues

LOOK AT A world map, and it is obvious that the Americas comprise two landmasses: North America extending from Alaska to Panama and South America from Colombia to Argentina. But here we are reminded that continents and geographic realms do not usually coincide. In Chapter 3 we discussed a North American realm whose southern boundary is the U.S.-Mexican border and the Gulf of Mexico. Between North America and South America lies a small but important geographic realm known as Middle America. Consisting of a mainland corridor and numerous Caribbean islands, Middle America is a highly fragmented realm.

From Figure 4-1 it is clear that Middle America is much wider longitudinally than it is long latitudinally. The distance from Baja California to Barbados is about 6000 kilometers (3800 mi), but from the latitude of Tijuana to Panama City is only half that distance. In terms of total area, Middle America is the second-smallest of the world's geographic realms. As the map shows, the dominant state of this realm is Mexico, larger than all its other countries and territories combined.

Middle America may be a small geographic realm by global standards, but comparatively it is densely peopled. Its population in 2010 was a bit less than 200 million, over half of it in Mexico alone, and the rate of natural increase (at 1.6 percent) remains above the world average. Figures 4-2 and 4-3 show Mexico's populous interior core area quite clearly, but note that population in Guatemala and Nicaragua tends to cluster toward the Pacific rather than the Caribbean coast. Among the islands, the maps reveal how crowded Hispaniola (containing Haiti and the Dominican Republic) is, but at these scales we cannot clearly see the pattern on the smaller islands of the eastern Caribbean, some of which are also densely populated.

What Middle America lacks in size it makes up in physiographic and cultural diversity. This is a realm of soaring volcanoes and spectacular shorelines, of tropical forests and barren deserts, of windswept plateaus and scenic islands. It holds the architectural and technological legacies of ancient indigenous civilizations. Today it is a mosaic of immigrant cultures from Africa, Europe, and elsewhere, richly reflected in music and the visual arts. Material poverty, however, is endemic: island Haiti is the poorest country in the Americas; Nicaragua, on the mainland, is almost as badly off. As we will find, a combination of factors has produced a distinctive but troubled realm between North and South America.

MAJOR GEOGRAPHIC QUALITIES OF

Middle America

1. Middle America is a fragmented realm that consists of all the mainland countries from Mexico to Panama and all the islands of the Caribbean Basin to the east.

2. Middle America's mainland constitutes a crucial barrier between Atlantic and Pacific waters. In physiographic terms, this is a land bridge that connects the continental landmasses of North and South America.

3. Middle America is a realm of intense cultural and political fragmentation. The political geography defies unification efforts, but countries and regions are beginning to work together to solve mutual problems.

4. Middle America's cultural geography is complex. African influences dominate the Caribbean, whereas Spanish and Amerindian traditions survive on the mainland.

5. The realm contains the Americas' least-developed territories. New economic opportunities may help alleviate Middle America's endemic poverty.

6. In terms of area, population, and economic potential, Mexico dominates the realm.

7. Mexico is reforming its economy and has experienced major industrial growth. Its hopes for continuing this development are tied to overcoming its remaining economic problems and to expanding trade with the United States and Canada under the North American Free Trade Agreement (NAFTA).

Defining the Realm

Sometimes you will see Middle and South America referred to, in combination, as "Latin" America, alluding to their prevailing Spanish-Portuguese heritage. This is an inappropriate regional name, just as Anglo America, a term once commonly used for North America, was also improper. Such culturally based terminologies reflect historic power and dominance, and they tend to make outsiders out of those people they do not represent.

In North America, the term *Anglo* (as a geographic appellation) was offensive to many Native Americans, to African Americans, to Hispanics, to Quebecers, and to others. In Middle (and South) America, millions of people of indigenous American, African, Asian, and European ancestries do not fit under the "Latin" rubric. You will not find the cultural landscape very "Latin" in the Bahamas, Barbados, Jamaica, Belize, or large parts of Guatemala and Mexico. So let us adopt the geographic neutrality of North, Middle, and South.

But is Middle America sufficiently different from either North or South America to merit regional distinction? Certainly, in this age of globalization and migration, border areas are becoming transition zones, as is happening along the U.S.-Mexican boundary. Still, consider this: North America encompasses only a pair of states, and the entire continent of South America only 13 (even including France's dependency on the northeast coast). But far smaller Middle America, as we define it, incorporates more than three dozen political

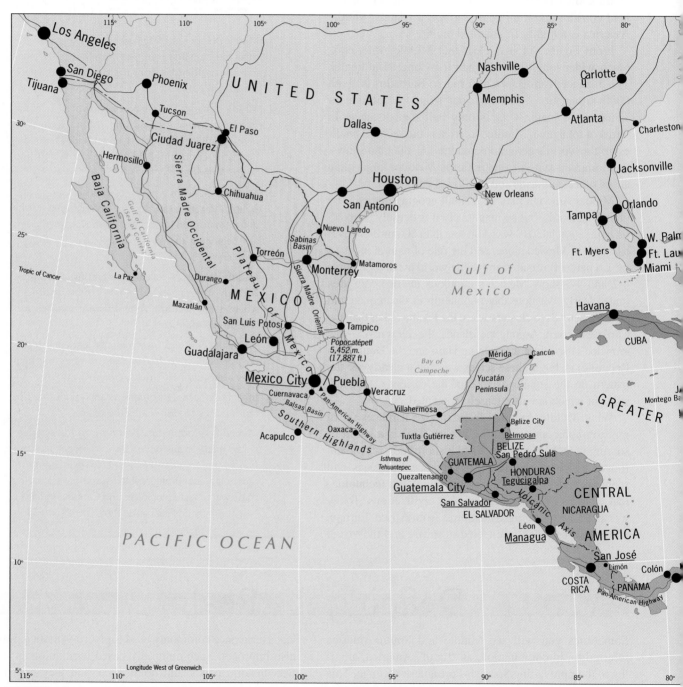

FIGURE 4-2

entities, including several dependencies (or quasi-dependencies) of the Netherlands, the United Kingdom, and France, but not counting constituent territories of the United States. Therefore, unlike South America, Middle America is a multilingual patchwork of independent states, territories in political transition, and residual colonial dependencies, with strong continuing ties to the United States and non-Iberian Europe. Middle America is defined in large measure by its vivid cultural-geographic pluralism.

The Regions

As Figure 4-2 indicates, Middle America contains four distinct regions. Dominant *Mexico* occupies the largest part of the collective territory, with much of its northern border defined by the Rio Bravo (Rio Grande) from El Paso to the Gulf of Mexico. To the southeast, Mexico yields to the region called *Central America*, consisting of the seven republics of Guatemala, Belize, Honduras, El Salvador, Nicaragua, Costa Rica, and Panama. Sometimes the entire realm is referred to as Central America, but only these seven countries constitute this region. The *Greater Antilles* is the regional name referring to the four large islands in the northern sector of the Caribbean Sea: Cuba, Jamaica, Hispaniola, and Puerto Rico; two countries, Haiti and the Dominican Republic, share the island of Hispaniola. And the *Lesser Antilles* form an extensive arc of smaller islands from the Virgin Islands off Puerto Rico to the Netherlands Antilles near the northwestern coast of Venezuela. In the Caribbean islands large and small, moist winds sweep in from the east, watering windward (wind-facing) coasts while leaving leeward (wind-protected) areas dry. On the map, the "Leeward" and "Windward" Islands actually are not "dry" and "wet," these terms having navigational rather than environmental implications.

PHYSIOGRAPHY

As Figure 4-1 shows, fragmented Middle America is a realm of high relief, studded with active volcanoes. Figure G-3 reminds us why: the Caribbean, North American, South American, and Cocos tectonic plates converge here, creating dangerous landscapes subject to earthquakes and landslides. Add to this the realm's exposure to Atlantic hurricanes, and it amounts to some of the highest-risk real estate on Earth.

A Land Bridge

The funnel-shaped mainland, a 4800-kilometer (3000-mi) connection between North and South America, is wide enough in the north to contain two major mountain chains and a vast interior plateau, but narrows to a slim 65-kilometer (40-mi) ribbon of land in Panama. Here this strip of land, or *isthmus*, bends eastward so that Panama's orientation is east-west. Thus mainland Middle America is what physical geographers call a **land bridge**, an isthmian link between continents.

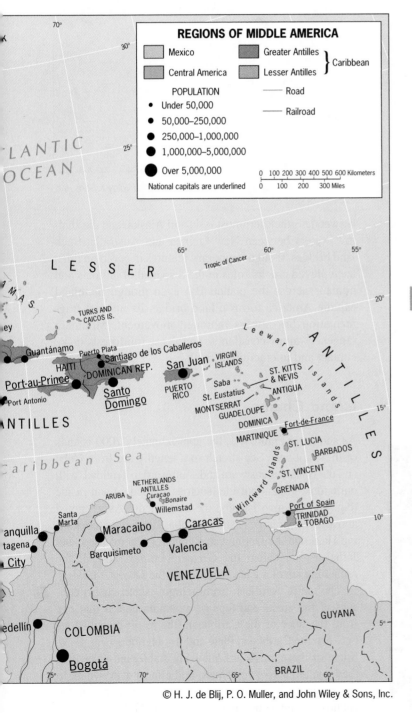

© H. J. de Blij, P. O. Muller, and John Wiley & Sons, Inc.

FIGURE 4-3

© H. J. de Blij, P. O. Muller, and John Wiley & Sons, Inc.

If you examine a globe, you can see other present and former land bridges: Egypt's Sinai Peninsula between Asia and Africa, the (now-broken) Bering land bridge between northeasternmost Asia and Alaska, and the shallow waters between New Guinea and Australia. Such land bridges, though temporary features in geologic time, have played crucial roles in the dispersal of animals and humans across the planet. But even though mainland Middle America forms a land bridge, its internal fragmentation has always inhibited movement. Mountain ranges, swampy coastlands, and dense rainforests make contact and interaction difficult.

Island Chains

As shown in Figure 4-1, the approximately 7000 islands of the Caribbean Sea stretch in a lengthy arc from Cuba and the Bahamas eastward and southward to Trinidad, with numerous outliers outside (such as Barbados) and inside (e.g., the Cayman Islands) the main chain. As we noted above, the four large islands—Cuba, Hispaniola (containing Haiti and the Dominican Republic), Puerto Rico, and Jamaica—are called the *Greater Antilles*, and all the remaining smaller islands constitute the *Lesser Antilles*.

The entire Antillean **archipelago** (island chain) consists of the crests and tops of mountain chains that rise from the floor of the Caribbean, the result of collisions between the Caribbean Plate and its neighbors (Fig. G-3). Some of these crests are relatively stable, but elsewhere they contain active volcanoes, and almost everywhere in this realm earthquakes are an ever-present danger—in the islands as well as on the mainland (Fig. G-4).

No part of Middle America is safe from environmental onslaught, but certain areas are riskier than others—and when disaster strikes, the poorest inhabitants, often living in the least desirable and most hazardous locales, suffer disproportionately. Haiti in 2008 was struck by three hurricanes and a tropical storm, repeatedly flooding cities and towns and dislocating some 800,000 people. Thick mud filled streets and houses, and for weeks after the waters receded, bodies continued to emerge from debris such as this in the city of Gonaïves. © AP/Wide World Photos

LEGACY OF MESOAMERICA

Mainland Middle America was the scene of the emergence of a major ancient civilization. Here lay one of the world's **3** true **culture hearths** (see Fig. 7-4), a source area from which new ideas radiated and whose population could expand and make significant material and intellectual progress. Agricultural specialization, urbanization, and transport networks developed, and writing, science, art, and other spheres of achievement saw major advances. Anthropologists refer to the Middle American culture hearth as *Mesoamerica*, which extended southeast from the vicinity of present-day Mexico City to central Nicaragua. Its development is especially remarkable because it occurred in very different geographic environments, each presenting obstacles that had to be overcome in order to unify and integrate large areas. First, in the low-lying tropical plains of what is now northern Guatemala, Belize, and Mexico's Yucatán Peninsula, and perhaps simultaneously in Guatemala's highlands to the south, the Maya civilization arose more than 3000 years ago. Later, far to the northwest on the high plateau in central Mexico, the Aztecs founded a major civilization centered on the largest city ever to exist in pre-Columbian times.

The Lowland Maya

The Maya civilization is the only major culture hearth in the world that arose in the lowland tropics. Its great cities, with their stone pyramids and massive temples, still yield archeological information today. Maya culture reached its zenith from the third to the tenth centuries AD.

The Maya civilization, anchored by a series of city-states, unified an area larger than any of the modern Middle American countries except Mexico. Its population probably totaled between 2 and 3 million; certain Maya languages are still used in the area to this day. The Maya city-states were marked by dynastic rule that functioned alongside a powerful religious hierarchy, and the great cities that now lie in ruins were primarily ceremonial centers. We also know that Maya culture produced skilled artists and scientists, and that these people achieved a great deal in agriculture and trade. They grew cotton, created a rudimentary textile industry, and exported cotton cloth by seagoing canoes to other parts of Middle America in return for valuable raw materials. They domesticated the turkey and grew cacao, developed writing systems, and studied astronomy.

The Highland Aztecs

In what is today the intermontane highland zone of Mexico, significant cultural developments were also taking place. Here, just north of present-day Mexico

FROM THE FIELD NOTES

"We spent Monday and Tuesday upriver at Lamanai, a huge, still mostly overgrown Maya site deep in the forest of Belize. On Wednesday we drove from Belize City to Altun Ha, which represents a very different picture. Settled around 200 BC, Altun Ha flourished as a Classic Period center between AD 300 and 900, when it was a thriving trade and redistribution center for the Caribbean merchant canoe traffic and served as an *entrepôt* for the interior land trails, some of them leading all the way to Teotihuacán. Altun Ha has an area of about 6.5 square kilometers (2.5 sq mi), with the main structures, one of which is shown here, arranged around two plazas at its core. I climbed to the top of this one to get a perspective, and sat down to have my sandwich lunch, imagining what this place must have looked like as a bustling trade and ceremonial center when the Roman Empire still thrived, but a more urgent matter intruded. A five-inch tarantula emerged from a wide crack in the sun-baked platform, and I noticed it only when it was about two feet away, apparently attracted by the crumbs and a small piece of salami. A somewhat hurried departure put an end to my historical-geographical ruminations." © H. J. de Blij.

www.conceptcaching.com

City, lay Teotihuacán, the first true urban center in the Western Hemisphere, which prospered for nearly seven centuries after its founding around the beginning of the Christian era.

A Once and Future Core Area

The Aztec state, the pinnacle of organization and power in pre-Columbian Middle America, is thought to have originated in the early fourteenth century with the founding of a settlement on an island in a lake that lay in the *Valley of Mexico* (the area surrounding what is now Mexico City). This urban complex, a functioning city as well as a ceremonial center, named Tenochtitlán, was soon to become the greatest city in the Americas and the capital of a large powerful state. The Aztecs soon gained control over the entire Valley of Mexico, a pivotal 50-by-65-kilometer (30-by-40-mi) mountain-encircled basin that is still the heart of the modern state of Mexico. Both elevation and interior location moderate its climate; for a tropical area, it is quite dry and very cool. The basin's lakes formed a valuable means of internal communication, and the Aztecs built canals to connect several of them. This fostered a busy canoe traffic, bringing agricultural produce to the cities, and tribute was paid by their many subjects to the headquarters of the ruling nobility.

A Regional Empire

With its heartland consolidated, the new Aztec state soon began to conquer territories to the east and south. The Aztecs' expansion of their empire was driven by their desire to subjugate peoples and towns in order to extract taxes and tribute. As Aztec influence spread throughout Middle America, the state grew ever richer, its population mushroomed, and its cities thrived and expanded.

The Aztecs produced a wide range of impressive accomplishments, although they were better borrowers and refiners than they were innovators. They developed irrigation systems, and they built elaborate walls to terrace slopes where soil erosion threatened. Indeed, the greatest contributions of Mesoamerica's Amerindians surely came from the agricultural sphere. Corn (maize), the sweet potato, various kinds of beans, the tomato, squash, cacao beans (the raw material of chocolate), and tobacco are just a few of the crops that grew in Mesoamerica when the Europeans first made contact.

COLLISION OF CULTURES

We in the Western world all too often believe that history began when the Europeans arrived in some area of the world and that the Europeans brought such superior power to the other continents that whatever existed there previously had little significance. Middle America confirms this misperception: the great, feared Aztec state apparently fell before a relatively small band of Spanish invaders in an incredibly short period of time (1519–1521). But let us not lose sight of a few realities. At first, the Aztecs believed the Spaniards were "White Gods" whose arrival had been predicted by Aztec prophecy. Hernán Cortés, for all his 508 soldiers, did not singlehandedly overthrow this powerful empire: he ignited a rebellion by Amerindian peoples who had fallen under Aztec domination and had seen their relatives carried off for human sacrifice to Aztec gods. Led

by Cortés with his horses and guns, these peoples rose against their Aztec oppressors and joined the band of Spaniards headed toward Tenochtitlán (where thousands of them would perish in combat against Aztec defenders).

Effects of the Conquest

Spain's defeat of Middle America's dominant indigenous state opened the door to Spanish penetration and supremacy. Throughout the realm, the confrontation between Hispanic and native cultures spelled disaster for the Amerindians: a catastrophic decline in population (perhaps as high as 90 percent), rapid deforestation, pressure on vegetation from grazing animals, substitution of Spanish wheat for maize (corn) on cropland, and the concentration of Amerindians into newly built towns.

The Spaniards were ruthless colonizers but not more so than other European powers that subjugated other cultures. True, the Spaniards first enslaved the Amerindians and were determined to destroy the strength of indigenous society. But biology accomplished what ruthlessness could not have achieved in so short a time: diseases introduced by the Spaniards and slaves imported from Africa killed millions of Amerindians.

The Rural Impact

Middle America's cultural landscape—its great cities, its terraced fields, its dispersed aboriginal villages—was thus drastically modified. Unlike the Amerindians, who had used stone as their main building material, the Spaniards employed great quantities of wood and used charcoal for heating, cooking, and smelting metal. The onslaught on the forests was immediate, and rings of deforestation swiftly expanded around the colonizers' towns. The Spaniards also introduced large numbers of cattle and sheep, and people and livestock now had to compete for available food (requiring the opening of vast areas of marginal land that further disrupted the region's food-production balance). Moreover, the Spaniards introduced their own crops (notably wheat) and farming equipment, and soon large fields of wheat began to encroach upon the small plots of corn that the natives cultivated.

The New Urban Settlements

The Spaniards' most far-reaching cultural changes derived from their traditions as town-dwellers. To facilitate domination, the Amerindians were moved off their land into nucleated villages and towns that the Spaniards established and laid out. In these settlements, the Spaniards could exercise the kind of rule and admin-istration to which they were accus-tomed (Fig. 4-4). The internal focus of each Spanish town was the central *plaza* or market square, around which both the local church and government buildings were located. The surrounding street pattern was deliberately laid out in gridiron form, so that any insurrections by the resettled Amerindians could be contained by having a small military force seal off the affected blocks and then root out the troublemakers. Each town was located near what was thought to be good agricultural land (which was often not so good), so that the Amerindians could go out each day and work in the fields. Packed tightly into these towns and villages, they came face to face with Spanish culture. Here they learned

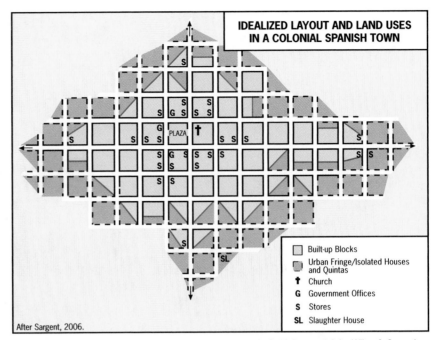

FIGURE 4-4

© H. J. de Blij, P. O. Muller, and John Wiley & Sons, Inc.

the Europeans' Roman Catholic religion and Spanish language, and they paid their taxes and tribute to a new master. Nonetheless, the nucleated indigenous village survived under colonial (and later postcolonial) admin-istration and is still a key feature of remote Amerindian areas in southeastern Mexico and inner Guatemala, where to this day native languages prevail over Spanish (Fig. 4-5).

Church and State

Once the indigenous population was conquered and resettled, the Spaniards were able to pursue another pri-mary goal in their New World territory: the exploitation of its wealth for their own benefit. Lucrative trade, com-mercial agriculture, livestock ranching, and especially mining were avenues to affluence. And wherever the Spaniards ruled—in towns, farms, mines, or indigenous villages—the Roman Catholic Church was the supreme cultural force transforming Amerindian society. With Jesuits and soldiers working together to advance the fron-tiers of New Spain, in the wake of conquest it was the church that controlled, pacified, organized, and accul-turated the Amerindian peoples.

MAINLAND AND RIMLAND

In Middle America outside Mexico, only Panama, with its twin attractions of interoceanic transit and gold de-posits, became an early focus of Spanish activity. From

there, following the Pacific side of the isthmus, Spanish influence radiated northwestward through Central Amer-ica and into Mexico. The major arena of international competition in Middle America, however, lay not on the Pacific side but on the islands and coasts of the Caribbean Sea. Here the British gained a foothold on the mainland, controlling a narrow coastal strip that extended southeast from Yucatán to what is now Costa Rica. As the colonial-era map (Fig. 4-6) shows, in the Caribbean the Spaniards faced not only the British but also the French and Dutch, all interested in the lucrative sugar trade, all searching for instant wealth, and all seeking to expand their empires.

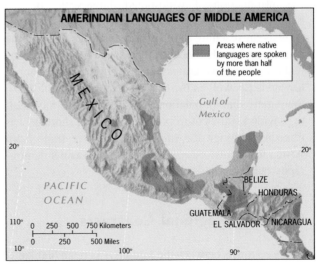

FIGURE 4-5 © H. J. de Blij, P. O. Muller, and John Wiley & Sons, Inc.

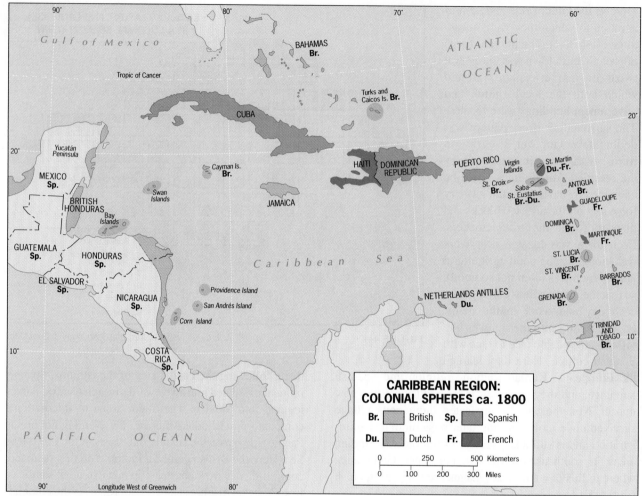

FIGURE 4-6

© H. J. de Blij, P. O. Muller, and John Wiley & Sons, Inc.

Much later, after centuries of European colonial rivalry in the Caribbean Basin, the United States entered the picture and made its influence felt in the coastal areas of the mainland, not through colonial conquest but through the introduction of widespread, large-scale, banana plantation agriculture. The effects of these plantations were as far-reaching as the impact of colonialism on the Caribbean islands. Because the diseases the Europeans had introduced were most rampant in these hot, humid lowlands (as well as the Caribbean islands to the east), the Amerindian population that survived was too small to provide a sufficient workforce. This labor shortage was remedied through the trans-Atlantic slave trade from Africa that transformed the Caribbean Basin's demography (see map in Chapter 6, p. 297).

Lingering Regional Contrasts

These contrasts between the Middle American highlands on the one hand and the coastal areas and Caribbean islands on the other were conceptualized by John Augel-li into the **Mainland-Rimland framework** (Fig. 4-7). Augelli recognized (1) a Euro-Amerindian **Mainland**, which consisted of continental Middle America from Mexico to Panama, excluding the Caribbean coastal belt from mid-Yucatán southeastward; and (2) a Euro-African **Rimland**, which included this coastal zone as well as the islands of the Caribbean. The terms *Euro-Amerindian* and *Euro-African* underscore the cultural heritage of each region. On the Mainland, European (Spanish) and Amerindian influences are paramount and also include **mestizo** sectors where the two ancestries mixed. In the Rimland, the heritage is European and African. As Figure 4-7 shows, the Mainland is subdivided into several areas based on the strength of the Amerindian legacy. The Rimland is also subdivided, with the most obvious division the one between the mainland-coastal plantation zone and the islands. Note, too, that the islands themselves are classified according to their cultural heritage (Figs. 4-6, 4-7).

Supplementing these contrasts are regional differences in outlook and orientation. The Rimland was an area of sugar and banana plantations, of high accessibil-

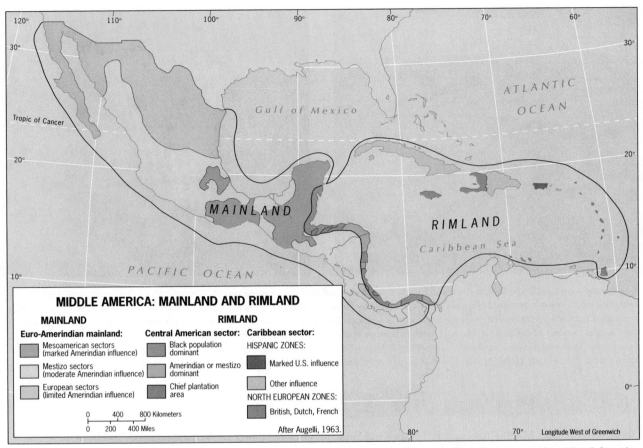

FIGURE 4-7

© H. J. de Blij, P. O. Muller, and John Wiley & Sons, Inc.

ity, of seaward exposure, and of maximum cultural contact and mixture. The Mainland, being farther removed from these contacts, was an area of greater isolation. The Rimland was the region of the great *plantation*, and its commercial economy was therefore susceptible to fluctuating world markets and tied to overseas investment capital. The Mainland was the region of the *hacienda*, which was more self-sufficient and less dependent on external markets.

The Hacienda

This contrast between plantation and hacienda land tenure in itself constitutes strong evidence for the Rimland-Mainland division. The hacienda was a Spanish institution, but the modern plantation, Augelli argued, was the concept of Europeans of more northerly origin. In the **hacienda**, Spanish landowners possessed a domain whose productivity they might never push to its limits: the very possession of such a vast estate brought with it social prestige and a comfortable lifestyle. Native workers lived on the land—which may once have been their land—and had plots where they could grow their own subsistence crops. All this is writ-

ten as though it is mostly in the past, but the legacy of the hacienda system, with its inefficient use of land and labor, still exists throughout mainland Middle America (see photos, p. 204).

The Plantation

The **plantation** was conceived as something entirely different from the hacienda. Robert West and John Augelli list five characteristics of Middle American plantations that illustrate the differences between hacienda and plantation: (1) plantations are located in the humid tropical coastal lowlands; (2) plantations produce for export almost exclusively—usually a single crop; (3) capital and skills are often imported so that foreign ownership and an outflow of profits occur; (4) labor is seasonal—needed in large numbers mainly during the harvest period—and such labor has been imported because of the scarcity of Amerindian workers; and (5) with its "factory-in-the-field" operation, the plantation is more efficient in its use of land and labor than the hacienda. The objective was not self-sufficiency but profit, and wealth rather than social prestige is a dominant motive for the plantation's establishment and operation.

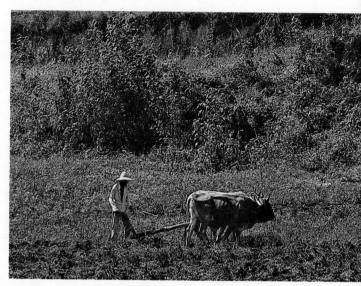

Contrasting land uses in the Middle American Rimland and Mainland give rise to some very different rural cultural landscapes. Huge stretches of the realm's best land continue to be controlled by (often absentee) landowners whose haciendas yield export or luxury crops, or foreign corporations that raise fruits for transport and sale on their home markets. The banana plantation shown here (*left*) lies near the Caribbean coast of Costa Rica. The vast fields of banana plants stand in strong contrast to the lone peasant who ekes out a bare subsistence from small cultivable plots of land, often in high-relief countryside where grazing some goats or other livestock is the only way to use most of the land. (Left): © John Coletti/Getty Images, Inc. (Right): © Jean-Gerard Sidaner/Photo Researchers.

POLITICAL FRAGMENTATION

Today continental Middle America is fragmented into eight countries, all but one of which have Hispanic origins. The largest of them all is Mexico—the giant of Middle America—whose 1,960,000 square kilometers (756,000 sq mi) constitute more than 70 percent of the realm's entire land area (the Caribbean region included) and whose 111 million people outnumber those of all the other countries and islands of Middle America combined.

The cultural variety in Caribbean Middle America is much greater. Here Hispanic-influenced Cuba dominates: its area is almost as large as that of all the other islands put together, and its population of 11.3 million is 10 percent greater than that of the next-ranking country—the Dominican Republic (10.3 million), also of Spanish heritage. As we noted, however, the Caribbean is hardly an arena of exclusive Hispanic cultural heritage: for example, Cuba's southern neighbor, Jamaica (population 2.8 million, mostly of African ancestry), has a legacy of British involvement, while to the east in Haiti (9.4 million, overwhelmingly of African ancestry) the strongest imprints have been African and French. The Lesser Antilles also exhibit great cultural diversity. There are the (once Danish) U.S. Virgin Islands; French Guadeloupe and Martinique; a group of British-influenced islands, including Barbados, St. Lucia, and Trinidad and Tobago; and the Dutch St. Maarten (shared with the French) and the A-B-C islands of the Netherlands Antilles—Aruba, Bonaire, and Curaçao—off the northwestern Venezuelan coast.

Independence

Independence movements stirred Middle America at an early stage. On the mainland, revolts against Spanish authority (beginning in 1810) achieved independence for Mexico by 1821 and for the Central American republics by the end of the 1820s. The United States, concerned over European designs in the realm, proclaimed the Monroe Doctrine in 1823 to deter any European power from reasserting its authority in the newly independent republics or from further expanding its existing domains.

POINTS TO PONDER

- Existing and projected border fences between the United States and Mexico spark anger on the Mexican side, but the issue is complicated by rising drug-related violence on both sides of the boundary.

- The Panama Canal is undergoing a major renovation. Watch for China to play a major role in the process.

- Puerto Rico is part of the United States, but nearly 50 percent of all Puerto Ricans live below the federal poverty line.

- The Netherlands Antilles were dissolved as a Caribbean political entity in 2007, its five islands accorded new status as parts of the Kingdom of the Netherlands.

By the end of the nineteenth century, the United States itself had become a major force in Middle America. The Spanish-American War of 1898 made Cuba independent and put Puerto Rico under the U.S. flag; soon afterward, the Americans were in Panama constructing the Panama Canal. Meanwhile, with U.S. corporations driving a boom based on huge banana plantations, the Central American republics had become colonies of the United States in all but name.

Independence came to the Caribbean Basin in fits and starts. Afro-Caribbean Jamaica as well as Trinidad and Tobago, where the British had brought a large South Asian population, attained full sovereignty from the United Kingdom in 1962; other British islands (among them Barbados, St. Vincent, and Dominica) became independent later. France, however, retains Martinique and Guadeloupe as overseas *départements* of the French Republic, and the Dutch islands are at various stages of autonomy.

Regions of the Realm

Middle America consists of four geographic regions: (1) Mexico, the giant of the realm in every respect; (2) Central America, the string of seven small republics occupying the land bridge to South America; (3) the four large islands that constitute the Greater Antilles of the Caribbean; and (4) the numerous islands of the Caribbean's Lesser Antilles (Fig. 4-2).

MEXICO

By virtue of its physical size, large population, cultural identity, resource base, economy, and relative location, the state of Mexico by itself constitutes a geographic region in this multifaceted realm. The U.S.-Mexican boundary on the map crosses the continent from the Pacific to the Gulf, but Mexican cultural influences penetrate deeply into the southwestern States and American impacts reach far into Mexico. To Mexicans the border is a reminder of territory lost to the United States in historic conflicts; to Americans it is a symbol of economic contrasts and illegal immigration. Along the Mexican side of this border the effects of NAFTA (the North American Free Trade Agreement between Canada, the United States, and Mexico [see p. 175]) are transforming Mexico's economic geography.

Physiography

The physiography of Mexico is reminiscent of that of the western United States, although its environments are more tropical. Figure 4-8 shows several prominent features: the elongated Baja (Lower) California Peninsula in the northwest, the far eastern Yucatán Peninsula, and the Isthmus of Tehuantepec in the southeast where the Mexican landmass tapers to its narrowest extent. Here in the southeast, Mexico most resembles

Central America physiographically; a mountain backbone forms the isthmus, curves southeastward into Guatemala, and extends northwestward toward Mexico City. Shortly before reaching the capital, this mountain range divides into two chains, the Sierra Madre Occidental in the west and the Sierra Madre Oriental in the east (Figs. 4-1, 4-8). These diverging ranges frame the funnel-shaped Mexican heartland, the center of which consists of the rugged, extensive Plateau of Mexico (the Valley of Mexico lies near its southeastern end). As Figure G-7 reveals, Mexico's climates are marked by dryness, particularly in the broad, mountain-flanked north. Most of the better-watered areas lie in the southern half of the country where the major population concentrations have developed.

MAJOR CITIES OF THE REALM

City	Population* (in millions)
Ciudad Juárez, Mexico	1.4
Guadalajara, Mexico	4.3
Guatemala City, Guatemala	1.2
Havana, Cuba	2.2
Managua, Nicaragua	1.3
Mexico City, Mexico	28.8
Monterrey, Mexico	3.9
Panama City, Panama	1.4
Port-au-Prince, Haiti	2.2
Puebla, Mexico	2.3
San José, Costa Rica	1.4
San Juan, Puerto Rico	2.7
San Salvador, El Salvador	1.5
Santo Domingo, Dominican Rep.	2.3
Tegucigalpa, Honduras	1.0
Tijuana, Mexico	1.7

*Based on 2010 estimates.

FIGURE 4-8

© H. J. de Blij, P. O. Muller, and John Wiley & Sons, Inc.

Population Patterns

Mexico's population grew rapidly during the last three decades of the twentieth century, doubling in just 28 years; but demographers have recently noted a sharp drop in fertility, and they are predicting that Mexico's population (currently 111.0 million) will stop growing altogether by about 2050. This will have enormous implications for the country's economy, and it will reduce the cross-border migration that is of so much concern at present.

Where the People Live

The distribution of Mexico's population is mapped in Figure 4-3, and its association with the country's 31 internal States is shown in Figure 4-9. The largest concentration, containing more than half the Mexican people, extends across the densely populated "waist" of the country from Veracruz State on the eastern Gulf coast to Jalisco State on the Pacific. The center of this corridor is dominated by the most populous State, Mexico (**3** on the map), at whose heart lies the Federal District of Mexico City (**9**). In the dry and rugged terrain to the north of

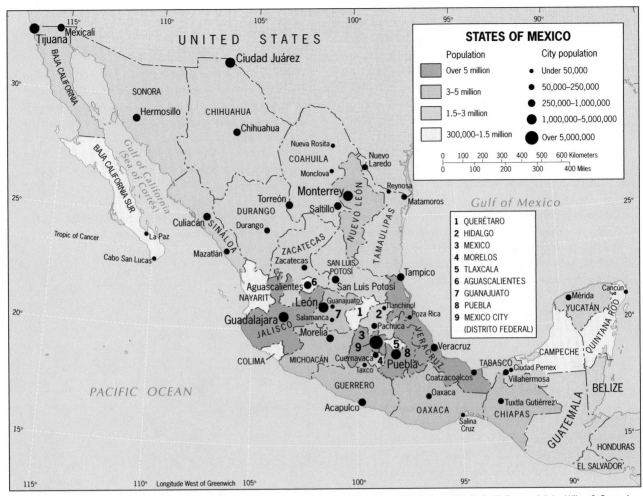

FIGURE 4-9

© H. J. de Blij, P. O. Muller, and John Wiley & Sons, Inc.

this central corridor lie Mexico's least-populated States. Southern Mexico also exhibits a sparsely peopled periphery in the hot and humid lowlands of the Yucatán Peninsula, but here most of the highlands of the continental spine contain sizeable populations.

Pull and Push

Another major feature of Mexico's population map is urbanization, driven by the *pull* of the cities (with their perceived opportunities for upward mobility) in tandem with the *push* of the economically stagnant countryside. Today, 76 percent of the Mexican people reside in towns and cities, a surprisingly high proportion for a developing country. Undoubtedly, these numbers are affected by the explosive recent growth of Mexico City, which now totals almost 29 million (making it the largest urban concentration on Earth) and is home to 26 percent of the national population. Among the other leading cities are Guadalajara, Puebla, and León in the central population

corridor, and Monterrey, Ciudad Juárez, and Tijuana in the northern U.S. border zone (Fig. 4-9). Urbanization rates at the other end of Mexico, however, are at their lowest in those remote uplands where Amerindian society has been least touched by modernization.

A Mix of Cultures

Nationally, the Amerindian imprint on Mexican culture remains strong. Today, 60 percent of all Mexicans are mestizos, 22 percent are predominantly Amerindian, and about 8 percent are full-blooded Amerindians; most of the remaining 10 percent are Europeans. Certainly the Mexican Amerindian has been Europeanized, but the Amerindianization of modern Mexican society is so powerful that it would be inappropriate here to speak of one-way, European-dominated **acculturation**. Instead, what took place in Mexico is **transculturation**—the two-way exchange of culture traits between societies in close contact. In the southeastern periphery (Fig. 4-5),

AMONG THE REALM'S GREAT CITIES . . . Mexico City

MIDDLE AMERICA HAS only one great metropolis: Mexico City. With 28.8 million inhabitants, Mexico City is home to just over one-fourth of Mexico's population and grows by about 300,000 each year. Even more significantly, Mexico City has surpassed Tokyo, Japan (current population: 26.7 million) to become the world's largest urban agglomeration.

Lakes and canals marked this site when the Aztecs built their city of Tenochtitlán here seven centuries ago. The conquering Spaniards made it their headquarters, and following independence the Mexicans made it their capital. Centrally positioned and well connected to the rest of the country, Mexico City, hub of the national core area, became the quintessential primate city.

Vivid social contrasts mark the cityscape. Historic plazas, magnificent palaces, churches, villas, superb museums, ultramodern skyscrapers, and luxury shops fill the city center. Beyond lies a zone of comfortable middle-class and struggling, but stable, working-class neighborhoods. Outside this belt, however, lies a ring of more than 500 slums and countless, even poorer *ciudades perdidas*—the "lost cities" where newly arrived peasants live in miserable poverty and squalor. (These squatter settlements contain no less than one-third of the metropolitan area's population.) Mexico City's more affluent residents have also been plagued by problems in recent years as the country's social and political order came close to unraveling. Rampant crime remains a serious concern, much of it associated with corrupt police as control of the country's long-time ruling party (the PRI) weakened and collapsed.

Environmental crises parallel the social problems. Local surface waters have long since dried up, and groundwater supplies are approaching depletion; to meet demand, the metropolis must now import a third of its water by pipeline from across the mountains (with about 40 percent lost through leakages in the city's crumbling waterpipe network). Air pollution here is among the world's worst as more than 4 million motor vehicles and 40,000 factories churn out smog that in Mexico City's thin, high-altitude air sometimes reaches

100 times the acceptably safe level. And add to all this a set of geologic hazards: severe land subsidence as underground water supplies are overdrawn; the ever-present threat of earthquakes that can wreak havoc on the city's unstable surface (the last big one occurred in 1985); and the risk of volcanic activity as nearby Mount Popocatépetl occasionally shows signs of ending centuries of dormancy.

In spite of it all, the great city continues to beckon, and every year more than 100,000 of the desperate and the dislocated arrive with hope—and little else.

© H. J. de Blij, P. O. Muller, and John Wiley & Sons, Inc.

several hundred thousand Mexicans still speak only an Amerindian language, and millions more still use these languages daily even though they also speak Mexican Spanish. The latter has been strongly shaped by Amerindian influences, as have Mexican modes of dress, foods and cuisine, artistic and architectural styles, and folkways. This fusion of heritages, which makes Mexico unique, is the product of an upheaval that began to reshape the country a century ago.

Revolution and Its Aftermath

Modern Mexico was forged in a revolution that began in 1910 and set into motion events that are still unfolding today. At its heart, this revolution was about the redistribution of land, an issue that had not been resolved after Mexico freed itself from Spanish colonial control in the early nineteenth century. As late as 1900, more than 8000 haciendas blanketed virtually all of Mexico's good farmland, and about 95 percent of all rural families owned no land whatsoever and toiled as *peones* (landless, constantly indebted serfs) on the haciendas. The triumphant revolution produced a new constitution in 1917 that launched a program of expropriation and parceling out of the haciendas to rural communities.

Land Reform

Mexico, alone among Middle America's countries with large Amerindian populations, has made significant progress toward land reform and has done so without major social upheavals. Since 1917, more than half the cultivated land in Mexico has been redistributed, mostly to peasant communities consisting of 20 families or more. On such farmlands, known as *ejidos*, the government holds title to the land, but the rights to use it are parceled out to village communities and then to individuals for cultivation. This system of land management is an Amerindian legacy, and not surprisingly most *ejidos* lie in central and southern Mexico, where Amerindian social and agricultural traditions are strongest. About half of Mexico's land continues to be held in such 'social landholdings,' but these tend to be excessively fragmented, causing both low farm yields and widespread rural poverty. In the 1990s the Mexican government tried to privatize *ejidos*, hoping to promote consolidation and increase productivity, but that effort did not work. To date, less than 10 percent of *ejidos* have been privatized.

States of Contrast

Countries with strong regional disparities face serious challenges that can be difficult to overcome and may worsen over time. Make a map of per-capita income for the States of Mexico (like the one on page 213), and you see at a glance that Mexico's southernmost States—Chiapas, Oaxaca, and Guerrero, all bordering the Pacific Ocean—are by far the poorest. States bordering the United States in the north, including Nuevo Léon, Chihuahua, and Baja California, have the highest incomes. Using rural poverty as a measure, only about 10 percent of people in the countryside in the north are in the poorest category, but nearly 50 percent in the south.

This is a serious problem for Mexico because the south is not only culturally the most Amerindian part of the country, but also the least well educated, the most isolated and remote, the least productive agriculturally, and lagging badly in terms of infrastructure investment and overall development. And the bad news is that the gap between north and south is widening, not closing. Successive Mexican government have pledged to do some-

thing about this, but nothing much happened—until 1994, when a radical group of Mayan peasant farmers in Chiapas calling themselves the Zapatista National Liberation Army (ZNLA) launched a guerrilla war with coordinated attacks on several towns. By taking the name of a legendary leader of the 1910 Revolution (Emiliano Zapata) and by timing their assault to coincide with the birth of NAFTA, the ZNLA achieved maximum impact and publicity.

Although the Zapatistas had substantial public support and even brought their campaign as a social movement to Mexico City itself, their demands of better treatment for all of Mexico's 33 million Amerindian citizens did not yield significant results. President Vicente Fox in 2001 launched something called *Plan Puebla-Panama*, designed to stimulate economic development in the south, but little came of it. Various job creation schemes and agricultural projects failed to reverse the south's fortunes, and the government quieted the Zapatista rebels, who still hold some villages in Chiapas, by providing water and electricity to them. Meanwhile, in 2006 tensions in the neighboring State of Oaxaca produced an uprising when public school teachers went on strike after their grievances went unheeded, leading to death and destruction and damaging the tourist industry. Mexico's cultural, economic, and political divide is a growing threat to its future.

Regions of Mexico

Physiographic, demographic, economic, historical, and cultural criteria combine to reveal a regionally diverse Mexico extending from the lengthy ridge of Baja California to the tropical lowlands of the Yucatán Peninsula, and from the economic frenzy of the NAFTA North to the Amerindian traditionalism of the Chiapas southeast (Fig. 4-10). In the Core Area, anchored by Mexico City, and the West, centered on Guadalajara, lies the transition zone from the more Hispanic-mestizo north to the more Amerindian-infused mestizo south. East of the Core Area lies the Gulf Coast, once dominated by major irrigation projects and huge livestock-raising schemes but now the mainland center of Mexico's petroleum industry. The dry, scrub-vegetated Balsas Lowland separates the Core Area from the rugged, Pacific-fronting Southern Highlands, where Acapulco's luxury hotels stand in stark contrast to the Amerindian villages and *ejidos* of the interior, scene of major land reform in the wake of the 1910 revolution.

The dry, vast north stands in sharp contrast to these southern regions: only in the Northwest was there significant sedentary Amerindian settlement when the Spanish

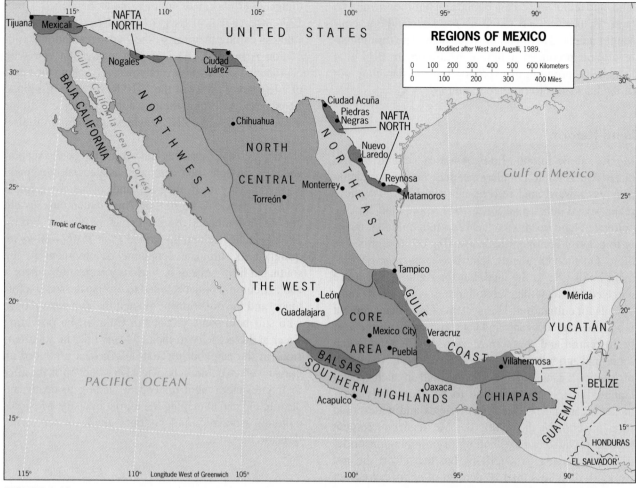

FIGURE 4-10

© H. J. de Blij, P. O. Muller, and John Wiley & Sons, Inc.

arrived. Huge haciendas, major irrigation projects, and some large cities separated by great distances mark this area, now galvanized by the impact of NAFTA. The NAFTA North region is still formative and discontinuous but is changing northern Mexico significantly. This is true even in Yucatán, where Mérida and environs are strongly affected by NAFTA development. To go from comparatively well-off northern Yucatán to poverty-stricken southern Chiapas is to see the whole range of Mexico's regional geography.

The Changing Geography of Economic Activity

During the last two decades of the twentieth century and in the first years of the twenty-first, Mexico's economic geography has changed, and in some respects progressed—though not without setbacks. During the early 1990s, the implementation of NAFTA led to an economic boom as Mexico became part of a free-trade zone and market comprising more than 400 million people. This

boom transformed urban landscapes along the 3115-kilometer (1936-mi) border between Mexico and the United States, but it could not, of course, close the economic gap between the two sides (see photo next page).

Today, Mexico is in progressive transition in many spheres: its democratic institutions are strengthening; its economy is more robust than it was at the time of NAFTA's inception; its once-rapid population growth is declining; its social fabric (notably relations with Amerindian minorities) is improving. The familiar problems of a country in transition, such as unchecked urbanization, inadequate infrastructure, corruption, and violent crime, will continue to afflict Mexico for decades to come. But, as the following discussion confirms, Mexico is a far stronger economy and society today than it was just one generation ago.

Agriculture

Although traditional subsistence agriculture and the output of the inefficient *ejidos* have not changed a great deal in the poorer areas of rural Mexico, larger-scale com-

The border between the United States and Mexico occasionally provides some stunning contrasts. Here, looking westward, we observe the opposing economic geographies that mark the border landscape of the Imperial Valley east of the urban area formed by Mexicali, Mexico and Calexico, California. The larger town of Mexicali (on the left) sprawls eastward along the border, with a cluster of maquiladora assembly plants, visible in the left foreground, surrounded by high-density, poor-quality housing. On the U.S. side, lush irrigated croplands blanket the otherwise dry countryside, fed by the All-American Canal that taps the waters of the Colorado River before they cross into Mexico. At this point, the canal swings northward to bypass border-hugging Calexico (right rear) before rejoining the international boundary beyond the town's western edge.
© Alex McLean/Landslides.

mercial agriculture has diversified during the past three decades and made major gains with respect to both domestic and export markets. The country's arid northern tier has led the way as major irrigation projects have been built on streams flowing down from the interior highlands. Along the booming northwest coast of the mainland, which lies within a day's drive of Southern California, mechanized large-scale cotton production now supplies an increasingly profitable export trade. Here, too, wheat and winter vegetables are grown, with fruit and vegetable cultivation attracting foreign investors.

But many Mexican small farmers are having a tough time of it, for two reasons. First, cheap corn from the United States, grown by American farmers heavily subsidized by the government, floods Mexico's markets, so that local farmers cannot make a profit. Second, the United States tried to create barriers against the import of low-priced Mexican produce such as avocados and tomatoes—violating the very rules NAFTA is supposed to stand for.

Energy and Industry

Until after 1990, Mexican industry was typical of a developing economy with state control over income-generating natural resources, among which oil and natural gas were (and are) the most productive. The government set up industries (such as the country's first steel plant, opened in Monterrey in 1903) without allowing competition. Employment in agriculture far exceeded that in manufacturing or mining, despite Mexico's substantial reserves of silver and other metals, high-quality coal (especially in Coahuila State), and an additional inventory ranging from antimony to zinc. Inefficiency prevailed, wages were low, inflation was rife, and Mexico's public debt was one of the highest in the world. The Mexican economy suffered from ups and downs directly related to world oil prices; when the price of oil dropped, federal expenditures had to be cut and deficits soared. In 2007, oil still accounted for more than one-third of total government revenues, a dangerous situation in a volatile world (the distribution of Mexican oil and gas reserves is mapped in Fig. 4-8).

During the 1980s, successive Mexican governments began trying to address these problems as part of the preparations for the momentous inception of NAFTA. Some of the inefficient mining and manufacturing operations were sold to private investors, and Mexico's financial systems (banking, tax collection) were strengthened. But with the inception of NAFTA in 1994, Mexico's industrial geography changed dramatically. Under the new agreement, factories based in Mexico could assemble imported, duty-free raw materials and components into finished products, which were then exported back to the United States market. Logically, these factories, called **maquiladoras**, would locate as close to the U.S. border as possible. As a result, manufacturing employment in the cities and towns along that border, from Tijuana in the west to Matamoros in the east, expanded rapidly. After only seven years of NAFTA's existence, there were 4000 factories with more than 1.2 million workers in the border zone and in northern Yucatán (where Mérida was part of the process), accounting for nearly one-third of Mexico's industrial jobs and 45 percent of its total exports.

NAFTA's impact on Mexico was far-reaching. Mexico benefited from foreign investment, job creation, tax receipts, and technology transfer. But Mexican employees worked long hours for low wages with few benefits and lived in the most basic shacks and slum dwellings encircling the burgeoning towns of the region we call the NAFTA North (Fig. 4-10). And there was no job security. Ten years after NAFTA's start, hundreds of American and other foreign corporations that had moved their factories to northern Mexico decided to relocate once again—to East and Southeast Asia where wages were even lower than the U.S. $2.00 per hour average paid by the maquiladoras. While factories assembling heavy and bulky items such as vehicles and refrigerators were still better off right across the border, others producing lighter and smaller goods such as electronic equipment and cameras moved to China, Vietnam, and other countries where wages were less than half of Mexico's. As a result, thousands of Mexican workers found themselves unemployed—and many crossed the border into the United States.

How can Mexico counter this trend? Here is one indicator: while low-wage maquiladora jobs were being lost to Asia (especially in textiles and apparel production), higher-wage jobs were being added in electronics, and Mexican company managers were soon in short supply. Education in high-tech and management fields is part of the answer. In the Northeast, the city of Monterrey in the high-income State of Nuevo Léon has a substantial international business community and modern industrial facilities that have attracted major multinational companies. The *Tecnológico* campus near Monterrey's airport, a degree-granting college supported by both government and private enterprise, lies at the heart of a network of more than 30 campuses throughout Mexico offering degrees in information technology, engineering, and administration. In the West, near Guadalajara, in the still-less-prosperous State of Jalisco, you can discern the beginnings of a "Silicon Valley of Mexico." In such places, the outlines of the landscape of the global economy are evident. Here lies hope for Mexico's future.

Geography of Inequality

The statistics of economic geography often hide the implications of the data for the people they purportedly represent. There is no doubt about it: NAFTA has enriched Mexico in general and northern Mexico in particular, and almost all the northern States bordering the United States have per-capita incomes above the national average. In fact, northern Mexico has historically been better off than

Hundreds of thousands of North American retirees, and many affluent purchasers of second homes, have sought the sun and low-cost residential opportunities of Middle America, converting some areas into virtual exclaves of the North. From Mexico to Panama, waterfront real estate is among the attractions for these permanent and seasonal migrants, as here in Cabo San Lucas at the southern tip of the Baja California peninsula.
© Liane Cory/Age Fotostock America, Inc.

the south—and NAFTA has widened the gap. We tend to think of NAFTA in terms of the industries it represents, but NAFTA also had an impact on Mexico's farmers. One stipulation of the agreement was that Mexico would lower and then drop its tariff barriers against U.S. and Canadian farm produce, including Mexico's staple, corn. But the United States generously subsidizes its corn farmers, who as a result can market corn in Mexico for prices even lower than Mexican farmers have to charge in order to make a small profit. In the process, hundreds of thousands of Mexican farmers are being put out of business. No wonder the U.S. government finds itself having to build fences to keep illegal immigrants out.

Mexico's north-south divide is starkly evident from the economic data (Fig. 4-11). In very general terms, the annual per-capita income in the northern States exceeds U.S. $10,000; in the southern States it falls below $5000. Economic growth in the northern States has averaged more than

1 QUERÉTARO
2 HIDALGO
3 MEXICO
4 MORELOS
5 TLAXCALA
6 AGUASCALIENTES
7 GUANAJUATO
8 MEXICO CITY
(DISTRITO FEDERAL)

MEXICO: ECONOMIC AND POLITICAL DISPARITIES

GDP PER CAPITA
National average=100

Above 149
125–149
100–124
75–99
50–74
Below 50

BORDER-ZONE MAQUILADORA EMPLOYMENT, 2006
100,000–300,000
50,000–100,000
15,000–50,000
5,000–15,000

Railroad
Road
NAFTA Highway

2006 PRESIDENTIAL VOTE BY STATE

WINNER OF EACH STATE
Calderón (Conservative)
Lopez Obrador (Populist)

FIGURE 4-11

© H. J. de Blij, P. O. Muller, and John Wiley & Sons, Inc.

4 percent in recent years; in southern States it is less than 2 percent. Far more people are poor in the south than in the north. Mexico's infrastructure, already inadequate, serves the south far less well than the north. Whatever the index—literacy, electricity use, water availability—the south lags.

These still-widening contrasts were thrown into sharp relief in 2006, when Mexico's most recent presidential election was contested by three candidates, of whom the

two leading ones were a conservative and a populist. When the ballots were counted, there was a near-tie, a result so close that the populist at first refused to concede. But on the map, it was clear that the north was won by the conservative candidate and the south by the populist (see inset map in Fig. 4-11). Mexico's electoral map thus showed that the country's regional economic disparities were having serious social consequences.

Mexico's Future

Mexico has accomplished much since the mid-1990s: it has strengthened its democratic institutions, brought greater stability to the economy, absorbed the impacts of NAFTA, and improved relations with its Amerindian minorities. Mexico's rate of population growth continues to decline, and while millions of Mexicans have emigrated to the United States since NAFTA's inception, fraying social fabrics, even this has a positive side: remittances from these workers in recent years are estimated to amount to between U.S. $25 and $30 billion, which is much greater than "official" foreign investment.

But, as noted above, the impact of globalization has fractured Mexico's economic and political geography, reflecting success in the north and failure in the south and deepening historic regional disparities. Northern Mexico is becoming more like North America; southern Mexico resembles Central America. The future of the country depends on the government's efforts to close this gap and on its plans to spread the positive effects of NAFTA from north to south. One such effort, a federal antipoverty program, requires families to keep their children in school, and it is having good effect. Another, the improvement of infrastructure to reduce the isolation of southern communities, has been less successful. This cannot be done by government alone, but Mexico's history of reliance on government (inefficient and corrupt as it has long been) is not easily overcome. Take a look at Figure 4-1: the dominance of Mexico City, the seat of power, is reflected by the national transport system. All roads (such as they are) seem to go to and through Mex-

ico City, creating a costly bottleneck that disadvantages the movement of southern products on their way to northern markets.

Mexico's future is inextricably bound up with the United States, in good times and bad. The good refers to the two countries' economic interaction: the United States is Mexico's most important trade partner. For all their differences over such issues as cross-border trucking, migration, and corporate practices in the maquiladora cities, the two governments have found ways to cooperate. The bad times were what NAFTA was supposed to remedy. But minor recessions in the American economy gave early warning that Mexico would quickly feel its effects, and when the major economic downturn began in 2008, the impact on NAFTA's Mexican operations—and thus the national economy—was staggering. At the same time, remittances from Mexican workers in the United States shrank as unemployment in construction and other industries rose sharply.

Clouds on Mexico's Horizon

Even as Mexico in 2009 suffered ever more serious economic woes, another crisis arose. Mexico had long been a waystation between producers of illegal drugs in South America and the world's richest single market for narcotics. In Chapter 5 we focus on the role of Colombian drug cartels in this illicit trade and the Colombian government's efforts (with major assistance from the United States) to eradicate the cocaine crop and arrest cartel leaders, many of whom were ensconced in cities such as Medellin and Cali. But over the past several years, the cartels that were centered in Colombia established new bases

It had been brewing for a long time, but the dramatic upsurge of drug-related violence in Mexico seemed sudden and unexpected when it happened in 2008. Drug cartels and their armed cadres committed murder, took hostages, tortured enemies, and created such mayhem, especially (and initially) in the northern maquiladora cities, that some American observers began to refer to Mexico as a 'failed state.' Certainly corruption and infiltration played their roles in Mexico's drug crisis, but the American market just across a porous border—and easy access to weapons in the United States—had much to do with Mexico's plight. In this photo, taken in April 2009 in the border city of Ciudad Juárez just across from El Paso, Texas, soldiers and local police guard forensic workers inspecting a car in which a prison guard was killed in a drive-by shooting. In 2008, cartel gunmen killed more than 6300 people in Mexico; the toll in 2009 already exceeded this number by early autumn. In June, a deadly four-hour gun battle in the Pacific tourist resort of Acapulco sent visitors scrambling for safety. Mexico's tourist industry, already afflicted by the global economic downturn, faced ruin. © Alejandro Bringas/Reuters/Landov in U.S.-border (and other) cities in northern Mexico and began a vicious battle for supremacy. They responded in part to the success of the antidrug campaign in Colombia, but they also saw new opportunities in Mexico directly across the border from their main market. So serious was the situation by mid-2009 that the drug barons and their henchmen were killing law-enforcement officers by the hundreds as well as competitors by the thousands, intimidating government and threatening all who stood in their way. Maps began appearing in the popular media to delineate which cartel prevailed in what part of Mexico, and in the United States some observers started referring to Mexico as a "failed state" because, in the words of its Director of National Intelligence, the Mexican government was not in control of parts of its own territory.

Fueling this rampage was the steady stream of weapons, bought with ease in the United States with the flood of drug money generated by that trade, which crossed the border and further empowered the cartels. The situation deteriorated so much that the once-busy day traffic between "twin" cities along the border (Fig. 4-11) dwindled to a trickle and American college students were warned against spring-break vacationing in certain Mexican resorts. Mexico's tourist industry was badly damaged; meanwhile, representatives of the U.S. government began referring to Mexico as a threat to national security. It surely was an ultimate geographic irony that, after reversing the devolutionary tide in its poverty-afflicted south, Mexico's government now faced what some called the state's "Colombianization" in the comparatively prosperous north.

Against this background, Mexico's longer-term problems, ranging from inadequate improvement of its infrastructure and inefficient management of its energy industry to a penchant for megaprojects and a neglect of regional integration, fade by comparison. Mexico is in a state of tumultuous transition, its progress impeded by a series of setbacks among which the current drug-related violence is only the latest. When times are good, Mexicans **11** dream of opportunities such as the so-called **dry canal** across the narrowest part of the country that would compete with the Panama Canal (see Fig. 4-8). When times are bad, the very fabric of the state seems to unravel, the cause not all of Mexico's making. For America's crucial neighbor, this again is such a time.

● THE CENTRAL AMERICAN REPUBLICS

Crowded onto the narrow segment of the Middle American land bridge between Mexico and the South American continent are the seven countries of Central America (Fig. 4-12). Territorially, they are all quite small; their population sizes range from Guatemala's nearly 15 million down to Belize's 300,000. Physiographically, the land bridge here consists of a highland belt flanked by coastal lowlands on both the Caribbean and Pacific sides (Fig. 4-1). These highlands are studded with volcanoes, and local areas of fertile volcanic soils are scattered throughout them. From earliest times, the region's inhabitants have been concentrated in this upland zone, where tropical temperatures are moderated by elevation and rainfall is sufficient to support a variety of crops.

Altitudinal Zonation of Environments

Continental Middle America and the western margin of South America are areas of high relief and strong environmental contrasts. Even though settlers have always favored temperate intermontane basins and valleys, people also cluster in hot tropical lowlands as well as high plateaus just below the snow line in South America's Andes Mountains. In each of these zones, distinct local climates, soils, vegetation, crops, domestic animals, and modes of life prevail. Such **altitudinal zones** (dia- **12** grammed in Fig. 4-13) are known by specific names as if they were regions with distinguishing properties—as, in reality, they are.

The lowest of these vertical zones, from sea level to 750 meters (2500 ft), is known as the *tierra* **13** *caliente*, the "hot land" of the coastal plains and low-lying interior basins where tropical agriculture predominates. Above this zone lie the tropical highlands containing Middle and South America's largest population clusters, the *tierra templada* of temperate **14** land reaching up to about 1800 meters (6000 ft). Temperatures here are cooler; prominent among the commercial crops is coffee, while corn (maize) and wheat are the staple grains. Still higher, from about 1800 to 3600 meters (6000 to nearly 12,000 ft), is the *tierra fría*, the cold country of the higher Andes where **15** hardy crops such as potatoes and barley are mainstays. Above the tree line, which marks the upper limit of the *tierra fría*, lies the *tierra helada*; this **16** fourth altitudinal zone, extending from about 3600 to 4500 meters (12,000 to 15,000 ft), is so cold and barren that it can support only the grazing of sheep and other hardy livestock. The highest zone of all is the *tierra nevada*, a zone of permanent snow and ice **17** associated with the loftiest Andean peaks. As we will see, the varied human geography of Middle and western South America is closely related to these diverse environments.

FIGURE 4-12

© H. J. de Blij, P. O. Muller, and John Wiley & Sons, Inc.

Population Patterns

Figure 4-3 shows that Central America's population tends to concentrate in the uplands of the *tierra templada* and that population densities are generally greater toward the Pacific than toward the Caribbean side. The most significant exception is El Salvador, whose political boundaries confine its people mostly to its tropical *tierra caliente*, tempered here by the somewhat cooler Pacific offshore. On the opposite side of the isthmus, Belize has the typically sparse population of the hot, wet Caribbean coastlands and their infertile soils. Panama is the only other exception to the rule, but for different reasons. Economic development has focused on the Panama Canal and the coasts, although new settlement is now moving onto the mountain slopes of the Pacific side.

Central America, we noted earlier, actually begins within Mexico, in Chiapas and in Yucatán, and the region's republics face many of the same problems as less-devel-

oped parts of Mexico. Population pressure is one of these problems. A population explosion began in the mid-twentieth century, increasing the region's human inhabitants from 9 million to almost 45 million by 2010. Unlike Mexico, Central America's population growth is not yet slowing down significantly except in Costa Rica and Panama, and population geographers talk of a demographic crisis in the making.

Emergence from a Turbulent Era

Devastating inequities, repressive governments, external interference, and the frequent unleashing of armed forces have destabilized Central America for much of its modern history. The roots of these upheavals are old and deep, and today the region continues its struggle to emerge from a period of turmoil that lasted through the 1980s into the mid-1990s.

Central America is not a large region, but because of its physiography it contains many isolated, comparatively inaccessible locales. Conflicts between Amerindian population clusters and mestizo groups are endemic to the region, and contrasts between the privileged and the poor are especially harsh. Dictatorial rule by local elites followed authoritarian rule by Spanish colonizers.

The Gang Problem

Almost everywhere in Central America today (and in parts of Mexico as well) a growing social problem is causing concern: the rising tide of gang violence. International criminal gangs of young and unemployed men, the most infamous of which currently is the *Mara Salvatrucha*, are committing random acts of violence (such as an attack on civilians in San Pedro Sula, Honduras, in late 2004 that cost 28 lives) and are provoking paramilitary death squads into combating them, resulting in increased murder rates in the region's cities. In El Salvador, for example, the homicide rate rose from 37 per 100,000 in 2002 to 48 in 2006. The problem reflects not only chronic unemployment, but also migration (including the return of deported criminals), corruption in law enforcement, and living conditions in the poorest neighborhoods.

The Promise of CAFTA

Political leaders and economic planners in the United States have long advocated a "free-trade" agreement to boost the economies of Central American countries. In 2005, this campaign resulted in the U.S. Congress's approval of the Central American Free Trade Agreement (CAFTA) linking the huge U.S. economy to five small Central American economies (Guatemala, El Salvador, Honduras, Nicaragua, and Costa Rica) and one Caribbean economy (the Dominican Republic). This is not the first time such an integrative effort has been made in Central America: the Central American Common Market, created in 1960, fell apart within a decade.

But today the region is more stable, the United States has stopped supporting dictators, democratic governments are holding their own, and trade between the republics and the United States, while still modest, is expanding. It is true, as critics of CAFTA argue, that many of its free-trade provisions are already in place, the result of earlier bilateral agreements, and that other tariffs will be reduced gradually over a period of two decades. But CAFTA will do more than enhance trade: it will signal to the world that the region has turned a corner. It will encourage investment, force governments to be more open and less corrupt, expand trade among these five Central American republics themselves, promote economic diversification, and thus reduce excessive dependence on farm products. It will take a long time for Central America's vital statistics (see the Data Table inside the back cover) to improve significantly, and CAFTA will have some negative as well as many positive consequences. But overall, its advantages outweigh its liabilities.

The Seven Republics

Guatemala, the westernmost of Central America's republics, has more land neighbors than any other. Straight-line boundaries across the tropical forest mark

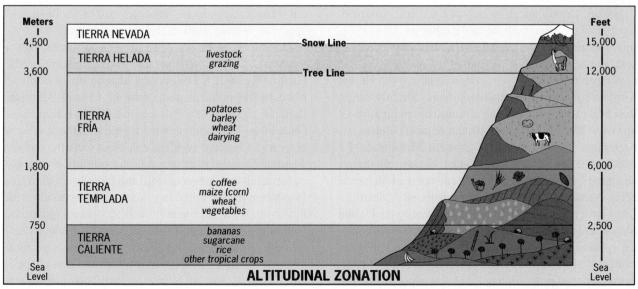

FIGURE 4-13   © H. J. de Blij, P. O. Muller, and John Wiley & Sons, Inc.

much of the border with Mexico, creating the box-like region of Petén between Chiapas State on the west and Belize on the east; also to the east lie Honduras and El Salvador (Fig. 4-12). This heart of the ancient Maya Empire, which remains strongly infused by Amerindian culture and tradition, has just a small window on the Caribbean but a longer Pacific coastline. Guatemala was still part of Mexico when the Mexicans threw off the Spanish yoke, and though independent from Spain after 1821, it did not become a separate republic until 1838. Mestizos, not the Amerindian majority, secured the country's independence.

Most populous of the seven republics with 14.5 million inhabitants (mestizos are in the majority with 55 percent, Amerindians 43 percent), Guatemala has seen much conflict. Repressive regimes made deals with U.S. and other foreign economic interests that stimulated development, but at a high social cost. Over the past half-century, military regimes have dominated political life. The deepening split between the wretchedly poor Amerindians and the better-off mestizos, who here call themselves *ladinos*, generated a civil war that started in 1960 and claimed more than 200,000 lives as well as 50,000 "disappearances" before it ended in 1996. An overwhelming number of the victims were of Mayan descent; the mestizos continue to control the government, army, and land-tenure system.

Guatemala's tragedy is that its economic geography has considerable potential but has long been shackled by the unending internal conflicts that have kept the income of 80 percent of the population below the poverty line. The country's mineral wealth includes nickel in the highlands and oil in the lower-lying north. Agriculturally, soils are fertile and moisture is ample over highland areas large enough to produce a wide range of crops including excellent coffee.

Belize, strictly speaking, is not a Central American republic in the same tradition as the other six. Until 1981, this country, a wedge of land between northern Guatemala, Mexico's Yucatán Peninsula, and the Caribbean (Fig. 4-12), was a dependency of the United Kingdom known as British Honduras. Slightly larger than Massachusetts and with a minuscule population of just over 300,000 (many of African descent), Belize has been more reminiscent of a Caribbean island than of a continental Middle American state. Today, all that is changing as the demographic complexion of Belize is being reshaped. Thousands of residents of African descent have recently emigrated (many went to the United States) and were replaced by tens of thousands of Spanish-speaking immigrants. Most of the latter are refugees from strife in nearby Guatemala, El Salvador, and Honduras, and their proportion of the Belizean population has risen from 33 to nearly 50 percent since 1980. Within the next few years the newcomers will be in the majority, Spanish will become the *lingua franca*, and Belize's cultural geography will exhibit an expansion of the Mainland at the expense of the Rimland.

The Belizean transformation extends to the economic sphere as well. No longer just an exporter of sugar and bananas, Belize is producing new commercial crops, and its seafood-processing and clothing industries have become major revenue earners. Also important is tourism, which annually lures more than 150,000 vacationers to the country's Mayan ruins, resorts, and newly legalized casinos; a growing speciality is ecotourism, based on the natural attractions of the country's near-pristine environment. Belize is also known as a center for *offshore banking*—a financial haven for foreign companies and individuals who want to avoid paying taxes in their home countries.

Honduras is a country still on hold as it struggles to rebuild its battered infrastructure and economy more than a decade after taking a direct hit from Category-5 Hurricane Mitch. The consequences of this 1998 tropical cyclone were catastrophic: massive floods and mudslides unleashed across the country killed 9200 people, demolished more than 150,000 homes, destroyed 21,000 miles of roadway and 335 bridges, and rendered at least 2 million homeless. Also devastated was the critical agricultural sector that employed two-thirds of Honduras's labor force, accounted for nearly a third of its gross domestic product, and earned more than 70 percent of its foreign revenues.

With 7.6 million inhabitants, about 90 percent mestizo, bedeviled Honduras still has years to go even to restore what was already the third-poorest economy in the Americas (after Haiti and Nicaragua). Agriculture, livestock, forestry, and limited mining formed the mainstays of the pre-1998 economy, with the familiar Central American products—bananas, coffee, shellfish, and apparel—earning most of the external income.

Honduras, in direct contrast to Guatemala, has a lengthy Caribbean coastline and a small window on the Pacific (Fig. 4-12). The country also occupies a critical place in the political geography of Central America, flanked as it is by Nicaragua, El Salvador, and Guatemala—all continuing to grapple with the aftermath of years of internal conflict and, most recently, natural disaster. The road back to economic viability is an arduous one, but once traversed will still leave four out of five Hondurans deeply mired in poverty and the country with little overall improvement in its development prospects.

El Salvador is Central America's smallest country territorially, smaller even than Belize, but with a population 25 times as large (7.5 million) it is the most densely peopled. With Belize, it is one of only two continental republics that lack coastlines on both the Caribbean and

Pacific sides (Fig. 4-12). El Salvador adjoins the Pacific in a narrow coastal plain backed by a chain of volcanic mountains, behind which lies the country's heartland. Unlike neighboring Guatemala, El Salvador has a quite homogeneous population (90 percent mestizo and just 1 percent Amerindian). Yet ethnic homogeneity has not translated into social or economic equality or even opportunity. Whereas other Central American countries were called banana republics, El Salvador was a coffee republic, and the coffee was produced on the huge landholdings of a few landowners and on the backs of a subjugated peasant labor force. The military supported this system and repeatedly suppressed violent and desperate peasant uprisings.

From 1980 to 1992, El Salvador was torn by a devastating civil war that was worsened by outside arms supplies from the United States (supporting the government) and Nicaragua (aiding the Marxist rebel forces). But ever since the negotiated end to that war, efforts have been under way to prevent a recurrence because El Salvador is having difficulty overcoming its legacy of searing inequality. The civil war did have one positive result: affluent citizens who left the country and did well in the United States and elsewhere send substantial funds back home, which now provide the largest single source of foreign revenues. This has helped stimulate such industries as apparel and footwear manufacturing as well as food processing. But a major stumbling block to revitalization of the agricultural sector has again been land reform, and El Salvador's future still hangs in the balance.

Nicaragua is best approached by reexamining the map (Fig. 4-12), which underscores the country's pivotal position in the heart of Central America. The Pacific coast follows a southeasterly direction, but the Caribbean coast is oriented north-south so that Nicaragua forms a triangle of land with its lakeside capital, Managua, located in a valley on the mountainous, earthquake-prone, Pacific side (the country's core area has always been located here). The Caribbean side, where the uplands yield to a coastal plain of rainforest, savanna, and swampland, has for centuries been home to Amerindian peoples such as the Miskito, who have been remote from the focus of national life.

Until the end of the 1970s, Nicaragua was the typical Central American republic, ruled by a dictatorial government and exploited by a wealthy land-owning minority, its export agriculture dominated by huge plantations owned by foreign corporations. It was a situation ripe for insurgency, and in 1979 leftist rebels overthrew the government. But the new regime quickly produced its own excesses, resulting in civil war through most of the 1980s, a conflict in which the United States got involved by arming the rebel *Contras*, who were trying to undermine the left-wing Sandinista regime. This strife ended in 1990, and since then more democratic governments have been voted into office. In 2006, Nicaraguans elected a former Sandinista leader, Daniel Ortega, back into office, to the consternation of the American government.

Nicaragua's economy has been a leading casualty of this turmoil, and for the past two decades it has ranked as continental Middle America's poorest. Hurricane Mitch struck here too, devastating the country's farms and driving tens of thousands into the impoverished towns just at a time when the agricultural sector was recovering and land reform promised a better life for some 200,000 peasant families.

Nicaragua's options are limited: none of the billions of aid dollars headed for Bosnia, Iraq, and Subsaharan Africa will be matched here. For years there has been talk of a *dry canal* transit role for this country, but other potential overland routes are more promising. Meanwhile, the mushrooming of Nicaragua's population continues (2010 total: 5.9 million), dooming hopes for a rise in standards of living.

Costa Rica underscores what was said about Middle America's endless variety and diversity because it differs significantly from its neighbors and from the norms of Central America as well. Bordered by two volatile countries (Nicaragua to the north and Panama to the east), Costa Rica is a nation with an old democratic tradition and, in this cauldron, no standing army for the past six decades! Although the country's Hispanic imprint is similar to that found elsewhere on the Mainland, its early independence, its good fortune to lie remote from regional strife, and its leisurely pace of settlement allowed Costa Rica the luxury of concentrating on its economic development. Perhaps most important, internal political stability has prevailed over much of the past 175 years.

Like its neighbors, Costa Rica is divided into environmental zones that parallel the coasts. The most densely settled is the central highland zone, lying in the cooler *tierra templada*, whose heartland is the *Valle Central* (Central Valley), a fertile basin that contains the country's main coffee-growing area and the leading population cluster focused on San José (Figs. 4-3; 4-12)—the most cosmopolitan urban center between Mexico City and the primate cities of northern South America. To the east of the highlands are the hot and rainy Caribbean lowlands, a sparsely populated segment of Rimland where many plantations have been abandoned and replaced by subsistence farming. Between 1930 and 1960, the U.S.-based United Fruit Company shifted most of the country's banana plantations from this crop-disease-ridden coastal plain to Costa Rica's third zone—the plains and gentle slopes of the Pacific coastlands. This move gave the Pacific zone a major boost in economic growth, and it is now an area of diversifying and expanding commercial agriculture.

Tropical Deforestation

18

BEFORE THE EUROPEANS arrived, two-thirds of continental Middle America was covered by tropical rainforests. The clearing and destruction of this precious woodland resource to make way for expanding settlement frontiers and the exploitation of new economic opportunities began in the sixteenth century during the Spanish colonial era, and the practice has continued systematically ever since. In recent decades, however, the pace of **tropical deforestation** in Central America has accelerated alarmingly, and since 1950 fully 90 percent of the region's forests have been decimated. Over the past decade, about 1.2 million hectares (3 million acres) of Central American and Mexican woodland disappeared annually, an area equivalent to one-third the size of Belgium. El Salvador has already lost nearly all of its forests, and most of the six other republics will soon approach that stage.

The causes of tropical deforestation are related to the persistent economic and demographic problems of disadvantaged countries. In Central America, the leading cause has been the need to clear rural lands for cattle pasture as many countries, especially Costa Rica, became meat producers and exporters. The price of this environmental degradation has been enormous, although some gains have been recorded. Because tropical soils are so nutrient-poor, newly deforested areas can function as pastures for only a few years at most. These fields are then abandoned for other freshly cut lands and quickly become the ravaged landscape seen in the photo in this box. Without the protection of trees, local soil erosion and flooding immediately become problems, affecting still-productive nearby areas (a sequence of events that reached catastrophic dimensions all across Honduras and northern Nicaragua when Hurricane Mitch struck in 1998). A second cause of deforestation is the rapid logging of tropical woodlands as the timber industry increasingly turns from the exhausted forests of the midlatitudes to harvest the rich tree resources of the equatorial zones, responding to accelerating global demands for housing, paper, and furniture. The third major contributing factor is related to the region's population explosion: as more and more peasants are required to extract a subsistence from inferior lands, they have no choice but to cut down the remaining forest for both firewood and additional crop-raising space, and their intrusion prevents the trees from regenerating (Haiti is the realm's extreme example of this denudation process).

Although deforestation is a depressing event, tropical pastoralists, farmers, and timber producers do not consider it to be life-threatening, and perhaps it even seems to offer some short-term economic advantages. Why, then, should there be such an outcry from the scientific community? And why should the World Resources Institute call this "the world's most pressing land-use problem"? The answer is that unless immediate large-scale action is taken, by the middle of this century the world's tropical rainforests will be reduced to two disappearing patches—the western Amazon Basin of northern South America and the middle Congo Basin of Equatorial Africa.

The tropical forest, therefore, must be a very important part of our natural world—and indeed it is. Biologically, the rainforest is by far the richest, most diversified arena of life on our planet: even though it covers only about (a shrinking) 3 percent of the Earth's land area, it contains about three-quarters of all plant and animal species. Its loss would cause not only the extinction of millions of species, but also what ecologist Norman Myers calls "the death of birth" because the evolutionary process that produces new species would be terminated. Because tropical rainforests already yield countless valuable medicinal, food, and industrial products, many potential disease-combating drugs or new crop varieties to feed undernourished millions will be irretrievably lost if they are allowed to disappear.

This scene in Costa Rica shows how badly the land can be scarred in the wake of deforestation. Without roots to bind the soil, tropical rains swiftly erode the unprotected topsoil. © Peter Poulides.

The long-term development of Costa Rica's economy has given it the region's highest standard of living, literacy rate, and life expectancy. Agriculture continues to dominate (with bananas, coffee, tropical fruits, and seafood the leading exports), and tourism is expanding steadily. Costa Rica is widely known for its superb scenery and for its efforts to protect what is left of its diverse tropical flora and fauna. But rainforest destruction is the regionwide price of human population growth and woodland exploitation, and even here more than 80 percent of the original forest has vanished (see box titled "Tropical Deforestation").

Still, the country's veneer of development cannot mask serious problems. In terms of social structure, about one-quarter of its population of 4.6 million is trapped in an unending cycle of poverty, and the huge gap between the poor and the affluent is constantly widening. With volatile neighbors and an economy that remains insufficiently diversified against risk, Costa Rica is only one step ahead of its regional partners.

Panama owes its existence to the idea of a canal connecting the Atlantic and Pacific oceans to avoid the lengthy circumnavigation of South America. In the 1880s, when Panama was still an extension of neighboring Colombia, a French company tried and failed to build such a waterway here. By the turn of the twentieth century, U.S. interest in a Panama canal rose sharply, and the United States in 1903 proposed a treaty that would permit a renewed effort at construction across Colombia's Panamanian isthmus. When the Colombian Senate refused to go along, Panamanians rebelled and the United States supported this uprising by preventing Colombian forces from intervening. The Panamanians, at the behest of the United States, declared their independence from Colombia, and the new republic immediately granted the United States rights to the Canal Zone, averaging about 15 kilometers (10 mi) in width and just over 80 kilometers (50 mi) in length.

Soon canal construction commenced, and this time the project succeeded as American technology and medical advances triumphed over a formidable set of obstacles. The Panama Canal (see the inset map in Fig. 4-12) was opened in 1914, a symbol of U.S. power and influence in Middle America. The Canal Zone was held by the United States under a treaty that granted it "all the rights, powers, and authority" in the area "as if it were the sovereign of the territory." Such language might suggest that the United States held rights over the Canal Zone in perpetuity, but the treaty nowhere stated specifically that Panama permanently yielded its own sovereignty in that transit corridor. In the 1970s, as the canal was transferring more than 14,000 ships per year (that number is now only slight-

FROM THE FIELD NOTES

"The Panama Canal remains an engineering marvel 90 years after it opened in August 1914. The parallel lock chambers each are 1000 feet long and 110 feet wide, permitting vessels as large as the Queen Elizabeth II to cross the isthmus. Ships are raised by a series of locks to Gatún Lake, 85 feet above sea level. We watched as tugs helped guide the QEII into the Gatún Locks, a series of three locks leading to Gatún Lake, on the Atlantic side. A container ship behind the QEII is sailing up the dredged channel leading from the Limón Bay entrance. The lock gates are 65 feet wide and 7 feet thick, and range in height from 47 to 82 feet. The motors that move them are recessed in the walls of the lock chambers. Once inside the locks, the ships are pulled by powerful locomotives called mules that ride on rails that ascend and descend the system. It was still early morning, and a major fire, probably a forest fire, was burning near the city of Colón, where land clearing was in progress. This was the beginning of one of the most fascinating days ever." © H. J. de Blij

www.conceptcaching.com

ly lower, but the cargo tonnage is up significantly) and generating hundreds of millions of dollars in tolls, Panama sought to terminate U.S. control in the Canal Zone. Delicate negotiations began. In 1977, an agreement was reached on a staged withdrawal by the United States from the territory, first from the Canal Zone and then from the Panama Canal itself (a process completed on December 31, 1999).

Panama today reflects some of the usual geographic features of the Central American republics. Its population of 3.5 million is about 70 percent mestizo and also contains substantial Amerindian, white, and black minorities. Spanish is the official language, but English is also widely used. Ribbon-like and oriented east-west, Panama's topography is mountainous and hilly. Eastern Panama, especially Darien Province adjoining Colombia, is densely forested, and here is the only remaining gap in the intercontinental Pan American Highway. Most of the rural population lives in the uplands west of the canal; there, Panama produces bananas, shrimps and other seafood, sugarcane, coffee, and rice. Much of the urban population is concentrated in the vicinity of the waterway, anchored by the cities at each end of the canal.

Near the northern end of the Panama Canal lies the city of Colón, site of the Colón Free Zone, a huge trading *entrepôt* designed to transfer and distribute goods bound for South America. It is augmented by the Manzanillo International Terminal, an ultramodern port facility capable of transshipping more than 1000 containers

a day. By 2002 China had become the third-largest user of the Canal (after the United States and Japan) and accounted for more than 20 percent of the cargo entering the Colón Free Zone. Near the southern end lies Panama City, the "Miami of the Caribbean" because of its waterfront location and skyscrapered skyline (see photo, p. 194). The capital is the financial center that handles the funds generated by the Canal, but its high-rise profile also reflects the proximity of Colombia's illicit drug industry and associated money-laundering and corruption. Panamanians' pervasive poverty presents a stark contrast to the city's modern image.

The Panama Canal has political as well as economic implications for Panama. For many years, Panama has recognized Taiwan as an independent entity and has sponsored resolutions to get Taiwan readmitted to the United Nations (where it was ousted in favor of China in 1971). In return, Taiwan has spent hundreds of millions of dollars in Panama in the form of investment and direct aid. But now China's growing presence in the Canal and the Colón Free Zone is causing a conundrum. Panama has awarded a contract to a Hong Kong-based (and thus Chinese) firm to operate and modernize the ports at both ends of the Canal, resulting in investments approaching U.S. $500 million. Not surprisingly, China has pressed Panama for a switch in recognition, using this commitment as leverage.

Meanwhile, many huge ships now sailing the oceans are too large to transit the Canal, and the Panama Canal Authority embarked on a plan to modernize the waterway by building new locks that will be able to handle vessels twice as large as the current maximum. Not only will this reverse the decline in the share of world cargo passing through the Canal, but it will also increase the Canal's efficiency. That will make potential competitors (such as Mexico with its dry-canal plan and Nicaragua with a scheme that would exploit Lake Managua in a mixed land-water venture) less eager to make huge investments in such projects.

In October 2006 Panamanian voters overwhelmingly approved the modernization plan, which will involve building a new set of locks that will be 40 percent longer and 60 percent wider than the existing ones (see diagram at left). The cost will be financed by charging ships more for passage and by international loans (China is sure to play its role), and the project is due to be completed in time for the Canal's centenary in 2014. To what extent it will help ordinary Panamanians escape poverty (which afflicts 40 percent of the population) remains uncertain.

Diagram showing the planned expansion and upgrading of the Panama Canal's lock system in order to accommodate ships that currently exceed the waterway's ("Panamax") capacity. The target date for completing the project is August 15, 2014, the one hundredth anniversary of the first interocean transit of the Canal by the passenger-cargo vessel, *S.S. Ancon*. © NG Image Collection.

THE CARIBBEAN BASIN

As Figure 4-2 reveals, the Caribbean Basin, Middle America's island region, consists of a broad arc of numerous islands extending from the western tip of Cuba to the southern coast of Trinidad. The four larger islands or *Greater Antilles* (Cuba, Hispaniola, Jamaica, and Puerto Rico) are clustered in the western half of this arc. The smaller islands or *Lesser Antilles* extend to the east in a crescent-shaped zone from the Virgin Islands to Trinidad and Tobago. (Breaking this tectonic plate-related regularity are the Bahamas and the Turks and Caicos, north of the Greater Antilles, and numerous other islands too small to appear on a map at the scale of Fig. 4-2.)

On these islands, whose combined land area constitutes only 9 percent of Middle America, lie 33 states and several other entities. (Europe's colonial flags have not totally disappeared from this region, and the U.S. flag flies over Puerto Rico.) The populations of these states and territories, however, comprise 21 percent of the entire geographic realm, making this the most densely peopled part of the Americas.

Economic and Social Patterns

The Caribbean is also one of the poorest regions of the world. Environmental, demographic, political, and economic circumstances combine to impede development in the region at almost every turn. The Caribbean Basin is scenically beautiful, but it is a difficult place to make a living, and poverty is the norm.

It was not always thus. Untold wealth flowed to the European colonists who invaded this region, enslaved and eventually obliterated the indigenous Amerindian (Carib) population, and brought to these islands in bondage the Africans to work on the sugar plantations that made the colonists rich. By the time competition from elsewhere ended the near-monopoly Caribbean sugar had enjoyed on European markets, the seeds of disaster had been sowed.

As the sugar trade collapsed, millions were pushed into a life of subsistence, malnutrition, and even hunger. Population growth now created ever-greater pressure on the land, but Caribbean islanders had few emigration options in their fragmented, mountainous habitats. The American market for sugar allowed some island countries to revive their sugar exports, including the Dominican Republic and Jamaica (Cuba's exports went to the Soviet Union, but when the USSR disintegrated Cuba's sugar industry was doomed). Some agricultural diversification occurred in the Lesser Antilles where bananas, spices, and sea-island cotton replaced sugar, but farming is a risky profession in the Caribbean. Foreign markets are undependable, and producers are trapped in a disadvantageous international economic system they cannot change.

Not surprisingly, many farmers and their families simply abandon their land and leave for towns and cities, hoping to do better there. Although the Caribbean region

Labadie, Haiti: Cruise ships from one of the richest countries in the world visit the shores of the Western Hemisphere's poorest. Because tourism provides jobs at coastal resorts favored by the visitors and generates revenues for local governments, the industry is welcomed. But most of the thousands of passengers sailing on ships like this never see the malnourished children, open sewage ditches, or fetid urban slums where UN peacekeeping forces struggle to keep order. At right, an appalling example of these most desperate of human environments is this hillside shacktown overlooking the capital, Port-au-Prince, where the most rudimentary structures are precariously crammed into every available space—on a steep slope fully exposed to the hazards of rainy-season downpours and intermittent tropical storms and hurricanes.   Left: © Ruth Fremson/The New York Times/Redux Pictures.   Right: © Charles Trainor Jr./Miami Herald Staff.

is less urbanized, overall, than other parts of the Americas, it is far more urbanized than, say, Africa or the Pacific Realm. Nearly two-thirds of the Caribbean's population now live in cities such as Santo Domingo in the Dominican Republic, Havana in Cuba, Port-au-Prince in Haiti, and San Juan in Puerto Rico. Many of the region's cities reflect the poverty of the citizens who were driven to seek refuge there. Port-au-Prince has some of the world's worst slums (see right photo, p. 223), and desolate squatter settlements ring cities such as Kingston, Jamaica and many smaller towns.

Ethnicity and Advantage

The human geography of the Caribbean region still carries imprints of the cultures of Subsaharan Africa. And the legacy of European domination also lingers. The historical geography of Cuba, Hispaniola, and Puerto Rico is suffused with Hispanic culture; Haiti and Jamaica carry stronger African legacies. But the reality of this ethnic diversity is that European lineages still hold the advantage. Hispanics tend to be in the best positions in the Greater Antilles; people who have mixed European-African ancestries, and who are described as **mulatto**, rank next. The largest part of this social pyramid is also the least advantaged: the Afro-Caribbean majority. In virtually all societies of the Caribbean, the minorities hold disproportionate power and exert overriding influence. In Haiti, the mulatto minority accounts for barely 5 percent of the population but has long held most of the power. In the adjacent Dominican Republic, the pyramid of power puts Hispanics (16 percent) at the top, the mixed sector (73 percent) in the middle, and the Afro-Caribbean minority (11 percent) at the bottom. Historic advantage has a way of perpetuating itself.

The composition of the population of the islands is further complicated by the presence of Asians from both China and India. During the nineteenth century, the emancipation of slaves and ensuing local labor shortages brought some far-reaching solutions. Some 100,000 Chinese emigrated to Cuba as indentured laborers, and Jamaica, Guadeloupe, and especially Trinidad saw nearly 250,000 South Asians arrive for similar purposes. To the African-modified forms of English and French heard in the Caribbean, therefore, can be added several Asian languages. The ethnic and cultural variety of the plural societies of Caribbean America is indeed endless.

Tourism: Promising Alternative?

Given the Caribbean Basin's limited economic options, does the tourist industry offer better opportunities? Opinions on this question are divided (see the Issue Box entitled "The Role of the Tourist Industry in Middle American Economies"). The resort areas, scenic treasures, and historic locales of Caribbean America attract well over 20 million visitors annually, with about half of these tourists traveling on Florida-based cruise ships. Certainly, Caribbean tourism is a prospective money-maker for many islands. In Jamaica alone, this industry now accounts for about one-sixth of the gross domestic product and employs more than one-third of the labor force.

But Caribbean tourism also has serious drawbacks. The invasion of poor communities by affluent tourists contributes to rising local resentment, which is further fueled by the glaring contrasts of shiny new hotels towering over substandard housing and luxury liners gliding past poverty-stricken villages. At the same time, tourism can have the effect of debasing local culture, which often is adapted to suit the visitors' tastes at hotel-staged "culture" shows. And while tourism does generate income in the Caribbean, the intervention of island governments and multinational corporations removes opportunities from local entrepreneurs in favor of large operators and major resorts.

THE GREATER ANTILLES

The four islands of the Greater Antilles contain five political entities: Cuba, Haiti, the Dominican Republic, Jamaica, and Puerto Rico (Fig. 4-2). Haiti and the Dominican Republic share the island of Hispaniola.

Cuba, the largest Caribbean island-state in terms of both territory (111,000 square kilometers/43,000 sq mi) and population (11.3 million), lies only 145 kilometers (90 mi) from the southern tip of Florida (Fig. 4-14). Havana, the now-dilapidated capital, lies almost directly across from the Florida Keys on the northwest coast of the elongated island. Cuba was a Spanish possession until the late 1890s when, with American help in the Spanish-American War, it attained independence. Fifty years later, a U.S.-backed dictator was in control, and by the 1950s Havana had become an American playground. The island was ripe for revolution, and in 1959 Fidel Castro's insurgents gained control, thereby converting Cuba into a communist dictatorship and a Soviet client. In the end, Castro's rule survived the collapse of the Soviet Empire despite the loss of subsidies and sugar markets on which it had long relied.

As the map shows, sugar was Cuba's economic mainstay for many years; the plantations, once the property of rich landowners, extend all across the territory. But as this map also shows, sugarcane is losing its position as the leading Cuban foreign exchange earner. Mills are being closed down, and the canefields are being cleared

The Role of the Tourist Industry in Middle American Economies

IN SUPPORT OF THE TOURIST INDUSTRY

"As the general manager of a small hotel in St. Maarten, on the Dutch side of the island, I can tell you that without tourists, we would be in deep trouble economically. Mass tourism, plain and simple, has come to the rescue in the Caribbean and in other countries of Middle America. Look at the numbers. Here it's the only industry that is growing, and it already is the largest worldwide industry. For some of the smaller countries of the Caribbean, this is not just the leading industry but the only one producing external revenues. Whether it's hotel patrons or cruise-ship passengers, tourists spend money, create jobs, fill airplanes that give us a link to the outside world, require infrastructure that's good not just for them but for locals too. We've got better roads, better telephone service, more items in our stores. All this comes from tourism. I employ 24 people, most of whom would be looking for nonexistent jobs if it weren't for the tourist industry.

"And it isn't just us here in the Caribbean. Look at Belize. I just read that tourism there, based on their coral reefs, Mayan ruins, and inland waterways, brought in U.S. $100 million last year, in a country with a total population of only about 300,000! That sure beats sugar and bananas. In Jamaica, they tell me, one in every three workers has a job in tourism. And the truth is, there's still plenty of room for the tourist industry to expand in Middle America. Those Americans and Europeans can close off their markets against our products, but they can't stop their citizens from getting away from their awful weather by coming to this tropical paradise.

"Here's another good thing about tourism. It's a clean industry. It digs no mine shafts, doesn't pollute the atmosphere, doesn't cause diseases, doesn't poison villagers, isn't subject to graft and corruption the way some other industries are.

"Last but not least, tourism is educational. Travel heightens knowledge and awareness. There's always a minority of tourists who just come to lie on the beach or spend all their time in some cruise-ship bar, but most of the travelers we see in my hotel are interested in the place they're visiting. They want to know why this island is divided between the Dutch and the French, they ask about coral reefs and volcanoes, and some even want to practice their French on the other side of the border (don't worry, no formalities, just drive across and start talking). Tourism's the best thing that happened to this part of the world, and other parts too, and I hope we'll never see a slowdown."

CRITICAL OF THE TOURIST INDUSTRY

"You won't get much support for tourism from some of us teaching at this college in Puerto Rico, no matter how important some economists say tourism is for the Caribbean. Yes, tourism is an important source of income for some countries, like Kenya with its wildlife and Nepal with its mountains, but for many countries that income from tourism does not constitute a real and fundamental benefit to the local economies. Much of it may in fact result from the diversion to tourist consumption of scarce commodities such as food, clean water, and electricity. More of it has to be reinvested in the construction of airport, cruise-port, overland transport, and other tourist-serving amenities. And as for items in demand by tourists, have you noticed that places with many tourists are also places where prices are high?

"Sure, our government people like tourism. Some of them have a stake in those gleaming hotels where they can share the pleasures of the wealthy. But what those glass-enclosed towers represent is globalization, powerful multinational corporations colluding with the government to limit the opportunities of local entrepreneurs. Planeloads and busloads of tourists come through on prearranged (and prepaid) tour promotions that isolate those visitors from local society.

"You geographers talk about cultural landscapes. Well, picture this: luxury liners sailing past poverty-stricken villages, luxury hotels towering over muddy slums, restaurants serving caviar when, down the street, children suffer from malnutrition. If the tourist industry offered real prospects for economic progress in poorer countries, such circumstances might be viewed as the temporary, unfortunate by-products of the upward struggle. Unfortunately, the evidence indicates otherwise. Name me a tourism-dependent economy where the gap between the rich and poor has narrowed.

"As for the educational effect of tourism, spare me the argument. Have you sat through any of those 'culture' shows staged by the big hotels? What you see there is the debasing of local culture as it adapts to visitors' tastes. Ask hotel workers how they really feel about their jobs, and you'll hear many say that they find their work dehumanizing because expatriate managers demand displays of friendliness and servitude that locals find insulting to sustain.

"I've heard it said that tourism doesn't pollute. Well, the Alaskans certainly don't agree—they sued a major cruise line on that issue and won. Not very long ago, cruise-ship crews routinely threw garbage-filled plastic bags overboard. That seems to have stopped, but I'm sure you've heard of the trash left by mountain-climbers in Nepal, the damage done by off-road vehicles in the wildlife parks of Kenya, the coral reefs injured by divers off Florida. Tourism is here to stay, but it is no panacea."

Regional ISSUE

Vote your opinion at www.wiley.com/college/deblij

for other crops and for pastures. Cuba has other economic opportunities, however, especially in its highlands. There are three mountainous areas, of which the south-

eastern chain, the Sierra Maestra, is the highest and most extensive. These highlands create considerable environmental diversity as reflected by extensive, timber-

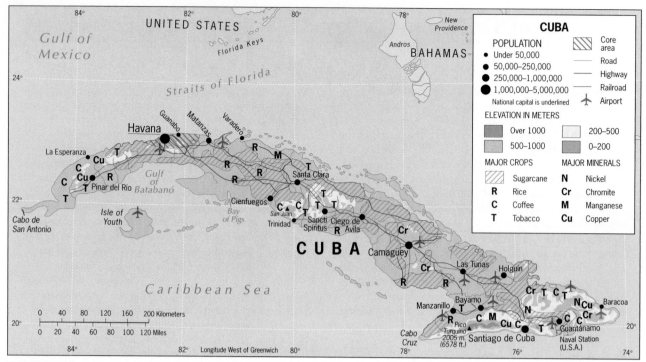

FIGURE 4-14

© H. J. de Blij, P. O. Muller, and John Wiley & Sons, Inc.

producing tropical forests and varied soils on which crops ranging from tobacco to coffee to rice to subtropical and tropical fruits are grown. Rice and beans are the staples, but Cuba cannot meet its dietary needs and so must import food. The savannas of the center and west support livestock. Although Cuba has only limited mineral reserves, its nickel deposits are extensive and have been mined for a century.

Early in the twenty-first century, Cuba found a crucial new supporter in Venezuela's leader, Hugo Chávez. Cuba has no domestic petroleum reserves, but Venezuela is oil-rich and, since 2003, has been providing all of the fuel that Cuba needs. In return, Castro sent 30,000 health workers and other professionals to Venezuela, where they aid the poor.

Poverty, crumbling infrastructure, crowded slums, and unemployment mark the Cuban cultural landscape, but Cuba's regime still has support among the general population. During the five-decade Castro period, much was done to bring the Afro-Cuban population into the mainstream through education and health provision. Cubans point to Guatemala, Nicaragua, and El Salvador and ask whether those countries are better off than they are under Castro. But in the United States, the view is different: Cuba, exiles and locals agree, could be the shining star of the Caribbean, its people free, its tourist economy booming, its products flowing to North American markets. Much will have to change for that vision to come true.

Jamaica lies across the deep Cayman Trench from southern Cuba, and a cultural gulf separates these two countries as well. Jamaica, a former British dependency, has an almost entirely Afro-Caribbean population. As a member of the British Commonwealth, Jamaica still recognizes the British monarch as the chief of state, represented by a governor-general. The effective head of government in this democratic country, however, is the prime minister. English remains the official language here, and British customs still linger.

Smaller than Connecticut and with 2.8 million people, Jamaica has experienced a steadily declining GNI over the past few decades despite its relatively slow population growth. Tourism has become the largest source of income, but the markets for bauxite (aluminum ore), of which Jamaica is a major exporter, have dwindled. And like other Caribbean countries, Jamaica has trouble making money from its sugar exports. Jamaican farmers also produce crops ranging from bananas to tobacco, but the country faces the disadvantages on world markets common to those in the global periphery. Meanwhile, Jamaica must import all of its oil and much of its food because the densely populated coastal flatlands suffer from overuse and shrinking harvests.

The capital, Kingston, on the southeastern coast, reflects Jamaica's economic struggle. Almost none of the hundreds of thousands of tourists who visit the country's beaches, explore its Cockpit Country of (karst) limestone towers and caverns, or populate the cruise ships calling at

Montego Bay or other points along the north coast get even a glimpse of what life is like for the ordinary Jamaican.

Haiti, the poorest state in the Western Hemisphere by virtually every measure, occupies the western part of the island of Hispaniola, directly across the Windward Passage from eastern Cuba (Fig. 4-15). Arawaks, not Caribs, formed the dominant indigenous population here, but the Spanish colonists who first took Hispaniola killed many, worked most of the survivors to death on their plantations, and left the others to die of the diseases they brought with them. French pirates established themselves in coastal coves along Hispaniola's west coast even as Spanish activity focused on the east, and just before the eighteenth century opened France formalized the colonial status of "Saint Dominique." During the following century, French colonists established vast plantations in the valleys of the northern mountains and in the central plain, laid out elaborate irrigation systems, built dams, and brought in a large number of Africans in bondage to work the fields. Prosperity made the colonists rich, but the African workers suffered terribly. They rebelled and, in a momentous victory over their European oppressors, established the independent republic of Haiti (resurrecting the original Arawak name for it) in 1804.

Strife among Haitian groups, American intervention, mismanagement, and dictatorship doomed the fortunes of this republic. In the 1990s, the United States attempted to help move Haiti toward more representative government, but the country was in economic and social collapse. By 2007, Haiti's GNI per capita had fallen to barely one-fifth of Jamaica's, a level lower than that of many poor African countries; foreign aid makes possible most of the country's limited public expenditures. Health conditions within the population of 9.4 million are dreadful: malnutrition is common, AIDS is rampant, diseases ranging from malaria to tuberculosis are rife, but hospital beds and doctors are in short supply. Conditions in and around the capital, Port-au-Prince, are among the worst in the world—less than 1000 kilometers (600 mi) from the United States.

The Dominican Republic has a larger share of the island of Hispaniola than Haiti (Fig. 4-15) in terms of both territory and population. Fly along the north-south border between the two countries, and you see a crucial difference: to the west, Haiti's hills and plains are treeless and gulleyed, its soils eroded and its streams silt-laden; to the east, forests drape the countryside and streams run clear (see photo, p. 228).

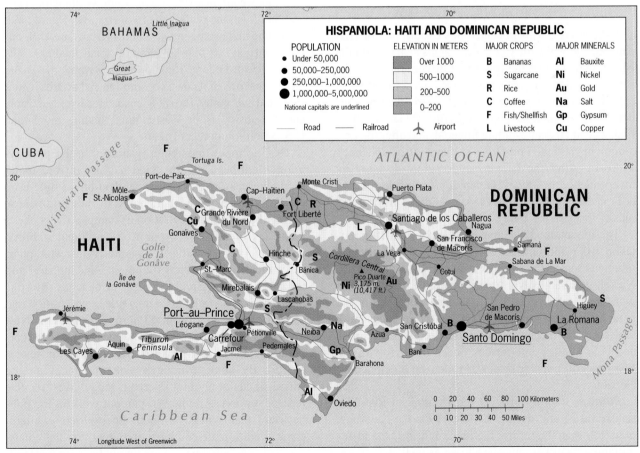

FIGURE 4-15   © H. J. de Blij, P. O. Muller, and John Wiley & Sons, Inc.

Indigenous Caribs inhabited this eastern part of Hispaniola, and when the European colonists arrived on the island, they were in the process of driving the Arawaks westward. Spanish colonists made this a prosperous colony, but then Mexico and Peru diverted Spanish attention from Hispaniola and the territory was ceded to France. But Hispanic culture lingered, and after the revolution in Haiti the French gave eastern Hispaniola back to Spain. After the Dominican Republic declared its independence in 1821, Haitian forces invaded it and occupied the republic until 1844, creating an historic animosity that persists today.

The mountainous Dominican Republic (2010 population: 10.3 million) has a wide range of natural environments and a far stronger resource base than Haiti. Nickel, gold, and silver have long been exported along with sugar, tobacco, coffee, and cocoa, but tourism (the great opportunity lost to Haiti) is the leading industry. A long period of dictatorial rule punctuated by revolutions and U.S. military intervention ended in 1978 with the first peaceful transfer of power following a democratic election.

Political stability brought the Dominican Republic rich rewards, and during the late 1990s the economy, based on manufacturing, high-tech industries, and remittances from Dominicans abroad as well as tourism, grew at an average 7 percent per year. But in the early 2000s the economy imploded, not only because of the downturn in the world economy but also because of bank fraud and corruption in government. Suddenly the Dominican peso collapsed, inflation skyrocketed, jobs were terminated, and blackouts prevailed. As the people protested, lives were lost and the self-enriched elite blamed foreign financial institutions that were unwilling to lend the government more money. Once again the hopes of ordinary citizens were dashed by greed and corruption among those in power.

Puerto Rico is the largest U.S. domain in Middle America, the easternmost and smallest island of the Greater Antilles (Fig. 4-16). This 9000-square-kilometer (3500-sq-mi) island, with a population of 4.0 million, is larger than Delaware and more populous than Oregon. It fell to the United States more than a century ago during the Spanish-American War of 1898. Since the Puerto Ricans had been struggling for some time to free themselves from Spanish control, this transfer of power was, in their view, only a change from one colonial power to another. As a result, the first half-century of U.S. administration was difficult, and it was not until 1948 that Puerto Ricans were permitted to elect their own governor.

When the island's voters approved the creation of a Commonwealth in a 1952 referendum, Washington, D.C. and San Juan, the two seats of government, entered into a complicated arrangement. Puerto Ricans are U.S. citizens but pay no federal taxes on local incomes. The Puerto Rican Federal Relations Act governs the island under the terms of its own constitution and awards it considerable autonomy. Puerto Rico also receives a sizeable annual subsidy from Washington, totaling more than U.S. $4 billion in recent years.

Despite these apparent advantages in the poverty-mired Caribbean, Puerto Rico has not thrived under U.S. administration. Long dependent on a single-crop economy (sugar), the island based its industrialization during the 1950s and 1960s on its comparatively cheap labor, tax breaks for corporations, political stability, and special access to the U.S. market. As a result, pharmaceuticals, electronic equipment, and apparel top today's list of exports, not sugar or bananas. But this industrialization failed to stem a tide of emigration that carried more than 1 million Puerto Ricans to the New York City area alone. The same wages that favored corporations kept many Puerto Ricans poor or unemployed. By some estimates, unemployment on the island stands at about 45 percent today. Most receive federal support. Another 30 percent work in the public sector, that is, in government. Puerto Rico's welfare-state condition discourages initiative, so that many people who could be working do not because their fed-

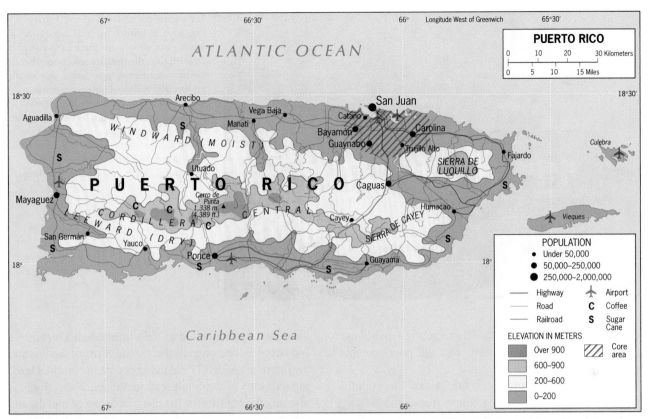

67°   66°30'   66°   Longitude West of Greenwich   65°30'

FIGURE 4-16

© H. J. de Blij, P. O. Muller, and John Wiley & Sons, Inc.

eral subsidy would decrease accordingly. Economists report that workers from the Dominican Republic even come to Puerto Rico to take low-wage jobs Puerto Ricans do not want.

Although Puerto Ricans are vocal in demanding political change, it is perhaps not surprising that successive referendums during the 1990s resulted in retention of the status quo—continuation of Commonwealth status rather than either Statehood or independence. The issue will no doubt continue to pose a formidable challenge to American statecraft in the years ahead.

THE LESSER ANTILLES

As Figure 4-2 shows, the Greater Antilles are flanked by two clusters of islands: the extensive Bahamas-Turks/Caicos archipelago to the north and the Lesser Antilles to the east and south. The Bahamas, the former British colony that is now the closest Caribbean neighbor to the United States, alone consists of nearly 3000 coral islands, most of them rocky, barren, and uninhabited, but about 700 carrying vegetation, of which some 30 are inhabited. Centrally positioned New Providence Island houses most of the country's 320,000

inhabitants and contains the capital, Nassau, a leading tourist attraction.

The Lesser Antilles are grouped geographically into the Leeward Islands and the Windward Islands, a (climatologically incorrect) reference to the prevailing airflows in this tropical area. The Leeward Islands extend from the U.S. Virgin Islands to the French dependencies of Guadeloupe and Martinique, and the Windward Islands from St. Lucia to the Netherlands Antilles off the Venezuelan coast (Fig. 4-2). It would be impractical to detail the individual geographic characteristics of each of the Lesser Antilles, but we should note that these countries and territories share the environmental risks of this region—earthquakes, volcanic eruptions, and hurricanes; that they face, to varying degrees, similar economic problems in the form of limited domestic resources, overpopulation, soil deterioration, land fragmentation, and market limitations; that tourism has become the leading industry for many; that political status ranges from complete sovereignty to continuing dependency; and that cultural diversity is strong not only between islands but within them as well. In economic terms, the GNI figures given (where available) in the Data Table at the end of the book may look encouraging, but these numbers conceal the social reality in virtually every

The capital city of Trinidad and Tobago may have an historic colonial name (Port of Spain), but after nearly three centuries of Spanish rule, the British took control here. English became the *lingua franca*, democratic government followed independence in 1962, and natural gas reserves propelled a thriving economy. As can be seen here, the harbor of Port of Spain is bursting at the seams as a car carrier delivers automobiles from Japan, and containers are stacked high on the docks.   © Boutin/Sipa Press.

entity: the gap between the fortunate few who are well-off and the great majority who are poor remains enormous.

Under such circumstances one looks for hopeful signs, and one of these signs comes from Trinidad and Tobago, the two-island republic at the southern end of the Lesser Antilles. This country (population: 1.3 million) has embarked on a natural-gas-driven industrialization boom that could turn it into an economic tiger.

Trinidad has long been an oil producer, but lower world prices and dwindling supplies in the 1990s forced a reexamination of its natural gas deposits to help counter the downturn. That quickly resulted in the discovery of major new supplies, and this cheap and abundant fuel has sparked a local gas-production boom as well as an influx of energy, chemical, and steel companies from Western Europe, Canada, and even India. (Meanwhile, Trinidad has become the largest supplier of liquefied natural gas for the United States.) Many of the new industrial facil-

ities have agglomerated at the ultramodern Point Lisas Industrial Estate outside the capital, Port of Spain, and they have propelled Trinidad to become the world's leading exporter of ammonia and methanol. Natural gas is also an efficient fuel for the manufacturing of metals, and steelmakers as well as aluminum refiners have been attracted to locate here. With Trinidad lying only a few kilometers from the Venezuelan coast of South America, it is also a sea-lane crossroads that is well connected to the vast, near-coastal supplies of iron ore and bauxite that are mined in nearby countries, particularly the Brazilian Amazon.

Middle America is a physically, culturally, and economically fragmented and diverse geographic realm that defies generalization. Given its strong Amerindian presence, its North American infusions, and its lingering Western European traditions, this certainly is not "Latin" America.

What You Can Do

RECOMMENDATION: Take a "serious" vacation! Hundreds of thousands of hard-working students take off for warmer weather between semesters or during spring break. Most go to relax in Florida or another "Sunbelt" State. But Middle America is nearby, and you could combine some sun and warmth with a valuable geographic field experience—if you prepare yourself and are ready for some challenges. Take some good, large-scale maps and a notebook and, when you spend a week in Puerto Rico, ask the people you meet how they feel about their island's political status. Visit Martinique and learn on the spot what happened to the "Paris of the Caribbean," St. Pierre, just over a century ago: the evidence is still visible. Go to Curaçao and find out how the island's economy survived the decline of its oil-refinery era. Or if you're in the West, cross the Mexican border and get your own take on the impact of NAFTA. The opportunities are right on our doorstep, and when you use your geographic perspective, you'll see the world in new ways of lasting value.

GEOGRAPHIC CONNECTIONS

1 When we studied the expansion of the European Union (EU), the issue of Turkey's application for membership was raised (Chapter 1). Turkish leaders want admission and are working to meet the conditions of membership, but Turkey still fails to meet these criteria in important respects. European Union members' views range across the spectrum, from enthusiastic to ambivalent to negative. Now imagine a North American Union of which Canada and the United States are members, and which Mexico wishes to join (NAFTA is a far less comprehensive union than the EU, in effect being just a first step toward it). Use your geographic insights to analyze the prospects of such a merger, and predict reactions in Canada and the United States. In what ways would Mexico qualify for membership, and in what respects would it not? Do you see any similarities between current Turkey-Europe and Mexico–North America relationships? In what ways are the two situations crucially different?

2 For many years, Caribbean islands such as St. Lucia have been exporting bananas to markets in North America and Europe, with these markets protected by preferential trade agreements. Now those agreements are ending as trade barriers fall all over the world, and the single-crop economies of the Caribbean are in trouble. What are the options for the farmers of St. Lucia and its neighbors? Relate your answer to core-periphery contrasts in the modern world (which can be reviewed on pp. 28-29).

Geographic Literature on Middle America: The key introductory works on this realm were authored by Blouet & Blouet, Clawson, James & Minkel, Kent, Preston, and West & Augelli et al. These works, together with a comprehensive listing of noteworthy books and recent articles on the geography of Mexico, Central America, and the Caribbean, can be found in the *References and Further Readings* section of this book's website at **www.wiley.com/college/deblij.**

BELIZE
GUAT.
HONDURAS
San Salvador
EL SALVADOR
Tegucigalpa
BELIZE
NICARAGUA
Managua
Lake Nicaragua
COSTA RICA
San José
ISTHMUS OF PANAMA
PANAMA
Panama City
Gulf of Panama
Cocos Island (COSTA RICA)

Caribbean Sea

DOMINICA
Martinique (FRANCE)
ST. LUCIA
ST. VINCENT & THE GRENADINES
BARBADOS
GRENADA
Lesser Antilles
Curaçao (NETH.)
Aruba (NETH.)
Bonaire (NETH.)
TRINIDAD & TOBAGO

ATLANTIC OCEAN

Santa Marta
BARRANQUILLA
CARTAGENA
MARACAIBO
Maracay CARACAS
Barquisimeto VALENCIA Barcelona
Ciudad Guayana Maturín
Boca Grande
Turbo
Cúcuta San Cristóbal
VENEZUELA
Bucaramanga
MEDELLÍN
Pereira
Buenaventura CALI
BOGOTÁ
COLOMBIA
Obando
ORINOCO
GUIANA
Angel Falls
HIGHLANDS
Georgetown
Paramaribo
Kourou
Cayenne
GUYANA
SURINAME
FRENCH GUIANA
San José del Guaviare
Mitú
Boa Vista
Pasto
Esmeraldas
QUITO
Cotopaxi (5,897 m, 19,347 ft.)
Equator
La Pedrera
Putumayo
Negro
Branco
Mouths of the Amazon
Macapá
Equator
Chimborazo (6,267 m, 20,561 ft.)
ECUADOR
GUAYAQUIL
Cuenca
Gulf of Guayaquil
Machala
Talara
Iquitos
Marañón
Tefé
Amazon
Porteira
Alenquer
Balbina Reservoir
Santarém
MANAUS
Ananindeua
Marajó Island
Marajó Bay
BELÉM
Parnaíba
Sobral FORTAL
Malpelo Island (COLOMBIA)
Archipiélago de Colón (Galápagos Islands) (ECUADOR)
Aguja Point
Piura
Chiclayo
Cajamarca
Pucallpa
Cruzeiro do Sul
Ucayali
Juruá
Purus
Madeira
Tapajós
Xingu
Telles Pires
Itaituba
Araras
Marabá
Imperatriz
Teresina
Mossoró
Nata
João Pessoa
RECIFE
Cape Bran
Trujillo
Chimbote
Huascarán (6,768 m, 22,205 ft.)
Huánuco
MONTAÑA
Rio Branco
Ji-Paraná (Rondônia)
BRAZIL
BRAZILIAN
SERTÃO
Petrolina
Maceió
Araca
SALVADOR
Cerro de Pasco
Callao
Huancayo
LIMA
Chincha Alta
Ica
Ayacucho
Cuzco
PERU
Reyes
Riberalta
Beni
Mamoré
Trinidad
Guaporé
Barreiras
Feira de Santana
Sobradinho Reservoir
São Francisco
Vitória da Conquista
La Paz
Illimani (6,462 m, 21,201 ft.)
BOLIVIA
Cochabamba
SANTA CRUZ
Corumbá
MATO GROSSO PLATEAU
Cuiabá
Rondonópolis
Rio Verde
Montes Claros
BRASÍLIA
GOIÂNIA
BELO HORIZONTE
Uberlândia
Juiz de Fora
HIGHLANDS
Arequipa
Tacna
Arica
Iquique
Sucre
Potosí
Tarija
Fortín Madrejón
Filadelfia
Campo Grande
Dourados
Londrina
CAMPINAS
RIO DE JANEIRO
Vila Velha (Espírito Santo)
Cape Frio
Tropic of Capricorn
Chuquicamata
ATACAMA DESERT
Calama
Antofagasta
Jujuy
Salta
PARAGUAY
GRAN CHACO
Asunción
Ciudad del Este
Fonte Grossa
Foz do Iguaçu
SÃO PAULO
Santos
São José dos Campos
CURITIBA
Joinville
Iguaçu Falls
Tropic of Capr
San Felix I. San Ambrosio I. (CHILE)
Copiapó
Mt. Ojos del Salado 22,572 ft.
Catamarca
Tucumán
Santiago del Estero
Resistencia
Corrientes
Posadas
Passo Fundo
Florianópolis
La Rioja
Reconquista
Santa Maria
Caxias do Sul
PORTO ALEGRE
Coquimbo
Córdoba
Santa Fe
Paraná
ENTRE RÍOS
Uruguaiana
Salto
URUGUAY
Patos Lagoon
Pelotas
SOUTH AMERICA
Viña del Mar
Valparaíso
SANTIAGO
Mendoza
San Luis
ROSARIO
Mt. Aconcagua (6,960 m, 22,834 ft.)
ARGENTINA
Paraná
Mercedes
Melo
MONTEVIDEO
Punta del Este
Juan Fernández Islands (CHILE)
Talca
Talcahuano Concepción
Chillán
Los Ángeles
Temuco
BUENOS AIRES
La Plata
Río de la Plata
Santa Rosa
PAMPAS
Río Colorado
Bahía Blanca
Mar del Plata
ATLANTIC OCEAN
Osorno
Neuquén
Puerto Montt
San Carlos de Bariloche
Chiloé Island
Corcovado Gulf
Viedma
San Matías Gulf
Valdés Peninsula (-40 m, -131 ft.)
Chonos Archipélago
Esquel
Puerto Madryn
Trelew
Coihaique
ANDES
PATAGONIA
Comodoro Rivadavia
Gulf of San Jorge
Gulf of Penas
Jaramillo
Wellington Island
Gran Bajo de San Julián (-105 m, -344 ft.)
Falkland Islands (U.K.)
Stanley
West Falkland
East Falkland
Puerto Natales
Puerto Santa Cruz
Río Gallegos
TIERRA DEL FUEGO
Punta Arenas
Strait of Magellan
Río Grande
Cape Horn
South Georgia Island (U.K.)

Elevation (m)
3000
1500
600
300
150
0
below sea level
0
-150
-1500
-3000
-6000
Roads
Railroads

0 km 400 800 1200 1600
0 miles 200 400 600 800 1000
Lambert Azimuthal Equal-Area Projection
Scale 1:30,000,000

CONCEPTS, IDEAS, AND TERMS

1 *Altiplano*
2 Land alienation
3 Plural society
4 Commercial agriculture
5 Subsistence agriculture
6 Free Trade Area of the Americas (FTAA)
7 Urbanization
8 Rural-to-urban migration
9 Megacity
10 "Latin" American City model
11 Informal sector
12 *Barrio (favela)*
13 Insurgent state
14 Failed state
15 Von Thünen model
16 El Niño
17 Forward capital
18 *Cerrado*
19 Growth-pole concept

REGIONS

● THE NORTH
● THE WEST
● THE SOUTH
● BRAZIL

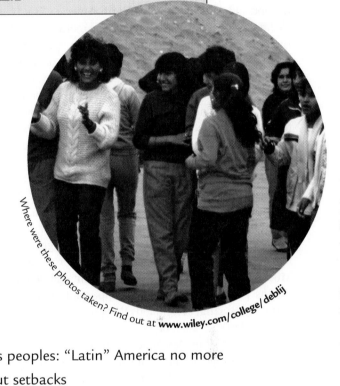

Where were these photos taken? Find out at **www.wiley.com/college/deblij**

In This Chapter

- The growing power of indigenous peoples: "Latin" America no more
- Democracy gains, but not without setbacks
- Tentative efforts toward economic integration: China lends a hand
- Chile: Star of the realm
- The poor performance of rich Argentina
- Brazil: Superpower in the making?

Photos: © H. J. de Blij

FIGURE 5-1   © H. J. de Blij, P. O. Muller, and John Wiley & Sons, Inc.

Of all the continents, South America has the most familiar shape—a giant triangle connected by mainland Middle America's tenuous land bridge to its neighbor in the north. South America also lies not only south but mostly east of its northern counterpart. Lima, the capital of Peru—one of the continent's westernmost cities—lies farther east than Miami, Florida. Thus South America juts out much more prominently into the Atlantic Ocean toward southern Europe and Africa than does North America. But lying so far eastward means that South America's western flank faces a much wider Pacific Ocean, with the distance from Peru to Australia nearly twice that from California to Japan.

As if to reaffirm South America's northward and eastward orientation, the western margins of the continent are rimmed by one of the world's longest and highest mountain ranges, the Andes, a gigantic wall that extends unbroken from Tierra del Fuego near the continent's southern tip in Chile to northeastern Venezuela in the far north (Fig. 5-1). The other major physiographic feature of South America dominates its central north—the Amazon Basin; this vast humid-tropical amphitheater is drained by the mighty Amazon River, which is fed by several major tributaries. Much of the remainder of the continent can be classified as plateau, with the most important components being the Brazilian Highlands that cover most of Brazil southeast of the Amazon Basin, the Guiana Highlands located north of the lower Amazon Basin, and the cold Patagonian Plateau that blankets the southern third of Argentina. Figure 5-1 also reveals two other noteworthy river basins beyond Amazonia: the Paraná-Paraguay Basin of south-central South America, and the Orinoco Basin in the far north that drains interior Colombia and Venezuela.

MAJOR GEOGRAPHIC QUALITIES OF
South America

1. South America's physiography is dominated by the Andes Mountains in the west and the Amazon Basin in the central north. Much of the remainder is plateau country.

2. Almost half of the realm's area and just over half of its total population are concentrated in one country—Brazil.

3. South America's population remains concentrated along the continent's periphery. Most of the interior is sparsely peopled, but sections of it are now undergoing significant development.

4. Interconnections among the states of the realm are improving rapidly. Economic integration has become a major force, particularly in southern South America.

5. Regional economic contrasts and disparities, both in the realm as a whole and within individual countries, remain strong.

6. Cultural pluralism exists in almost all of the realm's countries and is often expressed regionally.

7. Rapid urban growth continues to mark much of the South American realm, and urbanization overall is today on a par with the levels of the United States and Western Europe.

Defining the Realm

South America is a realm in dramatic transition, and it is not clear where this transition will lead. During much of the twentieth century, South American countries were in frequent political turmoil. Dictatorial regimes ruled from one end of the realm to the other; unstable governments fell with damaging frequency. Widespread poverty, harsh regional disparities, poor internal surface connections, limited international contact, and economic stagnation prevailed.

Entering the second decade of the twenty-first century, things were quite different. Democracy had taken hold almost everywhere. Long-isolated countries were becoming more interconnected through new transport routes and trade agreements. New settlement frontiers were being opened. Energy resources, some long exploited and others newly discovered, boosted national economies when world prices rose. Foreign states and corporations appeared on the scene to buy commodities and invest in infrastructure. The pace of globalization increased from Bogotá to Buenos Aires.

But these exciting developments must be seen against a backdrop of persistent problems. The realm's giant, Brazil, is embarked on a program of land reform, a campaign against poverty, and an effort to maintain financial rigor that have all run up against endemic corruption in government. The economy of Argentina is just recovering after an implosion that shook the country to its core. Whereas Chile is the realm's success story, emerging from the horrors of the Pinochet era as a stable and vibrant democracy with a thriving economy, neighboring Bolivia is in the grip of a social revolution arising in part from the realization of its energy riches. And on the north coast lies Venezuela, its oil reserves among the largest in the world (and by far the largest in the realm) and its political life dominated by a one-time coup leader whose closest ideological ally is Cuba's communist ruler and whose major adversary is the U.S. government.

Today, the United States hopes to foster democracy and encourage regional economic integration, but many South Americans remember past U.S. toleration of, and even support for, the realm's former dictators. Venezuela's populist leader champions the poor and uses oil income to counter American influence, campaigning vigorously against the notion of a Free Trade Area of the Americas (FTAA) and warning South American governments against capitalist plots. Such advice finds a ready market because the great majority of South Americans remain mired in poverty. By some measures, the disparity between rich and poor is wider in this realm than in any other, and wealth is disproportionately concentrated in the hands of a small minority (the richest 20 percent of the realm's inhabitants control 70 percent, while the poorest 20 percent own 2 percent). The question of the day is whether South America can sustain its progress against political, ideological, and economic odds.

STATES ANCIENT AND MODERN

Thousands of years before the first European invaders appeared on the shores of South America, peoples now referred to as *Amerindians* had migrated into the continent via North and Middle America and founded societies in coastal valleys, in river basins, on plateaus, and in mountainous locales. These societies achieved different and remarkable adaptations to their diverse natural environments, and by about one thousand years ago, a number of regional cultures thrived in the elongated valleys between mountain ranges of the Andes from present-day Colombia southward to Bolivia and Chile. These high-altitude valleys, called *altiplanos*, provided fertile soils, reliable water supplies, building materials, and natural protection to their inhabitants.

The Inca State

One of these *altiplanos*, at Cuzco in what is now Peru, became the core area of South America's greatest indigenous empire, that of the *Inca*. The Inca were expert

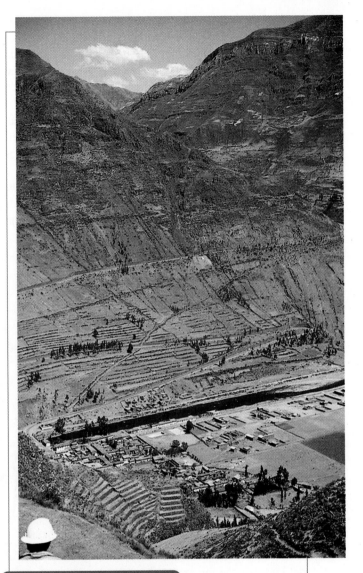

FROM THE FIELD NOTES

"From this high vantage point I got a good perspective of a valley near Pisac in the Peruvian Andes, not far from Cuzco. This was part of the Incan domain when the Spaniards arrived to overthrow the empire, but the terraces you can see actually predate the Inca period. Human occupation in these rugged mountains is very old, and undoubtedly the physiography here changed over time. Today these slopes are arid and barren, and only a few hardy trees survive; the stream in the valley bottom is all the water in sight. But when the terrace builders transformed these slopes, the climate may have been more moist, the countryside greener."
© Courtesy Philip L. Keating.

Concept Caching

www.conceptcaching.com

builders whose stone structures (among which Machu Picchu near Cuzco is the most famous), roads, and bridges helped unify their vast empire; they also proved themselves to be efficient administrators, successful farmers and herders, and skilled manufacturers; scholars studied the heavens, and physicians even experimented with brain surgery. Great military strategists, the Inca integrated the peoples they vanquished into a stable and well-functioning state, an amazing accomplishment given the high-relief terrain they had to contend with.

As a minority ruling elite in their far-flung empire, the Inca were at the pinnacle in their rigidly class-structured, highly centralized society. So centralized and authoritarian was their state that a takeover at the top was enough to gain immediate power over all of it—as a small army of Spanish invaders discovered in the 1530s. The European invasion brought a quick end to thousands of years of Amerindian cultural development and changed the map forever.

America, but the great majority of the settlers stayed on or near the coast, as is reflected in the current map (Fig. 5-3). Almost all of the realm's major cities have coastal or near-coastal locations, and the current population distribution map gives you the impression of a continent yet to be penetrated and inhabited. But look carefully at Figure 5-3, and you will see a swath of population well inland from the coastal settlements, most clearly in Peru but also northward into Ecuador and southward into Bolivia. That is the legacy of the Inca Empire and its incorporated peoples, surviving in their mountainous redoubt and still numbering in the millions.

The Population Map—Then and Now

If we were able to reconstruct a map of South America's population before the arrival of the Europeans (a "pre-Columbian" map, as it would be called), it would look quite different from the current map (Fig. 5-2). Indigenous Amerindian societies inhabited not only the Andes and adjacent lowlands but also riverbanks in the Amazon Basin, where settlements numbering in the thousands subsisted on fishing and farming. They did not shy away from harsh environments such as those of the island of Tierra del Fuego in the far south, where the fires they kept going against the bitter cold led the Europeans to name the place "land of fire."

Today the map looks quite different. Many of the indigenous societies succumbed to the European invaders, not just through warfare but also because of the diseases the Hispanic conquerors brought with them. Geographers estimate that 90 percent of native Amazonians died within a few years of contact, and the peoples of Tierra del Fuego also are no longer there to build their fires. From one end of South America to the other, the European arrival spelled disaster.

Spanish and Portuguese colonists penetrated the interior of South

FIGURE 5-2

© H. J. de Blij, P. O. Muller, and John Wiley & Sons, Inc.

FIGURE 5-3 © H. J. de Blij, P. O. Muller, and John Wiley & Sons, Inc.

SOUTH AMERICA: POPULATION
DISTRIBUTION, 2010
One dot represents 50,000 persons

Today, Amerindian political leaders are emerging to bring the plight of the realm's indigenous peoples not only to local but also to international attention. Amerindians were conquered, decimated by foreign diseases, robbed of their best lands, subjected to forced labor, denied the right to grow their traditional crops, socially discriminated against, and swindled out of their fair share of the revenues from resources in their traditional domains. They may still be the poorest of the realm's poor, but they are now asserting themselves. For some states in this realm, the consequences of this movement will be far-reaching.

The Iberian Invaders

The modern map of South America (Fig. 5-4) started to take shape when the Iberian colonists began to understand the location and economies of the Amerindian societies. The Inca, like Mexico's Maya and Aztec peoples, had accumulated gold and silver at their headquarters, possessed productive farmlands, and constituted a ready labor force. Not long after the defeat of the Aztecs in 1521, Francisco Pizarro sailed southward along the continent's northwestern coast, learned of the existence of the Inca Empire, and withdrew to Spain to organize its overthrow. He returned to the Peruvian coast in 1531 with 183 men and two dozen horses, and the events that followed are well known. In 1533, his party rode victorious into Cuzco.

At first, the Spaniards kept the Incan imperial structure intact by permitting the crowning of an emperor who was under their control. But soon the breakdown of the old order began. The new order that gradually emerged in western South America placed the indigenous peoples in serfdom to the Spaniards. Great haciendas were formed by **land alienation** (the takeover of indigenously held land by foreigners), taxes were instituted, and a forced-labor system was introduced to maximize the profits of exploitation.

Amerindian Reawakening

As we note in the regional discussion, South America's long-downtrodden Amerindians are staging a social, political, and economic awakening. They are not alone in this—in Chapter 4 we saw the Zapatista movement in southern Mexico's Chiapas State achieve national attention—but they have begun to realize their potential in several South American countries where their numbers translate into strength. Mexico's Amerindian population represents less than one-third of the total, but Peru's Amerindians constitute about 45 percent, and in Bolivia they are in the majority at 55 percent.

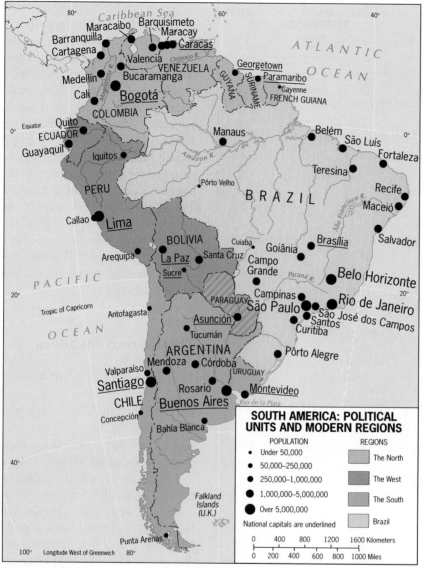

FIGURE 5-4   © H. J. de Blij, P. O. Muller, and John Wiley & Sons, Inc.

a Portuguese sphere of influence because Spain and Portugal had signed a treaty in 1494 to recognize a north-south line 370 leagues west of the Cape Verde Islands as the boundary between their New World spheres of influence. This border ran approximately along the meridian of 50°W longitude, thereby cutting off a sizeable triangle of eastern South America for Portugal's exploitation (Fig. 5-2). But a brief look at the political map of South America (Fig. 5-4) shows that this treaty did not limit Portuguese colonial territory to the east of the 50th meridian. Instead, Brazil's boundaries were bent far inland to include almost the entire Amazon Basin, and the country came to be only slightly smaller in territorial size than all the other South American countries combined. This westward thrust was the result of Portuguese and Brazilian penetration, particularly by the *Paulistas*, the settlers of São Paulo who needed Amerindian slave labor to run their plantations.

The Africans

As Figure 5-2 shows, the Spaniards initially got very much the better of the territorial partitioning of South America—not just in land quality but also in the size of the aboriginal labor force. When the Portuguese began to develop their territory, they turned to the same lucrative activity that their Spanish rivals had pursued in the Caribbean—the plantation cultivation of sugar for the European market. And they, too, found their labor force in the same source region, as millions of Africans (nearly half of all who came to the Americas) were brought in bondage to the tropical Brazilian coast north of Rio de Janeiro. Not surprisingly, Brazil now has South America's largest black population, which is still heavily concentrated in the country's poverty-stricken northeastern States. With Brazilians of direct or mixed African ancestry today accounting for nearly half of the population of 201 million, the Africans decidedly constitute the third major immigration of foreign peoples into South America.

Lima, the west-coast headquarters of the Spanish conquerors, soon became one of the richest cities in the world, its wealth based on the exploitation of vast Andean silver deposits. The city also served as the capital of the viceroyalty of Peru, as the Spanish authorities quickly integrated the new possession into their colonial empire (Fig. 5-2). Subsequently, when Colombia and Venezuela came under Spanish control and, later, when Spanish settlement expanded in what is now Argentina and Uruguay, two additional viceroyalties were added to the map: New Granada and La Plata.

Portuguese Brazil

Meanwhile, another vanguard of the Iberian invasion was penetrating the east-central part of the continent, the coastlands of present-day Brazil. This area had become

FROM THE FIELD NOTES

"Salvador is one of Brazil's oldest and most vibrant cities. Magnificent churches, public buildings, and mansions were built during the time when this was, by many measures, the most important city in the Southern Hemisphere. Long the capital of Brazil, Salvador was the point of entry for tens of thousands of Africans, and the fortune-making plantation economy, augmented by the whaling industry, concentrated enormous wealth here, some of which went into the construction of an opulent city center. But fortunes change, Salvador lost its political as well as economic advantages, and the city fell into disrepair. Walking the streets of the old town in 1982, I noted the state of decay of much of Salvador's architectural heritage and wondered about its survival: weathering in this tropical environment was destroying woodwork, façades, and roofs. But then the United Nations proclaimed Salvador's old town a World Heritage site, and massive restoration began (left). By the late 1990s, much of the district had been revived (right), and tourism's contribution to the local economy was on the rise." © H. J. de Blij.

Concept Caching

www.conceptcaching.com

Longstanding Isolation

Despite their adjacent location on the same continent, their common language and cultural heritage, and their shared national problems, the countries that arose out of South America's Spanish viceroyalties (along with Brazil) until quite recently existed in a considerable degree of isolation from one another. Distance and physiographic barriers reinforced this separation, and the realm's major population agglomerations still adjoin the coast, mainly the eastern and northern coasts (Fig. 5-3). The viceroyalties existed primarily to extract riches and fill Spanish coffers. In Iberia there was little interest in developing the American lands for their own sake. Only after those who had made Spanish and Portuguese America their home and who had a stake there rebelled against Iberian authority did things begin to change, and then very slowly. Thus South America was saddled with the values, economic outlook, and social attitudes of eighteenth-century Iberia—not the best tradition from which to begin the task of forging modern nation-states.

Independence

Certain isolating factors had their effect even during the wars for independence. Spanish military strength was always concentrated at Lima, and those territories that lay farthest from their center of power—Argentina and Chile—were the first to establish their independence from Spain (in 1816 and 1818, respectively). In the north Simón Bolívar led the burgeoning independence movement, and in 1824 two decisive military defeats there spelled the end of Spanish power in South America.

This joint struggle, however, did not produce unity because no fewer than nine countries emerged from the three former viceroyalties. It is not difficult to understand why this fragmentation took place. With the Andes intervening between Argentina and Chile and the Atacama Desert between Chile and Peru, overland distances seemed even greater than they really were, and these obstacles to contact proved quite effective. Hence, from their outset the new countries of South America began to grow apart amid friction and even wars. Only within the past two decades have the countries of this realm finally begun to recognize the mutual advantages of increasing cooperation and to make lasting efforts to steer their relationships in this direction.

CULTURAL FRAGMENTATION

When we speak of the interaction of South American countries, it is important to keep in mind just who does the interacting. The fragmentation of colonial South America into ten individual republics, and the subsequent postures of each of these states, was the work of a small minority that constituted the land-holding, upper-class elite. Thus in every country a vast majority—be they Amerindians in Peru or people of African descent in Brazil—could only watch as their European masters struggled with each other for supremacy.

Using the Land

3 South America, then, is a continent of **plural societies**, where Amerindians of different cultures, Europeans from Iberia and elsewhere, Africans mainly from western tropical Africa, and Asians from India, Japan, and Indonesia cluster in adjacent areas. The result is a cultural kaleidoscope of almost endless variety, whose internal divisions are also reflected in the realm's economic landscape. This is readily visible in the map of South America's dominant livelihood, agriculture **4** (Fig. 5-5). Here **commercial** or market (for-profit) and **5** **subsistence** (primarily for household use) agriculture exist side by side to a greater degree than anywhere else in the world. The geography of commercial agricultural systems (as shown by areas of soybean and non-soy grain production) was initially tied to the distribution of landholders of European background, while subsistence farming (such as highland mixed subsistence-market, agroforestry, and shifting cultivation) is associated with the spatial patterns of indigenous peoples as well as populations of African and Asian descent. Nonetheless, these patterns are changing in this era of globalization.

Cultural Landscapes

The cultural landscape of South America, similar to that of Middle America, is a layered one. Amerindians cultivated and crafted diverse landscapes throughout the continent, some producing greater impacts than others.

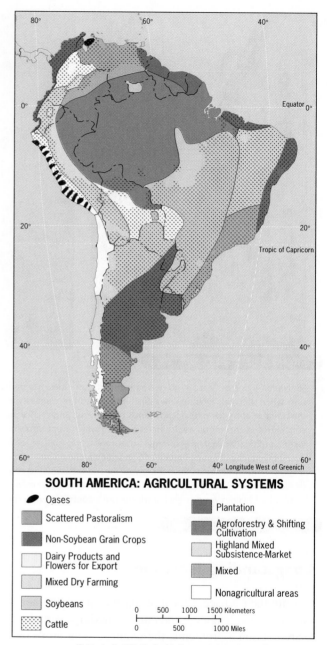

SOUTH AMERICA: AGRICULTURAL SYSTEMS

- Oases
- Scattered Pastoralism
- Non-Soybean Grain Crops
- Dairy Products and Flowers for Export
- Mixed Dry Farming
- Soybeans
- Cattle
- Plantation
- Agroforestry & Shifting Cultivation
- Highland Mixed Subsistence-Market
- Mixed
- Nonagricultural areas

0 500 1000 1500 Kilometers
0 500 1000 Miles

© H. J. de Blij, P. O. Muller, and John Wiley & Sons, Inc.

FIGURE 5-5

When the Europeans arrived, the cultural change that resulted from depopulation severely impacted the environment. Native peoples became minorities in their own lands, and Europeans introduced crops, animals, and ideas about land ownership and land use that changed South America irreversibly. They also brought in Africans from various parts of Subsaharan Africa. Europeans from non-Iberian Europe also started immigrating to South America, especially during the first half of the twentieth century. Japanese settlers arrived in Brazil and Peru during the same era. All of these elements have contributed to the present-day ethnic composition in this realm.

FIGURE 5-6 © H. J. de Blij, P. O. Muller, and John Wiley & Sons, Inc.

Figure 5-6 shows the distinct concentrations of Amerindian and African cultural dominance, as well as areas where these groups are hardly present and people of European ancestry dominate.

ECONOMIC INTEGRATION

As noted above, the separatism that has so long characterized international relations in this realm is giving way as South American countries discover the benefits of forging new partnerships with one another. With mutually advantageous trade the catalyst, a new continentwide spirit of cooperation is blossoming at every level. Periodic flareups of boundary disputes now rarely escalate into open conflict. Cross-border rail, road, and pipeline projects, stalled for years, are multiplying steadily. In southern South America, five formerly contentious nations are developing the *hidróvia* (water highway), a system of river locks that is opening most of the Paraná-Paraguay Basin to barge transport. Investments today flow freely from one country to another, particularly in the agricultural sector. Similar ideas have been proposed to connect the Paraná-Paraguay rivers to the Amazon River system.

Recognizing that free trade could solve many of the realm's economic-geographic problems, governments are now pursuing several avenues of economic supranationalism. In 2010, South America's republics were affiliating with the following major trading blocs:

- **Mercosur/l** (*Mercosur* in Spanish; *Mercosul* in Portuguese): Launched in 1995 by countries of the Southern Cone and Brazil, this Common Market established a free-trade zone and customs union linking Brazil, Argentina, Uruguay, Paraguay, and now Venezuela. Bolivia, Chile, Colombia, Ecuador, and Peru are associate members. This organization is becoming the dominant free-trade organization for South America.

- *Andean Community*: Formed as the Andean Pact in 1969 but restarted in 1995 as a customs union with common tariffs for imports, this bloc is made up of Colombia, Peru, Ecuador, and Bolivia. Venezuela was a member until it withdrew in 2006.

- *Union of South American Nations (UNASUR)*: Founded in 2008 in Brasília, Brazil, the 12 independent countries of South America signed a treaty to create a union envisioned as similar to the European Union (see Chapter 1) with the goal of a continental parliament, a coordinated defense effort, one passport for all its citizens, and greater cooperation on infrastructure development. However, significant disagreement between member-states about details still puts these efforts years into the future. UNASUR had been preceded by the South American Community of Nations.

- **Free Trade Area of the Americas (FTAA):** The United States and other NAFTA proponents have tried to move this hemispheric free-trade idea forward, but it has been resisted by peasants and workers in South America, and formally opposed by Mercosur/l. As long as the terms of trade remain set by the North, the Southern partners will be reluctant to participate in this initiative. **6**

URBANIZATION

As in most other realms, South Americans are leaving the land and moving to the cities. South America started relatively early in this **urbanization** process, which intensified **7** throughout the twentieth century. With South America's urban population now at 81 percent, the realm ranks with those of Europe and the United States. The urban population of South America has grown annually by about 5 percent since 1950, while the increase in rural areas was less than 2 percent. These numbers underscore not only the dimensions but also the durability of the **rural-to-urban** **8** **migration** from the countryside to the cities.

FROM THE FIELD NOTES

"Two unusual perspectives of Rio de Janeiro form a reminder that here the wealthy live near the water in luxury high-rises, such as these overlooking Ipanema Beach, while the poor have million-dollar views from their hillslope *favelas*, such as Rocinho."   © H. J. de Blij.

www.conceptcaching.com

Regional Patterns

The generalized spatial pattern of South America's urban transformation is displayed in Figure 5-7, which shows a *cartogram* of the continent's population.* Here we see not only the realm's countries in population-space relative to each other, but also the proportionate sizes of individual large cities within their total national populations.

Regionally, southern South America is the most highly urbanized. Today in Argentina, Chile, and Uruguay, almost all of the population resides in cities. Ranking next in urbanization is Brazil. The next highest group of countries borders the Caribbean in the north. Not surprisingly, the Andean countries constitute the realm's least urbanized zone. Figure 5-7 tells us a great deal about the relative positions of major metropolises in their countries. Three of them—Brazil's São Paulo and Rio de Janeiro, and Argentina's Buenos Aires—rank among the world's **megacities** (cities whose populations exceed 10 million). But even in the Amazon Basin the population is now 70 percent urban.

---

*As noted on page 20, a cartogram is a specially transformed map in which countries and cities are represented in proportion to their populations. Those containing large numbers are blown up in population space, while those containing lesser numbers are shrunk in size accordingly.

Causes and Challenges of Cityward Migration

In South America, as in Middle America, Africa, and Asia, people are attracted to the cities and driven from the poverty of the rural areas. Both *pull* and *push* factors are at work. Rural land reform has been slow in coming, and for this and other reasons every year tens of thousands of farmers simply give up and leave, seeing little or no possibility for economic advancement. The cities lure them because they are perceived to provide opportunity—the chance to earn a regular wage. Visions of education for their children, better medical care, upward social mobility, and the excitement of life in a big city draw hordes to places such as São Paulo and Caracas.

But the actual move can be traumatic. Cities in developing countries are surrounded and often invaded by squalid slums, and this is where the urban immigrant most often finds a first—and frequently permanent—abode in a makeshift shack without even the most basic amenities and sanitary facilities. Unemployment is persistently high, often exceeding 25 percent of the available labor force. But still the people come, hopeful for a better life, the overcrowding in the shantytowns worsens, and the threat of epidemic-scale disease (and other disasters) rises.

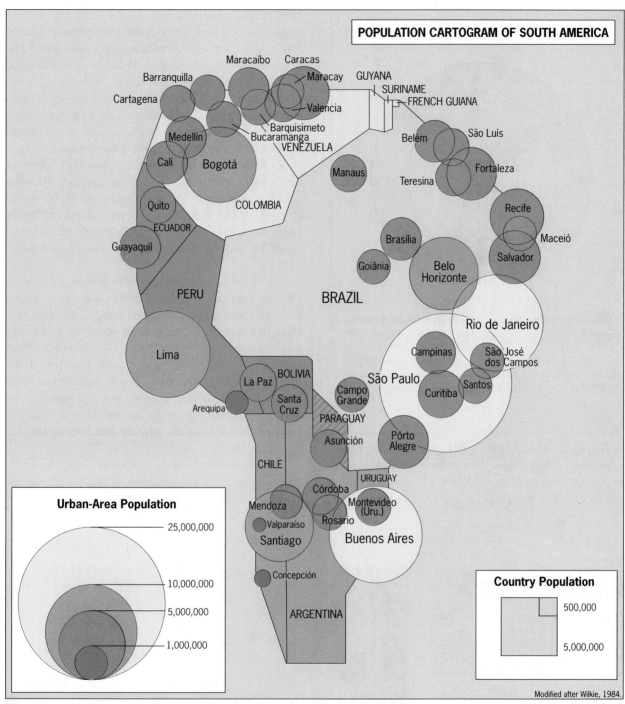

POPULATION CARTOGRAM OF SOUTH AMERICA

Urban-Area Population

25,000,000
10,000,000
5,000,000
1,000,000

Country Population

500,000
5,000,000

Modified after Wilkie, 1984.

FIGURE 5-7

© H. J. de Blij, P. O. Muller, and John Wiley & Sons, Inc.

The "Latin" American City Model

The urban experience in the South and Middle American realms varies because of diverse historical, cultural, and economic influences. Nonetheless, there are many common threads that have prompted geographers to search for useful generalizations. One is the model of the intraurban spatial structure of the **"Latin" American city** proposed by Ernst Griffin and Larry Ford (Fig. 5-8).

Model and Reality

As noted in Chapter 1, the idea behind a *model* is to create an idealized representation of reality, displaying as many key real-world elements as possible. In the case of South America's cities, the basic spatial framework of city structure, which blends traditional elements of South and Middle American culture with modernization forces now reshaping the urban scene, is a composite of radial sectors and concentric zones.

10

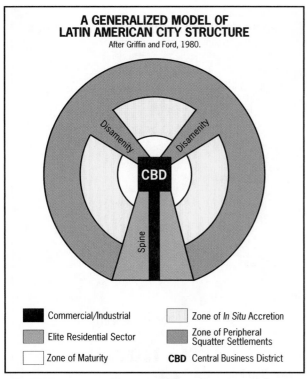

A GENERALIZED MODEL OF LATIN AMERICAN CITY STRUCTURE
After Griffin and Ford, 1980.

Legend:
- ■ Commercial/Industrial
- ▨ Elite Residential Sector
- □ Zone of Maturity
- ▢ Zone of *In Situ* Accretion
- ▨ Zone of Peripheral Squatter Settlements
- **CBD** Central Business District

FIGURE 5-8 © H. J. de Blij, P. O. Muller, and John Wiley & Sons, Inc.

Anchoring the model is the *CBD*, which is the primary business, employment, and entertainment focus of the surrounding metropolis. The CBD contains many modern high-rise buildings but also mirrors its colonial beginnings. As shown in Figure 4-4, by colonial law Spanish colonizers laid out their cities around a central square, or plaza, dominated by a church and government buildings. Santiago's *Plaza de Armas*, Bogotá's *Plaza Bolívar*, and Buenos Aires' *Plaza de Mayo* are classic examples. The plaza was the hub of the city, which later outgrew its old center as new commercial districts formed nearby; but to this day the plaza remains an important link with the past (photo below).

Radiating outward from the urban core along the city's most prestigious axis is the commercial *spine*, which is adjoined by the *elite residential sector* (shown in green in Fig. 5-8). This widening corridor is essentially an extension of the CBD, featuring offices, retail facilities, and housing for the upper and upper-middle classes.

The three remaining concentric zones are home to the less fortunate residents of the city, with income level and housing quality decreasing as distance from the CBD increases. The *zone of maturity* in the inner city contains housing for the middle class, who invest sufficiently to keep their aging dwellings from deteriorating. The adjacent *zone of in situ accretion* is one of much more modest housing interspersed with unkempt areas, representing a transition from inner-ring affluence to outer-ring poverty and taking on slum characteristics.

The outermost *zone of peripheral squatter settlements* is home to the millions of relatively poor and unskilled workers who have recently migrated to the

FROM THE FIELD NOTES

"At the heart of Buenos Aires lies the Plaza de San Martín, flanked by impressive buildings but, unusual for such squares in Iberian America, carpeted with extensive lawns shaded by century-old trees. Getting a perspective of the plaza was difficult until I realized that you could get to the top of the 'English Tower' across the avenue you see in the foreground. From there, one can observe the prominent location occupied by the monument to the approximately 700 Argentinian military casualties of the Falklands War of 1982 (center), where an eternal flame behind a brass map of the islands symbolizes Argentina's undiminished determination to wrest the islands from British control."
© H. J. de Blij.

Concept Caching

city. Here many newcomers earn their first cash income by becoming part of the **informal sector**, in which workers are undocumented and money transactions are beyond the control of government. The settlements consist mostly of self-help housing, vast shantytowns known as **barrios** in Spanish-speaking South America and **favelas** in Brazil. Some of their entrepreneurial inhabitants succeed more than others, transforming parts of these shantytowns into beehives of activity that can propel resourceful workers toward a middle-class existence.

A final structural element of many South American cities forms an inward, narrowing sectoral extension of the zone of peripheral squatter settlements and is known as the *zone of disamenity*. It consists of undesirable land along highways, rail corridors, riverbanks, and other low-lying areas; people are so poor that they are forced to live in the open. Thus this realm's cities present enormous contrasts between poverty and wealth, squalor and comfort—harsh contrasts all too frequently seen in the cityscape.

Regions of the Realm

South America divides geographically into four rather clearly defined regions (Fig. 5-4):

1. *The North* consists of five entities that display a combination of Caribbean and South American features: Colombia, Venezuela, and those that represent three historic colonial footholds by Britain (Guyana), the Netherlands (Suriname), and France (French Guiana).

2. *The West* is formed by four republics that share a strong Amerindian cultural heritage as well as powerful influences resulting from their Andean physiography: Ecuador, Peru, Bolivia, and, transitionally, Paraguay.

3. *The South*, often called the "Southern Cone," includes three countries that actually conform to the much-misused regional term "Latin" America: Argentina, Chile, and Uruguay (all with strong European imprints and little remaining Amerindian influence) plus aspects of Paraguay.

4. *Brazil* not only is South America's giant, accounting for just about half the realm's territory as well as population, but is fast developing into the Western Hemisphere's second superpower. In Brazil the dominant Iberian influence is Portuguese, not Spanish, and here Africans, not Amerindians, form a significant component of demography and culture. Brazil in some ways is a bridge between the Americas and Africa, which is why we discuss it last in this chapter, just before we turn to Subsaharan Africa.

MAJOR CITIES OF THE REALM

City	Population* (in millions)
Asunción, Paraguay	2.1
Belo Horizonte, Brazil	5.9
Bogotá, Colombia	8.1
Brasília, Brazil	3.8
Buenos Aires, Argentina	13.0
Caracas, Venezuela	3.2
Guayaquil, Ecuador	2.7
La Paz, Bolivia	1.7
Lima, Peru	8.3
Manaus, Brazil	1.9
Montevideo, Uruguay	1.5
Quito, Ecuador	1.8
Rio de Janeiro, Brazil	12.2
Santiago, Chile	5.9
São Paulo, Brazil	26.2

*Based on 2010 estimates.

THE NORTH:
FACING THE CARIBBEAN

As Figures 5-5 and 5-6 remind us, the countries of South America's northern tier have something in common besides their coastal location: each has a coastal tropical-plantation zone on the Caribbean colonial model. Especially in the three Guianas, early European plantation development entailed the forced immigration of African laborers and eventually the absorption of this element into the population matrix. Far fewer Africans were brought to South America's northern shores than to Brazil's Atlantic coasts, and tens of thousands of South Asians also arrived as contract laborers and stayed as settlers, so the overall situation here is not comparable to Brazil's. And it is also distinctly different from that of the rest of South America.

Today, Guyana, Suriname, and French Guiana still display the coastal orientation and plantation dependency with which the colonial period endowed them, although logging their tropical forests is penetrating and ravaging the interior. In Venezuela and Colombia, however, farming, ranching, and mining drew the population inland, overtaking the coastal-plantation economy and creating diversified economies.

Figure 5-9 shows that not only Venezuela but also Colombia is Caribbean in its orientation: in Venezuela, coastal oil and natural gas reserves have replaced plantations as the economic mainstay, and Colombia's Andean valleys open toward the north where roads, railways, and pipelines lead toward Caribbean ports such as Cartagena and Barranquilla. Colombia's Pacific coast is hardly a factor in the national economy despite the outlet at Buenaventura. But, as we will see, the North's locational advantages are countered by physical, economic, and political obstacles that continue to prevent the realization of their full potential.

Colombia

Imagine a country more than twice the size of France, not at all burdened by overpopulation, with an environmental geography so varied that it can produce crops ranging from the temperate to the tropical, possessing world-class oil reserves and other natural resources. This

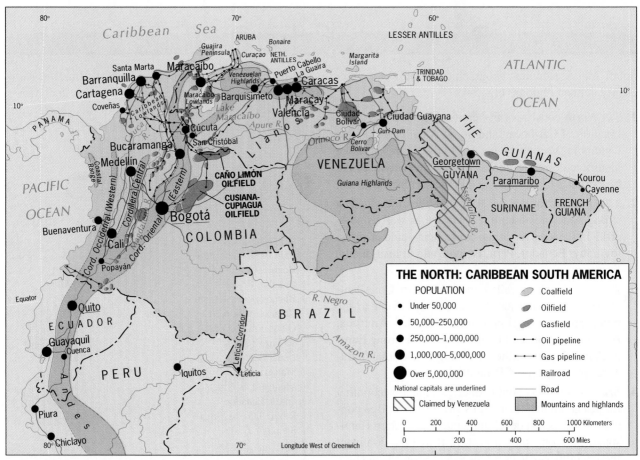

FIGURE 5-9

© H. J. de Blij, P. O. Muller, and John Wiley & Sons, Inc.

country is situated in the crucial northwest corner of South America, with 3200 kilometers (2000 mi) of coast-line on both Atlantic (Caribbean) and Pacific waters, closer than any of its neighbors to the markets of the north and sharing a border with giant Brazil to the south. Its nation uses a single language and adheres to one dom-inant religion. Wouldn't such a country thrive, near the very center of the burgeoning economic geography of its hemisphere?

The answer is no. Colombia has a history of strife and violence, its politics unstable, its economy damaged, its future clouded. Colombia's cultural uniformity did not produce social cohesion. Its spectacular, scenic physical geography also divides its population of 45.7 million into clusters not sufficiently interconnected to foster integra-tion; even today, this huge country has less than 800 kilo-meters (500 mi) of four-lane highways. Its proximity to U.S. markets is a curse as well as a blessing: at the root of Colombia's latest surge of internal conflict lies its role as one of the world's largest producers of illicit drugs.

History of Conflict

Colombia's current disorder is not its first. In the past, civil wars between conservatives and liberals (based on Roman Catholic religious issues) developed into con-flicts pitting rich against poor, elites against workers. In Colombia today, people still refer to the last of these wars as *La Violencia*, a decade of strife beginning in 1948 during which as many as 200,000 people died. In the 1970s, disaster struck again. In remote parts of the country, groups opposed to the political power structure began a campaign of terrorism, damaging the develop-ing infrastructure and destroying confidence in the future. Simultaneously, the U.S. market for narcotics expanded rapidly, and many Colombians got involved in the drug trade. Powerful and wealthy drug cartels formed in major cities such as Medellín and Cali, with networks that influenced all facets of Colombian life from the peasantry to the politicians. The fabric of Co-lombian society unraveled.

People and Resources

As Figures 5-1 and 5-9 show, Colombia's physiography is mountainous in its Andean west and north and com-paratively flat in its *Llanos* (Plains) interior. Look at Fig-ure 5-3, and you can see how Colombia's scattered population tends to cluster in the west and north, where the resources and the agricultural opportunities (including the coffee for which Colombia is famous) lie. On the map, the capital of Bogotá (elevation: 2700 meters [8500 ft]), forms the largest and most important of these clusters

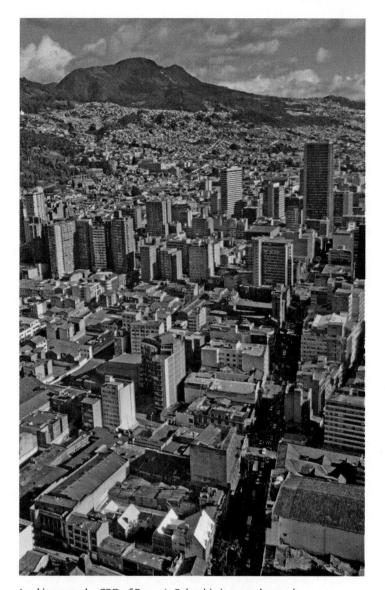

Looking over the CBD of Bogotá, Colombia in a southeasterly direction, where high relief curtails building, you are reminded of the formidable and divisive topography of this large country. The airport (Eldorado International!) lies behind us to the northwest, and the larger and less prosperous suburbs occupy the hilly area to the west (to the right of this view). With more than 8 million inhabitants, metropolitan Bogotá is larger than Colombia's next seven urban areas combined and is the country's primate city— but terrain, distance, and persistent insurgencies erode the capital's supremacy.   © Lonely Planet Images/Getty Images, Inc.

(see photo above) and seems to be centrally situated, but in fact it lies in the southeastern quadrant of Colombia's most densely peopled sector.

We noted earlier that Colombia's population clusters, of which there are about a dozen, are not well intercon-nected. Several of these lie on the Caribbean coast, cen-tered on Barranquilla, Cartagena, and Santa Marta, old colonial entry points. Others are anchored by major cities such as Medellín and Cali. What is especially interesting about Figure 5-9 in this context is how little development has taken place on the country's lengthy Pacific coast, where the port city of Buenaventura, across the mountains from Cali, is the only place of any size. The map suggests that Colombia's Pacific Rim era, already in full force in Chile, Peru, and Ecuador, has yet to arrive.

The map also shows that Colombia and neighboring Venezuela share the oil and gas reserves in and around the "Lake" Maracaibo area ("Lake" because this is really a gulf with a narrow opening to the sea), but Venezuela has the bigger portion. Recent discoveries, however, have boosted Colombia's production along the base of the easternmost cordillera, and our map shows a growing system of pipelines from the interior to the coast. Meanwhile, the vast and remote interior proved fertile ground for that other big money-maker in Colombia: drugs.

Cocaine's Cost

With its energy, mineral, and agricultural productivity, Colombia might have been on the fast track toward prosperity, but the rise of the narcotics industry, fueled by outside (especially U.S. but also Brazilian and European) demand, and coupled with its legacy of violence, crippled the state for decades and threatened its very survival. In a country as large and physiographically diverse as this, there are many opportunities to avoid detection and evade law enforcement, ranging from farming coca plants to smuggling weapons (see box titled "The Geography of Cocaine"). Drug cartels based in the cities controlled vast networks of producers and exporters; they infiltrated the political system, corrupted the army and police, and waged wars with each other that cost tens of

thousands of lives and destroyed Colombia's social order. The drug cartels organized their own armed forces to combat attempts by the Colombian government to control the illegal narcotics economy, and kidnapped and killed citizens at will even as the Colombian government appealed to the United States for help in its ineffective campaign. Meanwhile, the owners of large haciendas in the countryside hired private security guards to protect their properties, banding together to expand these units into what became, in effect, private armies. Colombia was in chaos as narcoterrorists committed appalling acts of violence in the cities, rebel forces and drug-financed armies of the political "left" fought paramilitaries of the political "right" in the countryside, and Colombia's legitimate economy, from coffee growing to tourism, suffered fatally. To further weaken the national government, the rebels even took to bombing oil pipelines.

State of Insurgency

What happened in Colombia beginning in the 1970s and escalating in the three decades following was not unique in the world—states have succumbed to chaos in the past—but this was an especially clear-cut case of a process long studied and modeled by political scientists. By the turn of this century, certain parts of Colombia were beyond the control of its government and armed forces. There, insurgents of various stripes created their own domains, successfully resisting interference and demanding to be left alone to pursue their goals, illegitimate though they might be. Leaders of some of these insurgent domains even sent emissaries to Bogotá to negotiate their terms for "independence."

When it comes to energy resources, Venezuela is in the lead in South America, but Colombia has formidable reserves as well, the most extensive of which lie in the central north on the inland side of the Andes' easternmost range. This example of the industry's environmental impact shows what is reputedly the largest oilfield in South America, based on reserves in the Cusiana river basin, about 250 kilometers (150 mi) east of Bogotá in the heart of the country. © Gamma Presse, Inc.

The Geography of Cocaine

ANY GEOGRAPHICAL DISCUSSION of northwestern South America today must take note of one of its most widespread activities: the production of illegal narcotics. Of the enormous flow of illicit drugs that enter the United States each year, the most widely used substance is cocaine—all of which comes from South America, mainly Colombia, Peru, and Bolivia. Within these three countries, cocaine annually brings in billions of (U.S.) dollars and "employs" tens of thousands of workers, constituting an industry that functions as a powerful economic force. Those who operate the industry have accumulated considerable power through bribery of politicians, kidnappings and other forms of intimidation, threats of terrorism, and alliances with guerrilla groups in outlying zones beyond governmental control. The cocaine industry itself is structured within a tightly organized network of territories that encompass the various stages of this drug's production.

The first stage of cocaine production is the extraction of coca paste from the coca plant, a raw-material-oriented activity that is located near the areas where the plant is grown. The coca plant was domesticated in the Andes by the Incas centuries ago; millions of their descendants still chew coca leaves for stimulation and brew them into coca tea, the leading beverage of high-altitude South America. The main zone of coca-plant cultivation is along the eastern slopes of the Andes and in adjacent tropical lowlands in Bolivia, Peru, and Colombia. Today, five areas dominate in the growing of coca leaves for narcotic production (see Fig. 5-11): Bolivia's Chaparé district in the marginal Amazon lowlands northeast of the city of Cochabamba; the Yungas Highlands north of the Bolivian capital, La Paz; north-central Peru's Huallaga and neighboring valleys; south-central Peru's Apurimac Valley, southeast of Huancayo; and the green-colored zones (in Fig. 5-10) around the guerrilla-controlled territories of southern Colombia. These areas, which can produce as many as nine crops per year, are especially conducive to high leaf yields thanks to favorable local climatic and soil environments that also allow plants to develop immunity to many diseases and insect ravages. Despite government efforts to aerially spray herbicides and manually eradicate plants in the field, all of these areas continue to thrive as coca cultivation has become a full-fledged cash crop. Operations in guerrilla-controlled portions of Colombia have reached the scale of plantations, inducing thousands of local subsistence farmers (sometimes at gunpoint) to join the more lucrative ranks of the field workforce. In Peru and Bolivia, the specialized coca-cultivation zones have lured an even larger peasant-farmer population to abandon the less profitable production of food crops (coca's per-acre income is now seven times greater than cocoa's), thereby further reducing the capability of these nutrition-poor countries to feed themselves.

(As [U.S.-supported] government efforts to eliminate coca-raising have intensified in recent years, the geography of production has made certain adjustments. The most aggressive pursuit occurred in Colombia, and the net effect was the displacement of coca cultivation to nearby Andean countries. This shift, however, actually appears to have strengthened rather than weakened the industry: recent press reports indicate that not only are coca-leaf prices—and overall productivity—rising steadily, but also that street prices for cocaine are falling worldwide.)

The coca leaves harvested in the source areas of the Andes and adjoining interior lowlands make their way to local collection centers, located at the convergence of rivers and trails, where coca paste is extracted and prepared. This begins the second stage of production, which involves the refining of that coca paste (about 40 percent pure cocaine) into cocaine hydrochloride (more than 90 percent pure), a lethal concentrate that is diluted with substances such as sugar or flour before being sold on the streets to consumers. Cocaine refining requires sophisticated chemicals, carefully controlled processes, and a labor force skilled in their supervision, and here Colombia has predominated. Most of this activity takes place in ultramodern processing centers located in the rebel-held territory of the lowland central-south and east, beyond the reach of the Bogotá government. Interior Colombia also possesses the geographic advantage of intermediate location, lying between the source areas to the south (as well as locally) and the U.S. market to the north.

The final stage of production entails the distribution of cocaine to the marketplace, which depends on an efficient, clandestine transportation network that leads into the United States. Private planes operating out of remote airstrips were the preferred "exporting" method until about a decade ago, but aggressive U.S. measures along the coasts of Florida and the Gulf of Mexico have effectively closed down trans-Caribbean flight paths. Most cocaine now travels overland to be smuggled by sea through northwestern South America's Pacific and Caribbean seaports. As noted in Chapter 4, its main destination is Mexico, which today supplies at least 90 percent of all the cocaine that enters the United States. This rerouting through Mexico has been aided by the greatly increased northward flow of goods into the U.S. under NAFTA—providing ever greater opportunities for smuggling via the millions of trucks that cross the border each year. It has also triggered the rapid rise of organized crime (see pp. 214–215) as several Mexican drug gangs evolved into sophisticated international cartels that control both the cocaine inflow and retail distribution within the United States, and increasingly dictate the terms of operation to their South American suppliers.

Viewing this process institutionally, political scientists recognized three evolutionary stages. During the first, *contention*, a rebellion erupts, sustains itself, and becomes based in some part of a country. During the second stage, *equilibrium*, the rebels gain effective control over a territorially defined sector of the state, establish a capital, impose rules of conduct, and have sufficient strength to bring the national government to the negotiating table. If they succeed, they may achieve secession from the state or even bring the government down and take control of all of it. But equilibrium may also be followed by a third stage, *counteroffensive*, in which the state, perhaps with the support of outside forces, resumes the conflict and ultimately defeats the insurgents.

We will encounter instances of all three stages in this model in later chapters of this book. The geographer Robert McColl looked at the sequence in spatial context and formulated the concept of an **insurgent state** to represent the second stage, that of equilibrium. He argued that this equilibrium implied the formation of more than just a rebel base: this is the moment of truth when the rebel domain takes on the elements of a nascent state complete with boundaries, a core area as well as a capital, a local government, and schools and other social ser-

vices that substitute for those the formal state may have provided previously.

Colombia in the 1990s and early 2000s contained several entities that were taking on the properties of insurgent states, and one of these, the red-striped zone shown in Figure 5-10, even acquired a name—"Farclandia"—after the initials of the Revolutionary Armed Forces of Colombia (FARC), one of the most brutal and by far the most powerful among Colombia's insurgent groups. In late 1999, FARC had the Colombian government on the ropes, forcing it to demilitarize its area south of Bogotá about the size of Switzerland and announcing plans for a second insurgent state centered on the remote southeastern town of Mitú. It appeared to be just a matter of time before the national government in Bogotá would lose control, equilibrium would turn into disintegration, and Colombia would devolve into a **failed state** (a country whose institutions have collapsed and in which anarchy prevails).

Counteroffensive?

During the first decade of the twenty-first century, Colombia mounted a counteroffensive whose effectiveness remains in doubt. In 2002, newly elected President Alvaro Uribe began a twin campaign to defeat the rebels militarily and to persuade them through legal means to give up their arms. He secured significant assistance from the United States and was able to increase the pressure on armed rebels as well as coca growers, invaded "Farclandia," and scored some noteworthy successes in killing and arresting leaders as well as freeing hostages held by the rebels. Colombia's nascent insurgent states weakened, and disarray marked the once-formidable rebel armies. Reelected in 2006 by a people weary of conflict and willing to accept the inequities inherent in the pacification effort, Uribe continued his campaign, although charges of links between government members and right-wing paramilitaries eroded public support.

The government matched its domestic counteroffensive with an international campaign to help revive the economy, promoting market-oriented, business-friendly policies. Under the political circumstances, it is remarkable that economic growth before the global downturn of 2008

FIGURE 5-10   © H. J. de Blij, P. O. Muller, and John Wiley & Sons, Inc.

was as high as 8 percent annually, led by revenues from energy, metals, and agricultural products; the comparative lull in violence boosted coffee exports, cut-flower exports (Colombia ranks second in the world in this business), and tourism.

But Colombia remains a nation in crisis, not all of it of its own making. About 70 percent of the cocaine consumed in the United States comes from Colombia; most of the remainder goes to Brazil or Europe. American support for Uribe's antidrug campaign (*Plan Colombia*), logistical as well as financial, raises political issues in Bogotá, where extraditions of Colombian druglords to the United States are not universally popular. Conversely, the U.S. Congress refused to approve a free-trade agreement with Colombia because of the Uribe government's alleged tilt toward right-wing paramilitaries in connection with the amnesty program. Colombians argue that the cocaine keeps flowing because Americans continue to buy it, and that Americans should look in the mirror before they accuse others of bad behavior. In 2010, Colombia's future still remained clouded. Sixty years—more than two generations—of violence have created propensities that cannot be cast off overnight.

Venezuela

A long and tortuous boundary separates Colombia from Venezuela, its neighbor to the east. Much of what is important in Venezuela is concentrated in the northern and western parts of the country, where the Venezuelan Highlands form the eastern spur of the north end of the Andes system. Most of Venezuela's 29.1 million people are concentrated in these uplands, which include the capital of Caracas, its early rival Valencia, and the commercial/industrial centers of Barquisimeto and Maracay.

The Venezuelan Highlands are flanked by the Maracaibo Lowlands and "Lake" Maracaibo to the northwest and by a vast plainland of savanna country, known as the *Llanos*, in the Orinoco Basin to the south and east (Fig. 5-9). The Maracaibo Lowlands, once a disease-infested, sparsely peopled coastland, today constitute one of the world's leading oil-producing areas; much of the oil is drawn from reserves that lie beneath the shallow waters of the lake itself. The country's third-largest city, Maracaibo, is the focus of the petroleum industry that transformed the Venezuelan economy in the 1970s; however, as we shall see, since then oil has been more of a curse than a blessing.

The *Llanos* on the southern side of the Venezuelan Highlands and the Guiana Highlands in the country's southeast, like much of Brazil's interior, are in an early stage of development. The 300- to 650-kilometer (180- to 400-mi)-long *Llanos* slope gently from the base of the Andean spur to the Orinoco River. Their mixture of savanna grasses and scrub woodland support cattle grazing on higher ground, but widespread wet-season flooding of the more fertile lower-lying areas has thus far inhibited the plainland's commercial farming potential (much of the development of the *Llanos* to date has been limited to the exploitation of its substantial oil reserves). Crop-raising conditions are more favorable in the *tierra templada* areas (see the discussion of the altitudinal zonation concept on p. 215) of the Guiana Highlands. Economic integration of this more remote interior zone with the rest of Venezuela has been spearheaded by the discovery of rich iron ores on the northern flanks of the Guiana Highlands southwest of Ciudad Guayana. Local railroads now connect with the Orinoco, and from there ores are shipped directly to foreign markets.

Oil and Politics

Despite these opportunities, Venezuela since 1998 has been in upheaval as longstanding economic and social problems finally intensified to the point where the electorate decided to push the country into a new era of radical political change. For nearly two decades since the euphoric 1970s, oil had not bettered the lives of most Venezuelans. A major reason was that the government unwisely acquired the habit of living off oil profits, forcing the country to suffer the consequences of the long global oil depression that began in the early 1980s. Venezuela found itself heavily burdened by a huge foreign debt it had incurred in borrowing against future oil revenues that were not materializing fast enough. By the mid-1990s, the government was required to sharply devalue the currency, and a political crisis ensued that resulted in a severe recession and widespread social unrest. With more and more Venezuelans enraged at the way their oil-rich country was approaching bankruptcy without making progress toward the more equitable distribution of the national wealth, voters resoundingly turned in an extreme direction in the 1998 presidential election. Expressing their disgust with Venezuela's ruling elite of both political parties, they elected Hugo Chávez, a former colonel who in 1992 had led a failed military coup. Reaffirming their decision in 2000, Venezuelans gave nearly 60 percent of their votes to Chávez, apparently providing him with a mandate to act as strongman on behalf of the urban poor and the increasingly penurious middle class.

Chávez has indeed pursued this course since entering office in 1999, sweeping aside Congress and the Supreme Court, supervising the rewriting of the Venezuelan constitution in his own image, and proclaiming himself the leader of a "peaceful leftist revolution" that will transform the country. Although he professes that social equality ranks highest on his agenda, Chávez has stirred up racial divisions by actively promoting mestizos (68 percent of the population) over people of European background (21 percent). And in the international arena, Chávez sparks controversy at every turn: angering the government of neighboring Colombia by expressing his "neutrality" in its confrontation with cocaine-producing insurgent forces; unsettling another neighbor, Guyana, by aggressively reviving a century-old territorial claim to the western zone of that country (the striped area in Fig. 5-9); embracing Fidel Castro and his successor, and providing communist Cuba with oil in exchange for the assignment to Venezuela of thousands of doctors and technicians; and obstructing the globalization plans of the United States throughout South America, using the windfall income from high-priced oil to reward allies in this effort with substantial subsidies.

In 2006, Chávez was reelected with an overwhelming majority over an opponent from the oil-rich western State of Zulia, where oil and cattle dominate the economy—the "Texas of Venezuela." Opposition to Chávez is so strong in this corner of the country (the State encircles Lake Maracaibo) that a pro-autonomy movement has arisen there, a devolutionary response to the current government's populist policies.

Undaunted, Chavez continued to raise Venezuela's profile in the Western Hemisphere and in the world beyond. He expounded his anti-American message in Moscow, proclaimed his solidarity with Iran's president, and hosted Russian warships in Venezuelan ports. In Middle America, he took up common cause not only with Cuba but also with Nicaragua, where he promised to build an oil refinery, fund social programs, and provide low-cost fuel. And in South America, Chavez used his financial resources to support Bolivian President Evo Morales, the realm's first national leader of Amerindian ancestry, who faced powerful opposition from the country's wealthy Hispanic minority.

Chávez has positioned himself as the champion of Venezuela's (and South America's) poor, trumpeting his "Bolivarian Revolution" as a local alternative to what he describes as the insidious encroachment of U.S. imperialism. He has taken Venezuela to center stage in the ideological contest in this realm, casting doubt on the virtues of democracy and free enterprise, castigating elites for not doing enough to reduce the gap between rich and poor, and urging the downtrodden to assert themselves. His message resonates across the continent.

The "Three Guianas"

Three small entities form the eastern flank of the realm's northern region: Guyana, Suriname, and French Guiana. They are good reminders of why the name "Latin" America is inappropriate: the first is a legacy of British colo-

nialism and has English as its official language; the second is a remnant of Dutch influence where Dutch is still official among its polyglot of tongues; and the third is still a dependency—of France. None has a population exceeding one million, and all three exhibit social indices and cultural landscapes more representative of Caribbean islands than South America. Here British, Dutch, and French colonial powers acquired possessions and established plantations, brought in African and Asian workers, and created economies similar to those of their Middle American domains (see Fig. 4-6).

Guyana, with 820,000 people, still has more inhabitants than Suriname and French Guiana combined. When Guyana became independent in 1966, its British rulers left behind an ethnically and culturally divided population in which people with South Asian (Indian) ancestry make up about 50 percent and those with African heritage (including African-European ancestry) 36 percent. This makes for contentious politics, given the religious mix, which is approximately 50 percent Christian and 45 percent Hindu and Muslim. Guyana remains dominantly rural, and plantation crops continue to figure strongly among exports (gold from the interior is the most valuable single product). Oil may soon become a factor in the economy, though, because a recently discovered reserve that lies offshore from Suriname extends westward beneath Guyana's waters. Still, Guyana is among the realm's poorest and least urbanized countries, and it is strongly affected by its neighbor's narcotics industry. Its thinly populated interior, beyond the reach of antidrug campaigns, has become a staging area for drug distribution to Brazil, North America, and even Europe. A recent official report suggests that drug money amounts to as much as one-fifth of Guyana's total economy.

Venezuela's land claim against Guyana may be in abeyance, but the United Nations in 2007 settled a dispute with Suriname over a potentially oil-rich maritime zone in Guyana's favor (Fig. 5-9). Some experts believe that this offshore basin may hold more oil than Europe's North Sea, which would—if properly managed—transform Guyana's economy. Exploratory drilling began in 2009.

Suriname actually progressed more rapidly than Guyana after it became independent in 1975, but persistent political instability soon ensued. The Dutch colonists brought South Asians, Indonesians, Africans, and even some Chinese to their colony, making for a fractious nation. More than 100,000 residents—about one-quarter of the entire population—emigrated to the Netherlands, and were it not for support from its former colonial ruler Suriname would have collapsed. Still, the rice farms laid out by the Dutch give Suriname self-sufficiency and even allow for some exporting, and again plantation crops continue to figure among the exports. Suriname's leading income producer, however, is its bauxite (aluminum ore) mined in a zone across the middle of the country. And the recent oil finds (Fig. 5-9) may provide important revenues in the years ahead.

Suriname, whose population totals 480,000, is one of South America's poorest countries, and its prospects remain bleak. Along with Guyana, Suriname is involved in an environmental controversy centering on its luxuriant tropical forests. Timber companies from Asia's Pacific Rim offer large rewards for the right to cut down the magnificent hardwood trees, but conservationists are trying to slow the destruction by buying up concessions before they can be opened to logging.

Suriname's cultural geography is enlivened by its languages. Dutch may be the official tongue, but other than by officials and in schools it is not heard much. A mixture of Dutch and English, *Sranan Tongo*, serves as a kind of common tongue, but you can also hear Amerindian, Hindi, Chinese, Indonesian, and even some French Creole in the streets. Suriname may have a Dutch past, but today it is a member of Caricom, the English-speaking, supranational Caribbean economic organization.

French Guiana, the easternmost outpost of Caribbean South America, is a dependency—mainland South America's only one. This territory is an anomaly in other ways as well. Consider this: it is not much smaller than South Korea (population: nearly 50 million) with a population of just over 200,000. Its status is an *Overseas Département* of France, and its official language is French. Nearly half the population resides in the immediate vicinity of the capital, Cayenne.

In 2009 there still was no prospect of independence for this severely underdeveloped relic of the former French Empire. Gold remains the most valuable export, and the small fishing industry sends some exports to France. But what really matters here in French Guiana is the European Space Agency's launch complex at Kourou on the coast, which accounts for more than half of the territory's entire economic activity. From plantation farming to spaceport . . . this must be the ultimate story of globalization.

● THE WEST: ANDEAN SOUTH AMERICA

The second regional grouping of South American states—the Andean West (Fig. 5-11)—is dominated physiographically by the great Andes Mountains and historically by Amerindian peoples. This region encompasses Peru, Ecuador, Bolivia, and transitional Paraguay, the last with one foot in the West and the other in the South (Fig. 5-4). Bolivia and Paraguay also constitute South America's only two landlocked countries.

Spanish conquerors overpowered the Amerindian nations, but they did not reduce them to small minorities as happened to so many indigenous peoples in other parts of the world. Today, roughly 45 percent of the people in Peru, the region's most populous country, are Amerindian; in Bolivia Amerindians are in the majority at 55 percent. About 25 percent of Ecuador's population identifies itself as Amerindian, and in Paraguay the ethnic mix, not regionally clustered as in the other three countries, is overwhelmingly weighted toward Amerindian ancestry.

As the Data Table inside the back cover reports, this is South America's poorest region economically, with lower incomes, higher numbers of subsistence farmers, and fewer opportunities for job-seekers. There is no oil-rich Venezuela or agriculturally bountiful Argentina in South America's West. For a very long time, the urbane lives of the land-owning elite have been worlds away from the hard-scrabble existence of the landless peonage (the word *peon* is an old Spanish term for an indebted day laborer). But today, this region, like the realm as a whole, is stirring, and oil and natural gas are part of the story. In Bolivia, the first elected president of Amerindian ancestry is trying to gain control over an energy industry not used to his aggressive tactics. In Ecuador's 2006 elections, a populist gained the presidency on a promise to divert more of the country's oil revenues toward domestic needs. In Peru, where an energy era seems to be opening as new reserves are discovered, the government is under pressure to protect Amazonian peoples and environments, limit foreign involvement, and put domestic needs and rights first.

In short, this is a crucial region in significant transition, where U.S. ambitions and intentions collide with domestic initiatives and where globalizers and locals find themselves face to face.

Peru

Peru straddles the Andean spine for more than 1600 kilometers (1000 mi) and is the largest of the region's four republics in both territory and population (28.7 million). Physiographically and culturally, Peru divides into three subregions: (1) the desert coast, the European-mestizo region; (2) the Andean highlands or *Sierra*, the Amerindian region; and (3) the eastern slopes, the sparsely populated Amerindian-mestizo interior (Fig. 5-11).

Three Subregions

Lima and its port, Callao, lie at the center of the desert *coastal strip*, and it is symptomatic of the cultural division prevailing in Peru that for nearly 500 years the

FIGURE 5-11

© H. J. de Blij, P. O. Muller, and John Wiley & Sons, Inc.

capital city has been positioned on the periphery, not in a central location in a basin of the Andes. From an economic point of view, however, the Spaniards' choice of a headquarters on the Pacific coast proved to be sound, for the coastal subregion has become commercially the most productive part of the country. A thriving fishing industry contributes significantly to the export trade; so do the products of irrigated agriculture in some 40 oases

distributed all along the arid coast, which include fruits such as citrus, olives, and avocados, and vegetables such as asparagus (a big money-maker) and lettuce.

The *Sierra* (Andean) subregion occupies about one-third of the country and is the ancestral home of the largest component in the total population, the Quechua-speakers, who had been subjugated by the Inca rulers when the Spanish conquerors arrived. Their survival dur-

AMONG THE REALM'S GREAT CITIES . . . **Lima**

EVEN IN A realm marked by an abundance of primate cities, Lima (8.3 million) stands out. Here reside 29 percent of the Peruvian population—who produce over 70 percent of the country's gross national income, 90 percent of its collected taxes, and 98 percent of its private investments. Economically, at least, Lima is Peru.

Lima began as a modest oasis in a narrow coastal desert squeezed between the cold waters of the Pacific and the soaring heights of the nearby Andes. Near here the Spanish conquistadors discovered one of the best natural harbors on the western shoreline of South America, where they founded the port of Callao; but they built their city on a site 11 kilometers (7 mi) inland where soils and water supplies proved more beneficial. This city was named Lima and quickly became the Spaniards' headquarters for all their South American territories.

Since independence, Lima has continued to dominate Peruvian national life. The city's population growth was manageable through the end of the 1970s, but the past three decades have witnessed a disastrous doubling in its size. As usual, the worst problems are localized in the peripheral squatter shantytowns that now house close to one-half of the metropolitan population. Lima's CBD,

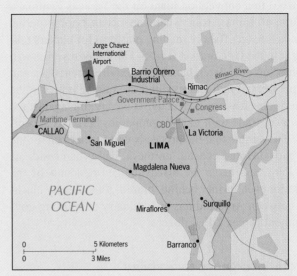

© H. J. de Blij, P. O. Muller, and John Wiley & Sons, Inc.

however, is worlds removed from this squalor, its historic landscape dotted with five-star hotels, new office towers, and prestigious shops as foreign investments have poured in since 1990.

---

ing the harsh colonial regime was made possible by their adaptation to the high-altitude environments they inhabited, but their social fabric was ripped apart by communalization, forced cropping, religious persecution, and outward migration to towns and haciendas where many became serfs. Another Amerindian people, the Aymara, live in the area of Lake Titicaca, making up about 6 percent of Peru's population but larger numbers in neighboring Bolivia (ca. 25 percent).

Although these Amerindian people make up almost half of the population of Peru, their political influence remains slight—even during the recent term of President Alejandro Toledo, who had strong support in the Andean districts. Nor is the Andean subregion a major factor in Peru's commercial economy—except, of course, for its mineral storehouse, which yields copper, zinc, and lead from mining centers, the largest of which is Cerro de Pasco. In the high valleys and intermontane basins, the Amerindian population is clustered either in isolated villages, around which people practice a precarious subsistence agriculture, or in the more favorably located and fertile areas where they are tenants, peons on white- or mestizo-owned haciendas. Most of these people never receive an adequate daily caloric intake or balanced diet of any sort. The wheat produced around Huancayo, for example, is mainly exported and would in any case be too expensive for the Amerindi-

ans themselves to buy. Potatoes, barley, and corn are among the subsistence crops grown here in the *tierra fría* zone, and in the *tierra helada* of the higher basins the Amerindians graze their llamas, alpacas, cattle, and sheep. The contrasts between the prosperous coast, booming Lima, and the thriving north on the one hand, and this poverty-stricken southern Andean zone on the other, constitute a threat to Peru's long-term stability.

Of Peru's three subregions, the *Oriente*, or East—the inland slopes of the Andes and the Amazon-drained, rainforest-covered *montaña*—is the most isolated. The focus of the eastern subregion, in fact, is Iquitos, a city that looks east rather than west and can be reached by oceangoing vessels sailing 3700 kilometers (2300 mi) up the Amazon River across northern Brazil. Iquitos grew rapidly during the Amazon wild-rubber boom of a century ago and then declined; now it is finally growing again and reflects Peruvian plans to open up the eastern interior.

Today, the *Oriente* subregion is on the threshold of a new energy era. Petroleum had been discovered west of Iquitos as long ago as the 1970s, when oil began to flow through a pipeline across the Andes to the Pacific port of Bayovar. But major new discoveries of oil and gas reserves since the turn of the millennium are promising (some say the verb should be *threatening*) to open a new energy era for Peru, with associated impacts on the Amerindian peoples still living an isolated existence in this remote, forested East. Already, pipelines carry natural gas from the Camisea reserve (north of Cuzco) to a conversion plant on the Paracas Peninsula south of Lima, from where an ocean-floor pipeline sends it to an offshore loading platform for tankers taking it to the U.S. market (Fig. 5-11). Other reserves will come "on line" later, but already environmentalists and supporters of indigenous rights are raising issues in the interior even as political activists are arguing against the terms of trade with the oil companies Peru has accepted. They argue that the proceeds will further benefit the already-favored coastal, northern, and urban residents of Peru, and leave the disadvantaged residents of the interior even further behind. In mid-2009, deadly clashes between Amerindians and industry-supporting government forces killed dozens.

From Insurgency to Stability

By the turn of this century, Peru appeared to have put behind it the lengthy period of instability during which its government was threatened by well-organized guerrilla movements. The most serious threat, which for a time during the 1980s seemed on the verge of achieving an insurgent state in its Andean base, was the so-called, Maoist-inspired *Sendero Luminoso* (Shining Path) movement. But forceful counteroffensive action under the later-disgraced president of Japanese ancestry, Alberto Fujimori, defeated the rebel movement and returned Peru to political stability and economic growth.

The social cost of the Fujimori regime, however, was high. Democratic principles were violated, and corruption was endemic. The Amerindian population made no significant gains, and the wealth gap increased. Nonetheless, the country held together during the next administration despite low public support, violent clashes between Andean mining companies and Amerindian villagers, strikes by coca farmers demanding legalization, and other major challenges. Then, in 2006, a presidential election evolved into an ideological contest between a former president, supported by the traditional Hispanic and mestizo sector, and a candidate who had significant Amerindian support as well as the endorsement of Venezuela's Hugo Chávez and Bolivia's Evo Morales. The former president, Alan Garcia, was narrowly elected, suggesting that Peru was not yet ready for the transition already underway in its neighbors. But developments in other parts of Andean South America, and in the realm as a whole, make it likely that Peru, too, will come to confront the implications of its cultural geography. Amerindians form a near-majority, they are restive, and they now have models of empowerment they have not previously seen. The question is what course Peru's transition will take.

Ecuador

On the map, Ecuador, smallest of the three Andean West republics, appears to be just a northern corner of Peru. But that would be a misrepresentation because Ecuador possesses a full range of regional contrasts (Fig. 5-11). It has a coastal belt; an Andean zone that may be narrow (under 250 kilometers [150 mi]) but by no means of lower elevation than elsewhere; and an *Oriente*—an eastern subregion that is as sparsely settled and as economically marginalized as that of Peru. As in Peru, nearly half of Ecuador's population (which totals 14.4 million) is concentrated in the Andean intermontane basins and valleys, and the most productive region is the coastal strip. Here, however, the similarities end.

Ecuador's Pacific coastal zone consists of a belt of hills interrupted by lowlands, of which the most important lies in the south between the hills and the Andes, drained by the Guayas River. Guayaquil—the country's largest city, main port, and leading commercial center—forms the focus of this subregion. Unlike Peru's coastal strip, Ecuador's is not a desert: it consists of fertile tropical plains not afflicted by excessive rainfall. Seafood (especially shrimp) is a leading product, and these lowlands support a thriving commercial agricultural economy built around bananas, cacao, cattle-raising, and coffee on the hillsides. Moreover, Ecuador's western subregion is also far less Europeanized than Peru's because its white component of the national population is only about 7 percent.

A greater proportion of whites are engaged in administration and hacienda ownership in the central Andean zone, where most of the Ecuadorians who are Amerindian also reside—and, not surprisingly, where land-tenure reform is an explosive issue. The differing interests of the Guayaquil-dominated coastal lowland and the Andean-highland subregion focused on the capital (Quito) have long fostered a deep regional cleavage between the two. This schism has intensified in recent years, and autonomy and other devolutionary remedies are now being openly discussed in the coastlands.

In the rainforests of the *Oriente* subregion, oil production is expanding as a result of the discovery of additional reserves. Some analysts are predicting that interior Ecuador as well as Peru will prove to contain reserves comparable to those of Venezuela and Colombia, and that an "oil era" will soon transform their economies. As it is, oil already tops Ecuador's export list, and the industry's infrastructure is being modernized. This is essential because much ecological damage has already been done: the trans-Andean pipeline to the port of Esmeraldas, constructed in 1972, was the source of numerous oil spills, and toxic waste was dumped along its route in a series of dreadful environmental disasters. A second, more modern pipeline went into operation in 2003, but like Peru, Ecuador faced grow-

"I can't remember being hotter anyplace on Earth, not in Kinshasa, not in Singapore . . . not only are you near the equator here in steamy Guayaquil, but the city lies in a swampy, riverine lowland too far from the Pacific to benefit from any cooling breezes and too far from the Andes foothills to enjoy the benefits of suburban elevation. But Guayaquil is not the disease-ridden backwater it used to be. Its port is modern, its city-center waterfront on the Guayas River has been renovated, its international airport is the hub of a commercial center, and it has grown into a metropolis of 2.7 million. Ecuador's oil revenues have made much of this possible, but from the White Hill (a hill that serves as a cemetery, with the most elaborate vaults near its base and the poorest at the feet of the giant statue of Jesus that tops it) you can see that globalization has not quite arrived here. High-rise development remains limited, international banks, hotels and other businesses remain comparatively few, and the 'middle zone' encircling the city shows little evidence of prosperity (top). Beating the heat and glare is an everyday priority: whole streets have been covered by makeshift tarpaulins and more permanent awnings to protect shoppers (bottom). Talk to the locals, though, and you find that there is another daily concern: the people 'up there in the mountains who rule this country always put us in second place.' Take the 45-minute flight from Guayaquil to cool and comfortable Quito, the capital, and you're in another world, and you quickly forget Guayaquil's problems. That's just what the locals here say the politicians do."   © H. J. de Blij.

Concept Caching

ing opposition from environmentalists and activists who, in 2005, shut it down for a week. With revenues from oil and gas exports exceeding 50 percent of all exports, Ecuador's leaders are hearing an increasingly familiar refrain from its people: demand more from the companies and give these funds to those who need it. But it is not as simple as that: oil and gas exploration and exploitation require huge investments, and foreign companies can afford to spend what the government's own state company, PetroEcuador, cannot. That leads to difficult choices, because the state needs income to cover its obligations. Clearly, energy riches are a double-edged sword.

There is more to Ecuador's economic geography than its energy sector, of course, but the country's other commercial opportunities (fish from its waters, fruits from the coast, flowers from the cooler slopes) do not reach their full potential because the country is in a difficult social transition. Ecuador's plural society is strained by a growing social movement, led by its indigenous peoples, that reflects what is happening in the Andean region generally.

Bolivia

Nowhere in this region are the problems faced by Ecuador and to a (hitherto) lesser extent by Peru more acute than they are in landlocked, volatile Bolivia. Before reading further, take a careful look at Bolivia's regional geography in both Figures 5-11 and 5-12. Bolivia is bounded by remote parts of Brazil and Argentina, mountainous Andean highlands and *altiplanos* of Peru, and coveted coastal zones of northern Chile. As the maps show, the Andes in this area broaden into a vast mountain complex some 700 kilometers (450 mi) wide. On the boundary between Peru and Bolivia, freshwater Lake Titicaca lies at 3700 meters (12,500 ft) above sea level and helps make the adjacent *Altiplano* livable by ameliorating the coldness in its vicinity, where the snow line lies just above the plateau surface. On the surrounding cultivable land, potatoes and grains have been raised for centuries dating back to pre-Inca times, and the Titicaca Basin still supports a major cluster of Aymara subsistence farmers. This

FIGURE 5-12

© H. J. de Blij, P. O. Muller, and John Wiley & Sons, Inc.

and the harsh exploitation of its labor force hangs heavily over a society where about two-thirds of the people, almost all Amerindian, live in dire poverty. In recent years, however, this underrepresented aboriginal majority has been making an impact on national affairs. In 2003, violent opposition to a government plan to export natural gas to the United States via a new pipeline to the Chilean coast led to chaos and the government's resignation. In 2005, Bolivian voters elected their first Aymara (Amerindian-ancestry) president, Evo Morales, who proceeded to nationalize the country's natural gas resources.

Departments and Capitals

Landlocked, physiographically bisected, culturally split, and economically divided, Bolivia is a state in trouble. The country's prospects are worsened by its political geography: look at Figure 5-12 and you can see that Bolivia's nine provinces are regionally divided between Amerindian-majority *Departments* (as subnational units are called here) in the west and those with mestizo majorities in the east. The capital, La Paz, lies in the Amerindian-majority mountains, but many mestizo Bolivians do not recognize it as such: historically, the functions of government have been split between La Paz (the administrative headquarters) and Sucre (which lost most government branches in 1899 during a civil war but retained the Supreme Court, calling itself the "constitutional capital"). And some mestizo Bolivians even suggest that the eastern city of Santa Cruz should be considered as a candidate for "compromise capital" of their country.

The Santa Cruz *Department*, like the others in the *Oriente*, stands in sharp contrast to those of the Andes in the west. Here the hacienda system (see p. 203) persists almost unchanged from colonial times, its profitable agriculture supporting a wealthy aristocracy. Nearly 90 percent of Bolivia's agriculturally productive land is still owned by about 50,000 families, much of it taken during the colonization period and the rest expropriated during subsequent military dictatorships. Now this eastern zone of Bolivia proves to contain rich energy resources as well, adding to its economic advantage. So there is talk (and there are frequent public demonstrations) in support of autonomy, even secession here, but neither the haciendas nor the energy industry could function without the Amerindian labor force.

portion of the *Altiplano* is the heart of modern Bolivia and also contains the capital city, La Paz.

The landlocked state of Bolivia is the product of the Hispanic impact, and the country's Amerindians (who still make up at least 55 percent of the national population of 10.4 million) no more escaped the loss of their land than did their Peruvian or Ecuadorian counterparts. What made the richest Europeans in Bolivia wealthy, however, was not land but minerals. The town of Potosí in the eastern cordillera became a legend for the immense deposits of silver in its vicinity; tin, zinc, copper, and several ferroalloys were also discovered there. Amerindian workers were forced to work in the mines under the most dreadful conditions.

Today natural gas and oil, exported to Argentina and Brazil, are major sources of foreign revenues, and zinc has replaced tin as the leading metal export. But Bolivia's economic prospects will always be impeded by the loss of its outlet to the Pacific Ocean during its war with Chile in the 1880s, despite its transit rights and dedicated port facilities at Antofagasta.

More critical than its economic limitations or its landlocked situation is Bolivia's social predicament. The government's history of mistreatment of indigenous people

Less than ten years ago Bolivia was in chaos, a failing state on the brink of all-out civil war. Venezuela's president, Hugo Chávez, openly involved himself in the country's politics through public opposition to the then-Bolivian leadership and by supporting the Amerindian side in the dispute. When Evo Morales was elected and began trying to gain control of the energy industry while making moves to alter the division of national wealth, the struggle between east and west took a new turn.

By the end of the first decade of this century, Bolivia still hung together, even as the president sought to change the constitution and the three eastern *Departments* tried to secure greater autonomy (their local newspapers often refer to Spain's Catalonia as their model for the future). When energy prices were high, there was money to spend on the social programs President Morales had promised to expand. But when, toward the end of the decade, the price of natural gas declined even as the Bolivian government got into trade disputes with its neighbors Brazil (the main market for Bolivia's gas) and Argentina, the specter of state failure rose again. The combination of rising expectations among Amerindian citizens and devolutionary hopes among the minority mestizos cast an ominous cloud over Bolivia's future.

Paraguay

Paraguay, Bolivia's landlocked neighbor to the southeast, is one of those transitional countries between regions and exhibits properties of each (see Fig. 5-4; additional examples on the world map [Fig. G-2] are Belarus, Kazakhstan, and Sudan). Certainly, Paraguay (population: 6.5 million) is not an "Andean" country: it has no highlands of consequence. Its well-watered eastern plains give way to the dry scrub of the Chaco in the west. Nor does it have clear, spatially entrenched ethnic divisions among Amerindians, mestizos, and others as do Bolivia and Peru. But Amerindian ancestry dominates the ethnic complexion of Paraguay, and Amerindian Guaraní is so widely spoken in the country that this is one of the world's most thoroughly bilingual societies. And continuing protests by landless peasants mirror those occurring elsewhere in South America's West. Looking south, there is little in Paraguay's economic geography to compare to Argentina, Uruguay, or Chile, as the back-cover Data Table confirms. In a sense, Paraguay is a bridge between West and South, but not a heavily traveled one.

Paraguay is transitional in another way: as many as 300,000 Brazilians have crossed the border to settle in eastern Paraguay, where they have created a thriving commercial agricultural economy that produces soybeans, livestock, and other farm products exported to or through Brazil. Brazil, of course, is the giant in the Mercosur/l free-trade zone, but Paraguay is in a geographically difficult position: Paraguay often complains that Brazil does not live up to its regional-trade obligations and creates unacceptable difficulties for Paraguayan exporters. Meanwhile, politicians raise fears that growing Brazilian immigration is creating a foreign enclave within Paraguay, where people speak Portuguese (including in the local schools), the Brazilian rather than the Paraguayan flag flies over public buildings, and a Brazilian cultural landscape is evolving. Like Bolivia, Paraguay pays heavily for its landlocked weakness.

Paraguay's human geography reflects this. As the Data Table inside the back cover shows, the country's low GNI resembles that of countries of the West, not the more advantaged South. This is also one of South America's least urbanized states, and poverty dominates the countryside as well as the slums encircling the capital, Asunción, and other towns (Fig. 5-11). In 2007, nearly two-thirds of the population lived at or below the official poverty level. Although records are inadequate, research suggests that 1 percent of the population owns about 75 percent of the land. That may be a record for inequality in the South American realm.

In 2008, Paraguay ended its history of ruthless dictatorial rule with the election of a radical, pro-Guaraní priest, Fernando Lugo, who in his days as a missionary supported hacienda invasions by landless peasants and promised land reform and other remedies for the poor.

Money for these remedies will be hard to come by, but in July 2009 President Lugo scored a major success when he negotiated new terms with Brazil for the sale of Paraguay's share of electricity generated by Itaipu Dam located on the Paraná River between the two countries. When this huge dam's turbines began producing power in 1984, each country (then ruled by dictators) got 50 percent of the electricity, but severely underdeveloped Paraguay needed just a tiny fraction of that. So Paraguay's ruler agreed to sell to Brazil the remainder of his country's share at prices far below market value. Democratically elected Lugo persuaded Brazil's (also democratic) government to begin paying market rates for the 40 percent of Paraguay's Itaipu-generated power it sells every year, adding hundreds of millions of dollars to Paraguay's income. Furthermore, Brazil will assist Paraguay in building a modern transmission line from Itaipu to Asunción, to be ready in 2012 and mark the start of a crucial expansion of Paraguay's electrical-power infrastructure.

Another problem arising from Paraguay's long-term weakness and misrule lies in the southeast, where the borders of Brazil, Argentina, and Paraguay converge in a chaotic scene of smuggling, money laundering, political intrigue, and even terrorist activity, centered on the town of Ciudad del Este. Locals call this the *Triple Frontier*, and warning flags went up when reports of Hizbullah (Iranian-backed terrorist) activity were confirmed by the discovery of maps of the area in a Taliban safe house in Afghanistan. Paraguay's state system clearly needs strengthening, and not just for domestic reasons.

● THE SOUTH: MID-LATITUDE SOUTH AMERICA

South America's three southern countries—Argentina, Chile, and Uruguay—constitute a region sometimes referred to as the *Southern Cone* because of its tapered, ice-cream-cone shape (Fig. 5-13). As noted earlier, Paraguay has strong links to this region and in some

FIGURE 5-13   © H. J. de Blij, P. O. Muller, and John Wiley & Sons, Inc.

ways forms part of it, although social contrasts between Paraguay and those in the Southern Cone remain sharp.

Since 1995, the countries of this region have been drawing closer together in an economic union named *Mercosur/l*, the hemisphere's second-largest trading bloc after NAFTA. Despite setbacks and disputes, Mercosur/l has expanded and today encompasses Argentina, Uruguay, Paraguay, Brazil, and (pending ratification) Venezuela; Chile, Bolivia, Peru, Ecuador, and Colombia participate as associate members.

Argentina

The largest Southern Cone country by far is Argentina, whose territorial size ranks second only to Brazil in this geographic realm; its population of 40.6 million ranks third after Brazil and Colombia. Argentina exhibits a great deal of physical-environmental variety within its boundaries, and the vast majority of the Argentines are concentrated in the physiographic subregion known as the *Pampa* (a word meaning "plain"). Figure 5-3 underscores the degree of clustering of Argentina's inhabitants on the land and in the cities of the Pampa. It also shows the relative emptiness of the other six subregions (mapped in Fig. 5-13): the scrub-forest *Chaco* in the northwest; the mountainous *Andes* in the west, along whose crestline lies the boundary with Chile; the arid plateaus of *Patagonia* south of the Rio Colorado; and the undulating transitional terrain of intermediate *Cuyo, Entre Rios* (also known as "Mesopotamia" because it lies between the Paraná and Uruguay rivers), and the *North*.

The Argentine Pampa is the product of the past 150 years. During the second half of the nineteenth century, when the great grasslands of the world were being opened up (including those of the interior United States, Russia, and Australia), the economy of the long-dormant Pampa began to emerge. The food needs of industrializing Europe grew by leaps and bounds, and the advances of the Industrial Revolution—railroads, more efficient ocean transport, refrigerated ships, and agricultural machinery—helped make large-scale commercial meat and grain production in the Pampa not only feasible but also highly profitable. Large haciendas were laid out and farmed by tenant workers; railroads radiated ever farther outward from the booming capital of Buenos Aires and brought the entire Pampa into production.

AMONG THE REALM'S GREAT CITIES . . . **Buenos Aires**

ITS NAME MEANS "fair winds," which first attracted European mariners to the site of Buenos Aires alongside the broad estuary of the muddy Rio de la Plata. The shipping function has remained paramount, and to this day the city's residents are known throughout Argentina as the *porteños* (the "port dwellers"). Modern Buenos Aires was built on the back of the nearby Pampa's grain and beef industry. It is often likened to Chicago and the Corn Belt in the United States because both cities have thrived as interfaces between their immensely productive agricultural hinterlands and the rest of the world.

Buenos Aires (13.0 million) is yet another classic South American primate metropolis, housing nearly one-third of all Argentines, serving as the capital since 1880, and functioning as the country's economic core. Moreover, Buenos Aires is a cultural center of global standing, a monument-studded city that contains the world's widest boulevard (Avenida 9 de Julio).

During the half-century between 1890 and 1940, the city was known as the "Paris of the South" for its architecture, fashion leadership, book publishing, and performing arts activities (it still has the world's biggest opera house, the Teatro Colón). With the recent restoration of democracy, Buenos Aires is now trying to recapture its golden years. Besides reviving these cultural functions, the

© H. J. de Blij, P. O. Muller, and John Wiley & Sons, Inc.

city has added a new one: the leading base of the hemisphere's motion picture and television industry for Spanish-speaking audiences.

A Culture Urban and Urbane

Argentina once was one of the richest countries in the world. Its historic affluence is still reflected in its architecturally splendid cities whose plazas and avenues are flanked by ornate public buildings and private mansions. This is true not only of the capital, Buenos Aires, at the head of the Rio de la Plata estuary—it also applies to interior cities such as Mendoza and Córdoba. The cultural imprint is dominantly Spanish, but the cultural landscape was diversified by a massive influx of Italians and smaller but influential numbers of British, French, and German immigrants. A sizeable immigration from Lebanon resulted in the diffusion of Arab ancestry to more than 8 percent of the Argentinian population.

Argentina has long been one of the realm's most urbanized countries: 91 percent of its population is concentrated in cities and towns, a higher percentage even than Western Europe or the United States. Almost one-third of all Argentinians live in metropolitan Buenos Aires, by far the leading industrial complex where processing Pampa products dominates. Córdoba has become the second-ranking industrial center and was chosen by foreign automobile manufacturers as the car-assembly center for the expanding Mercosur/l market. One in three Argentinian wage-earners is engaged in manufacturing, another indication of the country's economic progress. But what concentrates the urban populations is the processing of products from the vast, sparsely peopled interior: Tucumán (sugar), Mendoza (wines), Santa Fe (forest products), and Salta (livestock). Argentina's product range is enormous. There is even an oil reserve near Comodoro Rivadávia on the coast of Patagonia.

Economic Boom and Bust

Despite all these riches, Argentina's economic history is one of boom and bust. With just over 40 million inhabitants, a vast territory with diverse natural resources, adequate infrastructure, and good international linkages, Argentina should still be one of the world's wealthiest countries, as it once was. But political infighting and economic mismanagement have combined to ruin a vibrant and varied economy. What began as a severe recession toward the end of the 1990s became an economic collapse in the first years of this century.

To understand how Argentina finds itself in this situation, a map of its administrative structure is useful. On paper, Argentina is a federal state consisting of the Buenos Aires Federal District and 23 provinces (Fig. 5-14). As would be expected from what we have just learned, the urbanized provinces are populous, while the mainly rural ones have smaller populations—but the gap between

© H. J. de Blij, P. O. Muller, and John Wiley & Sons, Inc.

FIGURE 5-14

Buenos Aires Province (whose capital is La Plata), with nearly 15 million, and Tierra del Fuego (capital: Ushuaia), with barely over 100,000, is wide indeed. Several other provinces contain under 500,000 inhabitants, so that the larger ones in addition to dominant Buenos Aires are also disproportionately influential in domestic politics, especially Córdoba and Santa Fe, both with capitals of the same name.

The never-ending problem for Argentina has been corrupt politics and associated mismanagement. Following an army coup in 1946, Juan Perón got himself elected president, bankrupted the country, and was suc-ceeded by a military junta that plunged the country into its darkest days, culminating in the "Dirty War" of 1976–1983 which saw more than 10 (and perhaps as many as 30) thousand Argentinians disappear without a trace. In 1982, this ruthless military clique launched an invasion of the British-held Malvinas (Falkland Islands), resulting in a major defeat for Argentina. By the time civilian government replaced the discredited junta, inflation was soaring and the national debt had become staggering (see the Issue Box entitled "Who Needs Democracy?"). Economic revival during the 1990s was followed by another severe downturn that exposed the flaws in Argentina's fiscal system, includ-ing scandalously inefficient tax collection and uncon-ditional federal handouts to the politically powerful provinces.

In 2003, a new administration took office led by Pres-ident Néstor Kirchner, who hails from a small Patagon-ian province, not from one of the country's traditional power centers. He won on the Peronist Party platform, and then, true to Argentinian political tradition, two years later took on the political bosses in Buenos Aires Province by supporting the Senate candidacy of his wife, Cristina Fernandez de Kirchner, against the spouse of the province's most powerful politician. Fernandez de Kirchner won decidedly, cementing her husband's pres-idential power and allowing him to pursue goals he had long embraced: combating foreign economic interven-tion, controlling domestic companies, reviving Argenti-na's export economy by keeping its currency artificially low, and ending military immunity from prosecution (in response to growing public demands for justice follow-ing the military dictatorship).

In another one of those operatic scenarios that makes locals say "Only in Argentina," President Kirchner announced in mid-2007 that he would not seek reelec-tion to a second four-year term and instead nominated his wife as his party's presidential candidate. Fernandez de Kirchner won decisively, beating 13 other candidates with 45 percent of the vote; but economic problems soon intruded into her post-election political honeymoon despite her party's comfortable majorities in both the Chamber and the Senate. As the Data Table at the back of the book shows, Argentina has the highest per-capita GNI in the realm, but rising inflation and growing labor troubles cost the president much of her public support. Foreign investors were reluctant to assume risks in a country whose government habitually ignored contract terms. Moreover, Argentina continued to cope with un-certainties as to its energy supplies. The end of the old boom-and-bust cycle is not yet in sight.

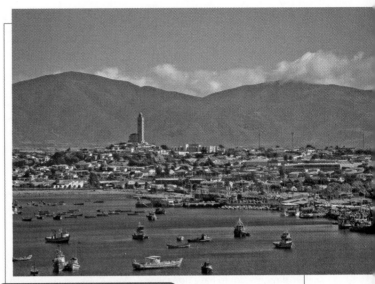

FROM THE FIELD NOTES

"You could see this (what I thought was a) church from many kilometers away on the fieldtrip to Coquimbo, a coastal town about 300 kilometers (200 mi) north of Valparaíso, Chile, where the climatic transition to the desert north is evident all around. But it was this urban scene that taught me a lesson: I glanced at the tower and assumed that this was an unusually prominent Roman Catholic church looming over the townscape, as you see all over the realm. But my colleague advised me to look closer. 'See any crosses?' he asked. I didn't. 'Anything unusual about the architecture?' I noticed the arched courtyard. 'Remind you of anything?' Well, yes, but surely not here? Here indeed—a mosque serving a widely scattered but significant minority, visible from afar and symbolizing change in a realm where evangelical movements are by no means the only challenge to Roman Catholicism's historic domination. I had seen the growing presence of Islam in the greater Lima area days earlier, but that was to be expected in a city of its dimensions and international linkages. But here in relatively remote Coquimbo?" © H. J. de Blij.

www.conceptcaching.com

Concept Caching

Chile

For 4000 kilometers (2500 mi) between the crestline of the Andes and the coastline of the Pacific lies the nar-row strip of land that is the Republic of Chile (Fig. 5-15). On average just 150 kilometers (90 mi) wide (and only rarely over 250 kilometers or 150 mi in width), Chile is the world's quintessential example of what *elon-gation* means to the functioning of a state. Accentuated by its north-south orientation, this severe territorial atten-uation not only results in Chile extending across numerous environmental zones; it has also contributed

Who Needs Democracy?

SOUTH AMERICA DOES!

"I'm old enough to remember the good as well as the bad old days here in Argentina. As a young man I worked for the national railroad company when that guy Perón came on the scene. I guess he learned his craft from the fascists in Italy, because he excelled in election by intimidation. He divided this country as never before, spending money on the workers and the poor while curbing freedoms and even abolishing freedoms guaranteed under our constitution. To tell you the truth, my salary actually went up a bit, and I even voted for him the second time around—largely, I think, because I wanted to do something for his beautiful wife Evita, who was loved by the whole country (except the military).

"Anyway, if you looked around South America in those days, we could have done worse. Wall-to-wall dictators. The military and police, secret and otherwise, put the fear of God into the populace, helped by the church. At least we had a guy who went through the motions of election. While those others divided the wealth among their cronies, Perón tried to spread it around. Even wealth we didn't have! Yes, he certainly put us into debt. And he had his cronies too.

"When they finally deposed Perón, I thought that we might see democratic elections here. (In retrospect, I don't know why I thought that—Spain and Portugal were ruled by two of the worst, Franco and Salazar; that's our 'Latin' heritage). Well, you already know how wrong I was. Instead of more democracy, we got military repression. We lived in terror, not just fear. You've probably heard of those nuns who asked questions about the fate of a baby whose mother had 'disappeared.' They loaded them into a helicopter, in their outfits, and threw them out over the Rio de la Plata. Later I heard some soldiers joke about 'the flying nuns.' Rumors of what could happen to you were rife. Everyone knew someone who knew someone who was a victim.

"Our cloud of military terror lifted after they made their mistake in the Malvinas, 25 years ago, and since then we've had a taste of real democracy. It hasn't been easy, but let me tell you, it's better than 'strongman' rule so many of us South Americans are familiar with. Most of all, there's openness. You can express your views without fear. Political parties can argue their positions without military intimidation. Corrupt public officials can be found out by reporters, and they can't send machine-gun-toting colonels to kill their pursuers. Yes, there are downsides—that openness applies to the economy too, and as we've discovered, outsiders can interfere in our financial affairs. But I'd rather have democracy and open disorder than dictatorship and isolated order, and I've known them both."

Regional **ISSUE**

SAY NO TO THE EXCESSES OF DEMOCRACY!

"You can't help noticing that the Great Democratic Revolution that was supposed to be sweeping Middle and South America is not exactly a success. Here in Brazil we've had democratic government of a sort since 1989, when our man Fernando Collor de Mello was allowed by the military to win an election. You'll remember what happened to him. He resigned three years later after being implicated in a corruption and influence-peddling scheme. I'm sure the poor benefited hugely from this return to democracy.

"South American countries have a history of being led by men who captured the imagination of the nation and whose vision forged the character of the state. These men knew that the state must serve the people. The state must ensure that its railroads and bus services are available at low cost to all citizens. The state must provide electricity, fuel, and clean water. The state must staff and maintain the schools. The state and the church are indivisible, and the networks of the church are in the service of the state. If the military is needed to maintain order in the interest of stability, so be it. Call them strongmen if you like, but they personified their nations and did so for centuries.

"Democracy is a luxury for rich countries. The Americans like to teach us about democracy, but as a geography high-school teacher here in Santa Catarina, let me ask you this: how can the U.S. call itself a democracy when, in their Senate, a few hundred thousand people in Wyoming have the same representation as over 30 million in California? Oh, I see. You call yourself a 'representative republic.' Well, then don't lecture us about democracy. Even Juan Perón was 'democratically' elected, you know. In fact, most of those allegedly dictatorial rulers you so harshly criticize would probably win at the ballot box.

"Anyway, look at what democracy has done for Argentina and for Venezuela before the Venezuelans had the good sense to throw the rascals out and elect Chávez. Look what it is doing for Peru, where democratically elected President Toledo was trying to sell off the electric company serving Arequipa before the citizens stopped him. That 'openness' you hear pro-democracy advocates talk about just means that outsiders can come in, put your country in debt, demand 'privatization' of public companies, buy up corporations and banks and haciendas, fire thousands of workers to increase profits and share prices back home, and make the poor even poorer. What we need in this part of the world is a strong hand, not only to guide the nation but to protect the state."

Vote your opinion at www.wiley.com/college/deblij

CHILE

POPULATION

- • Under 50,000
- • 50,000–250,000
- ● 250,000–1,000,000
- ● 1,000,000–5,000,000
- ⬤ Over 5,000,000

National capitals are underlined

—— Railroad

—— Road

0   100   200   300 Kilometers

0   100   200 Miles

© H. J. de Blij, P. O. Muller, and John Wiley & Sons, Inc.

FIGURE 5-15

to the country's external political, internal administrative, and general economic problems. Nonetheless, throughout most of their modern history, the Chileans have made the best of this potentially disastrous centrifugal force: from the beginning, the sea has constituted an avenue of longitudinal communication; the Andes Mountains continue to form a barrier to encroachment from the east; and when confrontations loomed at the far ends of the country, Chile proved to be quite capable of coping with its northern rivals, Bolivia and Peru, as well as Argentina in the extreme south.

Three Subregions

As Figures 5-13 and 5-15 indicate, Chile is a three-subregion country. About 90 percent of its 17.1 million people are concentrated in what is called Middle Chile, where Santiago, the capital and largest city, and Valparaíso, the chief port, are located. North of Middle Chile lies the Atacama Desert, wider, drier, and colder than the coastal desert of Peru. South of Middle Chile, the coast is broken by a plethora of fjords and islands, the topography is mountainous, and the climate—wet and cool near the Pacific—soon turns drier and colder against the Andean interior. South of the latitude of Chiloé Island, there are few permanent overland routes and hardly any settlements. These three subregions are also apparent on the realm's cultural map (Fig. 5-6), as well as the map of South America's agricultural systems (Fig. 5-5). In addition, a small Amerindian subsistence zone in northeasternmost Chile's Andes is shared with Argentina and Bolivia.

Some intraregional differences exist between northern and southern Middle Chile, the country's core area. Northern Middle Chile, the land of the hacienda and of Mediterranean climate with its dry summer season, is an area of (usually irrigated) crops that include wheat, corn, grapes, and other Mediterranean products. Livestock raising and fodder crops also take up much of the productive land, but continue to give way to the more efficient and profitable cultivation of fruits for export. Southern Middle Chile, into which immigrants from both the north and Europe (especially Germany) have pushed, is a better-watered area where raising cattle has predominated. But here, too, more lucrative fruits, vegetables, and grains are changing the area's agricultural specializations.

Prior to the 1990s, the arid Atacama region in the north accounted for more than half of Chile's foreign revenues. The Atacama Desert contains the world's largest exploitable deposits of nitrates, which was the country's economic mainstay before the discovery of methods of

FROM THE FIELD NOTES

"In a small, admission-charge private museum I found a treasure of historic information about this place in Chile—its environmental challenges ranging from droughts to earthquakes, its growth as a center for irrigated agriculture augmented by mining in the hinterland, its religious importance as the site of many churches and convents. But what matters today is tourism and retirement. Not only is La Serena (just inland from Coquimbo) on the international visitor circuit, but it lies at the center of a growing cluster of second-home and retirement complexes, benefiting from its equable climate and attractive scenery, and from the Southern Cone's comparative economic prosperity. So La Serena has become an amusement park, its stores converted to boutique shops, its historic homes on visitor tours, its plaza throbbing with the music of competing boomboxes, its streets clogged by buses. An older resident sitting on a park bench had his doubts. 'All this started when they built this Pan-American Highway right past us,' he said. 'Now you can get here by sea, by road, even by air. Is this your first visit? You came too late.' No doubt about it: what I saw here wasn't what I read in those faded pages in the museum."
© H. J. de Blij.

www.conceptcaching.com

Concept Caching

synthetic nitrate production a century ago. Subsequently, copper became the chief export (Chile again possesses the world's largest reserves, which in 2008 accounted for more than half of all export revenues). It is mined in several places, but the main concentration lies on the eastern margin of the Atacama near Chuquicamata, not far from the port of Antofagasta. For most of the past decade, commodity prices were high and Chile's export revenues soared; but prices plunged sharply as global recession deepened in 2009, underscoring the risks involved—and the continuing need for economic diversification.

Political and Economic Transformation

Chile today is emerging from a development boom that transformed its economic geography during the 1990s, a growth spurt that established its reputation as South America's greatest success story. Following the withdrawal of its brutal military dictatorship in 1990, Chile embarked on a program of free-market economic reform that brought stable growth, lowered inflation and unemployment, reduced poverty, and attracted massive foreign investment. The last is of particular significance because these new international connections enabled the export-led Chilean economy to diversify and develop in some badly needed new directions. Copper remains the single leading export, but many other mining ventures have been launched. In the agricultural sphere, fruit and vegetable production for export has soared because Chile's harvests coincide with the winter farming lull in the affluent countries of the Northern Hemisphere. Industrial expansion is occurring as well, though at a more leisurely pace, and new manufactures include a modest array of goods that range from basic chemicals to computer software.

Chile's increasingly globalized economy has propelled the country into a prominent role on the international economic scene. The United States, long Chile's leading trade partner, now is in second place compared to the Asian Pacific Rim, where China takes the bulk of Chile's exports followed by Japan and South Korea (Argentina remains Chile's leading source of imports, mostly energy, which is a potential problem given Argentina's own rising needs). Chile's regional commerce also is growing, and Chile has affiliated with Mercosur/l as an associate member.

A "Chilean Model"?

Chile's economic and democratic transformation has been widely touted as a "Chilean model" to be followed by other Middle and South American countries. But Chile benefits from an unusual combination of opportunities, and some economic geographers argue that Chile has in fact not taken sufficient advantage of all of them. Exports should be more diversified, and Chile's dependence on high copper prices entails a major risk. Despite reform efforts by President Michelle Bachelet, the education system remains inadequate. Income inequality—that realmwide affliction—has improved, but nearly 20 percent of Chileans still live in poverty. The plight of Chile's small number of indigenous people, brought to national attention by the Mapuche of the forested south, is of national concern. But compare

Chile to virtually any other country in this realm: the Chilean model may not always apply, but the Chilean example certainly provides hope for the future.

Uruguay

Uruguay, unlike Argentina or Chile, is compact, small, and rather densely populated. This *buffer state* of old became a fairly prosperous agricultural country, in effect a smaller-scale Pampa (though possessing less favorable soils and topography). Figures 5-1 and G-7 show the similarity of physical conditions on the two sides of the Plata estuary. Montevideo, the coastal capital, contains almost 40 percent of the country's population of 3.3 million; from here, railroads and roads radiate outward into the productive agricultural interior (Fig. 5-13). In the immediate vicinity of Montevideo lies Uruguay's major farming area, which produces vegetables and fruits for the metropolis as well as wheat and fodder crops. Most of the rest of the country is used for grazing cattle and sheep, with beef products, wool and textile manufactures, and hides dominating the export trade. Tourism is another major economic activity as Argentines, Brazilians, and other visitors increasingly flock to the Atlantic beaches at Punta del Este and other thriving resort towns.

Concentricity and Conflict

15 In Chapter 1, we noted the work of **von Thünen** and his **model** layout of agricultural zones (pp. 45–46). Here in Uruguay we can discern a real-world example of that scheme (inset map, Fig. 5-13). Market gardens and dairying cluster nearest the urban area of Montevideo, with increasingly extensive agriculture ringing this national market in concentric land-use zones.

Uruguay's 177,000 square kilometers (68,000 sq mi), about the size of Florida, evince a low physiologic density (see p. 414), and with the Montevideo urban area comprising 1.5 million people, there is plenty of agricultural potential here. But Uruguay's government seeks to diversify the economy, and one of its plans—the construction of two large cellulose (paper) factories on the Uruguayan side of the Uruguay River where it forms the border with Argentina—has caused a quarrel between the two Mercosur/l neighbors that has exposed some serious rifts. Montevideo is the administrative headquarters of Mercosur/l, but, like Paraguay, Uruguay has long felt neglected and even obstructed by its much larger neighbors. When Argentinians across the river began demonstrating, blocked a bridge, and stopped taking vacations on Uruguay's beaches, they cost the Uruguayan economy hundreds of millions of dollars.

The Argentinians, who dislike large foreign investments anyway (the factories were to be built by Finnish and Spanish companies), argued that the paper mills would accelerate deforestation, cause river pollution, create acid rain, and damage the farming, fishing, and tourist industries. When the presidents of the two countries met to negotiate a solution and Uruguay's leader agreed to halt construction for a "study period," Urugayan public opinion showed strong opposition to the compromise.

This issue reveals how quickly nationalist feelings can overwhelm the need for international cooperation. It also shows that Mercosur/l partners are far from united on economic matters, and indicates that the collaboration implied by Mercosur/l membership is still a distant reality.

BRAZIL: GIANT OF SOUTH AMERICA

The next time you board an airplane in the United States, don't be surprised if the aircraft you enter was built in Brazil. When you go to the supermarket to buy provisions, take a look at their sources. Chances are some of them will come from Brazil. When you listen to your car radio, some of the best music you hear is likely to have originated in Brazil. The emergence of Brazil as the regional superpower of South America and an economic superpower in the world at large is going to be the story of the twenty-first century. In 2003 Brazil's newspapers carried quotes from national leaders who asked why Brazil, for all its territorial size, population numbers, and energy needs, did not have a nuclear-energy program. Why, asked one editorial, should countries like North Korea and Pakistan have nuclear weapons when Brazil does not even possess nuclear technology? Certainly not because Brazil lacks the capacity.

Why is Brazil so upward-bound today? In large measure it is because, after a long period of dictatorial rule by a minority elite that used the military to stay in power, Brazil embraced democratic government in 1989 (its first democratically elected president was ousted on charges of corruption three years later) and has not looked back since. The era of military coups and repeated quashings of civil liberties is over, and in Brazil's presidential election of 2002 the leader of the Workers Party, Luiz Inácio Lula da Silva, popularly known as Lula, was elected in a runoff that marked a watershed in Brazilian democratic politics. (In a show of displeasure at the result, the United States did not send a single high-ranking representative to Lula's inauguration.) A champion of the poor and dispossessed, Lula came into office amid a rising tide of expectations but with the reality of a faltering economy confronting his government. In his first term, Lula skillfully managed Brazil's numerous countervailing forces, unleashing the country's regional and global potential. But toward its end his party faced bribery and corruption allegations that came close to the president's inner circle. Nonetheless, he was reelected to a second term in 2006 and continued his successful administration.

The Realm's Giant

By any measure, Brazil is South America's giant. It is so large that it has common boundaries with all the realm's other countries except Ecuador and Chile (Fig. 5-4). Its tropical and subtropical environments range from the equatorial rainforest of the Amazon Basin to the humid temperate climate of the far south. Territorially, Brazil occupies just under 50 percent of South America and is exceeded on the global stage only by Russia, Canada, the United States, and China. But in population size, it accounts for just over half the realm's total and is again the world's fifth-largest country (surpassed only by China, India, the United States, and Indonesia). The Brazilian economy, with its modern industrial base, is now the ninth largest on Earth, and is likely to continue advancing in the international ranks and become a world force as the twenty-first century unfolds.

Population Patterns

Brazil's population of 200.6 million is as diverse as that of the United States. In a pattern quite familiar in the Americas, the indigenous inhabitants of the country were decimated following the European invasion (fewer than 200,000 Amerindians now survive deep in the Amazonian interior). Africans came in great numbers, too, and today there are about 13 million Afro-Brazilians in Brazil. Significantly, however, there was also much racial mixing, and 86 million Brazilians have combined European, African, and minor Amerindian ancestries. The remaining 102 million—now barely in the majority at 51 percent—are mainly of European origin, the descendants of immigrants from Portugal, Italy, Germany, and Eastern Europe. The complexion of the population was further diversified by the arrival of Lebanese and Syrian Muslims following the collapse of the Ottoman Empire nearly a century ago. Estimates of the Muslim component in the Brazilian population suggest that about 10 million can claim to be of Arab descent, of whom about 1 million are nominally Muslims. But practicing Muslims are far fewer in number: in 2009 Brazil had only about 50 functioning mosques.

Another significant, though small, minority began arriving in Brazil in 1908: the Japanese, who today are concentrated in São Paulo State. The more than 1 million Japanese-Brazilians form the largest ethnic Japanese community outside Japan, and in their multicultural environment they have risen to the top ranks of Brazilian society as business leaders, urban professionals, farmers—and even as politicians in the city of São Paulo. Committed to their Brazilian homeland as they are, the Japanese community also retains its contacts with Japan, resulting in many a trade connection. And in recent years some ethnic Japanese have been emigrating to Japan, where young families are needed in a fast-aging population.

A National Culture

Brazilian society, to a greater degree than is true elsewhere in the Americas, has made progress in dealing with its racial divisions. To be sure, blacks are still the least advantaged among the country's major population groups, and community leaders continue to complain about discrimination. But ethnic mixing in Brazil is so pervasive that hardly any group is unaffected, and official census statistics about "blacks" and "Europeans" are meaningless. What the Brazilians do have is a true national culture, expressed in an historic adherence to the Catholic faith (now eroding under Protestant-evangelical and secular pressures), in the universal use of a modified form of Portuguese as the common language ("Brazilian"), and in a set of lifestyles in which soccer, "beach culture," distinctive music and dance, and a growing national consciousness and pride are fundamental ingredients.

Changing Demography

Brazil's population grew rapidly during the world's twentieth-century population explosion. But over the past three decades, the rate of natural increase has slowed from nearly 3.0 percent to 1.4 percent, and the average number of children born to a Brazilian woman has just about been halved from 4.4 in 1980 to 2.3 in 2008. These dramatic reductions occurred in the absence of any active family-planning policies by Brazil's federal or State governments and in direct contradiction to the teachings of the Catholic Church, re-

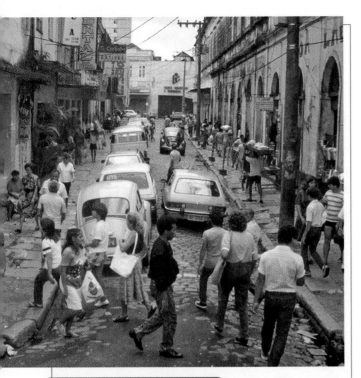

flecting religion's changing role in this former bastion of Roman Catholicism. Brazil's rapid urbanization, its persisting economic uncertainties, and the widespread use of contraceptives are among factors influencing this significant change.

African Heritage

As we noted earlier, Afro-Brazilians still suffer disproportionately from the social ills of this relatively integrated nation. And yet Brazil's culture is infused with African themes, a quality that has marked it from the very beginning. Three centuries ago, the Afro-Brazilian sculptor and architect affectionately called Aleijadinho was Brazil's most famous artist. The world-famous composer Heitor Villa-Lobos used numerous Afro-Brazilian folk themes in his music. So many Africans were brought in bondage to the city and hinterland of Salvador (Bahia State) from what is today Benin (formerly Dahomey) in West Africa that Bahia has become a veritable outpost of African culture. Candomblé, one of the Macumba religious sects arising from the arrival of African cultures in Brazil, is concentrated in Salvador and Bahia (Fig. 5-16). Its presence there has led to a reconnection with modern Benin as Afro-Brazilians travel to West Africa to search for their roots and West Africans come to Bahia to experience the cultural landscape to which their ancestors gave rise.

In some parts of Bahia, the rural Northeast, and urban areas in Brazil's core you can almost imagine being in Africa. As a nation, Brazil is taking an increased interest in African linkages, not least because the Lusitanian world incorporates not only Portugal and Brazil but also Angola and Moçambique, where the Portuguese colonial experiment failed but where the Lusitanian imprint also survives.

Inequality in Brazil

For all its accomplishments in multiculturalism, Brazil remains a country of stark, appalling social inequalities (Fig. 5-17). Although such inequality is hard to measure precisely, South America is often cited as the geographic realm exhibiting the world's sharpest division between affluence and poverty. And in South America, Brazil is reputed to have the widest gap of all.

The Scourge of Poverty

Data that underscore this situation include the following. Today, the richest 10 percent of the population own two-thirds of all the land and control more than half of Brazil's wealth. The poorest 20 percent of the people live in the most squalid conditions prevailing anywhere on the planet, even including the megacities of Africa and Asia. According to UN estimates, in this age of adequate available (but not everywhere affordable) food, about half the population of Brazil suffers some form of malnutrition and, in the northeastern States, even hunger. Packs of young orphans and abandoned children roam the cities, sleeping where they can, stealing or robbing when they must. Some of the world's most magnificent central cities are ringed and sectored by wretched *favelas*

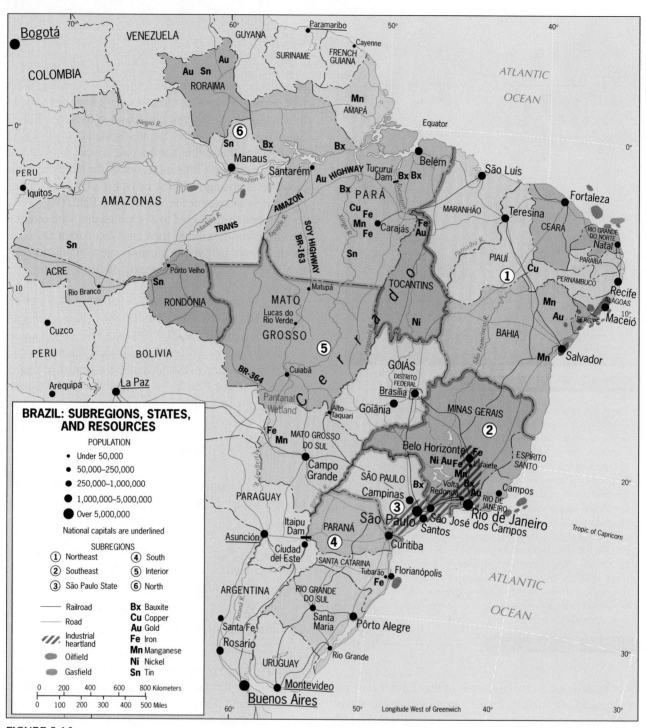

BRAZIL: SUBREGIONS, STATES, AND RESOURCES

POPULATION

● Under 50,000
● 50,000–250,000
● 250,000–1,000,000
● 1,000,000–5,000,000
● Over 5,000,000

National capitals are underlined

SUBREGIONS

① Northeast   ④ South
② Southeast   ⑤ Interior
③ São Paulo State   ⑥ North

— Railroad
— Road
▨ Industrial heartland
▨ Oilfield
▨ Gasfield

Bx Bauxite
Cu Copper
Au Gold
Fe Iron
Mn Manganese
Ni Nickel
Sn Tin

0   200   400   600   800 Kilometers
0   100   200   300   400   500 Miles

FIGURE 5-16

where poverty, misery, and crime converge (see Fig. 5-8 and the right-hand photo on p. 242).

Most Brazilian voters had long believed that these conditions were worsening, that what economic development was taking place benefited those already well off, and that powerful corporations and foreign investors were to blame for the spiral of rising and growing unemployment. When the latest official reports announced that the number of Brazilians afflicted by poverty had increased by 50 percent since 1980, the people demanded change, and, as we will note, they achieved that in 2002.

FIGURE 5-17   © H. J. de Blij, P. O. Muller, and John Wiley & Sons, Inc.

Brazilian manufacturers attain world-class quality. During the mid-1990s, the revenues from industrial exports surpassed those from agriculture. Commerce with Argentina made that member of *Mercosul* (as the Portuguese-speaking Brazilians call Mercosur) one of Brazil's leading trade partners. On the global stage, Brazil became a formidable presence in other ways. The country's enormous and easily accessible iron ore deposits, the relatively low wages of its workers, and the mechanized efficiency of its steelmakers enable Brazil to produce that commodity at half the cost of steel made in the United States. This causes American steel producers to demand government protection through tariffs, which goes against purported U.S. principles of free trade.

A New Era

Even though Brazil has become a major player on the world's economic stage, it has not escaped the cycles of boom and bust that affect transforming economies. Needing loans from international lending agencies to weather the downturns, Brazil's government had to agree to terms that included the privatization of public companies ranging from telephones to utilities. This often meant layoffs, rising prices to consumers, and even civil unrest. Brazil's currency, the *real*, was devalued in 1999, resulting in the outflow of billions of dollars withdrawn from Brazil's banks by panicked depositors. Strikes and marches against increased gas prices, highway tolls, and utility charges paralyzed the country for days. Newspapers reported scandals and corruption involving leading public figures and hundreds of millions of *reais*. In the political campaign that led up to the presidential election of 2002, these issues guaranteed the defeat of the right-of-center government held responsible for it all. In that election a leftist candidate, Luiz Inácio Lula da Silva, won in a landslide victory. He faced the daunting task of adhering to Brazil's existing financial commitments while steering a new course toward more fairness, openness, and honesty in government.

Continuing Challenges

The new left-of-center government in 2002 inherited Brazil's massive social problems without adequate resources to address them. The incoming administration

Development Prospects

Brazil is richly endowed with mineral resources, including enormous iron and aluminum ore reserves, extensive tin and manganese deposits, and sizeable oil- and gasfields (Fig. 5-16). Other significant energy developments involve massive new hydroelectric facilities and the successful substitution of sugarcane-based alcohol (gasohol) for gasoline—allowing well over half of Brazil's cars to use this fuel instead of costly imported petroleum. Besides these natural endowments, Brazilian soils sustain a bountiful agricultural output that makes the country a global leader in the production and export of coffee, soybeans, and orange juice concentrate. Commercial agriculture, in fact, is now the fastest growing economic sector, driven by mechanization and the opening of a major new farming frontier in the fertile grasslands of southwestern Brazil (see pp. 276–277).

Industrialization has propelled Brazil's rise as a global economic power. Much of the momentum for this continuing development was unleashed in the early 1990s after the government opened the country's long-protected industries to international competition and foreign investment. These new policies are proving to be effective because productivity has risen by over a third since 1990, as focused on two key areas: land reform and poverty reduction. At the turn of this century, 1 percent of Brazilians owned 50 percent of all productive land. But the government was able to settle only about 60,000 landless families per year, far below what was needed. This resulted in protests and disruptions, which in turn scared off investors, affecting the jobless rate. By 2006, the government had found the resources to increase the rate of settlement, but the voters who elected President Lula da Silva had expected much more. And in the poverty arena, the administration began a *Fome Zero* (Zero Hunger) program, but in the process created a huge, corruption-riddled bureaucracy. Officially, Brazil counts some 40 million people who live on half the minimum wage of about U.S. $80 per month or less, and the question was how to aid these poorest of the poor with government handouts that had to be modest because of budgetary constraints. By 2006, this muddled plan had been replaced by a much more efficient subsidy program that had significant results in the poorest States. This *Bolsa Familia* (Family Fund) plan gives families small amounts of cash to keep their children in school and to ensure their vaccinations against diseases that especially afflict the poor. In just a few years this program has become so successful that it now serves as a model for antipoverty campaigns in many other parts of the world.

Oil in Brazil's Future?

Brazil has been paying for its improvements with revenues mainly derived from commodities—iron ore, soybeans, coffee, orange juice, beef, sugar—that, during the first decade of this century, commanded high prices on international markets. Economic geographers note that Brazil's dependence on commodities is risky and that the country is linked ever more strongly to the world economy. This means that a global economic downturn could hurt Brazil more than was true in the past, when Brazil imported less and domestic products were protected by tariff barriers. According to this assessment, Brazil does not invest enough of its national income in technology, education, public services, infrastructure, and other needs, and when a downturn comes it could damage the country's prospects (it was too early to tell if this occurred in 2009).

In the past, a leading concern for Brazil was energy and its costs. Recessions were deepened by high oil and gas prices, and other than some small domestic reserves and those in neighboring Bolivia there were few nearby sources from which to acquire what was needed. But in 2008 Brazil's National Petroleum Agency announced the discovery of what it described as an "enormous" oil reserve, possibly the third-largest in the world, much larger than the nearby Tupi Reserve found the previous year in the same area off Rio de Janeiro beneath Atlantic waters (Fig. 5-16). The new-found reserve, appropriately called Carioca-Sugar Loaf, is being explored as you read this. Drilling deep below the sea through thick rock layers will be very costly, and if an energy era is in prospect for Brazil, it will be years off and require foreign investment and technology to which the state-owned oil company, Petrobras, is not accustomed—but the stakes could hardly be higher. A decade from now, optimistic projections suggest, Brazil may rank among the world's top ten oil producers.

Brazil's Subregions

Brazil is a federal republic consisting of 26 States and the federal district of the capital, Brasília (Fig. 5-16). As in the United States, the smallest States lie in the northeast and the larger ones farther west; their populations range from about 400,000 in the northernmost, peripheral Amazon State of Roraima to more than 40 million in burgeoning São Paulo State. Although Brazil is about as large as the 48 contiguous United States, it does not exhibit a clear physiographic regionalism. Even the Amazon Basin, which covers almost 60 percent of the country, is not entirely a plain: between the tributaries of the great river lie low but extensive tablelands. Given this physiographic ambiguity, the six Brazilian subregions discussed next have no absolute or even generally accepted boundaries. In Figure 5-16, those boundaries have been drawn to coincide with the borders of States, making identifications easier.

The Northeast ① was Brazil's source area, its culture hearth. The plantation economy took root here at an early date, attracting Portuguese planters, who soon imported the country's largest group of African slaves to work in the sugar fields. But the ample rainfall occurring along the coast soon gives way to lower and more variable patterns in the interior, which is home to about half of the region's 50-plus million people. This drier inland backcountry—called the *sertão*—is not only seriously overpopulated but also contains some of the worst poverty to be found anywhere in the Americas. The Northeast produces less than one-sixth of Brazil's gross domestic product, but its inhabitants constitute more than one-fourth of the national population. Given this staggering imbalance, it is not surprising that the region contains half of the country's poor, a literacy rate 20 percent below Brazil's mean, and an infant mortality rate twice the national average.

Much of the Northeast's misery is rooted in its un-equal system of land tenure. Farms must be at least 100 hectares (250 acres) to be profitable in the hard-scrabble *sertão*, a size that only large landowners can afford. Moreover, the Northeast is plagued by a monumental environmental problem: the recurrence of devastating droughts at least partly attributable to **El Niño** (periodic sea-surface-warming events off the continent's north-western coast that skew weather patterns).

Understandably, these conditions propel substantial emigration toward the coastal cities and, increasingly, out of the Northeast to the more prosperous Brazilian sub-regions that line the Atlantic seaboard to the south. To stem that human tide, Brazil's government has pursued the purchase of underutilized farmland for the settling of landless peasants; although tens of thousands of families have benefited, overall barely a dent has been made in the Northeast's massive rural poverty crisis.

The Northeast today is Brazil's great contradiction. In cities such as Recife and Salvador, hordes of peasants driven from the land constantly arrive to expand the sur-rounding shantytowns. As yet, few of the generalizations about emerging Brazil apply here, but there are some bright spots. A petrochemical complex has been built near Salvador, creating thousands of jobs and luring foreign investment. Irrigation projects have nurtured a number of productive new commercial agricultural ventures. Tourism is booming along the entire Northeast coast, whose thriving beachside resorts attract thousands of vacationing Europeans. Recife has spawned a budding software industry and a major medical complex. And For-taleza is the center of new clothing and shoe industries that have already put the city on the global economic map.

The Southeast ② has been modern Brazil's *core area*, with its major cities and leading population clus-ters. Gold first drew many thousands of settlers, and other

AMONG THE REALM'S GREAT CITIES . . . Rio de Janeiro

SAY "SOUTH AMERICA" and the first image most peo-ple conjure up is Sugar Loaf Mountain, Rio de Janeiro's landmark sentinel that guards the entrance to beautiful Guanabara Bay. Nicknamed the "magnificent city" be-cause of its breathtaking natural setting, Rio replaced Sal-vador as Brazil's capital in 1763 and held that position for almost two centuries until the federal government shifted its headquarters to interior Brasília in 1960. Rio de Janeiro's primacy suffered yet another blow in the late 1950s: São Paulo, its hated urban rival 400 kilometers (250 mi) to the southwest, surpassed Rio to become Brazil's largest city—a gap that has been widening ever since. Although these events triggered economic decline, Rio (12.2 million) remains a major *entrepôt*, air-travel and tourist hub, and center of international business. As a focus of cultural life, however, Rio de Janeiro is unques-tionably Brazil's leader, and the *cariocas* (as Rio's residents call themselves) continue to set the national pace with their entertainment industries, universities, museums, and libraries.

On the darker side, this city's reputation is increasing-ly being tarnished by the widening abyss between Rio's affluent and poor populations—symbolizing inequities that rank among the world's most extreme (see photos, p. 242). All great cities experience problems, and Rio de Janeiro is currently bedeviled by the drug use and crime waves emanating from its most desperate hillside *favelas* (slums) that continue to grow explosively.

Nonetheless, Rio's planners recently launched a major project (known as "Rio-City") to improve urban

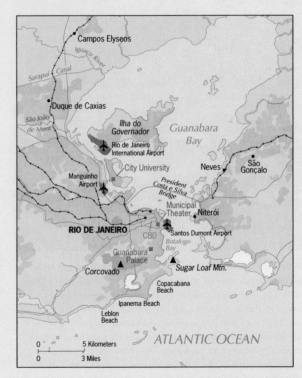

life for all residents. This ambitious scheme is designed to reshape nearly two dozen of the aging city's neigh-borhoods, introduce an ultramodern crosstown expressway to relieve nightmarish traffic congestion, and—most importantly—bring electrical power, paved streets, and a sewage-disposal network to the belea-guered *favelas*.

São Paulo may not have an imposing skyline to match New York or a skyscraper to challenge the Sears (now Willis) Tower of Chicago, but what this Brazilian megacity does have is mass. São Paulo is more than just another city—it is a vast conurbation of numerous cities and towns containing more than 26 million inhabitants who constitute the third-largest human agglomeration on Earth. This view shows part of the razor-sharp edge of the CBD, a concrete jungle with little architectural distinction but vibrant urban cultures. In the foreground, juxtaposed against the affluence of downtown, the teeming inner city reflects the opposite end of the social spectrum—a chaotic jumble of *favelas* that mainly house the newest urban migrants, especially from Brazil's hard-pressed Northeast. São Paulo, demographers say, gives us a glimpse of the urban world of the future.   © AP/Wide World Photos.

mineral finds also contributed to the influx—with Rio de Janeiro itself serving as the terminus of the "Gold Trail" and as the long-term capital of Brazil until 1960. Rio de Janeiro became the cultural capital as well, the country's most international center, *entrepôt*, and tourist hub. The third quarter of the twentieth century brought another mineral age to the Southeast, based on the iron ores around Lafaiete carried to the steelmaking complex at Volta Redonda (Fig. 5-16).

The surrounding State of Minas Gerais (the name means "General Mines") formed the base from which industrial diversification in the Southeast has steadily expanded. The booming metallurgical center of Belo Horizonte paved the way and is now the endpoint of a rapidly developing, ultramodern manufacturing corridor that stretches 500 kilometers (300 mi) southwest to metropolitan São Paulo (Fig. 5-16).

São Paulo State ③ is the leading industrial producer and primary focus of ongoing Brazilian development. This economic-geographic powerhouse accounts for nearly half of the country's gross domestic product, with an economy that today matches Argentina's in overall size. Not surprisingly, this subregion is growing phenomenally (it already contains 20 percent of Brazil's population) as a magnet for migrants, especially from the Northeast.

The wealth of São Paulo State was built on its coffee plantations (known as *fazendas*), and Brazil is still the world's leading producer. But coffee today has been eclipsed by other farm commodities. One of them is orange juice concentrate (here, too, Brazil leads the world). São Paulo State now produces more than double the annual output of Florida, thanks to a climate all but devoid of winter freezes, to ultramodern processing plants, and to a fleet of specially equipped tankers that ship the concentrate to foreign markets. Another leading pursuit is soybeans, in which Brazil ranks second among the world's producers.

Matching this agricultural prowess is the State's industrial strength. The revenues derived from the coffee plantations provided the necessary investment capital, ores from Minas Gerais supplied the vital raw materials, the nearby outport of Santos facilitated access to the ocean, and immigration from Europe, Japan, and other parts of Brazil contributed the increasingly skilled labor force. As the capacity of the domestic market grew, the advantages of central location and agglomeration secured São Paulo's primacy. This also resulted in metropolitan São Paulo becoming the country's—and South America's—leading industrial complex and megacity (population: 26.2 million).

AMONG THE REALM'S GREAT CITIES . . . São Paulo

SÃO PAULO, WHICH lies on a plateau 50 kilometers (30 mi) inland from its Atlantic outport of Santos, possesses no obvious locational advantages. Yet here on this site we find the third-largest metropolis on Earth, whose population has multiplied so uncontrollably that São Paulo has more than doubled in size (from 11 to just over 26 million) over the past three decades.

Founded in 1554 as a Jesuit mission, modern São Paulo was built on the nineteenth-century coffee boom. It has since grown steadily as both an agricultural processing center (soybeans, orange juice concentrate, and sugar besides coffee) and a manufacturing complex (accounting for about half of all of Brazil's industrial jobs). It has also become Brazil's primary focus of commercial and financial activity. Today São Paulo's bustling, high-rise CBD (see photo, p. 274) is the very symbol of urban South America and attracts the continent's largest flow of foreign investment as well as the trade-related activities that befit the city's rise as the business capital of Mercosul (Mercosur).

Nonetheless, even for this biggest industrial metropolis of the Southern Hemisphere, the increasingly global tide of postindustrialism is rolling in and São Paulo is being forced to adapt. To avoid becoming a Detroit-style Rustbelt, the aging automobile-dominated manufacturing zone on the central city's southern fringes is today attracting new industries. Internet companies, in particular, have flocked here in such numbers that they already constitute Brazil's largest cluster of e-commerce activity—and prompt many to now call this area Silicon Village.

Elsewhere in São Paulo's vast urban constellation—whose suburbs now sprawl outward up to 100 kilometers (60 mi) from the CBD—additional opportunities are being exploited. In the outer northeastern sector, new research facilities as well as computer and telecommunications-equipment factories are springing up. And to the west of central São Paulo, lining the ring road that follows the Pinheiros River, is South America's largest suburban office complex replete with a skyline of ultramodern high-rises.

© H. J. de Blij, P. O. Muller, and John Wiley & Sons, Inc.

The huge recent growth of this megacity has also been accompanied by massive problems of overcrowding, pollution, and congestion. Traffic jams here are among the world's worst, with more than twice as many gridlocked motor vehicles on any given day than in Manhattan. Staggering poverty—on a par with Mexico City's—is the most pressing crisis as the ever-expanding belt of shantytowns tightens its grip on much of the metropolis. With its prodigious growth rate expected to persist through the foreseeable future, can anything prevent Greater São Paulo from approaching an unimaginable population of 50 million barely 20 years from today?

The South ④ consists of three States, whose combined population exceeds 27 million: Paraná, Santa Catarina, and Rio Grande do Sul (Fig. 5-16). Southernmost Brazil's excellent agricultural potential has long attracted large numbers of European immigrants. Here the newcomers introduced their advanced farming methods to several areas. Portuguese rice farmers clustered in the valleys of Rio Grande do Sul, where tobacco production has now propelled Brazil to become the world's leading exporter. The Germans, specialists in raising grain and cattle, occupied the somewhat higher areas to the north and in Santa Catarina. The Italians selected the highest slopes, where they established thriving vineyards. All of these fertile lands proved highly productive, and with growing markets in the large urban areas to the north, this tristate subregion became Brazil's most affluent corner.

With the South firmly rooted in the European/commercial agricultural sphere (Figs. 5-5, 5-6), European-style standards of living match the diverse Old World heritage that is reflected in the towns and countryside (where German and Italian are spoken alongside Portuguese). This has led to hostility against non-European Brazilians, and many communities actively discourage poor, job-seeking migrants from the North by offering to pay return bus fares or even blocking their household-goods-laden vehicles. Moreover, extremist groups have arisen to openly espouse the secession of the South from Brazil.

Economic development in the South is not limited to the agricultural sector. Coal from Santa Catarina and Rio Grande do Sul is shipped north to the steel plants of Minas Gerais. Local manufacturing is growing as well, especially in Pôrto Alegre and Tubarão. During the 1990s, a major center of the computer software industry was established in Florianópolis, the island city and State capital just off Santa Catarina's coast. Known as *Tecnópolis*, this budding technopole continues to grow by capitalizing on its seaside amenities, skilled labor force, superior air-travel and global communications linkages, and government and private-sector incentives to support new companies. Only the sparsely populated interior portion of the South has lagged behind the rest of the region, despite the recent completion of massive Itaipu Dam in westernmost Paraná State.

The Interior ⑤ subregion—constituted by the States of Goiás, Mato Grosso, and Mato Grosso do Sul—is also known as the *Central-West*. This is the region that Brazil's developers have long sought to make a part of the country's productive heartland, and in 1960 the new capital of Brasília was deliberately situated on its margins (Fig. 5-16).

By locating the new capital city in the wilderness 650 kilometers (400 mi) inland from its predecessor, Rio de Janeiro, the nation's leaders dramatically signaled the opening of Brazil's development thrust toward the west. Brasília is noteworthy in another regard because it represents what political geographers call a **forward capital**. A state will sometimes relocate its capital to a sensitive area, perhaps near a peripheral zone under dispute with an unfriendly neighbor, in part to confirm its determination to sustain its position in that contested zone. Brasília does not lie near a contested area, but Brazil's interior was an internal frontier to be conquered by a growing nation. Spearheading that drive, the new capital occupied a decidedly forward position.

Despite the subsequent growth of Brasília to 3.8 million inhabitants today (which includes a sizeable ring of peripheral squatter settlements), it was not until the 1990s that the Interior began its economic integration

This is not the U.S. Great Plains but the heart of Brazil's *cerrado* in eastern Mato Grosso State, where the flat terrain of the fertile, well-watered savanna is ideally suited for large-scale, highly mechanized agriculture. As Figure 5-5 shows, farm production here is dominated by soybeans, the cheapest and most efficient source of protein available, and in high demand across the world. This crop has expanded like wildfire in the Central-West subregion, and the soy frontier is now pushing northward into the Amazon rainforest itself. Brazil's acreage has more than doubled since 2000, propelling the country into the topmost ranks of soybean producers; in 2008, Brazil accounted for no less than 28 percent of the global output, second only to the United States' 33 percent. © D. Donne Bryant/Stock Photography.

18 with the rest of Brazil. The catalyst was the exploitation of the vast *cerrado*—the fertile savannas that blanket the Central-West and make it one of the world's most promising agricultural frontiers (at least two-thirds of its arable land still awaits development). As with the U.S. Great Plains, the flat terrain of the *cerrado* is one of its main advantages because it facilitates the large-scale mechanization of farming with a minimal labor force (see photo, p. 276). Another advantage is rainfall, which is more prevalent there than in the Great Plains or Argentine Pampa (Fig. G-6).

The leading crop is soybeans, whose output per hectare here exceeds even that of the U.S. Corn Belt. Other grains and cotton are also expanding across the farmscape of the *cerrado*, but the current pace of regional development is inhibited by a serious accessibility problem. Unlike the Great Plains and Pampa, the growth of an efficient transportation network did not accompany the opening of this farming frontier. Thus the Interior's products must travel along poor roads and intermittent railroads to reach the markets and ports of the Atlantic seaboard. Today several projects are finally underway to alleviate these bottlenecks, including the privately financed *Ferronorte* railway that links Santos to the southeastern corner of Mato Grosso State, and the improved, so-called *Soy Highway* north to the Amazon River port city of Santarém.

The North ⑥ is Brazil's territorially largest and most rapidly developing subregion, which consists of the seven States of the Amazon Basin (Fig. 5-16). This was the scene of the great rubber boom a century ago, when the wild rubber trees in the *selvas* (tropical rainforests) produced huge profits and the central Amazon city of Manaus enjoyed a brief period of wealth and splendor. But the rubber boom ended in 1910, and for most of the seven decades that followed, Amazonia was a stagnant hinterland lying remote from the centers of Brazilian settlement. All that changed dramatically during the 1980s as new development began to stir throughout this awakening region, which currently is the scene of the world's largest migration into virgin territory as more than 200,000 new settlers arrive each year. Most of this influx is occurring south of the Amazon River, in the tablelands between the major waterways and along the Basin's wide rim.

Development projects abound in the Amazonian North. One of the most durable is the *Grande Carajás Project* in southeastern Pará State, a huge multi-faceted scheme centered on one of the world's largest known deposits of iron ore in the hills around Carajás (Fig. 5-16). In addition to a vast mining complex, other new construction here includes the Tucuruí Dam

on the nearby Tocantins River and an 850-kilometer (535-mi) railroad to the Atlantic port of São Luis. This ambitious development project also emphasizes the exploitation of additional minerals, cattle raising, crop farming, and forestry. What is occurring here is a manifestation of the **growth-pole concept**. A growth pole **19** is a location where a set of activities, given a start, will expand and generate widening ripples of development in the surrounding area. In this case, the stimulated hinterland could one day cover one-sixth of all Amazonia.

Understandably, tens of thousands of settlers have descended on this part of the Amazon Basin. Those seeking business opportunities have been in the vanguard, but they have been followed by masses of lower-income laborers and peasant farmers in search of jobs and land ownership. The initial stage of this colossal enterprise has boosted the fortunes of many towns, particularly Manaus northwest of Carajás. Here, a thriving industrial complex (specializing in the production of electronic goods) has emerged in the free-trade zone adjoining the city thanks to the outstanding air-freight operations at Manaus's ultramodern airport. But many problems have also arisen as the tide of pioneers rolled across central Amazonia. One of the most tragic involved the Yanomami people, whose homeland in Roraima State was overrun by thousands of claim-stakers (in search of newly discovered gold), who triggered violent confrontations that ravaged the fragile aboriginal way of life.

Another leading development scheme, known as the *Polonoroeste Plan*, is located about 1600 kilometers (1000 mi) to the southwest of Grande Carajás in the 2400-kilometer (1500-mile)-long Highway BR-364 corridor that parallels the Bolivian border and connects the western Brazilian towns of Cuiabá, Pôrto Velho, and Rio Branco (Fig. 5-16). Although the government had planned for the penetration of western Amazonia to proceed via the east-west Trans-Amazon Highway, the migrants of the 1980s and 1990s preferred to follow BR-364 and settle within the Basin's southwestern rim zone, mostly in Rondônia State. Agriculture has been the dominant activity here, but in the quest for land, bitter conflicts continue to break out between peasants and landholders as the Brazilian government pursues the volatile issue of land reform.

This region of Brazil is well known for the *deforestation* occurring there (an environmental issue of global significance that is discussed in the box on p. 220 in Chapter 4). Cutting the forest results directly from logging operations, but more of it is a matter of land occupation and use by settlers. The usual pattern of settlement

This is what the Amazon's equatorial rainforest looks like from an orbiting satellite after the human onslaught in preparation for settlement. The colors on this Landsat image emphasize the destruction of the trees, with the dark green of the natural forest contrasted against the pale green and pinks of the leveled forest. The linear branching pattern of deforestation here in Rondônia State's Highway BR-364 corridor is explained in the text. But farming here is not likely to succeed for very long, and much of the cleared land is likely to be abandoned. Then the onslaught will resume to clear additional land—as an entire ecosystem comes ever closer to failing forever. © NRSC LTS/Photo Researchers.

in this part of Brazil goes like this. As main and branch highways are cut through the forest, settlers follow (often due to government-sponsored land-occupation schemes), and they move out laterally to clear land for farming. Subsistence crops are planted, but within a few years weed infestation and declining soil fertility make the plots unproductive. As soil fertility continues to decline, and with few markets for commercial crops, the settlers plant grasses and then sell their land to cattle ranchers. The peasant farmers then move on to newly opened areas, clear more land for planting, and the cycle repeats itself. Unfortunately, this not only entrenches low-grade land uses across widespread areas, but also entails the burning and clearing of vast stands of tropical woodland (see photo above). Since the 1980s, an area of rainforest almost the size of Ohio has been lost *annually* in Amazonia—accounting for over half of all tropical deforestation on the planet and worsening an environmental crisis of global proportions.

Brazil is the cornerstone of South America, the dominant economic force in Mercosur/Mercosul, the only dimensional counterweight to the United States in the Western Hemisphere, a maturing democracy, and an emerging global giant. The future of the entire South American realm depends on Brazil's stability and social as well as economic progress.

What You Can Do

RECOMMENDATION: Take up a Region! As you encounter the world's regions in this book, is there one that interests you more than others, one that arouses your curiosity so that you might want to learn more about it than this course can offer? Well, regional specialists are ever fewer and farther between in our increasingly computer-screen-dominated world, and regional knowledge could benefit you for a lifetime. We recommend that you follow your interest in a realm or a region that forms part of it—South America's "Southern Cone" or Southern Africa or Mainland Southeast Asia—and pursue it in the news, in your reading, in other courses. English-language newspapers (with corresponding websites) are published in almost every major city, from Jakarta to Buenos Aires to Beijing, and it's fascinating to see what concerns people there on a daily basis—the local news and the letters to the editor can give you insights you'll never find on the networks. Who knows, you may even travel there and start a specialization that could become a hobby, if not a professional sideline!

GEOGRAPHIC CONNECTIONS

1 South America is a realm of diverse cultural landscapes and numerous immigrant sources. Some countries that are relatively homogeneous culturally, such as Colombia and Venezuela, are as divided politically and economically as are other countries with strong and obvious cultural-geographic divisions such as Peru and Bolivia. In virtually all South American countries, such divisions have slowed development and hampered national integration. What do you see as the best solution(s) for the realm and its states as they seek to overcome these historic obstacles? What evidence is there on the continent of "what works" and what does not?

2 Brazil is the giant of South America, and it has the largest and most productive economy. But Brazil is unlike other South American states in a number of ways (for example, there is no megacity on a scale with São Paulo anywhere else in the realm), so that what locals call the "Brazilian model" of development may not work in other countries. Still, Brazil does share key qualities with its many neighbors, and its experiences can help these neighbors in their efforts to make progress. Discuss the important similarities and differences between Brazil and the rest of South America that have a bearing on the still-evolving relationships among countries in this realm.

Geographic Literature On South America: The key introductory works on this realm were authored by Blouet & Blouet, Bromley & Bromley, Caviedes & Knapp, Clawson, James & Minkel, Kent, Morris, and Preston. These works, together with a comprehensive listing of noteworthy books and recent articles on the geography of South America, can be found in the *References and Further Readings* section of this book's website at **www.wiley.com/college/deblij**

SUBSAHARAN AFRICA

6

Where was this photo taken? Find out at **www.wiley.com/college/deblij**

In This Chapter

- Heart of the world, cradle of humanity
- Africa's rich and varied cultures
- Why Africa remains in the grip of poverty
- South Africa: Engine of its region, beacon for the realm
- Nigeria: Cracking cornerstone of West Africa
- The Islamic Front looms over the north

Photos: © H. J. de Blij

FIGURE 6-1 © H. J. de Blij, P. O. Muller, and John Wiley & Sons, Inc.

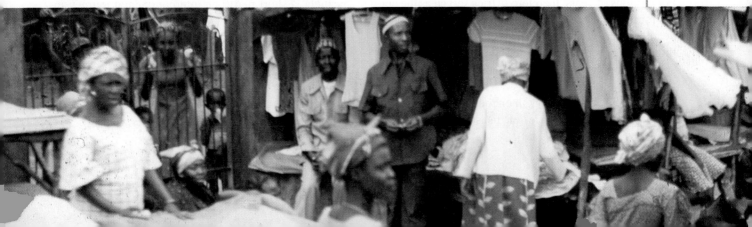

LOOK AT A GLOBE, or at a map of the physical world (such as Fig. G-1 in this book), and it is clear that Africa is the core while the rest of world is the periphery. Africa lies centrally positioned in the Land Hemisphere, with the Americas to the west, Eurasia to the north, Australia to the east, and Antarctica to the south.

In the economically globalizing world of today Africa may not be the *functional* core, but make no mistake: for most of human history, Africa was indeed the core, the very source of humanity. This is where the **1** great saga of **human evolution** began. Here in Africa we formed our first communities, created our first art, and fashioned our first weapons. After tens of thousands of years of adaptation to constantly changing environments, we followed our ancestor hominids out of Africa in fateful *migrations* that were to change the world.

This was in effect the first great wave of globalization. Modern humans crossed the narrow strait at the southern end of the Red Sea, skirted the South Asian coast, traversed the Indonesian archipelago, and reached Australia more than 40,000 years ago. Others went north, invading Europe and confronting the Neanderthals who had preceded them. The wide Pacific Ocean delayed their arrival in the Americas, but eventually, apparently less than 15,000 years ago, the migration's vanguard crossed the Bering Strait and started southward along North America's west coast, making South America the last continent they reached.

We tend to forget that this globalizing wave was also the *Africanization* of the world. Wherever humans migrated, their ancestors had started from Africa and carried their genetic and cultural baggage with them. Time, distance, and environment diversified the human map of the world, but at the source we are all Africans.

Cradle and Cauldron

For millions of years, therefore, Africa served as the cradle for humanity's emergence. For tens of thousands of years, Africa was the source of human cultures. For thousands of years, Africa led the world in countless spheres ranging from tool manufacture to plant domestication.

But in this chapter we will encounter an Africa that has been struck by a series of disasters ranging from environmental deterioration to human dislocation on a scale unmatched anywhere in the world. When we assess Africa's misfortunes, though, we should remember that these have lasted hundreds, not thousands, or millions, of years. Africa's catastrophic interlude will end, and Africa's time and turn will come again.

The focus in this chapter will be on Africa south of the Sahara, for which the unsatisfactory but convenient name *Subsaharan Africa* has come into use to signify not physically "under" the great desert but directionally "below" it. The African continent contains two geographic realms: (1) the African, extending from the southern margins of the Sahara to the Cape of Good Hope; and (2) the Northern, consisting of the western flank of the realm dominated by the Muslim faith and Islamic culture, whose heartland lies in the Middle East and the Arabian Peninsula. The great desert forms a formidable barrier between the two, but the powerful influences of Islam crossed it centuries before the first Europeans set foot in West Africa. By that time, the African kingdoms in what is known today as the Sahel had been converted, creating an Islamic foothold all along the northern periphery of the African realm (see Fig. G-2, p. 7). As we note later, this cultural and ideological penetration had momentous consequences for Subsaharan Africa.

Peril of Proximity

In the three previous chapters on the Americas, we made frequent reference to the forced migration of Africans to Brazil, the Caribbean region, and the United States. The slave trade was one of those African disasters alluded to above, and it was facilitated in part by what we may call the peril of proximity. The northeastern tip of Brazil, by far the largest single destination for the millions of Africans forced from their homes in bondage, lies about as far from the nearest West African coast as South Carolina lies from Venezuela, a short maritime intercontinental journey indeed (it is more than twice as far from West Africa to South Carolina). That proximity facilitated the forced migration of millions of West Africans to Brazil, which in turn contributed to the emergence of an African cultural diaspora in Brazil that is without equal in the New World. It is therefore logical to focus next on the African realm.

MAJOR GEOGRAPHIC QUALITIES OF

Subsaharan Africa

1. Physiographically, Africa is a plateau continent without a linear mountain backbone, with a set of Great Lakes, variable rainfall, generally low-fertility soils, and mainly savanna and steppe vegetation.

2. Dozens of nations, hundreds of ethnic groups, and many smaller entities make up Subsaharan Africa's culturally rich and varied population.

3. Most of Subsaharan Africa's peoples depend on farming for their livelihood.

4. Health and nutritional conditions in Subsaharan Africa need improvement as the incidence of disease remains high and diets are often unbalanced. The AIDS pandemic began in Africa and has become a major health crisis in this realm.

5. Africa's boundary framework is a colonial legacy; many boundaries were drawn without adequate knowledge of or regard for the human and physical geography they divided.

6. The realm is rich in raw materials vital to industrialized countries, but much of Subsaharan Africa's population has little access to the goods and services of the world economy.

7. Patterns of raw-material exploitation and export routes set up in the colonial period still prevail in most of Subsaharan Africa. Interregional and international connections are poor.

8. During the Cold War, great-power competition magnified conflicts in several Subsaharan African countries, with results that will be felt for generations.

9. Severe dislocation affects many Subsaharan African countries, from Liberia to Rwanda. This realm has the largest refugee population in the world today.

10. Government mismanagement and poor leadership afflict the economies of many Subsaharan African countries.

Defining the Realm

The African continent may be partitioned into two human-geographic realms, but the landmass is indivisible. Before we investigate the human geography of Subsaharan Africa, therefore, we should take note of the entire continent's unique physical geography (Fig. 6-1). We have already noted Africa's situation at the center of the planet's Land Hemisphere; moreover, no other landmass is positioned so squarely astride the equator, reaching almost as far to the south as to the north. This location has much to do with the distribution of Africa's climates, soils, vegetation, agricultural potential, and human population.

AFRICA'S PHYSIOGRAPHY

Africa accounts for about one-fifth of the Earth's entire land surface. The north coast of Tunisia lies 7700 kilometers (4800 mi) from the south coast of South Africa. Coastal Senegal, on the extreme western *Bulge* of Africa, lies 7200 kilometers (4500 mi) from the tip of the *Horn* in easternmost Somalia. These distances have environmental implications. Much of Africa is far from maritime

sources of moisture. In addition, as Figure G-7 shows, large parts of the landmass lie in latitudes where global atmospheric circulation systems produce arid conditions. The Sahara in the north and the Kalahari in the south form part of this globe-girdling desert zone. Water supply is one of Africa's great problems.

Rifts and Rivers

Africa's topography reveals several properties that are not replicated on other landmasses. Alone among the continents, Africa does not have an Andes-like linear mountain backbone; neither the northern Atlas nor the southern Cape Ranges are in the same league. Where Africa does have high mountains, as in Ethiopia and South Africa, these are deeply eroded plateaus or, as in East Africa, high, snowcapped volcanoes. Furthermore, Africa is one of only two continents containing a cluster of Great Lakes, and the only one whose lakes result from powerful tectonic forces in the Earth's crust. These lakes (with the exception of Lake Victoria) lie in deep trenches called

FIGURE 6-2

© H. J. de Blij, P. O. Muller, and John Wiley & Sons, Inc.

2 **rift valleys**, which form when huge parallel cracks or faults appear in the Earth's crust and the strips of crust between them sink, or are pushed down, to form great, steep-sided, linear valleys. In Figure 6-2 these rift valleys, which stretch almost 10,000 kilometers (6300 mi) from the Red Sea to Swaziland, are indicated by red lines.

Africa's rivers, too, are unusual: their upper courses often bear landward, seemingly unrelated to the coast toward which they eventually flow. Several rivers, such as the Nile and the Niger, have inland as well as coastal

deltas. Major waterfalls, notably Victoria Falls on the Zambezi, or lengthy systems of cataracts, separate the upper from the lower courses.

Finally, Africa may be described as the "plateau continent." Except for some comparatively limited coastal plains, almost the entire continent lies above 300 meters (1000 ft) in elevation, and fully half of it lies over 800 meters (2500 ft) high. As Figure 6-2 shows, the plateau surface has sagged under the weight of accumulating sediments into a half dozen major basins (three of them in the Sahara).

The margins of Africa's plateau are marked by escarpments, often steep and step-like. Most notable among these is the Great Escarpment of South Africa, marking the eastern edge of the Drakensberg Mountains.

Continental Drift and Plate Tectonics

Africa's remarkable and unusual physiography was one piece of evidence that geographer Alfred Wegener used to construct his hypothesis of **continental drift**. The present continents, Wegener reasoned, lay assembled as one giant landmass called *Pangaea* not very long ago geologically (220 million years ago). The southern part of this supercontinent was *Gondwana*, of which Africa formed the core (Fig. 6-3). When, about 200 million years ago, tectonic forces began to split Pangaea apart, Africa (and the other landmasses) acquired their present configurations. That process, now known as *plate tectonics*, continues, marked by earthquakes and volcanic eruptions. By the time it started, however, Africa's land surface had begun to acquire some of the features that mark it today—and make it unique. The rift valleys, for example, demarcate the zones where plate movement continues—hence the linear shape of the Red Sea, where the Arabian Plate is separating from the African Plate (Fig. G-3). And yes, the rift valleys of East Africa probably mark the further fragmentation of the African Plate (some geophysicists have already referred to a "Somali Plate," which in the future will separate, Madagascar-like, from the rest of Africa).

So Africa's ring of escarpments, its rifts, its river systems, its interior basins, and its lack of Andes-like mountains, all relate to the continent's central location in Pangaea, all pieces of the puzzle that led to the plate-tectonics solution. Once again, geography was the key.

FIGURE 6-3   © H. J. de Blij, P. O. Muller, and John Wiley & Sons, Inc.

NATURAL ENVIRONMENTS

Only the southernmost tip of Subsaharan Africa lies outside the tropics. Although African elevations are comparatively high, they are not high enough to ward off the heat that comes with tropical location except in especially favored locales such as the Kenya Highlands and parts of Ethiopia. And, as we have noted, Africa's bulky shape means that much of the continent lies far from maritime moisture sources. Variable weather and frequent droughts are among Africa's environmental problems.

It is useful at this point to refer back to Figure G-7 on page 17. As that map shows, Africa's climatic regions are distributed almost symmetrically about the equator, though more so in the center of the landmass than in the east, where elevation changes the picture. The hot, rainy climate of the Congo Basin merges gradually, both northward and southward, into climates with distinct winter-dry seasons. "Winter," however, is characterized more by drought than by cold. In parts of the area mapped *Aw* (savanna), the annual seasonal cycle produces two rainy seasons, often referred to locally as the "long rains" and the "short rains," separated by two "winter" dry periods. As you go farther north and south, away from the moist Congo Basin, the dry season(s) grow longer and the rainfall diminishes and becomes less and less dependable.

Crops and Animals

Most Africans still make their living by farming, and many grow crops in marginal areas where rainfall variability can have catastrophic consequences. Compare Figures G-6 and G-7 and note the steep decline in annual rainfall from more than 200 centimeters (80 in) around the equator in the Congo Basin to a mere 10 centimeters (4 in) in parts of Chad to the north and Namibia to the south. Millions of farmers till the often unproductive soils of the *savanna*, and many mix livestock herding with crop-raising to reduce their risk in this difficult environment. But the savanna's wildlife carries diseases that also infect livestock, which makes herding a risky proposition, too. Even where the savanna gives way to the still drier *steppe*, human pressure continues to grow, and people as well as animals trample fragile, desert-margin ecologies.

End of an Era

Africa's shrinking rainforests and vast savannas form the world's last refuges for wildlife ranging from primates to wildebeests. Gorillas and chimpanzees survive in dwindling numbers in threatened forest habitats, while millions of herbivores range in great herds across the savanna plains where people compete with them for space. European colonizers, who introduced hunting as a "sport" (a practice that was not part of African cultural traditions) and who brought their capacities for mass destruction to animals as well as people in Africa, helped clear vast areas of wildlife and push species to near extinction. Later they laid out game preserves and other types of conservation areas, but these were not sufficiently large or well enough connected to allow herd animals to follow their seasonal and annual migration routes. The same climatic variability that affects farmers also affects wildlife, and when the rangelands wither, the animals seek better pastures. When the fences of a game preserve wall them off, they cannot survive. When there are no fences, the wildlife invades neighboring farmlands and destroys crops, and the farmers retaliate. After thousands of years of equilibrium, the competition between humans and animals in Africa has taken a new turn. It is the end of an era.

Farmers' Problems

And yet it would seem that there could be room for humans as well as wildlife in Subsaharan Africa. As the Data Table inside the back cover confirms, all the countries of this realm *combined* have a population little more than half that of China alone. Indeed, Africa does have some areas of good soils, ample water, and high productivity: the volcanic soils of Mount Kilimanjaro and those of the highlands around the Western Rift Valley, the soils in the Ethiopian Highlands, and those in the moister areas of higher-latitude South Africa and parts of West Africa yield good crops when social conditions do not disrupt the farming communities. But such areas are small in the vastness of Africa. Ominously, a 2006 study by the International Center for Soil Fertility and Agricultural Development reports that the overall situation is deteriorating, and that it is worst in East Africa. Soils in Subsaharan Africa are losing nutrients at the highest rate in the world through erosion, exhaustion, and lack of fertilizer. The Center cites an example: in southern Somalia, soils are losing 88 kilograms (200 lb) of nutrients per hectare (2.5 acres) per year, compared to 9 kilograms (20 lb) in overstressed Egypt. And according to the journal *Science*, "African farmers desperate for fresh soil are clearing fragile forestlands and wildlife habitats."

The map of Subsaharan Africa's population distribution (Fig. 6-4) shows few major clusters and many sparsely peopled areas, so that the destruction of natural habitats still has decades, perhaps generations, to go. Nigeria, the realm's most populous state, has the largest numbers in

FIGURE 6-4

© H. J. de Blij, P. O. Muller, and John Wiley & Sons, Inc.

West Africa; a second large cluster occupies the Great Lakes region in East Africa; and the third well-defined population concentration centers on the Ethiopian Highlands in the northeast. Referring once again to Figure 6-4, compare these prominent population clusters to the sparsely peopled expanses of the interior from Namibia in the far southwest to the Central African Republic on the Saharan margin in the north: Subsaharan Africa has just one-tenth the population density of South Asia.

One major reason for this is that Subsaharan Africa has nothing to compare to the huge river basins of India or China and their fertile soils. In fact, this realm does not even have anything comparable to Egypt's teeming lower Nile Valley and Delta, where relatively small areas of fertile, irrigated soils can support tens of millions of people. Except for limited, often experimental patches of rice and wheat, Africa is the land of corn (maize), millet, and root crops, far less able to provide high per-hectare yields. Africa's natural environment poses a formidable challenge to the millions who depend directly on it.

ENVIRONMENT AND HEALTH

From birth, Africans (especially rural people living in *A* climates) are exposed to a wide range of diseases spread by insects and other organisms. The study of human health in spatial context is the field of **medical geography**, and medical geographers employ modern methods of analysis (including geographic information systems) to track disease outbreaks, identify their sources, detect their carriers, and prevent their repetition. Alliances between doctors and geographers have already yielded significant results. Doctors understand how a disease ravages the body; geographers understand how climatic conditions such as wind direction or variations in river flow can affect the distribution and effectiveness of disease carriers. This collaboration helps protect vulnerable populations.

Tropical Africa, the source of many serious illnesses, is the focus of much of medical geography's work. Not only the carriers (*vectors*) of infectious diseases but also

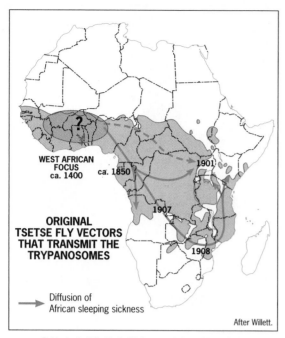

WEST AFRICAN FOCUS
ca. 1400

ca. 1850

1901

1907

1908

ORIGINAL TSETSE FLY VECTORS THAT TRANSMIT THE TRYPANOSOMES

→ Diffusion of African sleeping sickness

After Willett.

© H. J. de Blij, P. O. Muller, and John Wiley & Sons, Inc.

FIGURE 6-5

cultural traditions that facilitate transmission, such as sexual practices, food selection and preparation, and personal hygiene, play their role—and all can be mapped. Comparing medical, environmental, and cultural maps can lead to crucial evidence that helps combat the scourge.

In Africa today, hundreds of millions of people carry one or more maladies, often without knowing exactly what ails them. A disease that infects many people (the *hosts*) in a kind of equilibrium, without causing rapid and **5** widespread deaths, is said to be **endemic** to the population. People affected may not die suddenly or dramatically, but their health deteriorates, energy levels fall, and the quality of life declines. In tropical Africa, hepatitis, venereal diseases, and hookworm are among the public health threats in this category.

Epidemics and Pandemics

When a disease outbreak has local or regional dimen-**6** sions, it is called **epidemic**. It may claim thousands, even tens of thousands of lives, but it remains confined to a certain area, perhaps one defined by the range of its vector. In tropical Africa, trypanosomiasis, the disease known as sleeping sickness and vectored by the tsetse fly, has regional proportions. The great herds of savanna wildlife form the *reservoir* of this disease, and the tsetse fly transmits it to livestock and people. It is endemic to the wildlife, but it kills cattle, so Africa's herders try

to keep their animals in tsetse-free zones. African sleeping sickness appears to have originated in a West African source area during the fifteenth century, and it spread throughout much of tropical Africa (Fig. 6-5). Its epidemic range was limited by that of the tsetse fly: where there are no tsetse flies, there is no sleeping sickness. More than anything else, the tsetse fly has kept Subsaharan Africa's savannalands largely free of livestock and open to wildlife. Should a remedy be found, livestock would replace the great herds on the grasslands.

When a disease spreads worldwide, it is described as **pandemic**. Africa's and the world's most deadly vec-**7** tored disease is malaria, transmitted by a mosquito and killer of as many as one million children each year. Whether malaria has an African origin is not known, but it is an ancient affliction. Hippocrates, the Greek physician of the fifth century BC, mentions it in his writings. Apes, monkeys, and several other species also suffer from it. Recurrent fever attacks, anemia, and enlargement of the spleen are its symptoms.

Malaria spread around the world, not only in tropical but also in temperate areas (the United States did not conquer this disease until the mid-1950s), but Africa was hardest hit. Even today, 90 percent of tropical Africa's approximately 700 million inhabitants live with the threat or effects of malaria. Medical geographers Melinda Meade and Robert Earickson in 2005 wrote that malaria's "mosquito vectors are resistant to all the major insecticides, the agent is resistant to all the major drugs, and the ancient scourge is upon us again." For those at risk, this means that protection is the best course of action: millions of African children sleep in dwellings without windows or screens, and their best hope is insecticide-treated mosquito netting. Many organizations, governmental and private, are engaged in a massive campaign to distribute mosquito netting to even the most remote villages, but even its widespread use can only mitigate, not eradicate, the scourge. And now there are reports that global warming is causing malaria to again encroach on higher altitudes as well as higher latitudes, lending even greater urgency to this battle for children's lives. When you see the shockingly low life expectancies for tropical African countries in the Data Table at the end of the book, malaria is one of the leading causes, dramatically reflecting infant and child mortality.

Another mosquito-vectored pandemic disease with African origins, yellow fever, is also making a comeback in tropical Africa as well as in South America, again because its mosquito vector seems to be reinvigorated. And keep an eye out, too, for reports on the spread of dengue fever, still another growing threat. Africa's inventory of serious disease seems to have no limit.

Living with wildlife has its dangers as well, and millions of Africans still share their habitat with wild animals. "Bush meat" is common fare in such areas, and many an African epidemic had its origin in the forest. During the 1970s, AIDS appears to have erupted in the equatorial rainforest in western Equatorial Africa, and rapidly evolved from an epidemic into a global pandemic (see box titled "AIDS in Subsaharan Africa"). Monkeys may have activated this terrible malady in humans, but whatever the source, Africa was and remains most severely afflicted. In 2009, there were an estimated 34.3 million cases of AIDS in the world, at least 24.5 million of them in Subsaharan Africa, further reducing the life expectancies of this realm's inhabitants.

Challenges to Improving Health Conditions

The great incubator of vectored and nonvectored diseases in humid equatorial Africa continues to threaten local populations. Sudden outbreaks of Ebola fever in Sudan during the 1970s, in The Congo* in the 1990s, and in Uganda in 2000–2001 projected the risks that many Africans live with onto the world's television screens. But Africa's woes are soon forgotten. Only the fear that an African epidemic might evolve into a global pandemic mobilizes world attention. Africans cope with the planet's most difficult environments, but they are least capable of combating its hazards because their resources are so limited.

Even when international efforts are made to assist Africa, the unintended results may be catastrophic. Among Africa's endemic diseases is schistosomiasis, also called bilharzia. The vector is a snail, and the transmitted parasites enter via body openings when people swim or wash in slow-moving water where the snails thrive (fast-moving streams are relatively safe). When development projects designed to help African farmers dammed rivers and streams and sent water into irrigation ditches, farm production rose—but so did the incidence of schistosomiasis, which causes internal bleeding and fatigue, although it is rarely fatal. Medical geographers' maps showed the spread of schistosomiasis around the dam projects, and soon the cause was clear: by slowing down the water flow, the engineers had created an ideal environment for the snail vector.

Poverty is a powerful barrier to improving health, and major interventions in the natural environment carry costs as well as benefits. Africa needs improved medical

*Two countries in Africa have the same short-form name: Congo. In this book, we use *The Congo* for the larger Democratic Republic of the Congo (also called Congo DRC), and *Congo* for the smaller Republic of Congo.

services, more effective child-immunization programs, mobile clinics for remote areas, and similar remedies. While two-thirds of the realm's people continue to live in rural areas, the natural environment will weigh heavily on a vulnerable population.

LAND AND FARMING

Environmental conditions including rainfall and soil quality have much to do with the way Africans farm their land, but cultural, political, and economic factors also matter. Traditional land tenure (whether mainly subsistence or primarily commercial), farming techniques and equipment, markets, government intervention (in the form of promoting one crop over another through subsidies), and attitudes toward innovation such as irrigation or mechanization all play a role.

In their penetrating book, *Geography of Sub-Saharan Africa*, editor Samuel Aryeetey-Attoh and his colleagues focus on the issue of land tenure, crucial in Africa because most Africans are farmers. **Land tenure** refers to the way people own, occupy, and use land. African traditions of land tenure are different from those of Europe or the Americas. In most of Subsaharan Africa, communities, not individuals, customarily hold land. Occupants of the land have temporary, custodial rights to it and cannot sell it. Land may be held by large (extended) families, by a village community, or even by a traditional chief who holds the land in trust for the people. His subjects may house themselves on it and farm it, but in return they must follow his rules.

Stolen Lands

When the European colonizers took control of much of Subsaharan Africa, their land ownership practices clashed head-on with those of Africa. Africans believed that their land belonged to their ancestors, the living, and the yet-unborn; Europeans saw unclaimed space and felt justified in claiming it. What Africans called **land alienation**— the expropriation of (often the best) land by Europeans— changed the pattern of land tenure in Africa. By the time the Europeans withdrew, private land ownership was widely dispersed and could not be reversed. Postcolonial African states tried to deal with this legacy by nationalizing all the land and doing away with private ownership, reverting in theory to the role of the traditional chief. But this policy has not worked well. In the rural areas, the government in the capital is a remote authority often seen as unsympathetic to the plight of farmers. Governments keep the price of farm products low on urban markets,

AIDS in Subsaharan Africa

ACQUIRED IMMUNE DEFICIENCY Syndrome—AIDS—spread in Africa during the 1980s and became a pandemic during the 1990s. Its impact on Africa has been devastating. In 2007, more than 2 million people died of AIDS in Subsaharan Africa, 75 percent of the global toll.

Persons infected by HIV (human immunodeficiency virus) do not immediately or even soon display symptoms of AIDS. In the early stages, only a blood test will reveal infection, and then only by indicating that the body is mobilizing antibodies to fight HIV. People can carry the virus for years without being aware of it; during that period they can unwittingly transmit it to others. For that reason, official statistics of AIDS cases lag far behind the reservoir of those infected, especially in countries where populations cannot be thoroughly tested. Since AIDS was identified in the early 1980s, nearly 30 million people have died of it.

By 2009, more than 34 million people were known to be infected by AIDS worldwide, and of these about 25 million lived in 34 tropical African countries according to United Nations sources. In the early 1990s, the worst-hit countries were in Equatorial Africa in what was called the AIDS Belt from (then) Zaïre to Kenya. But a decade later, the most severely afflicted countries lay in Southern Africa. In Zimbabwe, Botswana, Lesotho, and Swaziland, more than 25 percent of all persons 15–49 were infected with HIV; in South Africa, the region's most populous country, the proportion was 22 percent (6.1 million). These are the official numbers; medical geographers estimate that 20 to 25 percent of the entire population of several tropical African countries are infected.

As the number of deaths from AIDS rises, African countries' vital statistics are showing the demographic impact. In Botswana, life expectancy in 1994 was 60 years; by 2006 it had bottomed out at 34. In South Africa it declined from 66 to 47 over the same time period. In Zimbabwe, population growth was 3.3 percent per year in the early 1980s; today it is 1.0 percent. Meanwhile, AIDS is killing parents at such a rate that it is already responsible for 20 million African orphans.

No part of tropical Africa has been spared. West African countries, too, are reporting growing numbers of AIDS cases. Ivory Coast, reporting that nearly 7 percent of adults are infected there, may provide the most accurate data. In Nigeria, where the full onslaught may not yet be known, AIDS incidence among adults is more than 5 percent (which, in that populous country, means that as many as 8.5 million people are ill with AIDS).

The misery the AIDS pandemic has created is beyond measure, but its economic impact is beyond doubt. Workers are sick on the job; but jobs are scarce and medical help is unavailable or unaffordable for many, so the workers stay on the job until they collapse. In a realm where unemployment is high, new workers immediately fill the vacancies, but they, too, are unwell. Corporations report skyrocketing AIDS-related costs. In Botswana, whose economy has been prospering due to increased diamond production, highly trained diamond sorters are being lost to AIDS, and training replacement workers is depressing profits. Medical expenses, multiplying death benefits, funeral payments, and recruiting and training costs are reducing corporate incomes from Kenya's sugar plantations to South Africa's gold mines. Thus the impact of AIDS is recorded in the continent's economic statistics. The pattern shown in Figure G-11, where Africa is concerned, is due in substantial part to the AIDS pandemic.

In countries of the global core, progress has been made in containing AIDS and prolonging the lives of those afflicted. This has taken AIDS off the medical center stage, but such a turnaround is not in prospect for Africa. The cost of the medications that combat AIDS is too high, and private drug companies will not dispense their products at prices affordable to African patients. Thus the best short-term strategy is for governments to attempt to alter public attitudes and behavior, notably by making condoms available as widely and freely as possible. Uganda, once one of tropical Africa's hardest-hit countries, managed to slow the rate of AIDS dispersal by these methods—and through a vigorous advertising campaign on billboards, in newspapers, and on radio and television.

But what Africa needs is a coordinated attack on AIDS through the provision of medicines. One such initiative was launched in 2003 when the United States provided Subsaharan Africa with the financial means to combat the disease: a U.S. $10 billion fund to facilitate treatment as well as prevention. The urgency of the matter is underscored by estimates that, in the absence of a massive campaign, some African countries may experience population declines of as much as 10 to 20 percent, a disaster comparable only to Europe's fourteenth-century bubonic plague and the collapse of Native American populations after the introduction of smallpox by the European invaders in the sixteenth. Once again environmental, cultural, and historical factors combine to afflict Africa's peoples.

pleasing their supporters but frustrating farmers. Large landholdings once owned by Europeans seem now to be occupied by officials or those the government favors. The colonial period's approach to land tenure left Africa with a huge social problem.

Rapid population growth makes the problem worse. Traditional systems of land tenure, which involve subsistence farming in various forms ranging from shifting cultivation to pastoralism, work best when population is fairly stable. Land must be left fallow to recover from cultivation, and pastures must be kept free of livestock so the grasses can revive. A population explosion of the kind Africa experienced during the mid-twentieth century destroys this equilibrium. Soils cannot rest, and pastures are overgrazed. As land becomes degraded, yields decline.

This is the situation in much of Subsaharan Africa, where two-thirds of the people depend on farming for their livelihood. That percentage is now declining, but it remains among the highest for any realm in the global periphery. Subsaharan Africa is sometimes thought of as a storehouse of minerals and fuels and great underground riches, but farming is the key to its economy. And not just subsistence farming: agricultural exports also produce foreign exchange revenues needed to pay for essential imports. The European colonizers identified suitable environments and laid out plantations to grow cocoa, coffee, tea, and other luxury crops for the markets of the global core. Irrigation schemes enabled the export of cotton, groundnuts (peanuts), and sesame seeds. Today, national governments and some private owners continue to operate these projects.

Persistent Subsistence

But most African farmers are subsistence farmers who grow grain crops (corn, millet, sorghum) in the drier areas and root crops (yams, cassava, sweet potatoes) where moisture is ample. Shifting cultivation still occurs in remote parts of The Congo and neighboring locales in Equatorial Africa, but this practice is declining as the forest shrinks. Others are pastoralists, driving their cattle and goats along migratory routes to find water and pasture, sometimes clashing with sedentary farmers whose fields they invade. Whatever their principal mode of farming, all African farmers today feel the effects of population pressure and must cope with dwindling space and damaged ecologies. By *intercropping*, planting several types of crops in one cleared field, some of which resemble those in the forest, shifting cultivators extend the life of their plots. In *com-*

Pyrethrum, a flowering plant native to Southeast Asia, does well in East Africa, where its flower heads are picked and crushed into a powder that forms the source of a useful insecticide of the same name. Take a look at the listed contents of many household bug sprays, and you are likely to see pyrethrum among them; it is also used in "dusts" for edible plants because pyrethrum is not toxic to humans. Detailed lists of some African states' products sometimes show pyrethrum as an export, and in Kenya the government has set up a Pyrethrum Board to cultivate and harvest the flower. These workers, seen picking the petaled pyrethrum flower, are government employees.   © Sven Torfinn/Panos Pictures.

pound farming, usually near a market center, farmers combine the use of compound (village) and household waste as fertilizer with intensive care of the crops to produce vegetables, fruits, and root crops as well as eggs and chickens for urban consumers. Other subsistence farmers use *diversification* to cope with such problems; they combine their cultivation of subsistence crops with nonfarm work, including full-time jobs in nearby towns.

Notwithstanding all these adaptations, African farm production is declining for many reasons, including government mismanagement and corruption. When Nigeria became independent in 1960, its government embarked on costly industrialization, and agriculture was given a lower priority. In that year, Nigeria was the world's largest producer and leading exporter of palm oil; today, Nigeria must import palm oil to meet its domestic demand. Inadequate rural infrastructures create another obstacle. Roads from the rural areas to the market centers are in bad and worsening shape, often impassable during the rainy season and rutted when dry. Ineffective links between producers and consumers create economic losses. Storage facilities in the villages tend to be inadequate, so that untold quantities of grains, roots, and vegetables are

A Green Revolution for Africa?

10

THE GREEN REVOLUTION—the development of more productive, higher-yielding types of grains—has narrowed the gap between world population and food production. Where people depend mainly on rice and wheat for their staples, the Green Revolution has pushed back the specter of hunger. The Green Revolution has had less impact in Subsaharan Africa, however. In part, this relates to the realm's high rate of population growth (which is substantially higher than that of India or China). Other reasons have to do with Africa's staples: rice and wheat support only a small part of the realm's population. Corn (maize) supports many more, along with sorghum, millet, and other grains. In moister areas, root crops, such as the yam and cassava, as well as the plantain (similar to the banana) supply most calories. These crops were not priorities in Green Revolution research.

Lately there have been a few signs of hope. Even as terrible famines struck several parts of the continent and people went hungry in places as widely scattered as Moçambique and Mali, scientists worked toward two goals: first, to develop strains of corn and other crops that would be more resistant to Africa's virulent crop diseases, and second, to increase the productivity of those strains. But these efforts faced serious problems. An average African hectare (2.5 acres) planted with corn, for example, yields only about half a ton of corn, whereas the world average is 1.3 tons. When a virus-resistant variety was developed and distributed throughout Subsaharan Africa in the 1980s, yields rose significantly where farmers could afford to buy the new strain as well as the fertilizers it requires. Nigeria saw a near-doubling of its production before it leveled off, but population growth negated the advantage. Hardier types of root crops also raised yields in some areas.

But Africa needs more than this to reverse the trend toward food deficiency. (Nor is the Green Revolution an unqualified remedy: the poorest farmers, who need the help most, can least afford the more expensive higher-yielding seeds and also cannot pay for the pesticides that may be required.) It has been estimated that, for the realm as a whole, food production has fallen about 1 percent annually even as population grew by more than 2 percent. Lack of capital, inefficient farming methods, inadequate equipment, soil exhaustion, male dominance, apathy, and devastating droughts contributed to this decline. The seemingly endless series of civil conflicts (The Congo, Sudan, Somalia, Rwanda, Ivory Coast, Chad) also reduced farm output. The Green Revolution may narrow the gap between production and need, but the battle for food sufficiency in Africa is far from won.

---

lost to rot, pests, and theft. Electricity supply in rural areas also requires upgrading: without power, pumps cannot be run for water supply, and the processing of produce cannot be mechanized. Rural farmers have inadequate access to credit: banks would rather lend to urban homeowners than to distant villagers. This is especially true for women who, according to current estimates, produce 75 percent of the food in Subsaharan Africa.

Along the banks of Africa's rivers and on the shores of the Great Lakes fishing augments local food supplies, but Africa's fishing industries are still underdeveloped and contribute only modestly to overall nutrition. During the postcolonial period, Subsaharan Africa's population has more than doubled; food production has declined. Agricultural exports have also decreased, but food imports have mushroomed. The situation is especially clear in terms of per-capita output, which is lower today than it was 20 years ago. Even advances in biotechnology that closed the global gap between food production and demand largely passed Africa by (see box titled "A Green Revolution for Africa?")

AFRICA'S HISTORICAL GEOGRAPHY

Africa is the cradle of humankind. Archeological research has chronicled 7 million years of transition from australopithecenes to hominids to *Homo sapiens*. It is therefore ironic that we know comparatively little about Subsaharan Africa from 5000 to 500 years ago—that is, before the onset of European colonialism. This is partly due to the colonial period itself, during which African history was neglected, many African traditions and artifacts were destroyed, and many misconceptions about African cultures and institutions became entrenched. It is also a result of the absence of a written history over most of Africa south of the Sahara until the sixteenth century—and over a large part of it until much later than that. The best records are those of the savanna belt immediately south of the Sahara, where contact with North African peoples was greatest and where Islam achieved a major penetration.

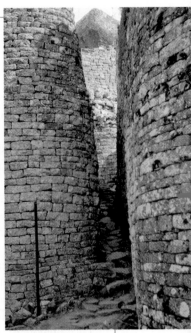

"Got up before dawn this date in the hope of being the first visitor to the Great Zimbabwe ruins, a place I have wanted to see ever since I learned about it in a historical geography class. Climbed the hill and watched the sun rise over the great elliptical 'temple' below, then explored the maze of structures of the so-called fortification on the hilltop above. Next, for an incredible 90 minutes I was alone in the interior of the great oval walls of the temple. What history this site has seen—it was settled for perhaps six centuries before the first stones were hewn and laid here around AD 750, and from the 11th to the 15th centuries this may have been a ceremonial or religious center of a vast empire whose citizens smelted gold and copper and traded them via Sofala and other ports on the Indian Ocean for goods from South Asia and even China. No cement was used: these walls are built with stones cut to fit closely together, and in the valley they rise over 30 feet (nearly 10 m) high. And what secrets Great Zimbabwe still conceals—where are the plans, the tools, the quarries? The remnants of an elaborate water-supply system suggest a practical function for this center, but a decorated conical tower and adjacent platform imply a religious role. Whatever the answers, history hangs heavily, almost tangibly, over this African site."
© H. J. de Blij

The absence of a written record does not mean, as some scholars have suggested, that Africa does not have a history as such prior to the coming of Islam and Christianity. Nor does it mean that there were no rules of social behavior, no codes of law, no organized economies. Modern historians, encouraged by the intense interest shown by Africans generally, are now trying to reconstruct the African past, not only from the meager written record but also from folklore, poetry, art objects, buildings, and other such sources. Much has been lost forever, though. Almost nothing is known of the farming peoples who built well-laid terraces on the hillsides of northeastern Nigeria and East Africa or of the communities that laid irrigation canals and constructed stone-lined wells in Kenya; and very little is known about the people who, perhaps a thousand years ago, built the great walls of Zimbabwe (see photos above). Porcelain and coins from China, beads from India, and other goods from distant sources have been found in Zimbabwe and other points in East and Southern Africa, but the trade routes within Africa itself—let alone the products that circulated and the people who handled them—still remain the subject of guesswork.

African Genesis

Africa on the eve of the colonial period was a continent in transition. For several centuries, the habitat in and near one of the continent's most culturally and economically productive areas—West Africa—had been changing. For 2000 years, probably more, Africa had been innovating as well as adopting ideas from outside. In West Africa, cities were developing on an impressive scale; in central and Southern Africa, peoples were moving, readjusting, sometimes struggling with each other for territorial supremacy. The Romans had penetrated to southern Sudan, North African peoples were trading with West Africans, and Arab *dhows* were sailing the waters along the eastern coasts, bringing Asian goods in exchange for gold, copper, and a comparatively small number of slaves.

Consider the environmental situation in West Africa as it relates to the past. As Figures G-6 and G-7 indicate, the environmental regions in this part of the continent exhibit a decidedly east-west orientation. The *isohyets* (lines of equal rainfall totals) run parallel to the southern coast (Fig. G-6); the climatic regions, now positioned somewhat differently from where they were two millennia ago, still trend strongly east-west (Fig. G-7); the vegetation pattern also reflects this situation, with a coastal forest belt yielding to savanna (tall grass with scattered trees in the south, shorter grass in the north) that gives way, in turn, to steppe and desert.

We know that African cultures had been established in all these environmental settings for thousands of years. One of these, the Nok culture, endured for over eight centuries on the Benue Plateau (north of the Niger-Benue confluence in modern Nigeria) from about 500 BC to the third century AD. The Nok people made stone as well as iron tools, and they left behind a treasure of art in the form of clay figurines representing humans and animals. But we have no evidence that they traded with distant peoples. The opportunities created by environments and technologies still lay ahead.

Early Trade in West Africa

West Africa, over a north-south span of a few hundred kilometers, displayed an enormous contrast in environments, economic opportunities, modes of life, and products. The peoples of the tropical forest produced and needed goods that were different from the products and requirements of the peoples of the dry, distant north. For example, salt is a prized commodity in the forest, where the humidity precludes its formation, but it is plentiful in the desert and steppe. This enabled the desert peoples to sell salt to the forest peoples in exchange for ivory, spices, and dried foods. Thus there evolved a degree of *regional complementarity* between the peoples of the forest and those of the drylands. And the savanna peoples—those located in between—found themselves in a position to channel and handle the trade (which is always economically profitable).

The markets in which these goods were exchanged prospered and grew, and cities arose in the savanna belt of West Africa. One of these old cities, now an epitome of isolation, was once a thriving center of commerce and learning and one of the leading urban places in the world—Timbuktu. Others, predecessors as well as successors of Timbuktu, have declined, some of them into oblivion. Still other savanna cities, such as Kano in the northern part of Nigeria, remain important.

Early States

Strong and durable states arose in the West African culture hearth (see Fig. 7-4). The oldest state we know anything about is Ghana. Ancient Ghana was located to the northwest of the modern country of Ghana, and covered parts of present-day Mali, Mauritania, and adjacent territory. Ghana lay astride the upper Niger River and included gold-rich streams flowing off the Futa Jallon Highlands, where the Niger has its origins. For a thousand years, perhaps longer, old Ghana managed to weld various groups of people into a stable state. The country had a large capital city complete with markets, suburbs for foreign merchants, religious shrines, and, some distance from the city center, a fortified royal retreat. Taxes were collected from the citizens, and tribute was extracted from subjugated peoples on Ghana's periphery; tolls were levied on goods entering Ghana, and an army maintained control. Muslims from the northern drylands invaded Ghana in about AD 1062, when it may already have been in decline. Even so, the capital was protected for 14 years. However, the invaders had ruined the farmlands and destroyed the trade links with the north. Ghana could not survive. It finally broke into smaller units.

Eastward Shift

In the centuries that followed, the focus of politico-territorial organization in the West African culture hearth shifted almost continuously eastward—first to ancient Ghana's successor state of Mali, which was centered on Timbuktu and the middle Niger River Valley, and then to the state of Songhai, whose focus was Gao, a city on the Niger that still exists. This eastward movement may have been the result of the growing influence and power of Islam. Traditional religions prevailed in ancient Ghana, but Mali and its successor states sent huge, gold-laden pilgrimages to Mecca along the savanna corridor south of the Sahara, passing through present-day Khartoum and Cairo. Of the tens of thousands who participated in these pilgrimages, some remained behind. Today, many Sudanese trace their ancestry to the West Africa savanna kingdoms.

Beyond the West

West Africa's savanna region undoubtedly witnessed momentous cultural, technological, and economic developments, but other parts of Africa also progressed. Early states emerged in present-day Sudan, Eritrea, and Ethiopia. Influenced by innovations from the Egyptian culture hearth, these kingdoms were stable and durable: the oldest, Kush, lasted 23 centuries (Fig. 6-6). The Kushites built elaborate irrigation systems, forged iron

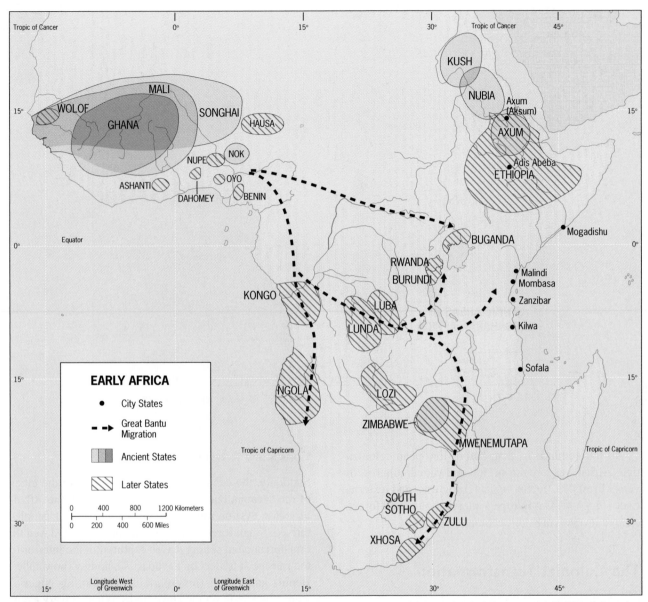

FIGURE 6-6

© H. J. de Blij, P. O. Muller, and John Wiley & Sons, Inc.

tools, and built impressive structures as the ruins of their long-term capital and industrial center, Meroe, reveal. Nubia, to the southeast of Kush, was Christianized until the Muslim wave overtook it in the eighth century. And Axum was the richest market in northeastern Africa, a powerful kingdom that controlled Red Sea trade and endured for six centuries. Axum, too, was a Christian state that confronted Islam, but Axum's rulers deflected the Muslim advance and gave rise to the Christian dynasty that eventually shaped modern Ethiopia.

The process of **state formation** spread throughout Africa and was still in progress when the first European contacts occurred in the late fifteenth century. Large and effectively organized states developed on the equatorial

west coast (notably Kongo) and on the southern plateau from the southern part of The Congo to Zimbabwe. East Africa had several city-states, including Mogadishu, Kilwa, Mombasa, and Sofala.

Bantu Migration

A crucial event affected virtually all of Equatorial, West, and Southern Africa: the great Bantu migration from present-day Nigeria–Cameroon southward and eastward across the continent. This migration appears to have occurred in waves starting as long as 5000 years ago, populating the Great Lakes area and penetrating South Africa, where it resulted in the formation of the powerful Zulu Empire in the nineteenth century (Fig. 6-6).

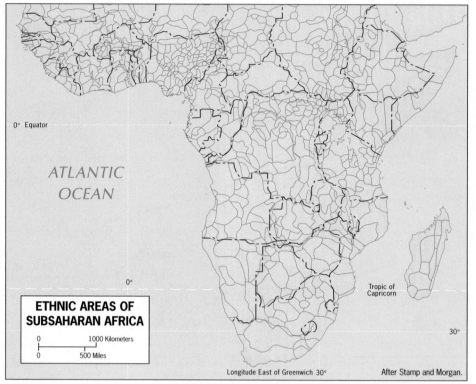

FIGURE 6-7

© H. J. de Blij, P. O. Muller, and John Wiley & Sons, Inc.

All this reminds us that, before European colonization, Africa was a realm of rich and varied cultures, diverse lifestyles, technological progress, and external trade (Fig. 6-7). But Europe's intervention would forever change its evolving political map.

The Colonial Transformation

European involvement in Subsaharan Africa began in the fifteenth century. It would interrupt the path of indigenous African development and irreversibly alter the entire cultural, economic, political, and social makeup of the continent. It started quietly in the late fifteenth century, with Portuguese ships groping their way along the west coast and rounding the Cape of Good Hope. Their goal was to find a sea route to the spices and riches of the Orient. Soon other European countries were sending their vessels to African waters, and a string of coastal stations and forts sprang up. In West Africa, the nearest part of the continent to European spheres in Middle and South America, the initial impact was strongest. At their coastal control points, the Europeans traded with African middlemen for the slaves who were needed to work New World plantations, for the gold that had been flowing northward across the desert, and for ivory and spices.

Coastward Reorientation

Suddenly, the centers of activity lay not with the cities of the savanna but in the foreign stations on the Atlantic coast. As the interior declined, the coastal peoples thrived. Small forest states gained unprecedented wealth, transferring and selling slaves captured in the interior to the European traders on the coast. Dahomey (now called Benin) and Benin (now part of neighboring Nigeria) were states built on the slave trade. When slavery eventually came under attack in Europe, those who had inherited the power and riches it had brought opposed abolition vigorously in both continents.

Horrors of the Slave Trade

Although slavery was not new to West Africa, the *kind* of slave raiding and trading the Europeans introduced certainly was. In the savanna, kings, chiefs, and prominent families traditionally took a few slaves, but the status of those slaves was unlike anything that lay in store for those who were shipped across the Atlantic. In fact, large-scale slave trading had been introduced in East Africa long before the Europeans brought it to West Africa. African middlemen from the coast raided the interior for able-bodied men and women and marched them in chains to the Arab markets on the coast (Zan-

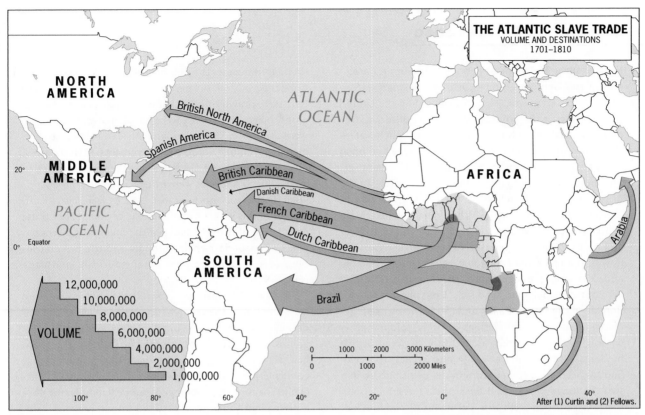

THE ATLANTIC SLAVE TRADE
VOLUME AND DESTINATIONS
1701–1810

After (1) Curtin and (2) Fellows.

FIGURE 6-8

© H. J. de Blij, P. O. Muller, and John Wiley & Sons, Inc.

zibar was a notorious market). There, packed in specially built *dhows*, they were carried off to Arabia, Persia, and India. When the European slave trade took hold in West Africa, however, its volume was far greater. Europeans, Arabs, and collaborating Africans ravaged the continent, forcing perhaps as many as 30 million persons away from their homelands in bondage (Fig. 6-8). Families were destroyed, as were entire villages and cultures; and those who survived their exile suffered unfathomable misery.

The European presence on the West African coast completely reoriented its trade routes, for it initiated the decline of the interior savanna states and strengthened the coastal forest states. Moreover, the Europeans' insatiable demand for slaves ravaged the population of the interior. But it did not lead to any major European thrust toward the interior or produce colonies overnight. The African middlemen were well organized and strong, and they held off their European competitors, not just for decades but for centuries. Although the Europeans first appeared in the fifteenth century, they did not carve West Africa up until nearly 400 years later, and in many areas not until after 1900.

In all of Subsaharan Africa, the only area where European penetration was both early and substantial was at its southernmost end, at the Cape of Good Hope.

There the Dutch founded Cape Town, a waystation to their developing empire in the East Indies. They settled in the town's hinterland and migrated deeper into the interior. They brought thousands of slaves from Southeast Asia to the Cape, and intermarriage was common. Today, people of mixed ancestry form the largest part of Cape Town's population. Later the British took control not only of the Cape but also of the expanding frontier and brought in tens of thousands of South Asians to work on their plantations. Multiracial South Africa was in the making.

Elsewhere in the realm, the European presence remained confined almost entirely to the coastal trading stations, whose economic influence was strong. No real frontiers of penetration developed. Individual travelers, missionaries, explorers, and traders went into the interior, but nowhere else in Africa south of the Sahara was there an invasion of white settlers comparable to that of southernmost Africa's.

Colonization

In the second half of the nineteenth century, after more than four centuries of contact, the European powers finally laid claim to virtually all of Africa. Parts of the continent had been "explored," but now representatives of European

The Berlin Conference

IN NOVEMBER 1884, the imperial chancellor and architect of the German Empire, Otto von Bismarck, convened a conference of 14 states (including the United States) to settle the political partitioning of Africa. Bismarck wanted not only to expand German spheres of influence in Africa but also to play off Germany's colonial rivals against one another to the Germans' advantage. The major colonial contestants in Africa were: (1) the British, who held beachheads along the West, South, and East African coasts; (2) the French, whose main sphere of activity was in the area of the Senegal River and north of the Congo Basin; (3) the Portuguese, who now desired to extend their coastal stations in Angola and Moçambique deep into the interior; (4) King Leopold II of Belgium, who was amassing a personal domain in the Congo Basin; and (5) Germany itself, active in areas where the designs of other colonial powers might be obstructed, as in Togo (between British holdings), Cameroon (a wedge into French spheres), South West Africa (taken from under British noses in a swift strategic move), and East Africa (where German Tanganyika broke the British design for a solid block of territory from the Cape north to Cairo).

When the conference convened in Berlin, more than 80 percent of Africa was still under traditional African rule. Nonetheless, the colonial powers' representatives drew their boundary lines across the entire map. These lines were drawn through known as well as unknown regions, pieces of territory were haggled over, boundaries were erased and redrawn, and African real estate was exchanged among European governments. In the process, African peoples were divided, unified regions were ripped apart, hostile societies were thrown together, hinterlands were disrupted, and migration routes were closed off. Not all of this was felt immediately, of course, but these were some of the effects when the colonial powers began to consolidate their holdings and the boundaries on paper became barriers on the African landscape (Fig. 6-9).

The Berlin Conference was Africa's undoing in more ways than one. The colonial powers superimposed their domains on the African continent. By the time Africa regained its independence after the late 1950s, the realm had acquired a legacy of political fragmentation that could neither be eliminated nor made to operate satisfactorily. The African politico-geographical map is therefore a permanent liability that resulted from three months of ignorant, greedy acquisitiveness during a period when Europe's search for minerals and markets had become insatiable.

---

governments and rulers arrived to create or expand African spheres of influence for their patrons. Competition was intense. Spheres of influence began to crowd each other. It was time for negotiation, and in late 1884 a conference was convened in Berlin to sort things out. This conference laid the groundwork for the now familiar politico-geographical map of Africa (see box titled "The Berlin Conference").

Figure 6-9 shows the result. The French dominated most of West Africa, and the British East and Southern Africa. The Belgians acquired the vast territory that became The Congo. The Germans held four colonies, one in each of the realm's regions. The Portuguese held a small colony in West Africa and two large ones in Southern Africa (see the map dated 1910).

The European colonial powers shared one objective in their African colonies: exploitation. But in the way they governed their dependencies, they reflected their differences. Some colonial powers were themselves democracies (the United Kingdom and France); others were dictatorships (Portugal, Spain). The British established a system of indirect rule over much of their domain, leaving indigenous power structures in place and making local rulers representatives of the British Crown. This was unthinkable in the Portuguese colonies, where harsh, direct control was the rule. The French sought to create culturally assimilated elites that would represent French ideals in the colonies.

In the Belgian Congo, however, King Leopold II, who had financed the expeditions that staked Belgium's claim in Berlin, embarked on a campaign of ruthless exploitation. His enforcers mobilized almost the entire Congolese population to gather rubber, kill elephants for their ivory, and build public works to improve export routes. For failing to meet production quotas, entire communities were massacred. Killing and maiming became routine in a colony in which horror was the only common denominator. After the impact of the slave trade, King Leopold's reign of terror was Africa's most severe demographic disaster. By the time it ended, after a growing outcry around the world, as many as 10 million Congolese had been murdered. In 1908 the Belgian government took over, and slowly its Congo began to mirror Belgium's own internal divisions: corporations, government administrators, and the Roman Catholic Church each pursued their sometimes competing interests. But no one thought to change the name of the colonial cap-

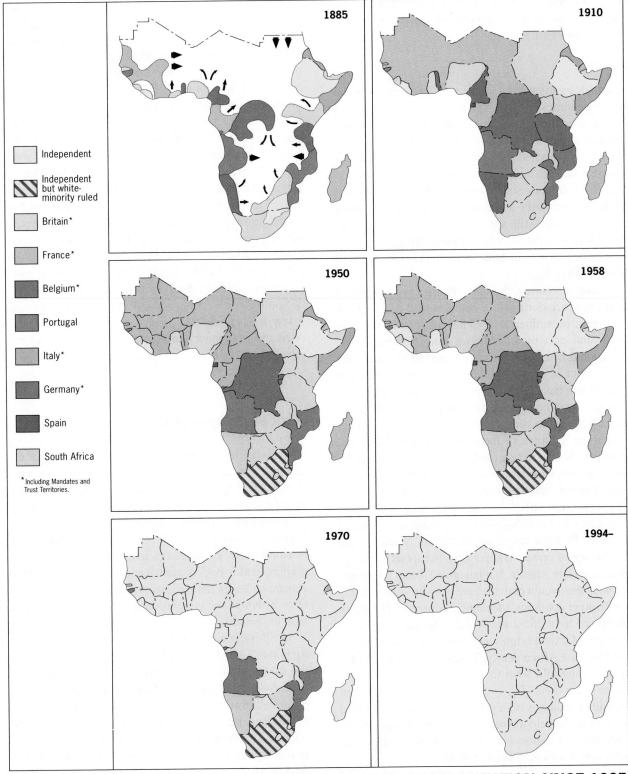

FIGURE 6-9

© H. J. de Blij, P. O. Muller, and
John Wiley & Sons, Inc.

COLONIZATION AND DECOLONIZATION SINCE 1885

ital: it was Leopoldville until the Belgian Congo achieved independence in 1960.

12   **Colonialism** transformed Africa, but in its post-Berlin form it lasted less than a century. In Ghana, for

example, the Ashanti (Asante) Kingdom still was fighting the British into the early twentieth century; by 1957, Ghana was independent again. A few years from now, much of Subsaharan Africa will have been independent for half a century, and the colonial period is becoming an interlude rather than a paramount chapter in modern African history.

Legacies

Nevertheless, some colonial legacies will remain on the map for centuries to come. The boundary framework has been modified in a few places, but it is a politico-geographical axiom that boundaries, once established, tend to become entrenched and immovable. The transport systems were laid out to facilitate the movement of raw materials from interior sources to coastal outlets. Internal circulation was only a secondary objective, and most African countries still are not well connected to each other. Moreover, the colonizers founded many of Subsaharan Africa's cities or built them on the sites of small towns or villages.

In many countries, the elites that gained advantage and prominence during the colonial period also have retained their preeminence. This has led to authoritarianism in some postcolonial African states (for details, see the regional discussion to come), and to violence and civil conflict in others. Military takeovers of national governments have been a byproduct of decolonization, and hopes for democracy in the struggle against colonialism have too often been dashed. Yet a growing number of African countries, including Ghana, Tanzania, Botswana, South Africa, and, most recently, Nigeria, have overcome the odds and achieved representative government under dauntingly difficult circumstances.

Although not strictly a colonial legacy, the impact of the Cold War (1945–1990) on African states should be noted. In three countries—Ethiopia, Somalia, and Angola—great-power competition, weapons, military advisers, and, in Angola, foreign armed forces magnified civil wars that would perhaps have been inevitable in any case. In other countries, ideological adherence to foreign dogmas led to political and social experiments (such as one-party Marxist regimes and costly farm collectivization) that cost Africa dearly. Today, debt-ridden Africa is again being told what to do, this time by foreign financial institutions. The cycle of poverty that followed colonialism exacts a high price from African societies. As for future development prospects, the disadvantages of peripheral location vis-à-vis the world's core areas continue to handicap Subsaharan Africa.

CULTURAL PATTERNS

We may tend to think of Africa in terms of its prominent countries and famous cities, its development problems and political dilemmas, but Africans themselves have another perspective. The colonial period created states and capitals, introduced foreign languages to serve as the *lingua franca*, and brought railroads and roads. The colonizers stimulated labor movements to the mines they opened, and they disrupted other migrations that had been part of African life for many centuries. But they did not change the ways of life of most of the people. No less than two-thirds of the realm's population still live in, and work near, Africa's hundreds of thousands of villages. They speak one of more than a thousand languages in use in the realm. The villagers' concerns are local; they focus on subsistence, health, and safety. They worry that the conflicts over regional power or political ideology will engulf them, as has happened to millions in Liberia, Sierra Leone, Ethiopia, Rwanda, The Congo, Moçambique, and Angola since the 1970s. Africa's largest peoples are major nations, such as the Yoruba of Nigeria and the Zulu of South Africa. Africa's smallest peoples number just a few thousand. As a geographic realm, Subsaharan Africa has the most complex cultural mosaic on Earth (see Fig. 6-7).

African Languages

Africa's linguistic geography is a key component of that cultural intricacy. Most of Subsaharan Africa's more than 1000 languages do not have a written tradition, making classification and mapping difficult. Scholars have attempted to delimit an African language map, and Figure 6-10 is a composite of their efforts. One feature is common to all language maps of Africa: the geographic realm begins approximately where the Afro-Asiatic language family (mapped in yellow in Fig. 6-10) ends, although the correlation is sharper in West Africa than to the east.

In Subsaharan Africa, the dominant language family is the Niger-Congo family, of which the Kordofanian subfamily is a small, historic northeastern outlier (Fig. 6-10), and the Niger-Congo languages carry the other subfamily's name. The latter subfamily extends across the realm from West to East and Southern Africa. The Bantu language forms the largest branch in this subfamily, but Niger-Congo languages in West Africa, such as Yoruba and Akan, also have millions of speakers. Another important language family is the Nilo-Saharan family, extending from Maasai in Kenya northwest to Teda in Chad. No other language families are of similar extent or importance: the Khoisan family, of ancient origins, now survives among the dwindling Khoi and San peoples of the Kalahari; the small white minority in South Africa speak Indo-European languages; and Malay–Polynesian languages prevail in Madagascar, which was peopled from Southeast Asia before Africans reached it.

The Most Widely Used Languages

About 40 African languages are spoken by 1 million people or more, and a half-dozen by about 10 million or more: Yoruba (28 million), Hausa (24 million), Ibo, Swahili, Lingala, and Zulu. Although English and French have become important *linguae francae* in multilingual countries such as Nigeria and Côte d'Ivoire (where officials even insist on spelling the name of their country—Ivory Coast—in the Francophone way), African languages also serve this purpose. Hausa is a common language across the West African savanna; Swahili is widely used in East Africa. And pidgin languages, mixtures of African and European tongues, are spreading along West Africa's coast. Millions of Pidgin English (called *Wes Kos*) speakers use this medium in Nigeria and Ghana.

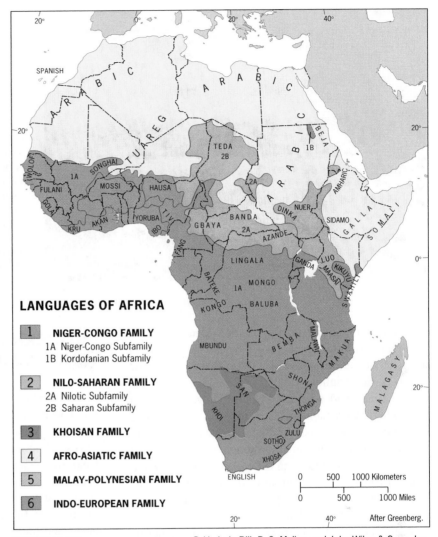

LANGUAGES OF AFRICA

1 **NIGER-CONGO FAMILY**
1A Niger-Congo Subfamily
1B Kordofanian Subfamily

2 **NILO-SAHARAN FAMILY**
2A Nilotic Subfamily
2B Saharan Subfamily

3 **KHOISAN FAMILY**

4 **AFRO-ASIATIC FAMILY**

5 **MALAY-POLYNESIAN FAMILY**

6 **INDO-EUROPEAN FAMILY**

After Greenberg.

FIGURE 6-10   © H. J. de Blij, P. O. Muller, and John Wiley & Sons, Inc.

Language and Culture

13 **Multilingualism** can be a powerful centrifugal force in society, and African governments have tried with varying success to establish "national" alongside local languages. Nigeria, for example, made English its official language because none of its 250 languages, not even Hausa, had sufficient internal interregional use. But using a European, colonial language as an official medium invites criticism, and Nigeria remains divided on the issue. On the other hand, making a dominant local language official would invite negative reactions from ethnic minorities. Language remains a potent force in Africa's cultural life.

Religion in Africa

Africans had their own religious belief systems long before Christians and Muslims arrived to convert them.

And for all of Subsaharan Africa's cultural diversity, Africans had a consistent view of their place in nature. Spiritual forces, according to African tradition, are manifest everywhere in the natural environment, not in a supreme deity that exists in some remote place. Thus gods and spirits affect people's daily lives, witnessing every move, rewarding the virtuous, and punishing (through injury or crop failure, for example) those who misbehave. Ancestral spirits can inflict misfortune on the living. They are everywhere: in the forest, rivers, mountains.

As with land tenure, the religious views of Africans clashed fundamentally with those of outsiders. Monotheistic Christianity first touched Africa in the northeast when Nubia and Axum were converted, and Ethiopia has been a Coptic Christian stronghold since the fourth century AD. But the Christian churches' real invasion did not begin until the onset of colonialism after the turn of the

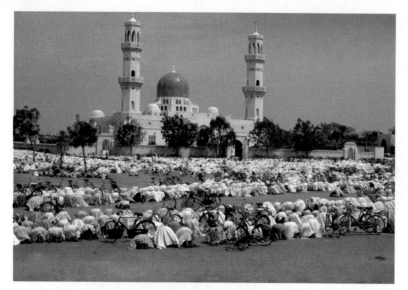

The faithful kneel during Friday prayers at a mosque in Kano, northern Nigeria. The survival of Nigeria as a unified state is an African success story; the Nigerians have overcome strong centrifugal forces in a multi-ethnic country that is dominantly Muslim in the north, Christian in the south. In the 1990s, some Muslim clerics began calling for an Islamic Republic in Nigeria, and after the death of the dictator Abacha and the election of a non-Muslim president, the Islamic drive intensified. A number of Nigeria's northern States adopted Sharia (strict Islamic) law, which led to destructive riots between the majority Muslims and minority Christians who felt threatened by this turn of events. Can Nigeria avoid the fate of Sudan (see pp. 366-367)?   © M. & E. Bernheim/Woodfin Camp & Associates.

sixteenth century. Christianity's various denominations made inroads in different areas: Roman Catholicism in much of Equatorial Africa mainly at the behest of the Belgians, the Anglican Church in British colonies, and Presbyterians and others elsewhere. However, almost everywhere, Christianity's penetration led to a blending of traditional and Christian beliefs, so that much of Subsaharan Africa is nominally, though not exclusively, Christian. Go to a church in Gabon or Uganda or Zambia, and you may hear drums instead of church bells, sing African music rather than hymns, and see African carvings alongside the usual statuary.

Islam had a different arrival and impact. Long before the colonial invasion, Islam advanced out of Arabia, across the desert, and down the east coast. Muslim clerics converted the rulers of African states and commanded them to convert their subjects. They Islamized the savanna states and penetrated into present-day northern Nigeria, Ghana, and Ivory Coast. They encircled and isolated Ethiopia's Coptic Christians and Islamized the Somali people in Africa's Horn. They established beachheads on the Kenya coast and took over Zanzibar. On the map, the African Transition Zone defines the Islamic Front (see Fig. 6-11). In the field, Arabizing Islam and European Christianity competed for African minds, and Islam proved to be a far more pervasive force. From Senegal to Somalia, the population is virtually 100 percent Muslim, and Islam's rules dominate everyday life (photo above). The Sunni *mullahs* would never allow the kind of marriage between traditional and Christian beliefs seen in much of formerly colonial Africa. This fundamental contradiction between Islamic dogma and Christian accommodation creates a potential for conflict in countries where both religions have adherents.

MODERN MAP AND TRADITIONAL SOCIETY

The political map of Subsaharan Africa contains almost 50 states but no nation-states (apart from some microstates and ministates in the islands and in the far south). Centrifugal forces are powerful, and outside interventions during the Cold War, when communist and anticommunist foreigners took sides in local civil wars, worsened conflict within African states. Colonialism's economic legacy was not much better. In tropical Africa, core areas, capitals, port cities, and transport systems were laid out to maximize profit and facilitate exploitation of minerals and soils; the colonial mosaic inhibited interregional communications except where cooperation enhanced efficiency. Colonial Zambia and Zimbabwe, for example (then called Northern and Southern Rhodesia), were landlocked and needed outlets, so railroads were built to Portuguese-owned ports. But such routes did little to create intra-African linkages. The modern map reveals the results: in West Africa you can travel from the coast into the interior of all the coastal states along railways or adequate roads. But no high-standard roadway was ever built to link these coastal neighbors to each other.

Supranationalism

To overcome such disadvantages, African states must cooperate internationally, continentwide as well as regionally. The Organization of African Unity (OAU) was established for this purpose in 1963 and in 2001 was superseded by the African Union. In 1975 the Economic Community of West African States (ECOWAS) was founded by 15 countries to promote trade, transportation, industry, and social affairs in the region. And in the early 1990s, another important step was taken when 12 countries joined in the Southern African Development Community (SADC), organized to facilitate regional commerce, intercountry transport networks, and political interaction.

Population and Urbanization

Subsaharan Africa is the second least urbanized major world realm, but it is urbanizing at a fast pace. The percentage of urban dwellers today stands at 35 percent. This means that almost 270 million people now live in cities and towns, many of which were founded and developed by the colonial powers.

African cities became centers of embryonic national core areas, and of course they served as government headquarters. This *formal sector* of the city used to be the dominant one, with government control and regulations affecting civil service, business, industry, and workers. Today, however, African cities look different. From a distance, the skyline still resembles that of a modern center. But in the streets, on the sidewalks right below the shopwindows, there are hawkers, basket weavers, jewelry sellers, garment makers, wood carvers—a second economy, most of it beyond government control. This *informal sector* now dominates many African cities. It is peopled by the rural immigrants, who also work as servants, apprentices, construction workers, and in countless other menial jobs.

Millions of urban immigrants, however, cannot find work, at least not for months or even years at a time. They live in squalid circumstances, in desperate poverty, and governments cannot assist them. As a result, the squatter rings around (and also within) many of Africa's cities are unsafe—uncomfortable, unhealthy slums without adequate shelter, water supply, or basic sanitation. Garbage-strewn (no solid-waste removal here), muddy and insect-infested during the rainy season, and stifling and smelly during the dry period, they are incubators of disease. Yet very few residents return to their villages. Every new day brings hope.

In our regional discussion we focus on some of Subsaharan Africa's cities, all of which, to varying degrees, are stressed by the rate of population influx. Despite the plight of the urban poor and the poverty of Africa's rural areas, some of Africa's capitals remain the strongholds of privileged elites who, dominant in governments, fail to address the needs of other ethnic groups. Discriminatory policies and artificially low food prices disadvantage farmers and create even greater urban-rural disparities than the colonial period saw. But today the prospect of democracy brings hope that Africa's rural majorities will be heard and heeded in the capitals.

ECONOMIC PROBLEMS

As the world map of economic development (Fig. G-11) shows, Subsaharan Africa as a geographic realm is the weakest link in the international economy. Proportionately more countries are poor and debt-ridden in Subsaharan Africa than anywhere else. Annual income per person is low, infrastructure is inadequate, linkages with the rest of the world are weak, farmers face obstacles to profit at home as well as abroad. Champions of globalization say that Africa's poverty is proof that a lack of globalizing interaction spells disadvantage for any country. Opponents of globalization argue that Africa's condition is proof that globalization is making the poor poorer and the already-rich richer.

From Mauritania to Madagascar, Africa's low-income economies reflect a combination of difficulties so daunting that overcoming them should be a global, not just an African, objective. If the world is a village of neighborhoods, as is so often said, this is the neighborhood that needs the most help—ranging from medical assistance against AIDS to fair market conditions for farmers. Lifting hundreds of millions of subsistence farmers from the vagaries of individual survival to the security of collective prosperity will take more than a Green Revolution. Africa's vast expanse of savanna soil may not be any less fertile than that of Brazil's booming *cerrado*, but if there is no capital to provide fertilizers and no incentive to produce, no secure financial system and no roads to markets, no government policies to secure fair domestic prices and no equal access to the international marketplace, Africa's key industry will not make progress. That is the crucial problem: farming will form Africa's dominant economic activity for decades to come. True, Africa continues to sell commodities ranging from asbestos to zinc to foreign buyers in a pattern established during the colonial era, but the gauge of Africa's well-being lies in its agriculture.

Figure G-11 may appear to contradict this prospect, but consider how few the exceptions are. Of Subsaharan Africa's nearly 50 countries, all but nine have low-income economies. The only well-balanced economy among these nine is South Africa, for reasons we discuss in the regional section of this chapter: South Africa combines industrial production, commodity sales, and commercial farming. Among the others, Botswana's prosperity is based on commodity (diamond) sales, Namibia's on metals, and Gabon's on oil and forest exploitation. But even the export of oil or metals cannot lift Nigeria, The Congo, Zimbabwe, or any other African economy out of its low-income status.

The remnants of colonial economic ties are mostly gone, and profits from commodities, from gold to manganese, have dropped as production costs rise and market prices fluctuate. But a new buyer has arrived on the African scene: China. The Chinese demand for metals, especially iron ore but also ferroalloys and copper, is burgeoning, and Africa has what China wants—except that Subsaharan Africa's railroad system, even in comparatively efficient South Africa, cannot get the cargoes to the coast in sufficient quantity or, in many areas, at all. China is offering to pay for infrastructure improvements, and there is hardly a country in Africa today without Chinese business representatives negotiating with the government. If China's interest in commodities were to help link farmers as well as mines to cities and ports, Africa's economic condition might improve.

POINTS TO PONDER

● More than one million infants and children die of malaria each year in Africa, many times more than in the rest of the susceptible countries of the world combined.

● China is building a dam on Ethiopia's Tekeze River that will be the largest in Africa, higher even than China's Three Gorges project, to generate electricity and facilitate irrigation.

● Liberia in 2005 became the first country in Africa to be governed by a woman when President Ellen Johnson-Sirleaf was elected with 59 percent of the vote.

● The staggering death toll in the wars in The Congo—more than 4 million since 1998—is the worst in the world, but international assistance, including UN peacekeeping efforts, has not been enough to end it.

Regions of the Realm

On the face of it, Africa seems to be so massive, compact, and unbroken that any attempt to justify a contemporary regional breakdown is doomed to fail. No deeply penetrating bays or seas create peninsular fragments as they do in Europe. No major islands (other than Madagascar) provide the broad regional contrasts we see in Middle America. Nor does Africa really taper southward to the peninsular proportions of South America. And

Africa is not cut by an Andean or a Himalayan mountain barrier. Given Africa's colonial fragmentation and cultural mosaic, is regionalization possible? Indeed it is.

Maps of environmental distributions, ethnic patterns, cultural landscapes, historic culture hearths, and other spatial data yield a four-region structure complicated by a fifth, overlapping zone as shown in Figure 6-11. Beginning in the south, we identify the following regions:

1. *Southern Africa*, extending from the southern tip of the continent to the northern borders of Angola, Zambia, Malawi, and Moçambique. Ten countries constitute this region, which extends beyond the tropics and whose giant is South Africa. The island state of Madagascar, with Southeast Asian influences, is neither Southern nor East African.

2. *East Africa*, where natural (equatorial) environments are moderated by elevation and where plateaus, lakes, and mountains, some carrying permanent snow, define the countryside. Six countries, including the highland part of Ethiopia, comprise this region.

3. *Equatorial Africa*, much of it defined by the basin of the Congo River, where elevations are lower than in East Africa, temperatures are higher and moisture more ample, and where most of Africa's surviving rainforests remain. Among the eight countries that form this region, The Congo dominates territorially and demographically, but others are much better off economically and politically.

MAJOR CITIES OF THE REALM

City	Population* (in millions)
Abidjan, Ivory Coast	4.3
Accra, Ghana	2.4
Adis Abeba, Ethiopia	3.6
Cape Town, South Africa	3.3
Dakar, Senegal	2.9
Dar es Salaam, Tanzania	3.4
Durban, South Africa	2.8
Harare, Zimbabwe	1.7
Ibadan, Nigeria	2.9
Johannesburg, South Africa	3.6
Kinshasa, The Congo	9.2
Lagos, Nigeria	10.6
Lusaka, Zambia	1.5
Mombasa, Kenya	1.0
Nairobi, Kenya	3.5

*Based on 2010 estimates.

4. *West Africa*, which includes the countries of the western coast and those on the margins of the Sahara in the interior, a populous region anchored in the southeast by Africa's demographic giant, Nigeria. Fifteen countries form this crucial African region.

5. *The African Transition Zone*, the complicating factor on the regional map of Africa. In Figure 6-11, note that this zone of increasing Islamic influence completely dominates some countries (e.g., Somalia in the east and Senegal in the west) while cutting across others, thereby creating Islamized northern areas and non-Islamic southern zones (Nigeria, Chad, Sudan).

● SOUTHERN AFRICA

Southern Africa, as a geographic region, consists of all the countries and territories lying south of Equatorial Africa's The Congo and East Africa's Tanzania (Fig. 6-12). Thus

defined, the region extends from Angola and Moçambique (on the Atlantic and Indian Ocean coasts, respectively) to South Africa and includes a half-dozen landlocked states. Also marking the northern limit of the region are Zambia and Malawi. Zambia is nearly cut in half by a long land extension from The Congo, and Malawi penetrates deeply into Moçambique. The colonial boundary framework, here as elsewhere, produced many liabilities.

Africa's Richest Region

Southern Africa constitutes a geographic region in both physiographic and human terms. Its northern zone marks the southern limit of the Congo Basin in a broad upland that stretches across Angola and into Zambia (the tan corridor extending eastward from the Bihe Plateau in Fig. 6-2). Lake Malawi is the southernmost of the East African rift-valley lakes; Southern Africa has none of East Africa's volcanic

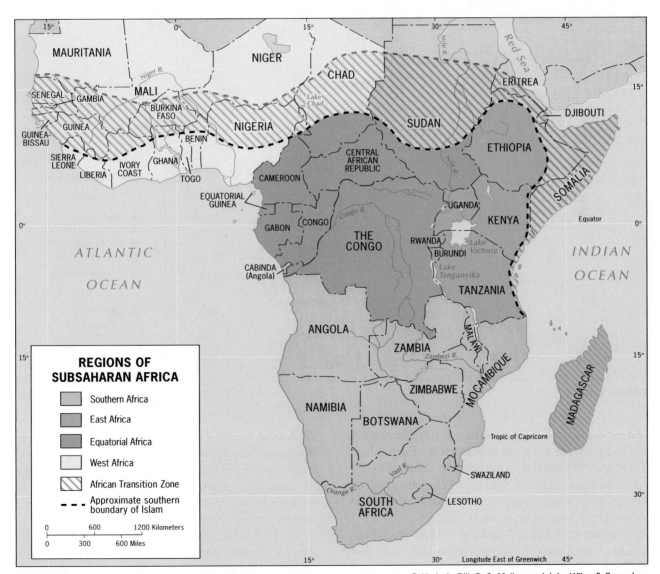

FIGURE 6-11

© H. J. de Blij, P. O. Muller, and John Wiley & Sons, Inc.

FIGURE 6-12

© H. J. de Blij, P. O. Muller, and John Wiley & Sons, Inc.

and earthquake activity. Most of the region is plateau country, and the Great Escarpment is much in evidence here. There are two pivotal river systems: the Zambezi (which forms the border between Zambia and Zimbabwe) and the Orange-Vaal (South African rivers that combine to demarcate southern Namibia from South Africa).

Southern Africa is the continent's richest region materially. A great zone of mineral deposits extends through the heart of the region from Zambia's Copperbelt through

Zimbabwe's Great Dyke and South Africa's Bushveld Basin and Witwatersrand to the goldfields and diamond mines of the Orange Free State and northern Cape Province in the heart of South Africa. Ever since these minerals began to be exploited in colonial times, many migrant laborers have come to work in the mines.

Southern Africa's agricultural diversity matches its mineral wealth. Vineyards drape the slopes of South Africa's Cape Ranges; tea plantations hug the eastern escarpment slopes of Zimbabwe. Before civil war destroyed its economy, Angola was one of the world's leading coffee producers. South Africa's relatively high latitudes and its range of altitudes create environments for apple orchards, citrus groves, banana plantations, pineapple farms, and many other crops.

Despite this considerable wealth and potential, not all of the ten countries of Southern Africa have prospered. As Figure G-11 shows, four remain mired in the low-income category (Malawi, Moçambique, Zambia, and Zimbabwe); although South Africa, Angola, Lesotho, Swaziland, Namibia, and Botswana fall into the two middle-income ranks, the (desert-dominated) last two rank among the realm's three most sparsely populated countries. A period of rapid population growth, followed by the devastating onslaught of AIDS, civil conflict, political instability, incompetent government, widespread corruption, unfair competition on foreign markets, and environmental problems have combined to constrain regional development.

Still, with the tragic exception of Zimbabwe (the realm's basket case in 2009), Southern Africa as a region is better off than any other in Subsaharan Africa: as Figure G-11 shows, six countries here have risen above the lowest-income rank among world states. Some international cooperation including a regional association and a customs union are emerging. And South Africa, the realm's most important country by many measures, gives hope for a better future.

South Africa

The Republic of South Africa is the giant of Southern Africa, an African country at the center of world attention, a bright ray of hope not only for Africa but for all humankind.

Long in the grip of one of the world's most notorious racial policies (**apartheid**, or "apartness," and its derivative, **separate development**), South Africa today is shedding its past and is building a new future. That virtually all parties to the earlier debacle worked cooperatively to restructure the country under a new flag, a new national anthem, and a new leadership was one of the great events of the late twentieth century. Now, more than 15 years after Nelson Mandela became South Africa's first democratically elected president, and with his successor reelected to a second term, South Africa is maturing politically. In 2008, a significant development occurred: the political party that represented the coalition of anti-apartheid forces, the African National Congress (ANC), broke apart through the secession of a group of dissidents calling themselves the Congress of the People (COPE), a new party that might form the core of a new opposition. This was exciting as well as ominous: exciting because it was a vigorous expression of democracy at work, but ominous because the political system could break down under the tensions of controversy.

An African—and Global—Magnet

South Africa stretches from the warm subtropics in the north to Antarctic-chilled waters in the south. With a land area of more than 1.2 million square kilometers (470,000 sq mi) and a multicultural population of just under 50 million, South Africa contains the bulk of its region's minerals, most of its good farmlands, its largest cities and best ports, its most productive factories, and best-developed transport networks. Commodities from as far away as Zambia move through South African ports. Workers from distant Malawi and nearby Moçambique arrive to find work in South Africa's mines and factories.

South Africa's location has much to do with its human geography. On the African continent, peoples migrated southward—first the Khoisan-speakers and then the Bantu peoples—into the South African cul-de-sac. On the oceans, the Europeans arrived to claim the southernmost Cape as one of the most strategic places on Earth, the gateway from the Atlantic to the Indian Ocean, a waystation on the route to Asia's riches. The Dutch East India Company founded Cape Town as early as 1652, and soon the Hollanders began to bring Southeast Asians to the Cape to serve as domestics and laborers, thereby creating a foothold for Islam here. By the time the British took over about 150 years later, Cape Town had a substantial population of mixed ancestry, the source of today's so-called Coloured sector of the country's citizenry.

The British also altered the demographic mosaic by bringing tens of thousands of indentured laborers from their South Asian domain to work on the sugar plantations of east-coast Natal. Most of these laborers stayed after their period of indenture was over, and today South Africa counts more than 1 million Indians among its people. Most are still concentrated in Natal and prominently so in metropolitan Durban, adhering to their Hindu faith.

British and Boers

As we noted above, South Africa had been occupied by Europeans long before the colonial "scramble for Africa" gained momentum. The Dutch, after the British took control of the Cape, trekked into the South African interior and, on the plateau they called the *highveld*, founded their own republics. When diamonds and gold were discovered there, the British challenged the *Boers* (descendants of the Dutch settlers) for these prizes. In 1899–1902, the British and the Boers fought the Boer War. The British won, and British capitalists took control of South Africa's economic and political life. But the Boers negotiated a power-sharing arrangement and eventually achieved supremacy. Having long since shed their European links, they now called themselves *Afrikaners*, their word for Africans, and proceeded to erect the system known as apartheid.

Zulu and Xhosa

These foreign immigrations and struggles took place on lands that Africans had already entered and fought over. When the Europeans reached the Cape, Bantu nations were driving the weaker Khoi and San peoples into less hospitable territory or forcing them to work in bondage. One great contest was taking place in the east and southeast, below the Great Escarpment. The Xhosa nation was moving toward the Cape along this natural corridor. Behind them, in Natal (Fig. 6-12), the Zulu Empire became the region's most powerful entity in the nineteenth century. On the highveld, the North and South Sotho, the Tswana, and other peoples could not stem the tide of European aggrandizement, but their numbers ensured survival.

In the process, South Africa became Africa's most pluralistic and heterogeneous society. People had converged on the country from Western Europe, Southeast Asia, South Asia, and other parts of Africa itself. At the end of the first decade of this century, Africans outnumbered non-Africans by just about 4 to 1 (Table 6-1).

TABLE 6-1

Demographic Data for South Africa

Population Groups	2010 Estimated Population (in millions)
African nations	**38.6**
Zulu	11.6
Xhosa	8.6
Tswana	4.0
Sotho (N and S)	3.9
Others (6)	10.5
Whites	**4.7**
Afrikaners	2.9
English-speakers	1.7
Others	0.1
Mixed (Coloureds)	**4.5**
African/White	4.1
Malayan	0.4
South Asian	**1.3**
Hindus	0.9
Muslims	0.4
TOTAL	**49.1**

The Ethnic Mosaic

Heterogeneity also marks the spatial demography of South Africa. Despite centuries of migration and (at the Cape) intermarriage, labor movement (to the mines, farms, and factories), and sustained urbanization, regionalism pervades the human mosaic. The Zulu nation still is largely concentrated in the province the Europeans called Natal. The Xhosa still cluster in the Eastern Cape, from the city of East London to the Natal border and below the Great Escarpment. The Tswana still occupy ancestral lands along the Botswana border. Cape Town still is the core area of the Coloured population; Durban still has the strongest Indian imprint. Travel through South Africa, and you will recognize the diversity of rural cultural landscapes as they change from Swazi to Ndebele to Venda.

This historic regionalism was among the factors that led the Afrikaner government to institute its "separate development" scheme, but it could not stem the tide of urbanization. Millions of workers, job-seekers, and illegal migrants converged on the cities, creating vast shantytowns on their margins. In the legal African "townships" such as Johannesburg's Soweto ("SOuth WEstern TOwnships") and in these squatter settlements the antiapartheid movement burgeoned, and the strength of the African National Congress movement grew. In February 1990 Nelson Mandela, imprisoned for 28 years on Robben Island, South Africa's Alcatraz, became a free man, and following the momentous first democratic election of 1994 he became president of an ANC-dominated government in Cape Town.

South Africa's New Map

Prior to 1994, South Africa had been administratively organized into four provinces, two of them based on British coastal holdings and two interior ones roughly representative of Boer "republics." Even before the momentous 1994 election, these were replaced by a framework of nine provinces carefully laid out to reflect South Africa's heterogeneous subregions (Fig. 6-13). During whiteminority rule, South Africa had two capital cities: Pretoria, the administrative headquarters, in the interior; and Cape Town, the legislative capital, on the coast. Pretoria and the country's largest city—the mining, industrial, and financial center of Johannesburg—were made part of the new Province of Gauteng, which approximately corresponds to South Africa's core area. Cape Town retained its legislative functions and simultaneously became the capital of the new Western Cape Province. The old Province of Natal was renamed Kwazulu-Natal in recognition of its dominant ethnic sector.

This politico-geographical reorganization proved to be a stroke of genius. In South Africa's first democratic

FIGURE 6-13

© H. J. de Blij, P. O. Muller, and John Wiley & Sons, Inc.

election ever, the African National Congress won seven of the new provinces, failing to win only in Kwazulu-Natal where the Zulu vote went to a local movement and in the Western Cape where an alliance of Coloured and white voters outpolled the ANC. This demonstration of democracy and the resulting representation of minorities in the new South African Parliament was a triumph. The ANC attracted support from some voters in the white, Coloured, and Asian communities, and President Mandela's cabinet was, indeed, a rainbow coalition.

In 1999, ANC leader Thabo Mbeki, who had served as Mandela's deputy, became the country's second pop-

ularly elected president. In many other African countries, the succession from heroic founder-of-the-nation to political inheritor-of-the-presidency has not gone well, but South Africa's constitution proved its worth. Still, the new president faced some formidable problems: inevitable comparisons to the incomparable Mandela; the rising tide of AIDS in South Africa, which Mbeki and his health minister attributed to causes other than HIV; the collapse of order and the rise of ruthless dictatorship in neighboring Zimbabwe, on which Mbeki's leadership failed; and impatience at home over slow progress in the betterment of living conditions for the poor. This last and

SUBSAHARAN AFRICA IS urbanizing rapidly, but it still has only one true conurbation, and Johannesburg, South Africa lies at the heart of it. Little more than a century ago, Johannesburg was a small (though rapidly growing) mining town based on the newly discovered gold reserves of the Witwatersrand.

Today Johannesburg forms the focus of a conurbation of more than 7 million, extending from Pretoria in the north to Vereeniging in the south, and from Springs in the east to Krugersdorp in the west. The population of metropolitan Johannesburg itself passed the 3 million mark in 2003 (it now totals 3.6 million), thereby overtaking Cape Town to become South Africa's largest metropolis.

Johannesburg's skyline is the most impressive in all of Africa, a forest of skyscrapers reflecting the wealth generated here over the past hundred years. Look southward from a high vantage point, and you see the huge mounds of yellowish-white slag from the mines of the "Rand," the so-called mine dumps, partly overgrown today, interspersed with suburbs and townships. In a general way, Johannesburg developed as a white city in the north and a black city in the south. Soweto, the black township, lies to the southwest; Houghton and other spacious, upper-class suburbs, formerly exclusively white residential areas, lie to the north.

Johannesburg has neither the scenery of Cape Town nor the climate and beaches of Durban. The city lies 1.5 kilometers (nearly 1 mi) above sea level, and its thin air often is polluted from smog created by automobiles, factories, mine-dump dust, and countless cooking fires in the townships and shantytowns that ring the metropolis.

Over the past century, the Johannesburg area produced nearly one-half of all the world's gold by value. But

today, Johannesburg lies at the heart of an industrial, commercial, and financial complex whose name on the new map of South Africa is **Gauteng**.

© H. J. de Blij, P. O. Muller, and John Wiley & Sons, Inc.

inevitable development is not unique to South Africa or any other country in the global periphery: the rise to power by a local leader results in a "revolution of rising expectations" that must be met by persuasive evidence of progress. Mbeki had to steer a course that kept South Africa's national economy on track while finding ways to fund social programs as well as foreign initiatives (he sent the South African Army to help quell the crisis in The Congo, for example). Mbeki was reelected in 2004, but in 2008 his presidency was cut short by disputes within the ANC that eventually led to his premature resignation.

Jacob Zuma of KwaZulu-Natal was elected in the important presidential vote of 2009. Mr. Zuma had positioned himself as a "man of the people," skillfully mixing populist campaign promises with reassurances to business and in-

dustry, firing up South Africa's long-simmering revolution of rising expectations. President Zuma also faced legal problems even as he ran for the presidency, having been charged in court with corruption and fraud, acquitted in a rape case in another trial, and facing criticism for other aspects of his personal life (he has three wives in a country where monogamy is the norm if not the law). The economic downturn starting in 2008 immediately challenged Zuma, not only nationally but also within his ANC party, whose left wing expects the results Mbeki allegedly failed to deliver.

South Africa's Economic Geography

South Africa has by far the largest and most diverse economy in all of Africa, but it is an economy facing major challenges. Ever since diamonds were discovered at

Kimberley in the 1860s, South Africa has been synonymous with minerals. The Kimberley finds, made in a remote corner of what was then the Boer "Republic" of Orange Free State (the British soon annexed it to the Cape), set into motion a new economic geography. Rail lines were laid from the coast to the "diamond capital" even as fortune seekers, capitalists, and tens of thousands of African workers, many from as far afield as Lesotho, streamed to the site. One of the capitalists was Cecil Rhodes, of Rhodes Scholarship fame, who used his fortune to help Britain dominate Southern Africa.

Just 25 years after the diamond discoveries, prospectors found what was long to be the world's greatest goldfield on a ridge called the Witwatersrand (Fig. 6-13). This time the site lay in the so-called South African Republic (the Transvaal), and again the Boers were unable to hold the prize. Johannesburg became the gold capital of the world, and a new and even larger stream of foreigners arrived, along with a huge influx of African workers. Cheap labor enlarged the profits. Johannesburg grew explosively, satellite towns developed, and black townships mushroomed. The Boer War was merely an interlude here on the mineral-rich Witwatersrand.

During the twentieth century, South Africa proved to be richer than had been foreseen. Additional goldfields were discovered in the Orange Free State. Coal and iron ore were found in abundance, which gave rise to a major iron and steel industry. Other metallic minerals, including chromium and platinum, yielded large revenues on world markets. Asbestos, manganese, copper, nickel, antimony, and tin were mined and sold; a thriving metallurgical industry developed in South Africa itself. Capital flowed into the country, white immigration grew, farms and ranches were laid out, and markets multiplied.

Infrastructural Gains

South Africa's cities grew apace. Johannesburg was no longer just a mining town: it became an industrial complex and a financial center as well. The old Boer capital, Pretoria, just 50 kilometers (30 mi) north of the Witwatersrand, became the country's administrative center during apartheid's days. In the Orange Free State, major industrial growth matched the expansion of mining. While the core area developed megalopolitan characteristics, coastal cities expanded as well. Durban's port served not only the Witwatersrand but also a wider regional hinterland. Cape Town was becoming South Africa's largest central city; its port, industries, and productive agricultural hinterland gave it primacy over a wide area.

The labor force for all this development, from mines to railroads and from farms to highways, came from the African peoples of South Africa (and, indeed, from beyond its borders as well). Draconian apartheid policies notwithstanding, workers from Moçambique, Zimbabwe, Botswana, and other neighboring countries sought jobs in South Africa even as millions of South African villagers also left their homes for the towns and cities. In the process they built the best infrastructure of any African country. But apartheid ruined South Africa's prospects. The cost of the separate development program was astronomical. Social unrest during the decade preceding the

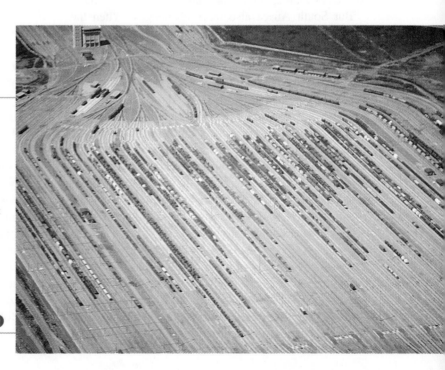

FROM THE FIELD NOTES

"Looking down on this enormous railroad complex, we were reminded of the fact that almost an entire continent was turned into a wellspring of raw materials carried from interior to coast and shipped to Europe and other parts of the world. This complex lies near Witbank in the eastern Rand, a huge inventory of freight trains ready to transport ores from the plateau to Durban and Maputo. But at least South Africa acquired a true transport network in the process, ensuring regional interconnections; in most African countries, railroads serve almost entirely to link resources to coastal outlets."
© H. J. de Blij

www.conceptcaching.com

end of apartheid created a vast educational gap among youngsters. International sanctions against the race-obsessed regime damaged the economy.

The Sputtering Engine of Africa

During the upheavals of the last decade of the twentieth century—think of the Soviet Union, Yugoslavia, the Middle East—was any country more fortunate than South Africa? Surely a massive revolution would sweep apartheid away and bring an all-African regime to power? Had that happened, South Africa today would still be repairing the damage. But instead, largely through the unmatched statesmanship of Nelson Mandela, South Africa is governed by a multiracial party, the erstwhile architects of apartheid are in parliament as a legitimate opposition, and its economy is the largest and healthiest in the realm. Check the Data Table inside the back cover and you will see that, on the African mainland, South Africa's per-capita GNI is practically in a class by itself. By some calculations, South Africa produces 45 percent of all of Subsaharan Africa's GDP.

In many respects South Africa is the most important country in Subsaharan Africa, and the entire realm's fortunes are bound up with it. No African country attracts more foreign investment or (still) foreign workers. None has the universities, hospitals, and research facilities. No other has the military forces capable of intervening in African trouble spots. Few have the free press, effective trade unions, independent courts, or financial institutions to match South Africa's. And with a population of 49.1 million (79 percent black, just under 10 percent white, 9 percent Coloured, 2.6 percent Asian), South Africa has a large, multiracial, and growing middle class.

But South Africa also faces challenges. When the African National Congress (ANC) won the 1994 elections (and subsequent democratic elections as well, another beacon for Africa), the country's African peoples harbored expectations not only of post-apartheid freedom but of better living conditions as well. And indeed the government made major progress: it began an old-age pension scheme, built nearly two million brick homes for the poor, provided free water to nearly 30 million people in remote areas—all this without destabilizing the economy. And through a series of laws, the ANC has made it mandatory for firms to hire more blacks and women. The growing African middle class strongly supports the ANC.

This progress notwithstanding, South Africa confronts long-range problems. The economy continues to depend far too much on the export of minerals and metals (diamonds, platinum, gold, iron, and steel) at a time when this entails risks at home and abroad. At home, those union-friendly labor laws, including rising wages, are making mining less profitable. Abroad, commodity prices are unreliable. In the

apartheid year of 1970, South Africa produced nearly 70 percent of the world's gold; today it produces less than 15 percent. And the manufacturing sector of the economy remains too weak. Even while the black middle class is expanding, unemployment among blacks may be as high as 50 percent—and the gap between rich and poor is growing. Land reform, an urgent matter in a country where land alienation reached huge proportions, is too slow in the view of many. Add to this the scourge of AIDS and the government's failure to address this crisis effectively, and South Africa is shadowed by serious problems.

Core-Periphery Contrasts

On maps of development indices, South Africa is portrayed as a middle-income economy, but averages mean little in this country of strong internal core-periphery contrasts. In its great cities, industrial complexes, mechanized farms, and huge ranches, South Africa resembles a high-income economy, much like Australia or Canada. But outside the primary core area (centered on Johannesburg) and beyond the secondary cores and their linking corridors lies a different South Africa, where conditions are more like those of rural Zambia or Zimbabwe. In terms of such indices as life expectancy, infant and child mortality, overall health, nutrition, education, and many others, a wide range marks the country's population sectors. South Africa has been described as a microcosm of the world, exhibiting in a single state not only a diversity of cultures but also a wide array of human conditions. If South Africa can keep on course, this one-time pariah of apartheid will become a guidepost to a better world.

The Middle Tier

Between South Africa's northern border and the region's northern limit lie two groups of states: those with borders with South Africa and those beyond.

Five countries constitute the middle tier, neighboring South Africa: Zimbabwe, Namibia, Botswana, and the ministates of Lesotho and Swaziland (Fig. 6-12). As the map shows, four of these five countries are landlocked. Botswana occupies the heart of the Kalahari Desert and surrounding steppe; it has an upper-middle-income economy boosted by diamond exports, tourism, and a democratic government, but the fate of its rural inhabitants continues to roil its politics. The government succeeded in transferring the sales and distribution of the country's diamonds from London back to the capital of Gaborone, where the De Beers Company built a major sorting center. Botswana's health services and budgets are strained by the high incidence of AIDS (one of the highest in all of tropical Africa). Lesotho, the landlocked, high-mountain ministate encircled by South Africa and a traditional kingdom that used to depend on revenues from the mine workers it sent to South Africa, now survives precariously on a textile industry built by Taiwanese investors that sells the bulk of its production in the United States. Swaziland, another ministate and nearly surrounded by South Africa, is also a remaining traditional kingdom whose absolute ruler resists democratic reforms while engaging in personal excesses that regularly make the world news. Its citrus-fruit-based export economy is essentially irrelevant to the king's subjects, more than 70 percent of whom live in extreme poverty. Both Lesotho and Swaziland have seen surges in AIDS incidence that put them among Africa's worst-afflicted.

As Figure 6-12 shows, Botswana is not the only desert country in Southern Africa. To its west, neighboring Namibia is named after what is known as the driest desert on Earth, the Namib. Oddly shaped by its narrow Caprivi Strip linking it to the Zambezi River, this former German colony was administered by South Africa from 1919 to 1990. Territorially the size of Texas and Oklahoma combined, vast Namibia with just 2.1 million inhabitants has one of the lowest population densities in the world (see Figure 6-4). Only in the far north is there enough water for sedentary agriculture; most of the rest of the country lies under huge livestock ranches. Beef ranks among Namibia's exports, but the export economy is mainly based on minerals: diamonds, gold, copper, zinc, lead, uranium, and other commodities all attract foreign investment. The interior capital, Windhoek, the major port, Walvis Bay, and the main mining town of Tsumeb are urban centers in a country whose population remains 65 percent rural. A major challenge for Namibians will be orderly land reform: the government plans to purchase and transfer commercial estates to local farmers, and by 2010 more than 1 million hectares (2.5 million acres) had been reallocated. The goal, however, is 15 times larger, for which Namibia seeks international assistance because the cost involves not just the price of the land, but also the subsequent help to the new owners. In 2010, some 250,000 applicants were awaiting their turn to qualify.

The Tragedy of Zimbabwe

Botswana's other neighbor, Zimbabwe, may be said to lie at the heart of Southern Africa, between the Zambezi River in the north and the Limpopo River in the south, between the escarpment to the east and the desert to the west. Landlocked but endowed with good farmlands, cool uplands, a wide range of mineral resources, and varied natural environments, Zimbabwe (named after historic stone ruins in its interior [see photo page 293]) had

one of Southern Africa's most vibrant economies and seemed to have a bright future.

Zimbabwe's core area is defined by the mineral-rich "Great Dyke" that extends across the heart of the country from the vicinity of its capital, Harare, in the north to the second city, Bulawayo, in the south. Gold, copper, asbestos, chromium, and platinum are among its major exports in normal times. But Zimbabwe is not just an ore-exporting country. Its farms produce tobacco, tea, sugar, cotton, and even cut flowers in addition to staples for the local market.

But these are not normal times for the two nations that constitute most of Zimbabwe's population, the Shona (82 percent) and the Ndebele (14 percent). Following their successful joint campaign to end white-minority rule, the Shona turned on the Ndebele in a bitter ethnic conflict that established Shona dominance. A leader named Robert Mugabe rose to prominence during this early period, and it seemed that relatively prosperous Zimbabwe would put its conflicts behind it and take its place among African success stories.

During the colonial period, the tiny minority of whites that controlled what was then called Southern Rhodesia (after Cecil Rhodes, the British capitalist of diamond fame and scholarship honors), took the most productive farmlands and organized the agricultural economy. After independence their descendants continued to hold huge estates, only parts of which were being farmed. What was needed was a comprehensive land reform program, but in the 1980s now President Mugabe began to encourage squatters to invade these white farms and a number of owners were killed. The agricultural economy began to collapse as some white farmers held out while others abandoned their lands. Corruption in the Mugabe government involved the transfer of land, not to needy squatters but to friends of top officials. Foreign investment in other Zimbabwean enterprises, including the mining industry, started to dry up. By the mid-1990s Zimbabwe was in an economic free fall.

Many Africans within and outside Zimbabwe had little sympathy for the white farmers or corporate powers controlling the means of production, but Mugabe had no compassion for his own people either. He now turned on the informal sector of the economy, which was all most Zimbabweans had left when jobs on farms and in factories and mines disappeared. Shantytown dwellings that were once confined to the outskirts of the cities now appeared in city centers, as did shacks where women repaired garments and men fixed bicycles. Mugabe ordered his henchmen to destroy the dwellings and shacks of some 700,000 of these "informal" urban residents, leaving them in the streets without shelter or livelihood.

As conditions worsened, people by the hundreds of thousands streamed out of Zimbabwe; by 2009 an estimated 4

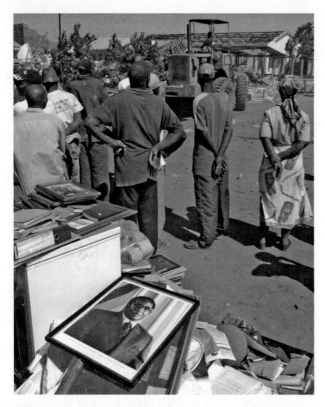

African cities are a jumble of "formal" and "informal" sectors and neighborhoods ranging from modern downtowns and leafy suburbs to shantytowns and slums. Many families that have moved from the countryside to the city have no choice but to rent (or build) a ramshackle dwelling lacking basic amenities. These poorest urban areas nevertheless have a vibrant economy as their occupants try to make a living as local vendors, repairers, builders, traders, or artisans. Often they find jobs in service or construction industries nearby without earning enough to allow them to move out of their abode. Most governments tolerate these neighborhoods, but not the president of Zimbabwe, who in 2005 ordered his capital city "cleaned up." Here, a group of residents of a shantytown in Harare watch the destruction, having hurriedly piled their belongings on the dirt road outside their dwellings—including a framed photograph of President Robert Mugabe, once a hero of Africa's liberation struggle and inheritor of a thriving economy, now the symbol of administrative failure and social repression.
© AP/World Wide Photos.

million of the country's 13.8 million people were refugees. About 80 percent of the people were jobless. A whole generation of children suffered from malnutrition. But still the Saddam Hussein of Africa, as Mugabe came to be called, would not yield what had become his dictatorial powers. In late 2008 a major outbreak of cholera killed thousands and sent another wave of desperate Zimbabweans across the border, but Mugabe did not express any sympathies; none of the victims was from his elite inner circle. A rapacious ruler had converted an African breadbasket into a basket case.

This terrible tragedy was compounded by the failure of Zimbabwe's key neighbors to intervene. President Thabo Mbeki of South Africa proved a weak leader in a time of need; it fell to former President Mandela to express the sentiments of outrage felt by many. Some international food assistance reached desperate Zimbabweans, but the international community did not prove a deterrent to Mugabe. In 2008 he was defeated in legislative elections by Morgan Tsvangirai, but through intimidation and violence Mugabe succeeded in preventing a legitimate runoff. Eventually Tsvangirai agreed to a power-sharing arrangement, but Zimbawe still does not have the representative, competent governement it needs. Meanwhile, the people suffer.

The human catastrophe of Zimbabwe stands in sharp contrast to the social achievement of South Africa. The question is whether South Africa's example will ultimately prevail in Zimbabwe, or if Zimbabwe's affliction will eventually infect its southern neighbor.

The Northern Tier

In the four countries that extend across the northern tier of the region—Angola, Zambia, Malawi, and Moçambique—problems abound. Angola (17.7 million), formerly a Portuguese dependency, with its exclave of Cabinda (Fig. 6-12) had a thriving economy based on a wide range of mineral and agricultural exports at the time of independence in 1975. But then Angola fell victim to the Cold War, with northern peoples choosing a communist course and southerners falling under the sway of a rebel movement backed by South Africa and the United States. The results included a devastated infrastructure, idle farms, looting of diamonds, hundreds of thousands of casualties, and millions of landmines that continue to kill and maim. But Angola's oil wealth yields about U.S. $3 billion per year, and stability, if sustained, may attract investors to begin rebuilding this ruined country.

By 2009, a building boom in Angola's capital, once-picturesque and then-ruined Luanda, signified what seemed to be a new era. As the second-largest oil exporter in Subsaharan Africa, Angola joined the Organization of Petroleum Exporting Countries (OPEC) and became the largest supplier of crude oil to China; the Chinese planned to help Angola in the reconstruction of its highway and railroad infrastructure. In 2008, an election contested by three major parties (including those representing the old north-south division) proved that democracy was making gains in this embattled country. But Angola was vulnerable to fluctuating world prices for the oil that brings more than 90 percent of its export revenues, and its economic growth rate, which had reached an unprecedented 15 percent, could not be sustained. Thus it will take a long time

Distinctive Madagascar

MAPS SHOWING THE geographic regions of the Subsaharan Africa realm often omit Madagascar. And for good reason: Madagascar differs strongly from Southern Africa, the African region to which it is nearest—just 400 kilometers (250 mi) away. It also differs from East Africa, from which it has received some of its cultural infusions. Madagascar is the world's fourth-largest island, a huge block of Africa that separated from the main landmass 160 million years ago. About 2000 years ago, the first settlers arrived—not from Africa (although perhaps via Africa) but from Southeast Asia. Malay communities flourished in the interior highlands of the island, which resembles Africa in having a prominent eastern escarpment and a central plateau (Fig. 6-14). Here was formed a powerful kingdom, the empire of the Merina. Its language, Malagasy, of Malay-Polynesian origin, became the indigenous tongue of the entire island (Fig. 6-10).

The Malay and Indonesian immigrants brought Africans to the island as wives and slaves, and from this forced immigration evolved the African component in Madagascar's population of 20.0 million. In all, nearly 20 discrete ethnic groups coexist in Madagascar, among which the Merina (about 5 million) and Betsimisaraka (about 3 million) are the most numerous. Like mainland Africa, Madagascar experienced colonial invasion and competition. Portuguese, British, and French colonists appeared after 1500, but the Merina were well organized and resisted colonial conquest. Eventually Madagascar became part of France's empire, and French became the *lingua franca* of the educated elite.

Because of its Southeast Asian imprint, Madagascar's staple food is rice, not corn. It has some minerals, including chromite, iron ore, and bauxite, but the economy is weak, damaged by long-term political turmoil and burdened by rapid population growth. The infrastructure has crumbled; the "main road" from the capital to the nearest port (Fig. 6-14) is now a potholed 250 kilometers (150 mi) that takes 10 hours for a truck to navigate.

Meanwhile, Madagascar's unique flora and fauna retreated before the human onslaught. Madagascar's long-term isolation kept evolution here so distinct that the island is a discrete zoogeographic realm. Primates living on the island are found nowhere else; 33 varieties of lemurs are unique to Madagascar. Many species of birds, amphibians, and reptiles are also exclusive to this island. Their home, the rainforest, covered 168,000 square kilometers (65,000 sq mi) in 1950, but today only about one-third of it is left. Logging, introduced by the colonists, damaged it; slash-and-burn agriculture is destroying it; and severe droughts since 1980 have intensified the impact. Obviously, Madagascar should be a global conservation priority, but funds are limited and the needs are enormous. Malnutrition and

poverty are powerful forces when survival is at stake for villages and families.

Madagascar's cultural landscape retains its Southeast Asian imprints, in the towns as well as the paddies. The capital, Antananarivo, is the country's primate city, its architecture and atmosphere combining traces of Asia and Africa. Poverty dominates the townscape here, too, and there is little to attract in-migrants (Madagascar is only 30 percent urbanized). But perhaps the most ominous statistic is the high natural increase rate (2.8 percent) of Madagascar's population.

© H. J. de Blij, P. O. Muller, and John Wiley & Sons, Inc.

FIGURE 6-14

for the peoples in Angola's countryside to believe that better times have truly arrived.

On the opposite coast, the other major former Portuguese colony, Moçambique (21.3 million), fared poorly in a different way. Without Angola's mineral base and with limited commercial agriculture, Moçambique's chief asset was its relative location. Its two major ports, Maputo and Beira, handled large volumes of exports and imports for South Africa, Zimbabwe, and Zambia. But upon independence Moçambique, too, chose a Marxist course with dire economic and dreadful political consequences. Another rebel movement supported by South Africa caused civil conflict, created famines, and generated a stream of more than a million refugees toward Malawi. Rail and port facilities lay idle, and Moçambique at one time was ranked by the United Nations as the world's poorest country. In recent years, the port traffic has been somewhat revived, and Moçambique and South Africa are working on a joint Johannesburg-Maputo Development Corridor (Fig. 6-12), but it will take generations for Moçambique to recover.

Landlocked Zambia (12.7 million) is an African success story: it remained stable even during economically difficult times and avoided the turbulence engulfing its neighbors, and is now reaping some rewards as a result. Its boundaries a product of British imperialism, Zambia shares the minerals of the Copperbelt with The Congo's Katanga Province (Fig. 6-12). In recent years, high copper prices have brought substantial external income; refugee farmers from Zimbabwe boosted the tobacco industry; and stability and relatively open government were rewarded when donors forgave nearly U.S. $4 billion in foreign debt. Zambia still remains a very poor country; it is too dependent on a few commodities, its access to world markets remains difficult, and its farming sector needs irrigation systems. But Zambia exemplifies what is possible when an African state is run responsibly.

Neighboring Malawi (14.5 million), by comparison, has suffered from political instability and infighting ever since the death of its founding autocratic ruler, Hastings Banda. Malawi's dependence on corn as its food staple, its variable climate, and its severely fragmented land-use pattern combine to create cycles of boom and bust from which the country has yet to escape, and its political representatives spend more time scheming and conniving than they do in cooperative planning. When the terrible drought of 2005 broke in 2006, producing a bumper crop that was matched by another in 2007, Malawi was actually able to export some corn to needy Zimbabwe. But when still-troubled local subsistence farmers asked for help, they were told by the government to increase their efficiency by forming cooperatives. Members of the National Assembly might consider the principle of cooperation as a means of improving their own efficiency.

Before leaving Southern Africa we should take note of Madagascar, the island-state that lies off the southeastern African coast. Despite its location, this country is neither Southern nor East African. Madagascar has historic ties to Southeast Asia and cultural links to both Asia and Africa (see box titled "Distinctive Madagascar").

● EAST AFRICA

In only two parts of the world do local people refer to the "Great Lakes" as a geographic landmark: North America and East Africa. A string of elongated lakes, some with colonial-era names, marks the western border of the region we define as East Africa (Fig. 6-15). Going to school there, you learn these names early: Albert, Edward, Kivu, Tanganyika, Malawi. But in some ways the most important lake of all lies at the heart of this region, not on its border: Lake Victoria. And as the map shows, Lake Victoria is anything but elongated. It is a large body of water lying in a huge basin—but the basin lies high on a plateau.

The equator crosses East Africa, even Lake Victoria itself, but you would hardly think so on the ground here. We tend to associate equatorial environments with rainforests, heat, and moisture, but in East Africa elevation changes everything (as we also note in Chapter 5 for low-latitude, Andean South America). Drought, not incessant rain, is the frequent problem in this region. Nights can be downright cold. Some volcanic peaks carry snow. Highland East Africa is as different from lowland Equatorial Africa west of the lakes as Amazonia is from the Andes.

East Africa, as the map shows, extends from the Ethiopian Highlands in the north to Tanzania in the south, and from the Indian Ocean to the eastern margins of the Congo Basin. This is a region of spectacular scenery. Volcanoes, some of them active, rise above the high plains. Canyons carved by rivers rival the one we call Grand; Ethiopia was once called Abyssinia, and not for nothing because steep scarps define *rift valleys* partly filled with water; views from the top are all the more breathtaking when you recall that this is the landscape seen by our earliest ancestors. Herds of wildlife that attract millions of tourists roam vast savannas whose names (Serengeti; Maasai Mara) have become world renowned.

Too many visitors come here to see the wildlife without also learning about East Africa's diversity of cultures. In addition to highland Ethiopia, five other countries form this East African region: Kenya, Tanzania, Uganda, Rwanda, and Burundi. Here the Bantu peoples that make up most of the population met Nilotic peoples from the north, including the Maasai. In the hills of Rwanda and Bu-

FIGURE 6-15   © H. J. de Blij, P. O. Muller, and John Wiley & Sons, Inc.

rundi a stratified society developed in which the minority Tutsi, in their cattle-owning kingdoms, dominated the Hutu peasantry. The Indian Ocean coast was the scene of many historic events: the arrival of Islam from Ara-

bia, the visit of Ming Dynasty Chinese fleets in the 1400s, the quest for power by the Turks, the Arab slave trade, the European colonial competition. Here, too, developed the East African *lingua franca*, Swahili.

AMONG THE REALM'S GREAT CITIES . . .   Nairobi

NAIROBI IS THE quintessential colonial legacy: there was no African settlement on this site when, in 1899, the railroad the British were building from the port of Mombasa to the shores of Lake Victoria reached it. However, it had something even more important: water. The fresh stream that crossed the railway line was known to the Maasai cattle herders as Enkare Nairobi (Cold Water). The railroad was extended farther into the interior, but Nairobi grew. Indian traders set up shop. The British established their administrative headquarters here. When Kenya became independent in 1963, Nairobi naturally was the national capital.

Nairobi owes its primacy to its governmental functions, which ensured its priority through the colonial and independence periods, and to its favorable situation. To the north and northwest lie the Kenya Highlands, the country's leading agricultural area and the historic base of the dominant nation in Kenya, the Kikuyu. Beyond the rift valley to the west lie the productive lands of the Luo in the Lake Victoria Basin. To the east, elevations drop rapidly from Nairobi's 1600 meters (5000 ft), so that highland environs make a swift transition to tropical savanna that, in turn, yields to semiarid steppe.

A moderate climate, a modern city center, several major visitor attractions (including Nairobi National Park, on the city's doorstep), and a state-of-the-art airport have boosted Nairobi's fortunes as a major tourist destination, though wildlife destruction, security concerns, and political conditions have damaged the industry in recent years.

Nairobi is Kenya's principal commercial, industrial, and educational center. But its growth (to 3.5 million today) has come at a price: its modern central business district stands in stark contrast to the squalor in the shantytowns that house the countless migrants its apparent opportunities attract.

© H. J. de Blij, P. O. Muller, and John Wiley & Sons, Inc.

Kenya

Kenya is neither the largest nor the most populous country in East Africa, but over the past half-century it has been the dominant state in the region. Its skyscrapered capital at the heart of its core area, Nairobi, is the region's largest city; its port, Mombasa, is the region's busiest. During the 1950s, the Kikuyu nation led a vicious rebellion that hastened the departure of the British from the region.

After independence, Kenya chose a capitalist path of development, aligning itself with Western interests. Without major known mineral deposits, Kenya depended on coffee and tea exports and on a tourist industry based on its magnificent national parks (Fig. 6-15). Tourism became its largest single earner of foreign exchange, and Kenya prospered, apparently proving the wisdom of its capitalist course.

But serious problems arose. During the 1980s, Kenya had the highest rate of population growth in the world, and population pressure on farmlands and on the fringes of the wildlife reserves mounted. Poaching became worrisome, and tourism declined. During the late 1990s, violent weather buffeted Kenya, causing landslides and washing away large segments of the crucial Nairobi-Mombasa Highway. This disaster was followed by a severe drought lasting several years, bringing famine to the interior. Meanwhile, government corruption siphoned off funds that should have been invested. Democratic principles were violated, and relationships with Western allies were strained. The AIDS pandemic brought yet another setback to a country that, in the early 1970s, had appeared headed for an economic takeoff.

Kenya is one of those African states in which leadership and representative government faltered after independence. Jomo Kenyatta was the father of the nation, the leader who personified the struggle for sovereignty. As the map shows, geography and history had placed his people, the Kikuyu (22 percent of the 2010 population of 40.2 million), at the center of this struggle and in a position of power. But even before independence, the Luo (13 percent today), whose domain borders Lake Vic-

Raila Odinga (the son of Kenya's first vice-president). The announcement of the outcome, a narrow victory for the incumbent, was followed by riots and ethnic clashes in which hundreds were killed and the country seemed to teeter on the brink of civil war. Eventually, the government and the opposition reached a mediated agreement that divided the powers, giving the presidency to the incumbent and the prime minister's position to the challenger. But this result failed to address the serious problems Kenya had been facing even before the election campaign started.

As Figure 6-15 shows, Kenya confronts an external problem as well. In its northeast, Kenya's neighbor is Islamic, chaotic, and violent Somalia, a failed state whose troubles—and people—sometimes spill over into Kenyan territory. Kenya already has been a victim of Islamic terrorism, which seriously impacted its tourist industry; since 2008, ships headed into and out of the port of Mombasa have fallen prey to Somali pirates, further damaging the economy. Kenya has long been the leading state in East Africa, but its problems are daunting.

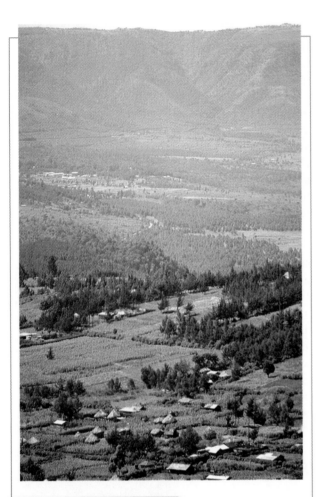

FROM THE FIELD NOTES

"Visiting the village in the Kenya Highlands where a graduate student was doing fieldwork on land reform, I took the long way and drove along the top of the eastern wall of the Eastern Rift Valley. Often the valley wall is not sheer but terraced, and soils on those terraces are quite fertile; also, the west-facing slopes tend to be well-watered. Here African farmers built villages and laid out communal plots, farming these lands in a well-organized way long before the European intrusion."
© H. J. de Blij

www.conceptcaching.com

Kenya has become one of the world's major exporters of flowers, but not without controversy. Here, workers are picking roses in a huge, 2-hectare (5-acre) greenhouse, part of an industry that currently concentrates around Lake Naivasha in the Eastern Rift Valley. European companies have set up these enterprises, buying large swaths of land and attracting workers far and wide. Locals see wildlife and waters threatened, but the government appreciates the revenues from an industry that, in 2009, accounted for about 20 percent of Kenya's agricultural export income. This is a revealing core-periphery issue: companies growing and exporting the flowers to European markets use pesticides not tolerated in Europe itself, so that the ecological damage prevented by regulations in the global core falls on the periphery, where rules are less stringent and the environment suffers. The burgeoning flower industry around Lake Naivasha also has created social problems as thousands of workers, attracted by rumored job opportunities, have converged on the area and find themselves living in squalid conditions.
© Marta Nascimento/REA.

toria in the west, posed a challenge to Kikuyu leadership. Other peoples, including the Luhya, Kalenjin, and Kamba, constitute an additional 37 percent. And still others in smaller numbers, such as the Maasai, Turkana, Boran, and Galla, live on Kenya's periphery, along with Muslim communities in the east that are not well integrated into Kenya's political fabric (Fig. 6-15).

During the increasingly autocratic rule of Kenyatta's successor, the seeds of serious political trouble were sown as corruption soared and efficiency shrank. When the time came for an election in 2007, a Kikuyu incumbent, Mwai Kibaki, faced a Luo challenger with a famous last name,

Tanzania

Tanzania (a name derived from Tanganyika plus Zanzibar) is the biggest and most populous East African country outside of Ethiopia. Its total area exceeds that of the region's four other countries combined. Tanzania has been described as a country without a core because its clusters of population and zones of productive capacity lie dispersed—mostly on its margins on the east coast (where the capital, Dar es Salaam, is located), near the shores of Lake Victoria in the northwest, near Lake Tanganyika in the far west, and near Lake Malawi in the interior south (Figs. 6-4, 6-15). This is in sharp contrast to Kenya, which has a well-defined core area in the Kenya Highlands (centered on Nairobi in the heart of the country). Moreover, Tanzania's population of 42.1 million consists of many peoples, none numerous enough to dominate the country. About 100 ethnic groups coexist; 35 percent of the population, mainly those on the coast, are Muslims. (Tanzania has the distinction of having had both a Roman Catholic and a Muslim president during its period of independence.)

In contrast to Kenya, following independence Tanzania embarked on a socialist course toward development, including a massive farm collectivization program that was imposed without adequate planning. Communist China helped Tanzania construct a railroad, the Tan-Zam Railway, from Dar es Salaam to Zambia, but the project failed, as did an effort to move the capital from colonial Dar es Salaam to Dodoma in the interior (Fig. 6-15). The country's limited tourist infrastructure was allowed to degenerate, giving Kenya virtually the entire tourist market.

But Tanzania did achieve what several other East African countries could not: political stability and a degree of democracy. Thus Tanzania changed economic direction in the late 1980s without social turmoil and embarked on a market-oriented "recovery" program. The government, aided by foreign donors and investors, encouraged commercial agriculture, but although Tanzanian coffee, cotton, tobacco, and other farm products sold increasingly well on foreign markets, the great majority of Tanzanians remained subsistence farmers and the country stayed wretchedly poor. However, its transparent government, relatively low corruption level, and effective leadership (President Mkapa and his successor, President Jakaya Kikwete, set the highest standards in the region) were rewarded with a growing and dependable stream of subsidies and investments. Today the economy is growing robustly, and the government is able to plan an expansion of commercial farming through investments in highways, irrigation, fertilizers, market mechanisms, and other needs. Meanwhile, Tanzania's tourist industry is back on track, its mining industry is expanding (gold is a major income earner, and Tanzania may possess some of the world's largest nickel reserves), and this one-time basket case of East Africa may yet rise to lead the region.

Uganda

Uganda (2010 population: 31.1 million) was the site of the most important African political entity in this region when the British arrived in the 1890s. This was the kingdom of Buganda (shown in dark brown in Fig. 6-15), which faced

FROM THE FIELD NOTES

"It was 108 in the shade, but the narrow alleys of Zanzibar's Stone City felt even hotter than that. I spent some time here when I was working on my monograph on Dar es Salaam in the 1960s and had not been back. In those days, African socialism and *uhuru* were the watchwords; since then Tanzania has not done well economically. But here were signs of a new era: a People's Bank in a former government building, and on the old fort's tower a poster saying 'Think Digital Go Tritel.' Nearby, on the sandy beach where I relaxed 35 years ago, was evidence that Zanzibar had not escaped the ravages of AIDS. Now the sand served as a refuge for the sick, who were resting there. 'It's better, bwana, than the corridor of the clinic,' said a young man who could walk only a few steps at a time with the help of a cane, and who breathed with difficulty as he spoke. Here as everywhere in Subsharan Africa, AIDS has severely strained already-limited facilities." © H. J. de Blij

www.conceptcaching.com

the north shore of Lake Victoria, had an impressive capital at Kampala, and was stable—as well as ideally suited for indirect rule over a large hinterland. The British established their headquarters at nearby Entebbe on the lake (thus adding to the status of the kingdom) and proceeded to organize their Uganda protectorate in accordance with the principles of indirect rule. The Baganda (the people of Buganda) became the dominant people in Uganda, and when the British departed, they bequeathed a complicated federal system to perpetuate Baganda supremacy.

16 **Landlocked states** tend to face particular problems, and Uganda is no exception. Dependent on Kenya for an outlet to the ocean, Uganda at independence in 1962 had better economic prospects than many other African countries. It was the largest producer of coffee in the British Commonwealth. It also exported cotton, tea, sugar, and other farm products. Copper was mined in the southwest, and an Asian immigrant population of about 75,000 dominated the country's commerce. Nevertheless, political disaster struck. Resentment over Baganda overlordship fueled revolutionary change, and a brutal dictator, Idi Amin, took control in 1971. He ousted the Asians, exterminated his opponents, and destroyed the economy. Eventually, in 1979, an invasion supported by neighboring Tanzania drove Amin from power, but by then Uganda lay in ruins. Recovery was slow, complicated by the AIDS pandemic, which struck Uganda with particular severity. President Yoweri Museveni became an international hero of the anti-AIDS campaign through his advocacy of condom use, responsible personal behavior, and medical intervention. Impressive results showed the effectiveness of this combination of countermeasures as Uganda's rate of infection fell dramatically.

In the 1990s, Uganda's interior location again afflicted its future: during a period of civil war in its northern neighbor, Sudan, Uganda was accused of supporting African rebels against their Muslim oppressors, and a costly conflict ensued. This destabilized the country's north, home of the Acholi people, and led to a second nightmare—the rise of a nominally Christian movement called the Lord's Resistance Army (LRA), which proclaimed that it wanted to rule Uganda according to the Ten Commandments. Known for unmatched brutality even in this zone of violent conflict, the LRA captured, conscripted, and enslaved children, and drove some 2 million Ugandans, most of them Acholi, into squalid refugee camps during the government's counterinsurgency campaign. Yet despite the government's best efforts in the form of military intervention combined with offers of negotiation, more than a million Acholi remained in these camps in 2009 as the LRA rebellion continued.

As if this were not enough, Uganda also found itself embroiled in the collapse of still another landlocked neighbor, Rwanda (Fig. 6-15), and in a civil war in The Congo. In

both neighbors, Uganda took the side of Tutsi minorities under threat from powerful enemies, putting further strain on its limited financial resources. Situated at one of Africa's most volatile crossroads, Uganda faces a difficult future.

Rwanda and Burundi

Rwanda and Burundi would seem to occupy Tanzania's northwest corner (Fig. 6-15), and indeed they were part of the German colonial domain conquered before World War I. But during that war Belgian forces attacked the Germans from their Congo bases and were awarded these territories when the conflict ended in 1918. The Belgians used them as labor sources for their Katanga mines.

Rwanda (10.1 million) and Burundi (9.4 million) are physiographically part of East Africa, but their cultural geography is linked to the north and west. Here, Tutsi pastoralists from the north subjugated Hutu farmers (who had themselves made serfs of the local Twa [pygmy] population), setting up a conflict that was originally ethnic but became cultural. Certain Hutu were able to advance in the Tutsi-dominated society, becoming to some extent converted to Tutsi ways, leaving subsistence farming behind, and rising in the social hierarchy. These so-called moderate Hutu were—and are—often targeted by other Hutus, who resent their position in society even as they despise the Tutsi. This longstanding discord, worsened by colonial policies, had repeatedly devastated both countries and, in the 1990s, spilled over into The Congo, generating the first interregional war in Subsaharan Africa (see pp. 322-324).

An estimated 4 million people have perished as Hutu, Tutsi, Ugandan, and Congolese rebel forces have fought for control over areas of the eastern Congo, unleashing longstanding local animosities that worsened the death toll. Only massive international intervention could stabilize the situation, but the world essentially turned a blind eye to the region's woes—again.

Highland Ethiopia

As Figure 6-15 shows, the East African region also encompasses the highland zone of Ethiopia, including the capital, Adis Abeba; the source of the Blue Nile, Lake Tana; and the Amharic core area that was the base of the empire that lost its independence only from 1935 to 1941. Ethiopia, mountain fortress of the Coptic Christians who held their own here, eventually became a colonizer itself. Its forces came down the slopes of the highlands and conquered much of the Islamic part of Africa's Horn, including present-day Eritrea and the Ogaden area, a Somali territory. (Geographically, these are parts of what we have mapped as the African Transition Zone [Fig. 6-11], which will be discussed in the final section of this chapter.)

Physiographically and culturally, highland Ethiopia is part of the East African region. But because Ethiopia was not colonized and because its natural outlets are to the Red Sea, not southward to Mombasa, effective interconnections between former British East Africa and highland Ethiopia never developed. Yet the Amhara and Oromo peoples of Ethiopia are Africans, not Arabs; nor have they been Arabized or Islamized as in northern Sudan and Somalia. The independence and secession of Eritrea in 1993 effectively landlocked Ethiopia, but for a few years there was cooperation and Ethiopia used Eritrean Red Sea ports (see Fig. 6-20). In 1998, however, a boundary dispute led to a bitter and costly war, and Ethiopia was forced to turn to Djibouti for a maritime outlet. This border conflict cost some 70,000 lives and did not end until UN mediation in 2000, when an agreement signed in Algiers gave Eritrea—in the view of Ethiopians—more than it deserved. By mid-decade, the arbitration still had not been completely accepted in Adis Abeba, but now a new problem loomed and Ethiopian troops were on (and across) the border with southern Somalia, where an Islamist militia was in the process of taking over the government. No matter how these issues turn out, Highland Ethiopia has adversaries on three sides and is likely to turn increasingly toward East Africa, however tenuous its surface links to the south may be today.

● EQUATORIAL AFRICA

The term *equatorial* is not just locational but also environmental. The equator bisects Africa, but as we noted above only the western part of central Africa features the conditions associated with the lowland tropics: intense heat, high rainfall and extreme humidity, little seasonal variation, rainforest and monsoon-forest vegetation, and enormous biodiversity. To the east, beyond the Western Rift Valley, elevations rise, and cooler, more seasonal climatic regimes prevail. As a result, we recognize two regions in these lowest latitudes: (1) Equatorial Africa to the west; and (2) just-discussed East Africa to the east.

Equatorial Africa is physiographically dominated by the gigantic Congo Basin. The Adamawa Highlands separate this region from West Africa; rising elevations and climatic change mark its southern limits (see the *Cwa* boundary in Fig. G-7). Its political geography consists of eight states, of which The Congo (officially the Democratic Republic of the Congo) is by far the largest in both territory (49 percent) and population (63 percent) (Fig. 6-16).

Five of the other seven states—Gabon, Cameroon, São Tomé and Príncipe, Congo, and Equatorial Guinea—all have coastlines on the Atlantic Ocean. The Central African Republic and Chad, the south of which is part of this region, are landlocked. In certain respects, the physical and human characteristics of Equatorial Africa extend even into southern Sudan. This vast and complex region is in many ways the most troubled in the entire Subsaharan Africa realm.

The Congo

As the map shows, The Congo has but a tiny window (37 km/23 mi) on the Atlantic Ocean, just enough to accommodate the mouth of the Congo River. Oceangoing ships can reach the port of Matadi, inland from which falls and rapids make it necessary to move goods by road or rail to the capital, Kinshasa. This is not the only place where the Congo River fails as a transport route. Follow it upstream in Figure 6-16, and you note that other transshipments are necessary between Kisangani and Ubundu, and at Kindu. Follow the railroad south from Kindu, and you reach another narrow corridor of territory at the city of Lubumbashi. That vital part of Katanga Province contains most of The Congo's major mineral resources, including copper and cobalt.

With a territory not much smaller than the United States east of the Mississippi, a population of just over 70 million, a rich and varied mineral base, and much good agricultural land, The Congo would seem to have all the ingredients needed to lead this region and, indeed, Africa. But strong centrifugal forces, arising from its physiography and cultural geography, pull The Congo apart. The immense forested heart of the basin-shaped country creates communication barriers between east and west, north and south. Many of The Congo's productive areas lie along its periphery, separated by enormous distances. These areas tend to look across the border, to one or more of The Congo's nine neighbors, for outlets, markets, and often ethnic kinship as well.

Crisis in the Interior

The Congo's civil wars of the 1990s started in one such neighbor, Rwanda, and spilled over into what was then still known as Zaïre. As noted above, Rwanda, Africa's most densely populated country, has for centuries been the scene of conflict between sedentary Hutu farmers and invading Tutsi pastoralists. Colonial borders and practices worsened the situation (see the Issue Box titled "The Impact of Colonialism on Subsaharan Africa"), and after independence a series of terrible

EQUATORIAL AFRICA
POPULATION
- Under 50,000
- 50,000–250,000
- 250,000–1,000,000
- 1,000,000–5,000,000
- Over 5,000,000

National capitals are underlined

- Area most affected by rebel activity, 2009
- Oilfields
- Potential oilfields
- Railroad
- Road
- Projected oil pipeline

0   200   400 Kilometers
0   100   200 Miles

Longitude East of Greenwich

FIGURE 6-16

© H. J. de Blij, P. O. Muller, and John Wiley & Sons, Inc.

crises followed. In the mid-1990s, the latest of these crises generated one of the largest refugee streams ever seen in the world, and the conflict engulfed eastern (and later northern and western) Congo. The death toll will never be known, but estimates in mid-2009 exceeded 5 million, a toll rivaling that of World War II. Although the United Nations Security Council authorized the largest and most expensive UN peacekeeping operation in its history to counter the crisis, only concerted action by the entire international community could break the cycle of violence. But this is Africa, not Yugoslavia or the Middle East. There is no threat to European stability. There is no oil to forfeit.

By 2005, stable government in The Congo's capital, Kinshasa, negotiations among the rebel groups and African states involved in the conflict in various ways,

For decades the easternmost areas of The Congo, especially the provinces of North Kivu and South Kivu (see Fig. 6-16), have endured recurrent violence as internal conflicts are worsened by civil strife in neighboring countries spilling across the borders. Government forces, rebel groups, militiamen and their followers, warlords, and thieves fight each other, often at the behest of outsiders, driving millions from their homes and sowing death and destruction in the countryside. This photo, taken late in 2008, shows thousands of refugees walking along a road north of the town of Goma, hoping to return to their looted homes near Kibumba, scene of a battle among militias seeking control but driven out by Congolese government forces.   © AP/Wide World Photos.

the UN peacekeeping effort and related assistance, and sheer exhaustion from endless strife had produced a semblance of security in all but a few areas of the eastern Congo, and there was hope for longer-term stability. In 2006, a democratic election in The Congo gave a second term to President Joseph Kabila, the son of the assassinated ruler whose forces had invaded and captured the capital region during the civil war, ushering in what seemed to be a new era of progress for this troubled state.

But as before, those hopes were dashed. In late 2008, a Tutsi "General" who also proclaimed himself to be a Christian pastor began a campaign of conquest in The Congo's North Kivu Province bordering Uganda and Rwanda, creating hundreds of thousands of refugees and demanding an autonomous enclave he would rule. Laying siege to the city of Goma, he put The Congo's President Kabila in a difficult position, insisting that Kabila's forces disarm Hutu militias that, he said, had attacked Tutsis in North Kivu. Meanwhile, President Paul Kagame of Rwanda was accused of supporting the "General," and in 2009 it appeared that another stage in this ongoing catastrophe was in the making. Conflict is endemic here in The Congo's east, and no end of it is in sight.

Across the River

To the west and north of the Congo and Ubangi rivers lie Equatorial Africa's other seven countries (Fig. 6-16). Two of them are landlocked. Chad, straddling the African Transition Zone as well as the regional boundary with West Africa, is one of Africa's most remote countries, although recent oil discoveries in the south are now changing this. Not only oil, but also the crisis in neighboring Darfur in Sudan is affecting Chad, with a major refugee influx and a rebel movement destabilizing this vulnerable state. The Central African Republic, chronically unstable and poverty-stricken, never was able to convert its agricultural potential and mineral resources (diamonds, uranium) into real progress. And one country consists of two small, densely forested volcanic islands: São Tomé and Príncipe, a ministate with a population of less than 250,000 whose economy is about to be transformed by recent oil discoveries.

The four coastal states present a different picture. All four possess oil reserves and share the Congo Basin's equatorial forests; oil and timber, therefore, rank prominently among their exports. In Gabon, this combination has produced Equatorial Africa's only upper-middle-income economy. Of the four coastal states, Gabon also has the largest proven mineral resources, including manganese, uranium, and iron ore. Its capital, Libreville (the only coastal capital in the region), reflects all this in its high-rise downtown, bustling port, and rapidly expanding squatter settlements.

Cameroon, less well endowed with oil or other raw materials, has the region's strongest agricultural sector by virtue of its higher-latitude location and high-relief topography. Western Cameroon is one of the more developed parts of Equatorial Africa and includes the capital, Yaoundé, and the port of Douala.

With five neighbors, Congo could be a major transit hub for this region, especially for The Congo if it recovers from civil war. Its capital, Brazzaville, lies across the Congo River from Kinshasa and is linked to the port of Pointe Noire by road and rail. But devastating power struggles have negated Congo's geographic advantages.

As Figure 6-16 shows, Equatorial Guinea consists of a rectangle of mainland territory and the island of Bioko, where the capital of Malabo is located. A former Spanish colony that remained one of Africa's least-developed territories, Equatorial Guinea, too, has been affected by the oil business in this area. Petroleum products now dominate its exports; but, as in so many other oil-rich countries, this bounty has not significantly raised incomes for most of the people.

The Impact of Colonialism on Subsaharan Africa

COLONIALISM IS THE CULPRIT!

"If you want to see and feel the way colonialism ruined Africa, you don't have to go beyond the borders of my own country, the so-called Democratic Republic of the Congo, for the evidence. I'm a school teacher here in Kikwit, east of Kinshasa, and my whole life I've lived under the dictatorship left us by the Belgians and in the chaos before and after it. All you have to do is look at the map. Who can say that the political map of Africa isn't a terrible burden? Imagine some outside power, say the Chinese, coming into Europe and throwing the Germans and the French together in one country. There would never be peace! Well, that's what they did here. All over Africa they put old enemies together and parceled up our great nations. And now they tell us to get over our animosities and 'tribalism' and live in democratic peace.

"And let me tell you something about tribalism. Before the conquest, we had ethnic groups and clans like all other peoples have, from the Scots to the Siamese. And we certainly fought with each other. But mostly we got along, and there was lots of gray area between and among us. You weren't immediately labeled a Hutu or a Luba or a Bemba. The colonial powers changed all that. Suddenly we were tribalized. Whether you liked it or not, whether you filled the 'profile' or not, you were designated a Tutsi or a Hutu, a Ganda or a Toro. This enabled the European rulers to use the strong to dominate the weak, to use the rich to tax the poor, to police migrants and assemble labor. The tribalization of Africa was a colonial invention. It worsened our divisions and destroyed our common ground. And now we just have to get over it?

"Of course the colonialists exploited our mineral resources—I tell my pupils how The Congo enriched Europeans and other 'civilized' peoples outside of Africa. But they also grabbed our best land, turning food-producing areas into commercial plantations. Are you surprised that in Zimbabwe Mugabe wants to chase those white farmers off 'their' land? I'm not, and I'm with him all the way. If the colonialists hadn't ruined traditional agriculture, there wouldn't be any food shortages today.

"Look at my country. The war between the Hutu and the Tutsi that started during colonial times has now spread into The Congo. We've got so used to taking sides in 'tribal' conflicts that much of the east is out of control. And who are the losers? The poorest of the poor, who get robbed and killed by the Mai, the Interahamwe, and the Tutsis and their friends. What are the ex-colonial powers doing now to help us with the mess they left behind? Nothing! So don't complain to me about poor African leadership and our failure to carry out economic reforms dictated by some foreign bank. The poisonous legacy of colonialism will hold Africa back for generations to come."

COLONIALISM IS A SCAPEGOAT!

"We Africans have many factors to blame for our troubles, but colonialism isn't one of them. Many African countries have been independent for nearly two generations; most Africans were born long after the colonial era ended. Are we better off today than we were at the end of the colonial period? Are those military dictatorships that ran so many countries into the ground any better than the colonizers who preceded them? Are the ghosts of colonialists past rigging our elections, stifling the media? Is our endemic corruption the fault of lingering colonialism? Did the colonial governors set up Swiss bank accounts to which to divert development funds? Can extinct colonial regimes be blamed for the environmental problems we face? Did the bandits of interior Sierra Leone or eastern areas of The Congo learn their mutilation practices from colonialism? As a nurse in a Uganda hospital for AIDS victims, I say that it's time to stop blaming ancient history for our current failures.

"It's not that we Africans aren't at a disadvantage in this unforgiving world. Our postcolonial population explosion was followed by an HIV-induced implosion. Apartheid South Africa spread its malign influence far beyond its borders. The Cold War pitted regimes against rebels and, to use a colonial term, tribe against tribe. Droughts and other plagues of biblical proportions ravaged our continent when we were more vulnerable than ever. Prices for our raw materials on international markets fell when we most needed a boost. Foreign governments protected their farmers against competition from African producers. Those are legitimate complaints. But colonialism?

"Forget about blaming boundaries, white farmers, tribalism, religion, and other residues of colonialism for our present troubles. Look at what happened here in Uganda. A decade ago we were at the heart of the 'AIDS Belt' in tropical Africa, with one of the worst infection rates on the continent. Then came President Museveni, whose government launched a vigorous self-help campaign, educating people to the risks, urging restraint, distributing condoms, encouraging women to resist unwanted sex. Now Uganda's AIDS incidence is dropping, and we are held up as an example for all of Africa to follow. While South Africa's President Mbeki reveals his confusion about the causes of AIDS, Zimbabwe's Mugabe is too busy chasing white farmers, and Kenya's Moi is trying to rig his succession, we in Uganda have leadership in an area where we need it desperately. That certainly beats blaming colonialism for everything.

"We have an African Union. If those colonial boundaries are so bad, let's change them, or do what the European Union is doing and begin to erase them. But first, let's isolate tyrants, nurture open democracy, and combat corruption—and stop blaming colonialism."

Regional ISSUE

Vote your opinion at www.wiley.com/college/deblij

One other territory would seem to be a part of Equatorial Africa: *Cabinda*, wedged between the pair of Congos just to the north of the Congo River's mouth. But Cabinda is one of those colonial legacies on the African map—it belonged to the Portuguese and was administered as part of Angola. Today it is an **exclave** of independent Angola, and a valuable one: it contains major oil reserves.

● WEST AFRICA

West Africa occupies most of Africa's Bulge, extending south from the margins of the Sahara to the Gulf of Guinea coast and from Lake Chad west to Senegal (Fig. 6-17). Politically, the broadest definition of this region includes all those states that lie to the south of Western

Sahara, Algeria, and Libya and to the west of Chad (itself sometimes included) and Cameroon. Within West Africa, a rough division is sometimes made between the large, mostly steppe and desert states that extend across the southern Sahara (Chad could also be included here) and the smaller, better-watered coastal states.

Apart from once-Portuguese Guinea-Bissau and long-independent Liberia, West Africa comprises four former British and nine former French dependencies. The British-influenced countries (Nigeria, Ghana, Sierra Leone, and Gambia) lie separated from one another, whereas Francophone West Africa is contiguous. As Figure 6-17 shows, political boundaries extend from the coast into the interior, so that from Mauritania to Nigeria, the West African habitat is parceled out among parallel, coast-oriented states. Across these boundaries, especially across those between former British and former French territories,

FIGURE 6-17

© H. J. de Blij, P. O. Muller, and John Wiley & Sons, Inc.

there is only limited interaction. For example, in terms of value, Nigeria's trade with Britain is about 100 times as great as its trade with nearby Ghana. The countries of West Africa are not interdependent economically, and their incomes are largely derived from the sale of their products on the non-African international market.

Given these cross-currents of subdivision within West Africa, why are we justified in speaking of a single West African region? First, this part of the realm has remarkable cultural and historical momentum. The colonial interlude failed to extinguish West African vitality, expressed not only by the old states and empires of the savanna and the cities of the forest, but also by the vigor and entrepreneurship, the achievements in sculpture, music, and dance, of peoples from Senegal to Nigeria's southeastern Iboland. Second, West Africa contains a set of parallel east-west ecological belts, clearly reflected in Figures G-6 and G-7, whose role in the development of the region is pervasive. As the transport-route pattern on the map of West Africa indicates, overland connections within each of these belts, from country to country, are poor; no coastal or interior railroad ever connected this tier of countries. Yet spatial interaction is stronger across these belts, and some north-south economic exchange does take place, notably in the coastal consumption of meat from cattle raised in the northern savannas. And third, West Africa received an early and crucial imprint from European colonialism, which—with its maritime commerce and slave trade—transformed the region from one end to the other. This impact reached into the heart of the Sahara and set the stage for the reorientation of the whole area, from which emerged the present patchwork of states.

Despite the effects of the slave trade, West Africa today is Subsaharan Africa's most populous region (Fig. 6-4). In these terms, Nigeria (whose census results are in doubt, but with an estimated population of 156 million) is Africa's largest state; Ghana (25.0 million) and Ivory Coast (21.7 million) rank prominently as well. The southern half of the region, understandably, is home to most of the people. Mauritania, Mali, and Niger include too much of the unproductive Sahel's steppe and the arid Sahara to sustain populations comparable to those of Nigeria, Ghana, or Ivory Coast.

The peoples along the coast reflect the modern era that the colonial powers introduced: they prospered in their newfound roles as middlemen in the coastward trade. Later, they experienced the changes of the colonial period; in education, religion, urbanization, agriculture, politics, health, and many other endeavors, they adopted new ways. In contrast, the peoples of the interior retained their ties with a different era in African history. Distant and aloof from the main theater of European colonial activity and often drawn into the Islamic orbit, they experienced a significantly different kind of change. But the map reminds us that Africa's boundaries were not drawn to accommodate such contrasts. Both Nigeria and Ghana possess population clusters representing the interior as well as the coastal peoples, and in both countries the wide cultural gap between north and south has produced political problems.

Nigeria: West Africa's Cornerstone

Nigeria, the region's cornerstone, is home to more than 155 million people, by far the largest population total of any African country. When Nigeria achieved full

Ghana's coastal town of Elmina may be too small to show up on our regional map, but this was the site of the first European settlement in West Africa. It was a key node in Portuguese maritime trade and later became notorious during the slave trade era. Today it is a fishing port of about 20,000, but the industry is not doing well. The local fishers must compete with the larger foreign trawlers that harvest the waters off the coast of West Africa, and cold-storage facilities remain minimal, so that almost all of the local catch is smoked and sold on the local market. This keeps prices low, so that Elmina's economy has changed very little even after the government in 1983 introduced economy-boosting policies. Talk to the locals (fishing and related work are the livelihoods of over half the population), and you realize that it was the capital of Accra that benefited most from the new economic rules, not places like this.
© Richard J. Grant.

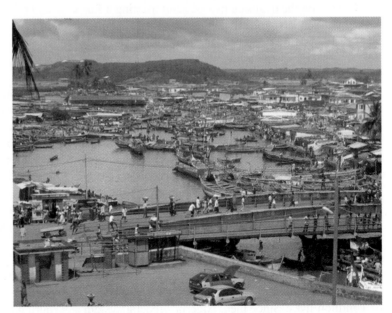

AMONG THE REALM'S GREAT CITIES . . . Lagos

IN A REALM that is only 35 percent urbanized, Lagos, former capital of federal Nigeria, is the exception: a teeming megacity of 10.6 million, sometimes called the Calcutta of Africa.

Lagos evolved over the past three centuries from a Yoruba fishing village, Portuguese slaving center, and British colonial headquarters into Nigeria's largest city, major port, leading industrial center, and first capital. Situated on the country's southwestern coast, it consists of a group of low-lying islands and sand spits between the swampy shoreline and Lagos Lagoon. The center of the city still lies on Lagos Island, where the high-rises adjoining the Marina overlook Lagos Harbor and, across the water, Apapa Wharf and the Apapa industrial area. The city expanded southeastward onto Ikoyi Island and Victoria Island, but after the 1970s most urban sprawl took place to the north, on the western side of Lagos Lagoon.

Lagos's cityscape is a mixture of modern high-rises, dilapidated residential areas, and squalid slums. From the top of a high-rise one sees a seemingly endless vista of rusting corrugated roofs, the houses built of cement or mud in irregular blocks separated by narrow alleys. On the outskirts lie the shantytowns of the less fortunate, where shelters are made of plywood and cardboard and lack even the most basic facilities.

By world standards, Lagos ranks among the most severely polluted, congested, and disorderly cities. Mismanagement and official corruption are endemic. Laws, rules, and regulations, from zoning to traffic, are flouted. The international airport is notorious for its inadequate security and for extortion by immigration and customs officers. In many ways, Lagos is a metropolis out of control.

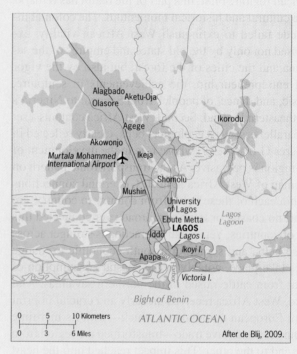

After de Blij, 2009.

© H. J. de Blij, P. O. Muller, and John Wiley & Sons, Inc.

independence from Britain in 1960, its new government was faced with the daunting task of administering a European political creation containing three major nations and nearly 250 other peoples ranging from several million to a few thousand in number.

For reasons obvious from the map, Britain's colonial imprint always was stronger in the south than in the north. Christianity became the dominant faith in the south, and later southerners, especially the Yoruba, took a lead role in the transition from colony to independent state. The choice of Lagos, the port of the Yoruba-dominated southwest, as the capital of a federal Nigeria (and not one of the cities in the more populous north) reflected British desires for the country's future. A three-region federation, two of which lay in the south, would ensure the primacy of the non-Islamic part of the state. But this framework did not last long. In 1967, the Ibo-dominated Eastern Region declared its independence as the Republic of Biafra, leading to a three-year civil war at a cost of 1 million lives. Since then, Nigeria's federal system has been modified repeatedly; today there are 36 States, and the capital has been moved from Lagos to centrally located Abuja (Fig. 6-18).

Fateful Oil

Large oilfields were discovered beneath the Niger Delta during the 1950s, when Nigeria's agricultural sector produced most of its exports (peanuts, palm oil, cocoa, cotton) and farming still had priority in national and State development plans. Soon, revenues from oil production dwarfed all other sources, bringing the country a brief period of prosperity and promise. But before long Nigeria's oil wealth brought more bust than boom. Misguided development plans now focused on grand, ill-founded industrial schemes and costly luxuries such as a national airline; the continuing mainstay of the vast majority of Nigerians, agriculture, fell into neglect. Worse, poor management, corruption, outright theft of oil revenues during military misrule, and excessive borrowing against

THE STATES OF FEDERAL NIGERIA

	POPULATION
Road	Under 50,000
Railroad	50,000–250,000
Pipeline	250,000–1,000,000
Oilfield	1,000,000–5,000,000
Sharia law proclaimed	Over 5,000,000
Core area	National capitals are underlined
	TIV Major ethnic group

FIGURE 6-18

© H. J. de Blij, P. O. Muller, and John Wiley & Sons, Inc.

future oil income led to economic disaster. The country's infrastructure collapsed. In the cities, basic services broke down. In the rural areas, clinics, schools, water supplies, and roads to markets crumbled. In the Niger Delta area, local people beneath whose land the oil was being exploited demanded a share of the revenues and reparations for ecological damage; the military regime under General Abacha responded by arresting and executing nine of their leaders. By the time the ruthless Abacha was succeeded by democratically elected President Olusegun Obasanjo, the situation in the Niger Delta had generated an armed resistance group named the Movement for the Emancipation of the Niger Delta

(MEND), which carried out attacks on pipelines, kidnapped oil-company workers, and committed other acts of sabotage. The delta has become one of the most dangerous oil-producing areas in the world for locals and workers alike.

On global indices of national well-being, Nigeria during the 1990s sank to the lowest rungs even as its oil production ranked it tenth in the world in that category, with the United States its chief customer. After 2000 some improvement occurred, but mismanagement and corruption slowed all progress, and so did rapid population growth (which had eased up by 2007, but Nigeria remains a fast-growing country). Take a careful look at the data in the

To the people living in Nigeria's "middle belt," the Islamic Front (p. 333) is more than a geographic concept: it is like living in an earthquake zone where things can fall apart in a moment. In Figure 6-18, you can see Plateau State bordering Islamic-law Kaduna and Bauchi States, with a land extension northward separating these two. In that land extension is where Jos, the State capital, is situated, and in that city Muslim and Christian tensions are never far from the surface. In late November 2008 local elections led to rumors that the mainly Christian Peoples Democratic Party had won over the mainly Muslim All Nigerian Peoples Party. Riots broke out, and in a few days hundreds died and much property destruction took place. Mosques and churches, as well as more than 3000 shops in one market alone, were destroyed before the government could restore order. This photo, taken in early December, shows the Nigerian military separating an advancing group of protesters from their targets by barricading a road in central Jos while local elders try to calm the people. It took the national army and State law enforcement several more days to control the violence, but it will not be the last time such strife will erupt along Nigeria's Islamic Front.   © AFP/Getty Images, Inc.

table at the end of the book and you will see that this African as well as regional cornerstone state has a long way to go.

Islam Ascendant

Nigeria faces problems beyond the economic and demographic spheres. The free election of a president in 1999 raised the people's hopes, but trouble arose in another context: its historic cultural schism. Soon after the return to democracy, federal Nigeria's northern States, beginning with Zamfara, decided to proclaim *Sharia* (strict Islamic) law. When Kaduna State followed suit, riots between Christians and Muslims devastated the old capital city of Kaduna. There, and in 11 other northern States (Fig. 6-18), the imposition of Sharia law led to the departure of thousands of Christians, intensifying the cultural fault line that threatens the cohesion of the country.

While the national government worked to ease the tensions, Muslim-Christian conflict in the northern cities is never far from the surface. In 2008 in the city of Jos, where Christians and Muslims had retreated to separate neighborhoods after earlier strife, local elections between parties that had split along religious lines produced violent clashes in which rampaging gangs killed nearly 1000 people (photo at left). As Figure 6-18 shows, Jos is not located in a Sharia-proclaiming State, but lies wedged between two States that did—Kaduna and Bauchi. Here as elsewhere the Muslim-Christian divide does not correspond to State boundaries, and Nigerians refer to this transition zone as their country's *Middle Belt*. Here, too, the devolutionary forces arising from Nigeria's cultural diversity and relative location are strongest, threatening the cohesion of Africa's most populous country, the cornerstone of West Africa.

Coast and Interior

Nigeria is one of 17 countries (counting Chad and offshore Cape Verde [not shown in Fig. 6-17]) that constitute the region of West Africa. Four of these states, comprising a huge territory on the Sahara's margins but containing small populations, are landlocked: Mali, Burkina Faso, Niger, and Chad. Figure G-7 shows clearly how steppe and desert conditions dominate the natural environments of these four interior states. Figure 6-4 reveals the concentration of population in the steppe zone and along the ribbon of water provided by the Niger River. Scattered oases form the remaining settlements and anchor regional trade. But even the coastal states do not escape the dominance of the desert over West Africa. Mauritania's environment is almost entirely desert. Senegal, as Figure 6-17 shows, is mostly a semiarid Sahel country; and not only northern Nigeria but also northern Benin, Togo, and Ghana have interior steppe zones. The loss of pastures to **desertification** (human-induced desert expansion) is a constant worry for the livestock herders there.

West Africa's states share the effects of the environmental zonation depicted in Figures G-6 and G-7, but they also exhibit distinct regional geographies. Benin, Nigeria's neighbor with a population of 9.9 million, has a growing cultural and economic link with the Brazilian State of Bahía, where many of its people were taken in bondage and where elements of West African culture have survived.

Ghana, once known as the Gold Coast, was the first West African state to achieve independence (1957), with a sound economy based on cocoa exports. Two grandiose post-independence schemes can be seen on

Even in the poorer countries of the world, you see something that has become a phenomenon of globalization: gated communities. Widening wealth differences, security concerns, and real estate markets in societies formerly characterized by traditional forms of land ownership combined to produce this new element in the cultural landscape. This is the entrance to Golden Gate, the first private gated community development in Accra, Ghana, started in 1993 as a joint venture between a Texas-based construction company and a Ghanaian industrial partner. © Richard J. Grant.

its map: the port of Tema, which was to serve a vast West African hinterland, and Lake Volta, which resulted from the region's largest dam project. When neither fulfilled expectations, Ghana's economy collapsed. But in the 1990s, Ghana's military regime was replaced by democratic and stable government in the capital, Accra, and over the past decade Ghana has become a model for Africa. Even ethnic strife in the northern region in 2002 did not disrupt the country's progress, and its agriculture-dominated export economy, based on cocoa, received an unexpected boost when chaos descended on its chief West African competitor, neighboring Ivory Coast. In 2006, Ghana authorized the world's largest mining conglomerate to begin exploiting gold reserves northwest of Kumasi, the historic Ashanti capital, under strict compensatory and environmental conditions that set an example to the rest of the continent and, indeed, the world.

In 2007 Ghana celebrated 50 years of independence, its democracy maturing, its economy robust, corruption persistent but declining, and its international stature rising. The same year also brought news of a major discovery of oil reserves off Ghana's coast, and caused a surge

of optimism. This came only a few months after (former) President Kufuor had signed a nearly $550 million aid package from the U.S. Millennium Challenge Corporation to expand commercial agriculture, further improve infrastructure, and combat poverty, a grant made in recognition of Ghana's achievements that, in late 2008, included a democratic, closely contested, and peaceful presidential election.

Ivory Coast (which, as we have stated, officially still goes by its French name, *Côte d'Ivoire*) translated three decades of autocratic but stable rule into economic progress that gave it lower-middle-income status based mainly on cocoa and coffee sales. Continued French involvement in the country's affairs contributed to this prosperity; the capital, Abidjan, reflected its comparative well-being. But then Ivory Coast's president-for-life engineered the transfer of the capital to his home village, Yamoussoukro, spending tens of millions of dollars to build a Roman Catholic basilica there to rival that of St. Peter's in Rome. It was dedicated just as the country's economy was slowing and its social conditions worsened.

By the late 1990s, Ivory Coast's economy had reverted to the low-income category (Fig. G-11). Worse, regional strife intensified during the run-up to a presidential election in 2000, when southern politicians objected to the ethnic origins and alleged Muslim sympathies of a northern candidate. In September 2002, a series of coordinated attacks by northern rebels failed to take the largest city and former capital, Abidjan, but succeeded in capturing major northern centers. The attacks cost hundreds of lives and confirmed the worst fears of citizens in this long-stable and comparatively prosperous country, where immigrants have long contributed importantly to the economy and society.

At the turn of this century, the population included about 2 million Burkinabes (immigrants from Burkina Faso to the north), as many as 2 million Nigerians, about one million Malians, half a million Senegalese, and smaller groups of Ghanaians, Guineans, and Liberians. This cultural mix had been kept stable by strong central government and a growing economy, but when both of these mainstays failed, disorder followed. In 2004 French troops, sent to Ivory Coast to stabilize the situation and to facilitate negotiations as well as protect the French expatriate community, suffered a deadly attack by the Ivoirian Air Force. This resulted in retaliatory action and led to assaults on French citizens in Abidjan and elsewhere. Most of the expatriate community departed, leaving Ivory Coast, its cocoa-based economy devastated, in dire straits, a West African success story with a tragic turn.

All of West Africa has been affected by what has happened in Liberia and Sierra Leone. Liberia, a country founded in 1822 by freed slaves who returned to Africa with the help of American colonization societies, was ruled by their descendants for more than six generations. Rubber plantations and iron mines made life comfortable for the "Americo-Liberians," but among the local peoples, resentment simmered. A military coup in 1980 was followed in 1989 by full-scale civil war that pitted ethnic groups against each other and drove hundreds of thousands of refugees into neighboring countries, including Ivory Coast, Guinea, and Sierra Leone. Monrovia, the capital (named after U.S. president James Monroe), was devastated; an estimated 230,000 people, almost 10 percent of the population, perished. In 1997, one of the rebel leaders, Charles Taylor, became president, but he was indicted by a UN war-crimes tribunal in 2003 just as he was negotiating an end to the civil war with his opponents. Nigerian forces and UN peacekeeping troops entered Monrovia in 2005 and a semblance of stability returned to the country. Later that year, Liberia became the first African state to elect a female president when Ellen Johnson Sirleaf won the first post-civil war contest for that office.

Among the countries in the West African region most severely affected by Liberia's turmoil was Sierra Leone, its coastal neighbor to the northwest. Like Liberia, Sierra Leone also was founded as a haven for freed slaves, in this case by the British in 1787. Independent since 1961, Sierra Leone went the all-too-familiar route from self-governing Commonwealth member to republic to one-party state to military dictatorship. But in the 1990s a civil war brought untold horror to this small country even as refugees from Liberia's war were arriving. Rebels enabled by "blood" diamond sales fought supporters of the legitimate government in a struggle that devastated town and countryside alike and killed and mutilated tens of thousands. Eventually a combination of West African, United Nations, and British forces intervened in the conflict and resurrected a semblance of representative government, but not before Sierra Leone had sunk to dead last on the world's list of national well-being.

As Figure 6-17 shows, Guinea borders all three of the troubled countries just identified, and it has involved itself in the affairs of all of them—receiving in return a stream of refugees in its border areas. Guinea, with a population of 10.9 million, has been under dictatorial rule ever since its founding president, Sekou Touré, turned from father of the nation to usurper of the people. As Figure G-11 shows, Guinea is one of Africa's poorer states, but representative government and less corrupt administration could have made it a far better place.

Guinea has significant gold deposits, may possess as much as one-third of the world's bauxite (aluminum ore) reserves, and can produce far more coffee and cotton than it does; offshore lie productive fishing grounds. As the map shows, this is no ministate like neighboring Guinea-Bissau, the former Portguese colony. Guinea's capital, Conakry, has a substantial hinterland reaching well into West Africa's interior, and, as a result, the country has a wide range of environments.

As Figure 6-17 shows, Senegal has a wedge-shaped territory situated between forested Guinea to the south and desert Mauritania to the north. Senegal's environmental problems notwithstanding, this country is far ahead of its neighbors in almost every respect. Its great advantage, a legacy from colonial times, is the city of Dakar, headquarters for France's West African empire and now the oversized capital of a state with 13.4 million people, more than half of them farmers. Dakar anchors an unusually well-defined core area, but its surface connections with a much larger West African hinterland are not as good as its large commercial and fishing port would suggest. Senegal's economy remains largely agricultural, but the export sector now depends mostly on fish, fertilizers, converted petroleum products, and some iron ore. Peanuts, once a big revenue earner, now figure less prominently.

In many ways Senegal's most valuable asset is its tradition of representative government. In 2010 this country could not only celebrate 50 years of independence (from France) but also more than four decades of democracy. Although Senegal is 94 percent Muslim, its cultural mosaic is varied, with the Wolof, concentrated in and around the capital, constituting nearly half the population, the Fulani farmers of the interior about one-quarter, and the Serer about one-seventh. In general, Senegal's leaders have maintained close relationships with France, which remains the country's chief financial supporter in times of need. Without significant oil reserves, diamonds, or other riches, and with a mainly subsistence farming population, Senegal is still a low-income economy but with GNI levels that exceed the region's average (see the Data Table inside the back cover). Here is evidence that reasonably representative government and stability are greater assets than gold, liquid or otherwise.

A closer look at Figure 6-17 reveals that Senegal's territory surrounds that of another state, the English-speaking enclave of Gambia. Efforts to achieve unification have failed, and Senegal has even faced a secession movement in the Casamance District lying south of Gambia. Yet none of this has threatened a state whose strong cultural traditions and steady democracy are a model for Francophone West Africa.

Saharan Shadows

The huge West African bulge is environmentally dominated by the Sahara, the great desert whose dust is blown by *Harmattan* "winter" winds across the entire region and out to sea. Under nature's regime, the Sahara expands and contracts over time, its southern margin or *Sahel* (an Arabic word meaning "border") shifting north and south. Figure 6-17 shows the Sahel in an average position, but people and their animals are increasingly affecting its configuration. Herds of livestock occupy the desert-adjoining steppes when they turn temporarily green, leaving the vegetation trampled and the thin soil disturbed when drought returns. The desert advances and the people retreat into the moister savannas to the south, invading farmlands and confronting settled inhabitants. It is an ancient cycle intensified by population growth and political partitioning, and it challenges virtually every state in this part of Africa.

Much of West Africa's countryside retains its traditional African forms of agriculture and herding which, as noted earlier, are skillful adaptations to environmental limitations and social constraints. Visit villages south of the Sahel, and you may find that there is not enough commerce to justify daily markets everywhere. This has **19** led to an economic tradition called the **periodic market**, involving a system whereby particular village markets are not open every day, but only every third or fourth day or in some other rotation adjusted to factors of product supply, distance, and population distribution. This system ensures that all villages of a certain size participate in the exchange network. Cultural traditions of this kind endure throughout West Africa, even as the towns and cities beckon the farmers and burst at the seams.

Look to the north of the Sahel, and even the vast Sahara is compartmentalized into states, from Mauritania in the west to Chad in the east. As we will see in the next chapter, Islam dominates in these sparsely peopled desert countries, but here too population pressures and environmental issues intensify. Within the West African region and beyond, the great challenges are economic survival and nation-building, complicated by a boundary framework that is as burdensome as any in this realm.

● THE AFRICAN TRANSITION ZONE

From our discussion of East, Equatorial, and West Africa, it is obvious that the northern margin of the realm we define as Subsaharan Africa is in turmoil. Along a zone from Ethiopia in the east to Guinea in the west, cultural and ethnic tensions tend to erupt into conflict, and often the conflict spreads to engulf large areas across international borders. It is also clear that much of this turmoil has to do with the religious transition from Islamic and Arabized Africa to the Africa where Christianity and traditional beliefs prevail. This zone, called the *African Transition Zone*, epitomizes the way one geographic realm yields to another. But in terms of its instability, variability, and volatility, this particular transition between realms is unique.

It is useful to look again at Figure 6-11, which shows that some entire countries lie within the African Transition Zone, such as Senegal, Guinea, Burkina Faso, Eritrea, and Somalia. Others are bisected by it, including Ivory Coast, Nigeria, Chad, Sudan, and Ethiopia. Africa's colonial boundary framework was laid out even as Islam was on the march and paid little heed to the consequences. But it is Islam that defines the southern periphery of the African Transition Zone along a religious frontier sometimes referred to as the **Islamic Front** (Fig. 6-19). **20**

As we noted earlier in this chapter, Islam arrived across the Sahara by caravan and up the Nile by boat, thus reaching and converting the peoples of the interior steppes and savannas on the southern side of the great desert. Then came Europe's colonial powers, and the modern politico-geographical map of Subsaharan Africa took shape. Muslims and non-Muslims were thrown together in countries not of their making. Vigorous Christian proselytism slowed the march of Islam, resulting in the distribution shown in Figure 7-3. The north-to-south transition is especially clear in the heart of the continent, where Libya is 96 percent Muslim, Chad 57, the Central African Republic 15, and The Congo 9.

Conflict is the hallmark of much of the Islamic Front across Africa. In Sudan, a 30-year war between the Arab, Islamic north and the African, Christian-animist south, costing hundreds of thousands of lives and displacing millions, appeared to be finally settled in late 2004. Even so, a new and deathly conflict had arisen in Sudan's far-western Darfur Province (see p. 367) even as the southern war ended. In Ivory Coast, livestock owned by Muslim cattle herders from Burkina Faso had been trampling African farms for years before the north-south schism broke into open conflict in 2002. In Nigeria, Islamic revivalism in the North is clouding the entire country's prospects. But perhaps the most conflict-prone part of the African Transition Zone lies in Africa's *Horn*, involving the historic Christian state of Ethiopia and its neighbors (Fig. 6-20).

Earlier we noted that Highland Ethiopia where the country's core area, capital (Adis Abeba), and Christian heartland lie, is virtually encircled by dominantly Muslim societies. One poor-quality road, along which

FIGURE 6-19

© H. J. de Blij, P. O. Muller, and John Wiley & Sons, Inc.

bandits rob bus passengers and law enforcement is virtually nonexistent, forms the only direct surface link between Adis Abeba and Nairobi, the capital of Kenya to the south (Fig. 6-20). But, as Figure 6-19 reminds us, Ethiopia's eastern Ogaden is traditional Somali country and almost entirely Islamic, a vestige of a time when the rulers in Adis Abeba extended their power from their highland fortress over the plains below. Today, 34 percent of Ethiopia's population of 83.1 million is Muslim, and the Islamic Front is quite sharply defined (Fig. 6-19).

With the realm's second-largest population, Ethiopia is an important African cornerstone state, but its economy remains weak, its politics unrepresentative, and its relations with neighbors adversarial. Landlocked and

fragmented by the African Transition Zone, Ethiopia forfeited international support when its disputed parliamentary election in 2005 was followed by the imprisonment of opposition leaders and journalists. The World Bank, for example, withheld budgetary subsidies while human-rights organizations criticized Ethiopian actions in its southern Oromo and eastern Somali areas. In 2007, an attack by the Ogaden National Liberation Front (a terrorist organization) on an oil rig in eastern Ethiopia killed 68 Ethiopians and 9 Chinese. Ethiopian forces, with U.S. approval, tried to intervene in the chaotic situation prevailing in southern Somalia, but the campaign proved beyond their capabilities.

Ethiopia's diverse environments, spectacular scenery, and archeological fame sustain an economy that depends

HORN OF AFRICA
POPULATION

- Under 50,000
- 50,000–250,000
- 250,000–1,000,000
- 1,000,000–5,000,000

National capitals are underlined

── Railroad   ── Road

OROMO Major ethnic group

0   200   400 Kilometers
0   100   200 Miles

FIGURE 6-20

© H. J. de Blij, P. O. Muller, and John Wiley & Sons, Inc.

primarily on farm products (coffee, tea, and spices) and tourism. But the overwhelming majority of Ethiopian citizens are subsistence farmers, and despite its historic abundance the country now requires food imports from relief agencies.

As Figure 6-20 shows, Ethiopia is landlocked not only by Somalia but also by the critically-positioned ministate of Djibouti and Ethiopia's former federal partner, Eritrea.

Once an Italian colony, Eritrea was under British administration during World War II and was made part of an Ethiopian "Federation" in 1952. This was one of those in-name-only federations, and when Ethiopia formally annexed Eritrea in 1962, Eritreans—Christian as well as Muslim—took up arms to fight a war of secession. By 1993 they had achieved their goal when an independent state of Eritrea gained international recognition.

The new state had the effect of landlocking Ethiopia, which had made Assab its key Red Sea port while it controlled Eritrea. As the Muslim sector of Eritrea's population of 5.3 million grew faster than the Christian sector, relations between Ethiopia and Eritrea worsened, and in 1998 the two countries launched a seemingly senseless war over their common border. This war flared up repeatedly, cost billions of dollars, and required foreign mediation to resolve, but a negotiated agreement reached in 2000 was not enough to stop it altogether. By 2009, more than 100,000 soldiers on both sides had died in old-style combat over a strip of rocky land and a few abandoned villages. But by then, Eritrea had become a magnet for Islamic militants eager to join a campaign against Christian Ethiopia. As millions went hungry and the once-promising Eritrean economy collapsed, it was a reminder that the Islamic Front is far more than a line on a map.

In 2002, a ministate in this part of the African Transition Zone, Djibouti, came to international attention as a result of the so-called War on Terror declared by the United States. Formerly a French colony and still host to a French military base, coastal Djibouti, ethnically divided between ethnic Afars and Issas but politically stable, proved to have a favorable relative location between troubled Somalia, vulnerable Eritrea, and crucial Ethiopia—and directly across the narrow Bab el Mandeb Strait (which separates the Red Sea and the Gulf of Adan) from Yemen, the ancestral home of Usama bin Laden (Fig. 6-20).

But the key component in the eastern sector of the African Transition Zone is Somalia, where 9.5 million people, virtually all Muslim, live at the mercy of a desert-dominated climate that enforces cross-border migration into Ethiopia's Ogaden area in pursuit of seasonal pastures. As many as three to five million Somalis live permanently on the Ethiopian side of the border, but this is not the only division the Somali "nation" faces. In reality, the Somali people is an assemblage of five major ethnic groups fragmented into hundreds of clans engaged in an endless contest for power as well as survival.

Early in this new century's first decade, Somalia's condition as a **failed state** led to the country's fragmentation into three parts (Fig. 6-20). In the north, the sector called *Somaliland*, which had proclaimed its independence in the 1990s and which remains by far the most stable of the three, functioned essentially as an African state, although the "international community" will not recognize it as such. In the east, a conclave of local chiefs declared their territory to be separate from the rest of Somalia, called it *Puntland*, and asserted an unspecified degree of autonomy. In the south, where the official capital, Mogadishu, is located on the Indian Ocean coast, local secular warlords and Islamic militias continued their struggle for supremacy, the warlords supported by U.S. funding. In 2006, Islamic militias stormed into the capital and took control, ousting the warlords and proclaiming their determination to create an Islamic state.

21

As noted above, Ethiopian forces intervened with the unheralded support of the United States. They were joined by an African Union peacekeeping force, but before long both came under continuous and increasing attack from Somali militias. Hundreds of thousands of refugees streamed out of Mogadishu, where the fighting was worst, and many others left their villages in the embattled zone. By 2008, the UN was estimating that more than a million Somalis were stranded between the Kenya border (where many were turned back) and the sea, and starvation was widespread.

Meanwhile, Indian Ocean waters off Somalia became an arena of action for a growing number of pirates based along the country's lengthy coastline and with increasing capabilities. Piracy had long been occurring here, but most of the incidents involved smaller boats and did not make the news. But then Somali pirates began to target larger ships; they captured a Ukrainian freighter loaded with tanks and other weapons, a

The waters off the Horn of Africa have become the scene of modern-day piracy as vessels ranging from oil tankers to cruise ships are being attacked by teams of armed criminals using all kinds of small boats to stalk their prey. The Gulf of Adan (Aden), flanked by failed-state Somalia and terrorist-base Yemen, is especially dangerous as ships converge on the choke-point entrance to the Red Sea. Several countries, including the United States, have deployed warships to combat the pirates, but the problem persists. This photo was taken in May 2009 as members of a boarding team from the cruiser *USS Gettysburg* and U.S. Coast Guard law enforcement officers approached a suspected pirate boat in the Gulf. © AP/Wide World Photos.

300,000-ton oil tanker, and even tried to board cruise ships. Their sphere of activity expanded from the Gulf of Adan to waters south of the latitude of Kenya's port of Mombasa, and their impact was global. Warships from Russia, India, and the United States were among vessels trying to secure the vast maritime region off Somalia, but the "industry" could not easily be closed down. The combination of a failed state in Africa's Horn, Islamic fundamentalism driven by local militants, al-Qaeda in the mix, and pirates operating off the coast makes this segment of the African Transition Zone a focus of international concern.

As this chapter has underscored, Africa is a continent of infinite social diversity, and Subsaharan Africa is a realm of matchless cultural history. Comparatively recent environmental change widened the Sahara and separated north from south; the arrival of Islam turned the north to Mecca and left the south to Christian proselytism. Millions of Africans were carried away in bondage; the cruelties of colonialism subjugated those who were left behind in a political straitjacket forged in Berlin. Africa's equatorial heart is an incubator for debilitating diseases second to none on Earth, but Africa's health problems never were a top priority in the medical world. When the richer countries finally abandoned their African empires, African peoples were left a task of reconstruction in which they not only got little help: they found their products uncompetitive on markets where rich-country producers basked in subsidies. We said at the beginning of this chapter that Africa's time and turn will come again, but it will take far more than is being done today to rectify half a millennium's malefactions.

What You Can Do

RECOMMENDATION: Write an op-ed piece! This is not merely a matter of writing a longer letter to the editor. An op-ed (opinion-editorial) essay is more challenging because you cannot assume the reader's familiarity with the topic. Therefore, you will want to write an introductory paragraph setting the stage and explaining what issue you plan to address, leading into a concisely argued position on the matter citing the necessary evidence, followed by a logical and judicious conclusion. Give yourself 1000 to 1200 words, and make every word count! It's great practice, and if the newspaper or magazine editor accepts your article, it will have much more impact than a letter to the editor. And, unlike a letter, you are able to title your op-ed piece. For example, you might try one under the headline Give African Farmers a Break! or another asking What Has Globalization Done for Africa? And here's a switch: readers are likely to write letters to the editor responding to your argument. They may even ask the editor to invite you to write again on another topic.

GEOGRAPHIC CONNECTIONS

1 Tourism is one of Africa's important industries, in some countries contributing substantially to the government's income and creating jobs where unemployment is high. But if you were a member of an organized tour group, you would soon discover that visitors to Subsaharan Africa tend to be more interested in the realm's remaining wildlife than in the people and their cultures. How would you go about adding some human interest to the itinerary? What aspects of the cultural geography of Subsaharan Africa would you like to see and why? In what ways are villages and small towns usually more evocative of regional cultures than large cities?

2 You have just been appointed to a key U.S.-government commission charged with the following responsibility: sufficient funds are available to give substantial assistance to just four countries in Subsaharan Africa, and the government hopes that the beneficial effects of this assistance will radiate outward into neighboring states. The first task of the commission of which you are a member is to identify the four countries. If you had this decision to make, which four African states would you nominate, and what criteria, including geographic factors, would you employ to arrive at that selection?

Geographic Literature on Subsaharan Africa: The key introductory works on this realm were authored by Aryeety-Attoh, Bohannon & Curtin, Cole & de Blij, de Blij, Grove, and Stock. These works, together with a comprehensive listing of noteworthy books and recent articles on the geography of Africa south of the Sahara, can be found in the *References and Further Readings* section of this book's website at **www.wiley.com/college/deblij.**

NORTH AFRICA/ SOUTHWEST ASIA

ATLANTIC OCEAN

ATLANTIC OCEAN

Mediterranean Sea

North Sea

Baltic Sea

Adriatic Sea

Tyrrhenian Sea

Ionian Sea

Bay of Biscay

Gulf of Sidra

Gulf of Guinea

Bight of Benin

Bight of Biafra

NORWAY SWEDEN

Shetland Islands (U.K.)

Bergen Oslo Stockholm Uppsala
Stavanger Göteborg Gotland (SWEDEN)
Orkney Islands (U.K.)
DENMARK Aalborg Malmö Kalininingrad
Aarhus COPENHAGEN Odense Gdansk Bydgoszcz
Glasgow Edinburgh Newcastle Bremen HAMBURG Szczecin **POLAND**
British Isles Leeds Hannover Berlin Poznan Wroclaw
IRELAND Belfast Liverpool Manchester Amsterdam Essen Leipzig Dresden
Dublin Birmingham UNITED The Hague Cologne GERMANY PRAGUE Krakow
Cork KINGDOM Antwerp BRUSSELS BELG. Frankfurt CZECH REP. Brno
Plymouth LONDON Rotterdam NETH. LUX. Stuttgart Munich SLOVAKIA
English Channel Le Havre Luxembourg LIECH. VIENNA BRATISLAVA
Channel Islands (U.K.) Rouen PARIS Strasbourg Zürich AUS. BUDAPEST
Brest Tours Bern SWITZ. Innsbruck Graz HUNGARY
FRANCE Geneva Lyon MILAN Venice Ljubljana Zagreb
Nantes Turin Genoa Rijeka SLOV. CROATIA BOSNIA SERBIA
La Coruña Gijón Bordeaux MASSIF Nice Bologna Sarajevo BELG.
Vigo CENTRAL Marseille MONACO Florence Podgorica Pristina
Porto Bilbao Pico de Aneto Toulouse SAN MARINO MONTENEGRO KOS. Skop.
Valladolid (3,404 m, Corsica (FRANCE) ROME Bari MACED.
PORTUGAL CANTABRIAN MOUNTAINS 11,168 ft.) Elba Tirane ALBANIA
Lisbon MADRID PYRENEES ANDORRA Sardinia Naples GREECE
IBERIAN BARCELONA (ITALY) ITALY Athen
SPAIN Zaragoza Balearic Sea
PENINSULA Valencia Palma de Majorca Cagliari Palermo Crete
Seville Murcia Balearic Is. Sicily (GREECE)
(SPAIN) ALGIERS Valletta
Madeira Is. Funchal Strait of Gibraltar Tangier Gibraltar (U.K.) Oran Mostaganem Bejaia Annaba Tunis MALTA
(PORTUGAL) RABAT Alboran Sea Setif Constantine Sousse
Salvage Is. CASABLANCA Fez Oujda Tlemcen Batna TUNISIA Qairouan
(PORTUGAL) Safi Taza Djelfa Biskra Tebessa Sfax
Canary Islands Khouribga Beni Mellal Laghouat Gafsa Gabes TRIPOLI Derna
(SPAIN) Marrakech Ouargla Zuwarah CYRENA
La Palma Santa Cruz de Tenerife Jebel Toubkal Bechar Ghardaia Gulf of Sidra Benghazi Ajdabiya
Hierro (4,165 m, 13,665 ft.) Ghadames Sidra
Tenerife Gran Agadir GREAT ERG OCCIDENTAL TADEMAIT TRIPOLITANIA
Canaria Las Lanzarote PLATEAU GREAT ERG ORIENTAL
Palmas Fuerteventura Tan Tan Adrar HAMADA DE TINRHERT
El Aaiún In-Salah ALGERIA LIBYA
Smara Tindouf I-n-Salah
Western Sahara ERGIGUIDI HAMADA DE TINRHERT
Ad Dakhla ERG CHECH Sabha
Fderik TANEZROUFT TASSILI-N-AJJER FEZZAN
Nouadhibou EL DJOUF S Al Birkah MARZUQ
Cape Blanc Akjoujt Atar A DESERT Al Birkah
Cape Verde Is. Nouakchott MAURITANIA H Mt. Greboun TIBESTI
Santo Mindelo Tidjikja (1,944 m, 6,378 ft.) AIR TENERE MTS.
Antão Nicolau CAPE VERDE Aleg A Arlit Mt. Koussi
São Boa Vista Kiffa R (3,415 m, 11,204 ft.)
Fogo Praia Saint Louis Nema Bilma BORKOU Ouianga Kebir
Santiago Rosso Tombouctou (Timbuktu) Agadez ENNEDI Faya-Largeau
Thies Aleg Kayes MALI Gao S NIGER
DAKAR Diourbel Nioro Tahoua
Cape Verde SENEGAL Kaolack Niger Segou Zinder CHAD
GAMBIA Banjul Nara BURKINA FASO Niamey Dosso Sokoto Maradi Biltine
Ziguinchor GUINEA- Koutiala Ouahigouya Diffa N'Djamena Mao Ati Abéché
Bissau BISSAU Bamako Sikasso Koudougou Katsina KANO Lake Chad Kousséri MARRA
Bafata Kankan Bobo OUAGADOUGOU Dori Kaduna Zaria Bauchi DAR
Boke Labe Siguiri Dioulasso Korhogo Parakou Minna Jos Maiduguri CHAINE DES
Kindia GUINEA FUTA-JALLON Tamale NIGERIA Abuja Birao BONGOS
CONAKRY HIGHLANDS Bouake IVORY GHANA Sokode BENIN Ilorin Oshogbo Makurdi N'Djamena Mokolo CENTRAL AFRICAN
SIERRA Man COAST Kumasi Lake Volta TOGO Abeokuta IBADAN BENIN CITY Enugu Garoua Maroua REPUBLIC Bria
LEONE Yamoussoukro Koforidua LOMÉ LAGOS ADAMAWA Moundou Bambari
Freetown Bo Kenema Divo ABIDJAN ACCRA Porto Novo PORT Umuahia Aba HIGHLANDS Doba Bangassou
Monrovia Buchanan Nzerekore LIBERIA San Pedro Sekondi Cotonou HARCOURT CAMEROON Ngaoundéré Sarh Ndélé CHAINE DES
River Cess Harper Cape Takoradi Cape Coast Bight of Benin Benue YAOUNDÉ Bertoua Bangui
Cape Palmas Three Points Malabo Bioko DOUALA Bafoussam Bouar
EQUATORIAL Bata Berberati
GUINEA SÃO TOMÉ & PRINCIPE Príncipe Libreville GABON Oyem Bumba
São Tomé São Tomé Port-Gentil Ouesso THE
Annobón (EQ. GUINEA) Lambaréné CONGO
Cape Lopez Mbandaka
Equator Kaulamoutou Impfondo (D
Francevile Mbandaka
Tropic of Cancer
Longitude West of Greenwich Longitude East of Greenwich

7

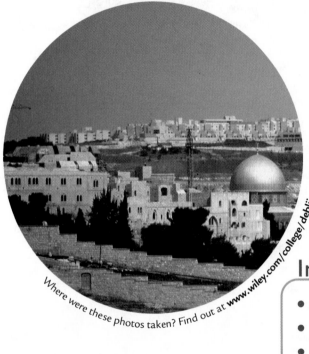

Where were these photos taken? Find out at www.wiley.com/college/deblij

In This Chapter

- How Islam transformed a realm
- Where the oil is—and is not
- Foreign intervention, Muslim reaction
- Political theory and practical reality in Iraq
- The predicament of Israel
- Afghanistan and the Taliban

Photos: © H. J. de Blij

FIGURE 7-1 *Map:* © H. J. de Blij, P. O. Muller, and John Wiley & Sons, Inc.

FROM MOROCCO ON the shores of the Atlantic to the mountains of Afghanistan, and from the Horn of Africa to the steppes of inner Asia, lies a vast geographic realm of enormous cultural complexity. It stands at the crossroads where Europe, Asia, and Africa meet, and it is part of all three (Fig. 7-1). Throughout history, its influences have radiated to these continents and to practically every other part of the world as well. This is one of humankind's primary source areas. On the Mesopotamian Plain between the Tigris and Euphrates rivers (in modern-day Iraq) and on the banks of the Egyptian Nile arose several of the world's earliest civilizations. In its soils, plants were domesticated that are now grown from the Americas to Australia. Along its paths walked prophets whose religious teachings are still followed by billions. And at the opening of the second decade of the twenty-first century, the heart of this realm is beset by some of the most bitter and dangerous conflicts on Earth.

Defining the Realm

So sprawling, vast, and diverse is this geographic realm that it is difficult just to assign a name to it. We identify it as North Africa/Southwest Asia, for which some geographers use the shorthand *NASWA*. Whatever it is called, there is no doubt that it lies at the heart of the inhabited world, and contains the geographic center of the Land Hemisphere. This *centrality* has historic consequences: peoples have converged on this realm, crossed it, left it, reentered it. In the process it acquired a cultural tapestry of endless variety, and when the political boundaries of the world were drawn, this geographic realm found itself with more neighbors than any other.

Given the diverse geographies of North Africa/Southwest Asia, it is not surprising that this realm is sometimes characterized in a few words that purport to summarize one or more of its dominant features. None of these is satisfactory, but all hold a kernel of truth. Four of them follow.

A "Dry World"?

Compare Figures 7-1 and G-7, and it appears that NASWA's expanse matches almost exactly that of the largest concentration of *BW* and *BS* climates, containing as it does the Sahara as well as the Arabian Desert. But add Figure 7-2 to the picture, and it is clear that most of the realm's nearly 600 million people live where there is surface water—in the Nile Valley and Delta, along

MAJOR GEOGRAPHIC QUALITIES OF

North Africa/Southwest Asia

1. North Africa and Southwest Asia were the scene of several of the world's great ancient civilizations, based in its river valleys and basins.

2. From this realm's culture hearths diffused ideas, innovations, and technologies that changed the world.

3. The North Africa/Southwest Asia realm is the source of three world religions: Judaism, Christianity, and Islam.

4. Islam, the last of the major religions to arise in this realm, transformed, unified, and energized a vast domain extending from Europe to Southeast Asia and from Russia to East Africa.

5. Drought and unreliable precipitation dominate natural environments in this realm. Population clusters exist where water supply is adequate to marginal.

6. Certain countries of this realm have enormous reserves of oil and natural gas, creating great wealth for some but doing little to raise the living standards of the majority.

7. The boundaries of the North Africa/Southwest Asia realm consist of volatile transition zones in several places in Africa and Asia.

8. Conflict over water sources and supplies is a constant threat in this realm, where population growth rates are high by world standards.

9. The Middle East, as a region, lies at the heart of this realm; and Israel lies at the center of the Middle East conflict.

10. Religious, ethnic, and cultural discord frequently cause instability and strife in this realm.

FIGURE 7-2

© H. J. de Blij, P. O. Muller, and John Wiley & Sons, Inc.

northwestern Africa's hilly Mediterranean coast (the *tell*, meaning "mound" in Arabic), along the eastern and northeastern shores of the Mediterranean Sea, in the basin of the Tigris and Euphrates rivers, in the uplands of western Iran and eastern Turkestan, and in far-flung oases throughout the realm. It is true that, in this realm, water is almost always at a premium (it is, for example, a major issue between the Israelis, who use far more of it per person, and their neighbors, the Palestinians), and that millions of peasants struggle to make soil and mois-ture yield a small harvest. In one of humanity's ancient adaptations, nomadic peoples and their animals still cir-culate across dust-blown flatlands, stopping at oases that form islands of farming and trade in the vast drylands, but they are now a minuscule—and dwindling—minority. Ancient as well as modern technologies have enabled nearly 10 percent of the planet's population to defy these environmental limitations. Dams, canals, and wells gather and bring water from rivers, snowy moun-taintops, and aquifers (natural underground reservoirs) to

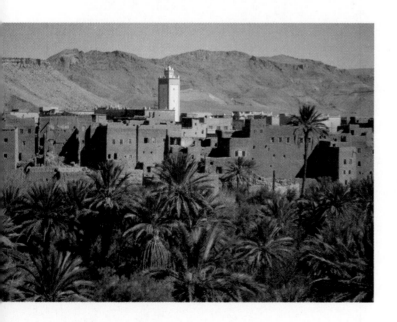

Compare Figures G-7 and 7-10, and you will see why this realm is often referred to as the "dry world." Desert (**BW**) and steppe (**BS**) dominate landscape and life here, and water is com-paratively scarce, but people manage to make the most of a resource locals will tell you is more valuable than oil. Irrigation schemes, lengthy canals, wells and wind-driven pumps, desalinization plants, qanats (see p. 389), and oases sustain hundreds of millions in cities, towns, villages, and farms. The realm's mountains yield water not only by wresting it from the moisture in the air but also by releasing it from underground aquifers that reach the surface in valleys such as the one shown here, the Todra Valley on the Saharan side of the High Atlas in Morocco, where a stand of palm trees in the barren countryside signifies the presence of the essence of life. People have clustered in this remote settlement named Tinerhir for centuries, building flat-roofed, mud-brick dwellings, wearing clothing to beat the heat, accommodating travelers in need of replenishment, their lives and occupations made possible by a trickle of water on the very margin of the Sahara. © Jose Fuste Raga/Corbis Images.

hundreds of millions of people on the land as well as in the great cities. And in some countries desalinization plants paid for with oil money are making life in the "dry world" possible in ways unimagined just decades ago.

The "Middle East"?

This realm is also frequently called the Middle East. That must sound odd to someone in, say, India, who might think of a Middle West rather than a Middle East! The name, of course, reflects the biases of its source: the "Western" world, which saw a "Near" East in Turkey, a "Middle" East in Egypt, Arabia, and Iraq, and a "Far" East in China and Japan. Still, the term has taken hold, and it can be seen and heard in everyday usage by scholars, journalists, and members of the United Nations. Even so, it should be applied to only one of the regions of this vast realm, not to the realm as a whole. It is likely that the entire realm's centrality, referred to earlier, contributes to this misuse: look again at Figure G-13 and you will note that not only does NASWA lie at the heart of the inhabited world, but the *Middle East* region lies at the core of NASWA.

An "Arab World"?

North Africa/Southwest Asia is often referred to by another appellation: the Arab World. This term implies a uniformity that does not actually exist. First, the name *Arab* is applied loosely to the peoples of this area who speak Arabic and related languages, but ethnologists normally restrict it to certain occupants of the Arabian Peninsula—the Arab "source." In any case, the Turks are not Arabs, and neither are most Iranians or Israelis. Moreover, although versions of the Arabic language prevail from Mauritania in the west across all of North Africa to the Arabian Peninsula, Syria, and Iraq in the east, it is not spoken in other parts of this realm (see Fig. G-10). In Turkey, for example, Turkish is the major language, and it has Ural-Altaic rather than Arabic's Semitic or Hamitic roots. The Iranian language belongs to the Indo-European linguistic family. Other "Arab World" languages that have separate ethnological identities are spoken by the Jews of Israel, the Tuareg people of the Sahara, the Berbers of northwestern Africa, and the peoples of the transition zone between North Africa and Subsaharan Africa to the south.

An "Islamic World"?

Finally, and perhaps most significantly, this realm is routinely referred to as the Islamic World. Indeed, the prophet Muhammad was born on the Arabian Peninsula in AD 571, and Islam, the religion to which he gave rise, spread across this realm and beyond after his death in 632. This great saga of Arab conquest saw Islamic armies, caravans, and fleets carry Muhammad's teachings across deserts, mountains, and oceans, penetrating Europe, converting the kings of West African states, threatening the Christian strongholds of Ethiopia, invading Central Asia, conquering most of what is now India, and even reaching the peninsulas and islands of Southeast Asia. Today, the world's largest Muslim state is Indonesia, and the Islamic faith extends far outside the realm under discussion (Fig. 7-3). In this context, the term *Islamic World* is also misleading in that it suggests that there is no Islam beyond NASWA's borders.

In any case, this World of Islam is not entirely Muslim either. Christian minorities continue to survive in all the regions and many of the countries of the North Africa/Southwest Asia realm. Judaism has its base in the Middle East region; and smaller religious communities, such as Lebanon's Druze and Iran's Bahais, many of them under pressure from Islamic regimes, diversify the religious mosaic.

Nonetheless, Islam's impact on this realm's cultural geography is pervasive. It ranges from dominant in such countries as Iran, Saudi Arabia, and Sudan to laminate (that is, more of a veneer) in several of the countries of Turkestan, where the former Soviet rulers discouraged the faith and where its comeback is still in progress—one of the criteria that allow us to identify Turkestan as a discrete region within this realm. The rise and impact of Islam lie at the heart of the story of this chapter.

States and Nations

Although Islam and its cultural expressions dominate this geographic realm, its political and social geographies are divided and often fractious. We will encounter states with strong internal divisions (Lebanon, Iraq), nations as yet without their own states (Palestinians, Kurds, Berbers), and territories still in the process of integration (Western Sahara, Palestine). No single state dominates this realm as Mexico eclipses the rest of Middle America or Brazil looms over South America. Of its nearly 30 states and territories, the three largest and in many ways the most important are Egypt in North Africa, Turkey on the threshold of Europe, and Iran at the margins of Turkestan. All three have populations between 70 and 80 million, modest by world standards. But at the other end of the continuum there are ministates such as Bahrain, Qatar, and Djibouti, each with fewer than one million inhabitants. It is important to study Figure 7-1 attentively, because the boundary framework—much of it inherited from the colonial era—lies at the root of many of the realm's current problems.

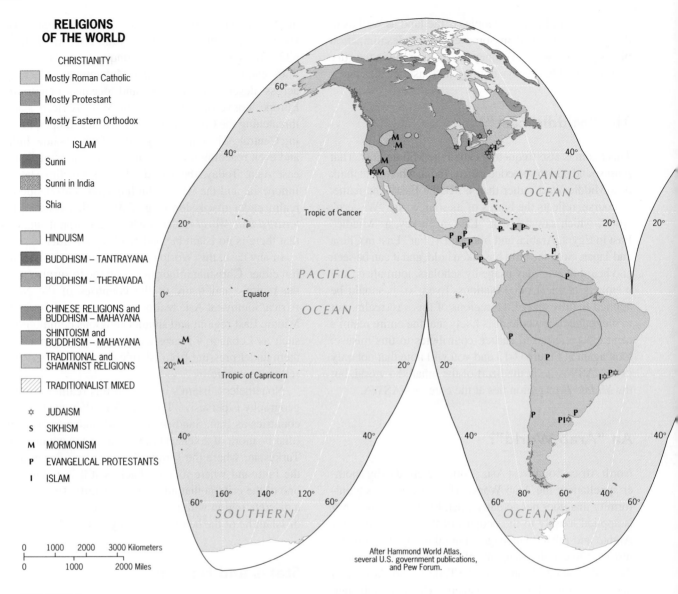

RELIGIONS OF THE WORLD

CHRISTIANITY
- Mostly Roman Catholic
- Mostly Protestant
- Mostly Eastern Orthodox

ISLAM
- Sunni
- Sunni in India
- Shia

- HINDUISM
- BUDDHISM – TANTRAYANA
- BUDDHISM – THERAVADA
- CHINESE RELIGIONS and BUDDHISM – MAHAYANA
- SHINTOISM and BUDDHISM – MAHAYANA
- TRADITIONAL and SHAMANIST RELIGIONS
- TRADITIONALIST MIXED

- ☼ JUDAISM
- S SIKHISM
- M MORMONISM
- P EVANGELICAL PROTESTANTS
- I ISLAM

0   1000   2000   3000 Kilometers
0   1000   2000 Miles

After Hammond World Atlas,
several U.S. government publications,
and Pew Forum.

FIGURE 7-3

It is also useful to take a careful look at Figure 7-2, because it clearly shows how differently populations are distributed in the countries we will examine. Compare the three largest countries just mentioned: Egypt's population forms a river-valley ribbon and delta cluster; Turkey's, between the Black Sea and the Mediterranean Sea, is far more evenly spread, although its west is more populous than the east; and in Iran you can draw a line down its middle from the Caspian Sea to the Persian Gulf that separates the comparatively crowded west, where the capital is located, from the drier and far more sparsely peopled east. In North Africa, almost everyone seems to live along or near the Mediterranean coast, but no such coastal concentrations mark the Arabian Peninsula except in its far south, where Yemen's people cluster near the Red Sea. Certainly the historic role of water can be seen in the population patterns shown in Figure 7-2.

HEARTHS OF CULTURE

This geographic realm occupies a pivotal part of the world: here Eurasia, crucible of human cultures, meets Africa, source of humanity itself. Two million years ago, the ancestors of our species walked from East Africa into North Africa and Arabia and spread from one end of Asia to the other. Less than one hundred thousand years ago, *Homo sapiens* crossed these lands on the way to Europe, Australia, and, eventually, the Americas. Ten thousand years ago, human communities in what we now call the Middle East began to domesticate plants and animals, learned to irrigate their fields, enlarged their settlements into towns, and formed the earliest states. One thousand years ago, the heart of the realm was stirred and mobilized by the teachings of Muhammad and the Quran (Koran), and Islam was on the march from North Africa

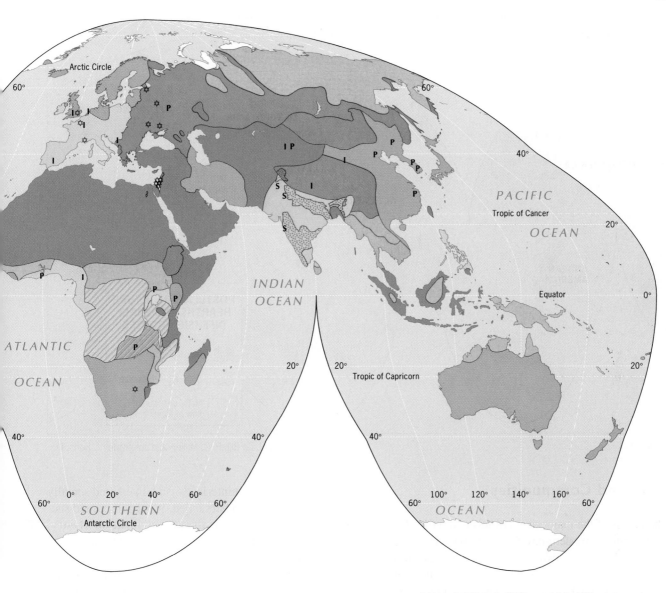

© H. J. de Blij, P. O. Muller, and John Wiley & Sons, Inc.

to India. Today this realm is a cauldron of religious and political activity, weakened by conflict but empowered by oil, plagued by poverty but fired by a wave of religious fundamentalism (Muslims prefer the term *religious revivalism*).

Dimensions of Culture

In the Introduction (pp. 20–23) we discussed the concept of culture and its regional expression in the cultural landscape. **Cultural geography**, we noted, is a wide-ranging and comprehensive field that studies spatial aspects of human cultures, focusing not only on cultural landscapes but also on **culture hearths**—the crucibles of civilization, the sources of ideas, innovations, and ideologies that changed regions and realms. Those ideas and innovations spread far and wide through a set of processes that we study under the rubric of **cultural diffusion**. Because we understand these processes better today, we can reconstruct ancient routes by which the knowledge and achievements of culture hearths spread (that is, *diffused*) to other areas.

Another topic of cultural geography, also relevant in the context of the NASWA realm, is the **cultural environment** that a dominant culture creates. Human cultures exist in long-term accommodation with (and adaptation to) their natural environments, exploiting opportunities that these environments present and coping with the extremes they can impose. The study of the relationship between human societies and natural environments has become a separate branch of cultural geography called **cultural ecology**. As we will see, the North Africa/Southwest Asia realm presents many opportunities to investigate cultural geography in regional settings.

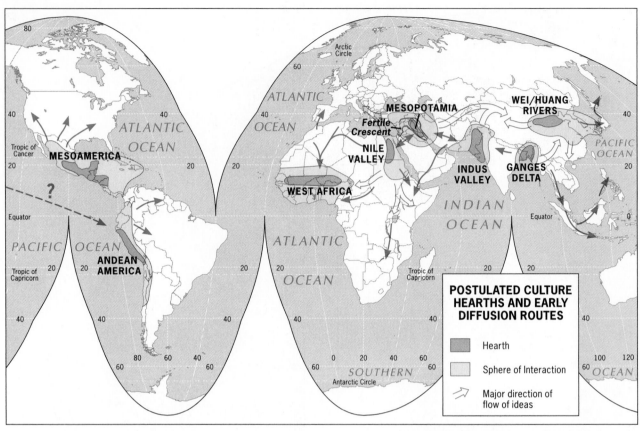

FIGURE 7-4

© H. J. de Blij, P. O. Muller, and John Wiley & Sons, Inc.

Rivers and Communities

In the basins of the major rivers of this realm (the Tigris and Euphrates of modern-day Turkey, Syria, and Iraq; and the Nile of Egypt) lay two of the world's earliest culture hearths (Fig. 7-4). *Mesopotamia*, "land amidst the rivers," had fertile alluvial soils, abundant sunshine, ample water, and animals and plants that could be domesticated. Here, in the Tigris-Euphrates lowland between the Persian Gulf and the uplands of present-day Turkey, arose one of humanity's first culture hearths, a cluster of communities that grew into larger societies and, eventually, into the world's first states. (Early state development probably was going on simultaneously in East Asia's river basins as well.) Mesopotamians were innovative farmers who knew when to sow and harvest crops, water their fields, and store their surplus. Their knowledge diffused to villages near and far, and a *Fertile Crescent* evolved extending from Mesopotamia across southern Turkey into Syria and the Mediterranean coast beyond (Fig. 7-4).

Mesopotamia

Irrigation was the key to prosperity and power in Mesopotamia, and urbanization was its reward. Among many settlements in the Fertile Crescent, some thrived,

grew, enlarged their hinterlands, and diversified socially and occupationally; others failed. What determined success? One theory, the **hydraulic civilization theory**, holds that cities that could control irrigated farming over large hinterlands held power over others, used food as a weapon, and thrived. One such city, Babylon on the Euphrates River, endured for nearly 4000 years (from 4100 BC). A busy port, its walled and fortified center endowed with temples, towers, and palaces, Babylon for a time was the world's largest city.

Egypt and the Nile

Egypt's cultural evolution may have started even earlier than Mesopotamia's, and its focus lay upstream from (south of) the Nile Delta and downstream from (north of) the first of the Nile's series of rapids, or cataracts (Fig. 7-4). This part of the Nile Valley is surrounded by inhospitable desert, and unlike Mesopotamia (which lay open to all comers), the Nile provided a natural fortress here. The ancient Egyptians converted their security into progress. The Nile was their highway of trade and interaction; it also supported agriculture through irrigation. The Nile's cyclical ebb and flow was much more predictable than that of the Tigris-Euphrates river system, adding to the certainties that stabilized life in its valley.

By the time Egypt finally fell victim to outside invaders (about 1700 BC), a full-scale urban civilization had emerged. Ancient Egypt's artist-engineers left a magnificent legacy of massive stone monuments, some of them containing treasure-filled crypts of god-kings called pharaohs. These tombs have enabled archeologists to reconstruct the ancient history of this culture hearth.

The Indus Valley

To the east, separated from Mesopotamia by more than 1900 kilometers (1200 mi) of mountain and desert, lies the Indus Valley (Fig. 7-4). By modern criteria, this eastern hearth lies outside the realm under discussion here; but in ancient times it had cultural and commercial ties with the Tigris-Euphrates region. Mesopotamian innovations reached the Indus region early, and eventually the cities of the Indus became power centers of a civilization that extended far into present-day northern India.

Today, the world continues to benefit from the accomplishments of the ancient Mesopotamians and Egyptians. They domesticated cereals (wheat, rye, barley), vegetables (peas, beans), fruits (grapes, apples, peaches), and many animals (horses, pigs, sheep). They also advanced the study of the calendar, mathematics, astronomy, government, engineering, metallurgy, and a host of other skills and technologies. In time, many of their innovations were adopted and then modified by other cultures in the Old World and eventually in the New World as well. Europe was the greatest beneficiary of these legacies of Mesopotamia and ancient Egypt, whose achievements constituted the foundations of "Western" civilization.

Decline and Decay

As Figure G-7 reminds us, many of the early cities of this culture realm lay in what is today desert territory. It is unlikely that our distant ancestors built large settlements in the middle of deserts, so it is reasonable to hypothesize that **climate change** overtook many of these ancient cities. This leads to the further hypothesis that the populations of some cities, confronting changing climatic conditions, were better able than others to cope with them. This suggests that knowledge of irrigation was not the only advantage that some ancient settlements had over others.

Indeed, climate change, which entails a shift in ecological zones in times of environmental transition, was happening quite rapidly during the most recent recession of Pleistocene ice we referred to in the Introduction (see page 12). In only a few thousand years, entire regions that were dry turned moist, but others that had been well-watered lost their indispensable lease on life. This may have destroyed many of the old city-centered states, some

suddenly and disastrously, others more slowly but inevitably. Thus climate change altered the map and made the "Fertile Crescent" history.

If this is what occurred, it is likely that overpopulation became a problem as the drying land could support ever-fewer people; moreover, destruction of natural vegetation reduced its carrying capacity for livestock. The question is: how did those early societies with rudimentary technologies (by today's standards) respond? Some cultural geographers suggest that the momentous innovations in agricultural planning and irrigation technology of the time were not "taught" by the seasonal flooding of rivers but were *forced* on the inhabitants as they faced this threat to their survival.

As has been true throughout human history, some societies found themselves favored over others. To the least fortunate, the encroaching desert was a fateful enemy: they abandoned their fields and gathered in the towns, then abandoned the towns and migrated—if they were able—to areas reputed to be better off. There, the migrants were not always welcome, stoking conflict of a kind that continues to this very day, as inhabitants of Africa's Sahel know all too well.

As old societies disintegrated, power emerged elsewhere. First the Persians, then the Greeks, and later the Romans imposed their imperial designs on the tenuous lands and disconnected peoples of North Africa/Southwest Asia. Roman technicians converted North Africa's farmlands into irrigated plantations whose products went by the boatload to Roman Mediterranean shores. Thousands of people were carried off as slaves to the cities of the new conquerors. Egypt was quickly colonized, as was the area we now call the Middle East. One region that lay distant, and therefore remote from these invasions, was the Arabian Peninsula, where no major culture hearth or large cities had emerged and where the turmoil had not affected Arab settlements and nomadic routes.

STAGE FOR ISLAM

In a remote place on the Arabian Peninsula, where the foreign invasions of the Middle East had had little effect on the Arab communities, an event occurred early in the seventh century that was to change history and affect the destinies of people in many parts of the world. In a town called Mecca (Makkah), about 70 kilometers (45 mi) from the Red Sea coast in the Jabal Mountains, a man named Muhammad in the year AD 611 began to receive revelations from Allah (God). Muhammad (571–632) was then in his early forties and had barely 20 years to live. Convinced after some initial self-doubt that he was indeed chosen to be a prophet, Muhammad committed his life to fulfilling the divine commands he believed he had received. Arab society was in social and cultural disarray, but Muhammad forcefully taught Allah's lessons and began to transform his culture. His personal power soon attracted enemies, and in 622 he fled from Mecca to the safer haven of Medina (Al Madinah), where he continued his work. This moment, the *hejira* ("migration"), marks the starting date of the Muslim era, Year 1 on Islam's calendar. Mecca, of course, later became Islam's holiest place.

The Faith

The precepts of Islam in many ways constituted a revision and embellishment of Judaic and Christian beliefs and traditions. All of these faiths have but one god, who occasionally communicates with humankind through prophets; Islam acknowledges that Moses and Jesus were such prophets but considers Muhammad to be the final and greatest prophet. What is Earthly and worldly is profane; only Allah is pure. Allah's will is absolute; Allah is omnipotent and omniscient. All humans live in a world that Allah created for their use but only to await a final judgment day.

Islam brought to the Arab World not only the unifying religious faith it had lacked but also a new set of values, a new way of life, a new individual and collective dignity. Islam dictated observance of the Five Pillars: (1) repeated expressions of the basic creed, (2) the daily prayer, (3) one month each year of daytime fasting (Ramadan), (4) the giving of alms, and (5) at least one pilgrimage in each Muslim's lifetime to Mecca (the *hajj*). And Islam prescribed and proscribed in other spheres of life as well. It forbade alcohol, smoking, and gambling. It tolerated polygamy, although it acknowledged the virtues of monogamy. Mosques appeared in Arab settlements, not only for the (Friday) sabbath prayer, but also as social gathering places to knit communities closer together. Mecca became the spiritual center for a divided, widely dispersed people for whom a collective focus was something new.

The Arab-Islamic Empire

Muhammad provided such a powerful stimulus that Arab society was mobilized almost overnight. The prophet died in 632, but his faith and fame spread like wildfire. Arab armies carrying the banner of Islam formed, invaded, conquered, and converted wherever they went. As Figure 7-5 shows, by AD 700 Islam had reached far into North Africa, into Transcaucasia, and into most of Southwest Asia. In the centuries that followed, it penetrated

southern and eastern Europe, Central Asia's Turkestan, West Africa, East Africa, and South and Southeast Asia, even reaching China by AD 1000.

Routes of Diffusion

The spread of Islam provides a good illustration of a series of processes called **spatial diffusion**, focusing on **8** the way ideas, inventions, and cultural practices propagate through a population in space and time. In 1952, Torsten Hägerstrand, a Swedish geographer, published a fundamental study of spatial diffusion entitled *The Propagation of Diffusion Waves*. He reported that diffusion takes place in two forms: **expansion diffusion**, when **9** propagation waves originate in a strong and durable source area and spread outward, affecting an ever larger region and population; and **relocation diffusion**, in **10** which an innovation, idea, or (for example) a virus is carried by migrants from the source to distant locations and diffuses from there. The recent global spread of AIDS is a case of relocation diffusion.

Islam on the March

Both expansion and relocation diffusion include several types of processes. The spread of Islam, as Figure 7-5 shows, initially proceeded by a form of expansion diffusion called **contagious diffusion** as the faith moved from **11** village to village across the Arabian Peninsula and the Middle East. But Islam got powerful boosts when kings, chiefs, and other high officials were converted, who in turn propagated the faith downward through their bureaucracies to far-flung subjects. This form of expansion diffusion, called **hierarchical diffusion**, served Islam well. **12**

But the map leaves no doubt that Islam later spread by relocation diffusion as well, notably to the Ganges Delta in South Asia, to present-day Indonesia, and to East Africa. That process continues to this day, but the heart of Islam remains in Southwest Asia. There, Islam became the cornerstone of an Arab Empire with Medina as its first capital. As the empire grew by expansion diffusion, its headquarters were relocated from Medina to Damascus (in present-day Syria) and later to Baghdad on the Tigris River (Iraq). And it prospered. In architecture, mathematics, and science, the Arabs overshadowed their European contemporaries. The Arabs established institutions of higher learning in many cities including Baghdad, Cairo, and Toledo (Spain), and their distinctive cultural landscapes united their vast domain (see box titled "The Flowering of Islamic Culture"). Non-Arab societies in the path of the Muslim drive were not only Islamized, but also Arabized, adopting other Arab traditions as well. Islam had spawned a culture; it still lies at the heart of that culture today.

The Flowering of Islamic Culture

THE CONVERSION OF a vast realm to Islam was not only a religious conquest: it was also accompanied by a glorious explosion of Arab culture. In science, the arts, architecture, and other fields, Arab society far outshone European society. While the Western European remnants of the Roman Empire languished, Arab energies soared.

When the wave of Islamic diffusion reached the Maghreb (western North Africa), the Arabs saw, on the other side of the narrow Strait of Gibraltar, an Iberia ripe for conquest and ready for renewal. An Arab-Berber alliance, called the Moors, invaded Spain in 711 and controlled all but northern Castile and Catalonia before the end of the eighth century.

It took seven centuries for Catholic armies to recapture all of the Arabs' Iberian holdings, but by then the Muslims had made an indelible imprint on the Spanish and Portuguese cultural landscape. The Arabs brought unity and imposed the rule of Baghdad, and their works soon overshadowed what the Romans had wrought.

Al-Andalus, as this westernmost outpost of Islam was called, was endowed with thousands of magnificent castles, mosques, schools, gardens, and public buildings. The ultimately victorious Christians destroyed most of the less durable art (pottery, textiles, furniture, sculpture) and burned the contents of Islamic libraries, but the great Islamic structures survived, including the Alhambra in Granada, the Giralda in Seville, and the Great Mosque of Córdoba, three of the world's greatest architectural achievements. While Spanish and Portuguese culture became Hispanic-Islamic culture, the Muslims were transforming their cities from Turkestan to the Maghreb in the image of Baghdad. The lost greatness of a past era still graces those townscapes today.

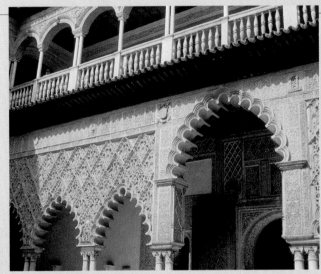

FROM THE FIELD NOTES

"Seville was a Muslim capital of Iberia, and walking the streets of the old city center is an adventure in cultural and historical geography. Seville fell to the Muslim invaders in 711, and the Muslims built a large mosque in the heart of town. Later the Christians, who ousted the Muslims in 1248, destroyed most of the mosque and built a huge cathedral on the site, but they saved the ornate minaret, made it the bell tower, and called it the Giralda (left). This minaret had been built by the Muslims to match the 67-meter (220-ft) minaret of the Koutoubia Mosque in Marrakech (photo p. 369). But the finest Islamic legacy in Seville surely is the Alcazar Palace (above), begun in the late twelfth century and embellished and finished by the Christians after their victory. Its arched façades and intricately carved walls remain a monument to the Muslim architects and artists who designed and created them." © H. J. de Blij

www.conceptcaching.com

Concept Caching

FIGURE 7-5

© H. J. de Blij, P. O. Muller, and John Wiley & Sons, Inc.

As we noted, Islam's expansion eventually was checked in Europe, Russia, and elsewhere. But a map showing the total area under Muslim sway in Eurasia and Africa reveals the enormous dimensions of the domain affected **13** by **Islamization** at one time or another (Fig. 7-6). Islam continues to expand, now mainly by relocation diffusion. There are Islamic communities in cities as widely scattered as Vienna, Singapore, and Cape Town, South Africa; Islam is also growing rapidly in the United States. With more than 1.3 billion adherents today, Islam is a vigorous and burgeoning cultural force around the world.

ISLAM DIVIDED

For all its vigor and success, Islam still fragmented into sects. The earliest and most consequential division arose after Muhammad's death. Who should be his legitimate successor? Some believed that only a blood relative should follow the prophet as leader of Islam. Others, a majority, felt that any devout follower of Muhammad was qualified.

The first chosen successor was the father of Muhammad's wife (and thus not a blood relative). But this did not satisfy those who wanted to see a man named Ali, a cousin of Muhammad, made *caliph* (successor). When Ali's turn came, his followers, the *Shi'ites*, proclaimed that Muhammad finally had a legitimate successor. This offended the *Sunnis*, those who did not see a blood relationship as necessary for the succession. From the beginning of this disagreement, the numbers of Muslims who took the Sunni side far exceeded those who regarded themselves as Shia (followers) of Ali. The great expansion of Islam was largely propelled by Sunnis; the Shi'ites survived as small minorities scattered throughout the realm. Today, about 85 percent of all Muslims are Sunnis.

The Strength of Shi'ism

But the Shi'ites vigorously promoted their version of the faith. In the early sixteenth century their work paid off: the royal house of Persia (modern-day Iran) made Shi'ism

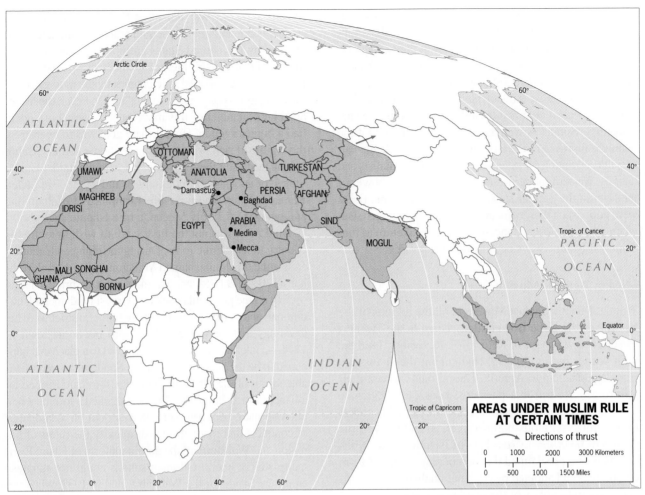

FIGURE 7-6

© H. J. de Blij, P. O. Muller, and John Wiley & Sons, Inc.

the only legal religion throughout its vast empire. That domain extended from Persia into lower Mesopotamia (modern Iraq), into Azerbaijan, and into western Afghanistan and Pakistan. As the map of religions (Fig. 7-3) shows, this created for Shi'ism a large **culture region** and gave the faith unprecedented strength. Iran remains the bastion of Shi'ism in the realm today, and the appeal of Shi'ism continues to radiate into neighboring countries and even farther afield.

During the late twentieth century, Shi'ism gained unprecedented influence in the realm. In its heartland, Iran, a *shah* (king) tried to secularize the country and to limit the power of the *imams* (mosque officials); he provoked a revolution that cost him the throne and made Iran a Shi'ite Islamic republic. Before long, Iran was at war with neighboring, Sunni-ruled Iraq, and Shi'ite parties and communities elsewhere were invigorated by the newfound power of Shi'ism. From Arabia to Africa's northwestern corner, Sunni-ruled countries warily watched their Shi'ite minorities, newly imbued with religious fervor. Mecca, the holy place for both Sunnis and Shi'ites, became a battle-

ground during the week of the annual pilgrimage, and for a time the (Sunni) Saudi Arabian government denied entry to Shi'ite pilgrims. Recently, that schism has healed somewhat, but intra-Islamic sectarian differences run deep.

Smaller Islamic sects further diversify this realm's religious landscape. Some of these sects play a disproportionately large role in the societies and countries of which they are a part, as we will see in our regional discussion.

Religious Revivalism in the Realm

Another cause of intra-Islamic conflict lies in the resurgence of religious fundamentalism, or as Muslims refer to it, **religious revivalism**. In the 1970s, the imams in Shi'ite Iran wanted to reverse the shah's moves toward liberalization and secularization: they wanted to (and did) recast society in traditional, revivalist Islamic molds. An *ayatollah* (leader under Allah) replaced the shah in 1979; Islamic rules and punishments were instituted. Urban women, many of whom had been considerably liberated and educated during the shah's regime, resumed more traditional Islamic roles. Vestiges of Westernization, encouraged by the shah, disappeared. Under these new conditions, even the war against Iraq (1980–1990), which began as a conflict over territory, became a holy war that cost more than a million lives.

Islamic revivalist fundamentalism did not rise in Iran alone, nor was it confined to Shi'ite communities. Many Muslims—Sunnis as well as Shi'ites—in all parts of the realm disapproved of the erosion of traditional Islamic values, the corruption of society by European colonialists and later by Western modernizers, and the declining power of the faith in the secular state. As long as economic times were good, such dissatisfaction remained submerged. But when jobs were lost and incomes declined, a return to fundamental Islamic ways became more appealing.

Muslim Against Muslim

This set Muslim against Muslim in all the regions of the realm (see the Issue Box entitled "Islam in the Twenty-First Century: Revive or Reform?"). Revivalists fired the faith with a new militancy, challenging the status quo from Afghanistan to Algeria. The militants forced their governments to ban "blasphemous" books, to resegregate the sexes in schools, to enforce traditional dress codes, to legitimize religious-political parties, and to heed the wishes of the *mullahs* (teachers of Islamic ways). Militant Muslims proclaimed that democracy inherited from colonialists and adopted by Arab nationalists was incompatible with the rules of the Quran (Koran).

The rift between moderate Muslims and militant revivalists began to spill over into other parts of the world decades ago, from Pakistan to the African Transition Zone and from the Caucasus to the Philippines. Even while Islamic countries such as Iran and Algeria struggled to overcome the instability and damage arising from their internal conflicts, militant activists planned and carried out attacks against the allies of the moderates—oil-guzzling Western countries guilty of military intervention, propping up unrepresentative regimes, mistreatment of immigrant Muslims, and cultural corruption. Before the worst of these attacks (through mid-2009) destroyed the World Trade Center in New York and severely damaged the Pentagon outside Washington, D.C., the French had foiled a similar assault on the Eiffel Tower in Paris, the Russians had suffered hundreds of casualties in Moscow and Transcaucasia, and U.S. targets were hit in the Middle East, the Arabian Peninsula, East Africa, and elsewhere.

Back to Basics

Although it has been argued that these attacks do not represent Islam as a religion in conflict with the West, they do have a religious context. During the 1990s, fol-

lowing the Cold War defeat of Soviet armed forces in Afghanistan and the subsequent abandonment of that country by the United States and its allies, an organization named al-Qaeda emerged, dedicated to punishing the perceived enemies of Islamic peoples by the most effective means at its disposal: terrorist attacks. However, the leader and chief financier of this movement, Usama bin Laden, had a bigger agenda. His ultimate goal was to overthrow the regime of rich princes ruling his home country, Saudi Arabia, accusing them of collusion with the infidel United States, of apostasy (faithlessness), and worse. Bin Laden wanted to make Saudi Arabia, birthplace of Muhammad and guarantor of holy Mecca, a true Islamic state.

Bin Laden was not the first Saudi to wish to return his country to its Islamic roots. During the eighteenth century an Islamic theologian named Muhammad ibn Abd al-Wahhab (1703–1792) founded a movement to bring Islamic society on the peninsula back to the most fundamental, traditional, and puritanical form of the faith. So strict were his teachings that he was expelled from his home town in 1744—a fateful exile because he moved to a place that happened to be the headquarters of Ibn Saud, ruler over a sizeable swath of the Arabian Peninsula. Ibn Saud formed an alliance with al-Wahhab, who had inherited a fortune when his wife died; between them, they set out on a campaign of conquest that made the Saudi dynasty the rulers of the region and **Wahhabism** the dominant form of Islam. Even the relatively liberal Ottoman sultans could not suppress either the Saudis or the Wahhabis, and in 1932, when the modern Kingdom of Saudi Arabia was established, Wahhabism became its official and dominant form of Islam.

Despite its reference to the name of the founder, Wahhabism today is a term used mostly by non-Muslims and outsiders. Followers of Wahhabi doctrines call themselves *Muwahhidun* or "Unitarians" to signify the strict and fundamental nature of their beliefs. These are the doctrines Usama bin Laden accuses the Saudi princes of having violated, and these are the harsh teachings so often cited in the Western media as emanating from the pulpits of Saudi Arabia's revivalist mosques.

Islam and Other Religions

Two additional major faiths had their sources in the *Levant* (the area extending from Greece eastward around the Mediterranean coast to northern Egypt), and both were older than Islam. Islam's rise submerged many smaller Jewish communities, but the Christians, not the Jews, waged centuries of holy war against the Muslims, seeking, through the Crusades, not only to drive Islam back but to reestablish Christian communities where they had dominated before the Islamic expansion. The aftermath of that

Islam in the Twenty-First Century: Revive or Reform?

ISLAMIC REVIVAL IS THE ONLY WAY

"We Muslims must return to, and protect, the fundamental tenets of our faith. Too many of us have strayed, lured and seduced by the vulgarities of Western and other foreign ways. Consider our history. The Prophet set us on course, the Quran was and is the source of all truth, and the message took root from Mecca to Morocco to Malaya. Don't forget: Allah's revelations came many centuries after other religions—Hinduism, Buddhism, Christianity—had captured the minds of millions. But many saw that Islam is the only way, and today practicing Muslims (which means every believer) outnumber Christians, millions of whom never see the inside of a church and do not live pious lives.

"As a devout Muslim living in Egypt and having suffered the consequences of my devotion, let me say this: we Muslims have tried to live harmoniously with other faiths. We not only brought science and enlightenment to the world we converted, but we lived in peace with non-Arabs. In Islamic Andalusia, Jewish communities were safe and secure. It was the Christian Catholics who invented the horrors of the Inquisition. From West Africa to Central Asia, we proved our willingness to live in peace with infidels. So what did we get in return? The 'three c's'—the crusades, colonialism, and commercial exploitation! From my American friends I understand that there's an expression, 'three strikes and you're out.' Well, you non-Muslims are out. And so are those Muslims who still cooperate with foreigners, including the president of my own country and the entire Saudi royal family. I agree with Sheikh Hamoud: whoever supports the infidel against Muslims is himself an infidel. When you sell your soul for oil dollars or foreign aid, you're subject to a fatwa.

"Yes, I'm a revivalist, or what you people call a fundamentalist. Our salvation lies in a return to the strictest rules of Islam. Every Muslim must adhere to every tenet of the faith. Youngsters should learn the Quran by heart: it is the source of all truth beyond which nothing else matters. Women should know and keep their place as it is accorded to them by the Quran. Sharia law should be imposed by all Islamic countries, so that the punishment of transgressors will serve as an example to others. We should dress conservatively, resist the degenerate temptations of other cultures, and follow the teachings of our religious leaders. Let me ask you this: are we better off today than when Sheikh al-Wahhab and Ibn Saud were forging the Islamic state of Arabia? I would say not. We have to go back to our roots to undo the damage that has been done to us.

"Ayatollah Khomeini's Islamic revolution failed in Iran, and our Algerian brothers were prevented from making Algeria an Islamic state, but these were just harbingers of things to come. Our goal is Saudi Arabia, where it all started fourteen centuries ago—and where we will make the new start that will transform the world."

ISLAM NEEDS A REFORMATION

"How is it that we Muslims, with our rich cultural and scientific heritage, wealth of oil, history of interaction with other societies, and central location in the world are held back today by autocratic, corrupt, and incompetent governments, interfering and medieval clerics, stultifying traditions, and internal discord? As a practicing Muslim who is also a realist, I can tell you from my perspective as a Tunisian working in the tourist industry that we need Islamic reform, not the revivalism about which I keep hearing.

"Let's be objective about this. Look at the world's social and economic statistics, and you'll see that countries where Islam is the dominant religion tend to rank near the bottom—not always, but far too often for this to be a coincidence. The world has been swept by a wave of democratization, but not here. The globe is in the grip of a revolution in information and communication technology, but not here. The planet has a mosaic of religions, each professing to be the real path to salvation, and each bedeviled by a fundamentalist fringe, but not here. Here the fundamentalists are a constituency, not a fringe. And they are using tactics ranging from intimidation to terrorism to maintain and increase their clout.

"It all boils down to this: freedom. We need the freedom to debate the issues that confront us without fear of fatwas or imprisonment. In no other religious society can you be sentenced to death for expressing an opinion about God or scriptures. It shouldn't happen here. We need freedom of the press and other media, which are more restricted here than anywhere else on Earth. We need freedom of association without fear of persecution. But most of all, we need the right to debate the role of religion in our societies. There is nothing incompatible between Islamic societies and others around the world, but you cannot designate every non-Muslim as an infidel and then expect to have cordial and trusting relations.

"Islam needs a reformation, a reinterpretation of its inspired writings in light of modern times. And the mullahs need to get out of the business of controlling government, which is their constant aim. The separation of mosque and state is indispensable to our progress.

"Our religion-suffused educational system produces graduates who are unable to use what they learned to enter the globalizing job market. I can visualize those kids in the Taliban's madrassas doing only one thing, day after day—'studying' to recite the Quran by heart. Even where schools don't go to that extreme, they tend to produce unemployable youngsters whose joblessness and frustration cause them to turn to the mullahs—who are all too happy to direct their anger toward the West.

"And our reformation will have to address the plight of women. As a woman, it infuriates me that half of all women in the Arab world can neither read nor write, that they have unequal citizenship, and that they play virtually no role in government. Nowhere in the world is the domination of men more absolute, and I refuse to believe that the Prophet would approve of this. We need a reformation, not a so-called revival."

Regional ISSUE

Vote your opinion at www.wiley.com/college/deblij

campaign still marks the realm's cultural landscape today. A substantial Christian minority (about 30 percent of the population) remains in Lebanon, and Christian minorities also survive in Israel, Syria, Egypt, and Jordan. Strained relations between the long-dominant Christian minority in Lebanon and the Muslim majority (itself divided into five sects) contributed to the disastrous armed struggle that engulfed this country in the 1970s and 1980s.

But the most intense conflict in modern times has pitted the Jewish state, Israel, against its Islamic neighbors near and far. Israel's United Nations-sponsored creation in 1948 precipitated more than six decades of intermittent strife and attempts at mediation; it also caused friction among Islamic states in the region. Jerusalem—holy city for Judaism, Christianity, and Islam—lies in the crucible of this confrontation.

The Ottoman Aftermath

It is a geographic twist of fate that Islam's last great advance into Europe resulted in the European occupation of Islam's very heartland. The Ottomans (named after their leader, Osman I), based in what is today Turkey, conquered Constantinople (now Istanbul) in 1453 and then pushed into Eastern Europe. Soon Ottoman forces were on the doorstep of Vienna; they also invaded Persia, Mesopotamia, and North Africa (Fig. 7-7). The Ottoman Empire under Suleyman the Magnificent, who ruled from 1522 to 1560, was the most powerful state in western Eurasia. As Figure 7-7 shows, its armies even advanced toward Moscow, Kazan, and Krakow from their base at Azov (near present-day Rostov in Russia). The Turks also launched marine attacks on Sicily, Spain, and France.

The Ottoman Empire survived for more than four centuries, but it lost territory as time went on, first to the Hungarians, then to the Russians, and later to the Greeks and Serbs until, after World War I, the European powers took over its provinces and turned them into colonies—colonies we now know by the names of Syria, Iraq, Lebanon, and Yemen (Fig. 7-8). As the map shows, the French and the British took large possessions; even the Italians annexed part of the Ottoman domain.

The boundary framework that the colonial powers created to delimit their holdings was not satisfactory. As we

FIGURE 7-7

© H. J. de Blij, P. O. Muller, and John Wiley & Sons, Inc.

noted earlier, this realm's population of 592 million is clustered, fragmented, and strung out in river valleys, coastal zones, and crowded oases (Fig. 7-2). The colonial powers laid out long stretches of boundary as ruler-straight lines across uninhabited territory; they saw no need to adjust these boundaries to cultural or physical features in the landscape. Other boundaries, even some in desert zones, were poorly defined and never marked on the ground. One product of this process was Iraq, delimited by (then) British Colonial Secretary Winston Churchill in 1921 as a kingdom with Baghdad as its capital; another was Jordan, centered on Amman. Later, when the colonies had become independent states, such boundaries led to quarrels, even armed conflicts, among neighboring Muslim states.

THE POWER AND PERIL OF OIL

Travel through the cities and towns of North Africa and Southwest Asia, talk to students, shopkeepers, taxi drivers, and migrants, and you will hear the same refrain:

"Leave us in peace, let us do things our traditional way. Our problems—with each other, with the world—seem always to result from outside interference. The stronger countries of the world exploit our weaknesses and magnify our quarrels. We want to be left alone."

That wish might be closer to fulfillment were it not for two relatively recent events: the creation of the state of Israel and the discovery of some of the world's largest oil reserves. We will discuss the founding of Israel in the regional section of this chapter. Here we focus on the realm's most valuable export product: oil.

Location and Dimensions of Known Reserves

About 30 of the world's countries have significant oil reserves. But five of these countries, all located in the North Africa/Southwest Asia realm, in combination possess larger reserves (ca. 77 percent of the world total) than the rest combined. The estimated reserves of this Big Five, averaged from several sources including the

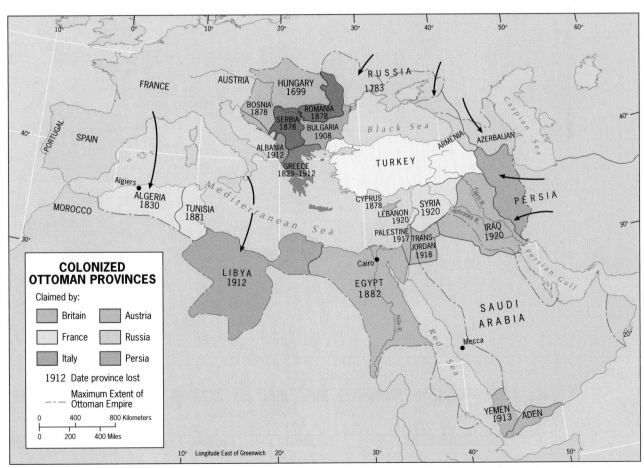

FIGURE 7-8

© H. J. de Blij, P. O. Muller, and John Wiley & Sons, Inc.

governments of the countries themselves and reported in billions of barrels, are as follows: (1) Saudi Arabia, 267; (2) Iraq, 114; (3) United Arab Emirates, 101; (4) Kuwait, 97; and (5) Iran, 91. The next-ranking country, Venezuela, had known reserves below 80 billion barrels; Russia below 60; the United States below 40; and Libya, Mexico, Nigeria, China, and Ecuador below 25. No other country's known reserves approached 20 billion barrels as of 2009. It should also be noted that Canada may pos-

sess the largest oil reserves of all—but most of this supply is contained in Alberta's *tar sands* (see pp. 159, 189), which requires complex and expensive recovery and refining processes.

In general terms, the NASWA realm's oil (and associated natural gas) lies in three discontinuous zones (Fig. 7-9). The most productive of these zones extends from the southern and southeastern part of the Arabian Peninsula northwestward around the rim of the Persian Gulf, reach-

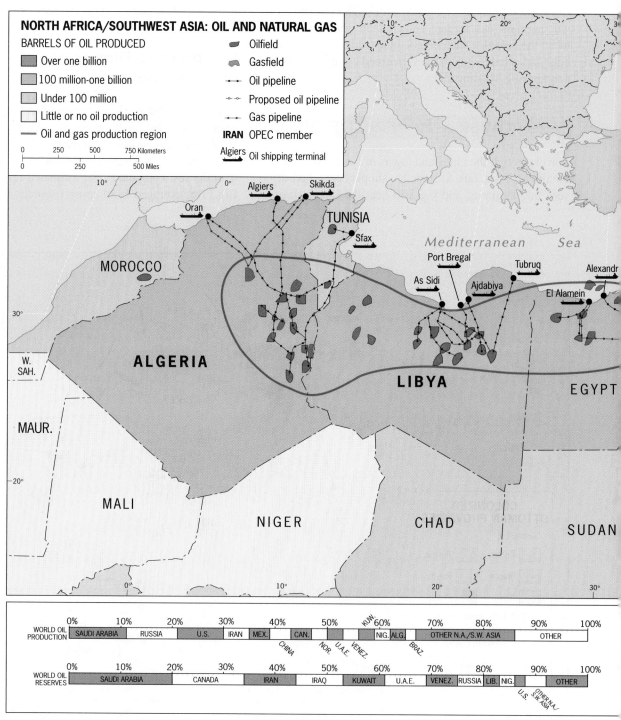

FIGURE 7-9

THE POWER AND PERIL OF OIL

ing into Iran and continuing northward into Iraq, Syria, and southeastern Turkey, where it peters out. The second zone lies across North Africa and extends from north-central Algeria eastward across northern Libya to Egypt's Sinai Peninsula, where it ends. The third zone begins on the margins of the realm in eastern Azerbaijan, continues eastward under the Caspian Sea into Turkmenistan and Kazakhstan, and also reaches into Uzbekistan, Tajikistan, Kyrgyzstan, and Afghanistan.

Producers and Consumers

Saudi Arabia is the world's largest oil exporter, but in recent years Russia has risen to second place. As we note in Chapter 2, oil and natural gas are by far Russia's most valuable commodities, and the Russian state direly needs the revenues they produce. Thus Russia, with modest (though expanding) known reserves, is vigorously exporting to international markets and is now a leading player in the

© H. J. de Blij, P. O. Muller, and John Wiley & Sons, Inc.

global energy arena. In combination, however, production by the countries of the North Africa/Southwest Asia realm far exceeds that from all other sources combined. The United States, not surprisingly, is the world's leading importer even while it consumes almost all of its own production.

As Figure G-11 indicates, the production and export of oil and gas has elevated several of this realm's countries into the higher-income categories. But petroleum wealth also has enmeshed these Islamic societies and their governments in global strategic affairs. When regional conflicts create instability in producing countries that have the potential to disrupt supply lines, powerful consumers are tempted to intervene—and have done so.

Colonial Legacy

When the colonial powers laid down the boundaries that partitioned this realm among themselves, no one knew about the riches that lay beneath the ground. A few wells had been drilled, and production in Iran had begun as early as 1908 and in Egypt's Sinai Peninsula in 1913. But the major discoveries came later, in some cases after the colonial powers had already withdrawn. Some of the newly independent countries, such as Libya, Iraq, and Kuwait, found themselves with wealth undreamed of when the Turkish Ottoman Empire collapsed. However, as Figure 7-9 shows, others were less fortunate. A few countries had (and still have) potential. The smaller, weaker emirates and sheikdoms on the Arabian Peninsula always feared that powerful neighbors would try to annex them (Kuwait faced this prospect in 1990 when Iraq invaded it). The unevenly distributed oil wealth, therefore, created another source of division and distrust among Islamic neighbors.

A Foreign Invasion

The oil-rich countries of the realm found themselves with a coveted energy source but without the skills, capital, or equipment to exploit it. These had to come from the Western world and entailed what many tradition-bound Muslims feared most: a strong foreign presence on Islamic soil, foreign intervention in political as well as economic affairs, and penetration of Islamic societies by the vulgarities of Western ways.

Many Muslims' worst fears came true in Iran during the 1950s, when the government of Iran tried to control British oil exploitation centered on Abadan and the Persian Gulf. The imperious ways of the British, the luxurious life of the European expatriates compared to the abject poverty of Iranian workers, and the unfair terms of the concession under which the oil was exported all contributed to the rise of nationalist sentiment in Iran, whose leaders appealed to the United States for help in negotiating a better deal with the British. American President Harry Truman had some sympathy for the Iranians, but his successor Dwight Eisenhower worried about Iran turning toward communism. Capitalizing on political disputes within Iran, CIA operatives engineered the overthrow of elected Premier Mohammad Mosaddeq, restoring to power the young shah and setting into motion a sequence of events that would eventually lead to the demise of Iran's monarchy and the proclamation of an Islamic Republic in 1979.

Even when foreign intervention was more subtle, it intensified clashes between traditional and modern. Oil revenues created cultural landscapes in which gleaming skyscrapers towered over ornate mosques and historic neighborhoods. The social chasms between rich, well-connected elites and less fortunate citizens bred resentment. Involvement of Western armies in regional conflicts produced the unthinkable: foreign armies on Arabia's hallowed soil. To many Muslims, all this violated the most basic tenets of their faith, and some retaliated with violence of their own. That violence, in the form we define as terrorism, has become one of the challenges of our time.

Oil's Geographic Impact

As is true of every natural resource found across the planet, oil has completely transformed cultural landscapes in certain parts of NASWA and left others virtually unchanged. For hundreds of millions of this realm's inhabitants, a patch of tillable soil and a source of water still mean more to daily life than all the oil in OPEC.* Countrysides in Arab as well as non-Arab regions in this part of the world continue to carry the imprints of centuries of cultural tradition, not decades of oil-driven modernization. Nevertheless, oil and natural gas—their location, production, transportation, and sale—have produced massive changes, including the following:

1. *Urban Transformation.* Undoubtedly the most visible manifestation of oil wealth is the modernization (and, as some call it, the "futurization") of cities. By far the tallest building in the world rises above Dubai in a state called the United Arab Emirates, but it is only one in a forest of glass-encased skyscrapers, many of which test the limits of design and engineering. Not only the capitals (such as Saudi Arabia's Riyadh) but also other cities reflect the riches

---

*OPEC, the Organization of Petroleum Exporting Countries, is the international oil cartel (syndicate) formed by 12 producing countries to promote their common economic interests through the setting of joint pricing policies and the limitation of market options for consumers. Its 8 NASWA members are Algeria, Iran, Iraq, Kuwait, Libya, Qatar, Saudi Arabia, and the United Arab Emirates.

oil has brought—yes, even Mecca. And, as we will see in the regional section, entirely new cities are springing up from the deserts and along the coasts.

2. *Variable Incomes.* Oil and natural gas prices fluctuate on world markets (during 2008, oil sold for more than [U.S.] $140 per barrel before plunging to less than $40). When energy prices are high, several states in this realm rank among the highest-income societies in the world. Even when they decline, many petroleum-exporting countries manage to remain at least in the upper-middle-income category (Fig. G-11).

3. *Infrastructure.* Massive spending on airports, seaports, bridges, tunnels, four-lane highways, public buildings, shopping malls, recreational facilities, and other components of national infrastructures creates an image of comfort and affluence quite unlike that prevailing in countries without major oil or gas revenues. Saudi Arabia is now engaged in a vast modernization project extending from coast to coast.

4. *Industrialization.* Some far-sighted governments among those with oil wealth, realizing that reserves will not last forever, are investing some of their income in industries that will outlast the era of massive oil exporting. Petrochemical manufacturing using domestic supplies, aluminum, steel, and fertilizers are among these industries, although others potentially more promising, for example in high-technology fields, are not yet a significant part of this important initiative.

5. *Regional Disparities.* Oil wealth, like other high-value resources, tends to create strong regional contrasts both within and between countries. Saudi Arabia's ultramodern east coast is a world apart from most of its interior, which remains a land of desert, oasis, vast distances, isolated settlements, and slow change. This is not unique to countries in this realm: sharp regional disparities mark oil-rich countries from Algeria to Azerbaijan.

6. *Foreign Investment.* Governments and private entrepreneurs have invested massive quantities of oil-generated wealth in foreign countries, buying stock, and acquiring prestigious hotels, famous stores, and other high-cachet properties. These investments have created a network of international linkages that not only connects NASWA states and individuals to the economies of non-NASWA countries but also links them to the growing Islamic communities in those states.

7. *Foreign Involvement.* To many inhabitants of this realm, especially those with strong Islamic-revivalist convictions, the inevitable presence of foreigners (including businesspeople, politicians, architects, engineers, and even armed forces) on Islamic soil is an unwelcome byproduct of the energy era. Public opinion in Saudi Arabia forced its ruling regime to negotiate the departure of American troops from the country's territory.

8. *Intra-Realm Migration.* Oil wealth allows governments, industrialists, and private individuals to hire workers from less-favored parts of the realm to work in the oilfields, ports, and in other mainly menial capacities. This has brought many Shi'ites to the countries of the eastern zone of the Arabian Peninsula, where they now form significant sectors of national populations; hundreds of thousands of Palestinian Arabs also sought temporary work in that region's industries. In 2010 it was estimated that Saudi Arabia, with a population of just under 30 million, hosted more than 5 million foreign workers from within the realm.

9. *Migration from Other Realms.* The labor market in such places as Dubai and Abu Dhabi in the United Arab Emirates has also attracted workers from beyond this realm's borders during recent periods of rapid growth. Wages in such countries as Pakistan, India, Sri Lanka, and Bangladesh are even lower than those paid by the building industry or private

This photo, taken during June 2009 in Dubai, suggests a booming construction industry in a city that is already a modern legend complete with the world's tallest building, the ultimate in luxury hotels, superb golf courses and other sports facilities, a huge indoor ski slope—all in the desert of the Arabian Peninsula. But take a closer look and you will see no workers on the roofs or scaffolding, no supplies being hoisted up, no activity. The global economic recession had its impact even here in this playground of the wealthy, and foreign construction workers (who have built Dubai's infrastructure) were being sent home by the thousands, some without wages owed to them. Dubai achieved with commerce and services what its fellow emirate in the UAE, Abu Dhabi, built with oil. When the economic downturn came, it was Dubai that suffered most.
© Joern Pollex/Getty Images, Inc.

employers in oil-boom-driven NASWA states. These workers serve mainly as domestics, gardeners, trash collectors, and the like; at the beginning of 2009, about 80 percent of the total population of Dubai consisted of foreign employees. Over the past few years, working conditions and wage issues also created protests that exposed the harsh circumstances faced by many guest workers throughout this realm.

10. *Diffusion of Revivalism.* Oil and gas revenues are used by Islamist regimes to support Islamic communities as well as their mosques and cultural centers throughout the world. No NASWA state has spent more money on such causes than Saudi Arabia, and thousands of mosques from England to Indonesia prosper as a result. This example of relocation diffusion creates nodes of recruitment for the faith—and ensures the dissemination of revivalist principles.

Oil brought this realm into contact with the outside world in ways unforeseen just a century ago. Oil has strengthened and empowered some of its peoples, but has dislocated and imperiled others. It has truly been a double-edged sword. So before we study the regions of this realm we should remind ourselves that the great majority of NASWA's inhabitants are not, in their daily lives, directly affected by the changes the energy era has brought. Most Moroccans, Tunisians, Egyptians, Jordanians, Yemenis, Turks, and many millions of others—Kurds, Palestinians, Berbers, Tuaregs—make ends meet by trading or farming or working at jobs their parents and grandparents performed. Take the case of Iran, which as a country earns about two-thirds of its income from oil and gas. Only half of 1 percent of that country's workforce (just over 100,000 out of more than 20 million workers) earn a salary from energy or energy-related work. By far the largest number in any single occupation are the farmers (5 million). And for all their oil, Iranians in 2007 earned a mere (U.S.) $10,800 per person—less than the average Turk and one-third of the average earned per capita in Singapore. As we will discover, it is cultural geography rather than economic geography that dominates the regionalization of the North Africa/Southwest Asia realm.

POINTS TO PONDER

● The world's tallest building by far, four-fifths of a kilometer (one half-mile) high, is rising over the Emirate of Dubai on the Arabian Peninsula.

● As a result of the war started by the United States in 2003, as many as 3 million Iraqis had fled to neighboring countries and more than 2 million had become internal refugees. By mid-2009, only a few hundred thousand had fully resettled in Iraq.

● The Obama Administration, while continuing America's pro-Israeli policies, has been more reserved than previous U.S. governments in its acceptance of Israel's conduct in Gaza and the West Bank.

● Iran's government claims that its pursuit of nuclear capability is an energy-supply matter; the United States and other countries fear that Iran is developing nuclear weapons capacity.

Regions of the Realm

Identifying and delimiting regions in this vast geographic realm is quite a challenge. Population clusters are widely scattered in some areas, highly concentrated in others. Cultural transitions and landscapes—internal as well as peripheral—make it difficult to discern a regional framework. Furthermore, this is a highly changeable realm and always has been. Several centuries ago it extended into Eastern Europe; now it reaches into Central Asia, where

17 an Islamic **cultural revival** (the regeneration of a long-dormant culture through internal renewal and external infusion) is well under way.

The following are the regional components of this far-flung realm today (Fig. 7-10):

1. *Egypt and the Lower Nile Basin.* This region in many ways constitutes the heart of the realm as a whole. Egypt (together with Iran and Turkey) is one of the realm's three most populous countries. It is the historic focus of this part of the world and a major polit-

MAJOR CITIES OF THE REALM

City	Population* (in millions)
Algiers, Algeria	3.5
Almaty, Kazakhstan	1.2
Baghdad, Iraq	5.7
Beirut, Lebanon	1.9
Cairo, Egypt	12.5
Casablanca, Morocco	3.3
Damascus, Syria	2.7
Istanbul, Turkey	10.5
Jerusalem, Israel	0.7
Khartoum, Sudan	5.4
Riyadh, Saudi Arabia	4.9
Tashkent, Uzbekistan	2.3
Tehran, Iran	8.2
Tel Aviv-Jaffa, Israel	3.2
Tunis, Tunisia	2.3

*Based on 2010 estimates.

ical and cultural force. It shares with its southern neighbor, Sudan, the waters of the lower Nile River.

2. *The Maghreb and Its Neighbors.* Western North Africa (the *Maghreb*) and the areas that border it also form a region, consisting of Algeria, Tunisia, and Morocco at the center, and Libya, Chad, Niger, Mali, and Mauritania along the broad periphery. The last four of these countries also lie astride or adjacent to the broad transition zone where the Arab-Islamic realm of northern Africa merges into Subsaharan Africa.

3. *The Middle East.* This region includes Israel, Jordan, Lebanon, Syria, and Iraq. In effect, it is the crescent-like zone of countries that extends from the eastern Mediterranean coast to the head of the Persian Gulf.

4. *The Arabian Peninsula.* Dominated by the large territory of Saudi Arabia, the Arabian Peninsula also includes the United Arab Emirates (UAE), Kuwait, Bahrain, Qatar, Oman, and Yemen. Here lies the source and focus of Islam, the holy city of Mecca; here, too, lie many of the world's largest oil reserves.

5. *The Empire States.* We refer to this region as the Empire States because two of the realm's giants, both the centers of historic empires, dominate its geography. Turkey, heart of the former Ottoman Empire, now is the realm's most secular state and aspires to join the European Union. Shi'ite Iran, the core of the former Persian Empire, has become an Islamic republic. To the north, Azerbaijan, once a part of the Persian Empire, lies in the turbulent, Muslim-infused Transcaucasian Transition Zone. To the southwest, the island of Cyprus is divided between Turkish and Greek spheres, the Turkish sector being another remnant of imperial times.

6. *Turkestan.* Turkish influence ranged far and wide in southwestern Asia, and following the Soviet collapse that influence proved to be durable and strong. In the five former Soviet Central Asian republics the strength and potency of Islam vary, and the new governments deal warily (and sometimes forcefully) with Islamic revivalists. Here, Russian influence continues, democratic institutions remain weak, and Western (chiefly American) involvement has increased, especially in Afghanistan (also part of this region).

● EGYPT AND THE LOWER NILE BASIN

No country in the vast NASWA realm has a more pivotal position than does Egypt. Here the great Nile River, whose twin origins respectively lie in Uganda and Ethiopia, forms a fertile valley and creates a vast and populous delta. Here the Red Sea meets the Mediterranean Sea through the Suez Canal, vital link between the Atlantic and Indian oceans. Here lies Cairo, Egypt's capital, the realm's largest city, and a leading center of Islamic civilization. With 78.1 million people, Egypt is also the realm's largest state.

With its ruler-straight north-south and east-west borders in the northeastern corner of the African landmass, Egypt is one of the world's most recognizable states (Fig. 7-10). But its relative location also is noteworthy: across the Red Sea lie Saudi Arabia and Mecca (Makkah); and directly across the Mediterranean Sea from the Nile Delta lies Turkey, the second-largest country in the realm. In the east, Egypt's Sinai Peninsula is bounded by Israel, and Egypt also has a short border with Gaza, recently vacated by Israel and now ruled by a militant Palestinian faction engaged in a low-grade war with the Israelis. In the west, Egypt adjoins Libya, an energy-rich former Italian dependency with a history of militant autocracy. And to the south lies Sudan, currently entering an oil era, its regime engaged in an internationally condemned war in Darfur, and its southern provinces poised to secede.

What happens in Sudan does not stay in Sudan: not only the waters of the Nile but also the cultures and politics of Sudan infiltrate Egypt, and have done so for as long as the Nile has been a North African corridor. We hardly need to further justify the designation of Egypt and its southern neighbor as a discrete and crucial region of this realm.

Gift of the Nile

Egypt's Nile is the aggregate of two great branches upstream: the White Nile, which originates in the streams that feed Lake Victoria in East Africa, and the Blue Nile, whose source lies in Lake Tana in Ethiopia's highlands. The two Niles converge at Khartoum, capital of modern-day Sudan.

About 95 percent of Egypt's 78 million people live within 20 kilometers (12 mi) of the great river's banks or in its delta (Figs. 7-11, 7-2). It has always been this way: the Nile rises and falls seasonally, watering and replenishing soils and crops on its banks.

The ancient Egyptians used *basin irrigation*, building fields with earthen ridges and trapping the floodwaters with their fertile silt, to grow their crops. That practice continued for thousands of years until, during the nineteenth century, the construction of permanent dams made it possible to irrigate Egypt's farmlands year round. These dams, with locks for navigation, controlled floods, and expanded the country's cultivable

POLITICAL UNITS AND GEOGRAPHIC REGIONS OF NORTH AFRICA/SOUTHWEST ASIA

Egypt and the Lower Nile Basin
The Maghreb and Its Neighbors
Middle East
Arabian Peninsula
The Empire States
Turkestan
African Transition Zone

POPULATION

● Under 50,000
● 50,000–250,000
● 250,000–1,000,000
● 1,000,000–5,000,000
● Over 5,000,000

National capitals are underlined

—— Railroad
—— Road
++++ Canal

0   400   800   1200   1600   2000 Kilometers
0   400   800   1000 Miles

NORWAY   FINLA
Oslo   Stockholm   Helsinki
SWEDEN   ESTON
North Sea   DENMARK   LATV
UNITED   RUSSIA
KINGDOM   LITHUANIA
London   NETH.   Berlin   Warsaw   Men
BELG.   GERMANY   POLAND
LUX.   CZECH
Paris   REPUBLIC   SLOVAKIA
FRANCE   Vienna   Budapest
SWITZ.   AUSTRIA   SLOVAKIA   ROMANI
SLO.   HUNGARY   Bucha
Milan   CRO.   Belgrade
ITALY   BOSNIA-   SERBIA   BULGAR
Barcelona   HERZEGOVINA
Rome   MONTENEGRO   MAC.   ALBANIA
Naples   GREECE
MALTA   Athens   Izm

ATLANTIC OCEAN

Gibraltar (U.K.)   SPAIN   Algiers   Annaba   Tunis
Tangier   Oran   Constantine
Rabat   Mediterranean Se
Casablanca   MAGHREB   TUNISIA   Tripoli   Benghazi
Marrakech   Atlas Mountains   Sfax
MOROCCO   Bechar

Canary Is. (Sp.)

El Aaiún

Tropic of Cancer

WESTERN SAHARA   ALGERIA   LIBYA   Libyan Desert

Sahara

Nouakchott

MAURITANIA

Dakar
SENEGAL   MALI
GAMBIA   NIGER   CHAD
GUINEA-BISSAU   Ségou   Mopti   N'Djamena
GUINEA   Bamako   Niamey   AFRICAN   TRANSITION
Conakry   BURKINA FASO   Kano
SIERRA LEONE   BENIN   Kaduna
Freetown   IVORY   GHANA   TOGO   Niger R.   NIGERIA   N'Djamena
LIBERIA   COAST   Abuja
Monrovia   Yamoussoukro   Ibadan   Benue R.
Abidjan   Accra   Lagos   CENTRAL AFRICAN REPUBLIC
            Yaoundé   Bangui
EQ. GUINEA   CAMEROON
SÃO TOME & PRÍNCIPE
ATLANTIC OCEAN   GABON   CONGO   Congo R.   THE CONGO
Equator

Brazzaville   Kinshasa

Longitude West of Greenwich

FIGURE 7-10

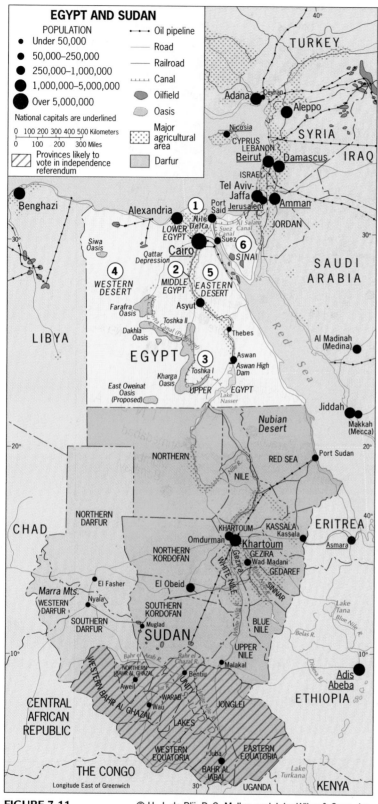

FIGURE 7-11   © H. J. de Blij, P. O. Muller, and John Wiley & Sons, Inc.

The greatest of all Nile projects, the Aswan High Dam (which began operating in 1968), creates Lake Nasser, one of the world's largest artificial lakes (Fig. 7-11). As the map shows, the lake extends into Sudan, where 50,000 people had to be resettled to make way for it. The Aswan High Dam increased Egypt's irrigable land by nearly 50 percent and today provides the country with about 40 percent of its electricity. But, as is so often the case with megaprojects of this kind, the dam also produced serious problems. Snail-carried schistosomiasis and mosquito-transmitted malaria thrived in the dam's standing water, afflicting hundreds of thousands of people living nearby. By blocking most of the natural fertilizers in the Nile's annual floodwaters, the dam necessitated the widespread use of artificial fertilizers, proving very costly to small farmers and damaging to the natural environment. And the now fertilizer- and pesticide-laden Nile no longer supports the fish fauna offshore, reducing the catch and depriving coastal populations of badly needed proteins.

Valley and Delta

Egypt's elongated oasis along the Nile, just 5 to 25 kilometers (3 to 15 mi) wide, broadens north of Cairo across a delta anchored in the west by the great city of Alexandria and in the east by Port Said, gateway to the Suez Canal. The delta contains extensive farmlands, but it is a troubled area today. The ever more intensive use of the Nile's water and silt upstream is depriving the delta of much needed replenishment. And the low-lying delta is geologically subsiding, raising fears of salt-water intrusion from the Mediterranean Sea that would damage soils here.

Egypt's millions of subsistence farmers, the *fellaheen*, still struggle to make their living off the land, as did the peasants of the Egypt of five millennia past. Rural landscapes seem barely to have changed; ancient tools are still used, and dwellings remain rudimentary. Poverty, disease, high child mortality rates, and low incomes prevail. The Egyptian government is proceeding with grandiose plans to expand its irrigated acreage along a series of depressions west of the Nile (Fig. 7-11), but without modernization and greater efficiency in the existing farmlands the future is dim.

area, allowing farmers to harvest more than one crop per year on the same field. In a single century, all of Egypt's farmland was brought under this kind of *perennial irrigation*.

AMONG THE REALM'S GREAT CITIES . . . Cairo

STAND ON THE roof of one of the high-rise hotels in the heart of Cairo, and in the distance you can see the great pyramids, monumental proof of the longevity of human settlement in this area. But the present city was not founded until Muslim Arabs chose the site as the center of their new empire in AD 969. Cairo became and remains Egypt's primate city, situated where the Nile River opens into its delta, home today to nearly one-sixth of the entire country's population.

Cairo at 12.5 million ranks among the world's 15 largest urban agglomerations, and it shares with other cities of the poorer world the staggering problems of crowding, inadequate sanitation, crumbling infrastructure, and substandard housing. But even among such cities, Cairo is noteworthy for its stunning social contrasts. Along the Nile waterfront, elegant skyscrapers rise above carefully manicured, Parisian-looking surroundings. But look eastward, and the urban landscape extends gray, dusty, almost featureless as far as the eye can see. Not far away, more than a million squatters live in the sprawling cemetery known as the City of the Dead (see photo below). On the outskirts, millions more survive in overcrowded shantytowns of mud huts and hovels.

And yet Cairo is the dominant city not only of Egypt but for a wider sphere; it is the cultural capital of the Arab World, with centers of higher learning, splendid museums, world-class theater and music, and magnificent mosques and Islamic learning centers. Although Cairo always has been primarily a center of government, administration, and religion, it also is a river port and an industrial complex, a commercial center, and, as it sometimes seems, one giant bazaar. Cairo is the heart of the Arab World, a creation of its geography, and a repository of its history.

© H. J. de Blij, P. O. Muller, and John Wiley & Sons, Inc.

FROM THE FIELD NOTES

"On the eastern edge of central Cairo we saw what looked like a combination of miniature mosques and elaborate memorials. Here lie buried the rich and the prominent of times past in what locals call the 'City of the Dead.' But we found it to be anything but a dead part of the city. Many of the tombs here are so large and spacious that squatters have occupied them. Thus the City of the Dead is now an inhabited graveyard, home to at least one million people. The exact numbers are impossible to determine; indeed, whereas metropolitan Cairo in 2010 has an official population of 12.5 million, many knowledgeable observers believe that 17 million is closer to the mark."
© H. J. de Blij

www.conceptcaching.com

Subregions of Egypt

Egypt has six subregions, which are mapped in Figure 7-11. Most Egyptians live and work in Lower (i.e., northern) and Middle Egypt (regions ① and ②), the country's core area anchored by Cairo and flanked by the leading port and second industrial center, Alexandria. The economy has benefited from further oil discoveries in the Sinai (region ⑥) and in the Western Desert (region ④), so that Egypt now is self-sufficient and even exports some petroleum. Cotton and textiles are the other major source of external income, but the important tourist industry has repeatedly been hurt by Islamic extremists. As the population mushrooms, the gap between food supply and demand widens, and Egypt must import grain. Since the late 1970s, Egypt has been a major recipient of U.S. foreign aid.

Egypt today is at a crossroads in more ways than one. Its planners know that reducing the high birth rate would improve the demographic situation, but revivalist Muslims object to any programs that promote family planning. Its accommodation with Israel helps ensure foreign aid but divides the people. Its government faces challenges from pro-democracy as well as fundamentalist sides. Egypt's future, in this crucial corner of the realm, is uncertain.

Divided, Turbulent Sudan

As noted above, Egypt's Nile results from the confluence of the White and Blue Niles in Sudan (Fig. 7-11). Only two roads—one interrupted by a ferry across Lake Nasser and the other along the coast—link the two countries of this region, but the historical geographies of these fractious neighbors have long been interconnected.

The map reminds us how much larger Sudan is, although its population is barely half that of Egypt. This whole region fell under British administration during colonial times, when the British decided to combine Arabized and Islamized northern Sudan with the African and Christianized South. Unlike the North, where virtually everyone was Muslim, many southerners not converted to Christian faiths continued to adhere to traditional African religions.

Contrasts between North and South were not just cultural. The North is desert; the South is swamp and savanna. Figure 7-1 reveals this transition effectively: in the South, the White Nile is just one of many streams in a vast tropical wetland called the **Sudd**; but in the North, the Nile is confined to a single, desert-flanked channel. Sudan is a geographically divided country whose capital, Khartoum, lies in the northern sector where the two

Nile Rivers meet. In the inverted V-shaped zone between the converging rivers near the capital (known as the Gezira), the British colonists laid out a large irrigated agricultural project where cotton was the cash crop.

After the British departed, the regime in Khartoum wanted to impose its Islamic rule on the African South, resulting in a bitter and protracted civil war lasting more than three decades and costing an estimated 2 million lives. Millions more faced starvation as refugees in their own country, their government obstructing international relief efforts. It was one of the worst manifestations of the impact of the African Transition Zone (see pp. 333-337) on local peoples. Not until 2005 did international mediation produce a peace agreement that includes the stipulation that the South will be able to vote on secession—and independence—in 2011.

Misery in the Periphery

Whatever the outcome of this referendum, if indeed it ever happens, the future for Sudan's South is not bright. Look at the landlocked, poverty-stricken territory whose voters would decide either to remain the disadvantaged, marginalized periphery of their Muslim-dominated state or become an isolated, economically desperate country bordered by areas of endemic war in northern Uganda and The Congo, collapse in the Central African Republic and Chad, repudiation from the North of Sudan, and disconnects from Ethiopia and Kenya, with which there are no effective surface connections (Fig. 7-11). Among the tier of landlocked states along the African Transition Zone that extends from Mali in the west to Ethiopia in the east (Fig. 7-10), this as-yet unnamed country, if the voters approve secession, would still have the poorest prospects of all.

Following independence in 1956, Sudan itself epitomized the global periphery: its economy relied on the sale of cotton, sugar, sheep, and other farm products for revenues with which to buy essentials such as oil. Sudan has a 500-kilometer (300-mi) coastline on the Red Sea, where the limited facilities of Port Sudan handled what little trade there was, most of it from and to Saudi Arabia. The pursuit of the war against its southern provinces impoverished the Islamic regime in Khartoum, and in the 1970s and 1980s Sudan's per capita income was one of the lowest in the world.

Oil in the Desert

Fly into Khartoum today, and you are in for a surprise: a cluster of gleaming high-rise buildings tower over a low-rise townscape that has not changed for centuries.

The "international community" has been slow to respond to the crisis in Darfur, arguing over the question of whether the assault by Arabized Muslim militias on sedentary African farmers in this western province of Sudan technically constitutes genocide, and relying on an undermanned, poorly equipped, and inadequately mandated African Union force to stop the violence. The result can be seen in this photograph of the northern Darfur village of Bandago, where virtually every dwelling has been burned by *janjaweed* militias empowered by Khartoum's Islamic regime. By the middle of 2009, more than 2.5 million people had been driven from their homes, leaving behind charred remains like this; more than 300,000 had died; and this part of the African Transition Zone faced famine and dislocation for many years to come, whatever the result of the power struggles afflicting them. © AP/Wide World Photos

Only one of two commodities could be responsible for so rapid a transformation: oil or natural gas. In Sudan's case, it was oil.

During the 1980s it had become clear that Sudan's midsection contained significant oil reserves, and discoveries during the 1990s in Kordofan Province and even in some northern areas of the embattled South proved that an energy era was about to open here (Fig. 7-11). This complicated the political as well as the economic situation, and in short order Sudan became an exporter rather than an importer of oil. Foreign companies from Canada to China arrived to explore and exploit. The government saw its income rise dramatically, allowing it to buy more weaponry. Local peoples beneath whose lands the oil was found were ruthlessly dispossessed and pushed away. Those ultramodern high-rises in the Khartoum-Omdurman area reflected a new era of wealth and privilege.

Some leaders in the South viewed the oil bonanza as a ticket to self-sufficiency (the peace agreement stipulated that the South would receive half of local oil revenues, but the records are kept by the North, and trust has always been in short supply). Full control by the South over its share of oil exports would entail construction of a pipeline that avoids Sudan, through Kenya to the Indian Ocean coast, probably an unrealistic prospect. Furthermore, oil discoveries elsewhere in Africa (and, indeed, worldwide) have rarely had the favorable consequences anticipated by those most immediately affected. The South's relative location spells disadvantage either way.

Tragedy in Darfur

What oil can do for an authoritarian regime was tragically demonstrated by events in another part of Sudan that has become a byword for state-sponsored terrorism: Darfur. As Figure 7-11 shows (also see Fig. 6-19), three western provinces of Sudan are named Darfur (Arabic for "House of the Fur"). Southern and part of Western Darfur are inhabited by the Islamized but ethnically African Fur people. The north is predominantly Arab. Apparently, the centuries-old Islamization of the African Fur was not enough for the regime in Khartoum, which blamed these sedentary farmers for sympathizing with anti-Khartoum rebels. Soon the province's camel- and cattle-herding Arabs were invading the Africans' lands, supported by the army and by an informal militia named the *janjaweed* that burned villages, destroyed farms, and killed at will (see photo above).

By mid-2009, more than 2.5 million people had been driven from their lands, over 300,000 had been killed, the United Nations was trying to feed hundreds of thousands at risk of starvation, the conflict had spread into neighboring Chad and the Central African Republic, and the Khartoum regime, flush with money and influence with governments interested in its oil, could ignore the world's appeals. The tragedy of Sudan is a reminder of the capacity of a rapacious, oil-empowered regime to destroy the lives and hopes of millions under its territorial sway.

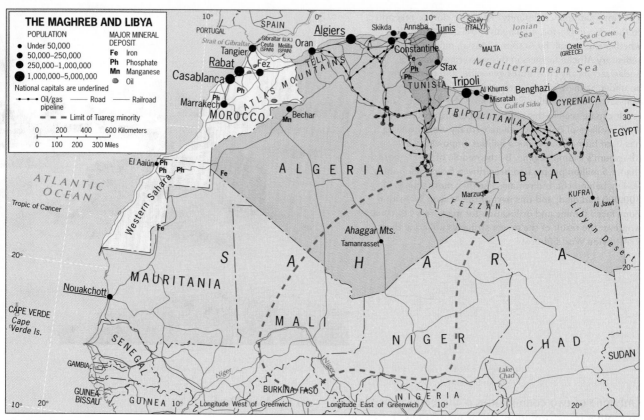

FIGURE 7-12

© H. J. de Blij, P. O. Muller, and John Wiley & Sons, Inc.

● THE MAGHREB AND ITS NEIGHBORS

The countries of northwestern Africa are collectively called the *Maghreb*, but the Arab name for them is more elaborate than that: *Jezira-al-Maghreb*, or "Isle of the West," in recognition of the great Atlas Mountain range rising like a huge island from the Mediterranean Sea to the north and over the sandy flatlands of the immense Sahara to the south.

The countries of the Maghreb (which is sometimes spelled *Maghrib*) are Morocco, last of the North African kingdoms; Algeria, a secular republic beset by the religious-political problems we noted earlier; and Tunisia, smallest and most Westernized of the three (Fig. 7-12). Neighboring Libya, facing the Mediterranean between the Maghreb and Egypt, is unlike any other North African country: an oil-rich desert state whose population is almost entirely clustered in settlements along the coast.

Atlas Mountains

Whereas Egypt is the gift of the Nile, the Atlas Mountains facilitate settlement of the Maghreb. These high ranges wrest from the rising air enough orographic rain-

fall to sustain life in the intervening valleys, where good soils support productive farming. From the vicinity of Algiers eastward along the coast into Tunisia, annual rainfall averages more than 75 centimeters (30 in), a total more than three times as high as that recorded for Alexandria in Egypt's delta. Even 250 kilometers (150 mi) inland, the slopes of the Atlas still receive over 25 centimeters (10 in) of rainfall. The effect of the topography can be read on the world map of precipitation (Fig. G-6): where the highlands of the Atlas terminate, desert conditions immediately begin.

The Atlas Mountains trend southwest-northeast and begin in Morocco as the High Atlas, with elevations close to 4000 meters (13,000 ft). Eastward, two major ranges dominate the landscapes of Algeria proper: the Tell Atlas to the north, facing the Mediterranean, and the Saharan Atlas to the south, overlooking the great desert. Between these two mountain chains, each consisting of several parallel ranges and foothills, lies a series of intermontane basins (analogous to South America's Andean *altiplanos* but at lower elevations), markedly drier than the northward-facing slopes of the Tell Atlas. In these valleys, the *rain shadow* effect of the Tell Atlas is reflected not only in the steppe-like natural vegetation but also in land-use patterns: pastoralism replaces cultivation, and stands of short grass and bushes blanket the countryside.

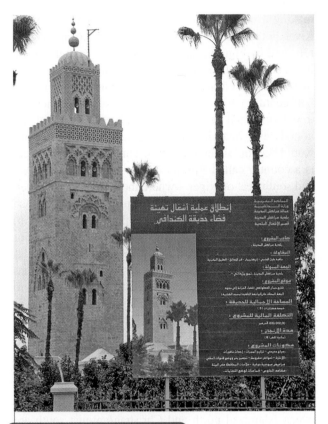

"Having seen the Giralda in Seville, Spain (photo p. 349), I was interested to view the tower of the Koutoubia Mosque in Marrakech, Morocco, also built in the twelfth century and closely resembling its Andalusian counterpart. Some 67 meters (220 ft) tall, the Koutoubia minaret was built by Spanish slaves and is a monument to the heyday of Islamic culture of the time. I soon discovered that it is not as easy as it looks to get a clear view of the tower, and a boy offered his help, guiding me around the block on which the mosque is situated. He was well informed on the building's history and on the varying fortunes of his home town. It was European mispronunciation of the name 'Marrakech' that led to Morocco being called Morocco, he reported. And Marrakech was Morocco's capital for a very long time, and still should be. 'But back to the mosque,' I said. 'What does the sign say?' He smiled. 'You'll just have to learn some Arabic, just as we know French,' he said. Obviously a teacher in the making, I thought."
© H. J. de Blij

Colonial Impact

During the colonial era, which began in Algeria in 1830 and lasted until the early 1960s, well over a million Europeans came to settle in North Africa—most of them French, and a large majority bound for Algeria—and these immigrants soon dominated commercial life. They stimulated the renewed development of the region's towns;

Casablanca, Algiers, Oran, and Tunis became the urban foci of the colonized territories. Although the Europeans dominated trade and commerce and integrated the North African countries with France and the European Mediterranean world, they did not confine themselves to the cities and towns. They recognized the agricultural possibilities of the favored parts of the *tell* (the lower Tell Atlas slopes and narrow coastal plains that face the Mediterranean) and established thriving farms. Not surprisingly, agriculture here is Mediterranean. Algeria soon became known for its vineyards and wines, citrus groves, and dates; Tunisia has long been one of the world's leading exporters of olive oil; and Moroccan oranges went to many European markets.

The Maghreb Countries and Libya

Between desert and sea, the Maghreb states display considerable geographic diversity. Morocco, a conservative kingdom in a revolutionary region, is tradition-bound and weak economically. Its core area lies in the north, anchored by four major cities; but the Moroccans' attention is focused on the south, where the government is seeking to absorb its neighbor, *Western Sahara* (a former Spanish dependency with less than 500,000 inhabitants, many of them immigrants from Morocco). Even if this campaign is successful, it will do little to improve the lives of most of Morocco's 32.1 million people. Hundreds of thousands have emigrated to Europe, many of them via Spain's two tiny exclaves on the Moroccan coast, Ceuta and Melilla (Fig. 7-12).

Algeria, rich in oil and natural gas, fought a bitter war of liberation against the French colonizers and has been in turmoil virtually ever since, with devastating economic and social consequences. Algiers, Algeria's primate city, is centrally situated along its Mediterranean coast and is home to 3.5 million of the country's 36 million citizens—but there are more Algerians in France than in the capital today. Conflict between those wanting to make Algeria an Islamic republic and the military-backed regime cost an estimated 100,000 lives; in 2005, the government sought to put this episode to rest by holding a referendum on the question of a general amnesty (it was approved). In recent years, new strife has threatened in the Kabylia region east of Algiers involving Algeria's Berber minority.

The smallest of the Maghreb states, Tunisia, lies at the eastern end of the region. As the Data Table inside the back cover shows, Tunisia in many ways outranks surrounding countries. Stability, national cohesion, and the maximization of limited natural resources (no petroleum reserves here comparable to Algeria or Libya) have yielded a higher GNI, higher social indicators, and a lower population growth rate (among its total of 10.5

million) than elsewhere in the Maghreb. Most of the country's productive capacity lies in the north, in the hinterland of its large, historic capital, Tunis. As the authoritarian national government slowly loosens its grip, Tunisia's relations with Europe—already the strongest in all of North Africa—continue to improve.

Almost rectangular in shape, Libya (6.6 million) lies on the Mediterranean Sea between the Maghreb states and Egypt (Fig. 7-12). What limited agricultural possibilities exist lie in the district known as Tripolitania in the northwest, centered on the capital, Tripoli, and in the northeast in Cyrenaica, where Benghazi is the urban focus. But it is oil, not farming, that drives the economy of this highly urbanized country. The oilfields lie well inland from the Gulf of Sidra, linked by pipelines to coastal terminals. Libya's two interior corners, the desert Fezzan district in the southwest and the Kufra oasis in the southeast, are connected to the coast by two-lane roads subject to sandstorms.

Between Maghreb and Sahel

Between the coastal states of the Maghreb/Libya and the southern margin of the Sahara lies a tier of five states, all but one of them landlocked, dominated by the desert, sustained by ribbons of water, and under the sway of Islam. Figure 7-1 best displays this chain of countries whose populations are not insignificant by African standards, totaling some 60 million.

Only coastal Mauritania has a minuscule population even in this company: less than 4 million in a territory the size of Texas and New Mexico combined, half of it concentrated in and around the coastal capital, Nouakchott, base of a small fishing fleet. Overwhelmingly Muslim and experiencing its first democratic presidential election as recently as 2007, Mauritania has been infamous for tolerating slavery despite its official ban in 1981; following the 2007 election, its parliament voted to impose prison terms on slave-owners continuing the practice. By contrast, neighboring Mali (13.6 million), is a multiparty, democratic state with a multicultural population that includes a sizeable minority adhering to traditional beliefs as well as a small Christian minority. Mali depends on the waters of the upper Niger (pronounced *NYE-jer*), the river on which its capital and irrigated farming zone lie, concentrating the population in the southwest—within the African Transition Zone. The fabled city of Timbuktu, today a mere shadow of its former greatness, lies downstream from the capital, Bamako. Mali shares with several neighbors an increasingly assertive minority of Tuaregs, a nomadic people of the interior Sahara (see Fig. 7-12) who are demanding a share of local mineral revenues, including uranium.

Uranium is now the leading export of Niger (pronounced *nee-ZHAIR*), at a time when nuclear power is making a worldwide comeback. As in Mali, the Niger River is the lifeline of this country of 15.6 million, as reflected by Figure 7-2, and Niamey, the capital, lies on its banks in the far southwest. We should remember that these landlocked and low-income countries emerged a half-century ago as independent states from French colonial rule, and when their boundaries were laid out nobody ever imagined that these colonial dependencies would become sovereign states. This explains why there is a country called Burkina Faso—lying south of Mali and southwest of Niger—where there is no Niger River, no coastline, and little economic opportunity (this is one of the world's poorest countries, with nearly half the population of 16.1 million subsisting on one dollar a day or less). Environmental variability here is marked by destructive floods alternating with severe droughts, the latter driving people and their livestock southward into Ivory Coast, resulting in deadly conflicts. Despite all this, a representative government in the capital of Ouagadougou manages to play a regional and international role in this part of the continent.

The fifth member of this tier of desert-margin states, Chad, lies east of Niger and, as Figure 7-10 shows, also astride the African Transition Zone. Here, divisions between Islamized north and Christian/animist south are strong, and Figures 6-19 and 7-10 show the capital, N'Djamena, lies directly on the Islamic Front. Since 2000, Chad (10.7 million) has been in turmoil, the crisis in Sudan's Darfur provinces spilling over into its eastern territorial margin, the south experiencing an oil boom (see Fig. 6-16), and the capital the focus of a struggle among rebel factions trying to overthrow the government.

Islamic militancy, commodities of growing value, and restive minorities are drawing the international attention of global powers to a periphery of North Africa long in the shadow of more prominent coastal neighbors to the north and south.

● THE MIDDLE EAST: CRUCIBLE OF CONFLICT

The regional term *Middle East*, we noted earlier, is not satisfactory, but it is so common and generally used that avoiding it creates more problems than it solves. It originated when Europe was the world's dominant realm and when places were "near," "middle," and "far" from Europe: hence a Near East (Turkey), a Far East (China, Japan, Korea, and other countries of East Asia), and a Middle East (Egypt, Arabia, Iraq). If you check definitions used in the past, you will see that the terms were applied inconsistently: Syria, Lebanon, Palestine, even Jordan sometimes were included in the "Near" East, and Persia and Afghanistan in the "Middle" East.

FIGURE 7-13

© H. J. de Blij, P. O. Muller, and John Wiley & Sons, Inc.

Today, the geographic designation *Middle East* has a more specific meaning. And at least half of it has merit: this region, more than any other, lies at the middle of the vast Islamic realm (Fig. 7-10). To the north and east of it, respectively, lie Turkey and Iran, with Muslim Turkestan beyond the latter. To the south lies the Arabian Peninsula. And to the west lie the Mediterranean Sea and Egypt, and the rest of North Africa. This, then, is the pivotal region of the realm, its very heart.

Five countries form the Middle East (Fig. 7-13): Iraq, largest in population and territorial size, facing the Persian Gulf; Syria, next in both categories and fronting the Mediterranean; Jordan, linked by the narrow Gulf of Aqaba to the Red Sea; Lebanon, whose survival as a unified state has come into question; and Israel, Jewish nation in the crucible of the Muslim world. Because of the extraordinary importance of this region in world affairs, we focus in some detail on issues of cultural, economic, and political geography in the discussion that follows.

Iraq: War and Its Consequences

In the immediate aftermath of the 9/11 terrorist attacks in the United States in 2001, American forces overthrew the ruling revivalist Taliban regime in Afghanistan that had given refuge to Usama bin Laden, who plotted the

A Future Kurdistan?

POLITICAL MAPS OF the region do not show it, but where Turkey, Iraq, and Iran meet, the cultural landscape is not Turkish, Iraqi, or Iranian. Here live the Kurds, a fractious and fragmented nation of at least 25 million (their numbers are uncertain). More Kurds live in Turkey than in any other country (perhaps as many as 15 million); possibly as many as 8 million in Iran; about half that number in Iraq; and smaller numbers in Syria, Armenia, and even Azerbaijan (see Fig. 7-13).

The Kurds have occupied this isolated, mountainous, frontier zone for over 3000 years. They are a nation, but they have no state; nor do they enjoy the international attention that peoples of other *stateless nations* (such as the Palestinians) receive. Turkish and Iraqi repression of the Kurds, and Iranian betrayal of their aspirations, briefly make the news but are soon forgotten. Relative location has much to do with this: spatial remoteness and the obstacles created by the ruling regimes inhibit access to their landlocked domain.

Many Kurds dream of a day when their fractured homeland will become a nation-state. Most would agree that the city of Diyarbakir, now in southeastern Turkey, would become the capital. It is the closest any Kurdish town comes to a primate city, although the largest urban concentration of Kurds today is in the shantytowns of Istanbul, where more than 3 million have migrated. In their heartland, meanwhile, the Kurds, their shared goals notwithstanding, are a divided people whose intense disunity has thwarted their objectives; and, as in the case of Europe's Basques, a small minority of extremists has tended to use violence in pursuit of their aims.

The Kurds—without oil reserves, without a seacoast, without a powerful patron, without a global public relations machine—are victims of a conspiracy between history and geography. One of the world's largest stateless nations keeps hoping for a deliverance that is unlikely to come.

assault. Soon, however, Afghanistan lost priority among U.S. concerns as Iraq was designated a security risk based on intelligence reports of its terrorist links, its alleged arsenal of weapons of mass destruction (WMD), its murderous regime, and the regional threat it posed. In March 2003 the United States attacked and invaded Iraq and following a brief struggle ended the rule of the Sunni-Muslim clique led by Saddam Hussein. While the search for WMD continued, the American government now faced the task of stabilizing Iraq, reconstructing its damaged infrastructure, and reconstituting its government.

It is therefore important to understand key aspects of the regional geography of California-sized Iraq and its official population of 30.9 million (which may in fact be significantly less because of unregistered emigration). The country constitutes nearly 60 percent of the total area of the Middle East as we define it and has more than 40 percent of its population. With its world-class oilfields, major gas reserves, and large areas of irrigable farmland, Iraq also is best endowed of all the region's countries with natural resources. Iraq is heir to the early Mesopotamian states and empires that emerged in the Tigris-Euphrates Basin, and the country is studded with matchless archeological sites and museum collections, to which disastrous damage was done during and after the 2003 invasion from combat and looting.

The War's Wider Regional Impact

Please take a careful look at Figure 7-13, showing the location of Iraq in the region we have defined as the Middle East, which reveals that Iraq has six immediate neighbors. Perhaps the most obvious feature of this map is the lavender-striped area that not only covers northeastern Iraq but also large parts of Turkey and Iran as well as smaller parts of adjacent countries. This is the area where the majority of the inhabitants are Kurds, who have been discriminated against by all the countries in which they live. About 15 million Kurds are in Turkey, some 8 million in Iran, and about 4 million in Iraq, and together they form one of the world's largest **stateless nations**, 18 divided and often exploited by those who rule over them (see box titled "A Future Kurdistan?").

The Kurdish population not only in Iraq but also in several of Iraq's neighbors is only one reason why it is important to study Figure 7-13 closely. Turkey is just one of the countries bordering war-torn Iraq that are directly concerned over the outcome of the struggle that began in 2003. Also anxious is Iran to the east, which fought a bitter war with Iraq in the 1980s but which shares Shia Islam with a majority of Iraqis. To the west, Syria was Iraq's ally during Saddam's despotic rule, and like Iraq under Saddam, Syria itself still is ruled by a strongman and his minority clique. To the south, Iraq is bordered by Kuwait, the country it invaded in 1990, triggering the first Gulf War, and by Saudi Arabia, separated from populated Iraq by a wide swath of desert but troubled by the impact any resolution of Iraq's problems may have at home. And to the southwest lies Jordan, which gave Iraq some help during the first Gulf War but later turned its back not only on Saddam but also on the Sunni insurgents who waged war on American and allied forces following Saddam's ouster.

With so many land neighbors, it is no surprise that Iraq is very nearly landlocked, and Figure 7-14 shows how short and congested its single outlet to the Persian Gulf is (that outlet was a factor in Saddam's 1990 decision to try to conquer and annex Kuwait). As a result, a network of pipelines across Iraq's neighbors must carry much of Iraq's oil export volume to coastal terminals in other countries—another reason these countries have an interest in what happens to Iraq in the years ahead.

Cultural and Political Geography in Discord

Let us now carefully examine the regional map of Iraq (Fig. 7-14). Note that Iraq's two vital rivers, the Tigris and Euphrates, rise in Turkey. The map shows that Iraq is divided into three cultural domains: that of the Shia majority in the south, the Sunni minority in the northwest, and the Kurdish minority in the hilly northeast. Not far from the geographic center of Iraq lies the capital, Baghdad, astride the Tigris River at the heart of the country's core area. The Baghdad urbanized area has a population today of between 6 and 7 million, a vast and often violent multicultural agglomeration of about one-quarter of the entire country's population (Fig. 7-15). Only the Shia south, which extends from the eastern neighborhoods of Baghdad all the way south to Basra and the shores of the Persian Gulf, has more people than Baghdad. The Sunni northwest, including the "Sunni Triangle" where Saddam Hussein's clan had its base and the turbulent province of Anbar (see inset map in Fig. 7-14) has between 5 and 6 million, and the Kurdish sector about 4 million. Of these three cultural domains, the Kurdish sector, the most remote of the three from the worst of Iraq's violent postwar transition, shows what might have been in the rest of the country: comparative stability and security have produced progress and prosperity.

The Outcome in Sight?

In the middle of 2009 the outlines of Iraq's postwar future seemed to be in prospect, although serious obstacles left the overall outcome of the military campaign begun in 2003 in doubt. Although the notion of a formal three-way partitioning of Iraq had been discarded, there could be no doubt that the concept of three spheres of dominant influence would be imprinted on Iraq's future map. The most difficult problem in this context was not so much the relatively blurred Sunni-Shia frontier but

rather the boundary between the Sunni and Kurdish spheres, which had become a key issue involving ethnic, historical, and economic factors.

As a postwar political and administrative framework takes shape, Shia and Sunni interests are being contested, mostly, in the evolving Iraqi parliament. A border between their respective cultural domains is not an overriding issue. But when it comes to the boundaries of a future Kurdish region, things get more complicated. The frontier between Kurdish and non-Kurdish domains in the northeast involves not only the Kurds themselves but also Arabs and the Turkmen minority concentrated here. It involves the history of Saddam's dreadful ethnic cleansing and "Arabization" of Kurdish areas. And, most crucial of all, it involves oil.

In 2009, the officially recognized Kurdish region still consisted only of the three northern provinces of Dahuk (Dohuk), Arbil (Erbil), and Sulaymaniyah (see the inset map on page 374). But as the main map in Figure 7-14 shows, the Kurdish domain—that is, the region where Kurds are effectively in control—is much larger. What the Kurds want is recognition of that larger domain, including not only significant oil reserves but also the city they call Kirkuk and the Arabs call Tamim. The dashed line in Figure 7-14 reflects approximately what the Kurds want. However, not only the (Sunni) Arabs but also the local Turkmen object to this. And the government of Turkey next door makes it clear that it disapproves of anything that further empowers the Kurds in Iraq.

Iraq's future as a stable and well-functioning state depends in large measure on the resolution of this complicated problem. The Kurdish region has been by far the most successful of Iraq's four regions (including Baghdad), making economic progress, working well administratively, improving its social services including schools and hospitals, and mostly avoiding the mayhem that engulfed the rest of Iraq. But now the danger is that Kirkuk could become the new Baghdad, the adversaries not being Shi'ites and Sunnis but Arabs and Kurds. If the parties succeed in reaching a consensus, Iraq's future will brighten. If they fail, the costs would be incalculable, not only to Iraq but to all the countries of the region.

Syria

One of those countries is Syria which, like Iraq in the past, is ruled by a minority. Although Syria's population of 20.9 million is about 75 percent Sunni Muslim, the ruling elite comes from a smaller Shi'ite sect based there, the Alawites. Leaders of this powerful minority have retained control over the country for decades, at times by ruthless suppression of dissent. In 2000, president-for-life

IRAQ

POPULATION
- Under 50,000
- 50,000–250,000
- 250,000–1,000,000
- 1,000,000–5,000,000
- Over 5,000,000

- Oilfield
- Road
- Railroad
- Oil pipeline
- Province boundary
- **T** Turkmen Minority
- **Su** Sunni Minority

Limit of strong Kurdish regional influence

National capitals are underlined

Domains depict prevailing majorities; minorities (Turkmen, Assyrians) not shown. Borderlands between domains have mixed populations.

0 40 80 120 160 200 Kilometers
0 20 40 60 80 100 Miles

TURKEY

Elazig
Malatya
Adiyaman
Gaziantep
Kilis
Sanliurfa
Aleppo
Diyarbakir
Batman
Lake Van
Tatvan
Van
Hakkari
Salmas
Naxcivan
Orumiyeh
Lake Orumiyeh
Zanjan
Qazvin

SYRIA
Qamishli
Zakhu
Dayr az Zawr
Lake Asad

Mosul
T Irbil
Kurdish Domain
Kirkuk (Tamim) **T**
Sulaymaniyah
T
Sanandaj
Hamadan
IRAN
Arak
Borujerd
Kermanshah

Mesopotamian
Sunni Domain

Tikrit
Samarra
Tigris River
Euphrates River
Qadisiyah Lake
Tharthar Lake
"Sunni Triangle"
Fallujah
Ramadi
Baqubah
Mandali
Khorramabad

Baghdad

Zagros Mts.
Dezful

JORDAN
Rutbah

D E S E R T
Nukhayb
Badanah

I R A Q
Karbala
Hillah
Najaf
Kut
Diwaniyah
Shia Domain
Amarah
Ahvaz
KHUZESTAN PROVINCE

Rafha

Nasiriyah
Su
Su
Basra
Abadan
Su

SAUDI ARABIA
Hafar al Batin

Rumaylah Oilfield
Kuwait
Persian Gulf
KUWAIT
Subayhiyah

PROVINCES OF IRAQ

0 100 200 Kilometers
0 50 100 Miles

TURKEY

Dahuk
Arbil
SYRIA
Ninawa (Nineveh)
Tamim (Kirkuk)
Sulaymaniyah
Salah ad Din (Saladin)
IRAN
Diyala
Baghdad
Anbar
Babil
Wasit
Karbala
Qadisyah
Maysan
Najaf
Dhi Qar
JORDAN
Basra
Muthanna

SAUDI ARABIA
KUWAIT

Longitude East of Greenwich

Hafez al-Assad died and was succeeded by his son Bashar, signaling a continuation of the political status quo. For 25 years, part of this status quo was the occupation and control of neighboring Lebanon, but in 2005 that came to an end.

Like Lebanon and Israel, Syria has a Mediterranean coastline where crops can be raised without irrigation. Behind this densely populated coastal zone, Syria has a much larger interior than its neighbors, but its areas of productive capacity are widely dispersed. Damascus, in the southwest corner of the country, was built on an oasis and is considered to be the world's oldest continuously inhabited city. It is now the capital of Syria, with a population of 2.7 million. The far northwest is anchored by Aleppo, the focus of cotton- and wheat-growing areas in the shadow of the Turkish border. Here the Orontes River is the chief source of irrigation water, but in the eastern part of the country the Euphrates Valley is the crucial lifeline. It is in Syria's interest to develop its eastern provinces, and recent discoveries of oil there will contribute to this.

As Figure 7-13 shows, and other larger-scaled maps later in this regional discussion emphasize, Syria's pending geographic issue involves a small southwestern corner of the country—the Golan Heights. Taken from Syria by Israel during a war in 1967 that changed the map of this part of the Middle East, this strategic territory remains under Israeli control despite occasional and tentative steps toward a resolution (for a close-up view, see the right inset map on page 377).

© H. J. de Blij, P. O. Muller, and John Wiley & Sons, Inc.

FIGURE 7-15

Jordan

Jordan, Syria's southern neighbor, is a classic case (and victim) of changing relative location. This desert kingdom was a product of the Ottoman collapse, but when Israel was created it lost its window on the Mediterranean Sea, the (now Israeli) port of Haifa, previously in the British-administered Mandate of Palestine. Following its independence in 1946 with about 400,000 inhabitants, newly-created Israel two years later bequeathed Jordan some 500,000 West Bank Palestinians and, later, a huge inflow of refugees. Today Palestinians outnumber original residents by more than two to one in the population of 6.1 million. It may be said that the 47-year rule by King Hussein, ending in 1999, was the key centripetal force that held the country together.

With an impoverished capital, Amman, without oil reserves, and possessing only a small and remote outlet to the Gulf of Aqaba, Jordan has survived with U.S., British, and other aid. It lost its West Bank territory in the 1967 war with Israel, including its sector of Jerusalem (then the kingdom's second-largest city). No third country has a greater stake in a settlement between Israel and the Palestinians than Jordan.

Lebanon

The Middle East map suggests that Lebanon has significant geographic advantages in this region: a lengthy coastline on the Mediterranean Sea; a world-class capital, Beirut, on its shoreline; oil terminals on its seafront; and a major capital (Syria's Damascus) in its hinterland. The map at the scale of Figure 7-13 cannot reveal still another asset: the fertile, agriculturally productive Bekaa Valley in the eastern interior.

French colonialism in the post-Ottoman era created a territory that in 1930 was about equally Muslim and Christian. Beirut became known as the "Paris of the Middle East." After independence in 1946, Lebanon functioned as a democracy and did well economically as the region's leading banking and commercial center. But the

FIGURE 7-14   © H. J. de Blij, P. O. Muller, and John Wiley & Sons, Inc.

Muslim sector of the population grew much faster than the Christian one, and in the late 1950s the Muslims launched their first rebellion against the established, Christian-dominated order. Further upheaval resulted from the arrival of several massive waves of Palestinian refugees, and in 1975 Lebanon fell apart in a civil war that wrecked Beirut, devastated the economy, and left the country at the mercy of Syrian forces that entered Lebanon to take control. In its wake came the emergence of the Iran-sponsored terrorist movement known as Hizbollah which, based in the dominantly Shi'ite south of Lebanon facing the border with Israel, also became a political force that gained representation in Lebanon's parliament. For so small a country (4.1 million, including more than 400,000 naturalized as well as noncitizen Palestinians), Lebanon is riven with religious and ethnic factions. After the Syrian forces were ousted in 2005 under United Nations auspices, a Sunni-led government, supported by a Christian faction, began the task of reconstruction. But in mid-2006 a Hizbollah team kidnapped two Israeli soldiers and killed several others, provoking massive Israeli retaliation that destroyed much of Lebanon's infrastructure and left the south devastated, though not incapacitated either politically or logistically: Iranian money ensured that Hizbollah would continue as a force in the state. Too many external powers are involved in Lebanon's affairs, it seems, for the country to have a chance at achieving stability and progress.

Israel and the Palestinian Territories

Israel, the Jewish state, lies at the heart of the Arab World (Fig. 7-10). Since 1948, when Israel was created as a homeland for the Jewish people on the recommendation of a United Nations commission, the Arab-Israeli conflict has overshadowed all else in the region.

Figure 7-16 helps us understand the complex issues involved here. In 1946 the British, who had administered this area in post-Ottoman times, granted independence to what was then called "Transjordan," the kingdom east of the Jordan River. In 1948, the orange-colored area became the U.N.-sponsored state of Israel—including, of course, land that had long belonged to Arabs in this territory called Palestine.

As soon as Israel proclaimed its independence, neighboring Arab states attacked it. Israel, however, not only held its own but pushed the Arab forces back beyond its borders, gaining the green areas shown in Figure 7-16. Meanwhile, Transjordanian armies crossed the Jordan River and annexed the yellow-colored area named the West Bank, including part of the city of Jerusalem. The king called his newly enlarged country Jordan.

More conflict followed. In 1967 a week-long war produced a major Israeli victory: Israel took the Golan Heights from Syria, the West Bank from Jordan, and the Gaza Strip from Egypt, and conquered the entire Sinai Peninsula all the way to the Suez Canal. In later peace agreements, Israel returned the Sinai but not the Gaza Strip. In 2005, however, the Israeli government decided to remove all Jewish settlements from Gaza, yielding this territory to the Palestinian Authority (Fig. 7-16, left inset map).

All this strife produced a huge outflow of Palestinian Arab refugees and displaced persons. The Palestinian Arabs constitute another of the region's stateless nations; about 1.3 million continue to live as Israeli citizens within the borders of Israel, but more than 2.3 million are in the West Bank and another 1.4 million in the Gaza Strip (the Golan Heights population is comparatively insignificant). Others live in neighboring and nearby countries including Jordan (2.7 million), Syria (435,000), Lebanon (405,000), and Saudi Arabia (325,000); another 200,000 or so reside in Iraq, Egypt, Kuwait, and Libya. Many have been assimilated into the local societies, but tens of thousands continue to live in refugee camps. In 2009, the scattered Palestinian population was estimated to total about 10 million.

Israel is about the size of Massachusetts and has a population of 7.7 million (including its 1.3 million Arab citizens), but because of its location amid adversaries and its strong international links, these data do not reflect Israel's importance. Israel has built a powerful military even as its regional Arab neighbors have also grown stronger; its policies arouse Arab passions; and Palestinian aspirations for a territorial state have yet to bear fruit. As a democracy with strong Western ties, Israel has been a recipient of massive U.S. financial aid, and U.S. foreign policy has been to seek a permanent accommodation between the Jews and Palestinians, recently in the form of a "two-state solution" that would create a Palestinian state alongside Israel. In wider context, the United States seeks a general accommodation between Israel and its Arab neighbors as well as between Jews and Arabs overall.

The geographic obstacles to such an accommodation include the following:

1. *The West Bank.* Even after its capture by Israel in 1967, the West Bank might have become a Palestinian homeland (and possibly a state), but Jewish immigration to the area made such a future difficult. In 1977 only 5000 Jews lived in the West Bank; by 2010 there were over 500,000, making up more than 18 percent of the population, claiming about 40 percent of the land, and creating a seemingly inextricable jigsaw of Jewish and Arab settlements (Fig. 7-17). This, in the view of many analysts, has created a situation that may render a two-state solution unattainable, with troubling implications for the future.

FIGURE 7-16

© H. J. de Blij, P. O. Muller, and John Wiley & Sons, Inc.

2. *The Gaza Strip.* The two-state solution was always going to be difficult for many reasons, including the separation of its territory between the West Bank and the Gaza Strip. Gaza is the region's most densely populated territory, a rectangular box of land on the Mediterranean coast of less than 380 sq km (147 sq mi) previously under Egyptian military rule but taken by Israel in the l967 war. Now with a population of almost 1.5 million, many living as refugees in squalid circumstances, Gaza was vacated in 2005 by the few Jewish settlers living there, its administration turned over to the Palestinian Authority. In 2006, Palestinian elections resulted in a significant split: the mili-

tant Islamist Hamas movement won in Gaza, and the more secular and moderate Fatah party prevailed in the West Bank. In 2007 the two factions fought a bitter battle in Gaza, where Hamas took complete control, ousting pro-Fatah moderates from all positions of influence. The Hamas leadership would not recognize Israel's right to exist nor participate in two-state negotiations. Rockets fired into Israel from Gaza created an atmosphere of insecurity that impelled Israel to launch a massive air and ground assault into Gaza in December 2008, which lasted nearly a month until just before the Obama administration was sworn into office on January 20, 2009.

LEBANON

GOLAN
HEIGHTS
(Occupied
by Israel)

SYRIA

33°

35° 35°30'

Jordan R.

Haifa

Sea of
Galilee
(Lake
Kinneret)

Nazareth

Afula

32°30'

Irbid

32°30'

Jenin

Hadera

Netanya

Tulkarm

Mediterranean Sea

Qalqiya

Nablus

JORDAN

WEST

BANK

Jordan R.

32°

Tel Aviv-
Jaffa

32°

Amman

Ramla

Ramallah

Jericho

Ashdod

Jerusalem

ISRAEL

Bethlehem

31°30'

Hebron

Dead
Sea

31°30'

Beersheba

35° Longitude East of Greenwich 35°30'

THE WEST BANK

Jewish settlement areas
inside Security Barrier

Jewish settlement areas
outside Security Barrier

Israeli settlements

Palestinian areas

National water carrier

Road

Railroad

Security Barrier,
completed and planned
Source: Israeli Ministry of Defence

National capitals are underlined

0   5   10   15 Kilometers

0   5   10 Miles

FIGURE 7-17   © H. J. de Blij, P. O. Muller, and John Wiley & Sons, Inc.

3. *The Security Fence.* Suicide attacks on Israel from the West Bank induced the Jewish state to construct a 700-km (440-mi) Security Barrier along its borders consisting of walls, fences, and intervening no-go zones (Fig. 7-17; see photo at right). This barrier does not conform either to Israel's 1948 UN boundary nor to its *de facto* postwar borders, cutting off an additional 10 percent of the West Bank's territory and in effect annexing it (and its inhabitants) to Israel. This reinforcement of the West Bank border imposes hardships on Palestinians: for some it runs between their homes and their farmlands, and for others it creates problems of transit to and from jobs through gated entry points where long delays and uncivil processing are common. Palestinians demand that the barriers be taken down; Israelis reply that the Palestinian Authority must control the terrorists, and point to the reduction in terrorist incidents as proof that the fence is necessary.

4. *Jerusalem.* The United Nations intended Tel Aviv to be Israel's capital, and Jerusalem an international city. But the Arab attack and the 1948–1949 war allowed Israel to drive toward Jerusalem (see Fig. 7-16, the green wedge into the center of the West Bank). By the time a cease-fire was arranged, Israel held the western part of the city, and Arab forces the eastern sector. But in this eastern sector lay major Jewish historic sites, including the Western Wall. Still, in 1950 Israel declared the western sector of Jerusalem its capital, making this, in effect, a *forward capital.* Figure 7-18 shows the position of the (black) armistice line, leaving most of the Old City in Jordanian hands. But then, in the 1967 war, Israel conquered all of the West Bank, including East Jerusalem; in 1980 the Jewish state reaffirmed Jerusalem's status as capital, calling on all nations to move their embassies from Tel Aviv. Meanwhile, the government redrew the map of the ancient city, building Jewish settlements in a ring around East Jerusalem that would end the old distinction between Jewish west and Arab east. This enraged Palestinian leaders, who still view Jerusalem as the eventual headquarters of a hoped-for Palestinian state.

5. *The Golan Heights.* The right inset map in Figure 7-16 suggests how difficult the Golan Heights issue is. The "heights" overlook a large area of northern Israel, and they flank the Jordan River and crucial Lake Kinneret (the Sea of Galilee), the water reservoir for Israel. Relations with Syria are not likely to become normalized until the Golan Heights are returned, but in democratic Israel the political climate may make ceding this territory impossible.

Israel's Security Barrier takes several forms – a concrete wall, an iron fence – and its effect also has multiple dimensions. This section of the barrier separates the village of Abu Dis on the outskirts of Jerusalem (left) from the West Bank (right). While Israel defines this demarcation as a security issue the Arabs, citing its location inside Palestinian territory, argue that it is motivated primarily by politics.   © AP/Wide World Photos

Israel lies at the center of a fast-moving geopolitical storm. The five issues raised above are only part of the overall problem: others include water rights, compensation for land expropriation, and the Palestinians' "right to return" to pre-refugee abodes. Now, in the age of nuclear, chemical, and biological weapons and longer-range missiles, Israel's search for an accommodation with its Arab neighbors is a race against time.

FIGURE 7-18

© H. J. de Blij, P. O. Muller, and John Wiley & Sons, Inc.

● THE ARABIAN PENINSULA

The Arabian Peninsula is one of the world's most clearly defined regions (Fig. 7-19). Flanked on three sides by water and on the fourth by only two countries (Jordan and Iraq), this expanse of deserts and mountains contains the birthplace of Islam, the largest known concentration of oil reserves in the world, and the country whose name is synonymous with oil and revivalist Islam, Saudi Arabia. Six other countries share the peninsula with the Saudi state. Clockwise from the north, these are Kuwait, Bahrain, Qatar, the United Arab Emirates (UAE), Oman, and Yemen. All of these names should be familiar to Americans. The United States led a war to liberate Kuwait from Iraq's invaders in the 1991 Gulf War. Bahrain and Qatar provided indispensable assistance following the U.S.-led invasion of Iraq in 2003. The United Arab Emirates is home to the "Hong Kong of Arabia," Dubai. Oman overlooks one of the world's narrowest and most dangerous strategic straits at the entrance to the Persian Gulf. Yemen, where a U.S. warship, the *U.S.S. Cole*, was incapacitated by a terror-

ist attack in 2000, has a key location alongside another narrow strait (at the mouth of the Red Sea) just across from Africa's violent Transition Zone.

Saudi Arabia

Saudi Arabia's planners think big. This vast desert country with only 29.6 million inhabitants (including about 7 million expatriate workers) for two decades has been spending its oil revenues on the planning and execution of gigantic projects, ranging from planned-from-scratch "economic cities" to bridges and tunnels such as the 50-kilometer (30-mi), U.S. $3 billion Aqaba Bridge that will link Egypt's waterfront resort town of Sharm-el-Sheikh at the tip of the Sinai Peninsula and Saudi Arabia's Ras Humayd via the island of Tiran. New industrial plants are being built to produce plastics, chemicals, and fertilizers as well as aluminum and steel for the building boom. There are blueprints for oil refineries and even more water desalinization plants than those currently operating.

Saudi Arabia owns the lion's share of what may be the largest concentration of oil reserves on Earth, and as Fig-

FIGURE 7-19

© H. J. de Blij, P. O. Muller, and John Wiley & Sons, Inc.

ure 7-19 shows, most of these reserves lie in the eastern part of the peninsula near (and under) the Persian Gulf and beneath the desert of the Rub al Khali (Empty Quarter). The oil goes to market in huge tankers that must negotiate the narrow entrance to the Persian Gulf or by cross-country pipelines to terminals on the Red Sea at Yanbu. Look at the map, and you will see that most of Saudi Arabia's economic development has occurred in a wide belt extending westward from the oil center of Dhahran on the

Persian Gulf through the national capital of Riyadh in the interior to the cities on and near the Red Sea coast that include the ports of Jiddah and Yanbu as well as the historic, fast-modernizing Mecca (Makkah) and Medina.

The Saudis have good reason to invest in their economic future. The population of 30 million citizens is growing fast—the rate of natural increase is twice the global average—and 38 percent of that population today is 15 years of age or less. Jobs by the millions will be needed. So the Saudi royal rulers are spending their country's oil money on six "economic cities" that by 2020 will create 1.3 million new jobs. By 2009, the sites of four of these cities had been selected, and construction was under way (see Fig. 7-19). King Abdullah Economic City (KAEC), at Rabigh on the Red Sea, is projected to be completed by 2016, when it will have a population of more than 2 million and an industrial complex of more than 2000 factories.

According to the grand design, the various economic cities will have diversified economic bases but will focus on particular combinations of industries (Ha'il, in the desert north, will concentrate on agricultural research, equipment, and innovation; Medina will become the site of the "Knowledge Economic City," with a new graduate-level university). In anticipation of the end of the oil era, the ultimate goal is to transform Saudi Arabia from a petrodollar-based economy into a global-scale industrial power.

Disparities and Doubts

This planned economic modernization, however, should be seen against a backdrop of cultural impediments that cast doubt on the entire enterprise. Skyscrapers and superhighways are no substitute for modern social institutions, and Saudi Arabia's conservative monarchism and ultraconservative religious bureaucracy create difficulties for the modernizers. Inevitable social contrasts arising from rapid economic growth tend to generate political opposition, but no adequate channels for the expression of such opposition exist in the absence of representative government. The country's legal system is in effect controlled by a cadre of puritanical judges whose rulings are not based on a body of codified law but on their interpretation of the will of Allah as revealed in religious texts. Wahhabist revivalism (see page 352) reinforces the strict segregation of women and men, and perpetuates archaic rules that include the prohibition against women driving motor vehicles, for which Saudi Arabia is castigated even in other quarters of the Islamic world. But that rule is only part of the "male guardianship" that inhibits women's advancement in this supposedly modernizing country. Women outnumber men in Saudi universities and colleges, but women constitute only a tiny fraction of the workforce. Women can graduate in their gender-segregated classes as lawyers, but the all-male justice ministry does not give them licenses to practice and, in any case, no judge would allow them into his courtroom.

Whether the Saudis' grandiose plans can succeed amidst such contradictions is therefore in question. The status of women is a crucial index of a society's modernity. Even as housing, health care, and education have improved markedly, regional disparities in the kingdom reveal the uneven allocation of its enormous revenues. In the more remote areas of the interior, where Bedouin nomads still follow their ancient caravan routes across the desert, Saudi Arabia presents a different aspect. Perhaps more significantly, most of the 9/11 terrorists were middle-class Saudis; Usama bin Laden's extremism arose from his distaste for the Saudi royalty and its Western links; significant numbers of Saudis enlisted as suicide terrorists during the U.S.-led war in Iraq; and al-Qaeda "leaders" in their remote mountain redoubts regularly rail against the corruption of Islamic "values" in the kingdom.

Meanwhile, the Arabian Peninsula is changing dramatically in all directions from the Saudis' conservative bastion. And beyond, a transformed Iraq and an increasingly militant Iran pose regional challenges that will affect long-term stability in this, the birthplace of the Prophet Muhammad.

On the Peninsula's Periphery

Saudi Arabia's neighbors on the peninsula it dominates may be small, but they are not inconsequential. They are easily remembered in two groups of three (Fig. 7-19). Clockwise from the head of the Persian Gulf, the first three, all small traditional Islamic monarchies, are Kuwait, Bahrain, and Qatar. The second group, larger and more diverse, consists of the United Arab Emirates (UAE), Oman, and Yemen. All but one (Yemen) are members, with Saudi Arabia, of the *Gulf Cooperation Council (GCC)*, formed in 1981 to promote the joint interests of states in this region. As Figure G-11 shows, four of the six are high-income economies.

At the very head of the Persian Gulf, Kuwait (2.8 million) almost cuts Iraq off from the open sea, which is one reason Saddam Hussein launched his war of annexation in 1990. Ministate Kuwait's oil reserves are among the world's five largest, but even Kuwait feels the need to diversify its economy. A major oil refinery, the largest of its kind in the realm, is being built near the capital. Like other wealthy countries in the GCC, Kuwait has attracted a large foreign workforce; in this mostly Sunni emirate, where 80 percent of the workers in private industry are foreigners, this has produced a 30-percent Shia minority. And, as in other GCC states, the working conditions for foreigners in this Arab country are often dismal, resulting in labor unrest and strikes. On the other hand, women were allowed to register and vote for the first time in the 2007 parliamentary elections.

With Shi'ite Iran directly across the Persian Gulf, Iran's nuclear ambitions and growing militancy worry the GCC countries, but in different ways. Bahrain, a group of tiny islands containing a population of about 800,000, already has a 70-percent Shia majority but is ruled by a Sunni-dominated regime. Iran has from time to time claimed Bahrain, most recently in 2007 when a high-ranking Iranian official renewed the notion in a government-endorsed newspaper. The clouds on Bahrain's horizon include its social divisions and recurrent (and at times violent) demonstrations as well as its diminishing reserves of oil and gas. The latter still support its high-income economy, but required reforms are not yet taking place.

Nearby Qatar lies on a small peninsula jutting into the Persian Gulf, a sandy wasteland made habitable by oil (now declining in importance) and gas (rising simultaneously and rapidly), with growing exports worldwide. With a small population (ca. 900,000) but a strong Sunni majority, this emirate is trying to establish itself as not only an Arab World focus for international economic, social, and political reform, but as a global center for financial and economic deliberation. Its capital, Doha, is the site of annual "Doha Round" meetings of the world's leading powers.

The second triad of GCC members occupies the large southern periphery of the Arabian Peninsula where they exhibit a very different geography. The United Arab Emirates (UAE), a federation of seven emirates with a total population of 4.6 million, faces the Persian Gulf between Qatar and Oman. In each emirate the reigning sheik is an absolute monarch, and the seven sheiks together form the Supreme Council of Rulers with a rotating presidency. Abu Dhabi has most of the oil in this federation and the largest cash reserves, but all eyes are on *Dubai* (Dubayy) and its huge port, which is increasingly called the Hong Kong of the Arabian Peninsula. A burgeoning financial center, high-technology hub, globally-linked airport, shipping base, ultramodern city, residential magnet, and tourist haven, Dubai became the synonym for modernization and comparative liberalism in the region. But the economic downturn starting in 2008 exposed globalizing Dubai's financial and social vulnerabilities, resulting in stagnation, labor unrest, and uncertainty (see photo p. 359).

The eastern corner of the Arabian Peninsula is occupied by the Sultanate of Oman (population: 2.8 million), another absolute monarchy, centered on the capital, Muscat. Figure 7-19 shows that Oman consists of two parts: the large eastern corner of the peninsula and a small but critical cape to the north, the Musandam Peninsula, that protrudes into the Persian Gulf to form a narrow *choke point* (see box titled "Choke Points: Danger on the Sea Lanes"). Here the narrow Hormuz Strait separates the Arabian Peninsula from Iran. Tankers that leave the other Gulf states must negotiate this narrow channel at slow speed, and during politically tense times warships have had to protect them. Iran's claim to several small islands near the Strait that are

FROM THE FIELD NOTES

"The port of Mutrah, Oman, like the capital of Muscat nearby, lies wedged between water and rock, the former encroaching by erosion, the latter crumbling as a result of tectonic plate movement. From across the bay one can see how limited Mutrah's living space is, and one of the dangers here is the frequent falling and downhill sliding of large pieces of rock. It took about five hours to walk from Mutrah to Muscat; it was extremely hot under the desert sun but the cultural landscape was fascinating. Oil also drives Oman's economy, but here you do not find the total transformation seen in Kuwait or Dubai. Townscapes (as in Mutrah) retain their Arab-Islamic qualities; modern highways, hotels, and residential areas have been built, but not at the cost of the older and the traditional. Oman's authoritarian government is slowly opening the country to the outside world after long-term isolation." © H. J. de Blij.

www.conceptcaching.com

Concept Caching

owned by the UAE is a potential source of dispute. Oman, with only limited oil resources that are likely to run out by 2030, has used its income frugally to improve infrastructure (a good road system and clean water supplies for all outlying areas) and to prepare for an oil-less future. A giant container port and free-trade zone have been built at Salalah, and major investments have been made in luxury tourist facilities, capitalizing on the country's sunshine and beaches. Oman also has gone further than most of the region's countries in attracting foreign investment and privatizing facilities such as its international airport.

At the southern tip of the Arabian Peninsula lies a key state on this region's periphery: Yemen (Fig. 7-19). Directly across the funnel-shaped Gulf of Adan from failed-state Somalia (where pirates prey on international shipping) and bordered by Saudi Arabia to the north, Yemen is the product of a merger between two neighbors

Choke Points: Danger on the Sea Lanes

19

THE HORMUZ STRAIT between the Arabian Peninsula and Iran is only one of dozens of maritime choke points that have taken on added significance during this time of terrorist challenges. A **choke point** is defined as the narrowing of an international waterway causing marine traffic congestion, requiring reduced speeds and/or sharp turns, and increasing the risk of collision as well as vulnerability to attack.

For many years, Southeast Asia's Strait of Malacca between the Malay Peninsula and the Indonesian island of Sumatera (Sumatra) has been a notorious choke point because reduced speeds in the Strait give pirates the opportunity to board vessels and plunder them or worse, kill their crews, and take them over. Since the 1980s, hundreds of acts of piracy have made this one of the world's least safe waterways, especially for smaller ships.

One of the world's busiest choke points is the Strait of Gibraltar between Spain and Morocco, which was in the news a few years ago when an al-Qaeda document was found that referred to plans for an attack on Western ships slowing down through this 60-kilometer (35-mi)-long funnel that narrows to 13 kilometers (8 mi) in width and is subject to high winds and fast currents. Another particularly risky choke point today is the Gulf of Adan leading to the narrow Bab el Mandeb Strait between Africa and the Arabian Peninsula, or, more specifically, between Djibouti

and Yemen, at the southern entrance to the Red Sea. Barim (Perim) Island partially blocks this 30-kilometer (18-mi)-wide strait. Hundreds of acts of piracy by Somali bandits, based on the beaches of the nearby Horn of Africa, have imperiled shipping in these sea lanes (see photo page 336). In 2008 they seized, among other ships, a freighter loaded with heavy weapons and a fully loaded 300,000-ton oil tanker; but their attempt to hijack the *Maersk Alabama* and its American crew in April 2009 triggered a successful counterattack by the U.S. Navy and strengthened international resolve to put an end to this scourge.

The English Channel between Britain and France, Turkey's Bosporus and Dardanelles between the Black Sea and the Mediterranean Sea, Indonesia's Sunda Strait between the Java Sea and the Indian Ocean, the Strait of Magellan for ships rounding the southern end of South America, and the Hainan Strait between southernmost China's Leizhou Peninsula and Hainan Island are just some of the many other choke points on the world map.

And some choke points are artificial. The Suez Canal between the Red Sea and the Mediterranean Sea and the Panama Canal between Atlantic (Caribbean) and Pacific waters are the busiest and most prominent, but many other route-shortening canals are vulnerable choke points at a time when the risk of hostile action looms larger than the risk of collision.

---

dating from 1990. San'a, the north's interior capital, became the new country's seat of government. Adan (Aden), major port and former capital, continues to anchor the south. For some time, it seemed that united Yemen was on a path toward representative government and progressive social policies; the minister for human rights, a woman named Wahiba Fare, campaigned for greater freedoms. But Yemen also was a haven for terrorists and kidnappers, and is the ancestral home of Usama bin Laden, whose family moved to Saudi Arabia before he was born.

With a complex cultural geography and a population almost as large as giant Saudi Arabia's (23.6 million, 55 percent Sunni and 45 percent Shi'ite), Yemen's future is in the balance. Its strategic location at the Bab el Mandeb ("Gate of Grief"), a choke point between the Red Sea and the Indian Ocean, attracts opportunists af all kinds, from pirates to terrorists. The possibility of significant oil reserves in the recently delimited border area with Saudi Arabia entices other interests. Meanwhile, the government has difficulty asserting its authority over the tribal

interior, and its early momentum toward democracy has been lost. Yemen may not be a well-known country, but its crucial location and turbulent neighborhood make it an important corner of the Arabian Peninsula.

● THE EMPIRE STATES

Two major states, both with imperial histories, dominate the region that lies immediately to the north of the Middle East and Persian Gulf: Turkey and Iran (Fig. 7-10). Although they share a short border and are both Islamic countries, they display significant differences as well. Even their versions of Islam are different: Turkey is an officially secular but dominantly Sunni state, whereas Iran is the heartland of Shi'ism. Turkey's leaders have established satisfactory relations with Israel; Iran's president promised to "wipe Israel off the map." In recent years, Turkey has been trying to negotiate entry into the European Union. Iran's goals have been quite different as the country has defied international efforts to constrain its nuclear ambitions. Two smaller countries are inextricably bound up with these two regional powers: island *Cyprus* to the southwest, divided today between Turks and Greeks in a way that threatens Turkey's European ambitions (see pp. 91-93); and oil-rich, Caspian Sea-bordering *Azerbaijan* to the north, with ethnic and religious ties to Iran but economic links elsewhere (see p. 144).

Turkey

Earlier in this chapter we chronicled the historical geography of the Ottoman Empire, its expansion, cultural domination, and collapse. By the beginning of the twentieth century, the country we now know as Turkey lay at the center of this decaying and corrupt state, ripe for revolution and renewal. This occurred in the 1920s and thrust into prominence a leader who became known as the father of modern Turkey: Mustafa Kemal, known after 1933 as Atatürk, meaning "Father of the Turks."

Capitals Old and New

The ancient capital of Turkey was Constantinople (now Istanbul), located on the Bosporus, part of the strategic straits connecting the Black and Mediterranean seas. But the struggle for Turkey's survival had been waged from the heart of the country, the Anatolian Plateau, and it was here that Atatürk decided to place his seat of government. Ankara, the new capital, possessed certain advantages: it would remind the Turks that they were (as Atatürk always said) Anatolians; it lay nearer the center of the country than Istanbul; and it could therefore act as a stronger unifier (Fig. 7-20). Istanbul lies on the threshold of Europe, with the minarets and mosques of this largest and most varied Turkish city rising above a townscape that resembles Eastern Europe.

Although Atatürk moved the capital eastward and inward, his philosophical orientation was westward and outward. To implement his plans for Turkey's modernization, he initiated reforms in almost every sphere of life within the country. Islam, formerly the state religion, lost its official status. The state took over most of the religious

FIGURE 7-20

© H. J. de Blij, P. O. Muller, and John Wiley & Sons, Inc.

AMONG THE REALM'S GREAT CITIES . . .   Istanbul

FROM BOTH SHORES of the Sea of Marmara, northward along the narrow Bosporus toward the Black Sea, sprawls the fabulous city of Istanbul, known for centuries as Constantinople, headquarters of the Byzantine Empire, capital of the Ottoman Empire, and, until Atatürk moved the seat of government to Ankara in 1923, capital of the modern Republic of Turkey as well.

Istanbul's site and situation are incomparable. Situated where Europe meets Asia and where the Black Sea joins the Mediterranean, the city was built on the requisite seven hills (as Rome's successor), rising over a deep waterway that enters the Bosporus from the west, the famous Golden Horn. A sequence of empires and religions endowed the city with a host of architectural marvels that give it, when approached from the water, an almost surreal appearance—and what is seen today is a mere remnant of history's accretion, survivals of earthquakes, fires, and combat.

Turkey's political capital may have moved to Ankara, but Istanbul remains its cultural and commercial headquarters. It also is the country's leading urban magnet, luring millions from the poverty-stricken countryside. In the heart of the city known as Stamboul and in the "foreign" area north of the Golden Horn called Beyoglu, modern buildings vie for space and harbor views, blocking the vistas that long made Istanbul's cultural landscape

unique. On the outskirts, shantytowns emerge so rapidly that Istanbul's population is doubling every 15 years at current rates, having reached 10.5 million in 2010. Istanbul's infrastructure is crumbling under this unprecedented influx, its growth is virtually without planning controls, and its legacy of two millennia of cultural landscapes is threatened.

© H. J. de Blij, P. O. Muller, and John Wiley & Sons, Inc.

schools that had controlled education. The Roman alphabet replaced the Arabic. A modified Western code supplemented Islamic law. Symbols of old—growing beards, wearing the fez—were prohibited. Monogamy was made law, and the emancipation of women was begun. The new government emphasized Turkey's separateness from the Arab World, and it has remained aloof from the affairs that engage other Islamic states.

As Figure 7-20 indicates, Turkey is a mountainous country of generally moderate relief; it also exhibits considerable environmental diversity ranging from steppe to highland (Fig. G-7). On the dry Anatolian Plateau villages are small, and subsistence farmers grow cereals and raise livestock. Coastal plains are not large, but they are productive and densely populated. Textiles (from home-grown cotton) and farm products dominate the export economy, but Turkey also has substantial mineral reserves, some oil in the southeast, massive dam-building projects on the Tigris and Euphrates rivers, and a small steel industry based on domestic raw materials. Normally self-sufficient in staples, Turkey imports oil and gas, chemical

products, automobiles, and machinery—a pattern that is changing as its economy records rapid growth in most (recent) years. Germany, Turkey's leading trade partner and home to as many as 3 million Turkish (many of them Kurdish) immigrants, has not been among Europe's most enthusiastic advocates of Turkish admission to the EU, a political complication that reflects current tensions in Germany over immigration issues. Meanwhile, Turkey's tourist industry is thriving as the country's incomparable archeological, architectural, and scenic attractions are enhanced by improving infrastructure. Turkey is an international, globalizing state in many ways far ahead of any and all other countries in this realm.

Armenians, Kurds, and Religious Minorities

But old issues still cloud Turkey's horizons. From before Atatürk's time, the Turks had a history of mistreating minorities. Soon after the outbreak of World War I, the (pre-Atatürk) regime decided to expel all the Armenians, concentrated at the time in Turkey's northeast. Nearly two million Turkish Armenians were uprooted and brutally forced out. An estimated 600,000 died in a campaign that still arouses anti-Turkish emotions among Armenians today.

In more recent times, Turkey has been criticized for its treatment of its large and regionally concentrated Kurdish population (a nation discussed in the box on p. 372). About one-fifth of Turkey's population of 76.8 million is Kurdish, and successive Turkish governments have mishandled relationships with this minority nation, even prohibiting the use of Kurdish speech and music in public places during one especially repressive period. The historic Kurdish homeland lies in southeastern Turkey, centered on Diyarbakir, but millions of Kurds have moved to the shantytowns around Istanbul—and to jobs in the countries of the European Union. The Kurdish issue has been an obstacle in Turkey's aspirations to join the EU, whose regulations relating to human rights and minority protections are strict. The Turks have made substantial progress: the death penalty has been abolished, and Kurds have been given freedom to express their culture in ways long forbidden. But some Kurds still resort to violence to redress old (or new) grievances, and Turkish forces still pursue some of them across international borders. Reports of mistreatment in jails and abrogations of due process continue to surface. In a related human rights arena, protections for women subject to Islamic family traditions have been strengthened, although these are more difficult to ensure in remote rural areas than in urban communities.

Another cultural issue in modern Turkey involves one of Islam's many sects, in this case the Alevis. Most of Turkey's Muslims are Sunnis, and some are orthodox and strict. The Alevis, who number about 12 million (more than 15 percent of the population), are Shi'ites—but they practice a rather nonchalant version of the faith. While orthodox Sunnis worship in mosques, live by Islam's daily rules, and fast during the daytime during Ramadan, Alevis do not gather in mosques and do not follow the rules of the Ramadan fast. This has led to serious cultural conflict, including murder and arson. The historic center of Turkey's Alevi sect lies in Sivas, in the east-central part of the country northwest of the Kurdish zone. Like the Kurds, many Alevis have left their cultural base, where they are under pressure from orthodox Sunnis, for the more cosmopolitan cities.

Still another, and growing, cultural issue in Turkey involves the country's Christian minority, numbering about 350,000. Religious freedom is a requirement for EU admission, and Turkey has repealed its laws prohibiting Christians from repairing damage to their churches, for example after local earthquakes. But reports of physical attacks on Christians are becoming more frequent, and Turkey will not acknowledge the religious title of the spiritual leader of more than 200 million Orthodox Chris-

tians—who is a Turkish citizen. As visitors to Istanbul can attest, the Greek Orthodox Church in Istanbul is in disrepair as its membership dwindles. The state may be secular, but some faiths have it easier than others.

A New Era?

The first decade of the twenty-first century was marked by a series of important developments in Turkey, many of them related to the prospect that Turkey would eventually achieve membership in the European Union. In the general election of 2002, an Islamist party led the voting, and its leader, Recep Tayyip Erdogan, became prime minister. In previous elections, Islamist parties were either prevented from running or were barred by the secular-minded army from taking office. The 2002 election was therefore a milestone and satisfied EU requirements that prospective members' elections must be free and fair. Erdogan, who had proclaimed his intention to maintain Turkey's secular status, won again in 2007, and nominated for president a politician whose wife insisted on wearing the Muslim head scarf, which is banned by law from all official functions in Turkey. Despite protest rallies and angry demonstrations, President Abdullah Gul was inaugurated. Meanwhile, Erdogan managed to persuade the Turkish Parliament to approve a referendum on the question of a constitutional amendment to henceforth elect the country's president by popular vote rather than parliamentary majority. Voters approved that amendment, ushering in a new and more representative era in Turkish politics. Furthermore, in an unprecedented and complicated development, Kurdish candidates in the 2007 election whose party had no chance of reaching the minimum threshold for admission to parliament sought election as independents; 26 were elected and of these 20 rejoined their party, thereby creating a formal parliamentary presence for Kurdish nationalists in the Turkish Parliament for the first time ever. This development was greeted with approval in the EU.

All these events bear witness to a crucial country in challenging transition. Turkey seeks to steer an independent and singular course in one volatile realm while accommodating the demands of another, and many Turks are having second thoughts. Around the turn of this century, support for joining the EU among adult Turks was about 75 percent, but reports of opposition in some EU members (notably Germany) and criticism of Turkey's cultural norms took their toll on public opinion. By 2009, support had dwindled to about 25 percent. If a modernizing Islamic country as secular and sophisticated as Turkey cannot be accepted by a Europe that was willing to accept Christian Romania and Bulgaria with all their shortcomings, Turks argue, then it is Islam, not human

Tehran is a city of murals, the most famous of which probably is the "Down with America" slogan on a high-rise at a busy downtown intersection. The face of Iran's "supreme leader," the Ayatollah Ali Khamenei, looms over parks and squares, in this case Enghelab Square, also in central Tehran. Its painter, Mehdi Qadyanloo, would prefer to replace slogan-inspired murals with new ones reflecting Iranian culture and expressing messages of peace and hope. One of these hopes lies in the increasingly assertive role played by women in Iran: women's advancement was one of the positive legacies of Shah Reza Pahlavi, who, however, tolerated no political opposition. During the June 2009 presidential election and its violent aftermath, women not only voted in large numbers but also risked their lives by joining prohibited rallies on behalf of the candidate of their choice.   © Behrouz Mehri/AFP/Getty Images, Inc.

Iran

Iran, Turkey's neighbor to the east, has its own history of conquest and empire, collapse and revival. Oil-rich and vulnerable, Iran already was just a remnant of a once-vast regional empire that waxed and waned, from the Danube to the Indus, over more than two millennia. Long known to the outside world as *Persia* (although the people had called themselves Iranians for centuries), the state was renamed *Iran* by the reigning shah (king) in 1935. His intention was to stress his country's Indo-European heritage as opposed to its Arab and Mongol infusions.

As we noted earlier, democracy was stirring in oil-rich Iran during the 1950s when the United States undermined a popular prime minister and helped reinstate the degenerating monarchy. In 1971, the then-ruling shah and his family celebrated the 2500th anniversary of Persia's royal rule with unmatched splendor. By 1979 revolution had engulfed Iran, and Shi'ite fundamentalists, with widespread public support, drove the shah from power. The monarchy was replaced by an Islamic republic ruled by an ayatollah, and a frightful wave of retribution followed. Today, despite a veneer of democracy, Islamists continue to control the state. The disputed 2009 presidential election led to a wave of street protests, violence, reprisals, and murder that dramatically exposed these schisms.

Crucial Location, Dangerous Terrain

As Figure 7-10 shows, Iran, again like Turkey, occupies a critical area in this turbulent realm. It controls the entire corridor between the Caspian Sea and the Persian Gulf. To the west it adjoins Turkey and Iraq, both historic enemies. To the north (west of the Caspian Sea) Iran borders Azerbaijan and Armenia, where once again Muslim confronts Christian. To the east Iran meets Pakistan and Afghanistan, and east of the Caspian Sea lies volatile Turkmenistan.

Iran, as Figure 7-21 demonstrates, is a country of mountains and deserts. The heart of the country is an upland, the Iranian Plateau, surrounded by even higher mountains, including the Zagros in the west, the Elburz in the north along the Caspian Sea coast, and the mountains of Khorasan to the northeast. The Iranian Plateau therefore is actually a huge highland basin marked by salt flats and wide expanses of sand and rock. The highlands wrest some moisture from the air, but elsewhere only oases break the arid monotony—oases that for countless centuries have been stops on the area's caravan routes.

In eastern Iran, neighboring Pakistan, and Afghanistan, people still move with their camels, goats, and other livestock along routes that are almost as old as the human history of this realm. Usually they follow a seasonal and an annual cycle, visiting the same pastures year after year, pitching their tents near the same stream. It is a lifestyle

or minority rights or any other factor that is making the real difference. So, ten years later, it appeared that Turkey might turn its back on Europe before Europe shuts the door to Turkey. Either way, such a diplomatic failure would have enormous consequences.

FIGURE 7-21

© H. J. de Blij, P. O. Muller, and John Wiley & Sons, Inc.

20 especially associated with this realm: **nomadism**. In Iran as elsewhere, nomads are not aimless wanderers across boundless plains. They know their terrain intimately, and they carefully judge how long to linger and when to depart based on many years of experience along the route.

City and Countryside

In ancient times, Persepolis in southern Iran (located near the modern city of Shiraz) was the focus of a powerful Persian kingdom, a city dependent on *qanats*, underground tunnels carrying water from moist mountain slopes to dry flatland sites many kilometers away. Today, Iran's population of 74.4 million is 67 percent urban, and the capital, Tehran, lies far to the north, on the southern slopes of the Elburz Mountains. This mushrooming metropolis of 8.2 million, lying at the heart of modern Iran's core area, still depends in part on the same kinds of qanats that sustained Persepolis more than 2000 years ago. As such, Tehran symbolizes the internal contradictions of Iran: a country in which modernization has taken hold in the cities, but little has changed in the vast countryside, where the mullahs led their peasant followers in a revolution that overthrew a monarchy and installed a theocracy.

The Ethnic-Cultural Map

Iranians dominate Iran's affairs, but Iran is no more an ethnically unified state than Turkey is. Figure 7-21 shows that three corners of the country are inhabited by non-Iranians, two in the west and one in the southeast. Iran's northwestern corner is part of greater Kurdistan, the multistate region in which Kurds are in the majority and of which southeastern Turkey also is a part (Fig. 7-13). Between the Kurdish area and the Caspian Sea the majority population is Azeri, who may number as many as 18 million and whose kinspeople form the majority in neighboring Azerbaijan to the north.

In the southwest, the minority is Arab and is concentrated in Khuzestan Province. During the war of the 1980s, Iraq's Saddam Hussein had his eye on this part of Iran, whose Arab minority often complained of mistreatment by the Iranian regime, and where a large part of Iran's huge oil reserve is located (note the situation of the export terminal of Abadan in Fig. 7-21). The Arab minority here is among Iran's poorest and most restive, in a province whose stability is key to the country's economy.

The third non-Iranian corner of Iran lies in the southeast, where a coastal Arab minority centered on the key port of Bandar-e-Abbas on the Strait of Hormuz gives way eastward to a scattered and still-nomadic population of Baluchis who have ties to fellow ethnics in neighboring Pakistan (see Fig. 7-23). On the Iranian side of the border, this is a remote and presently rather inconsequential area, but in Pakistani Baluchistan the government faces a persistent rebellion by tribespeople who feel slighted by official neglect (see pp. 423-424). Iran's ethnic-cultural map, therefore, is anything but simple.

Ideology, Energy, and Strategy

Iran possesses the third-largest share of the known oil reserves in the world and the second-largest natural gas reserves, with political as well as economic implications. Nearly 90 percent of the country's export revenues come from oil and gas, and although this huge income is not enough to lift this state into the high-income category, it enables Iran to acquire sophisticated weapons technologies (including ballistic missiles and warships) and to support movements and activities elsewhere in the world that promote its ideological goals. Plentiful energy availability translates into strategic advantage: in 2009 Iran and Turkey were negotiating an agreement whereby Iran would send large supplies of natural gas via Turkey to Europe. The United States opposed this

deal, as well as another initiative to send natural gas from Iran to India via Afghanistan or Pakistan. Meanwhile, the Chinese were working on a terminal in extreme southwestern Pakistan, at Gwadar, from where Iranian oil and gas will be shipped to China. When a country has ample oil and gas to sell, it will find friendly customers.

As Figure 7-9 shows, Iran's oil and gas reserves lie in a broad zone along its southwestern and southern periphery, where Abadan long ago became the country's "oil capital." In the days of the shahs, Iran's rulers tried to use oil revenues to modernize the country, and had the last shah succeeded in such efforts the 1979 revolution might have been forestalled. But Iran is a very large and populous country where traditions are strong and change comes slowly. The shah's modernization remained just a veneer: in the villages far from Tehran's polluted air, the holy men who saw the shah's innovations as so many threats continued to hold sway. There, as elsewhere in much of the Muslim world, urbanites, villagers, and nomads remained enmeshed in a web of production, profiteering, serfdom, and indebtedness that was a world away from the islands of modernity seen by visitors and tourists.

The revolution swept this pattern away, but it did not improve the lot of Iran's millions any more than oil could. A devastating war with Iraq (1980–1990), into which Iran ruthlessly poured hundreds of thousands of its young men, sapped both the coffers and energies of the state. When it was over, Iran was left poorer, weaker, and aimless, its revolution spent on unproductive pursuits.

In the early years of the twenty-first century, evidence abounded that the people of Iran remain divided between conservatives determined to protect the power of the mullahs and reformers intent on modernizing and liberalizing Iranian society. In 2002, a conservative court sentenced a university professor to death for publishing a proposal for an Islamic "enlightenment"; thousands of students took to the streets in protest. That such a sentence could be handed down at all is indicative of the gulf between the dogmatic and the rational in this historic and civilized society. (In the end, the death sentence was not carried out.)

Iran has long been accused of supporting terrorist activities and organizations, and even of using its embassies in planning and executing terrorist attacks (Argentina has accused Tehran of such operations in the bombing of Jewish targets in Buenos Aires in 1992 and 1994). In 2002, former President Bush designated Iran as part of his "axis of evil." Shortly thereafter, it became evident that Iran was developing nuclear-energy capacity with the potential to enrich uranium that could become part of a nuclear-weapons program. Iran's huge oil reserves would seem to make nuclear energy redundant, and when Iran also tested longer-range missiles in 2004, concern over its intentions grew in Europe as well as the United States (Russia, citing its commercial interests, provided technical assistance for the construction of the nuclear plant at Bushehr, shown in the photo on this page). This concern was heightened by the geography of Iran's nuclear development program: unlike Iraq, which in the 1980s had concentrated all of its nuclear operations in a single facility at Osirak (which was completely destroyed by one Israeli air strike in 1983), the Iranians dispersed their facilities throughout the country (Fig. 7-21) and buried much of their nuclear infrastructure deep underground.

Iran's rise as a potential nuclear power should not surprise its neighbors or negotiators. Although revolutionary Iran following the 1979 fall of the shah disavowed its Persian imperial memories and ambitions, this does not mean that its regime's interests now end at its borders. Iran already has one nuclear neighbor, Pakistan, and it has another in near-neighbor Israel. Iran remains a regional force with cultural links to minorities in Pakistan, Afghanistan, Iraq, and Azerbaijan. It is flanked on two sides by countries under the influence of the "Great Satan," the United States. Tehran has economic interests in oil reserves and pipelines in the Caspian Basin. Iran's revolution ended a monarchy, but it did not extinguish all ties to its imperial past.

● TURKESTAN

For centuries Turkish (Turkic) peoples held sway over a vast central Asian domain that extended from Mongolia and Siberia to the Black Sea. Propelled by population growth and energized by Islam, they penetrated Iran, defeated the Byzantine Empire, and colonized much of Eastern Europe. Eventually, their power declined as Mongols, Chinese, and Russians invaded their strongholds. But these conquerors could not expunge them, as the names on the modern map prove (Fig. 7-22). The latest conquerors, the Russian czars and their communist successors, created Soviet Socialist Republics named after the majority peoples within their borders. Thus the Kazakhs, Turkmen, Kyrgyz, and other Turkic peoples retained some geographic identity in what was Soviet Central Asia.

Central Asia—*Turkestan*—is a still-changing region. In some areas, the cultural landscapes of neighboring realms extend into it, for example, in northern (Russian) Kazakhstan. In other areas, Turkestan extends into adjacent realms, as in Uyghur-influenced western China (Xinjiang). Some areas once penetrated by Turkic peoples are no longer dominated by them, for instance, Afghanistan. And certain peoples now living

During the first decade of the twenty-first century, Iraq and Iran lay at the center of world attention and concern—Iraq because of America's armed invasion, the overthrow of dictator Saddam Hussein, and the Sunni-based insurgency that followed; and Iran because of its determination to achieve nuclear power capacity, the suspicion that its nuclear objectives went beyond electricity generation, and the election of President Mahmoud Ahmadinejad. Iran's nuclear infrastructure already is formidable, as shown by this satellite image of just one facility near Bushehr on the Persian Gulf; other installations are deep underground. The prospect of a nuclear-armed Iran at the heart of Eurasia troubles many members of the "international community," but it has not been enough to generate a unified and sufficiently firm response. © DIGITAL GLOBE/Reuters/Corbis

in Turkestan are not of Turkic ancestry, notably the Tajiks. This is a fractious region in sometimes turbulent transition.

The Ethnic Mosaic

The underlying cultural-geographic reason for this contentiousness—not only in Turkestan but in neighboring regions in Southwest and South Asia as well—is illustrated in Figure 7-23. This detailed map of ethnolinguistic groups actually is a generalization of an even more complex mosaic of peoples and cultures. Every cultural domain on the map also includes minorities that cannot be mapped at this scale, so that people of different faiths, languages, and ways of life rub shoulders everywhere. Often this results in friction, and at times such friction escalates into ethnic conflict.

FIGURE 7-22

© H. J. de Blij, P. O. Muller, and John Wiley & Sons, Inc.

Regional Framework

As we define it, Turkestan includes most or all of six states: (1) *Kazakhstan*, territorially larger than the other five combined but situated astride an ethnic transition zone; (2) *Turkmenistan*, with important frontage on the

Caspian Sea and bordering Iran and Afghanistan; (3) *Uzbekistan*, the most populous state and situated at the heart of the region; (4) *Kyrgyzstan*, wedged between powerful neighbors and chronically unstable; (5) *Tajikistan*, regionally and culturally divided as well as strife-torn; and (6) *Afghanistan*, engulfed in warfare almost

SOUTH-CENTRAL EURASIA

FIGURE 7-23

After National Geographic.

© H. J. de Blij, P. O. Muller, and John Wiley & Sons, Inc.

MAJOR ETHNOLINGUISTIC GROUPS

ALTAIC
1 Azerbaijani
2 Kalmyk
3 Kazakh
4 Kyrgyz
5 Turk
6 Turkmen
7 Uzbek

INDO-EUROPEAN
12 Armenian
13 Bakhtiari
14 Baluchi
15 Chahar Aimak
16 Hazara
17 Kurdish
18 Lur
19 Persian
20 Punjabi
21 Pushtun
22 Russian
23 Sindhi
24 Tajik

CAUCASIAN
8 Chechen-Ingush
9 Dagestani
10 Georgian

AFRO-ASIATIC
11 Arabic

25 Other
26 Uninhabited or sparsely populated

REGIONAL CONFLICTS
A Abkhazia – Georgia
B South Ossetia – Georgia
C North Ossetiya–Alaniya – Ingushetiya
D Chechnya – Russia
E Turkey – Kurdish separatists
F Nagorno-Karabakh – Azerbaijan
G Afghan war
H Iraq war

0   200 km
0   200 mi

continuously since it was invaded by Soviet forces in 1979, and frequently throughout its history.

During their hegemony over Central Asia, the Soviets tried to suppress Islam and install secular regimes (this was their objective when they invaded Afghanistan as well), but today Islam's revival is one of the defining qualities of this region. From Almaty to Samarqand, mosques are being repaired and revived, and Islamic dress again is part of the cultural landscape. National leaders make high-profile visits to Mecca; most are sworn into office on the Quran. All of Central Asia's countries now observe Islamic holidays. In other ways, too, Turkestan reflects the norms of the NASWA realm: in its dry-world environments and the clustering of its population, its mountain-fed streams irrigating farms and fields, its sectarian conflicts, its oil-based economies. It also is a region where democratic government remains an elusive goal.

The two influences of resurgent Islam and residual Russian both contribute significantly to the shaping of post-Soviet Turkestan. In recent years, Russia's leaders have worked to revive Moscow's links with this part of the Near Abroad, and to exert Russian influence here. For example, the government of Kyrgyzstan, which had granted the United States permission to lease an air base in the country for the military operations in Afghanistan, came under Russian pressure in 2009 to end that agreement and to oust the Americans, which it ultimately resisted. But with more than 250,000 Kyrgyz citizens working in Russia and nearly 40 percent of the impoverished country's GDP consisting of remittances sent back by workers outside their homeland, Moscow held a strong hand. The leaders of Turkestan's states tend to have Russian or Russified names, not Turkic ones (Kyrgyzstan's Bakiyev, Tajikistan's Akilov, Turkmenistan's late Niyazov, Uzbekistan's Karimov, Kazakhstan's

Nazarbayev), and are inheritors of the power vested in pro-Russian elites during Soviet rule in the region. More importantly, the concentration of authority prevailing in communist times continues today: representative government has but a weak foothold in Turkestan. The personality cult and absolute power that marked Niyazov's two-decade rule in Turkmenistan was on a par with North Korea's Kim Il Sung/Kim Jong Il.

The States of Former Soviet Central Asia

Kazakhstan is the territorial giant of the region and borders two greater giants: Russia and China. During the Soviet period, northern Kazakhstan was heavily Russified and became, in effect, part of Russia's Eastern Frontier (see Fig. 2-11). Rail and road links crossed the area mapped in Figure 7-22 as the Kazakh-Russian Transition Zone, connecting the north to Russia. The Soviets made Almaty, in the heart of the Kazakh domain, the territory's capital. Today the Kazakhs are in control, and they in turn have relocated the capital to Astana, right in the heart of the Kazakh-Russian Transition Zone, where more than 4.3 million Russians (who constitute 27 percent of the total national population of 16.0 million) still live. Clearly, Astana is another *forward capital*.

Figure 7-22 also reveals Kazakhstan's situation as a corridor between the Caspian Basin's vast oil reserves and westernmost China. Oil and gas pipelines now cross Kazakhstan, and enable the Chinese to reduce their dependence on oil carried by tankers along distant sea lanes.

Uzbekistan, by far the most populous country in this region (28.1 million), occupies the heart of Turkestan and borders every other state in it. Uzbeks not only make up 80

If you were asked where this photo might have been taken, Russia would seem to be a good guess, or perhaps Belarus (although the English word "bar" might not be expected there). Certainly the bleak apartment building in the background seems to be of Soviet-socialist vintage. But this is the entrance to a casino on the main shopping street of Almaty, the largest city and former capital of Kazakhstan. The Soviets and their Kazakh allies administered Kazakhstan from Almaty (then known as Alma-Ata) for some 70 years, and the Russian imprint remains strong in this city. But, as Figure 7-22 shows, the Russian imprint is even stronger in the country's north, and in 1998 the president and his party, over the objections of the mainly Kazakh opposition, moved the national capital from scenic and warm Almaty to drab and cold Astana. Yet Almaty retains its attractions: more than 40 casinos as well as countless bars and clubs coexist with Sunni Islam, Kazakhstan's leading religion.
© Abbie Trayler-Smith/Panos Pictures

percent of this population, but also form substantial minorities in several neighboring states. The capital, Tashkent, lies in the far eastern core area of the country, where most of the people live in towns and farm villages, and the crowded Fergana Valley is the focus. In the west lies the shrunken, now somewhat recovering Aral Sea, whose feeder streams were diverted into cotton fields and croplands during the Soviet occupation; heavy use of pesticides contaminated the groundwater, and countless thousands of local people suffered severe medical problems as a result.

In the post-Soviet period, Islamic (Sunni) revivalism has become a problem for Uzbekistan. *Wahhabism*, the particularly virulent form of Sunni fundamentalism, took root in eastern Uzbekistan; government efforts to suppress it have only increased its strength in the countryside. In turn, the country's regime has taken on increasingly dictatorial habits, and public demonstrations have been put down with many casualties.

Turkmenistan, the autocratic desert republic that extends all the way from the Caspian Sea to the borders of Afghanistan, has a population of just 5.4 million, of which more than three-quarters are Turkmen. During Soviet times, the communist planners began work on a massive project: the Garagum (Kara Kum) Canal designed to bring water from Turkestan's eastern mountains into the heart of the desert. Today the canal is 1100 kilometers (700 mi) long, and it has enabled the cultivation of some 1.2 million hectares (3 million acres) of cotton, vegetables, and fruits. The plan is to extend the canal all the way to the Caspian Sea, but meanwhile Turkestan has hopes of greatly expanding its oil and gas output from Caspian Basin reserves. But look again at Figure 7-22: Turkmenistan's relative location is not advantageous for export routes.

Kyrgyzstan's topography and political geography are reminiscent of the Caucasus. Shown in yellow in Figure 7-22, Kyrgyzstan lies intertwined with Uzbekistan and Tajikistan to the point of exclaves and enclaves along its borders. The Kyrgyz, for whom the Soviets established this "republic," make up less than two-thirds of the population of 5.4 million. Uzbeks (13 percent) and other minorities form the rest of the complex cultural geography; mountains and valleys isolate communities and make nation-building difficult. The agricultural economy is weak, consisting of pastoralism in the mountains and farming in the valleys. About 70 percent of the people profess allegiance to Islam, and Wahhabism has gained a strong foothold here. The town of Osh is often referred to as the headquarters of the movement in Turkestan. As noted in Chapter 2, Kyrgyzstan's relative location yielded complications when its government authorized the United States to operate a military base in support of the allied campaign of intervention in Afghanistan, and

the Russians, wary of their loss of influence, sought to end that agreement. Eventually, the U.S. prevailed, but this incident served to illustrate the weakness of a Kyrgyzstan caught in a quarrel between major powers.

Tajikistan's mountainous scenery is even more spectacular than Kyrgyzstan's, and here, too, topography is a barrier to the integration of a multicultural society. The Tajiks, who constitute about 65 percent of the population of 7.6 million, are of Persian (Iranian), not Turkic, origin and speak a Persian (and thus Indo-European) language. Most Tajiks, despite their Persian affinities, are Sunni Muslims, not Shi'ites. Small though Tajikistan is, regionalism plagues the state: the government in Dushanbe often is at odds with the barely connected northern area (see Fig. 7-22), a hotbed not only of Islamic revivalism but also of anti-Tajik, Uzbek activism. As Figure 7-23 shows, Tajiks form a significant minority in neighboring Afghanistan, so that the Dushanbe government has a strong interest in what happens there.

Fractious Afghanistan

Afghanistan, the southernmost country in this region, exists because the British and Russians, competing for hegemony in this part of inner Asia during the nineteenth century, agreed to tolerate it as a cushion, or **buffer state,** between them. This is how Afghanistan acquired the narrow extension leading from its main territory eastward to the Chinese border—the Vakhan Corridor (Fig. 7-24). As the colonialists delimited it, Afghanistan adjoined the domains of the Turkmen, Uzbeks, and Tajiks to the north, Persia (now Iran) to the west, and the western flank of British India (now Pakistan) to the east.

Landlocked and Fractured

Geography and history seem to have conspired to divide Afghanistan. As Figure 7-24 shows, the towering Hindu Kush range dominates the center of the country, creating three broad environmental zones: the relatively well-watered and fertile northern plains and basins; the rugged, earthquake-prone central highlands; and the desert-dominated southern plateaus. Kabol (Kabul), the capital, lies on the southeastern slope of the Hindu Kush, linked by narrow passes to the northern plains and by the Khyber Pass to Pakistan. Across this variegated landscape moved countless peoples: Greeks, Turks, Arabs, Mongols, and others. Some settled here, their descendants today speaking Persian, Turkic, and other languages. Others left archeological remains or no trace at all. The present population of Afghanistan (34.4 million) has no ethnic majority. This is a country of minorities in which the Pushtuns (or Pathans) of the east are the most numerous but make

up barely 40 percent of the total. The second-largest minority are the Tajiks (ca. 25 percent), a world away across the Hindu Kush, concentrated in the zone near Afghanistan's border with Tajikistan. The Hazaras of the central highlands and the south, the Uzbeks and Turkmen in the northern border areas, the Baluchi of the southern deserts, and other, smaller groups scattered across this fragmented country create one of the world's most complex cultural mosaics (Fig. 7-23). Two major languages, Pushtun and Dari (the local variant of Persian), plus several others create a veritable Tower of Babel here.

Costs of Conflict

Episodes of conflict have marked the history of Afghanistan, but none was as costly as its involvement in the Cold War. Following the Soviet intervention of 1979, the United States supported the Muslim opposition, the *Mujahideen* ("strugglers"), with modern weapons and money, and the Soviets were forced to withdraw. Soon the factions that had been united during the anti-Soviet campaign were in conflict, delaying the return of some 4 million (mainly Pushtun) refugees who had fled to Pakistan and Iran. The situation resembled the pre-Soviet past: a feudal country with a weak and ineffectual government in Kabol.

In 1994, what at first seemed to be just another warring faction appeared on the scene: the so-called *Taliban* ("students of religion") from religious schools in Pakistan. Their avowed aim was to end Afghanistan's chronic factionalism by instituting strict Islamic law. Popular support in the war-weary country, especially among the Pushtuns, led to a series of successes, and by 1996 the Taliban had taken Kabol.

The Taliban's imposition of Islamic law was so strict and severe that Islamic as well as non-Islamic countries objected. Restrictions on the activities of women ended their professional education, employment, and freedom of movement, and had a devastating impact on children as well. Public amputations and stonings enforced the Taliban's code. In the process, Afghanistan became a haven for groups of revolutionaries whose goals went far beyond those of the Taliban: they plotted attacks on Western interests throughout the realm and threatened Arab regimes they deemed compliant with Western priorities. Taliban-ruled, cave-riddled, remote and isolated Afghanistan was an ideal locale for these outlaws. Already

in possession of arms and ammunition (Soviet as well as American) left over from the Cold War, they also benefited from Afghanistan's huge, rapidly expanding, illicit opium trade. By mid-2009, according to reliable estimates, Afghanistan was producing more than 92 percent of the world's opium. Much of the revenue had found its way into the coffers of the conspirators.

Failed State, Terrorist Base

With such resources, the militants were able to organize and launch attacks against several Western targets in the realm and elsewhere, but events took a fateful turn in 1996 that would empower them as never before. During the conflict with the Soviets, the Mujahideen cause had been supported not only by the United States but also by a devout and revivalist Saudi Muslim named Usama bin ("son of") Laden. The child of a construction billionaire who had more than 50 children including 22 sons, Usama graduated from a university in Jiddah in 1979 and headed for Afghanistan with an inherited fortune estimated at about $300 million to help the anti-Soviet campaign. Well-connected in Saudi Arabia and now a pivotal figure in Afghanistan, bin Laden

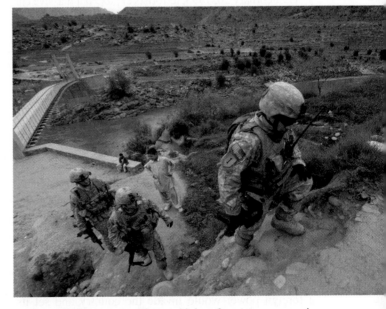

Even as United States armed forces withdraw from Iraq, a new and even greater challenge looms: the pacification and stabilization of Afghanistan and the defeat, in neighboring Pakistan as well as in Afghanistan, of the Taliban movement. This challenge is most urgent (and difficult) in Afghanistan's eastern and southern provinces, where high relief, distance, and poor surface communications favor the Taliban. This image shows U.S. soldiers on patrol near a bridge at Nishagam in Konar province (see Fig. 7-24) in mid-2009, not far from the border with the Federally Administered Tribal Areas of Pakistan, a Taliban stronghold.
© Liu Jin/AFP/Getty Images, Inc.

FIGURE 7-24
© H. J. de Blij, P. O. Muller, and John Wiley & Sons, Inc.

saw his fame as well as his war chest grow as the Soviet intervention collapsed. Following the withdrawal of the communist forces, he returned to Saudi Arabia, where he denounced his government for allowing U.S. troops on Saudi soil during the Gulf War. The Saudi regime responded by stripping him of his citizenship and expelling him, and bin Laden fled to Sudan. There he set up several legitimate businesses to facilitate his now-global financial transactions, but he also established terrorist training camps. Under international pressure, the Khartoum regime ousted him in 1996, and bin Laden returned to a country he knew would welcome him again: Afghanistan.

Bin Laden's fateful return to Afghanistan coincided with the Taliban's conquest of Kabol, and now he helped its forces push northward into the fertile and productive northern plains. Meanwhile, a terrorist organization named *al-Qaeda* took root in the country, a global network that would further the aims of the revolutionaries once loosely allied. Afghanistan became al-Qaeda's headquarters, and bin Laden its director; its exploits were funded by numerous Muslim sources and ranged from terrorist-training in local camps to lethal attacks on American targets, including a warship in Yemen's port of Adan and two U.S. embassies in East Africa.

On February 26, 1993, terrorists exploded a massive car bomb in the basement garage of the World Trade Center in New York, but their objective—to topple the 110-story tower—failed. Even as those responsible went on trial, al-Qaeda's leaders were planning the suicide attacks of September 11, 2001 that destroyed the buildings, killed thousands, and caused billions of dollars in damage. Several weeks later, United States and British forces, with the acquiescence of Pakistan, attacked both the Taliban regime and the al-Qaeda infrastructure in Afghanistan, the first stage of the Bush-era global "War on Terror" that would leave no country unaffected. Proof of bin Laden's and al-Qaeda's complicity in the September 11 assault was found in videotape and documentary form.

The Taliban and al-Qaeda leaderships may not have counted on Pakistan's compliance with Western demands, but they surely knew what the consequences of 9/11 would be for Afghanistan and its people. Once again a foreign power would invade the country and set its political course, the associated upheaval enabling rapacious warlords to exploit the peoples of border provinces far from the capital. But as it turned out, Afghanistan made significant progress toward representative government; most of the warlords were defeated or co-opted, and many of the refugees returned.

Nonetheless, even as the intensity of the war in Iraq declined and American (and Allied) troop reductions there became feasible, the situation in Afghanistan proved increasingly difficult for both its government and for the international forces trying to help the country achieve peace and stability. In a pattern recurring throughout Afghanistan's history, opponents of modernization in any form grew stronger and resumed attacking foreign forces and projects (e.g., burning girls' schools and threatening female students and teachers). The reformist government in Kabol endorsed the enlargement of the foreign forces on Afghanistan's soil; inevitable civilian casualties in the growing conflict created propaganda advantages for its revivalist opponents.

During 2009, its first year in office, the Obama Administration committed the United States to an expanding and multiform campaign in Afghanistan that required thousands of additional forces, stronger participation by allies, social and economic as well as military and strategic initiatives, and other moves designed to strengthen representative government in Afghanistan and to thwart a comeback by the Taliban. As the maps in this chapter as well as the next one show, this campaign inevitably and directly affects landlocked Afghanistan's neighbors, especially Pakistan. Even as the war in Afghanistan intensified, Taliban forces staged a major attack in northern Pakistan, briefly holding a large area including several sizeable towns where they conducted their hallmark public executions, destroyed girls' schools, and terrorized shopkeepers. A massive military response by Pakistani forces drove the Taliban out, but at a huge cost in terms of refugee dislocation and property damage. And as in Afghanistan, Taliban commanders and their henchmen disappeared into the cave-riddled mountains along the Pakistan-Afghanistan border to regroup for their next assault. Even as Afghans went to the polls for a violence-marred national election in August 2009, American casualties in Afghanistan had begun to exceed those in Iraq, and the question was how long a majority of the U.S. public would support an increasingly costly campaign in the aftermath of that misguided intervention.

CONCEPTS, IDEAS, AND TERMS

1 Wet monsoon
2 Social stratification
3 Refugees
4 Population geography
5 Population distribution
6 Population density
7 Physiologic density
8 Rate of natural population change
9 Demographic transition
10 Population explosion
11 Forward capital
12 Marchland
13 Caste system
14 Intervening opportunity
15 Natural hazards
16 Failed state
17 Insurgent state

REGIONS

- PAKISTAN
- INDIA
- BANGLADESH
- MOUNTAINOUS NORTH
- SOUTHERN ISLANDS

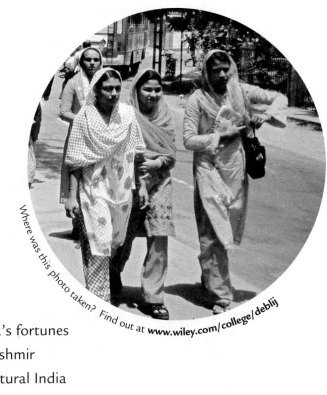

Where was this photo taken? Find out at www.wiley.com/college/deblij

In This Chapter

- The monsoon still is the key to India's fortunes
- Two nuclear powers quarrel over Kashmir
- Hindu nationalism rising in multicultural India
- Sri Lanka: End of the tragedy?

Photos: © H. J. de Blij.

FIGURE 8-1   *Map:* © H. J. de Blij, P. O. Muller, and John Wiley & Sons, Inc.

THE EURASIAN LANDMASS incorporates all or part of six of the world's geographic realms, and of these six, none is more clearly defined by nature than the one we call South Asia. Figure 8-1 shows us why: the huge triangle that divides the northern Indian Ocean between the Arabian Sea and the Bay of Bengal is so sharply demarcated by mountain walls and desert wastes that you could take a pen and mark its boundary, from the Naga Hills in the east and the great Himalayas in the north to the Hindu Kush and the Iran-bordering Baluchi wastelands in the west. Note how short are the distances over which the green of habitable lowlands turns to the dark brown of massive, snowcapped mountain ranges.

This is a realm of fabled geographic names: Mount Everest, Kashmir, the Khyber Pass, the Ganges River.

There was a time when South Asia was legendary, but remote from the affairs of the world—hungry, weak, exploited, an epitome of the global periphery. Today South Asia commands the world's attention. It is poised to become the most populous realm on Earth in 2011 (see the Data Table inside the back cover). Two of its states, India and Pakistan, are nuclear powers. In the remote mountain hideaways of Pakistan, a terrorist organization's leaders planned attacks that changed the skyline of New York and altered the battleground of Iraq. In the ports of India, a growing navy reflects that country's determination to become more than a regional power. We should all pay attention. Our daily lives will increasingly be affected by what happens in this crowded and restive part of the world.

Defining the Realm

South Asia's kaleidoscope of cultures may be the most diverse in the world, proving that neither formidable mountains nor forbidding deserts stopped foreign influ-

ences from further diversifying an already variegated realm. We will encounter many of these influences in this chapter, but South Asia also had one unifying force: the

MAJOR GEOGRAPHIC QUALITIES OF

South Asia

1. South Asia is clearly defined physiographically, and much of the realm's boundary is marked by mountains, deserts, and the Indian Ocean.

2. South Asia's great rivers, especially the Ganges, have for tens of thousands of years supported huge population clusters.

3. South Asia covers just over 3 percent of the Earth's land area but contains nearly 23 percent of the world's human population.

4. South Asia's population continues to grow at annual rates higher than the world average. It becomes the world's most populous geographic realm in 2011.

5. Poverty remains endemic in South Asia, despite pockets of progress. Hundreds of millions of people are plagued by inadequate nutrition and poor overall health.

6. British commercial penetration and colonial domination left strong geographic imprints on this realm, ranging from its boundary framework to its cultural landscapes.

7. South Asia's annual monsoon continues to dominate life for hundreds of millions of subsistence and commercial farmers. Failure of the monsoon cycle spells economic crisis.

8. Despite its physiographic demarcation, invading armies and cultures, from ancient Greeks to modern British and from Mongols to Muslims, diversified this realm's cultural mosaic.

9. Several of the world's great religions and many minor faiths have strong bases in South Asia, including Hinduism, Islam, and Buddhism. Religion remains a powerful force in political and economic life.

10. India, the world's largest democracy and a functioning federation, is the dominant state in this realm but has contentious relations with several of its neighbors.

11. Kashmir, the mountainous territory in the northern frontier zone between India and Pakistan, is a dangerous source of friction between these two nuclear powers.

British Empire, which in its nineteenth-century heyday held sway over all of it. When, in the mid-twentieth century, the British wanted to transfer their authority to one regional government, local objections nullified that notion. That regional government would have been Hindu-dominated, but Muslims in the realm's eastern and western flanks refused, as did a pair of kingdoms in the mountainous north, as well as Buddhists in the southern island then called Ceylon (now Sri Lanka). Negotiations and compromises produced the map in Figure 8-2. India, the realm's giant, is flanked by six countries (Pakistan, Nepal, Bhutan, Bangladesh, Sri Lanka, and the Maldives) and one remaining disputed territory, Kashmir.

Not immediately obvious from the map is just how comparatively small this realm is territorially, given its enormous human content. South Asia is smaller even than Europe and less than two-fifths the size of equally populous East Asia. Comparing the world's two giants,

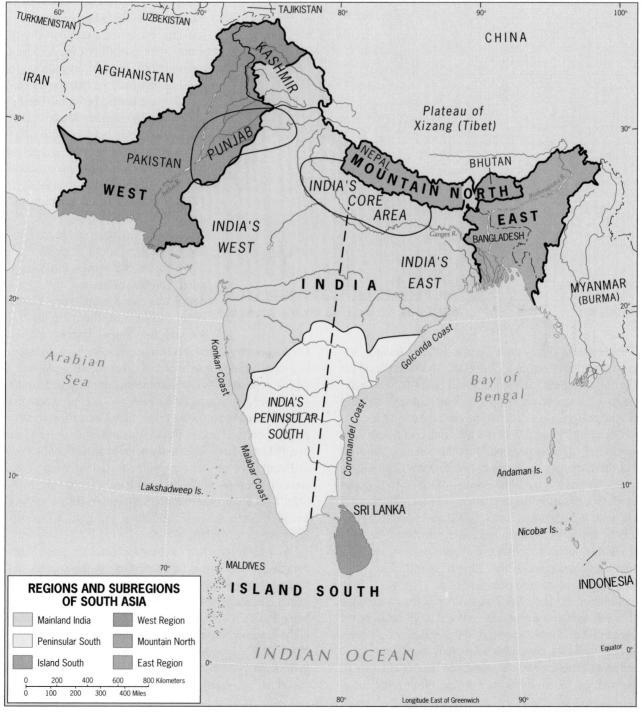

FIGURE 8-2

© H. J. de Blij, P. O. Muller, and John Wiley & Sons, Inc.

SOUTH ASIA POPULATION DISTRIBUTION: 2010
One dot represents 75,000 persons

FIGURE 8-3   © H. J. de Blij, P. O. Muller, and John Wiley & Sons, Inc.

China's area is almost three times as large as India's. Adjectives such as "teeming," "crowded," "jammed," and the like are often used to describe South Asia's habitable living space, and with reason. South Asia's intricate cultural mosaic is tightly packed, at times resulting in clashes.

Certainly Figure 8-3 shows how much of South Asia is relatively densely peopled. Only the deserts in the west and the mountain fringe in the north have large empty spaces, and you can see the outlines of the densely populated river basins in the dot pattern. Compare Figure 8-3 to Figure 9-2 (the map containing China's population distribution) and note how much more open space there is in China—more than half the country is very sparsely populated, and it seems that 90 percent of Chinese citizens live in the eastern one-third of it.

To return to Figure 8-2, note that India, the cornerstone of South Asia, has a core area centered in the basin of the Ganges River, the historic heart of this realm. India is a region as well as a state, with subregions based on cultural and other criteria to be discussed later in this chapter. To the west lies Islamic Pakistan, whose lifeline, the Indus

River and its tributaries, creates a core area in the Punjab. But note that the Punjab subregion extends across the border with India. What you see here is an aftermath of an enormous human tragedy. When independence approached, Muslim leaders in the west demanded the creation of an Islamic state. A boundary drawn hastily in 1947 between a future Muslim Pakistan and a dominantly Hindu India precipitated the sudden migration of millions of people in the border area— Muslims westward and Hindus eastward—with countless casualties and innumerable human tragedies. In South Asia's east, the British had earlier drawn a border between (Hindu) Bengal and (dominantly Muslim) East Bengal, and that border became the boundary of what became known as East Pakistan, the political connection based on shared Islamic values. In 1971, however, East Pakistan severed its link with West Pakistan (which renamed itself Pakistan) and became independent Bangladesh.

As Figure 8-2 shows, three separated entities make up the region defined as the Mountainous North. From east to west, these are the traditional kingdom of Bhutan, as isolated a country as the world has today; the troubled state of Nepal, where royal rule failed but the struggle to establish representative government continues; and, at the western end, the disputed territory of Kashmir, where the British boundary-making effort failed to resolve a complicated cultural and political situation, and where India and Pakistan have repeatedly been in armed conflict.

Finally, the region mapped as the Island South consists of two very different countries: dominantly Buddhist Sri Lanka, where a quarter-century-long civil war ended in 2009, and the Republic of Maldives, whose official religion is Islam.

We will view each of these regions and subregions in geographic perspective in the second half of this chapter, but here we should briefly address the question of Pakistan's designation as a regional component of South Asia. Since Islam is Pakistan's official religion (India has none) and Islam is the key criterion in defining the realm we designated as North Africa/Southwest Asia, should Pakistan be included in the latter? The answer lies in several aspects of Pakistan's historical geography. One criterion is ethnic continuity, which links Pakistan to India rather than to Afghanistan or Iran. Another factor involves language: although Urdu is Pakistan's official language, English is the *lingua franca*, as it is in India. Still another, of course, is Pakistan's evolution as part of the British South Asian Empire. Furthermore, the boundary between Pakistan and India does not signify the eastern frontier of Islam in Asia. As we shall see, more than 170 million of India's nearly 1.2 billion citizens are Muslims (which is just about as many as there are in all of Pakistan), and millions live very close to the Indian side of the border whose creation cost so many lives in 1947. And not only are Pakistan and India linked in the cultural-historical arena: they are locked in a deadly and dangerous embrace in embattled Kashmir. Pakistan today remains part of a realm that changes not in the Punjab, but at the Khyber Pass, the highland gateway to Afghanistan.

SOUTH ASIA'S PHYSIOGRAPHY: FROM FERTILE FARMLAND TO TERRORIST HIDEOUT

South Asia may be small as world geographic realms go, but its physiographic diversity is unmatched (Fig. 8-4). From snowcapped peaks to tropical forests and from bone-dry deserts to lush farmlands, this part of the world presents a virtually endless range of ecologies and environments—a range that is matched by its cultural mosaic.

As Figures G-3 and G-4 remind us, the spectacular relief in the north of this realm results initially from the collision of two of the Earth's great tectonic plates, pushing parts of the crust upward into elevations where permanent snow and ice make the landscape look polar. The swing of the seasons melts enough of this snow in spring and summer to sustain the great rivers below, providing water for farmlands that support hundreds of millions of people.

Thus the mountains give life—but they also bring death. This is one of the world's most dangerous areas, where people live in valleys between mountains that bear witness to the forces below, forces that frequently shake the ground. During 2005 a series of earthquakes along the collision zone struck northern Pakistan and Kashmir, opening huge fissures, destroying villages, killing more than 70,000 people, and making over 3 million homeless.

And for governments, the mountains cause other problems. In the remote, rugged uplands of Waziristan, along the border between Pakistan and Afghanistan, fugitive leaders of the terrorist al-Qaeda organization hide in caves, and members of the deposed extremist Islamic Taliban movement blend in with fiercely inde-

pendent tribespeople who resist government scrutiny. In the distant hills of desert Baluchistan, a separatist movement festers beyond the reach of Pakistan's government. And at the opposite corner of South Asia, the Indian government struggles to control various rebel movements in the rugged mountain spurs of Nagaland.

The Monsoon

Physical geography, therefore, is crucial here in South Asia—but not just on and below the ground. What happens in the atmosphere is critical as well. The name "South Asia" is almost synonymous with the term *monsoon*, because the annual rains that come with its onset, usually in June, are indispensable to subsistence as well as commercial agriculture in the realm's key country, India (see box titled "South Asia's Life-Giving Monsoon"). As Figure 8-4 shows, the wet monsoon brings life-giving rains to the southwestern (Malabar) coast, and a second arm from the Bay of Bengal turns and sweeps across northern India toward Pakistan, losing power (and moisture) as it proceeds. This means that amply watered eastern India and Bangladesh grow rice as their staple crop, but drier northwestern India and Pakistan raise wheat. Look again at Figure 8-2 and note the dashed line extending from India's core area southward. India's historic division between east and west has much to do with the paths of the monsoon.

Physiographic Regions

Figure 8-4 underscores South Asia's overall division into three zones: northern mountains, southern plateaus, and, in between, a tract of river lowlands. It is useful to be aware of the traditional names of the four great stretches of coastline—Konkan, Malabar, Coromandel, and Golconda—each of which has distinctive relief. Approximately where the Konkan and Malabar coasts meet lies Goa, once the sole non-British colony on the peninsula and today a thriving tourist mecca.

Northern Mountains

The northern mountains extend from the Hindu Kush and Karakoram ranges in the northwest through the Himalayas in the center (Mount Everest, the world's tallest peak, lies in Nepal) to the ranges of Bhutan and the Indian State of Arunachal Pradesh in the east. Dry and barren in the west on the Afghanistan border, the ranges become green and tree-studded in Kashmir, forested in the lower sections of Nepal, and even more densely

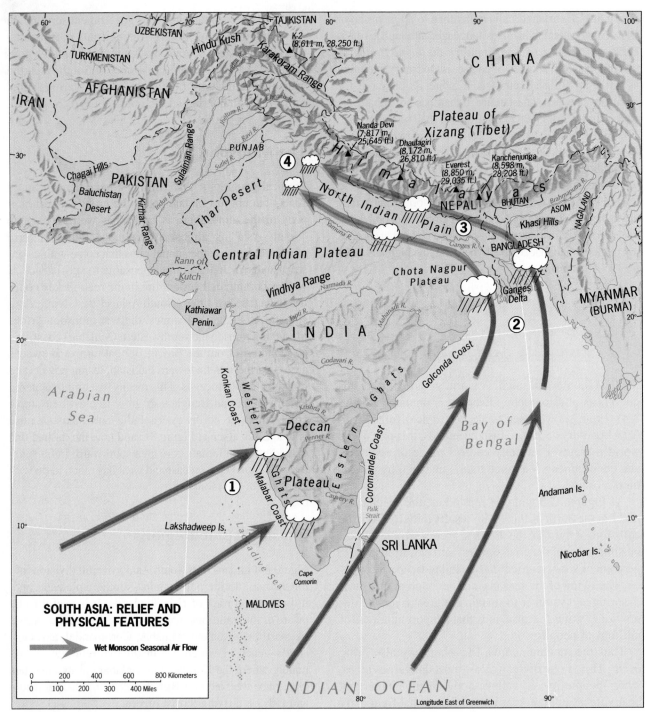

FIGURE 8-4

© H. J. de Blij, P. O. Muller, and John Wiley & Sons, Inc.

vegetated in Arunachal Pradesh. Transitional foothills, with many deeply eroded valleys cut by rushing melt-waters, lead to the river basins below.

River Lowlands

The belt of river lowlands extends eastward from Pakistan's lower Indus Valley (the area known as Sindh) through the wide plain of the Ganges Valley of India and on across the great double delta of the Ganges and Brahmaputra in Bangladesh (Fig. 8-1). In the east, this physiographic region often is called the North Indian Plain. To the west lies the lowland of the Indus River, which rises in Tibet, crosses Kashmir, and then bends southward to receive its major tributaries from the Punjab ("Land of Five Rivers").

Southern Plateaus

Peninsular India is mostly plateau country, dominated by the massive *Deccan*, a tableland built of basalt that

South Asia's Life-Giving Monsoon

THE DYNAMICS OF daily summer sea breezes are well known. The hot sun heats up the land, the air over this surface rises, and cool air from over the water blows in across the beach, replacing it. Teeming beaches and expensive seafront residential units attest to the efficacy of this natural air conditioning.

Under certain circumstances, this kind of circulation can affect whole regions. When an entire landmass heats up, a huge low-pressure system forms over it. This system can draw vast volumes of air from over the ocean onto the land. It is not just a local, daily phenomenon. It takes months to develop, but once it is in place, it also persists for months. When the inflow of moist oceanic air starts over South Asia, the **wet monsoon** has arrived. It may rain for 60 days or more. The countryside turns green, the paddies fill, and another dry season's dust and dirt are washed away. The region is reborn (see photo pair below).

Not all continents or coasts experience monsoons. A particular combination of topographic and atmospheric circumstances is needed. Figure 8-4 shows how it works in South Asia. It is June. For months, a low-pressure system has been building over northern India. Now, this cell has become so powerful that it draws air from over a large area of the warm, tropical Indian Ocean toward the interior. Some of that moisture-laden air is forced upward against the Western Ghats ①, cooling as it rises and condensing large amounts of rainfall. Other streams of air flow across the Bay of Bengal and get caught up in the convection over northeastern India and Bangladesh ②. Seemingly endless rain now inundates a much larger area, including the whole North Indian Plain. The mountain wall of the Himalayas stops the air from spreading and the rain from dissipating ③. The air moves westward, drying out as it flows toward Pakistan ④.

After persisting for weeks, the system finally breaks down and the wet monsoon gives way to periodic rains and, eventually, another dry season. Then the anxious wait begins for the next year's monsoon, for without it India would face disaster. In recent years, the monsoon has shown signs of irregularity and sometimes partial failure. In India, life hangs by a meteorological thread.

poured out when India separated from Africa during the breakup of Gondwana (see Fig. 6-3). The Deccan (meaning "South") tilts to the east, so that its highest areas are in the west and the major rivers flow into the Bay of Bengal. North of the Deccan lie two other plateaus, the Central Indian Plateau to the west and the Chota Nagpur Plateau to the east (Fig. 8-4). On the map, note the Eastern and Western Ghats ("hills") that descend from Deccan plateau elevations to the narrow coastal plains below. Onshore winds of the annual wet monsoon bring ample rain to the Western Ghats. As a result, here lies one of India's most productive farming areas and one of southern India's largest population concentrations.

The arrival of the annual rains of the wet monsoon transforms the Indian countryside. By the end of May, the paddies lie parched and brown, dust chokes the air, and it seems that nothing will revive the land. Then the rains begin, and blankets of dust turn into layers of mud. Soon the first patches of green appear on the soil, and by the time the monsoon ends all is green. The photograph on the left, taken just before the onset of the wet monsoon in the State of Goa, shows the paddies before the rains begin; three months later the countryside looks as on the right.
© Steve McCurry/ Magnum Photos, Inc.

LOCALS AND INVADERS

For tens of thousands of years, South Asia has seen peoples (and their cultural attributes and technologies) arrive, organize, compete, succeed, fail, move out, and change. Although this can be said of other world geographic realms, South Asia has a special place in the historical geography of humankind. Recent research indicates that the Ganges Basin was a crucial stop on the great out-of-Africa migration that carried our species from the shore of the Red Sea to the coast of Australia. Here, perhaps as long as 70,000 years ago, environmental conditions during the Wisconsinan Glaciation enabled groups of hunter-gatherers to stabilize and expand. From here, they went on to Southeast Asia and Australia and, scholars conclude, emigrated toward the Middle East and Europe as well. The Ganges Basin may have been a crucible of humanity.

Indus Valley Civilization

Whatever the distant past, we know that a complex and technologically advanced civilization had emerged in the Indus Valley by about 5000 years BC, encompassing a large portion of the lower Indus River Basin and, apparently, succeeding a still-earlier culture (Fig. 8-5). The Indus Valley Civilization (IVC) was centered on two major cities, Harappa and Mohenjo-Daro, which may have been capitals during different periods of its history; in addition, there were more than 100 smaller towns and villages. All this indicates that here was an early state in which political centralization had been achieved; the culture was literate, there appear to have been commercial contacts with Mesopotamia, and technologies such as irrigation, flood control, and well construction left their imprints. Today, it is known that IVC influences extended far to the west and eastward as far as the Delhi area, confirming the long-held suspicion that this was a foundation for the later Iron Age civilization that arose in what is now India. The locals apparently called their state *Sindhu* in the language that developed late in the IVC

period, and both *Indus* (for the river) and *India* (for the later state) may derive from this name.

A combination of environmental problems resulting from destructive floods and foreign invasion destroyed the IVC, apparently quite suddenly, about 3500 years ago. The invading *Aryans* (peoples speaking Indo-European languages based in what is today Iran) destroyed urban civilization in this northwestern corner of South Asia for centuries to come; but in the south, on and around the Kathiawar Peninsula you see in Figure 8-5, vestiges of the IVC survived. As the Iron Age dawned in India, acculturated Aryans began the process of welding the Ganges Basin's isolated tribes and villages into a new organized system, and urbanization made a comeback.

Incipient India

The Aryans brought their language (Sanskrit) and a new social order to the vast riverine flatlands of northern India, and they introduced their own religious belief sys-

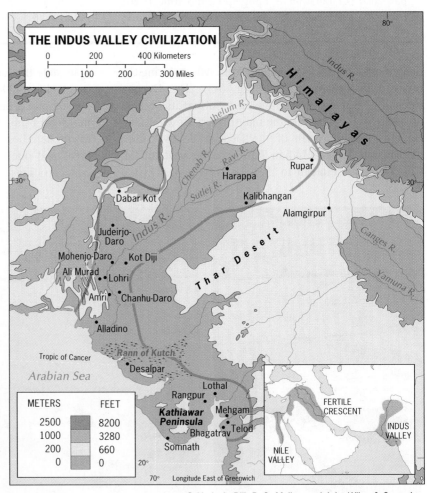

FIGURE 8-5 © H. J. de Blij, P. O. Muller, and John Wiley & Sons, Inc.

tem, *Vedism*. Out of the texts of Vedism and local creeds there arose a new belief system—*Hinduism*—and with it a new way of life. Starting about 3000 years ago, a combination of regional integration, the organization of villages into controlled networks, and the emergence of numerous small city-states produced a hierarchy of power among the people, a ranking from the very powerful (Brahmans—highest-order priests) to the weakest **2** (peasants). Thus a complex **social stratification** was part of Hinduism's *caste system*, in which soldiers, merchants, artists, toolmakers, and others all had their place. Those in the lowest castes, deemed to be there because they deserved it, were worst off, without hope of advancement and at the mercy of those higher up on the social ladder. Meanwhile, India's political geography evolved into a growing number of small states ruled by kings in consort with Brahmans.

Hinduism's restrictive and punitive qualities led others to seek a better way, and in the sixth century BC (thus more than 2500 years ago) a prince born into one of India's northern kingdoms proclaimed that he had found it. Prince Siddhartha, better known as Buddha,

gave up his royal rights to teach religious salvation through meditation, the rejection of Earthly desires, and a reverence for all forms of life. In truth, his teachings did not have a major impact on the Hindu-dominated society of his time, but he traveled widely, left followers in many parts of India, and was not forgotten. Centuries later, the ruler of a powerful Indian state decided to make *Buddhism* the state religion, and Buddhism spread far and wide.

Earlier Invaders

The cultural geography of South Asia bears witness to the numerous times outsiders invaded and put their stamp on entire regions of this realm. Invasions, we noted, have taken place almost throughout human history; the Persia-based Aryans were preceded by others about whom little or nothing is known. The map of one aspect of regional culture, the distribution of languages, reflects this conundrum (Fig. 8-6). When you compare this map to the world map of language families (Fig. G-10), it is clear that the languages of South Asia

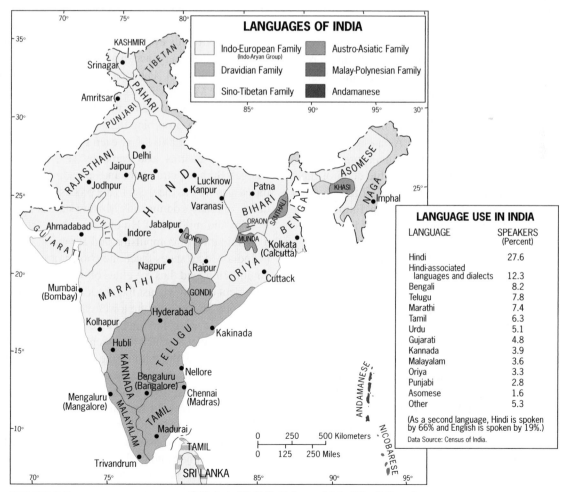

LANGUAGES OF INDIA

☐ Indo-European Family (Indo-Aryan Group)	Austro-Asiatic Family
Dravidian Family	Malay-Polynesian Family
Sino-Tibetan Family	Andamanese

LANGUAGE USE IN INDIA

LANGUAGE	SPEAKERS (Percent)
Hindi	27.6
Hindi-associated languages and dialects	12.3
Bengali	8.2
Telugu	7.8
Marathi	7.4
Tamil	6.3
Urdu	5.1
Gujarati	4.8
Kannada	3.9
Malayalam	3.6
Oriya	3.3
Punjabi	2.8
Asomese	1.6
Other	5.3

(As a second language, Hindi is spoken by 66% and English is spoken by 19%.)
Data Source: Census of India.

FIGURE 8-6   © H. J. de Blij, P. O. Muller, and John Wiley & Sons, Inc.

(shown in yellow in Fig. 8-6) lie at the eastern end of a language family—named *Indo-European*—that extends from the British Isles to the shores of the Bay of Bengal. That is why Hindi, the major domestic language of India, belongs to the same language family as Persian, Italian, and English. But look at the south: in what is now southern India, the indigenous languages are not Indo-European. They belong to a family named *Dravidian*, evidence of migrations that long preceded the heyday of the Indus Valley Civilization. But these are not fossil languages: they remain vibrant today and have long literary histories. Telugu, Tamil, Kanarese (Kannada), and Malayalam are spoken by some 250 million people, a basis for recognizing a southern subregion in peninsular India (Fig. 8-2).

Aśoka's Mauryan Empire

After the Aryans from Persia came the Greeks under Alexander the Great, who in the late fourth century BC made an incursion that would not change the realm but did demonstrate its vulnerability to invasion: part of his huge army came through the Khyber Pass, and other forces crossed the mountains and hills to the north. Alexander never made it to the Ganges Basin, however, and now, for the first time, a local state grew into a regional power. Starting in what is today the Indian State of Bihar, this expanding state came to control the whole of the North Indian Plain (see Fig. 8-4). When the emperor Aśoka took control of it, he headed what was to be the first empire in South Asia—reaching as far west as the Indus Valley, thereby incorporating the Punjab, and as far east as the great double delta of the Ganges and the Brahmaputra at the head of the Bay of Bengal. And to the south, this Mauryan Empire extended to the latitude of the modern city of Bengaluru (Bangalore).

Aśoka was no ordinary leader. For 40 years during the middle of the third century BC, he guided and expanded his vast empire *not* by conquest and subjugation, but through persuasion and teaching. A devout Buddhist, he deployed a veritable army of missionaries throughout his domain, all preaching the virtues of peace and stability. In the process, he disseminated Indian culture far and wide, and he did not stop at the frontiers of the Mauryan state. Aśoka's emissaries spread the faith as far afield as Southwest and Southeast Asia, ensuring Buddhism's survival after his death. Today, Sri Lanka, the island-state off the southern tip of India, is a dominantly Buddhist country, a legacy of its great benefactor—and Buddhism now has an estimated 400 million adherents around the globe (see Fig. 7-3).

Fracture and Fragmentation

As is so often the case when an inspiring leader has unified a state for a long time, the Mauryan Empire fell apart after Aśoka's death around 230 BC. In the more remote outposts, local princes took control and ceased to recognize the new regime, which itself was in dispute over matters of succession. Buddhism was in retreat, Hinduism in ascendance, and South Asia set on a course that was to mark it for many centuries to come: a course of disunity and strife. Only during the Gupta Empire (ca. AD 320–540), again centered in present-day Bihar, did a lengthy dynasty achieve regional unification and nurture a great flowering of Indian culture. Great Sanskrit epics, superb Hindu art, and major accomplishments in science and technology marked this formative period. But the Gupta period was the exception, not the rule, and thus South Asia was vulnerable to renewed invasions.

The Power of Islam

In the late tenth century, Islam came rolling like a giant tide across South Asia, spreading across Persia and Afghanistan, through the high mountain passes, into the Indus Valley, across the Punjab, and into the Ganges Basin, converting virtually everyone in the Indus Valley and foreshadowing the emergence, many centuries later, of the Islamic Republic of Pakistan. By the early thirteenth century, the Muslims had established a long-surviving and powerful sultanate at Delhi, which expanded across much of the peninsula, absorbing Indian states in the process and ruling over an empire larger even than the Mauryan Empire. The Muslims also came by sea, arriving at the Ganges-Brahmaputra Delta and spreading the faith from the east as well as the west, in the process laying the foundation of today's dominantly Muslim state of Bangladesh.

Why did millions of Hindus convert to Islam? It was a combination of compulsion and attraction. The princes of India's local states faced the choice of cooperation or annihilation, and thus the elites of the native states found it prudent to convert, just like the leaders of West Africa's states did. And among the lower castes of Hindu society, Islam was a welcome alternative to the rigid socioreligious hierarchy in which they were trapped at the bottom. Thus Islam was the faith of the rulers and of the disadvantaged, a powerful force in the heartland of Hinduism.

The Muslims, like other conquerors, also fought among themselves. In the early 1500s, a descendant of Genghis Khan named Babur put his forces in control of Kabol (Kabul) in Afghanistan, and from that base he penetrated the Punjab and challenged the Delhi Sultanate. In the 1520s, his Islamicized Mongol armies ousted the Delhi rulers and established the Mughal (Mongol) Empire.

By all accounts, Mughal rule was remarkably enlightened, especially under the leadership of Babur's grandson Akbar, who expanded the empire by force but adopted tolerant policies toward Hindus under his sway; Akbar's grandson, Shah Jahan, made his mark on India's cultural landscape through such magnificent architectural creations as the Taj Mahal and the Great Mosque of Delhi. But then religious bigotry took over, infighting intensified, and by the eighteenth century the Mughal Empire was in decline. Maratha, a Hindu state in the west, expanded not only into the peninsular south but also northward toward Delhi, capturing the allegiance of local rulers and weakening Islam's hold. Fractured India now lay open to still another foreign intrusion, this time from Europe.

Reflecting on more than seven centuries of Islamic rule in South Asia, it is remarkable that Islam never achieved proportional dominance over the realm as a whole. While Pakistan is more than 96 percent Muslim and Bangladesh 83 percent, India—where the Delhi Sultanate and the Mughal Empire were centered—remains less than 15 percent Muslim today; thus fewer than one-third of all South Asians are Muslims. By contrast, more than 60 percent of the realm's population is Hindu, with about 81 percent of all Indians adhering to that faith. Islam arrived like a giant tide, but Hinduism stayed afloat and outlasted the invasion.

The European Intrusion

Although the European invasion of South Asia gained momentum after 1800, European colonialists had been on South Asia's shores for a very long time—in fact, ever since the arrival of the Portuguese seafarer Vasco da Gama in 1498, even before the advent of the Delhi Sultanate. But the Portuguese, Dutch, French, and British were less interested in territorial acquisition than in control over trade, and eventually the British got the better of their competitors. Exploiting local rivalries and even defeating a combined Mughal-French force in 1757, the British East India Company (EIC) took control not only of the trade with Europe in spices, cotton, and silk goods, but also India's longstanding commerce with Southeast Asia, long in the hands of Indian, Arab, and Chinese merchants. In effect, the EIC became India's colonial administration.

In time, however, the EIC faced problems that made it increasingly difficult to combine commerce with administration. Eventually, a mutiny among Indian troops in the service of the EIC led to the abolition of the company. The British government took over in 1857 and maintained its rule (*raj*) for 90 years until 1947.

Colonial Transformation

When the British took power over South Asia, they controlled a realm with considerable industrial development (notably in metal goods and textiles) and an active trade with both Southwest and Southeast Asia. The colonialists saw this as competition, and soon India was exporting raw materials and importing manufactured goods—from Europe, of course. Local industries declined, and Indian merchants lost their markets.

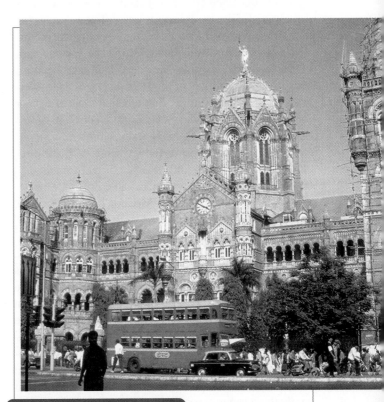

FROM THE FIELD NOTES

"More than a half-century after the end of British rule, the centers of India's great cities continue to be dominated by the Victorian-Gothic buildings the colonizers constructed here. Here is evidence of a previous era of globalization, when European imprints transformed urban landscapes. Walking the streets of Mumbai (the British called it Bombay) you can turn a corner and be forgiven for mistaking the scene for London, double-deckered buses and all. One of the British planners' major achievements was the construction of a nationwide railroad system, and railway stations were given great prominence in the urban architecture. I had walked up Naoroji Road, having learned to dodge the wild traffic around the circles in the Fort area, and watched the throngs passing through Victoria (now Chhatrapati Shivaji) Station. Inside, the facility is badly worn, but the trains continue to run, bulging with passengers hanging out of doors and windows." © H. J. de Blij.

www.conceptcaching.com

Concept Caching

Unifying their realm was a more difficult task for the British. In 1857, about 2 million square kilometers (750,000 sq mi) of South Asian territory still lay beyond British control, including hundreds of entities that had been guaranteed autonomy by the EIC during its administration. These "Native States," ranging in size from a few hectares to Hyderabad's 200,000 square kilometers (80,000 sq mi), were assigned British advisors, but in fact India was a near-chaotic amalgam of modern colonial and traditional feudal systems.

Colonialism did produce assets for India. The country was bequeathed one of the best transport networks of the colonial domain, especially the railroad system (although the network focused on interior-to-seaport linkages rather than fully interconnecting the various parts of the country). British engineers laid out irrigation canals through which millions of acres of land were brought into cultivation. Settlements that had been founded by Britain developed into major cities and bustling ports, led by Bombay (now Mumbai), Calcutta (now Kolkata), and Madras (now Chennai). These three cities are still three of India's largest urban centers, and their cityscapes bear the unmistakable imprint of colonialism (see photo, p. 411). Modern industrialization, too, was brought to India by the British on a limited scale. In education, an effort was made to combine English and Indian traditions; the Westernization of India's elite was supported through the education of numerous Indians in Britain. Modern practices of medicine were also introduced. Moreover, the British administration tried to eliminate features of Indian culture that were deemed undesirable by any standards—such as the burning alive of widows on their husbands' funeral pyres, female infanticide, child marriage, and the caste system. Obviously, the task was far too great to be achieved in barely three generations of colonial rule, but independent India itself has continued these efforts where necessary.

Partition

Even before the British government decided to yield to Indian demands for independence, it was clear that British India would not survive the coming of self-rule as a single political entity. As early as the 1930s, Muslim activists were promoting the idea of a separate state (that would become Pakistan). They circulated pamphlets arguing that British India's Muslims were a nation distinct from the Hindus and that a separate state consisting of Sindh, Punjab, Baluchistan, Kashmir, and a portion of Afghanistan should be created from the British South Asian Empire in this area. The first formal demand for such partitioning was made in 1940, and, as later elections proved, the idea had almost universal support among the realm's Muslims.

As the colony moved toward independence, a political crisis developed: India's majority Congress Party would not even consider partition, and the minority Muslims refused to participate in any future unitary government. But partition would not be a simple matter. True, Muslims were in the majority in the western and eastern sectors of British India, but Islamic clusters were scattered throughout the realm (Fig. 8-7A). Any new boundaries between Hindus and Muslims to create an Islamic Pakistan and a Hindu India would have to be drawn right through areas where both sides coexisted. People by the millions would be displaced.

The consequences of this migration for the social geography of India were far-reaching, especially in the northwest (Fig. 8-7B). Comparing the 1931 and 1951 distributions in Figure 8-7, you can see the dimensions of the losses in the Indian Punjab and in what is today the State of Rajasthan. (Because Kashmir was mapped as three entities before the partition and as one afterward, the change there represents an administrative, not a major numerical, alteration.) Even in the east a Muslim exodus occurred, as reflected on the map by the State of West Bengal, adjacent to Bangladesh, where the Muslim component in the population declined substantially.

The world has seen many **refugee** migrations but none involving so many people in so short a time as the one resulting from British India's partition (which occurred on independence day, August 15, 1947). Scholars who study the refugee phenomenon differentiate between "forced" and "voluntary" migrations, but, as this case underscores, it is not always possible to separate the two. Many Muslims, believing they had no choice, feared for their future in India and joined the stampede. Others had the means and the ability to make a decision to stay or leave, but even these better-off migrants undoubtedly sensed a threat. In any event, massive as the refugee movement was, it left far more Muslims in India than departed. As Figure 8-7 shows, the percentage of Muslims in India's total population (then about 360 million) declined even below what it had been in 1931, but India's Muslim minority has always had a high rate of population growth. Indeed, by 2010 it surpassed 170 million, more than 14 percent of the total population—the largest cultural minority in the world and almost as large as Pakistan's entire population (181 million).

And what about Hindus who found themselves on the "wrong" side of the border at the time of partition? The great majority of Hindus preferred to live in a secular democracy than under Islamic rule, so they, too, crossed the border (see photo at right). The Hindu component of present-day Pakistan may have neared 16 percent in 1947 but is only about 1 percent today; in Bangladesh (which was East Pakistan at the time of partition) it declined from 30 percent to about 16 percent today. The end of British rule in South Asia changed the realm's cultural geography forever.

FIGURE 8-7

© H. J. de Blij, P. O. Muller, and John Wiley & Sons, Inc.

SOUTH ASIA'S POPULATION DILEMMA

South Asia, as we noted earlier, is on course to become the most populous geographic realm on Earth in 2011, and India the most populous country (perhaps as early as the end of the next decade). That prospect has serious implications not only for this realm, but for the world as a whole. Over the past three decades, population growth in many parts of the world has been slowing, and some countries in Europe and Asia are even reporting that their populations are actually decreasing. The countries where this is happening tend to be the richer countries. This means that an ever-larger proportion of the new births in this century will be occurring in already poverty-afflicted regions. This has troubling ramifications, and nowhere is the trouble more serious than in South Asia.

4   All this is part of the field of **population geography**, in which research focuses on the dimensions, distribution, growth, and other aspects of human population in a country, region, or realm as this relates to soils, climates, land ownership, social conditions, economic development, and other factors. Population geography,

Flight was one response to the 1947 partition of what had been British India, resulting in one of the greatest mass population transfers in human history. Here, two trainloads of eastbound Hindu refugees fleeing (then) West Pakistan arrive at the station in Amritsar, the first city inside India.
© Corbis-Bettmann.

therefore, is the *geography of demography*. As usual, the map is a powerful ally in this geographic research.

5 In the Introduction chapter we referred to the various ways in which maps can reveal **population distribution**. Figure 8-3 shows how crowded South Asia is—despite the existence of sparsely peopled deserts and highlands even in this comparatively small realm. As we noted earlier, you can actually discern the courses of the great Ganges, Indus, and other rivers because of the immense concentrations of population that inhabit their basins. But **6** even more telling than distribution is **population density**, that is, the number of people per unit area (such as a square kilometer or square mile) in a country, province, or physiographic zone such as a river basin or mountain belt. Please take a careful look at the data displayed for South Asia in the table at the very end of the book, because you will see that population density appears in two columns. The first of these shows what we call the *arithmetic* population density for a country, which is simply the number of people per square kilometer (in this case) for each country. Truth be told, this statistic is not very meaningful despite its appearance in gazetteers and atlases, for obvious reasons. Take the case of Pakistan: when tens of millions of people cluster in river basins and valleys and almost nobody lives in deserts and high mountains, what does 227 people per square kilometer (783 per sq mi) really mean? Not much, which is why **7** geographers prefer to use the **physiologic density** measure that reports the number of people in a country per

unit of arable land area (suitable for agriculture). According to this measure, note that Bangladesh and Sri Lanka, not India, are the most densely populated countries of this realm (the very high Maldives number reflects the virtual absence of farmland on its tiny tropical islands).

Population Change

As noted above, geographers are especially interested in the **rate of natural population change** in a country, re- **8** gion, or realm, that is, the number of births minus the number of deaths, usually reported as a percentage (as we do in the Data Table inside the back cover) or as an index per thousand of the population. These data can reveal much about the social condition and prospects of a country. We have already seen how stagnant population growth produces problems of aging, how too-rapid population decline (such as Russia's) signals serious societal problems, and how a slower decline (as in South America) can be helpful in economic development.

Population geographers analyze how population growth rates change as a region's economy evolves. The higher-income economies have gone through the so-called **demo- 9 graphic transition**, a five-stage sequence that took them from high birth rates and high death rates in preindustrial times to low death rates and even lower birth rates today (Fig. 8-8). Stages 2 and 3 in this model, with high birth rates and low death rates, form the **population explosion** that **10** caused such great alarm during the twentieth century when death rates in industrializing and urbanizing countries dropped sharply, but birth rates took longer to follow suit. In 1930, the planet's human population was approximately 2 billion; by 1975, it had doubled to 4 billion. With the population explosion at full throttle, some scholars at mid-century predicted that the global population could exceed 10 billion by the year 2000.

What was not yet obvious at the time was that population growth might not only decrease quite rapidly (stage 4) but would even enter a negative stage (stage 5) as birth rates dipped below death rates, a phenomenon of the global core in high-income

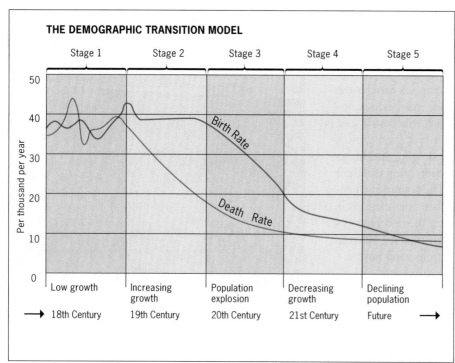

FIGURE 8-8

© H. J. de Blij, P. O. Muller, and John Wiley & Sons, Inc.

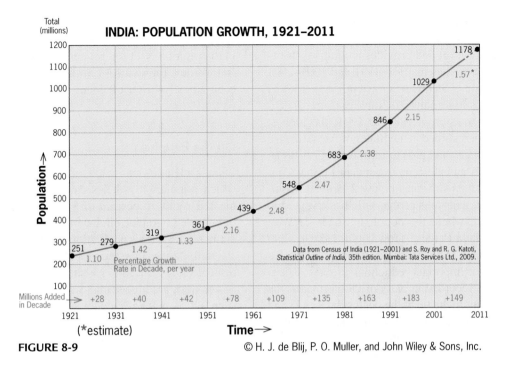

FIGURE 8-9 © H. J. de Blij, P. O. Muller, and John Wiley & Sons, Inc.

economies. In Chapter 1, we noted this is now happening in certain countries of Europe; it also affects Russia and Japan. The causes are not everywhere the same, but overall the global rate of population growth is in stage 4; the book-end Data Table shows it to be 1.3 percent today.

Does this mean that the global population crisis has dissipated? It does not. Even as growth in the global core declines to near zero, the rates of increase in many regions of the periphery remain high, in some countries nearly double the global average. Over the next nine decades the world may add between 3 and 4 billion inhabitants—most of them in the poorest parts of the periphery.

Prospects for South Asia

When the British ruled India during the nineteenth century, the country still was in the first stage, with high birth rates and high death rates; the high death rates were caused not only by a high incidence of infant and child mortality but also by famines and epidemics. As Figure 8-8 indicates, the population during stage 1 does not grow or decline much, but it is not stable. Famines and disease outbreaks kept erasing the gains made during better times. But then India entered the second stage. Birth rates remained high, but death rates declined because medical services improved (soap came into widespread use), food distribution networks became more effective, farm production expanded, and urbanization developed. In the

1920s, India's population still was growing at a rate of only 1.10 percent, but by the 1970s that rate had shot up to 2.47 percent per year (Fig. 8-9). Note that India gained 28 million people during the 1920s but a staggering 135 million during the 1970s.

Where does South Asia find itself today? Since India's component of the realm's population is the largest by far, how India fares is critical. And as Figure 8-9 shows, India's overall growth rate has been declining since the 1970s, when growth was still explosive but stage 3 had been reached. India's growth rate has declined from 2.47 percent in the 1970s to an estimated 1.57 percent (so far) during the decade 2001–2011—still above the global average but clearly and consistently declining.

But India's twentieth-century population explosion has left it with a huge burden of numbers. In Figure 8-9, note that even as the rate of growth decreased after 1971, the actual numbers of people added to the total per decade continued to increase until the decade that began in 2001. Even so, India is adding about 15 million people *per year* in the census period of 2001–2011. Thus after India enters stage 4 of the demographic transition cycle, it will still be burdened by the legacy of its stage 3 population explosion.

India: Internal Geographic Variation

Statistics for a country as large as India tend to lose their usefulness unless they are put in geographic context. In its demographic as in so many of its other dimensions, there is not just one India but several regionally different and distinct Indias. Figure 8-10 takes population growth down to the State level and provides a comparison between the census periods of 1981–1991 and 1991–2001 (the names of India's States are shown in Fig. 8-13, and their populations are displayed in Table 8-1 on p. 429).

During the period from 1981 to 1991, the highest growth rates were recorded in the States of the northeast, and only eight States had growth rates below 2 percent. But between 1991 and 2001, no fewer than 11 States, including Goa, had growth rates below 2 percent. Comparing these two maps tells us that little has changed in India's heartland, where the populous States of West Bengal and Madhya Pradesh show slight decreases but Uttar Pradesh and Bihar display similarly slight increases. The most important reductions in rates of natural increase are recorded in parts of the northeast, the east, and the south. In the tip of the peninsula, Kerala and neighboring Tamil Nadu have growth rates lower than that of the realm's leading country, Sri Lanka.

One reason for these contrasts lies in India's federal system: individual States pursued population-control policies to reduce population growth, including mass sterilizations. Another reason, as we note in more detail in the regional section, relates to spatial differences in India's development. As in the world at large, India's economically-better-off States have lower rates of population growth.

On the Periphery

India is a Hindu-dominated, officially secular country in which population policies can be debated and implemented. With 76 percent of South Asia's total population, its demographic circumstances obviously affect the entire realm. India's population growth rate remains substantially higher than the global rate (1.6 to 1.3 percent, respectively). But the growth rates of Pakistan and Bangladesh are higher than India's (in the case of Pakistan at 2.3, pratically double the world rate and one of the highest among the more populous countries on Earth). In both countries, the predominance of Islamic traditions, low levels of urbanization and modernization, and the inferior status of women (check the percentage

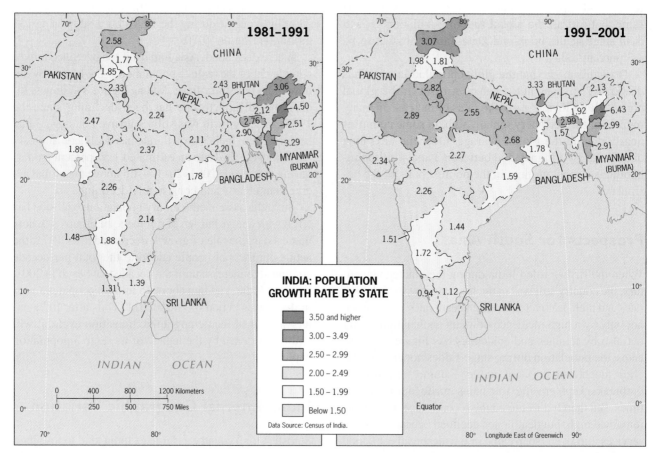

FIGURE 8-10

© H. J. de Blij, P. O. Muller, and John Wiley & Sons, Inc.

of female literacy in the Data Table at the end of the book!) are among the factors contributing to this troubling situation.

When the world's geographic realms are viewed in demographic context, South Asia and Subsaharan Africa appear to face similar problems. But we should remember that the total population of all of Subsaharan Africa's countries is less than half of South Asia's, in an area almost five times as large. South Asia's demographic condition is one of its key diagnostic properties.

SOUTH ASIA'S BURDEN OF POVERTY

In recent years, optimistic television news reports and articles in the popular press have been proclaiming a new era for South Asia, marked by rising growth rates for the realm's national economies, rewards from globalization and modernization, and increasing integration into the global economy. India, obviously the key to the realm, has been described as "India Shining" during this wave of enthusiasm.

And indeed, a combination of circumstances ranging from America's involvement with Pakistan in the campaign against terrorism to the real estate and stock market booms in India suggest that a new era has arrived. But consider this: more than two-thirds of India's nearly 1.2 billion people continue to live in poverty-stricken rural areas, their villages and lives virtually untouched by what is happening in the cities (where, by the way, tens of millions of urban dwellers inhabit some of the world's poorest slums). Fully a third of Pakistan's population lives in abject poverty; female literacy is below 30 percent. Half of the people of Bangladesh, and nearly half those in the realm as a whole, live on the equivalent of one U.S. dollar per day or less. It is estimated that half the children in South Asia are malnourished and underweight, a majority of them girls—this at a time when the world is able to provide adequate calories for all its inhabitants, if not adequately balanced daily meals.

Why is poverty so severe in this realm, comparable in this respect only to Subsaharan Africa? We have already noted more than one reason: rapid population growth in a realm with already high physiologic densities; a continued and direct dependence, by hundreds of millions of subsistence farmers, on the vagaries of climate and weather; cultural traditions that put females at a perilous disadvantage. But there are other reasons, including land ownership (in Pakistan's province of Sindh, for example, 7 percent of landowners

hold more than 40 percent of the land; in India, hundreds of millions of peasants have no control over the land they cultivate) plus government inefficiency, corruption, and venality. Even as children go hungry, powerful farm lobbies pressure the government to buy and store their grain, raising prices and thus increasing the market cost for poor villagers unable to purchase enough of it. India's nationwide subsidized food distribution system, set up nearly a half-century ago to deal with food crises, suffers from disorganization, corruption, and theft. And in this era of globalization, the government is also under pressure from international organizations that lend India money to limit subsidies to consumers. In the pages that follow, the specter of poverty haunts even the most optimistic projections for this troubled realm.

THE LATEST INVASIONS

Although it may be premature to forecast a dramatic economic rise of the sort that has thrust China onto the world stage, there can be no doubt that South Asia today commands the world's attention. Even before the events of 9/11, the end of the Cold War had altered the geopolitics of Eurasia. Before the 1990s India had tilted toward Moscow, but the post-Soviet era saw renewed linkages between India and the United States propelled by business as well as strategic motivations. Following the 2001 terrorist attacks in New York and Washington, the United States secured Pakistan's support in its War on Terror in return for massive financial assistance and support for Pakistan's authoritarian ruler (who had come to power through a coup in 1999). To get an idea of the consequences, consider this: in 2001, Pakistan's economy grew by just 2 percent (substantially below its rate of population growth); five years later, it was growing at 7 percent. But Pakistanis became increasingly resentful of American intrusions into the Tribal Areas along their border with Afghanistan and with autocratic rule, and demanded a return to democratic government. Following the assassination of presidential candidate Benazir Bhutto in 2007, an election did produce a somewhat representative government, but the American presence continued undiminished.

Meanwhile, India became a bastion for information technology and outsourcing industries, a factor in the fortunes of many large American companies. In 2005, the United States signed an agreement (finalized in 2008) to grant India full civil nuclear energy cooperation, encompassing fuel supplies and technology transfers, and easing energy-poor India's way toward nuclear energy development. Since both Pakistan and India possess nuclear weapons, the United States had withheld such cooperation because India never signed international nonproliferation agreements.

Thus South Asia today is witnessing still another of its historic invasions. But this time it is an invasion of investment, commerce, technology, custom, and connectivity—in short, *globalization*. As we noted, hundreds of millions of people are as yet virtually unaffected by this penetration, but from the textile sweatshops of Bangladesh to the telephone banks of Bengaluru (Bangalore) you can see the future. And in the streets, where shops carry ever more international goods, you see the increasing blend of the traditional and the external. It would be impractical to draw a map of all this, because the mosaic is far too intricate. India's teeming Mumbai (Bombay) has become a global metropolis, part of the international network of world cities. On the other hand, Pakistan's Lahore remains traditional and locally focused; chaotic, dangerous Karachi is somewhere in between. The ricefields of Bangladesh and the grinding rural poverty of India's adjoining West Bengal State are a world apart from the expansive wheatfields of the Punjab.

The latest—and current—invasion of South Asia is in the process of linking this realm to the wider world as never before, not even during colonial times. As we focus on the regions of this realm, we should remember that the fortunes of the twenty-first century world will be linked directly to the fate of these constituent parts.

POINTS TO PONDER

- More than 60 years after independence, India's changing of geographic names continues. Bangalore is now Bengaluru, Cochin is Kochi. Not everyone in India is happy about these name changes.

- In 2008 India and the United States finalized a landmark agreement allowing India to import civilian nuclear technology in exchange for opening key nuclear facilities to international inspections. India already has a nuclear weapons arsenal.

- Vegetarian Indians are occasionally eating meat in ever-larger numbers, a consequence of globalization. Loosening observance of religious strictures, cheaper available meat (especially chicken), and increasing foreign travel appear to be driving the change.

- Populous India contains the world's largest cultural minority, the 171-million-strong Muslim sector. But apart from a small, affluent elite, Muslims remain near the bottom of India's social ladder.

Regions of the Realm

As noted at the outset of this chapter, India forms the heart of the South Asian realm, its own core area constituting one of the greatest human agglomerations on Earth. Please refer again to Figure 8-2 on page 403, which shows the Indian coreland lying at the center of a ring of peripheral entities defined by cultural or natural borders. India is a multicultural society dominated by the Hindu faith; the border with Muslim Pakistan to the west and dominantly Islamic Bangladesh to the east have religious origins. Physiographically, India mainly consists of plateau-and-basin country, but its neighbors to the north, Nepal and Bhutan, straddle the mighty Himalaya range and its foothills. The waters of the Indian Ocean separate India from neighbors Sri Lanka and the Maldives in the south.

Thus there are good geographic reasons for using the political framework to designate the regions of this realm. Culture and nature constitute formidable regionalizing factors here, even within India itself. The cultural forces that broke British India apart at the time of independence still exert enormous power in still-evolving South Asia.

MAJOR CITIES OF THE REALM

City	Population* (in millions)
Ahmadabad, India	5.8
Bengaluru (Bangalore), India	7.2
Chennai (Madras), India	7.7
Colombo, Sri Lanka	0.7
Delhi–New Delhi, India	17.1
Dhaka, Bangladesh	14.6
Hyderabad, India	6.9
Karachi, Pakistan	13.3
Kathmandu, Nepal	1.0
Kolkata (Calcutta), India	15.8
Lahore, Pakistan	7.2
Mumbai (Bombay), India	20.2
Varanasi, India	1.5

*Based on 2010 estimates.

PAKISTAN: ON SOUTH ASIA'S WESTERN FLANK

If India is the dominant entity in South Asia, why focus first on Pakistan? There are several reasons, both historic and geographic. Here in Pakistan lay South Asia's earliest urban civilizations, whose innovations radiated southeastward into the great peninsula. Here lies South Asia's Muslim frontier, contiguous to the great Islamic realm to the west and irrevocably linked to the enormous Muslim minority to its east. Pakistan's cultural landscapes bear witness to its transitional location. Teeming, disorderly Karachi is the typical South Asian city; as in India, the largest urban center lies on the coast. Historic, architecturally Islamic Lahore is reminiscent of the scholarly centers of Muslim Southwest Asia. In Pakistan's east, the 1947 partition boundary divides a Punjab that is otherwise continuous, a land of villages, wheatfields, and irrigation ditches. In the northwest, Pakistan resembles Afghanistan in its huge migrant populations and its mountainous frontier. And in the far north, Pakistan and India are locked in a deadly conflict over Jammu and Kashmir. This western flank is clearly South Asia's most critical region.

This was perhaps never more true than it is today. Pakistan lies on the eastern periphery of the vast contiguous Islamic realm that extends from Casablanca to Kashmir, in the shadow of Afghanistan. Islamic Pakistan's coherence and stability have become crucial at a time when the global struggle against terrorism is entangling its leaders with Western power and priorities. Pakistan's role in this struggle, endorsed by its government, does not have the approval of tens of millions of Pakistanis, who express their distaste by voting for militant Islamic parties, voicing support for the resurgent Taliban movement, and giving refuge to fugitive al-Qaeda operatives in their homes. The future of Pakistan is once again in the balance, and with it the stability of South Asia.

Gift of the Indus

If, as is so often said, Egypt is the gift of the Nile, then Pakistan is the gift of the Indus. The Indus River and its principal tributary, the Sutlej, nourish the ribbons of life that form the heart of this populous country (Fig. 8-11). Territorially, Pakistan is not large by Asian standards; its area is about the same as that of Texas plus Louisiana. But Pakistan's population of 180.8 million makes it one of the world's ten most populous states. Among Muslim countries (officially, it is known as the Islamic Republic of Pakistan), only Southeast Asia's Indonesia is larger, but Indonesia's Islam has long been less militant than Pakistan's. Pakistan has been in the forefront of support for Islamic revivalism; it lies near the heart of central Asia; it has the nuclear bomb. Pakistan is a key factor in the world of Islam.

Pakistan lies like a gigantic wedge between Iran and Afghanistan to the west and India to the east. This wedge extends from the Arabian Sea in the south, where it is widest, to the glacier-studded mountains of the north. Here in the far north of Pakistan, the boundary framework is complex and jurisdictions are uncertain. India, China, Pakistan, and, until recently, the former Soviet Union have all vied for control over parts of this mountainous northern frontier. The Soviets, by virtue of their temporary rule over Afghanistan, controlled the Vakhan Corridor (the long narrow segment of land lying just south of Tajikistan in Fig. 8-11). The Chinese have claimed areas to the east. But the major territorial conflict

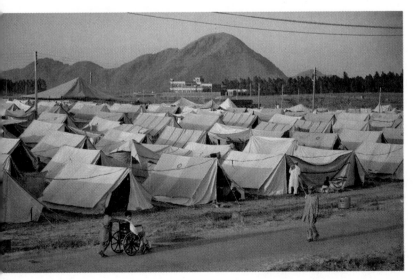

In the spring of 2009, the Pakistani government began a major military offensive against the Taliban in an area known as the Swat Valley, part of the Malakand Division of the North West Frontier Province located to the north of the city of Peshawar. The Taliban had established a virtual insurgent state in this far northern area of the country, and their armed incursions southward had overrun the regional center of Mingora and were threatening the divisional headquarters of Mardan (see Fig. 8-1). The offensive drove the Taliban back, but at a high cost to the civilian population. By mid-2009, a humanitarian crisis of massive proportions had dislocated as many as 2 million civilians, many of them surviving in refugee camps such as this at Swabi, where 18,000 people depend on help from the Red Cross/Red Crescent (ICRC).   © Jeroen Oerlemans/Panos Pictures.

FIGURE 8-11

© H. J. de Blij, P. O. Muller, and John Wiley & Sons, Inc.

has been between Pakistan and India over the areas of Jammu and Kashmir. This issue has plagued relations between these two countries for decades (see box titled "The Problem of Kashmir").

Postcolonial Reorientation

When Pakistan became an independent state following the partition of British India in 1947, its capital was Karachi on the south coast, near the western end of the Indus Delta. As the map shows, however, the present capital is Islamabad, near the larger city of Rawalpindi in the north, not far from Kashmir. By moving the cap-

ital from the "safe" coast to the embattled interior and by placing it on the doorstep of contested territory, Pakistan announced its intent to stake a claim to its northern frontiers. And by naming the city Islamabad, Pakistan proclaimed its Muslim foundation, here in the face of the Hindu challenge. This politico-geographical use of a national capital can be assertive, and Islamabad exemplifies the principle of the **forward** **11** **capital**.

At independence, Pakistan had a bounded national territory, a capital, a cultural core, and a population—but it had few centripetal forces to bind state and nation. The disparate regions of Pakistan shared the Islamic faith and an aversion for Hindu India, but little else.

Karachi and the coastal south, the desert of Baluchistan, the city of Lahore and the Punjab, the rugged northwest along Afghanistan's border, and the mountainous far north are worlds apart, and a Pakistani nationalism to match that of India at independence did not exist. Successive Pakistani governments, civilian as well as military, turned to Islam to provide the common bond that history and geography had denied the nation. In the process, Pakistan became one of the world's most theocratic states; its common law, based on the English model, was gradually transformed into a Quranic (Koranic) system with Islamic Sharia courts and associated punishments.

But even Islam itself is not unified in restive Pakistan. About 77 percent of the people are Sunni Muslims, and the Shia minority numbers about 20 percent. Sunni fanatics intermittently attack Shi'ites, leading to retaliation and creating grounds for subsequent revenge.

Despite the Islamization of Pakistan's plural society, it remains a strongly regionalized country in which Urdu is the official language and English is still the *lingua franca* of the elite. Yet several other major languages prevail in diverse parts, and lifeways vary from nomadism in Baluchistan to irrigation farming in the Punjab to pastoralism in the northern highlands.

A Long Way to Go

To govern so diverse and fractious a country would test any system, and so far Pakistan has failed the test. Democratically elected governments have repeatedly squandered their opportunities, only to be overthrown by military coups; a military regime was in power when the United States launched its War on Terror in late 2001, and Pakistan lay in the crosshairs. Becoming an American ally boosted the economy, but Pakistan's fundamental problems may in fact have worsened. The economic boom has not filtered down to the poor; literacy rates are not rising; health conditions are not improving significantly; national institutions are weak (for example, there are only about 2 million registered taxpayers in a country of more than 180 million); and one consequence of the War on Terror is that Pakistanis, who have always been overwhelmingly secular in their political choices, are joining Islamic parties in larger numbers.

Meanwhile, too little is being done to confront a growing water-supply crisis, an insurgency festers in Baluchistan, the army is incapable of establishing control over mountainous Waziristan (where al-Qaeda and Taliban organizations maintain hideouts), the issue of Kashmir costs Pakistan dearly, and relations with neighboring India, which if they were satisfactory would bring enormous benefits, remain conflicted. And looming over all this are the nuclear capabilities of both Pakistan and India and their potential to trigger disaster in this highly contentious part of the world.

Subregions

Punjab

Pakistan's core area is the Punjab (Fig. 8-11), the Muslim heartland across which the postcolonial boundary between Pakistan and India was superimposed. (As a result, India also has a region called Punjab, sometimes spelled Panjab there.) Pakistan's Punjab is home to more than 55 percent of the country's population. In the triangle formed by the Indus River and its tributary, the Sutlej, live almost 100 million people. Punjabi is the language here, and intensive farming of wheat is the mainstay.

Three cities anchor this core area: Lahore, the outstanding center of Islamic culture in the realm; Faisalabad; and Multan. Lahore, now home to 7.2 million people, lies close to the India-Pakistan border. Founded about 2000 years ago, Lahore was situated favorably to become a great Muslim center during the Mughal period, when the Punjab was a corridor into India. As a center of royalty, Lahore was adorned with numerous magnificent buildings, including an imposing fort, multiple palaces, and several mosques displaying superb stonework and marble embellishments. The site of an old university and many majestic gardens, Lahore is the cultural focus of Islam. The city received hundreds of thousands of refugees and grew rapidly after partition. Lahore did lose its eastern hinterland, but its new role in independent Pakistan sustained its growth.

Punjab's relationship with Pakistan's other three provinces is one of the country's weak points. Both the governments and peoples of the other three provinces feel uneasy about the dominance of Punjab, the populous, powerful core of the country, from which most of the army is drawn.

Sindh

The lower Indus River is the key to life in Sindh (Fig. 8-11), but the Punjab controls the waters upstream, which is one of the issues dividing this country. When the Punjab-dominated regime proposes to build dams across the Indus and its tributaries, Sindhis (who make up almost one-fourth of Pakistan's population) are reminded of their underrepresentation in government and talk of greater autonomy. Nationalist, anti-Pakistan rage swept Sindh following the assassination of presidential

The Problem of Kashmir

KASHMIR IS A territory of high mountains surrounded by Pakistan, India, China, and, along more than 50 kilometers (30 mi) in the far north, Afghanistan (Fig. 8-12). Although known simply as Kashmir, the area actually consists of several political divisions, including the State properly referred to as Jammu and Kashmir (one of the 562 Indian States at the time of independence) and the administrative areas of Gilgit in the northwest and Ladakh (including Baltistan) in the east. The main conflict between India and Pakistan over the final disposition of this

FIGURE 8-12   © H. J. de Blij, P. O. Muller, and John Wiley & Sons, Inc.

territory has focused on the southwest, where the State of Jammu and Kashmir is located.

When partition took place in 1947, the existing States of British India were asked to decide whether they wanted to be incorporated into India or Pakistan. In most of the States, the local ruler made this decision, but Kashmir was an unusual case. It had about 5 million inhabitants at the time, nearly half of them Muslims concentrated in the basin known as the Vale of Kashmir, where the capital, Srinagar, is located (Fig. 8-12). Another 45 percent of the people, dominantly Hindu, were concentrated in Jammu, which leads down the foothill slopes of the Himalayas to the edge of the Punjab. A small minority of Buddhists made the area called Ladakh their mountainous home.

But in the State of Jammu and Kashmir the ruler was Hindu, not Muslim, although the overall population was more than 75 percent Muslim. Thus the ruler had to make a difficult decision in 1947—to go with Pakistan and thereby exclude the State from Hindu India, or to go with India and thereby incur the wrath of most of the people. Hence the maharajah of Kashmir sought to remain outside both Pakistan and India and to retain autonomous status. This decision was followed, after the partitioning of India and Pakistan, by a Muslim uprising against Hindu rule in Kashmir. The maharajah asked for Indian help, and Pakistan's forces came to the aid of the Muslims. After more than

candidate Benazir Bhutto in 2007; Bhutto's prominence had always given Sindh a voice on Pakistan's national stage and had been a factor in diminishing the Sindhis' separatist movement. But her murder sparked a wave of riots aimed at government offices, police posts, railroad stations, banks, and other symbols of the Pakistani state—and fueled anti Punjab sentiment in the press that included open calls for secession.

The ribbon of fertile, irrigated alluvial land along the lower Indus, where the British laid out irrigation systems, makes Sindh a Pakistani breadbasket for wheat and rice. Commercially, cotton is king here, supplying textile factories in the cities and towns; textiles account for more than half of Pakistan's exports by value.

But the dominant presence in southern Sindh is the chaotic, crime-ridden megacity of Karachi, with its stock market, dangerous streets, crowded beaches, and poverty-stricken shantytowns, a place of searing contrasts under a

broiling sun. Karachi grew explosively during and after the partition of 1947, when refugee Muhajirs from across the new border with India streamed into this urban area, setting off riots and gang battles whose aftermath still simmers. With little effective law enforcement, Karachi later became a hotbed of terrorist activity, but somehow the city functions as Pakistan's (and Afghanistan's) major maritime outlet and seat of the provincial government.

North West Frontier

As Figure 8-11 shows, the North West Frontier Province lies wedged between the powerful Punjab to the east and troubled Afghanistan to the west, with the territory long known as the Tribal Areas intervening in the south (see box titled "Pakistan's Defiant Tribal Areas" on p. 425).

The province's mountainous physical geography reflects its remoteness and the isolation of many of its a year's fighting and through the intervention of the United Nations, a cease-fire line left Srinagar, the Vale, and most of Jammu and Kashmir (including nearly four-fifths of the territory's population) in Indian hands (Fig. 8-12). Eventually, this line began to appear on maps as the final boundary settlement, and Indian governments have proposed that it be so recognized.

Why should two countries, whose interests would be served by peaceful cooperation, allow a distant mountainland to trouble their relationship to the point of war? There is no single answer to this question, but there are several areas of concern for both sides. In the first place, Pakistan is wary of any situation whereby India would control vital irrigation waters needed in Pakistan. As the map shows, the Indus River, the country's lifeline, crosses Kashmir. Moreover, other tributary streams of the Indus originate in Kashmir, and it was in the Punjab that Pakistan learned the lessons of dealing with India for water supplies. Second, the situation in Kashmir is analogous to the one that led to the partition of the subcontinent: Muslims are under Hindu domination. The majority of Kashmir's people are Muslims, so Pakistan argues that free choice would deliver Kashmir to the Islamic Republic. A free plebiscite is what Pakistanis have sought—and the Indians have thwarted. Furthermore, Kashmir's connections with Pakistan before partition were a great deal stronger than those between Kashmir and India, although India has invested heavily in improving its links to Jammu and Kashmir since the military stalemate. In recent years, it did seem more likely that the cease-fire line (now called

the Line of Control) would indeed become a stable boundary between India and Pakistan. (The incorporation of the State of Jammu and Kashmir into the Indian federal union was accomplished as far back as 1975, when India was able to reach an agreement with the State's chief minister and political leader.)

The drift toward stability was reversed in the late 1980s when a new crisis engulfed Kashmir. Extremist Muslim groups demanding independence escalated a long-running insurgency in 1988. That brought a swift crackdown by the Indian military, whose harsh tactics prompted the citizenry to support the separatists far more strongly. In the early 1990s, Pakistan charged the Indians with widespread human rights violations, and the Indians accused the Pakistanis of inciting secession and supplying arms. Between 1990 and 1995, nearly 10,000 people were killed in the sporadic fighting that continues to afflict this scenic area. Among the casualties was the Kashmiri Pandit (Hindu) community of about 350,000 living in the Vale, which suffered as many as 1000 deaths and left the area en masse, many ending up in refugee camps in a more secure part of the territory. In 1995, Muslim extremists began a campaign that involved capturing and killing foreign visitors, a move that was condemned by representatives of both sides. Nevertheless, the memory of three wars between India and Pakistan defied every attempt at compromise. But all this was overshadowed by the implications of the 1998 test explosion of nuclear bombs by both India and Pakistan. This development transformed Kashmir from a problem frontier to a potentially calamitous flashpoint.

people. Mountain passes lead to Afghanistan; the Khyber Pass, already noted as the historic route of invaders, is legendary (see photo, p. 424). Coming from Afghanistan, the road takes you directly to the provincial capital, Peshawar, which lies in a broad, alluvium-filled and fertile valley where wheat and corn drape the countryside.

Here again Pakistan had to deal with a huge influx of immigrants: during the Soviet occupation of Afghanistan in the 1980s and later during the Taliban regime, several million Pushtun refugees streamed through the Khyber and other passes into refugee camps. There they found themselves subject to political pressure to demand autonomy within Pakistan, a case of *irredentism* that was relieved in part by United Nations assistance to the migrants. Following the ouster of the Taliban in late 2001, the great majority of Pushtuns went home. The aptly named North West Frontier Province remains a conser-

vative, deeply religious, militant territory, where Islamic political parties and movements are proportionately stronger than in any other part of the country and where obstruction of national policies (including the War on Terror) is a common goal.

Baluchistan

As Figure 8-11 indicates, Baluchistan is by far the largest of Pakistan's four provinces, accounting for (not including Kashmir) nearly half of the national territory but inhabited only by an estimated 9.5 million people or barely 5 percent of the country's population. For a sense of its terrain, take a look at Figure 8-1; much of this vast territory is desert, with mostly barren mountains that wrest some moisture from the air only in the northeast. Sheep raising is the leading livelihood here, and wool the primary export. In the far north, Baluchistan abuts the Tribal

Areas and Afghanistan. The provincial capital, Quetta, lies in this zone. This province could easily be called the "South West Frontier Province."

However, Baluchistan is not unimportant. Beneath its parched surface lie possibly substantial reserves of oil as well as coal, and already the province produces most of Pakistan's natural gas. Yet 90 percent of local residents have no energy supply themselves. And there is groundwater in several localities, but 8 out of 10 Baluchis do not have access to clean water. For decades. short-lived local rebellions have signaled dissatisfaction with the government, but recently a more serious insurgency, under the banner of the Baluchistan Liberation Army (BLA), has taken a heavier toll and appears more durable.

Part of the problem involves the construction of a major port and energy terminal at Gwadar on Baluchistan's southwestern coast (see photo, p. 425). Not only would Gwadar serve Baluchistan; it will become a transshipment point for oil and gas from remote sources, such as Iran and the Caspian Basin, destined for markets in East Asia. China dispatched engineers to plan and build this port with a loan to Pakistan, but Baluchis complained that they were left out of the process. In 2004 three Chinese engineers were killed and a dozen wounded in a terrorist attack at Gwadar. The future of this province remains very much in doubt.

Pakistan's Prospects

The Data Table inside the back cover tells the story of Pakistan's social circumstances, but not the entire story. What began as a corner of South Asia that would be a haven for Muslims in a realm of Hindus has become one of the world's ten most populous countries, the largest Islamic state on the Eurasian mainland, a nuclear power, and, officially, a partner against terrorism. It is a country of enormous cultural contrasts, where the modern and medieval exist side by side, where you hear people of one province call those of another (but not themselves) "Pakistanis," and where a sense of nationhood is still elusive among many people whose loyalties to family, clan, and village are stronger. Take another look at Figure 8-1 to be reminded how much regional geography has to do with this: Pakistan, with neighbors such as Iran, Afghanistan, disputed and divided Kashmir, and India has been what geographers call a **marchland**, an area of uncertain boundaries entered or crossed by so many armies, refugees, and migrants that its identity has become blurred. From the Muhajirs of Karachi to the Pushtuns of the Tribal Areas, from rebellious Baluchis to disaffected Sindhis, Pakistan carries the scars of its relative location.

12

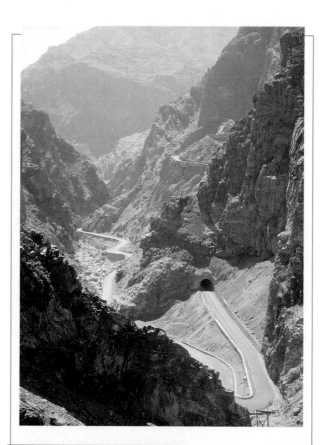

FROM THE FIELD NOTES

"Driving through the Khyber Pass that links Afghanistan and Pakistan was a riveting experience. From the dry and dusty foothills on the Afghanistan side, with numerous ruins marking historic battles, I traversed the rugged Hindu Kush via multiple hairpin turns and tunnels into Pakistan. This is one of the most strategic passes in the world, but for invading armies it was no easy passage: defensive forces had the advantage in these treacherous valleys. But now the roads and tunnels facilitate the movement of refugees, drugs, and arms, and they are used by militant separatist forces from both sides of the border. Pakistan's North West Frontier Province, and especially the city of Peshawar, are hotbeds of these activities."
© Barbara A. Weightman.

www.conceptcaching.com

Concept Caching

Location and Livelihoods

Nothing heals such scars as effectively as a thriving economy, representative government, dependable and transparent institutions, rising social indicators, and confidence in the future. But Pakistan has experienced so many cycles of progress and failure that confidence is in short supply. And yet there are areas of progress against all odds: a combination of expanded irrigation and Green Revolution farming techniques has allowed

Pakistan's Defiant Tribal Areas

WHY IS THERE a zone designated "Tribal Areas" (Fig. 8-11) in a country otherwise subdivided into provinces? It started during colonial times, when the British assigned a special status to the obstinate Pushtun villages and their chiefs living in the remote hills and mountains on the Afghan border. The "Tribal Agencies" were the responsibility of an official "agent" who exercised control over the local chiefs through ample reward and harsh punishment—and little or no accountability. There were seven of them, some very small, several large enough to appear on a relatively small-scale map such as Figure 8-11. North and South Waziristan were among the largest.

Today, the Tribal Areas (as they are now called) continue to resist authority. The Pakistan government left them pretty much alone; the army had never even entered the Tribal Areas until after 9/11, when the search for al-Qaeda and Taliban members hiding there required it. But neither Pakistan's military nor U.S. helicopters could root out the fugitive terrorists or the Taliban, and the Tribal Areas continue to be impenetrable and ungovernable.

Pakistan in recent years to export some rice (although wheat imports continue), and the country's legitimate economy has shown substantial growth. Exports include not only cotton-based textiles but also carpets, tapestries, and leather goods. Domestic manufacturing remains limited, but Pakistan has built its own steel mill near Karachi. Meanwhile, the authorities struggle to control Pakistan's growing production and trade in opium and hashish. Neighboring Afghanistan remains the world's dominant source of this illegal commerce, and the spillover effect continues; but Pakistan has many remote corners where poppy fields yield high returns and trade routes are well established. In this as in so many other spheres, Pakistan displays the contradictory symptoms of a state in transition.

Emerging Regional Power

A country's underdevelopment is not necessarily a barrier against its emergence as a power to be reckoned with in international affairs. When Pakistan became independent in 1947, the country was weak, disorganized, and divided; but only six decades later, Pakistan is a major military force and a nuclear power.

To say that Pakistan in 1947 was a divided country is no exaggeration. In fact, upon independence, present-day Pakistan was united with present-day Bangladesh, and the two countries were called West Pakistan and East Pakistan, respectively. As we have noted, the basis for this scheme was Islam: in Bangladesh, too, Islam is the state religion. Between the two Islamic wings of

Take a look at the map of Pakistan (Fig. 8-11) and you will see a small town named Gwadar on the coast of the Indian Ocean close to the border with Iran. This is an Arab corner of the South Asian realm, and it lies about as far away as you can get from the capital of Islamabad. Furthermore, this is Baluchistan Province, where insurgents frequently confront the Pakistani army. Nonetheless, Gwadar may loom large on the map of the future. Shown here is what is going on: this fishing town is being transformed by heavy equipment into a deep-water seaport, oil and gas terminal, and a gateway that is expected to serve much of landlocked central Asia. This is one of many Chinese-funded megaprojects that are in the process of transforming trade networks of the future. Among commodities in Gwadar's hinterland: Iran's vast reserves of natural gas.
© Aamir Qureshi/AFP/Getty Images, Inc.

Who Should Govern Kashmir?

KASHMIR SHOULD BE PART OF PAKISTAN!

"I don't know why we're even debating this. Kashmir should and would have been made part of Pakistan in 1947 if that colonial commission hadn't stopped mapping the Pakistan–India boundary before they got to the Chinese border. And the reason they stopped was clear to everybody then and there: instead of carrying on according to their own rules, separating Muslims from Hindus, they reverted to that old colonial habit of recognizing "traditional" States. And what was more traditional than some Hindu potentate and his minority clique ruling over a powerless majority of Muslims? It happened all over India, and when they saw it here in the mountains they couldn't bring themselves to do the right thing. So India gets Jammu and Kashmir and its several million Muslims, and Pakistan loses again. The whole boundary scheme was rigged in favor of the Hindus anyway, so what do you expect?

"Here's the key question the Indians won't answer. Why not have a referendum to test the will of all the people in Kashmir? India claims to be such a democratic example to the world. Doesn't that mean that the will of the majority prevails? But India has never allowed the will of the majority even to be expressed in Kashmir. We all know why. About two-thirds of the voters would favor union with Pakistan. Muslims want to live under an Islamic government. So people like me, a Muslim carpenter here in Srinagar, can vote for a Muslim collaborator in the Kashmir government, but we can't vote against the whole idea of Indian occupation.

"Life isn't easy here in Srinagar. It used to be a peaceful place with boats full of tourists floating on beautiful lakes. But now it's a violent place with shootings and bombings. Of course we Muslims get the blame, but what do you expect when the wishes of a religious majority are ignored? So don't be surprised at the support our cause gets from Pakistan across the border. The Indians call them terrorists and they accuse them of causing the 60,000 deaths this dispute has already cost, but here's a question: why does it take an Indian army of 600,000 to keep control of a territory in which they claim the people prefer Indian rule?

"Now this so-called War on Terror has made things even worse for us. Pakistan has been forced into the American camp, and of course you can't be against 'terrorism' in Afghanistan while supporting it in Kashmir. So our compatriots on the Pakistani side of the Line of Control have to stay quiet and bide their time. But don't underestimate the power of Islam. The people of Pakistan will free themselves of collaborators and infidels, and then they will be back to defend our cause in Kashmir."

Regional **ISSUE**

KASHMIR BELONGS TO INDIA!

"Let's get something straight. This stuff about that British boundary commission giving up and yielding to a maharajah is nonsense. Kashmir (all of it, the Pakistani as well as the Indian side) had been governed by a maharajah for a century prior to partition. What the maharajah in 1947 wanted was to be ruled by neither India nor Pakistan. He wanted independence, and he might have gotten it if Pakistanis hadn't invaded and forced him to join India in return for military help. As a matter of fact, our Prime Minister Nehru prevailed on the United Nations to call on Pakistan to withdraw its forces, which of course it never did. As to a referendum, let me remind you that a Kashmir-wide referendum was (and still is) contingent on Pakistan's withdrawal from the area of Kashmir it grabbed. And as for Muslim 'collaborators', in the 1950s the preeminent leader on the Indian side of Kashmir was Sheikh Muhammad Abdullah (get it?), a Muslim who disliked Pakistan's Muslim extremism even more than he disliked the maharajah's rule. What he wanted, and many on the Indian side still do, is autonomy for Kashmir, not incorporation.

"In any case, Muslim states do not do well by their minorities, and we in India generally do. As far as I am concerned, Pakistan is disqualified from ruling Kashmir by the failure of its democracy and the extremism of its Islamic ideology. Let me remind you that Indian Kashmir is not just a population of Hindus and Muslims. There are other minorities, for example, the Ladakh Buddhists, who are very satisfied with India's administration but who are terrified at the prospect of incorporation into Islamic Pakistan. You already know what Sunnis do to Shi'ites in Pakistan. You are aware of what happened to ancient Buddhist monuments in Taliban Afghanistan (and let's not forget where the Taliban came from). Can you imagine the takeover of multicultural Kashmir by Islamabad?

"To the Muslim citizens of Indian Kashmir, I, as a civil servant in the Srinagar government, say this: look around you, look at the country of which you are a subject. Muslims in India are more free, have more opportunities to participate in all spheres of life, are better educated, have more political power and influence than Muslims do in Islamic states. Traditional law in India accommodates Muslim needs. Women in Muslim-Indian society are far better off than they are in many Islamic states. Is it worth three wars, 60,000 lives, and a possible nuclear conflict to reject participation in one of the world's greatest democratic experiments?

"Kashmir belongs to India. All inhabitants of Kashmir benefit from Indian governance. What is good for all of India is good for Kashmir."

Vote your opinion at www.wiley.com/college/deblij

Pakistan lay Hindu India. But there was little else to unify the easterners and westerners, and their union lasted less than 25 years. In 1971, a costly war of secession, in which India supported East Pakistan, led to the collapse of this unusual arrangement. East Pakistan, upon its second independence in 1971, took the name *Bangladesh*; and since there was no longer any need for a "West" Pakistan, that qualifier was dropped and the name *Pakistan* remained on the map.

India's encouragement of independence for Bangladesh emphasized the continuing tension between Pakistan and India, which had already led to war in 1965, to further conflict in the 1970s over Jammu and Kashmir, and to flare-ups over other issues. (The continuing dispute over Kashmir is elaborated further in the Issues Box entitled "Who Should Govern Kashmir?") During the Cold War, India tilted toward Moscow, and Pakistan found favor in Washington because of its strategic location adjacent to Afghanistan. Still, armed conflict between the two South Asian countries seemed to be a regional matter—until the early 1990s, when their arms race took on ominous nuclear proportions. Both India and Pakistan had been developing nuclear weaponry, and in 1998 India exploded several nuclear bombs at its test site, a provocation that put intense pressure on Pakistan to follow suit. This took place only days later, and the notion of an "Islamic Bomb" had become reality. Since then, the specter of nuclear war has hung over the conflicts that continue to embroil Pakistan and India, a concern not just for the South Asian realm but for the world as a whole. No longer merely a decolonized, divided, and disadvantaged country trying to survive, Pakistan has taken a crucial place in the political geography of a geographic realm in turbulent transition.

● INDIA ASTIR

If you have been reading the press and watching television over the past few years, you have seen the growing attention being paid to India—not just in North America but around the world. Some examples include a *New York Times* series under the title "India Accelerating," a *Newsweek* cover story produced as a special feature under the banner headline "India Rising," and a prominent Survey of India (introduced by an editorial asking "Can India Fly?") published by the leading British newsmagazine, *The Economist*. Meanwhile, CNN and other news channels kept up a drumbeat of reportage about American jobs departing for India as U.S. corporations saved money by "outsourcing." And in the diplomatic sphere, state visits by an Indian prime minister to Washington, an American president to New Delhi, and a senior Chinese leader to India all followed in short order, reflecting India's growing global importance. Undoubtedly, India is on the move—but will it really become the next economic superpower?

Giant of the Realm

Certainly India has the dimensions to make the world pay attention. Not only does it occupy three-quarters of the great land triangle of South Asia: India also is poised to overtake China to become the world's most populous country. Already, India is the planet's largest democracy, a federation of 28 States and several additional Territories with a 2010 population of 1 billion, 186 million.

That India has endured as a unified country is a politico-geographical miracle. India is a cultural mosaic of immense ethnic, religious, linguistic, and economic diversity and contrast; it is a state of many nations. The period of British colonialism gave India the underpinnings of unity: a single capital, an interregional transport network, a *lingua franca*, a civil service. Upon independence in 1947, India adopted a federal system of government, giving regions and peoples some autonomy and identity, and allowing others to aspire to such status. Unlike Africa, where federal systems failed and where military dictatorships replaced them, India remained essentially democratic and retained a federal framework in which States have considerable local authority.

This political, democratic success has been achieved despite the presence of powerful centrifugal forces in this vast, culturally diverse country. Relations between the Hindu majority and the enormous Muslim minority, better in some States than in others, have at times threatened to destabilize the entire federation. Local rebellions, demands by some minorities for their own States, frontier wars, even involvement in a foreign but nearby civil war (in Sri Lanka) have buffeted the system—which has bent but not collapsed. India has succeeded where others in the postcolonial world have failed.

This success has not been matched in the field of economics, however. After more than 60 years of independence, India remains a very poor country, and not all of this can be blamed on the colonial period or on population growth, although overpopulation remains a strong impediment to improvement of living standards. Much of it results from poor and inconsistent economic planning, too much state ownership of inefficient industries, excessive government control over economic activities, bureaucratic suppression of initiative, corruption, and restraints against foreign investment. As we shall presently see, a few bright spots in some of India's States contrast sharply to the overwhelming poverty of hundreds of millions.

States and Peoples

The map of India's political geography shows a federation of 28 States, 6 Union Territories (UTs), and 1 National Capital Territory (NCT) (Fig. 8-13). The federal government retains direct authority over the UTs, all of which are small in both territory and population. The NCT, however, includes Delhi and the capital, New Delhi, and has more than 17 million inhabitants.

Postcolonial Restructuring

The political spatial organization shown in Figure 8-13 is mainly the product of India's restructuring following independence from Britain. Its State boundaries reflect the broad outlines of the country's cultural mosaic: as far as possible the system recognizes languages, religions, and cultural traditions. Indians speak 14 major and numerous minor languages, and while Hindi is the official

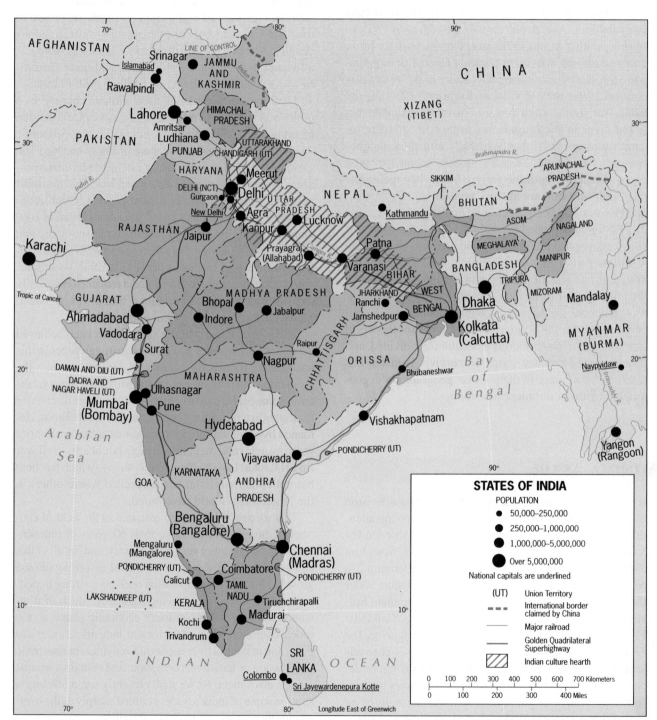

FIGURE 8-13

© H. J. de Blij, P. O. Muller, and John Wiley & Sons, Inc.

language (and English is the *lingua franca*), it is by no means universal. The map is the product of endless compromise—endless because demands for modifications of it continue to this day. As recently as 2000, the federal government authorized the creation of three new States. In the northeast lie very small States established to protect the local traditions of small populations; minority groups in the larger States ask why they should not receive similar recognition.

Staggering Numbers

With only 28 States for a national population of 1.186 billion, several of India's States contain more people than many countries of the world (Table 8-1). As Figure 8-13 shows, the (territorially) largest States lie in the heart of the country and on the great southward-pointing peninsula. Uttar Pradesh (197 million, according to the 2010 population estimate) and Bihar (96 million) constitute much of the Ganges River Basin and are the core area of modern India (see box titled "Solace and Sickness from the Holy Ganges"). Maharashtra (111 million), anchored by the great coastal city of Bombay (renamed Mumbai in 1996), also has a population larger than that of most countries. West Bengal, the State that adjoins Bangladesh, contains 89 million residents, about 16 million of whom live in its urban focus, Calcutta (renamed Kolkata in 2000).

These are staggering numbers, and they do not decline much toward the south. Southern India consists of four States linked by a discrete history and by their distinct Dravidian languages. Facing the Bay of Bengal are Andhra Pradesh (84 million) and Tamil Nadu (67 million), both part of the hinterland of the city of Madras (renamed Chennai in 1997) and located on the coast near their joint border. Facing the Arabian Sea are Karnataka (almost 59 million) and Kerala (34 million). Kerala, often at odds with the federal government in New Delhi, has long had the highest literacy rate in India and one of the lowest rates of population growth owing to strong local government and strictly enforced policies. "It's a matter of geography," explained a teacher in the Kerala city of Kochi. "We are here about as far away as you can get from the capital, and we make our own rules."

The Northern Peripheries

As Figure 8-13 shows, India's smaller States lie mainly in the northeast, on the far side of Bangladesh, and in the northwest, toward Jammu and Kashmir. North of Delhi, India is flanked by China and Pakistan, and physical as well as cultural landscapes change from the flatlands of the Ganges to the hills and mountains of spurs of the Himalayas. In the State of Himachal Pradesh, forests cover the hillslopes

TABLE 8-1

India: Population by State

State	1991 Census	2001 Census	Estimated 2010
Andhra Pradesh	66,508,000	75,728,000	84,075,000
Arunachal Pradesh	865,000	1,091,000	1,244,000
Asom (Assam)	22,414,000	26,638,000	30,187,000
Bihar	86,374,000	82,879,000	96,422,000
Chhattisgarh*		20,796,000	23,886,000
Goa	1,170,000	1,344,000	1,704,000
Gujarat	41,310,000	50,597,000	58,229,000
Haryana	16,464,000	21,083,000	24,963,000
Himachal Pradesh	5,171,000	6,077,000	6,692,000
Jammu and Kashmir	7,719,000	10,070,000	11,633,000
Jharkhand*		26,909,000	30,977,000
Karnataka	44,977,000	52,734,000	58,815,000
Kerala	29,099,000	31,839,000	34,256,000
Madhya Pradesh	66,181,000	60,385,000	71,141,000
Maharashtra	78,937,000	96,752,000	111,138,000
Manipur	1,837,000	2,167,000	2,447,000
Meghalaya	1,775,000	2,306,000	2,639,000
Mizoram	690,000	891,000	994,000
Nagaland	1,210,000	1,989,000	2,241,000
Orissa	31,660,000	36,707,000	40,433,000
Punjab	20,282,000	24,289,000	27,388,000
Rajasthan	44,006,000	56,473,000	66,795,000
Sikkim	406,000	541,000	604,000
Tamil Nadu	55,859,000	62,111,000	67,039,000
Tripura	2,757,000	3,191,000	3,536,000
Uttar Pradesh	139,112,000	166,053,000	197,319,000
Uttarakhand*		8,480,000	9,767,000
West Bengal	68,078,000	80,221,000	88,724,000
National Capital Territory**		13,783,000	17,116,000
Union Territories	1,998,000	2,670,000	3,512,000

*Established 2000
**Established 1993

Solace and Sickness from the Holy Ganges

STAND ON THE banks of the Ganges River in Varanasi, Hinduism's holiest city, and you will see people bathing in holy water, drinking it, and praying as they stand in it—while the city's sewage flows into it nearby, and the partially cremated corpses of people and animals float past. It is one of the world's most compelling—and disturbing—sights.

The Ganges (*Ganga*, as the Indians call it) is Hinduism's sacred river. Its ceaseless flow and spiritual healing power are Earthly manifestations of the Almighty. Therefore, tradition has it, the river's water is immaculate, and no amount of human (or other) waste can pollute it. On the contrary: just touching the water can wash away a believer's sins.

The stone steps leading into the Ganges' waters in Varanasi, India. Varanasi is India's holiest city, and millions descend these ghats every year. © David Zimmerman/Masterfile.

At Varanasi, Prayagraj (Allahabad), and other cities and towns along the Ganges, the riverbanks are lined with Hindu temples, decaying ornate palaces, and dozens of wide stone staircases called *ghats*. These stepped platforms lead down to the water, enabling thousands of bathers to enter the river. They come from the city and from afar, many of them pilgrims in need of the healing and spiritual powers of the water. It is estimated that more than a million people enter the river somewhere along its 2600-kilometer (1600-mi) course every day. During religious festivals, the number may be ten times as large.

By any standards, the Ganges is one of the world's most severely polluted streams, and thousands among those who enter it become ill with diarrhea or other diseases; many die. In 1986, then-Prime Minister Rajiv Gandhi launched a major scheme to reduce the level of pollution in the river, a decade-long construction program of sewage treatment plants and other facilities. In the mid-1980s, Gandhi was told, nearly 400 million gallons of sewage and other wastes were being disgorged into the Ganges every day. The plan called for the construction of nearly 40 sewage treatment plants in riverfront cities and towns.

Many Hindus, however, opposed this costly program to clean up the Ganges. To them, the holy river's spiritual purity is all that matters. Getting physically ill is merely incidental to the spiritual healing power contained in a drop of Ganga's water.

and relief reduces living space; less than 7 million people live here, many in small, comparatively isolated clusters. Before independence and political consolidation, the colonial government called this area the "Hill States."

The map becomes even more complex in the distant northeast, beyond the narrow corridor between Bhutan and Bangladesh. The dominant State here is Asom (Assam), home to 30.2 million, famed for its tea plantations and important because its petroleum and gas production amounts to more than 40 percent of India's total.

In the Brahmaputra Valley, Asom resembles the India of the Ganges. But in almost all directions from Asom, things change. To the north, in sparsely populated Aru-

nachal Pradesh (1.2 million), we are in the Himalayan offshoots again. To the east, in Nagaland (2.2 million), Manipur (2.4 million), and Mizoram (1 million), lie the forested and terraced hillslopes that separate India from Myanmar (Burma). This is an area of numerous ethnic groups (more than a dozen in Nagaland alone) and of frequent rebellion against Delhi's government. And to the south, the States of Meghalaya (2.6 million) and Tripura (3.5 million), hilly and still wooded, border the teeming floodplains of Bangladesh. Here in the country's northeast, where peoples are always restive and where population growth is still soaring, India faces one of its strongest regional challenges.

China's Latent Claims

Here, too, India faces a challenge from its powerful neighbor China. As Figure 8-13 shows, China's version of its southern border with India lies deep inside the Northeast, even coinciding with the Asom boundary part of the way. This dispute has its origins in the Simla Conference of 1913–1914, when the British negotiated a treaty with Tibet that defined their boundary in accordance with the terms laid down on the map by their chief negotiator, Sir Henry McMahon. Essentially, this *McMahon Line*, as it came to be known, ran along the most prominent crestline of the Himalayas. However, China's representatives at the conference refused to sign the agreement, arguing that Tibet was part of China and had no power to enter into treaties with foreign governments. In 1962, Chinese armed forces crossed the border and occupied Indian territory, but then withdrew. Today the matter remains unresolved.

India's Changing Map

As we noted earlier, 90 years of British rule in India—the *raj*—created an intricate mosaic of direct and indirect rule. Geographically, British authority prevailed in coastal areas, in the major river valleys where large populations clustered, and in strategic border zones where outside threats required direct rule. Most of the historic Princely States in the interior were controlled by British "advisors" to local princes whose positions depended on their acceptance of indirect rule. The British began experimenting with representative government at an early stage, allowing the formation of an Indian National Congress in 1885, transferring some political power to elected provincial officials in 1919, and approving elected provincial legislatures in 1935. Indian leaders, notably Jawaharlal Nehru, wanted to achieve independence through a united front, but this was thwarted by the Muslim League, which demanded the establishment of a separate Islamic state. When sovereignty came in 1947, there was chaos and conflict—and India's administrative map needed immediate restructuring.

After independence, the government initiated this process by phasing out the still-privileged Princely States. Next, the government reorganized the country on the basis of its major regional languages (see Fig. 8-6). Hindi, then spoken by more than one-third of the population, was designated the country's official language, but the Indian constitution gave national status to 13 other major languages, including the four Dravidian languages of the south. English, it was anticipated, would become India's common language, its *lingua franca* at government, administrative, and busi-

ness levels. Indeed, English not only remained the language of national administration but also became the chief medium of commerce in growing urban India. English was the key to better jobs, financial success, and personal advancement, and the language constituted a common ground in higher education.

Devolutionary Pressures

The newly devised framework based on the major regional languages, however, proved to be unsatisfactory to many communities in India. In the first place, many more languages are in use than the 14 that had been officially recognized. Demands for additional States soon arose. As early as 1960, the State of Bombay was divided into two language-based States, Gujarat and Maharashtra.

This devolutionary pressure has continued throughout India's existence as an independent country. In 2000, the three newest States were recognized: Jharkhand, carved from southern Bihar State on behalf of 18 poverty-stricken districts there; Chhattisgarh, where tribal peoples had been agitating since the 1930s for separation from the State of Madhya Pradesh; and Uttarakhand (originally named Uttaranchal), which split from India's most populous, Ganges Basin, core-area State of Uttar Pradesh on the basis of its highland character and lifeways (Fig. 8-13).

For many years India has faced quite a different set of cultural-geographic problems in its northeast, where numerous ethnic groups occupy their own niches in a varied, forest-clad topography. The Naga, a cluster of peoples whose domain had been incorporated into Asom (Assam) State, rebelled soon after India's independence. A protracted war brought federal troops into the area; after a truce and lengthy negotiations, Nagaland was proclaimed a State in 1961. That led the way for other politico-geographical changes in India's problematic northeastern wing.

The Sikhs

A further dilemma involved India's Sikh population. The Sikhs (the word means "disciples") adhere to a religion that was created about five centuries ago to unite warring Hindus and Muslims into a single faith. This faith's principles rejected the negative aspects of Hinduism and Islam, and it gained millions of followers in the Punjab and adjacent areas. During the colonial period, many Sikhs supported British administration of India, and by doing so they won the respect and trust of the British, who employed tens of thousands of Sikhs as soldiers and policemen. By 1947, there was a large Sikh middle class in the Punjab. When independence came, many left their rural homes and moved to the cities to enter urban professions. Today, they still exert a strong influence over Indian affairs, far in excess of the less than 2 percent of the population (about 23 million) they constitute.

For a time after India's independence, the Sikhs created India's most serious separatist problem, demanding the formation of an autonomous State they wanted to call *Khalistan*. This campaign entailed violence that not only rocked India but also reached into Sikh communities as far away as Canada. There, in the worst terrorist act in that country's history, Sikhs in 1985 placed bombs aboard an Air India jumbo jet, killing 329 as the plane exploded over the North Atlantic on its way back to India. But once again India's capacity to accommodate divisive forces eventually overcame the Sikhs' devolutionary drive. The government's decision to divide the original State of Panjab (Punjab) into a Sikh-dominated northwest—which retained the name Punjab—and a Hindu-dominated southeast (Haryana) created the Sikh stronghold this minority had wanted and that turned out to be crucial in defusing the separatist movement.

The Muslims

When India became a sovereign state and the great population shifts across its borders had ended, the country was left with a Muslim minority of about 35 million widely dispersed throughout the country. By 1991, that minority had grown to nearly 99 million, representing 11.7 percent of the total population (Fig. 8-14A). The 2001 Census of India reported a Muslim population of more than 138 million, constituting 13.4 percent and growing at a rate of nearly 4 million per year. This indicated a rate of increase in the Muslim sector of about 2.8 percent compared to India's overall rate of 1.6 percent, that is, three-quarters higher than the national average. As the Data Table at the back of the book indicates, this ranks among the world's fastest rates of natural increase; only a few countries are growing as rapidly, and none in South Asia ranks higher than 2.3 percent. The latest estimates for 2010 project a Muslim population total of 171 million (14.4 percent of the national total).

The map displaying the 2001 distribution of Muslims in India (Fig. 8-14B) reveals actual numbers as well as percentages and indicates that the core-area State of Uttar

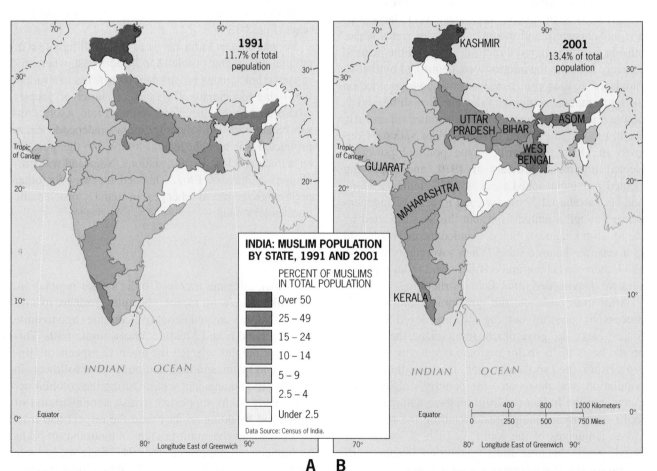

INDIA: MUSLIM POPULATION BY STATE, 1991 AND 2001

PERCENT OF MUSLIMS IN TOTAL POPULATION

- Over 50
- 25 – 49
- 15 – 24
- 10 – 14
- 5 – 9
- 2.5 – 4
- Under 2.5

Data Source: Census of India.

FIGURE 8-14

© H. J. de Blij, P. O. Muller, and John Wiley & Sons, Inc.

Pradesh and the Bangladesh-bordering State of West Bengal each contain more than 25 million Muslim inhabitants, while another two States, core-area Bihar and west-central Maharashtra, each contain more than 10 million. In terms of percentage of the total State population, however, Jammu and Kashmir leads with 67 percent, followed by Asom in northeastern India with 31 percent. Also noteworthy is Kerala in the far south, with nearly 25 percent of its population adherents of the Islamic faith. India's great geographic advantage is that its Muslim minority is not regionally concentrated, precluding the kind of secession movement that often develops when communities with different cultures are spatially divided (as we shall see, Sri Lanka in this realm suffers from exactly that kind of problem).

Relations between the Hindu majority and the Muslim minority are complex. What makes the news is conflict—for example, in 2002 when Muslims attacked a train carrying Hindus to a contested holy site in Ayodhya, killing dozens, which was followed by retaliation in several towns (but not others) in Gujarat State. What does not make the news is that a Muslim population of more than 170 million lives and participates in the kind of democracy that is all but unknown in the Islamic world itself. In the first decade of the twenty-first century, a Hindu nationalist party was the leading force in the federal government, but a Muslim scientist, the man who developed India's first nuclear weapons, was the country's president. On the other hand, many Muslims have been angered by the Hindu-inspired name changes appearing on India's map, most recently the substitution of Prayagraj for the venerable Islamic name of Allahabad, a key city on the Ganges in the State of Uttar Pradesh.

The most serious threat to Muslim integration into Indian society arises from the comparatively low level of education in Muslim communities. Even as small cadres of Muslim university graduates have become successful professionals, Muslims in some of India's States are falling behind even dalits (members of lowest-ranked Hindu castes, discussed in the following section) in terms of education and levels of poverty. According to official statistics, less than 4 percent of Muslims nationwide complete secondary school, so that India's Muslim minority consists of a huge mass of poor locals and a minuscule middle class. India's leaders worry that such data reflect a lack of aspiration among Muslim citizens of the country; a large majority of Muslims do not see secular education as a way to get ahead in this non-Islamic country. Bridging this growing gap is one of India's most urgent challenges, because a disaffected cultural minority forms a ready market for militants, foreign and domestic.

Already, concern over linkages between foreign (read Pakistani) terrorists and local Muslim subversives is changing Hindu attitudes and tactics, as shown by the growing cooperation between Indian and Israeli intelligence operations and official statements accusing Indian Muslim citizens of acting as collaborators in terrorist attacks (indeed, by 2010, India had become the most frequently targeted country in the world). India's leaders questioned Pakistan's prominent role as an American ally in the campaign against terrorism when Pakistan itself was involved in such acts in Kashmir.

Centrifugal Forces: From India to Hindustan?

In Chapter 1 we introduced the concept of centrifugal and centripetal forces affecting the fabric of the state. No country in the world exhibits greater cultural diversity than India, and variety in India comes on a scale unmatched anywhere else on Earth. Such diversity spells strong centrifugal forces, although, as we will see, India also has powerful consolidating bonds.

Class and Caste

Among the centrifugal forces, Hinduism's stratification of society into castes remains pervasive. Under Hindu dogma, *castes* are fixed layers in society whose ranks are based on ancestries, family ties, and occupations. The **caste system** may have its origins in the early social divisions into priests and warriors, merchants and farmers, craftspeople and servants; it may also have a racial basis, for the Sanskrit term for caste is color. Over the centuries, its complexity grew until India had thousands of castes, some with a few hundred members, others containing millions. Thus, in city as well as in village, communities were segregated according to caste, ranging from the highest (priests, princes) to the lowest (the untouchables). The term *untouchable* has such negative connotations that some scholars object to its use. Alternatives include *dalits* (oppressed), the common term in Maharashtra State but coming into general use; *harijans* (children of God), which was Gandhi's designation, still widely used in the State of Bihar; and *Scheduled Castes*, the official government label.

A person was born into a caste based on his or her actions in a previous existence. Hence, it would not be appropriate to counter such ordained caste assignments by permitting movement (or even contact) from a lower caste to a higher one. Persons of a particular caste could perform only certain jobs, wear only certain clothes, worship only in prescribed ways at particular places. They or their children could not eat, play, or even walk with people of a higher social status. The untouchables occupying the lowest tier were the most debased, wretched members of this rigidly structured social system. Although the British ended the worst excesses of the caste system, and postcolonial Indian leaders—including Mohandas (Mahatma) Gandhi (the great spiritual leader who sparked the independence movement) and Jawaharlal Nehru (the first prime minister)—worked to modify it, a few decades cannot erase centuries of class consciousness. In traditional India, caste provided stability and continuity; in modernizing India, it constitutes an often painful and difficult legacy.

Today we can discern a geography of caste—a degree of spatial variation in its severity. Cultural geographers estimate that about 15 percent of all Indians are of lower caste, about 40 percent of backward caste (one important rank above the lower caste), and some 18 percent of upper caste, at the top of which are the Brahmans, men in the priesthood. (The caste system does not extend to the Muslims or Sikhs, although converted Christians are affected, which is why these percentages do not add up to 100.) The colonial government and successive Indian governments have tried to help the lowest castes. This effort has had more effect in urban than in rural areas of India. In the isolated villages of the countryside, the untouchables often are made to sit on the floor of their classroom (if they go to school at all); they are not allowed to draw water from the village well because they might pollute it; and they must take off their shoes, if they wear any, when they pass higher-caste houses. But in the cities, untouchables have reserved for them places in the schools, a fixed percentage of State and federal government jobs, and a quota of seats in national and State legislatures. Gandhi, who took a special interest in the fate of the untouchables in Indian society, accomplished much of this reform.

The caste system remains a powerful centrifugal force, not only because it fragments society but also because efforts to weaken it often result in further division. Gandhi himself was killed, only a few months after independence, by a Hindu fanatic who opposed his work for the least fortunate in Indian society. But progress is being made, and while efforts to help the poorest are not always popular among the better-off, the future of India depends on it.

Hindutva

Another growing centrifugal force in India has to do with a concept known as *hindutva* or Hinduness—a desire to remake India as a society in which Hindu principles prevail. This concept has become the guiding agenda for a political party that became a powerful component of the federal government, and it is variously expressed as Hindu nationalism, Hindu patriotism, and Hindu heritage. This naturally worries Muslims and other minorities, but it also concerns those who understand that India's secularism, its separation of religion and state, is indispensable to the survival of its democracy. *Hindutva* enthusiasts want to impose a Hindu curriculum on schools, change the flexible family law in ways that would make it unacceptable to Muslims, inhibit the activities of non-Hindu religious proselytizers, and forge an India in which non-Hindus are essentially outsiders. Moderate Hindus and non-Hindus in India oppose such notions, which are as divisive as any India has faced. They nonetheless acknowledged the appeal of the Bharatiya Janata Party (BJP), which swept to power on a Hindu-nationalist platform, and realized that only the constraints of a coalition government kept the party from pursuing its more extreme goals. Recent State and national elections have seen a decline in the fortunes of the BJP, reflecting an internal party struggle between moderates and *hindutva* hardliners—and also confirming the continuing strength and vitality of Indian democracy.

The radicalization of Hinduism and the infusion of Hindu nationalism into federal politics might be seen by outsiders as threats to India's unity, but Indian voters have not rushed to embrace these initiatives. During the most recent national election campaign, the opposition Congress Party accused the BJP of polarizing India, citing the recent spate of name changes on India's map as one example of this divisive effect. The BJP narrowly lost its majority in the 2004 elections, and its prime minister resigned, yielding to a successor from the opposition Congress Party. But the Congress Party does not have a stellar record of governance, so the pendulum could swing back again. Increasingly, however, the key issues are likely to revolve around the management of India's diverse and growing economy, and the *hindutva* campaign may well take a back seat.

Just how far India's political pendulum can swing was shown in provincial legislative elections in Uttar Pradesh State in 2007, where polls had predicted that no party would come close to securing a majority vote—and where a dalit party, led by a woman named Mayawati Kumari, won an absolute majority of 208 of the 403 seats in the legislature. It was a stunning victory that made headlines throughout the country and shook the political establishment, revealing the growing power of the lowest castes and marking a turning point for India's representative government.

Communist Insurgency

Over the past several years there have been troubling signs of still another divisive force in India: communist-(avowedly Maoist-) inspired rebellions that appear to be growing into a coordinated revolutionary campaign (see photo to the right). It is known in India as the *Naxalite* movement, named after a town in the State of West

Bengal where it was founded in the 1960s to oppose, by force of arms, the still-new government in New Delhi. In the decades that followed, India's police managed to control and at times virtually wipe out the Naxalites, but in recent years they have been making a comeback. Based in some of India's most remote, poorest, and disaffected areas of States such as Bihar, Jharkhand, Chhattisgarh, and even Andhra Pradesh, the Maoist Naxalites have had notable successes against India's law enforcement, including liberating prisoners from jails, blowing up railroad tracks, laying landmines, and fighting pitched battles against police.

A look at Figure 8-13 shows that one of the Indian States where the Naxalites are having an impact lies adjacent to Nepal, where a Maoist insurgency grew over several decades, took control over a large part of the country's west, ruined the economy, and destabilized the national government (already weakened by a sclerotic royal family, corruption, and infighting) at a cost of many thousands of lives. In Nepal, the Maoists eventually saw their political opportunity when the monarchy imploded, and they joined the political system. In Nepal's 2008 elections, their Communist Party emerged as the country's largest and their leader became prime minister. This sequence of events raised alarm in India, where the Naxalites are active in 13 of the country's 28 States and, according to the government, in nearly 80 of its 604 Administrative Districts (the next level below that of a State), although nongovernment observers put the numbers much higher.

Why do the Maoists form a threat in economically developing India? They appeal to the lower-caste poorest of the poor, whose plight they blame on India's burgeoning economy and globalizing efforts that, for a long time to come, will leave tens of millions of families ever farther behind. Their tactics damage the country's infrastructure, target mines and factories, and intimidate corporations and contractors, and their successes against police and army units serve to expose weaknesses in India's security system. Stability is key to India's economic prospects; this is what the Naxalites try to undermine.

Islamic Extremism

Comparatively successful as the integration of India's Muslim communities into the fabric of the Indian state has been, the risk of Islamic violence, directed against Indian society in general, is also rising. For many years India has suffered from Islamic terrorism, attributable to the longstanding quarrel with Pakistan over Kashmir, to discord with Islamic Pakistan in general, and to occasional incidents of sectarian violence of the kind that followed the razing of a mosque in the town of Ayodhya to make way for a Hindu temple. At times such attacks have brought the two neighbors close to war, for example, in 2001 when Islamic terrorists with roots in Pakistan assaulted the Indian Parliament. In 2006, a series of simultaneous, coordinated attacks struck several trains along a commuter line in Mumbai (Bombay), killing about 200 people. While Indian investigators found the usual Pakistani links (some of the perpetrators had attended Pakistani terrorist-training camps), others were local Mumbai residents. Among those arrested were a doctor and a software engineer, middle-class Muslim residents of a multicultural city who had been drawn into the circle of Islamic terrorism. In late 2008, when terrorists struck a number of targets in Mumbai including two international hotels, some Indian officials were quick to suggest that the attackers, who belonged to a Pakistan-based terrorist group, must have had local accomplices to succeed as they did. No evidence of this, however, emerged.

The ghost of Mao Zedong may be dissipating in China, but it lives on in India. Shown in this remarkable 2006 photo is a brigade of Mao-inspired rebels marching during what they claimed was their ninth "convention," deep in the forest of the State of Chhattisgarh 1500 kilometers (930 mi) southeast of New Delhi. These Naxalites, as they are called in India, have a presence in nearly half of India's 28 States, and they prove capable of damaging infrastructure and defying law enforcement. On March 15, 2007, they attacked a police station in Orissa State, killing more than 50 officers and destroying the post. They claim to support the causes of India's dalits and demand land and jobs for the poor in India's rural east. ©AP/World Wide Photos.

Such developments have serious implications for India, where the overwhelming majority of Muslim citizens have remained uninvolved in extremist causes, even after provocations such as the deathly anti-Muslim riots of 2002 in the State of Gujarat. But now it seems that a few educated Muslims are joining local terrorist cells with links to Pakistan and perhaps other Islamic countries. Their terrorist acts spur investigations that offend ordinary and peaceful Indian Muslims, radicalizing a number of them and expanding the market for Islamic militancy. It is as yet too early to gauge the potential impact of this development on a country that has long and justly prided itself on its multicultural democracy and multisectarian stability, but the portents for India's political, social, and economic geography are obviously serious.

Centripetal Forces

In the face of all these divisive forces, what bonds have kept India unified for so long? Without question, the dominant binding force in India is the cultural strength of Hinduism, its sacred writings, holy rivers, and influence over Indian life. For most Indians, Hinduism is a way of life as much as it is a faith, and its diffusion over virtually the entire country (regardless of the Muslim, Sikh, and Christian minorities) brings with it a national coherence that constitutes a powerful antidote to regional divisiveness. Over the long term, however, the key ingredients of this Hinduism have been its gentility and introspection, radical outbursts notwithstanding.

Democracy

Another centripetal force lies in India's democratic institutions. In a country as culturally diverse and as populous as India, reliance on democratic institutions has been a birthright ever since independence, and democracy's survival—raucous, often corrupt, always free—has been a crucial unifier.

Furthermore, communications are better in much of India than in many other countries in the global periphery, and the continuous circulation of people, ideas, and goods helps bind the disparate state together. Before independence, opposition to British rule was a shared philosophy, a strong centripetal force. After independence, the preservation of the union was a common objective, and national planning made this possible.

Accommodation

India's capacity for accommodating major changes and its flexibility in the face of regional and local demands also constitute a centripetal force. Boundaries have been shifted; internal political entities have been created, relocated, or otherwise modified; and secessionist demands have been handled with a mixture of federal power and cooperative negotiation. Indians in South Asia have accomplished what Europeans in former Yugoslavia could not, and India's history of success is itself a centripetal force.

Education

Still another centripetal force in India is education. The country takes great pride in its high literacy rates, which in urban India exceed 96 percent for both males and females; these numbers are substantially above those of neighboring mainland countries and reflect Indians' determination to avail themselves of every educational opportunity. Private institutes teaching English abound in the cities, and when opportunities for the outsourcing of service jobs arose on the global economic scene, India had the educated workforce to seize them.

Leadership

Finally, no discussion of India's binding forces would be complete without mentioning the country's strong leadership. Gandhi, Nehru, and their successors did much to unify India by the strength of their compelling personalities. For many years, leadership was a family affair: Nehru's daughter, Indira Gandhi, twice took decisive control (in 1966 and 1980) after weak governments, and her son, Rajiv Gandhi (who, in 1991, like his mother seven years earlier, also was assassinated), was prime minister in the late 1980s. Since then, political leadership of India has been less dynastic—and also less cohesive.

Urbanization

India is famous for its great and teeming cities, but India is not yet an urbanized society. Only 28 percent of the population lived in cities and towns in 2010—but in terms of sheer numbers, that 28 percent amounts to 332 million people, more than the entire population of the United States.

India's rate of urbanization is on the upswing. People by the hundreds of thousands are arriving in the already overcrowded cities, swelling urban India by about 3 percent annually, twice as fast as the overall population growth. Not only do the cities attract as they do everywhere: many villagers are driven off the land by the desperate conditions in the countryside. As villagers manage to establish themselves in Mumbai or Kolkata or Chennai, they help their relatives and friends to join them in

"Searing social contrasts abound in India's overcrowded cities. Even in Mumbai (Bombay), India's most prosperous large city, hundreds of thousands of people live like this, in the shadow of modern apartment buildings. Within seconds we were surrounded by a crowd of people asking for help of any kind, their ages ranging from the very young to the very old. Somehow this scene was more troubling here in well-off Mumbai than in Kolkata (Calcutta) or Chennai (Madras), but it typified India's urban problems everywhere." © H. J. de Blij.

www.conceptcaching.com

squatter settlements that often are populated by newcomers from the same area, bringing their language and customs with them and cushioning the stress of the move.

As a result, India's cities display staggering social contrasts. Squatter shacks without any amenities at all crowd against the walls of modern high-rise apartments and condominiums (see photograph above). Hundreds of thousands of homeless roam the streets and sleep in parks, under bridges, on sidewalks. As crowding intensifies, social stresses multiply. Disorder never seems far from the surface; sporadic rioting, often attributable to rootless urban youths unable to find employment, has become commonplace in India's cities.

Urban Tradition and Evolution

India's modern urbanization has its roots in the colonial period, when the British selected Calcutta (Kolkata), Bombay (Mumbai), and Madras (Chennai) as regional trading centers and fortified ports. Madras was fortified as early as 1640; Bombay (1664) had the situational advantage of being the closest of all Indian ports to Britain; and Calcutta (1690) lay on the margin of India's largest population cluster and had the most productive hinterland, to which the Ganges Delta's countless channels connected it. This natural transport network made Calcutta an ideal colonial headquarters, but the population of Bengal was often rebellious. In 1912, the British moved their colonial government from Calcutta to the safer interior city of New Delhi, built adjacent to the old Mughal capital of Delhi.

Figure 8-13 displays the distribution of major urban centers in India. Except for Delhi–New Delhi, the largest cities have coastal locations: Kolkata dominates the east, Mumbai the west, and Chennai the south. But urbanization also has expanded in the interior, notably in the core area. In 2007, India had 41 metropolitan areas with populations exceeding 1 million, but road links among these (and other)

AMONG THE REALM'S GREAT CITIES . . . Mumbai (Bombay)

ANOTHER HISTORIC NAME is disappearing from the map: Bombay. In precolonial times, fishing folk living on the seven small islands at the entrance to this harbor named the place after their local Hindu goddess, Mumbai. The Portuguese, first to colonize it, called it Bom Bahia, "Beautiful Bay." The British, who came next, corrupted both to Bombay, and so it remained for more than three centuries. Now local politics has taken its turn. Leaders of a Hindu nationalist party that control the government of the State of which Bombay is the capital, Maharashtra, passed legislation to change its name back to Mumbai. In 1996, India's federal government approved this change.

Mumbai's 20.2 million people make this India's largest city. Maharashtra is India's economic powerhouse, the State that leads the country in virtually every respect. Locals dream of an Indian Ocean Rim of which Mumbai would be the anchor.

As such, Mumbai is a microcosm of India, a burgeoning, crowded, chaotic, fast-moving agglomeration of humanity (see photo, p. 437). The Victorian-Gothic architecture of the city center is a legacy of the British colonial period (see photo, p. 411). Shrines, mosques, temples, and churches evince the pervasive power of religion in this multicultural society. Street signs come in a bewildering variety of scripts and alphabets. Creaking double-decker buses compete with oxwagons and handcarts on the congested roadways. The throngs on the sidewalks—businesspeople, holy men, sari-clad women, beggars, clerks, homeless wanderers—spill over into the streets.

Mumbai is an urban agglomeration of city-sized neighborhoods, each with its own cultural landscape. The seven islands have been connected by bridges and causeways, and the resulting Fort District still is the center of the city, with many of its monuments and architectural landmarks. Marine Drive leads to the wealthy Malabar Hill area, across Back Bay. Northward lie the large Muslim districts, the Sikh neighborhoods, and other ethnic

© H. J. de Blij, P. O. Muller, and John Wiley & Sons, Inc.

and cultural enclaves. Beyond are some of the world's largest and poorest squatter settlements, including Dharavi of *Slumdog Millionaire* fame. It is situation, not site, that gave Bombay primacy in South Asia. The opening of the Suez Canal in 1869 made Bombay the nearest Indian port to Europe. Today, Mumbai is the focus of India's fastest-growing economic zone—not yet a tiger but on the move.

cities remain grossly inadequate. Now, in the second decade of the twenty-first century, India is constructing a nationwide four-lane expressway that will link the four anchors of this urban system (Delhi, Mumbai, Chennai, and Kolkata) and in the process connect 15 other major cities along this national route called the *Golden Quadrilateral* (Fig. 8-13). The impacts of this project are multiple: it is expanding urban hinterlands, commuters are using it to travel farther to work than ever before, once-remote villages now have a link to markets, and it is accelerating the rural-to-urban migration flow that will transform India in this century (see photo, p. 439). In this edition of our book we are

still able to say that India is a land of 600,000 villages in a country whose population is less than one-third urbanized. Future editions will undoubtedly chronicle the emergence of a new India with fewer villages, ever-larger metropolitan complexes, and an expanding number of megacities.

Economic Geography

Improving India's infrastructure, though, is not enough to overcome all serious impediments to the country's economic advancement. Take another look at Figure 8-13

AMONG THE REALM'S GREAT CITIES . . . Kolkata (Calcutta)

CALCUTTA IS SYNONYMOUS with all that can go wrong in large cities: poverty, dislocation, disease, pollution, crime, corruption. To call a city the Calcutta of its region is to summarize urban catastrophe. West Bengal's leaders had this in mind when they proposed that the city's local name, Kolkata, be restored. The city got its dreadful reputation (as well as its Anglicized name) during colonial times, when plague, malaria, and other diseases claimed countless thousands in legendary epidemics.

The British chose the site, 130 kilometers (80 mi) up the Hooghly River from the Bay of Bengal and less than 9 meters (30 ft) above sea level, not far from some unhealthy marshes but well placed for commerce and defense. When the British East India Company was granted freedom of trade in the populous hinterland, Calcutta's heyday began; when (in 1772) the British made it their colonial capital, the city prospered. The British sector of the city was drained and raised, and so much wealth accumulated here that Calcutta became known as the "city of palaces." Outside the British town, rich Indian merchants built magnificent mansions. Beyond lay neighborhoods that were often based on occupational caste, whose names are still on the map today (such as Kumartuli, the potters' district). Almost everywhere, on both banks of the Hooghly, lay the huts and hovels of the poorest of the poor. Searing social contrasts characterized Calcutta.

The twentieth century was not kind to Calcutta. In 1912 the British moved their colonial capital to New Delhi. The 1947 partition that created Pakistan also created then-East Pakistan (now Bangladesh), cutting off a large part of Calcutta's hinterland and burdening the city with a flood of refugees. The Indian part of the city had arisen virtually without any urban planning, and the in-

© H. J. de Blij, P. O. Muller, and John Wiley & Sons, Inc.

flux created almost unimaginable conditions. Today, Kolkata counts 15.8 million residents, including as many as 500,000 homeless. Beyond the façade of the downtown lies what may be the sickest city of all.

The Golden Quadrilateral Highway (see Fig. 8-13) is India's largest and most ambitious infrastructure project to date, a modern traffic artery that connects the capital, New Delhi in the north, the largest city, Mumbai on the northwest coast, Chennai on the far southeastern coast, and Kolkata in the northeast. Started in the mid-1990s, the project has taken over 15 years to complete; many kilometers of it had to be built on viaducts over congested local roads in urban areas such as Bengaluru, where construction is shown here in progress. Not only does the Golden Quadrilateral Highway connect the four major cities; it also links numerous smaller centers to make long-distance surface transport much more efficient than it has long been. What is needed now is improvement in the roads connecting second-tier cities such as Hyderabad, Kochi, Lucknow, and Patna to this artery. © Ed Kashi/Getty Images, Inc.

FLY DIRECTLY OVER the Delhi–New Delhi conurbation into its international airport, and you may not see the place at all. A combination of smog and dust creates an atmospheric soup that can limit visibility to a few hundred feet for weeks on end. Relief comes when the rains arrive, but Delhi's climate is mostly dry. The tail-end of the wet monsoon reaches here during late June or July, but altogether the city only gets about 60 centimeters (25 in) of rain a year. When the British colonial government decided to leave Calcutta and build a new capital city adjacent to Delhi, conditions were different. South of the old city lay a hill about 15 meters (50 ft) above the surrounding countryside, on the right bank of the southward-flowing Yamuna tributary of the Ganges. Compared to Calcutta's hot, swampy environment, Delhi's was agreeable. In 1912 it was not yet a megacity. Skies were mostly clear. Raisina Hill became the site of a New Delhi.

This was not the first time rulers chose Delhi as the seat of empire. Ruins of numerous palaces mark the passing of powerful kingdoms. But none brought to the Delhi area the transformation the British did. In 1947 the Indian government decided to keep its headquarters here. In 1970 the metropolitan-area population exceeded 4 million. By 2010 it was 17.1 million.

Delhi is popular as a seat of government for the same reason as its present expansion: the city has a fortuitous relative location. The regional topography creates a narrow corridor through which all land routes from north-

© H. J. de Blij, P. O. Muller, and John Wiley & Sons, Inc.

western India to the North Indian Plain must pass, and Delhi lies in this gateway. Thus the twin cities not only contain the government functions; they also anchor the core area of this populous country.

Old Delhi was once a small, traditional, homogeneous town. Today Old and New Delhi form a multicultural, multifunctional urban giant.

and note that the Golden Quadrilateral crosses numerous State boundaries. In the United States, we are used to seeing thousands of trucks on the interstate highways, crossing from one State into another without slowing down; there are truck-weighing stations along these expressways, but in the newest ones all the truck has to do is slow down enough for electronic surveillance to record its passing.

In India, truck drivers face a very different experience. It can take as long as nine days, including more than 30 hours waiting at State-border checkpoints and tollbooths, for a loaded truck to travel from Kolkata to Mumbai via Chennai, or less than the distance from Los Angeles to New Orleans on Interstate –10. Drivers are subject to daunting piles of paperwork and repeated demands for bribes, and when mechanical problems occur it may take days to get the truck back on the road. It will take more than expanding India's highway network to improve the circulation the country so desperately needs.

Globalization

When you arrive in any Indian city, you are struck by the number of small shops everywhere—tiny businesses wedged into every available space in virtually every non-public building along every street. Even the upper, walkup floors are occupied by shops, their advertising signs suspended from windows and balconies. According to a study by the Indian government's Department of Consumer Affairs, India has the highest density of retail outlets of any country in the world with 15 million shops (compared to well below 1 million in the United States, where the marketplace is 13 times richer). After farming, the retail sector is India's largest provider of jobs.

What keeps all these small stores in business? Most of them earn very little and can afford to stay open only because they are part of what economic geographers call the *informal sector*: they are essentially unregistered, pay no rent and probably no taxes, use family labor, have been handed down through generations, and survive because India's economic geography, bound by longstanding protec-

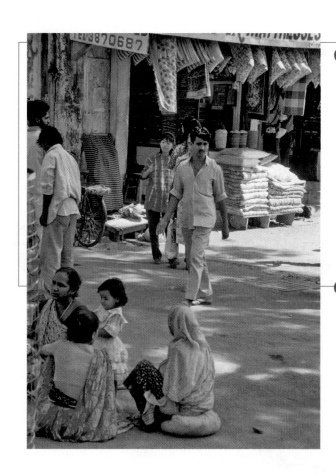

tive government regulation, has been slow to change. When you are in India you may wonder how so many shops can stay in business, but the answer is in the throngs of people on the sidewalks (spilling over into the clogged traffic in the streets). Not many of these people are wealthy, but all of them need basic goods and some can afford small luxuries, and so the shops tend to be busy all day.

Now imagine what would happen if India suddenly opened its doors to large international retailing companies like Wal-Mart. An invasion of foreign superstores would force millions of India's small shops to close, throw countless workers into unemployment, and destroy ways of life that have put most urban residents within walking distance of their daily needs. And yet this is what India faces. This is the era of globalization, and large-scale organized retailing, long held back by a combination of restraints (such as the nationwide distribution problems discussed above, but also legal and cultural obstacles), is emerging within India itself. This means that competition from foreign companies will somehow infiltrate the market, and in this respect the Indian government faces only one key question: how to keep control of the process through gradual deregulation. That issue is complicated, however, by India's chaotic democracy. Even if New Delhi approves appropriate legislation, individual States may counter it by imposing their own regulations and restrictions.

Still, the statistics show that the process is already well underway. India's middle class, now estimated to include some 300 million people (equal to the entire U.S. population), is growing rapidly, and something very new is beginning to appear in the cities: shopping malls. As recently as 2000, this vast country containing over one-sixth of humankind did not have a single shopping center! But by 2005 the 100th had opened its doors, and in mid-2009 more than 500 were in operation. You will see American fast-food restaurants among the establishments in these malls as well as the brand names of numerous other foreign companies, proving that globalization has already breached India's walls. The question remains: can India's economic transformation be achieved without severe social dislocation?

Farming's Enduring Importance

Agriculture, we noted earlier, still provides more jobs in India than any other economic sector, and India's fortunes (and misfortunes) remain strongly tied to farming. For all the emphasis on India's cities and middle-class growth, more than 70 percent of the people still live on (and from) the land, traditional farming methods persist, yields per hectare remain among the world's lowest, and hunger and malnutrition still afflict millions even as grain surpluses accrue. The relatively few areas of modernization, as in the wheatlands of the Punjab, are islands in a sea of stagnation. Thus the agricultural sector and indeed the entire country is directly vulnerable to environmental variations, with all the risks that entails. Furthermore, land reform has essentially failed; roughly one-quarter of India's entire cultivated area, including much of the best land, is still owned by less than 5 percent of the country's landholders, who have much political influence and obstruct redistribution. Perhaps half of all rural families own less than one hectare (2.5 acres) or no land at all: an estimated 175 million live and work as tenants, always uncertain of their fate.

Getting produce to markets—if there is any to sell—is yet another struggle for millions of farmers. In 2009, almost half of India's 600,000 villages could not be accessed by truck or car, and in this era of modern transportation animal-drawn carts still far outnumber motor vehicles nationwide.

Figure 8-15 maps India's agriculture and reflects the climatic distribution shown in Figure G-7, the rainfall pattern in Figure G-6, and the monsoonal cycle depicted in Figure 8-4. Rice dominates along the Arabian Sea-facing southwestern coast and in the monsoon-drenched peninsular northeast; where drier conditions develop, wheat and other grains prevail. Although, as the Data Table inside the back cover reminds us, physiologic densities in India are lower than in neighboring Bangladesh to the east, the comparison is deceptive because Indian farming, especially in the rice-growing zones and despite the effects of the Green Revolution, is so inefficient. For India's economic transformation to have real impact in the rural areas, nothing less than a technological revolution is required.

The Energy Problem

If you have a friend in (or from) India, you know someone who is familiar with power outages. They are a way of life in India, where electricity demand routinely exceeds available supply, where governments cannot bring themselves to require customers to pay for the actual cost

FIGURE 8-15   © H. J. de Blij, P. O. Muller, and John Wiley & Sons, Inc.

of the power they consume, and where power grids, generating equipment, and other related infrastructure are in bad shape. All this while hundreds of millions of villagers still have no electricity supply at all.

And yet electrical power is key to India's modernization. Already, foreign companies doing business in India sometimes import their own generators, but others are discouraged from investing in factories and other facilities because the power supply is so unreliable. Most of India's electricity is generated in thermal plants burning coal, oil, or natural gas; about 25 percent comes from hydroelectric sources, and about 3 percent from nuclear plants. The problems are many: India has substantial coal deposits, but the railroads cannot handle the transport to power plants. So India must import coal, but the ports do not have the required capacity. And India possesses only limited oil and natural gas reserves. Add to this an increasingly inadequate national power-supply grid in a country with a still-exploding population and even faster-growing demand, and you have trouble.

A key remedy, of course, lies in increased oil and gas imports, but here India runs into geopolitical problems. While the American government was eager to give India leeway in the nuclear arena, the United States made it clear that it does not like India's plan to buy Iranian natural gas via a pipeline across Pakistan. India's other options lie in interior Asia, but those sources are more distant and pipeline construction would involve further diplomatic complications. Once again, the other alternative—importing oil and gas via tankers and ports—is constrained by inadequate infrastructure.

Limitations on Manufacturing Growth

Given these problems, the geography of India's manufacturing (Fig. 8-16) is changing too slowly for the country's needs. The map is a legacy of colonial times, with coastal Mumbai, Kolkata, and Chennai anchoring major industrial zones and textile industries—the entry-level industry of disadvantaged countries—dominating the scene. India's information technology (IT) industries, centered in and around Bengaluru (Bangalore), Hyderabad, Mumbai, and Delhi, draw much international attention, and software and IT services

account for more than one-quarter of merchandise exports by value, but what India needs far more are manufacturing industries competitively selling goods on world markets, putting tens of millions to work, and transforming the economy (as has happened in China, which we discuss in Chapter 9). It is almost unbelievable that as many people worked in manufacturing in 1991 as in 2001 (the two most recent census years) while China's industrial workforce more than quadrupled during the same period.

Could India follow in China's footsteps? Some economic geographers suggest that India might leapfrog China and move quickly from an "underdeveloped" to a "postindustrial" services-led economy. Certainly India has the requisite intellectual clout: Indian corporations have set up hundreds of IT schools in China where tens of thousands of Chinese students are learning the business. But to reach India's hundreds of millions of potential wage-earners, India needs a vigorous expansion of its secondary industries, those that make goods (beyond textiles) and sell them at home and abroad. Here's how it may happen: when an economy churns along the way China's has over the past three decades, labor and production costs tend to rise. That causes manufacturers

FIGURE 8-16

© H. J. de Blij, P. O. Muller, and John Wiley & Sons, Inc.

to look for places where cheaper labor will reduce such costs. India, with its long history of local manufacturing, huge domestic market, and vast reservoir of capable labor, would then take its turn on the world stage.

As always in India, cultural issues also intrude when it comes to economic change. In 2008, a division of one of India's largest companies, Tata Motors, introduced a tiny automobile that would sell for little more than U.S. $2000. Experts internationally praised this vehicle as a marvel of Indian ingenuity and engineering, and when Tata proposed to build the factory to produce it in one of India's poorest States, West Bengal, it planned to boost the job market where this seemed to be most needed. But Tata's planners failed to reckon with the farmers in West Bengal and their political party. First, urged on by political leaders, these farmers refused to negotiate over the price of their land. Next, the company faced rallies and blockades as it tried to persuade locals that the thousands of prospective jobs would more than compensate for the approximately 400 hectares (1000 acres) of land it sought. Then, suddenly, the company withdrew its plans and began considering offers from other States. But the implications were clear: if the Tata Group, an Indian corporation, could not build a factory in an Indian State, would foreign investors looking for cheap labor want to risk similar problems? As one of our colleagues wrote us, "It would never have happened in China. Democracy has its costs."

Improving Prospects

Despite problems of the kind just discussed, India's prospects are brightening, if not yet in the dramatic terms used in popular-media hype. A middle class of more than 300 million demands goods ranging from cellphones (in recent years, 2.5 million new subscribers were signing up every month) to motor bikes (10,000 per day were being sold). Even if Tata's mini-car was not going to be built in West Bengal State, Chennai was already becoming an Indian automobile center: Korean and German cars are coming off assembly lines not only to be marketed in India but also to be sent to foreign consumers (that BMW you saw today may have been assembled in India and sold in Europe to an American buyer).

India's economy today ranks as the world's sixth-largest, and although the effect of the global economic crisis that began in 2008 cannot yet be gauged, many economists suggest that, whatever happens, India by 2020 may be in third place. Over the past several years, the economy has grown by an average of 7 percent—no match for China, but far ahead of the growth (and negative-growth) rates in most other parts of the world. As we will see in the next chapter, China's dramatic growth resulted from decisions at the top, a transformation planned and implemented in controlled detail. In India the economy is

growing from the bottom up, with all the traditional chaos that makes India a country like no other. As with China's provinces, some of India's States will advance ahead of others, and India's already incredible socioeconomic contrasts will intensify. But over time, India could achieve what China has—hitherto—not: an economic and cultural geography of consensus.

India East and West

The most commonly cited, and most clearly evident, regional division of India is between north and south. The north is India's heartland, the south its Dravidian appendage; the north speaks Hindi as its *lingua franca*, the south prefers English over Hindi; the north is bustling and testy, the south seems slower and less agitated.

But there is another, as yet less obvious, but potentially more significant divide across India. In Figure 8-16, draw a line from Lucknow, on the Ganges River, south to Madurai, near the southern tip of the peninsula (also review Figure 8-2 on page 403). To the west of this line, India is showing signs of economic progress, the kind of economic activity that has brought Pacific Rim countries such as

India's leading high-tech center is Bengaluru (formerly Bangalore: another of India's ongoing place-name changes), and the Indian company Infosys Technologies, based here, is India's leading software exporter. This Silicon Valley–like campus in suburban Electronics City is the center of the company's booming outsourcing business (Bank of America and Citigroup are among its American customers). Bengaluru, as Figure 8-13 shows, lies near the intersection of the states of Karnataka, Andhra Pradesh, and Tamil Nadu. While some skilled workers have emigrated to Europe, Australia, and America, where their skills earn them more than at home, many more choose to stay here and benefit from the higher local incomes the IT industries pay. Thus Bengaluru attracts India's best and brightest, whose children attend the city's superior schools. And this is a pleasant place to live: at 900 meters (3000 ft) above sea level, even the summer weather is tolerable. Bengaluru, however, is growing so rapidly that its infrastructure cannot keep up, so that congestion, traffic jams, and commuting times are all growing as well. Sound familiar? ©Reuters/Corbis.

Thailand and Indonesia a new life. To the east, India has more in common with less promising countries also facing the Bay of Bengal: Bangladesh and Myanmar (Burma).

As with other regional divides, there are exceptions to our east-west delineation. Indeed, our map seems to suggest that much of India's industrial strength lies in the east. But what Figure 8-16 cannot reveal is the profitability of those industries. True, the east is rich in iron and coal, but the heavy industries built by the state in the 1950s are now outdated, uncompetitive, and in decline. The hinterland of Kolkata contains India's Rustbelt. The government keeps many industries going, but at a high cost. Old industries, such as carpetmaking and cotton-weaving, continue to use child labor to remain viable. The State of Bihar represents the stagnation that afflicts much of India east of our line: by several measures it ranks among the poorest of the 28 States.

Compare this to western India. The State of Maharashtra, the hinterland of Mumbai, leads India in many categories, and Mumbai leads Maharashtra. Many smaller, private industries have emerged here, manufacturing goods ranging from umbrellas to satellite dishes and from toys to textiles. Across the Arabian Sea lie the oil-rich economies of the Arabian Peninsula. Hundreds of thousands of workers from western India have found jobs there, sending money back to families from Punjab to Kerala. (The western State of Gujarat offered Tata Motors the land it needed as soon as the mini-car producer gave up on West Bengal.) More importantly, many have used their foreign incomes to establish service industries back home. Outward-looking western India, in contrast to the inward-looking east, has begun to establish other ties to the outside world. Satellite links have enabled Bengaluru (Bangalore) to become the center of a growing software-producing complex reaching world markets.

The beaches of Goa, the small coastal State immediately to the south of Maharashtra, appeal to the tourist markets of Europe. This is, in fact, a classic case of **intervening opportunity** because resorts have sprung up along Goa's shore, and European tourists who once went to the more distant Maldives and Seychelles are coming to Goa. Maharashtra's economic success also has spilled over into Gujarat to the north, and even landlocked Rajasthan (the next State to the north) is experiencing the beginnings of what, by Indian standards, is a boom.

The boom has created political problems, however. Not only is Maharashtra State a rising economic power; it also is the base of a strong Hindu nationalist political movement whose leaders object to foreign intrusions and have blocked major development projects and other enterprises. They halted a huge industrial scheme about halfway through and closed a fast-food operation that they deemed incompatible with local culture. Such clashes between foreign interests and domestic traditions continue even as India's economy forges ahead.

Nevertheless, India's east-west divide shows a growing contrast that puts the west far ahead. The hope is that Maharashtra's success will spread northward and southward along the Arabian Sea coast and will ultimately diffuse eastward as well. But for this to happen, India will have to bring its population spiral under control.

BANGLADESH: CHALLENGES OLD AND NEW

On the map of South Asia, Bangladesh looks like another State of India: the country occupies the area of the double delta of India's great Ganges and Brahmaputra rivers, and India almost completely surrounds it on its landward

FIGURE 8-17   © H. J. de Blij, P. O. Muller, and John Wiley & Sons, Inc.

is the deltaic plain of the Ganges–Brahmaputra river system, combining fertile alluvial soils that attract farmers with the low elevations that endanger them when water rises. The shape of the Bay of Bengal forms a funnel that sends cyclones and their storm surges of wind-whipped water barreling into the delta coast. Without money to build seawalls, floodgates, elevated shelters in sufficient numbers, or adequate escape routes, hundreds of thousands of people are at continuous risk, with deadly consequences. And as if this is not enough, millions of people have been found to be exposed to excessive (natural) arsenic in the drinking water from their wells.

Limits to Opportunity

Bangladesh remains a nation of subsistence farmers. Urbanization is at only 24 percent; Dhaka, the megacity capital, and the cities of Chittagong, Rangpur, Khulna, and Rajshahi are the only urban centers of consequence. Moreover, Bangladesh has one of the highest physiologic densities in the world (1679 people per square kilometer/4349 per square mile), and only higher-yielding varieties of rice and the introduction of wheat in the crop rotation (where climate permits) have improved diets and food security. But diets remain poorly balanced, and, overall, the people's nutrition is unsatisfactory. The troubled textile industry provides most of Bangladesh's foreign revenues, but the once-thriving jute industry continues its decline. The discovery of a natural gas reserve is now the subject of a national debate: home consumption or money-making export?

Bangladesh is a dominantly Muslim society, but not one dominated by revivalists (for example, 30 seats in the national legislature are reserved for women). But Bangladesh's politics are chaotic, patronage-driven, corrupt, and seemingly beyond redemption. Foreign aid, essential to the country's survival, pours in, but the country remains mired among the world's lowest-income societies. Its relations with neighboring India have at times been strained over water resources (India controls the Ganges, which is Bangladesh's lifeline), cross-border migration (one-sixth of the population is Hindu), and transit

side (Fig. 8-17). But Bangladesh is an independent country, born in 1971 after its brief war for independence against Pakistan, with a territory about the size of Wisconsin. Today it remains one of the poorest and least developed countries on Earth, with a population of 152.4 million that is growing at an annual rate of 1.7 percent.

A Vulnerable Territory

Not only is Bangladesh a poor country; it also is highly **15** susceptible to damage from **natural hazards**. During the twentieth century, eight of the ten deadliest natural disasters in the entire world struck this single country. The most recent, a 1991 cyclone (as hurricanes are called in this part of the world), killed more than 150,000 people.

The reasons for Bangladesh's vulnerability can be deduced from Figures 8-17 and 8-1. Southern Bangladesh between parts of India across Bangladesh's north (refer to Figure 8-16 to see the reason). All the disadvantages of the global periphery afflict this populous, powerless country where survival is the leading industry and all else is luxury.

THE MOUNTAINOUS NORTH

As Figures 8-1 and 8-2 show, a tier of landlocked countries and territories lies across the mountainous zone that walls India off from China. One of them, Kashmir, is in a state of near-war. Another, Sikkim, was absorbed by India in 1975 and made into one of its federal States. But Nepal and Bhutan retain their independence.

Nepal, northeast of India's Hindu core, has a population of 28.1 million and is the size of Illinois. It has three geographic zones (Fig. 8-18): a southern, subtropical, fertile lowland called the *Terai*; a central belt of Himalayan foothills with swiftly flowing streams and deep valleys; and the spectacular high Himalayas themselves (topped by Mount Everest) in the north. The capital, Kathmandu, lies in the east-central part of the country in an open valley of the central hill zone.

Nepal, birthplace of Buddha, is materially poor but culturally rich. The Nepalese are a people of many sources, including India, Tibet, and interior Asia. About 85 percent are Hindu, and Hinduism is the country's official religion; but Nepal's Hinduism is a unique blend of Hindu and Buddhist ideals. Thousands of temples and pagodas ranging from the simple to the ornate grace the cultural landscape, especially in the valley of Kathmandu, the country's core area. Although over a dozen languages are spoken, 90 percent of the people also speak Nepali, a language related to Indian Hindi.

Nepal is a troubled country suffering from severe underdevelopment, with the lowest GNI in the realm. It also faces strong centrifugal social and political forces. Environmental degradation, crowded farmlands and soil erosion, and deforestation scar the countryside (see photo at right). The Himalayan peaks form a world-renowned tourist attraction, but tourist spending in Nepal, always relatively modest, has been cut back because of recurrent disorder associated with a Maoist-communist insurrection and the collapse of the country's monarchy.

Nepal's cultural and political geographies have long been in turmoil. Strong regionalism divides the country both north-south and east-west. The southern Terai with its tropical lowlands is a world apart from the Himalayan foothills in the central interior, and the peoples of the west have origins and traditions different from those in the east. An absolute (Hindu) monarchy held Nepal together until 1990, when antigovernment demonstrations

and resulting casualties persuaded the king to lift the ban on political parties. But chaos soon resumed, the king was assassinated, and by 2002 Maoist rebels controlled about 40 percent of the country, turning Nepal into a **failed state**. The ruling king was forced from office in 2006, the insurgents were brought into the political system, and in the following year Nepal became a secular state. Elections in 2008 confirmed the capacity of Nepal's new system to accommodate the still-powerful centrifugal forces in the country.

Human-induced environmental degradation is vividly displayed on Nepal's highest inhabitable slopes. Against the spectacular backdrop of the main Himalaya range north of Kathmandu, population pressures in the teeming valleys have forced farming and fuel-wood extraction to ever higher elevations. The inevitable result is deforestation and when the dense vegetation and binding roots are removed, copious summer-monsoon rainwater is free to cascade down the steep, denuded mountainsides. This massive runoff soon produces myriad gullies that further destabilize slopes by triggering mass movements of soil and rock. The lower half of this photo exhibits the evidence of this erosional process, which affects the more level terrain in the foreground as well. Ominously, we can also discern human dwellings creeping upslope, which portends a final onslaught on the single remaining forested ridge.
© Galen Rowell/Mountain Light Photography, Inc.

FIGURE 8-18

© H. J. de Blij, P. O. Muller, and John Wiley & Sons, Inc.

Mountainous Bhutan, with a population of 740,000, lies wedged between India and China's Tibet (Fig. 8-18), the only other buffer between Asia's giants. In landlocked, fortress-like Bhutan, time seems to have stood still. Bhutan long was a constitutional monarchy ruled by a king whose absolute power was unquestioned by his subjects. But in 2007 the reigning king, who had recently succeeded his father, and perhaps with an eye on what had happened to the monarchy in nearby Nepal, decided to order his subjects to vote for a political party in a newly created democracy. Thus in 2008 Bhutan went from absolute monarchy to multiparty democracy on the orders of its monarch, and Thimphu, the capital (population: ca. 80,000), became the seat of a new National Assembly.

In the mountainous countryside, the symbols of Buddhism, the state religion, dominate the cultural landscape. Social tensions persist through the presence of a Nepalese (Hindu) minority; and a Bhutanese refugee population remains housed in camps in eastern Nepal. Add to this a still-unresolved boundary issue with China, and newly democratic Bhutan faces challenges aplenty.

Forestry, hydroelectric power, and tourism all have potential here, and Bhutan has considerable mineral resources. But isolation and inaccessibility preserve traditional ways of life in this mountainous *buffer state*.

● THE SOUTHERN ISLANDS

As Figure 8-1 shows, South Asia's continental landmass is flanked by several sets of islands: Sri Lanka off the southern tip of India, the Maldives in the Indian Ocean to the southwest, and the Andaman Islands (belonging to India) marking the eastern edge of the Bay of Bengal.

The Maldives consists of more than a thousand tiny islands whose combined area is just 300 square kilometers (115 sq mi) and whose highest elevation is less than 2 meters (6 ft) above sea level. Its population of just over 300,000 from Dravidian and Sri Lankan sources is now virtually 100 percent Muslim, with one-quarter of it concentrated on the capital island named Maale. The Maldives might be unremarkable, except that, as the book-end Data Table shows, this country has the realm's highest GNI per capita. The locals have translated their palm-studded, beach-fringed islands into a tourist mecca that attracts tens of thousands of mainly European visitors annually. The key question is whether the Maldives' economy can withstand the global downturn that started in 2008, because tourism is a vulnerable industry.

The Maldives' low elevation has been mentioned repeatedly in assessments of the future impact of global warming, which would produce rising sea levels. Those risks became sudden reality on December 26, 2004 when the Indian Ocean tsunami generated off Indonesia swept over the islands, killing more than 100 residents as well as tourists and destroying resort facilities along the shore as well as inland.

The Maldives may remain the richest country per capita in South Asia, but there are concerns beyond those of low relief and global warming. A democratic election in 2008 was a first, but whether representative government can survive economic hard times is an open question. And sunbathing globals tend not to notice what is in effect a heroin epidemic, a soaring rate of unemployment among locals, rising poverty, and signs of Islamic extremism long unknown in this Indian Ocean paradise.

Sri Lanka: South Asian Tragedy

Sri Lanka (known as Ceylon before 1972), the compact, pear-shaped island located just 35 kilometers (22 mi) across the Palk Strait from southernmost India, became independent from Britain in 1948 (Fig. 8-19). There were good reasons to create a separate sovereignty for Sri Lanka. This is neither a Hindu nor a Muslim country: the

majority of its population of 20.8 million people—some 70 percent—are Buddhists. Furthermore, unlike India or Pakistan, Sri Lanka is a plantation country, and commercial farming still is the mainstay of the agricultural economy.

The great majority of Sri Lanka's people are descended from migrants who came to this island from northwestern India beginning about 2500 years ago. Those migrants introduced the advanced culture of their source area, building towns and irrigation systems and bringing Buddhism. Today, their descendants, known as the Sinhalese, speak a language (Sinhala) that belongs to the Indo-European language family of northern India.

The Dravidians who lived on the mainland, just across the Palk Strait, came later and in far smaller numbers—until the British colonialists intervened. During the nineteenth century the British brought hundreds of thousands of Tamils to work on their tea plantations, and soon a small minority became a substantial segment of Ceylonese society. The Tamils brought their Dravidian tongue to the island and introduced their Hindu faith. At the time of independence, they constituted more than 15 percent of the population; today they total about 18 percent.

Hope and Disaster

When Ceylon became independent, it was one of the great hopes of the postcolonial world. The country had a sound economy and a democratic government, and it was renowned for its tropical beauty. Its reputation soared when a massive effort succeeded in eradicating malaria and when family-planning campaigns reduced

FIGURE 8-19   © H. J. de Blij, P. O. Muller, and John Wiley & Sons, Inc.

"Sri Lanka is a dominantly Buddhist country, the religion of the majority Sinhalese. Most areas of the capital, Colombo, have numerous reminders of this in the form of architecture and statuary: shrines to the Buddha, large and small, abound. But walk into the Tamil parts of town, and the cultural landscape changes drastically. This might as well be a street in Chennai or Madurai: elaborate Hindu shrines vie for space with storefronts and Buddhist symbols are absent. The people here seemed to be less than enthused about the Tamil Tigers' campaign for an independent state. 'This would never be a part of it anyway,' said the fellow walking toward me as I took this photograph. 'We're here for better or worse, and for us the situation up north makes it worse.' But, he added, Sri Lankan governments of the past had helped create the situation by discriminating against Tamils."
© H. J. de Blij.

www.conceptcaching.com

Concept Caching

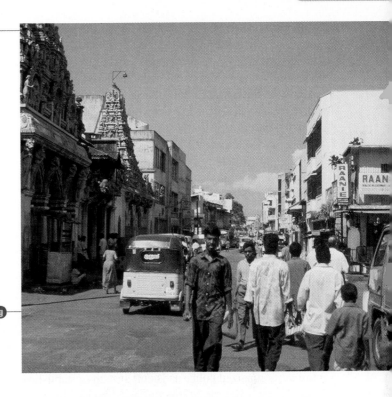

population growth while the rest of the realm was experiencing a population explosion. Rivers from the cool, forested interior highlands fed the paddies that provided ample rice; crops from the moist southwest paid the bills, and the capital, Colombo, grew to reflect the optimism that prevailed.

In the midst of this glowing scenario, the seeds of disaster were already being sown. Sri Lanka's Tamil minority soon began proclaiming its sense of exclusion, demanding better treatment from the Sinhalese majority. Although the government recognized Tamil as a "national language" in 1978, sporadic violence marked the Tamil campaign, and in 1983 full-scale civil war broke out. Now many in the Tamil community demanded a separate Tamil state to encompass the north and east of the country (see Fig. 8-19 inset map), and a rebel army called the Liberation Tigers of Tamil Eelam (LTTE) confronted Sri Lanka's national forces.

This is the tragic end of an insurgent state overpowered by a national counteroffensive: here in northeastern Sri Lanka the Tamil Tigers made their last stand before their defeat by the Sri Lankan army in May 2009. Thousands of civilians, used as human shields in the final stage of the battle, died or were wounded; as many as 200,000 were displaced by the fighting. In the aftermath of the civil war, the Sinhalese-dominated government must find ways to reintegrate Tamils and others swept up in the insurgents' campaign, in order to avoid a return to the social conditions that gave rise to the insurgency in the first place.
© AP/Wide World Photos.

17 In Chapter 5 (pp. 248-250) we discuss the concept of the **insurgent state** in the context of Colombia's struggle against armed revolutionaries seeking to establish a "parallel" state within its borders. As noted, such campaigns do not always reach the stage of *equilibrium*, when the rebels succeed in wresting a piece of territory from the government and set up their own administration. But in Sri Lanka, the LTTE did achieve this. First they took the Jaffna Peninsula (Fig. 8-19), next they extended their power over the Vanni Region, and soon they started publishing maps showing their ultimate demand: the entire northern and eastern periphery of Sri Lanka would become the secessionist state of *Eelam* (whose most recent rendering is shown by the striped zone on the inset map). They backed up these demands with terrorist attacks in the capital, Colombo, and unleashed combat in many parts of the state they were seeking to establish.

Eventually the town of Kilinochchi, on the mainland near the entry to the Jaffna Peninsula, became the LTTE "capital," with all the hallmarks of an insurgent state. The "Tigers" collected taxes, administered banks and issued currency, organized police and court systems, established schools, and even ran hotels and guest houses for visitors. By 2002 the government of Sri Lanka had agreed to a cease-fire, and negotiations leading to the country's partition—at least from the LTTE perspective—began.

But in 2007 the Colombo government decided to embark on the *counteroffensive* that marks the third and decisive stage in the success or failure of an insurgent state. After five years of intermittent fighting, cease-fires, and negotiations, it concluded that the only way to settle the conflict was through an all-out military campaign. On January 2, 2009, President Mahinda Rajapaksa announced the capture of Kilinochchi and proclaimed that the threat of partition of Sri Lanka had ended.

In the late spring of 2009, the final battle of the counteroffensive took place in the Vanni Region amid terrible losses among civilians, who were used as human shields by the LTTE even as the government army impeded their escape. The leader of the insurgency was killed, and the Tamil Tigers' insurgent state disappeared from the map. An international outcry over the tactics of the Sri Lankan army was followed by expressions of concern over the future of Sinhalese-Tamil relations in this still-fractured country. Reconciliation is direly needed, but while several million Tamils wanted nothing do with any armed insurgency, the Colombo government seemed to focus more on retribution than pacification. More than 30 years ago the insurgency arose in part because of Hindu-Tamil anger over the government's Buddhist-Sinhalese cultural dominance and social policies, and only more even-handed management of national affairs can stave off a resumption of the cycle that transformed South Asia's great success story into a dreadful tragedy whose aftereffects will be felt for decades.

What You Can Do

RECOMMENDATION: Seek an internship! If you are willing to trade your time for some experience and perhaps even education and training, consider looking for an internship where you can use your knowledge of geography while learning more about it. Should you decide to become a geography major or take a minor, you may be eligible for one of the National Geographic Society's internships in Washington, D.C. The Society places interns in many of its operations, from the Geography Bee to National Geographic Maps. Inquire about this and other internship opportunities at the office of the Geography Department at your college or university.

GEOGRAPHIC CONNECTIONS

1 One of the current issues in the United States involves the so-called outsourcing of jobs: the transfer of salaried employment from American to foreign locations. Many economists say that this is a logical outgrowth of globalization, that it involves comparatively few jobs, and that it helps U.S. companies survive competition to which they might otherwise succumb—costing more jobs than the outsourcing did in the first place. India has become a major player in the outsourcing of certain work. What are these jobs? Where in India do they cluster and why? Do language and/or time-zone location have anything to do with India's success, and if so, how?

2 Six decades after India and Pakistan ceased to be part of the British colonial empire, the contrasts between India and Pakistan—demographic, economic, social, political—could hardly be greater despite the fact that they are neighbors. What, in your view, are the key differences between the two South Asian states in these contexts? Can you identify crucial geographic factors that would account for such enormous dissimilarities? Given their shared histories, do you see any common ground between Pakistan and India today that might form a basis for their future convergence and cooperation?

Geographic Literature on South Asia: The key introductory works on this realm were authored by Farmer, Spate & Learmonth, Spencer & Thomas, Weightman, and Wolpert. These works, together with a comprehensive listing of noteworthy books and recent articles on the geography of South Asia, can be found in the *References and Further Readings* section of this book's website at **www.wiley.com/college/deblij.**

KAZAKH
UPLANDS
KAZAKHSTAN

Qaraghandy
70° 50°
80°
Rubtsovsk
Semey
Öskemen
Ob
Irtysh
Novokuznetsk
90°
100°
Angara
Lena
R U S
Irkutsk
Ula-Ude
Angarsk
Darhan
Erdenet
Ulaanb

MOYNQUM
Lake
Balqash
SARYESIK-ATYRAU
DESERT

Syr Darya

TASHKENT
UZBEKISTAN
ALAY MTS.
TAJIKISTAN
AFGHANISTAN
HINDU
KUSH
Khyber
Pass
Pamir

Bishkek
KYRGYZ RANGE
KYRGYZSTAN
T I A N   S H A N
ALMATY
Ysyk-Köl

Ismoil Somoni Peak
(7,495 m, 24,590 ft.)
KARAKORAM RANGE
K2
(8,611 m,
28,250 ft.)

Kashi
(Kashgar)
Shache
Yecheng
KUN LUN

Karamay
Yining
Shihezi
Changji
ÜRÜMQI
Baicheng
Aksu
Kuqa
Yanqi
Korla
TARIM BASIN
TAKLA MAKAN
DESERT

Hotan
Qiemo
Ruoqiang
Lop Nur

JUNGGAR
BASIN
Turpan
Turpan Depression
(-154 m, 505 ft.)
Hami
Dunhuang

Mt. Belukha
(4,506 m,
14,783 ft.)
ALTAY
Ölgiy
Altay
MOUNTAINS

SAYAN MOUNTAINS
Kyzyl
Yenisey
Mt. Munku Sardyk
(3,491 m, 11,453 ft.)

Ulaangom
Dund-Us
HANGAYN MTS.
Uliastay
MONGOL
Ulaan-Uul
Arvayheer
Altay

MONGO
PLATE
GOBI
Ejin Qi
Dalandzadgad

YIN
Linhe

PAKISTAN
KASHMIR
Islamabad
RAWALPINDI
LAHORE
FAISALABAD
LUDHIANA
Chandigarh

M
Wujiang
Shiquanhe
Qagcaka
Barga
Nyima
Zhongba
Brahmaputra
Xigaze
Gyangze

Kathmandu
Mt. Everest
(8,850 m, 29,035 ft.)

QINGHAI-XIZANG
PLATEAU
TANGGULA RANGE
NYAINQÊNTANGLHA SHAN
Nagqu
Salween
Lhasa
Nyingchi
Wenquan
Gyêga
Riwoka
Markam

ALTUN SHAN
Golmud
Da Qaidam
QAIDAM BASIN
Lenghu

QILIAN SHAN
Yumen
Zhangye
Jinchang
Dulan
Qinghai Hu
Xining
Nangdo
LANZHOU
Qabqa
Guyuan

C   H   I   N
Wudu
Tianshui
Baoji
XIANYA
QIN LIN
Tongchuan
Yar
Chang Jiang
Wei

Huang
Wuwei
Qingtongxia
Wuzhong
Yinchuan
Shizuishan
Wuda
ORD
DESE

Hanzhong
Guangyuan
Ankang
Mianyang
CHENGDU
(RED)
Nanchong
SICHUAN
BASIN
Qianning
Yaan
Neijiang
CHONGQING
Zigong
Yibin
Fulin
DAXUE MOUNTAINS
Chang Jiang

DELHI
MEERUT
New Delhi
Ghaziabad
FARIDABAD
JAIPUR
AGRA
KANPUR
LUCKNOW
GANGES PLAIN
Yamuna
Ganges
Prayagraj
(Allahabad)
VARANASI
PATNA

N E P A L
H   I   M   A   L   A   Y   A   S
Thimphu
BHUTAN
Brahmaputra
Guwahati

Mekong

Xichang
Zhaotong
Zunyi
Panzhihua
Liupanshui
Anshun
GUIYANG
Dali
YUNNAN
PLATEAU
KUNMING
Qujing

Hechi
Liuzhou

OPAL
INDYA RANGE
APTURA RANGE
NAGPUR
CHOTA-NAGPUR
PLATEAU
INDIA
Narmada
Tropic of Cancer

KHASI HILLS
Rajshahi
BANGLADESH
DHAKA
KHULNA
CHITTAGONG
Mouths of the Ganges

NAGA HILLS
Myitkyina

Baoshan

Yuxi
Gejiu

Nanning
Qinzhou
Beihai

VIETNAM
HANOI
Haiphong
Gulf of
Tonkin

Bhubaneshwar
Vishakhapatnam

Bay of
Bengal

Ramree Island
Cheduba
Island

ARAKAN MTNS.
MYANMAR
(BURMA)
MANDALAY

Irrawaddy

Salween

Nay Pyi Daw
YANGON
(RANGOON)

LAOS
Chiang Mai
Viangchan
THAILAND
ANNAMITE CORDILLERA
Sany
Mekong
Hongshui

Elevation (m)

| 3000 |
| 1500 |
| 600 |
| 300 |
| 150 |
| 0 |
| below sea level |
| -150 |
| -1500 |
| -3000 |

0 km   200   400   600   800   1000   1200
0 miles   200   400   600   800
Albers Equal-Area Projection
Scale 1:20,000,000

Roads
Railroads

EAST ASIA

9

Where was this photo taken? Find out at www.wiley.com/college/deblij

Photos: © H. J. de Blij

FIGURE 9-1 *Map:* © H. J. de Blij, P. O. Muller, and John Wiley & Sons, Inc.

In This Chapter

- From Xia to Xiamen: China's ancient roots, modern transformation
- China's global economic challenge
- Minority problems roil China
- Rebounding Japan: Still the region's powerhouse
- How geography could help reunite the Korean Peninsula
- Signs of progress over the issue of Taiwan

EAST ASIA IS a geographic realm like no other. At its heart lies the world's most populous country. On its periphery lies one of the globe's leading national economies. Along its coastline, on its peninsulas, and on its islands an economic boom has transformed cities and countrysides. Its interior contains the world's highest mountains and vast deserts. It is a storehouse of raw materials. The basins of its great rivers produce food that can sustain more than a billion people.

Defining the Realm

The East Asian geographic realm consists of six political entities: China, Japan, South Korea, North Korea, Mongolia, and Taiwan. Note that we refer here to political entities rather than states. In changing East Asia, the distinction is significant. The island of Taiwan, whose government calls this the Republic of China, functions as a democratic state but is regarded by mainland China (the communist People's Republic of China) as an out-of-control province that must one day be reunited with the motherland. Some foreign government officials do not always get this right. When the president of the People's Republic visited the White House in 2006 and the Chinese national anthem was about to be played during the welcoming ceremony, the announcer introduced it as the anthem of the Republic of China. The mistake was embarrassing, the guests were insulted, and the lesson—once again—was that geographic literacy is important. Fortunately, the band had the right music.

As we define it, East Asia lies between the sparsely peopled Siberian expanses of Russia to the north and the populous countries that make up South and Southeast Asia to the south. The realm extends from lofty snow-capped peaks and vast empty deserts in central Asia to earthquake-prone, crowded Pacific islands offshore (Fig. 9-1). Environmental diversity is one of its hallmarks, and this is reflected in the distribution of East Asia's population (Fig. 9-2), the largest of all the world's geographic realms.

MAJOR GEOGRAPHIC QUALITIES OF

East Asia

1. East Asia is encircled by snowcapped mountains, vast deserts, cold climates, and Pacific waters.

2. East Asia was one of the world's earliest culture hearths, and China is one of the world's oldest continuous civilizations.

3. East Asia is the world's most populous geographic realm, but its population remains strongly concentrated in its eastern regions.

4. China, the world's largest state demographically, is the current rendition of an empire that has expanded and contracted, fragmented, and unified many times during its long existence.

5. China today remains a strongly rural society, and its vast river basins feed hundreds of millions in a historic pattern that continues today.

6. China's sparsely peopled western regions are strategically important to the state, but they lie exposed to minority pressures and Islamic influences.

7. Along China's Pacific frontage an economic transformation is taking place, affecting all the coastal provinces and creating an emerging Pacific Rim region.

8. Increasing regional disparities and fast-changing cultural landscapes are straining East Asian societies.

9. Japan, one of the economic giants of the East Asian realm, has a history of colonial expansion and wartime conduct that still affects international relations here.

10. East Asia may witness the rise of the world's next superpower as China's economic and military strength and influence grow—and if China avoids the devolutionary forces that fractured the Soviet Union.

11. The political geography of East Asia contains a number of flashpoints that can generate conflict, including Taiwan, North Korea, and several island groups in the realm's seas.

FIGURE 9-2

© H. J. de Blij, P. O. Muller, and John Wiley & Sons, Inc.

Currently, East Asia is experiencing a massive geographic transformation that we will study from several perspectives because it affects our daily lives in numerous ways. Take a look at a geography textbook written 35 years ago, and you will find that neither text nor **1** index will refer to the regional name **Pacific Rim**. Today, that term signifies the fastest and most dramatic economic and social transfiguration in the history of humankind, and its continuing development is headline news. Travel through China's coastal provinces and cities from Dalian in the north to Hong Kong (Xianggang) in the south, and you will see what economic geographers describe as the greatest public works project on Earth (for Chinese spellings, see the box on the next page titled "Places and Names"). Whether in Beijing, Shanghai, Guangzhou, or dozens of other cities on or near the Pacific coast, forests of futuristic skyscrapers rise where traditional neighborhoods, villages, and farmers resisted change for countless centuries. China's economic juggernaut set in motion the greatest human

migration in the country's history as millions moved from countryside to city in search of a better life.

Before China's Pacific Rim became the geographic subregion it is today, Japan built a gigantic economy with global connections and was the only highly developed country in East Asia. Japanese capital fueled the emergence of other economies in East Asia (such as Hong Kong) and Southeast Asia (including Singapore), but Japanese investment in China's Pacific Rim played a key role in the process we will chronicle in this chapter. China and Japan have some serious issues to resolve, ranging from lingering arguments over what happened during World War II to growing disputes over Pacific islands, but there can be no doubt about their common interest in the economic miracle that has transformed East Asia in little more than one generation.

No matter how powerful Japan's influence was in the evolution of East Asia's Pacific Rim, China is the realm's colossus, with more than 10 times Japan's population and 25 times its area. One of the world's oldest (perhaps *the*

Places and Names

AS IN OTHER parts of the world, European colonists in China wrote down the names of places and people as they heard them—and often got them wrong. The Wade-Giles system put such place names as Peking, Canton, and Tientsin on the map, but that's not how the Chinese knew them. In 1958, the communist regime replaced the foreign version with the *pinyin* system, based on the pronunciation of Chinese characters in Northern Mandarin, which was to become the standard form of the Chinese language throughout China. The world now became familiar with these same three cities written as Beijing, Guangzhou, and Tianjin. Among prominent Chinese geographic names, only one has not gained universal recognition: Tibet, called Xizang by China but referred to by its old name on many maps for reasons that will become obvious later in this chapter.

Pinyin usage may be standard in China today, but China remains a country of many languages and dialects. Not only do minorities speak many different languages, but Mandarin-speakers use numerous dialects—so that when you sit down with some Chinese friends who have just met, it may take just one sentence for one to say to the other "ah, you're from Shanghai!"

Migration is changing this, of course. China's population is on the move, and many residents of Shanghai today were not born in the city. But China's language map is anything but simple. And not just place names, but personal names, too, were revised under pinyin rules. The communist leader who used to be called Mao Tse-tung in the Western media is now Mao Zedong. Also remember that the Chinese write their last name first: President Hu Jintao is Jintao to friends and Mr. Hu to others. To the Chinese, it is Obama Barack and Biden Joe, not the other way around.

oldest) continuous civilizations, China already formed a major culture hearth when Japan was still an isolated frontier inhabited by the Ainu. Migrations that originated in China advanced through the Korean Peninsula and reached the islands. Philosophies, religions, and cultural traditions (including, for example, urban design and architecture) diffused from China to Korea and Japan, and to other locales on the margins of the Chinese hearth. Over the past millennium, China repeatedly expanded to acquire empires, only to collapse in chaos when its center failed to maintain control. European colonial powers took such opportunities to partition China among themselves. The Russians and the Japanese pushed their empires deep into China. But always China recovered, and today China may again be poised to expand its sphere of influence and, indeed, to take its place as a world power in the twenty-first century.

NATURAL ENVIRONMENTS

Figure 9-1 dramatically illustrates the complex physical geography of the East Asian realm. In the southwest lie ice-covered mountains and plateaus, the Earth's crust in this region crumpled up like the folds of an accordion. A gigantic collision of tectonic plates is creating this landscape as the Indian Plate pushes northward into the underbelly of the Eurasian Plate (Fig. G-3). The result is some of the world's most spectacular scenery, but snow, ice, and cold are not the only dangers to human life here. Earthquakes and tremors occur almost contin-

uously, causing landslides, avalanches, and worse. In 2008, a magnitude-7.9 earthquake and a series of aftershocks devastated a large area of Sichuan Province, killing nearly 70,000 people and injuring many more. Have another look at Figure G-4 and you will see the zone of earthquake risk that penetrates China from the south. So close to the Three Gorges Dam, this is a major environmental concern.

Physiography and Population

As Figures 9-2 and G-8 show, much of this Asian interior is one of the world's most sparsely populated areas, but it is nevertheless critical to the lives of hundreds of millions of people. In this high mountainous terrain, fed by the melting ice and snow, rise the great rivers that flow eastward across China and southward across Southeast and South Asia. Throughout the Holocene, these rivers have been eroding the uplands and depositing their sediments in the lowlands, in effect creating the alluvium-filled basins that now sustain huge populations. Fertile alluvial soils and adequate growing seasons, combined with ample water and millions of hands to sow the wheat and plant the rice, have allowed the emergence of one of the great population concentrations on Earth.

Figure G-7 classifies this high-elevation zone of East Asia as category *H*, highland climate, which is by far the largest such area on Earth. Here one criterion overpowers the rules of climatic classification: altitude. The mountains wrest virtually all moisture from air masses

FIGURE 9-3

© H. J. de Blij, P. O. Muller, and John Wiley & Sons, Inc.

moving northward and block their path into interior Asia. As Figure G-7 shows, desert conditions prevail in East Asia's northern interior. Figure 9-3 names two of the more famous ones: the Takla Makan of far western China and the Gobi of Mongolia.

Physiography, therefore, has much to do with East Asia's population distribution, but even the more habitable and agriculturally productive east has its limitations. The northeast suffers from severe continentality, with long and bitterly cold winters. High relief encircles the river basins north of the Yellow Sea and dominates much of the northern part of the Korean Peninsula, creating strong local environmental contrasts. South Korea, as Figure G-7 shows, experiences relatively moderate conditions comparable to those of the U.S. Southeast; North Korea has a harsh continental climate like that of North Dakota.

In general, coastal, peninsular, and island East Asia possess more moderate climates than the interior. Like South Korea, southern Japan and southeastern China have humid temperate climates, and southernmost Taiwan and

Hainan Island even have areas of tropical (*A*) conditions. Proximity to the ocean tends to moderate climatic environments, and coastal East Asia proves the point.

Three Great Rivers

Before we focus on East Asia's human geography, it is useful to look at the map of this realm's complex physical stage (Fig. 9-3). From the high-relief interior come three major river systems that have played crucial roles in the human drama. In the north, the *Huang He* (Yellow River) arises deep in the high mountains, crosses the Ordos Desert and the Loess Plateau, and deposits its fertile sediments in the vast North China Plain, where East Asia's earliest states emerged. In the center, the *Chang Jiang* (Long River), called the *Yangzi* downstream, crosses the Sichuan Basin and the Three Gorges, where the world's largest dam project was recently completed, and waters extensive ricefields in the Lower Chang Basin. And in the south, the *Xi Jiang*

(West River) originates on the Yunnan Plateau and becomes the *Pearl River* in its lowest course. The Pearl estuary, flanked by one of China's leading urban-industrial complexes, has become a crucial hub in the still-evolving Pacific Rim.

Further scrutiny of Figure 9-3 indicates that a fourth river system plays a role in China: the *Liao River* in the northeast and its basin, the Northeast China Plain. As the map suggests, however, the Liao is not comparable to the great rivers to its south, its course being shorter and its basin, in this higher-latitude zone, much smaller. As we will see, China's Northeast was for some time the country's industrial, not agricultural, heartland.

Interior Environments

Looking again toward the interior, note the Loess Plateau south of the Ordos Desert, where the Huang He makes its giant loop. *Loess* is a fertile, windblown deposit composed of rock pulverized by glaciers. Add water (in this case the middle Huang and its tributaries) and an adequate growing season, and a sizeable population will arise. To the south, deep in the interior, lies the Sichuan (Red) Basin, crossed by the Chang Jiang. Despite its vulnerability to earthquakes, this basin has supported human communities for a long time, and you can readily discern its current population agglomeration in Figure 9-2. The Sichuan Basin, encircled as it is by mountains, is also one of the world's most sharply defined physiographic regions, and the concentration of its approximately 125 million inhabitants reflects that definition.

Still farther to the south lies the Yunnan Plateau, source of the tributaries that feed the Xi River. Much of southeastern China has comparatively high relief; it is hilly and in places even mountainous. This high relief has inhibited overland contacts between China and Southeast Asia.

Along the Coast

East Asia's Pacific margin is a jumble of peninsulas and islands. The Korean Peninsula looks like a near-bridge from Asia to Japan, and indeed it has served as such in the past. The Liaodong and Shandong peninsulas protrude into the Yellow Sea, which continues to silt up from the sediments of the Huang and Liao rivers. Off the mainland lie the islands that have played such a crucial role in the modern human geography of Asia and, indeed, the world: Japan, Taiwan, and Hainan. Japan's environmental range is expressed by cold northern Hokkaido and warm southern Kyushu, but Japan's core area lies on its main island, Honshu. As Figure 9-1 shows, myriad smaller islands flank the mainland and dot the East and South China seas. As we will discover, some of these smaller islands have major significance in the human geography of this realm.

FROM THE FIELD NOTES

"From the train, traveling from Beijing to Xian, we had a memorable view of the Loess Plateau. Loess is a fine-grained dust formed from rocks pulverized by glacial action, blown away by persistent winds, and deposited in sometimes well-defined locales. Loess covers much of the North China Plain, which is what makes it so fertile, but there it is not as thick as in the Loess Plateau in the middle basin of the Huang He (Yellow River). Here the loess averages 75 meters (250 ft) in thickness and in places reaches as much as 180 meters (600 ft). Because loess has some very distinctive physical properties, it tends to create unusual landscapes. Through a complicated physical process following deposition, loess develops the capacity to stand upright in walls and columns, and resists collapse when it is excavated. As a result the landscape looks terraced: streams cut deep and steep-sided valleys. The Loess Plateau is a physiographic region, but it is a cultural region too. Hundreds of thousands of people have literally dug their homes into vertical faces of the kind shown here, creating cave-like, multi-room dwellings with wooden exterior doors."
© H. J. de Blij.

www.conceptcaching.com

HISTORICAL GEOGRAPHY

Consider this: there is no evidence that *Homo sapiens*, modern humans, reached the Americas any earlier than 40,000 years ago, and many paleoanthropologists argue that the only available evidence indicates that humans crossed from Eurasia to Alaska a mere 14,000 to 13,000 years ago. On the opposite side of the Pacific, however, the story of hominid and human settlement spans hundreds of thousands of years—perhaps 1 million.

What took so long? If hominids like *Homo erectus* left their ancestral African homelands and migrated as far away as eastern Asia more than half a million years ago, and if modern humans followed them and reached Australia some 50,000 years ago, what kept the hominids from the Americas altogether, and what delayed the humans?

No satisfactory answers have yet been found to questions like these. The width of the Pacific Ocean may have doomed maritime migrants, although trans-Pacific migration may eventually have occurred. The frigid temperatures along the northern route from Siberia to Alaska may have stopped hominids and humans alike. But the Pleistocene was punctuated by warm interglaciations; why did it take until the early Holocene for humans to make the Bering Strait crossing?

All this implies that for hundreds of thousands of years East Asia was a cul-de-sac, a dead-end for migrants out of Africa. Many archeological sites in the realm have yielded evidence of *Homo erectus*, including what is perhaps the most famous of all: Peking Man. In a cave near the Chinese capital, Beijing, an archeologist in 1927 found a single tooth he recognized as representing a hominid. Later excavations proved him right as parts of more than three dozen skeletons were unearthed. They proved that Peking Man and his associates made stone and bone tools, had a communal culture, controlled the use of fire, hunted wildlife, and cooked their meat. Later arrivals undoubtedly challenged older communities, and the frontier of settlement must have expanded across the river basins of the east. But when *Homo sapiens*, modern humans, arrived in East Asia, the hominids could not compete. Current anthropological theory holds that the better-equipped humans eliminated the hominids, perhaps between 60,000 and 40,000 years ago, after which the human migration into the Americas could begin.

Early Cultural Geography

Whatever the outcome of the debate over human origins in East Asia, it is clear that humans have inhabited the plains and river basins, foothills, and islands of this realm for a very long time. Hunting sustained both the hominids and the early human communities; fishing drew them to the coasts and onto the islands. The first crossing into Japan may have occurred as long as 10,000 to 12,000 years ago, possibly much longer, when the Jomon people, a Caucasoid population of uncertain geographic origins, entered the islands; their modern descendants, the Ainu, spread throughout the Japanese archipelago. Today, only about 20,000 persons living in northernmost Hokkaido trace their ancestry to Ainu sources.

About 2300 years ago the Yayoi people, rice farmers who had settled in Korea, appear to have crossed by boat to Kyushu, Japan's southernmost island, from where they advanced northward. The Ainu, who subsisted by fishing, trapping, and hunting, were driven back, but gene-pool studies show that much mixing of the groups took place; they also show that the Yayoi invasion was followed by other incursions from the Asian mainland. By then, powerful dynastic states had already arisen in what is today China, and early Chinese culture traits thus found their way into Japan through the process we know as relocation diffusion.

Mainland Cultures

On the Asian mainland, plant and animal domestication had begun as early as anywhere on Earth. We in the Western world take it for granted that these momentous processes began in what we now call the Middle East and diffused from the Fertile Crescent to other parts of Eurasia and the rest of the world. But the taming of animals and the selective farming of plants may have begun as early, or earlier, here in East Asia. As in Southwest Asia, the fertile alluvial soils of the great river basins and the ebb and flow of stream water created an environment of opportunity, and millet and rice were being harvested between 7000 and 8000 years ago.

Even during this Neolithic period of increasingly sophisticated stone tools, East Asia was a mosaic of regional cultures. Their differences are revealed by the tools they made and the decorations on their bowls, pots, and other utensils. An especially important discovery of two 8000-year-old pots in the form of a silkworm cocoon, from China's Hebei Province, suggests a very ancient origin for one of the region's leading historic industries.

As noted earlier, plant and animal domestication produced surpluses and food storage, enabling population growth and requiring wider regional organization. Here as elsewhere during the Neolithic, settlements expanded, human communities grew more complex, and power became concentrated in a small group, an *elite*.

Formation of States and Dynasties

This process of **state formation** is known to have occurred in only a half-dozen regions of the world, and China was one of these. But evidence about China's earliest states has long been scarce. Today, however, archeologists are focusing on the lower Yi-Luo River Valley in the western part of Henan Province, where the first documented Chinese dynasty, the Xia Dynasty (2200–1770 BC), existed. The capital of this ancient state, Erlitou, has been found, and archeologists now refer to the Xia Dynasty as the Erlitou culture. Secondary centers are being discovered, and Erlitou tools and implements in a wider area prove that the Xia Dynasty represents a substantial state.

All early states were ruled by elites, but China's political history is chronicled in *dynasties* because here the succession of rulers came from the same line of descent, sometimes enduring for centuries. In the transfer of power, family ties counted for more than anything else. Dynasties were overthrown, but the victors did not change this system. Dynastic rule lasted into the twentieth century.

The Xia Dynasty may have been the earliest Chinese state, but it lay in the area where, later, more powerful dynastic states arose: the North China Plain. Here the tenets of what was to become Chinese society were implanted early and proved to be extremely durable. From this culture hearth ideas, innovations, and practices diffused far and wide. From agriculture to architecture, poetry to porcelain-making, influences radiated southward into Southeast Asia, westward into interior

Asia, and eastward into Korea and Japan. In the North China Plain lay the origins of what was to become the *Middle Kingdom*, which its citizens considered the center of the world.

Table 9-1 summarizes the sequence of dynasties that may have begun with Xia more than four millennia ago and ended in 1911 when the last emperor of the Qing (Manchu) Dynasty, a boy 6 years old, was forced to abdicate the Chinese throne. As the table shows, every dynasty contributed importantly to the development of Chinese society, sometimes progressively through enlightened policies and efficient administration, at other times regressively when capricious cruelty, authoritarian excesses, and xenophobia prevailed. During those 4000 years, China evolved into the world's most populous nation, always rebounding after disastrous famines and floods, its territory covering more than 80 percent of the East Asian realm (Fig. 9-4).

Dominant as China is in its regional sphere, there is more to East Asia than the Chinese giant. To the north, the landlocked state of Mongolia is what remains of the time when Mongol armies conquered much of Eurasia (as Table 9-1 shows, even China fell under the Mongol sway). To the east, the Korean Peninsula escaped incorporation into China. Offshore, Japan first surpassed China as a military power, colonizing Manchuria (as it was then called) and penetrating deep into China's heartland. Later, Japan outclassed China as an economic power, achieving world status while China languished under Maoist communism. And as the twenty-first century dawned, Taiwan remained a separate political entity, its detachment from China resulting from the struggle that brought the communists to power in Beijing.

**EVOLUTION OF
THE CHINESE EMPIRE**

—— The Great Wall

Earliest core area (1900 BC)

Shang Dynasty (1766–1080 BC)

Han Dynasty (206 BC–220 AD)

Qing (Manchu) Dynasty (1644–1911 AD)

0   500   1000 Kilometers
0   300   600 Miles

Modified after Tregear and others.

FIGURE 9-4 © H. J. de Blij, P. O. Muller, and John Wiley & Sons, Inc.

POINTS TO PONDER

- China has the world's deadliest mining industry. Accidents kill thousands of Chinese coal miners annually. Coal-fired power plants produce the bulk of China's electricity.

- China is the first state in the world whose population will age before it grows rich.

- Russia's leaders have promised Japan that it will build a 4000-kilometer (2600-mi) oil pipeline from Siberia to a terminal on the East Sea (Sea of Japan) directly across from Honshu that will deliver 1.6 million barrels of oil a day.

- Serious rebellions in Tibet and Xinjiang reveal China's problems in governing minority peoples.

TABLE 9-1

The Chinese Dynasties, 2200 BC to AD 1911

Dynasty Name(s)	Date(s)	Areas Governed	Major Features	Geographical Impact
Xia	c2200–c1770 BC	Small part of Huang River Basin, centered in Yi-Luo tributary.	First dynastic state (?); Neolithic technology, sophisticated stone tools.	Beginnings of stream diversion, irrigation.
Shang (latter part Yin)	c1766–c1080 BC	Huang-Wei River confluence across Shandong and Henan Provinces. Anyang capital.	Neolithic to Bronze Age transition. Timber houses with thatched roofs, beginnings of Chinese writing, moon-cycle-based calendar. Superb bronze vases and other implements.	Unifies large region in lower Huang River Basin.
Zhou	c1027–221 BC Two periods: Spring and Autumn c1027–481 BC; Warring States 475–221 BC	A people centered in Shaanxi Province to the west of the Shang area overthrow the Shang rulers and expand their domain eastward. Warring States is a period of feudalism.	Formative period for China. Irrigation systems, iron smelting, horses, expanded farm production. Major towns develop. Writing system is established. Chopsticks come into use. Daoism and Confucianism arise; Daoism centers on mysticism, individualism, the personal "way," and Confucianism on duties, community standards, respect for government.	Daoism diffuses widely throughout East Asia, reaches South Asia, and influences Buddhism. Confucianism becomes China's guiding philosophy for more than 2000 years. Building of *Great Wall* begins.
Qin (Ch'in)	221–206 BC	Most of North China Plain, south into Lower Chang Basin; first consolidation of large Chinese state.	Time of bureaucratic dictatorship after ferment of Zhou: Confucian Classics burned, Daoism combated. Much of Great Wall is built at huge cost in lives. Cruel despotism, excesses. Emperor's tomb discovered at Xian in 1976 with 6000 terracotta life-sized men and horses.	Ruler Ch'in (Qin)'s name immortalized as *China*.
Han	206 BC–AD 220	Major territorial expansion adds Xinjiang in interior Asia, Vietnam in Southeast Asia.	Second formative dynasty for China; restoration of Confucian principles and a flowering of culture. First use of eunuchs by authoritarian government, planting a long-term weakness in administration. Xian becomes one of the greatest cities of the ancient world. Chinese refer to themselves as the *People of Han*.	About the same time, same size as the Roman Empire. *Silk Road* carries goods from China across inner Asia to Syria and on to Rome.

Period from AD 220–580 witnessed disunion and division. Important developments included the arrival of Buddhism in East Asia and the diffusion of Chinese influence and ideas into Korea and across Korea into Japan.

Sui	AD 581–618	North and South China reunified, Xian rebuilt and greatly expanded. Conflict with Turks in Central Asia and with Koreans in east.	Brief but important dynasty with revival of Confucian rituals and practices in education. Brutal but modernizing regime conducts a census, establishes a penal code, and engages in massive public works.	Construction of key segments of Grand Canal links Huang and Yangzi Rivers and thus connects northern and southern economies.
Tang	AD 618–907	Defeat of Turks in Central Asia. Campaign to put down rebellions in the south succeeds, though northern frontier remains unstable.	A golden age for China and a third formative dynasty. Enormous and efficient bureaucracy dictates virtually all aspects of life. Xian now the cultural capital and largest city in the world. *Buddhism* thrives, pagodas arise everywhere. The arts, Chinese and foreign, flourish.	Arab and Persian seafarers visit Chinese ports, the Silk Route is loaded with trade, and China seems set for an international era. It is not to be: *Islam* arrives in Central Asia and Tang armies are defeated by Arabs in 751. The Silk Route breaks down. In the east, however, Chinese influences permeate Korea and penetrate Japan.

TABLE 9-1

The Chinese Dynasties, 2200 BC to AD 1911 (*Continued*)

Dynasty Name(s)	Date(s)	Areas Governed	Major Features	Geographical Impact
The 907–960 period is known as the Five Dynasties because five leaders tried to establish dynasties in northern China in quick succession, each failing in turn. In southern China it is known as the time of the Ten Kingdoms, for ten regimes ruled various parts of the region. Cultural continuity prevails despite political instability.				
Song (Northern (Southern	AD 960–1279 AD 960–1127) AD 1127–1279)	Rulers consolidate the Five (northern) Dynasties, then begin capturing the Ten (southern) Kingdoms. They are unable to control the Liao state north of the Great Wall. Mongols invade, drive Song rulers southward, and end dynasty in 1279.	Preoccupied with organization and administration, Song rulers create a competent, efficient *civil service*. A defensive rather than an expansionist period. An era of rich cultural and intellectual achievement in mathematics, astronomy, mapmaking. Paper and movable type are invented, as is *gunpowder*. Large ships are built. The arts, from porcelain-making to poetry, continue to thrive. Practice of footbinding begins. Improved rice varieties feed estimated 100 million inhabitants.	Capital is moved from northern Kaifeng on Huang River to Hangzhou in the south as Song rulers lose control over the north. Mongol invasion is followed by Kublai Khan's choice of Beijing as his capital (1272). Mongols use Chinese authoritarian and bureaucratic traditions to forge their own dynastic rule. Several cities have populations exceeding 1 million.
Yuan (Mongol)	AD 1264–1368	Controls eastern Asia from Siberia in the north to the Vietnam border in the south, including present-day Russian Far East and most of Korea.	Dynasty begins in north before Song ends in south. Repressive regime, social chasms between Mongol rulers and Chinese ruled. Deep and enduring hostility between Chinese and Mongols. Cultural *isolationism* results; Mongols are acculturated to Chinese norms. Marco Polo visits East Asia.	Most of present-day Mongolia is part of the Chinese sphere for the first time. Mongol effort to enter Japan fails. Mongol power over much of Eurasia stimulates trade and flow of information. Beijing grows into major city.
Ming	AD 1368–1644	Rules over all eastern China from Amur River in north to Red River (Vietnam) in south, as far west as Inner Mongolia and Yunnan. At various times incorporates North Korea, most of Mongolia, even Myanmar (Burma).	Chinese rule again after Mongol Yuan dynasty. Stable but autocratic and inward-looking regime. Major advances in science and technology. In first half of fifteenth century, Chinese oceangoing vessels, larger than any in Europe, explore Pacific and Indian Ocean waters and reach East Africa. Farming expands, silk and cotton industries improve, printing plants multiply. Defensive walls are built in the north and around cities, including the Forbidden City in Beijing, the Ming capital. But mismanagement, infighting, eunuch influence in the palaces, and anticommercialism weaken Ming rule.	Early Ming fleets herald an era of exploration and potential colonialism, but the impact of the Little Ice Age creates environmental crisis forcing massive grain redistribution from central to northern China. The great ships are burned, barges are built, and Grand Canal is expanded to cope with food crisis. Political system erodes and central authority falters.
Qing (Manchu)	AD 1644–1911	*Largest* China-centered empire ever incorporates Mongolia, much of Turkestan, Xizang (Tibet), Myanmar, Indochina, Korea, Taiwan.	Foreign rule again: Manchu, a people with Tatar links living in present-day Northeast China, seize an opportunity and take control of Beijing. Improbably, a group numbering about 1 million rule a nation of several hundred million. It is done by keeping Ming systems of administration and retaining (and rewarding) Ming officials. Expansion creates an empire. Population pressure and the concentration of land ownership in fewer hands create problems worsened by floods and famines. European powers and Japan force concessions on weakening Qing rulers; defeats in war lead to revolution and collapse.	Manchu (Qing) rulers create the map that today forms the justification for China's claims to Xizang (Tibet), Taiwan, and the central Eurasian interior; at the height of its power, the Qing empire forces the Russians to recognize Manchu authority over China's Northeast as far north as the Argun River. Latent claims to Russia's Far East are based on Manchu predominance there.

Regions of the Realm

East Asia presents us with an opportunity to illustrate the changeable nature of regional geography. Our regional delimitation is based on current circumstances, and it predicts ways the framework may change. It is anything but static. At the opening of the second decade of the twenty-first century, we can identify five geographic regions in the East Asian realm (Fig. 9-5). These are:

1. *China Proper*. Almost any map of China's human geography—population distribution, urban centers, surface communications, agriculture, or industry—emphasizes the strong concentration of Chinese activity in the country's eastern sector. This is the "real" China, where its great cities, populous farmlands, and historic sources are located. Long ago, scholars called this *China Proper*, and it is a good regional designation. But China is a large and complex country, and a number of subregions are nested within China Proper. Some of these, such as the North China Plain and the Sichuan Basin, are old and well-established geographic units. One in particular is new: China's Pacific Rim, still growing, yet poorly defined, and shown as "formative" in Figure 9-5.

2. *Xizang (Tibet)*. The high mountains and plateaus of Xizang, ruled by China but still widely known by its older name of Tibet, form a stark contrast to teeming China Proper. Here, next to one of the world's largest and most populous regions, lies one of the emptiest and, in terms of inhabited space, smallest regions.

3. *Xinjiang*. The vast desert basins and encircling mountains of Xinjiang form a third East Asian region. Again, physical as well as human geographic criteria come into play: here China meets Islamic Central Asia.

4. *Mongolia*. The desert state of Mongolia forms East Asia's fourth region. Like Tibet, landlocked Mongolia, vast but sparsely peopled, stands in stark contrast to populous China Proper.

5. *Jakota Triangle*. Just as the Pacific Rim is a formative region in changing China, so a group of three entities currently forms a discrete region in East Asia. Anchored by a *Ja*pan flanked by South *Ko*rea and *Ta*iwan, we call this the *Jakota Triangle* (the acronym "Jakota" derives from the first two letters of each). Economic and political factors define this region which, like the Pacific Rim, is changing fast.

● CHINA PROPER

When we in the Western world chronicle the rise of civilization, we tend to focus on the historical geography of Southwest Asia, the Mediterranean, and Western Europe. Ancient Greece and Rome were the crucibles of culture; Mediterranean and Atlantic waters were the avenues of its diffusion. China lay remote, so we believe, barely connected to this Western realm of achievement and progress. When an Italian adventurer named Marco Polo visited China during the thirteenth century and described the marvels he saw there, his work did little to change European minds. Europe was and would always be the center of civilization.

The Chinese, naturally, take a different view. Events on the western edge of the great Eurasian landmass were deemed irrelevant to theirs, the most advanced and refined culture on Earth. Roman emperors were rumored to be powerful, and Rome was a great city, but nothing could match the omnipotence of China's rulers. Certainly the Chinese city of Xian (which had a different name at the time) far eclipsed Rome as a center of sophistication. Chinese civilization existed long before ancient Greece and Rome emerged, and it was still there long after they collapsed. China, the Chinese teach themselves, is eternal. It was, and always will be, the center of the civilized world.

We should remember this notion when we study China's regional geography because 4000 years of Chi-

MAJOR CITIES OF THE REALM

City	Population* (in millions)
Beijing (Shi), China	11.7
Busan, South Korea	3.5
Chongqing (Shi), China	6.8
Guangzhou, China	9.4
Hong Kong (Xianggang) SAR, China	7.4
Macau SAR, China	0.5
Nanjing, China	4.0
Osaka, Japan	11.3
Seoul, South Korea	9.9
Shanghai (Shi), China	15.6
Shenyang, China	5.0
Shenzhen, China	8.0
Taipei, Taiwan	2.7
Tianjin (Shi), China	7.5
Tokyo, Japan	26.7
Wuhan, China	7.5
Xian, China	4.2

*Based on 2010 estimates.

FIGURE 9-5

© H. J. de Blij, P. O. Muller, and John Wiley & Sons, Inc.

nese culture and perception will not change overnight—not even in a generation. Time and again, China overcame the invasions and depredations of foreign intruders, and afterward the Chinese would close off their vast country against the outside world. Just 40 years ago, at the end of the 1960s, there were only a few *dozen* foreigners in the entire country with its (then) nearly 1 billion inhabitants. The institutionalization of communism required this insularity, and even the Soviet advisors had been thrown out. But by the early 1970s, China's rulers decided that an opening to the Western world would be advantageous, and so U.S. President Richard Nixon was invited to visit Beijing. That historic occasion, in 1972, ended this latest period of isolation—as always, on China's terms. Since then, China has been open to tourists and businesspeople, teachers, and investors. Hundreds of thousands of Chinese students have been sent to study at American and other Western institutions. Long-suppressed ideas flowed into

China, and a pro-democracy movement arose and climaxed in 1989. China's rulers knew that their violent repression of this movement would anger the world, but that did not matter because they deemed foreign condemnation irrelevant. Foreigners in China had done much worse. Moreover, Westerners had no business interfering in China's domestic affairs.

RELATIVE LOCATION

Throughout their nation's history, the Chinese have at times decided to close their country to any and all foreign influences; the most recent episode of exclusion occurred just a few decades ago. Exclusion is one of China's recurrent traditions, made possible by China's relative location and Asia's physiography. In other words, China's "splendid isolation" was made possible by geography.

Earlier we noted the role of relief and desert in encircling the culture hearth of East Asia, but equally telling is the factor of distance. Until recently, China lay far from the modern source areas of innovation and change. True, China—as the Chinese emphasize—was itself such a hearth, but China's contributions to the outside world remained limited, essentially, to finely made arts and crafts. China did interact with Korea, Japan, Taiwan, and parts of Southeast Asia, and eventually millions of Chinese emigrated to neighboring countries. But compare these regional links to those of the Arabs, who ranged worldwide and who brought their knowledge, religion, and political influence to areas from Mediterranean Europe to Bangladesh and from West Africa to Indonesia. Later, when Europe became the center of intellectual and material innovation, China found itself farther removed, by land or sea, than almost any other part of the world.

Today, modern communications notwithstanding, China still is distant from almost anywhere else on Earth. Going by rail from Beijing to Moscow, the capital of China's Eurasian neighbor, involves a tedious journey that takes the better part of a week. Direct surface connections with India are, for all practical purposes, nonexistent. Overland linkages with Southeast Asian countries, though improving, remain tenuous.

But for the first time in its history, China now lies near a world-class hearth of technological innovation and financial power: Japan. This proximity to an industrial and financial giant is critical for the momentous economic developments taking place in China's coastal provinces. Japanese investments and business partnerships have transformed the economic landscape of Pacific-coast China. American and European trade links also are important, but Japan's role has been crucial. Japan's economic success set the Pacific Rim engine in motion, and Japan's best financial years happened to coincide with China's reopening to foreigners in the 1970s. Geographic and economic circumstances combined to transform the map and made "Pacific Rim" a household word around the world.

EXTENT AND ENVIRONMENT

China's total area is slightly smaller than that of the United States including Alaska: each country has about 9.6 million square kilometers (3.7 million sq mi). As Figure 9-6 reveals, the longitudinal extent of China and the 48 contiguous U.S. States also is similar. Latitudinally, however, China is considerably wider. Miami, near the southern limit of the United States, lies halfway between Shanghai and Guangzhou. Thus China's lower-latitude

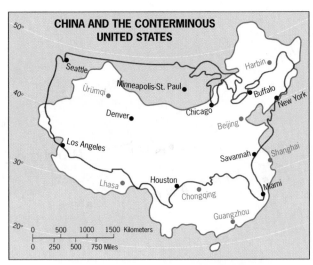

© H. J. de Blij, P. O. Muller, and John Wiley & Sons, Inc.

FIGURE 9-6

southern region takes on characteristics of tropical Asia. In the Northeast, too, China incorporates much of what in North America would be Quebec and Ontario. Westward, China's land area becomes narrower and physiographic similarities increase. But, of course, China has no west coast.

Now compare the climate maps of China and the United States in Figure 9-7 (which are enlargements of the appropriate portions of the world climate map in Fig. G-7). Note that both have a large southeastern climatic region marked *Cfa* (that is, humid, temperate, warm-summer), flanked in China by a zone of *Cwa* (where winters become drier). Westward in both countries, the *C* climates yield to colder, drier climes. In the United States, moderate *C* climates develop again along the Pacific coast. China, however, stays dry and cold as well as high in elevation at equivalent longitudes.

Note especially the comparative location of the U.S. and Chinese *Cfa* areas in Figure 9-7. China's lies much farther to the south. In the United States, the *Cfa* climate extends beyond 40° North latitude, but in China, cold and generally winter-dry *D* climates take over at the latitude of Virginia. Beijing has a warm summer but a bitterly cold and long winter. Northeast China, in the general latitudinal range of Canada's lower Quebec and Newfoundland, is much more severe than its North American equivalent. Harsh environments prevail over vast regions of China, but, as we will see, nature compensates in spectacular fashion. From the climatic zone marked *H* (for highlands) in the west come the great life-giving rivers whose wide basins contain enormous expanses of fertile soils. Without this water supply, China would not have a population more than four times that of the United States.

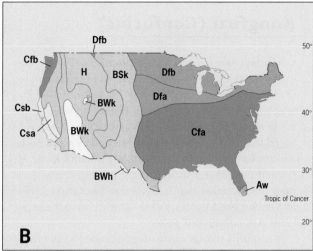

CLIMATES OF CHINA AND THE CONTERMINOUS UNITED STATES
After Köppen-Geiger

A HUMID EQUATORIAL CLIMATE	B DRY CLIMATE	C HUMID TEMPERATE CLIMATE	D HUMID COLD CLIMATE	H HIGHLAND CLIMATE
Am Short dry season	**BS** Semiarid	**Cf** No dry season	**Df** No dry season	**H** Unclassified highlands
Aw Dry winter	**BW** Arid	**Cw** Dry winter	**Dw** Dry winter	
	h=hot **k**=cold	**Cs** Dry summer **a**=hot summer **b**=cool summer **c**=short, cool summer		

FIGURE 9-7   © H. J. de Blij, P. O. Muller, and John Wiley & Sons, Inc.

EVOLVING CHINA

Even if China is not the world's longest continuous civilization (Egypt may claim this distinction), no other state on Earth can trace its cultural heritage as far back as China can. China's fortunes rose and fell, but over more than 40 centuries its people created a society with strong traditions, values, and philosophies. Kongfuzi (Confucius), whose teachings and writing still influence China, lived during the Zhou Dynasty, 2500 years ago (see box titled "Kongfuzi [Confucius]"). The Chinese still refer to themselves as the "People of Han," the dynasty that marks the breakdown of the old feudal order, the rise of military power, the unification of a large empire, the institution of property rights, the flourishing of architecture, the arts, and the sciences, and the development of trade along the Silk Route (Table 9-1). The Han Dynasty reigned 2000 years ago. The walled city of Xian, then called Ch'angan, was the early Han capital and one of the greatest cities of the ancient world.

As we try to gain a better understanding of China's present political, economic, and social geography, we should not lose sight of the China that endured for thousands of years under a sequence of imperial dynasties as a single entity. Larger and more populous than Europe, China had its divisive feudal periods, but always it came together again, ruled—mostly dictatorially—from a strong center as a unitary state. China was, and remains, a predominantly rural society, a powerless and subservient population controlled by an often ruthless bureaucracy. Retributions, famines, epidemics, and occasional uprisings took heavy tolls in the countryside. But in the cities, Chinese culture flourished, subsidized by the taxes and tribute extracted from the hinterlands. As China grew, it incorporated minorities ranging from Koreans and Mongols to Uyghurs and Tibetans; in the far south, it now includes several peoples with Southeast Asian affinities. Like the Chinese citizenry itself, these minorities have experienced both benevolent government and brutal subjugation. But **sinicization** has always been China's wish: to endow them with the elements of Chinese culture.

A Century of Convulsion

When the European colonialists appeared in East Asia, China long withstood them with a self-assured superiority based on the strength of its culture and the reassuring continuity of the state. There was no market for the British East India Company's rough textiles in a country long used to finely fabricated silks and cottons. There was little interest in the toys and trinkets the Europeans produced in the hope of barter for Chinese tea and porcelain. Even key European inventions, such as the

Kongfuzi (Confucius)

2 **Confucius** (*Kongfuzi* or *Kongzi* in pinyin) was China's most influential philosopher and teacher. His ideas dominated Chinese life and thought for over 20 centuries. Kongfuzi was born in 551 BC and died in 479 BC. Appalled at the suffering of ordinary people during the Zhou Dynasty, he urged the poor to assert themselves and demand explanations for their harsh treatment by the feudal lords. He tutored the indigent as well as the privileged, giving the poor an education that had hitherto been denied them and ending the aristocracy's exclusive access to the knowledge that constituted power.

Kongfuzi's revolutionary ideas extended to the rulers as well as the ruled. He abhorred supernatural mysticism and cast doubt on the divine ancestries of China's aristocratic rulers. Human virtues, not godly connections, should determine a person's place in society, he taught. Accordingly, he proposed that the dynastic rulers turn over the reins of state to ministers chosen for their competence and merit. This was another Kongzi heresy, but in time this idea came to be accepted and practiced.

His Earthly philosophies notwithstanding, Kongfuzi took on the mantle of a spiritual leader after his death. His thoughts, distilled from the mass of philosophical writing (including Daoism) that poured forth during his lifetime, became the guiding principles of the formative Han Dynasty. The state, he said, should not exist just for the power and pleasure of the elite; it should be a cooperative system for the well-being and happiness of the people.

With time, a mass of writings evolved, much of which Kongfuzi never wrote. At the heart of this body of literature lay the *Confucian Classics*, 13 texts that became the basis for education in China for 2000 years. From government to morality and from law to religion, the *Classics* were Chinese civilization's guide. The entire national system of education (including the state examinations through which everyone, poor or privileged, could enter the civil service and achieve political power) was based on the *Classics*. Kongfuzi championed the family as the foundation of Chinese culture, and the *Classics* prescribe a respect for the aged that was a hallmark of Chinese society.

But Kongfuzi's philosophies also were conservative and rigid, and when the colonial powers penetrated China, Kongzi's *Classics* came face to face with practical Western education. For the first time, some Chinese leaders began to call for reform and modernization, especially of teaching. Kongzi principles, they said, could guide an isolated China, but not China in the new age of competition. But the Manchu rulers resisted this call, and the Nationalists tried to combine Kongzi and Western knowledge into a neo-Kongzi philosophy, which was ultimately an unworkable plan.

The communists who took power in 1949 attacked Kongzi thought on all fronts. The *Classics* were abandoned, indoctrination pervaded education, and, for a time, even the family was viewed as an institution of the past. Here the communists miscalculated. It proved impossible to eradicate two millennia of cultural conditioning in a few decades. When China entered its post-Mao period, public interest in Kongfuzi surged, and the shelves of bookstores again sag under the weight of the *Analects* and other Confucian and post-Kongfuzi writings. The spirit of Kongfuzi will pervade physical and mental landscapes in China for generations to come.

FROM THE FIELD NOTES

"To pass through the gate from Tiananmen Square and enter the Manchu Emperors' Forbidden City was a riveting experience. Look back now toward the gate, and you see throngs of Chinese visiting what less than a century ago was the exclusive domain of the Qing Dynasty's absolute rulers—so absolute that unauthorized entry was punished by instant execution. The vast, walled complex at the heart of Beijing not only served as the imperial palace; it was a repository of an immense collection of Chinese technological and artistic achievements. Some of this is still here, but most of all you are awed by the lingering atmosphere, the sense of what happened here in these exquisitely constructed buildings that epitomized what high Chinese culture could accomplish. Inevitably you think of the misery of millions whose taxes and tribute paid for what stands here."   © H. J. de Blij (1980).

www.conceptcaching.com

mechanical clock, though considered amusing and entertaining, were ignored and even deprecated as irrelevant to Chinese culture.

The self-confident Ming emperors even beat the Europeans at their own game. In the fourteenth century, great oceangoing vessels sailed the South China Sea; Chinese fleets carrying as many as 20,000 men reached Southeast and South Asia and even East Africa. Several times as large and far more sophisticated than anything built in Europe, China's ships, together with their sailors and the goods they carried, demonstrated the potential of the Middle Kingdom. But suddenly China faced a crisis at home: the impact of the onset of the Little Ice Age that also affected Europe. Cold and drought decimated the wheat harvest on the North China Plain, and boats were needed by the thousands to carry rice from the Yangzi/Chang Basin to the hungry north. That spelled the end of China's long-range maritime explorations; it may have changed the historical geography of the world.

Even when Europe's sailing ships made way for steam-driven vessels and newer and better European products (including weapons) were offered in trade for China's tea and silk, China continued to reject European imports and resisted commerce in general. (When the Manchus took control of Beijing in 1644, their bows and arrows proved superior to Chinese-manufactured muskets, which were so heavy and hard to load that they were almost useless.) The Chinese kept the Europeans confined to small peninsular outposts, such as Macau, and minimized interaction with them. Long after India had succumbed to mercantilism and economic imperialism, China maintained its established order. This was no surprise to the Chinese. After all, they had held a position of undisputed superiority in their Celestial Kingdom as long as could be remembered, and they had dealt with foreign invaders before.

A (Lost) War on Drugs and Its Aftermath

All this confidence was shattered during the Manchu (Qing) Dynasty, China's last imperial regime. The Manchus, who had taken control in Beijing as a small minority and who had grafted their culture onto China's, had the misfortune of reigning when the standoff with Europeans ended and the balance of power shifted in favor of the colonialists.

On two fronts in particular, the economic and the political, the European powers destroyed China's invincibility. Economically, they succeeded in lowering the cost and improving the quality of manufactured goods, especially textiles, and the handicraft industries of China began to collapse in the face of unbeatable competition. Politically, the demands of the British merchants and the growing English presence in China led to conflicts. In the early part of the nineteenth century, the central issue was the importation into China from British India of opium, a dangerous and addictive intoxicant. Opium was destroying the very fabric of Chinese culture, weakening the society, and rendering China easy prey for colonial profiteers. As the Manchu government moved to stamp out the opium trade in 1839, armed hostilities broke out, and soon the Chinese found themselves losing a war on their own territory. The First Opium War (1839–1842) ended in disaster: China's rulers were forced to yield to British demands, and the breakdown of Chinese sovereignty was under way.

Drugs and Defeat

British forces penetrated up the Chang Jiang and controlled several areas south of it (Fig. 9-8); Beijing hurriedly sought a peace treaty by which it granted leases and concessions to foreign merchants. In addition, China ceded Hong Kong Island to the British and opened five ports, including Guangzhou (Canton) and Shanghai, to foreign commerce. No longer did the British have to accept a status that was inferior to the Chinese in order to do business; henceforth, negotiations would be pursued on equal terms. Opium now flooded into China, and its impact on Chinese society became even more devastating. Fifteen years after the First Opium War, the Chinese again tried to stem the disastrous narcotic tide, and the foreigners who had attached themselves to their country again defeated them. Now the government legalized cultivation of the opium poppy in China itself. Chinese society was disintegrating; the scourge of this drug abuse was not defeated until after the revival of Chinese power in the twentieth century.

Colonial Indignities

But before China could reassert itself, much of what remained of its independence steadily eroded (see box titled "Extraterritoriality"). In 1898 the Germans obtained a lease on Qingdao on the Shandong Peninsula, and the French acquired a sphere of influence in the far south at Zhanjiang (Fig. 9-8). The Portuguese took Macau; the Russians obtained a lease on Liaodong in the Northeast as well as railway concessions there; even Japan got into the act by annexing the Ryukyu Islands and, more importantly, Formosa (Taiwan) in 1895.

AMONG THE REALM'S GREAT CITIES . . . **Xian**

THE CITY KNOWN today as Xian is the site of one of the world's oldest urban centers. It may have been a settlement during the Shang-Yin Dynasty more than 3000 years ago; it was a town during the Zhou Dynasty, and the Qin emperor was buried here along with 6000 life-sized terra-cotta soldiers and horses, reflecting the city's importance. During the Han Dynasty the city, then called Ch'angan, was one of the greatest centers of the ancient world, the Rome of ancient China. Ch'angan formed the eastern terminus of the Silk Route, a storehouse of enormous wealth. Its architecture was unrivaled, from its ornamental defensive wall with elaborately sculpted gates to the magnificent public buildings and gardens at its center.

Situated on the fertile loess plain of the upper Wei River, Ch'angan was the focus of ancient China during crucial formative periods. After two centuries of Han rule, political strife led to a period of decline, but the Sui emperors rebuilt and expanded Ch'angan when they made it their capital. During the Tang Dynasty, Ch'angan again became a magnificent city with three districts: the ornate Palace City; the impressive Imperial City, which housed the national administration; and the busy Outer City containing the homes and markets of artisans and merchants.

After its Tang heyday the city again declined, although it remained a bustling trade center. During the Ming Dynasty it was endowed with some of its architectural landmarks, including the Great Mosque marking the arrival of Islam; the older Big Wild Goose Pagoda dates from the influx of Buddhism. After the Ming period, Ch'angan's name was changed to Xian (meaning "Western Peace"), then to Siking, and in 1943 back to Xian again.

Having been a gateway for Buddhism and Islam, Xian in the 1920s became a center of Soviet communist ide-

© H. J. de Blij, P. O. Muller, and John Wiley & Sons, Inc.

ology. The Nationalists, during the struggle against the Japanese, moved industries from the vulnerable east to Xian, and when the communists took power, they enlarged Xian's industrial base still further. The present city (population: 4.2 million) lies southwest of the famed tombs, its cultural landscape now dominated by a large industrial complex that includes a steel mill, textile factories, chemical plants, and machine-making facilities. Little remains (other than some prominent historic landmarks) of the splendor of times past, but Xian's location on the railroad to the vast western frontier of China sustains its long-term role as one of the country's key gateways.

After four millennia of recurrent cultural cohesion, economic security, and political continuity, the Chinese world lay open to the aggressions of foreigners whose innovative capacities China had negated to the end. Now the ships flying European flags lay in the ports of China's coasts and rivers, but China had not learned to manufacture the cannons to blast them out of the water. The smokestacks of foreign factories rose over the townscapes of its great cities, and no Chinese force could dislodge them. The Japanese, seeing China's weakness, invaded Korea. The Russians entered Manchuria, the home base of the Manchus. The foreign invaders even took to fighting among themselves, as Japan and Russia

did in Manchuria in 1904. (Today this part of China is called the **Northeast**, the name Manchuria having been—understandably—rejected.)

A New China Emerges

In the meantime, organized opposition to the foreign presence in China was gathering strength, and the twentieth century opened with a large-scale revolt against all outside elements. Bands of revolutionaries roamed both cities and countryside, attacking not only the hated foreigners but also Chinese who had adopted Western cul-

FIGURE 9-8

© H. J. de Blij, P. O. Muller, and John Wiley & Sons, Inc.

tural traits. Known as the Boxer Rebellion (after a loose translation of the Chinese name for these revolutionary groups), this 1900 uprising was put down with much bloodshed by an international force consisting of British, Russian, French, Italian, German, Japanese, and American soldiers. Simultaneously, another revolutionary movement (led by the nationalist, Sun Yat-sen) was gaining support, aimed against the Manchu leadership itself. In 1911, the emperor's garrisons were attacked all over China,

and in a few months the 267-year-old Qing Dynasty was overthrown. Indirectly, it too was a casualty of the foreign intrusion, and it left China divided and disorganized.

The fall of the Manchus and the proclamation of a republican government in China did little to improve the country's overall position. The Japanese captured Germany's holdings on the Shandong Peninsula, including the city of Qingdao, during World War I. When the victorious European powers met at Versailles in 1919 to

Extraterritoriality

4

DURING THE NINETEENTH century, as China weakened and European colonial invaders entered China's coastal cities and sailed up its rivers, the Europeans forced China to accept a European doctrine of international law—**extraterritoriality**. Under this doctrine, foreign states and their representatives are immune from the jurisdiction of the country in which they are based. Today, this applies to embassies and diplomatic personnel. But in Qing (Manchu) China, it went far beyond that.

The European, Russian, and Japanese invaders established as many as 90 treaty ports—extraterritorial enclaves in China's cities under unequal treaties enforced by gunboat diplomacy. In their "concessions," diplomats and traders were exempt from Chinese law. Not only port areas but also the best residential suburbs of large cities were declared to be "extraterritorial" and made inaccessible to Chinese citizens. In the city of Guangzhou (Canton in colonial times), Sha Mian Island in the Pearl River was a favorite extraterritorial enclave. A sign at the only bridge to the island stated, in English and Cantonese, "No Dogs or Chinese."

Christian missionaries fanned out into China, their residences and churches fortified with extraterritorial security. In many places, Chinese found themselves unable to enter parks and buildings without permission from foreigners. This involved a loss of face that contributed to bitter opposition to the presence of foreigners—a resentment that exploded in the Boxer Rebellion of 1900.

After the collapse of the Qing Dynasty in 1911, the Chinese Nationalists negotiated an end to all commercial extraterritoriality in China Proper; the Russians, however, would not yield in then-Manchuria. Only Hong Kong and Macau retained their status as colonies.

When China's government in 1980 embarked on a new economic policy that gave major privileges and exemptions to foreign firms in certain coastal areas and cities, opponents argued that this policy revived the practice of extraterritoriality in a new guise. This issue remains a sensitive one in a China that has not forgotten the indignities of the colonial era.

---

divide the territorial spoils, they affirmed Japan's rights in the area. This led to yet another Chinese effort to counter the foreign scourge. Nationwide protests and boycotts of Japanese goods were organized in what became known as the May Fourth Movement. One participant in these demonstrations was a charismatic young man named Mao Zedong.

Nationalists and Communists

During the chaotic 1920s, the Nationalists and the Communist Party at first cooperated, with the remaining foreign presence their joint target. After Sun Yat-sen's death in 1925, Chiang Kai-shek became the Nationalists' leader, and by 1927 the foreigners were on the run, escaping by boat and train or falling victim to rampaging Nationalist forces. But soon the Nationalists began purging communists even as they pursued foreigners, and in 1928, when Chiang established his Nationalist capital in the city of Nanjing, it appeared that the Nationalists would emerge victorious from their campaigns. They had driven the communists ever deeper into the interior, and by 1933 the Nationalist armies were on the verge of encircling the last communist stronghold in the area of Ruijin in Jiangxi Province.

This led to a momentous event in Chinese history: the *Long March*. Nearly 100,000 people—soldiers, peas-

ants, leaders—marched westward from Ruijin in 1934, a communist column that included Mao Zedong and Zhou Enlai. The Nationalist forces rained attack after attack on the marchers, and of the original 100,000, about three-quarters were killed. But new sympathizers joined along the way (see the route marked on Fig. 9-8), and the 20,000 survivors found a refuge in the mountainous interior of Shaanxi Province, 3200 kilometers (2000 miles) away. There, they prepared for a renewed campaign that would bring them to power.

Japan in China

Although many foreigners fled China during the 1920s and 1930s, others seized the opportunity presented by the contest between the Nationalists and communists. The Japanese took control over the Northeast, and when the Nationalists proved unable to dislodge them, they set up a puppet state there, appointed a Manchu ruler to represent them, and called their possession Manchukuo.

The inevitable full-scale war between the Chinese and the Japanese broke out in 1937, with the Nationalists bearing the brunt of it (which gave the communists an opportunity to regroup). The gray boundary in Figure 9-8 shows how much of China the Japanese conquered. The Nationalists moved their capital to Chongqing, and the communists controlled the area centered on Yanan. China had been broken into three pieces.

The Japanese committed unspeakable atrocities in their campaign in China. Millions of Chinese citizens were shot, burned, drowned, subjected to gruesome chemical and biological experiments, and otherwise wantonly victimized. The "Rape of Nanking," an orgy of murder, violation, torture, and pillage following the fall of Nanjing to the Japanese Army in late 1937, was perhaps the most heinous of Japan's war crimes during this period. Years later, when China's economic reforms of the 1980s and 1990s led to a renewed Japanese presence in China, the Chinese public and its leaders called for Japan to acknowledge and apologize for these wartime abuses. In Japan, this pitted apologists against strident nationalists, a dispute that continues today. The Chinese remain dissatisfied with Japan's failure to apologize unconditionally for its actions, and the book is not yet closed on this sensitive issue.

Communist China Arises

After the U.S.-led Western powers defeated Japan in 1945, the civil war in China quickly resumed. The United States, hoping for a stable and friendly government in China, sought to mediate the conflict but at the same time recognized the Nationalists as the legitimate government. The United States also aided the Nationalists militarily, destroying any chance of genuine and impartial mediation. By 1948, it was clear that Mao Zedong's well-organized militias would defeat Chiang Kai-shek. Chiang kept moving his capital—back to Guangzhou, seat of Sun Yat-sen's first Nationalist government, then back again to Chongqing. Late in 1949, after a series of disastrous defeats in which hundreds of thousands of Nationalist forces were killed, the remnants of Chiang's faction gathered Chinese treasures and valuables and fled to the island of Taiwan. There, they took control of the government and proclaimed their own Republic of China.

Meanwhile, on October 1, 1949, standing in front of the assembled masses at the Gate of Heavenly Peace on Beijing's Tiananmen Square, Mao Zedong proclaimed the birth of the People's Republic of China.

CHINA'S HUMAN GEOGRAPHY

After more than six decades of communist rule, China is a society transformed. It has been said that the year 1949 actually marked the beginning of a new dynasty not so different from the old, an autocratic system that dictated from the top. In that view, Mao Zedong simply bore

the mantle of his dynastic predecessors. Only the family lineage had fallen away; now communist "comrades" would succeed each other.

And certainly some of China's old traditions continued during the communist era, but in many other ways Chinese society was totally overhauled. Benevolent or otherwise, the dynastic rulers of old China headed a country in which—for all its splendor, strength, and cultural richness—the fate of landless people and of serfs often was undescribably miserable; in which floods, famines, and diseases could decimate the populations of entire regions without any help from the state; in which local lords could (and often did) repress the people with impunity; in which children were sold and brides were purchased. The European intrusion made things even worse, bringing slums, starvation, and deprivation to millions who had moved to the cities.

Leap and "Revolution"

The communist regime, dictatorial though it was, attacked China's weaknesses on many fronts, mobilizing virtually every able-bodied citizen in the process. Land was taken from the wealthy; farms were collectivized; dams and levees were built with the hands of thousands; the threat of hunger for millions receded; health conditions improved; child labor was reduced. But China's communist planners also made terrible mistakes. The *Great Leap Forward*, requiring the reorganization of the peasantry into communal brigade teams to speed industrialization and make farming more productive, had the opposite effect and so disrupted agriculture that between 20 and 30 million people died of starvation between 1958, when the program was implemented, and 1962, when it was abandoned.

Mao ruled China from 1949 to 1976, long enough to leave lasting marks on the state. Another of his communist dictums had to do with population. Like the Soviets (and influenced by a horde of Soviet advisors and planners), Mao refused to impose or even recommend any population policy, arguing that such a policy would represent a capitalist plot to constrain China's human resources. As a result, China's population grew explosively during his rule.

Yet another costly episode of Mao's rule was the so-called *Great Proletarian Cultural Revolution*, launched by Mao Zedong during his last decade in power (1966–1976). Fearful that Maoist communism was being contaminated by Soviet "deviationism" and concerned about his own stature as its revolutionary architect, Mao unleashed a campaign against what he viewed as emerging elitism in society. He mobilized young people living in cities and towns into cadres known as Red Guards and ordered them to attack "bourgeois" elements throughout China, criticize Communist Party officials, and root out "opponents" of the system. He shut down all of China's schools, persecuted untrustworthy intellectuals, and encouraged the Red Guards to engage in what he called a renewed "revolutionary experience." The results were disastrous: Red Guard factions took to fighting among themselves, and anarchy, terror, and economic paralysis followed. Thousands of China's leading intellectuals died, moderate leaders were purged, and teachers, elderly citizens, and older revolutionaries were tortured to make them confess to "crimes" they did not commit. As the economy suffered, food and industrial production declined. Violence and famine killed as many as 30 million people as the Cultural Revolution spun out of control. One of those who survived was a Communist Party leader who had himself been purged and then been reinstated—Deng Xiaoping. Deng was destined to lead the country in the post-Mao period of economic transformation.

Political and Administrative Subdivisions

Before we investigate the emerging human geography of contemporary China, we should acquaint ourselves with the country's political and administrative framework (Fig. 9-9). For administrative purposes, China is divided into the following units:

4 Central-Government-Controlled Municipalities (*Shi's*)

5 Autonomous Regions (A.R.'s)

22 Provinces

2 Special Administrative Regions (SAR's)

The four central-government-controlled *Municipalities* are the capital, Beijing; its nearby port city, Tianjin; China's largest metropolis, Shanghai; and the Chang River port of Chongqing, in the interior. These *Shi's* form the cores of China's most populous and important subregions, and direct control over them from the capital entrenches the central government's power.

We should note that the administrative map of China continues to change, posing problems for geographers. The city of Chongqing was made a *Shi* in 1996, and its "municipal" area was enlarged to incorporate not only the central urban area but a huge hinterland covering all of eastern Sichuan Province. As a result, the "urban" population of Chongqing is officially 30 million, making this the world's largest metropolis—but in truth, as shown in the cities population table on page 466, the cen-

tral urban area contains only 6.8 million inhabitants. And because Chongqing's population is officially *not* part of the province that borders it to the west (Fig. 9-9), the official population of Sichuan declined by 30 million when the *Chongqing Shi* was created. So despite what Table 9-2 suggests, the population within Sichuan Province's outer borders remains the largest of any of China's administrative divisions.

The five *Autonomous Regions* were established to recognize the non-Han minorities living there. Some laws that apply to Han Chinese do not apply to certain minorities. As we saw in the case of the former Soviet Union, however, demographic changes and population movements affect such regions, and the policies of the 1940s may not work in the twenty-first century. Han Chinese immigrants now outnumber several minorities in their own regions. The five Autonomous Regions (A.R.'s) are: (1) Nei Mongol A.R. (Inner Mongolia); (2) Ningxia Hui A.R. (adjacent to Inner Mongolia); (3) Xinjiang Uyghur A.R. (China's northwestern corner); (4) Guangxi Zhuang A.R. (far south, bordering Vietnam); and (5) Xizang A.R. (Tibet).

China's 22 *Provinces*, like U.S. States, tend to be smallest in the east and largest toward the west. The territorially smallest are the three easternmost provinces on China's coastal bulge: Zhejiang, Jiangsu, and Fujian. The two largest are Qinghai, flanked by Tibet, and Sichuan, China's Midwest.

As with all large countries, some provinces are more important than others. The Province of Hebei nearly surrounds Beijing and occupies much of the core of the country. The Province of Shaanxi is centered on the great ancient city of Xian. In the southeast, momentous economic developments are occurring in the Province of Guangdong, whose urban focus is Guangzhou. When, in the pages that follow, we refer to a particular province or region, Figure 9-9 is a useful locational guide.

In 1997, the British dependency of Hong Kong (Xianggang) was taken over by China and became the country's first **Special Administrative Region (SAR)**. In 1999, Portugal similarly transferred Macau, opposite Hong Kong on the Pearl River Estuary, to Chinese control, creating the second SAR under Beijing's administration.

Population Issues

Table 9-2 underscores the enormity of China's population: many Chinese provinces have more inhabitants than do most of the world's countries. With a 2010 population of 1 billion, 338 million, China remains the largest nation on Earth (Fig. 9-10). We should view these huge numbers in the context of China's population growth

FIGURE 9-9

© H. J. de Blij, P. O. Muller, and John Wiley & Sons, Inc.

rate. High rates of population growth dim national development prospects in many countries in the periphery, but China has a special problem. Even a modest growth rate—such as today's 0.5 percent annually—still adds 6.7 million to the total every year.

Mao Zedong, as we have noted, refused to institute policies that would reduce the rate of population increase because he believed (encouraged by Soviet advisors) that this would play into the hands of his capitalist enemies. Fast-growing populations, he said, constituted the only asset many poor, communist, or socialist countries had. If the capitalist world wanted to reduce growth rates, it must be an anticommunist plot.

The One-Child Policy and Its Impact

But other Chinese leaders, realizing that rapid population growth would stymie their country's progress, tried to stem the tide through local propaganda and education.

Some of these leaders lost their positions as "revisionists," but their point was not lost on China's future leaders. After Mao's death, China embarked on a vigorous population-control program. In the early 1970s, the annual rate of natural increase was about 3 percent; by the mid-1980s, it was down to 1.2 percent. Families were ordered to have one child only, and those who violated the policy were penalized by losing tax advantages, educational opportunities, and even housing privileges. As we noted, China's census bureau now officially reports a slowing growth rate of 0.5 percent.

The one-child policy was vigorously policed throughout China by Communist Party members, leading to serious excesses. The number of abortions, sometimes forced, rose sharply; female infanticide increased. During the 1980s, when reports of these byproducts of the policy reached the outside world, the regime began to institute certain exceptions (minorities were not subject to the one-child limit). Couples living in rural areas were allowed to

TABLE 9-2

China: Population by Major Administrative Divisions* (in millions)

Provinces	1990 Population	2000 Population	2008 Population**
Anhui	56.2	59.9	61.2
Fujian	30.0	34.7	35.8
Gansu	22.4	25.6	26.2
Guangdong	62.8	86.4	94.5
Guizhou	32.4	35.3	37.6
Hainan	6.6	7.9	8.5
Hebei	61.1	67.4	69.4
Heilongjiang	35.2	36.9	38.2
Henan	85.5	92.6	93.6
Hubei	54.0	56.3	57.0
Hunan	60.7	62.4	63.6
Jiangsu	67.1	74.4	76.3
Jiangxi	37.7	41.4	43.7
Jilin	24.7	26.9	27.3
Liaoning	39.5	41.9	43.0
Qinghai	4.5	5.2	5.5
Shaanxi	32.9	36.1	37.5
Shandong	84.4	90.8	93.7
Shanxi	28.8	33.0	33.9
Sichuan	107.2	83.3	81.3
Yunnan	37.0	42.9	45.1
Zhejiang	41.5	46.8	50.6
Autonomous Region			
Guangxi Zhuang	42.2	44.9	47.7
Nei Mongol	21.5	23.2	24.1
Ningxia Hui	4.7	5.6	6.1
Xinjiang Uyghur	15.2	19.3	21.0
Xizang	2.2	2.6	2.8

*Population data for the four centrally-administered municipalities and the Hong Kong (Xianggang) and Macau Special Administrative Regions are shown in the table on page 466.

**Source: National Sample Survey on Population Changes (as reported in *China Statistical Yearbook*).

have a second child if their firstborn was female. Certain farming and fishing families could apply for permission to raise a second child in view of future labor needs. Since 2001, urban couples could have two children if both the father and mother came from one-child families.

6 Chinese demographers report that these **restrictive population policies** have produced 300 million "non-births" over the past 30 years; that is, China's population would be more than 1.6 billion today if these policies had not been instituted. They estimate that the fertility rate declined from 2.3 children per woman in 1960 to 1.6 in 2008, which would mean that China's population will level off by mid-century and then begin a steady decline. This will entail a rapid increase in the proportion of the population over 65 years old, with consequences similar to Western Europe (a swift reduction in the number of workers supporting these old-age pensioners)—except that Europe's decline came after most of its countries had become wealthy, whereas China's is likely to start before its population achieves high-income status.

This worries China's official planners, especially because China has already encountered problems meeting its pension obligations. And there is another concern. China's one-child policy may have prevented a population explosion, but because families preferred male heirs over females, it also led to an excess of men over women in the population pyramid. According to the latest (2005) official data, the ratio at birth has averaged 120 boys for every 100 girls, a troubling statistic because it implies selective abortion, female infanticide, and child neglect. And this ratio also imperils China's future: millions of young men will be unable to find wives, and geographers and others who study this phenomenon, and its precedents elsewhere, report that such circumstances tend to induce governments to expand their armed forces and, sometimes, to find reasons to give those forces a practical purpose. More immediately, Chinese men are venturing abroad in search of potential spouses, and China's neighbors, notably Vietnam, have complained of their aggressive behavior, even of kidnapping tactics.

For several years the one-child policy has been debated in the media, a sign that the government itself is weighing these issues. In 2009 the policy was openly discussed in the National People's Congress, some of whose members favored a two-child policy, with no exceptions. This would raise China's fertility rate and keep the population growing beyond 2050, but the advantages would include a larger number of younger workers to support the aged and a reduction of the skewed male-female ratio that threatens to undermine this vast country's future social fabric.

The Minorities

We asserted earlier that China remains an empire; its government controls territories and peoples that are in effect colonized. The ethnolinguistic map (Fig. 9-11) apparently confirms this proposition. This map should be seen in context, however. It does not show local-area majority populations but instead reveals where minorities are concentrated. For example, the Mongolian population in China is shown to be clustered along the southeastern border of Mongolia in the Autonomous Region called Nei Mongol (see Fig. 9-9). But even in that A.R., the Han Chinese, not the Mongols, are now in the majority.

Nonetheless, the map gives definition to the term ***China Proper*** as the home of the People of Han, the ethnic (Man-

FIGURE 9-10

© H. J. de Blij, P. O. Muller, and John Wiley & Sons, Inc.

darin-speaking) Chinese depicted in tan and light orange in Figure 9-11. From the upper Northeast to the border with Vietnam and from the Pacific coast to the margins of Xinjiang, this Chinese majority predominates. When you compare this map to that of population distribution (to be discussed shortly), it will be clear that the minorities constitute only a small percentage of the country's total. The Han Chinese form the largest and densest clusters.

In any case, China controls non-Chinese areas that are vast, if not populous. The Tibetan group numbers under 3 million, but it extends across all of settled Xizang. Turkic peoples inhabit large areas of Xinjiang. Thai, Vietnamese, and Korean minorities also occupy areas on the margins of Han China. As we will see later, the Southeast Asian minorities in China have participated strongly in the Pacific Rim developments on their doorsteps. Hundreds of thousands have migrated from their Autonomous Regions to the economic opportunities along the coast.

Numerically, Chinese dominate in China to a far greater degree than Russians dominated their Soviet Empire. But

territorially, China's minorities extend over a proportionately larger area. The Ming and Manchu rulers bequeathed the People of Han an empire.

PEOPLE AND PLACES OF CHINA PROPER

The map of China's population density (Fig. 9-10) reveals the continuing relationship between the physical environment and its human occupants. In technologically advanced countries, we have noted, people shake off their dependence on what the land can provide; they cluster in cities and in other areas of economic opportunity. This depopulates rural areas that may once have been densely inhabited. In China, that stage has not yet been reached. Although China has a steadily growing number of large cities, a majority of the people (55 percent) still live on—and from—the land. Thus the population map reflects the livability and productivity of China's basins, lowlands, and plains. Compare Figures

FIGURE 9-11

© H. J. de Blij, P. O. Muller, and John Wiley & Sons, Inc.

9-7A and 9-10, and China's continuing dependence on soil, water, and warmth will be evident.

The population map also suggests that in certain areas environmental limitations are being overcome. Industrialization in the Northeast, irrigation in the Inner Mongolia Autonomous Region, and oil-well drilling in Xinjiang enabled millions of Chinese to migrate from China Proper into these frontier zones, where they now outnumber the indigenous minorities.

Nonetheless, physiography and demography remain closely related in China. To grasp this relationship, it is useful to compare Figures 9-3 (physiography) and 9-10 (population density) as our discussion proceeds. On the population map, the darker the color, the denser the population: in places we can follow the courses of major rivers through this scheme. Look, for example, at China's Northeast. A ribbon of population follows the Liao and the Songhua rivers. Also note the huge, nearly circular population concentration on the western edge of the red zone; this is the Sichuan Basin, in the upper course of the Chang Jiang. Four major river basins contain more than three-quarters of China's 1.3 billion people:

1. The Liao-Songhua Basin or the Northeast China Plain

2. The Lower Huang He (Yellow River) Basin, known as the North China Plain

3. The Upper and Lower Basins of the Chang Jiang (Long River; lower course known as the Yangzi)

4. The Basins of the Xi (West) and Pearl River

Not all the people living in these river plains are farmers, of course. China's great cities also have developed in these populous areas, from the industrial centers of the Northeast to the Pacific Rim upstarts of the South. Both

Beijing and Shanghai lie in major river basins. And, as the map shows, the hilly areas between the river basins are not exactly sparsely populated either. Note that the hill country south of the Chang Basin (opposite Taiwan) still has a density of 50 to 100 people per square kilometer (130 to 260 per sq mi).

Northeast China Plain

Let us look at the geography of China by going from north to south, being guided by the rivers that have formed the lifelines for the people over thousands of years. China's northernmost major river to reach the coast is also its least well known: the Liao River, which makes a pronounced bend before flowing into the Liaodong Gulf (Fig. 9-12). True, there is a river even to the north of the Liao named the Songhua, but it joins the Amur River along the border and eventually enters Russia.

The most populous part of the basin of the Liao River, the Northeast China Plain, forms the heartland of China's Northeast, ancestral home of the Manchus who founded China's last dynasty, battleground between Japanese and Russian invaders, Japanese colony, and later the heartland of communist industrial development. As the administrative map (Fig. 9-9) shows, there are three provinces here: Liaoning in the south, facing the Yellow Sea; Jilin in the center; and Heilongjiang, by far the largest, in the north.

When you are familiar with the vast farmlands of the North China Plain, the southernmost part of the Northeast China Plain looks so similar that it seems to be a continuation. Near the coast, farms look as they do near the mouth of the great Huang He; to the east, the Liaodong Peninsula looks like the Shandong Peninsula across the water. That impression soon disappears, however. Travel northward, and the Northeast China Plain reveals the effects of cold climate and thin soils. Farmlands are patchy; smokestacks seem to rise everywhere. This is industrial, not primarily farming, country.

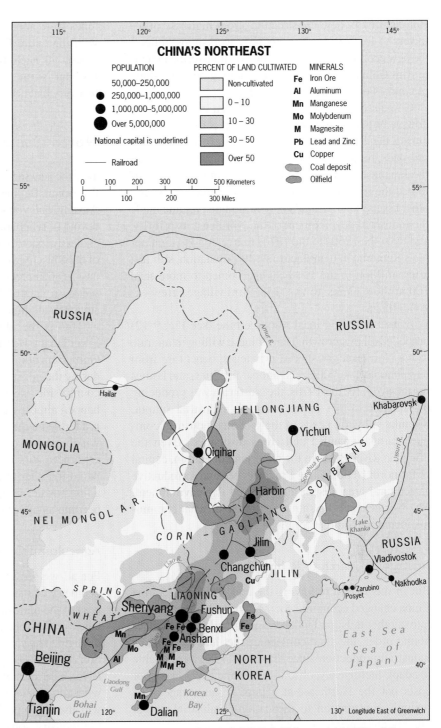

FIGURE 9-12

© H. J. de Blij, P. O. Muller, and John Wiley & Sons, Inc.

But if you look closely, you will see two stages in China's recent history imprinted in the human landscape. Still standing, from Dalian in the south to Yichun in the north, are the old, state-run factories of the communist period, many of them inactive and in disrepair, reminiscent of "Rustbelt" landscapes in former Soviet Eastern Europe—and of abandoned industrial plants in the more stressed districts of the American Manufacturing Belt.

But elsewhere, new construction, from high-rise office towers to low-rise high-technology workshops, reflects the new economic era that has already transformed Pacific Rim China. The Northeast is finally getting its share **7** of China's **economic restructuring**.

China's Northeast has gone through many ups and downs. When it fell under Japanese colonial control during the 1930s, this region was transformed by new railroads, roads, bridges, and other infrastructure. After the Japanese were ousted, the Soviets in the late 1940s looted the region of factory equipment, machinery, and other assets. Then China's new communist regime made the industrial development of the Northeast its priority, and from the 1950s to the 1970s this region led the People's Republic in manufacturing. Its population, still just a few million in the 1940s, mushroomed to more than 100 million. Cities, towns, even farm villages grew exponentially.

A well-endowed local resource base (see Fig. 9-12) propelled all this growth. Han Chinese willing to migrate to the Northeast got special privileges, benefiting from new schools, hospitals, recreational facilities, and homes for the aged. By the 1970s, the Northeast was producing one-quarter of the entire country's industrial output. Stimulated by the needs of the growing industrial complex, farms expanded and crops diversified. Even the Songhua River Basin in the hinterland of Harbin, capital of far-north Heilongjiang Province, became a productive agricultural area despite its cold and dry environment. The Northeast was a shining example of what communist planning could accomplish.

Rustbelt in Search of Rebound

But in the 1970s, following the end of Mao's regime, China changed economic course and the Northeast went into another tailspin. Under the new, market-driven economic order, the state companies that had led China's industrial production since the early 1950s were unable to adapt. Inefficiency, obsolescence, and excessive labor costs put the Northeast at a huge disadvantage. The region's contribution to the nation's industrial output quickly plummeted to about 7 percent.

But economic geographers know that the Northeast still is a storehouse of resources, has strong locational advantages, lies close to booming South Korea, and has a history of recovery. Thus it was not surprising that the port city of Dalian, the terminus of the railroad from the far north, began to take on the features of a Pacific Rim growth pole over the past decade, and that the cities in its hinterland showed signs of revitalization as well. Just before the global economic downturn that began in 2008, economists were predicting that the entire Northeast's for-

tunes were about to change for the better. In mid-2009, with the entire Chinese economy slowed by the world recession, such forecasts had to be revised. But the resilient Northeast had already shown its capacity for a recovery that is likely to resume when the recession eases.

North China Plain

Take the new superhighway between the port of Tianjin and the capital, Beijing, or the train from Beijing southward to Anyang, and you see a physical landscape almost entirely devoid of relief and a cultural landscape that repeats itself endlessly (photo at right). Villages, many showing evidence of the collectivization program of the early communist era, pass at such regular intervals that some kind of geometry seems to control the farmscape. Carefully diked canals stretch in all directions. Rows of trees mark most fields.

This is the physical and cultural landscape of the lower Huang He (Yellow River) as it flows into the Bohai Gulf. Depending on the season, the air is clear, as it is during the moist summer when the wheat crop covers the countryside like a carpet, or choked with dust, which can happen almost any other time of the year. The soil here is a fertile mixture of alluvium and loess (deposits of fine, windborne silt or dust of glacial origin), and the slightest breeze stirs it up. In the early spring the dust hangs like a fog over the landscape, covering everything with a fine powder. Add this to urban pollution and the cities become as unhealthy as any in the world.

Agricultural Colossus

For all we see, there are geographic explanations. The geometry of the settlement pattern results from the flatness of the uniformly fertile surface and the distance villagers could walk to neighboring villages and back during daylight. The canals form part of a vast system of irrigation channels designed to control the waters of the Huang River. The rows of trees serve as windbreaks. And while wheat dominates the crops of the North China Plain, you will also see millet, sorghum, soybeans, cotton for the textile industry, tobacco (the Chinese are heavy smokers), and fruits and vegetables for the urban markets.

The North China Plain is one of the world's most densely populated agricultural areas. Figure 9-10 shows that most of it has a density of more than 400 people per square kilometer (1000 people per sq mi), and in some parts that figure is twice as high. Here, the ultimate hope of the Beijing government lay less in land redistribution than in raising yields through improved fertilization, expanded irrigation facilities, and mechanization. A series of dams on the Huang, including the Xiaolangdi Dam upstream in

FROM THE FIELD NOTES

"We flew over the North China Plain for an hour, and one could not miss the remarkably regular spacing of the villages on this seemingly table-flat surface. Many of these villages became communes during the communist reorganization of China's social order. The elongated buildings (many added since the 1950s) were each designed to contain one or more extended families. After the demise of Mao and his clique, the collectivization effort was reversed and the emphasis returned to the small nuclear family. In this village and others we overflew, the long sides of the buildings almost always lay east-west, providing maximum sun exposure to windows and doors, and enhancing summer ventilation from longitudinal breezes. The more distant view looks hazy, because the North China Plain during the spring is wafted by loess-carrying breezes from the northwest. On the ground, a yellow-gray hue shrouds the atmosphere until a passing weather front temporarily clears the air."
© H. J. de Blij (1980).

Concept Caching

www.conceptcaching.com

Henan Province, now reduce the flood danger, but outside the irrigated areas the ever-present problem of rainfall variability and drought persists. The North China Plain has not produced any substantial food surplus even under normal circumstances; thus when the weather turns unfavorable, the situation soon becomes precarious. The specter of famine may have receded, but the food situation is still uncertain in this very critical part of China Proper.

Core-Area Cities

The North China Plain is an excellent example of the concept of a national **core area**. Not only is it a densely populated, highly productive agricultural zone, but it also is the site of the capital and other major cities, a sub-

stantial industrial complex, and several ports, among which Tianjin ranks as one of China's largest (Fig. 9-13). Tianjin, on the Bohai Gulf, is linked by rail and highway (less than a two-hour drive) to Beijing. Like many of China's harbors, that of Tianjin's river port is not particularly good; but Tianjin is well situated to serve not only the northern sector of the North China Plain and the capital, but also the Upper Huang Basin and Inner Mongolia beyond (see Fig. 9-1). Tianjin, again like several other Chinese ports, had its modern start as a treaty port, but the city's major growth awaited communist rule. For decades it was a center for light industry and a flood-prone harbor, but after 1949 the communists constructed a new artificial port and flood canals. They also chose Tianjin as a site for major industrial development and made substantial investments in the chemical industry (in which Tianjin still leads China), iron and steel production, heavy machine manufacturing, and textiles. Today, with a population of 7.5 million, Tianjin is China's fifth-largest metropolis and, as we note later, looms large in the grand design of China's planners.

Northeast of Tianjin lies the smaller city of Tangshan. In 1976, a devastating earthquake, with its epicenter between Tangshan and Tianjin, killed an estimated 750,000 people, the deadliest natural disaster of the twentieth century. So inept was the government's relief effort in nearby Beijing that the widespread resentment it provoked may have helped end the Maoist period of communist rule in China.

Beijing, unlike Tianjin, is China's political, cultural, and educational center. Its industrial development has not matched Tianjin's. The communist administration did, however, greatly expand the municipal area of Beijing, which (as we noted) is not controlled by the Province of Hebei but is directly under the central government's authority. In one direction, Beijing was enlarged all the way to the Great Wall—50 kilometers (30 mi) to the north—so that the "urban" area includes hundreds of thousands of farmers. Not surprisingly, this has circumscribed an enormous total population, enough to rank Beijing, with its 11.7 million inhabitants, among the world's megacities.

Inner Mongolia

Northwest of the core area as defined by the North China Plain, along the border with the state of Mongolia, lies Inner Mongolia, administratively defined as the Nei Mongol Autonomous Region (see Fig. 9-9). Originally established to protect the rights of the approximately 5 million Mongols who live outside the Mongolian state, Inner Mongolia has been the scene of massive immigration by Han Chinese. Today Chinese outnumber Mongols here by nearly four to one. Near the Mongolian

FIGURE 9-13

© H. J. de Blij, P. O. Muller, and John Wiley & Sons, Inc.

border, Mongols still traverse the steppes with their tents and herds, but elsewhere irrigation and industry have created an essentially Chinese landscape. The A.R.'s capital, Hohhot, has been eclipsed by Baotou on the Huang He, which supports a corridor of farm settlements as it crosses the dry land here on the margins of the Gobi and Ordos deserts. With just over 24 million inhabitants, Inner Mongolia cannot compare to the huge numbers that crowd the North China Plain, but recent mineral discoveries have boosted industry in Baotou and

livestock herding is expanding. Nei Mongol may retain its special administrative status, but it functions as part of Han China in all but name.

Basins of the Chang/Yangzi

We come now to the third of China's great rivers on our southward traverse, one that originates in the deep, icy interior of the mountainous west and is appropriately

AMONG THE REALM'S GREAT CITIES . . . **Beijing**

BEIJING (POPULATION: 11.7 MILLION), capital of China, lies at the northern apex of the roughly triangular North China Plain, just over 160 kilometers (100 mi) from its port, Tianjin, on the Bohai Gulf. Urban sprawl has reached the hills and mountains that bound the Plain to the north, a defensive barrier fortified by the builders of the Great Wall. You can reach the Great Wall from central Beijing by road in about an hour.

Although settlement on the site of Beijing began thousands of years ago, the city's rise to national prominence began during the Mongol (Yuan) Dynasty, more than seven centuries ago. The Mongols, preferring a capital close to their historic heartland, endowed the city with walls and palaces. Following the Mongols, later rulers at times moved their capital southward, but the government always returned to Beijing. From the third Ming emperor onward, Beijing was China's focus; it was ideally situated for the Manchus when they took over in 1644. During the twentieth century, China's Nationalists again chose a southern capital, but when the communists prevailed in 1949 they reestablished Beijing as their (and China's) headquarters.

Ruthless destruction of historic monuments carried out from the time of the Mongols to the communists (and now during China's modernization) has diminished, but not totally destroyed, Beijing's heritage. From the Forbidden City of the Manchu emperors (see photo on p. 470) to the fifteenth-century Temple of Heaven, Beijing remains an open-air museum and the cultural focus of

© H. J. de Blij, P. O. Muller, and John Wiley & Sons, Inc.

China. Successive emperors and aristocrats, moreover, bequeathed the city numerous parks and additional recreational spaces that other cities lack. Today Beijing is being transformed as a result of China's new economic policies. A forest of high-rises towers above the retreating traditional cityscape, avenues have been widened, and expressways have been built. A new era has arrived in this historic capital.

called the Chang Jiang—the Long River (see Fig. 9-1). But most of us know it as the *Yangzi*, the name it carries along its lower navigable course (Fig. 9-13). In contrast to the sediment-laden Huang River and its vast, flat alluvial plain, the Yangzi has one major channel and is flanked by higher relief. Near its mouth lies the country's largest city, Shanghai. Part of its middle course is now being transformed by a gigantic engineering project—the just-completed Three Gorges Dam. Farther upstream, the Chang crosses the populous, productive Sichuan Basin. And unlike the Huang, the Chang Jiang is navigable to oceangoing ships for over 1000 kilometers (600 mi) from the coast all the way to Wuhan (Fig. 9-13). Smaller ships can reach Chongqing, even after dam construction. Several of the Chang River's tributaries also are navigable, so that 30,000 kilometers (18,500 mi) of water transport routes serve its drainage basin. Thus the Chang Jiang constitutes one of China's leading transit corridors. With its tributaries it handles the trade of a vast area, including nearly all of middle China and sizeable parts of the north and south. The

North China Plain may be the core area of China, but in many ways the Lower Chang Basin is its heart.

Early in Chinese history, when the Chang's basin was being opened up and rice and wheat cultivation began, a canal was built to link this granary to the northern core of old China. Over 1600 kilometers (1000 mi) long, this was the longest artificial waterway in the world, but during the nineteenth century it fell into disrepair. Known as the Grand Canal, it was dredged and rebuilt when the Nationalists controlled eastern China. After 1949, the communist regime continued this restoration effort, and much of the canal is now again open to barge traffic, supplementing the huge fleet of vessels that hauls domestic interregional trade along the east coast (Fig. 9-13).

Pivotal Corridor

As Figure 9-14 shows, the Lower Chang Basin is an area of both rice and wheat farming, offering further proof of its pivotal situation between south and north in the heart of China Proper. Shanghai lies at the coastal gateway to this productive region on a small tributary of the Chang (now Yangzi) River, the Huangpu (see map p. 488). The city has an immediate hinterland of some 50,000 square kilometers (20,000 sq mi)—smaller than West Virginia—containing more than 50 million people. About two-thirds of this population are farmers who produce food, silk filaments, and cotton for the city's industries.

Travel upriver along the Yangzi/Chang Jiang, and you meet an unending stream of vessels large and small. The traffic between Shanghai and Wuhan—Wuhan is short for Wuchang, Hanyang, and Hankou, a coalescing conurbation of three cities—makes the Yangzi one of the world's busiest waterways. Smoke-belching boat engines create a more or less permanent plume of pollution here, worsening the regional smog created by the factories of Nanjing and others perched on the waterfront. Here China's Industrial Revolution is in full gear, with all its environmental consequences.

Three Gorges Dam

Above Wuhan the river traffic dwindles because the depth of the Chang reduces the size of vessels that can reach Yichang. River boats carry coal, rice, building materials, barrels of fuel, and many other items of trade. But the middle course of the Chang River is becoming more than a trade route. In the westward bend of the great river between Yichang and Chongqing, where the Chang has cut deep troughs on its way to the coast, one of the world's largest engineering projects was completed in 2009. It has several names: the Sanxia Project, the Chang Jiang Water Transfer Project, the Three Gorges Dam, and the New China Dam. It is most commonly called the Three Gorges Dam because it is here that the great river flows through a 240-kilometer (150-mi)-long series of steep-walled valleys less than 100 meters (330 ft) wide. Near the lower end of this natural trough, the new dam (shown in the photo to the right) now rises to over 180 meters (600 ft) above the valley floor to a width of 2 kilometers (1.3 mi), creating a reservoir that inundates the historic Three Gorges and extends more than 600 kilometers (over 400 mi) upstream. China's planners say this dam will end the river's rampaging flood cycle, enhance navigation, stimulate development along the new lake perimeter, and provide more than one-tenth of the country's electrical power supply. As such it will transform the heart of China, which is why many Chinese like to compare this gigantic project to the Great Wall and the Grand Canal, and call it the New China Dam.

FIGURE 9-14

© H. J. de Blij, P. O. Muller, and John Wiley & Sons, Inc.

The Sichuan Basin

As Figure 9-1 shows, all this development is taking place downstream from Chongqing, the city near the eastern margin of the great and populous Sichuan Basin that was made, along with its hinterland, a *Shi* in the hierarchy of China's administrative units. Obviously this city is strongly affected by the dam's construction, which disrupted its river trade (boats up to 1000 tons

could reach Chongqing previously). A diversion channel around the Three Gorges Dam is expected to restore and enhance Chongqing's port and transit functions, and the city figures prominently in an important economic

The Three Gorges Dam may not look so impressive in this aerial photograph taken over the Chang/Yangzi River, but the dam wall whose top you see here rises 180 meters (600 ft) above the now-inundated valley floor and is 2 kilometers (a mile and a quarter) wide. The dam lies upstream from the city of Yichang in Hubei Province, and its reservoir—the largest in the world—extends more than 600 kilometers (400 mi) upriver beyond the city of Chongqing to the margin of Sichuan Province. China today may be called the "home of the megaproject"—already the government is planning a massive water diversion scheme to reroute waters of the upper Chang River into the basin of the Huang River in the drier north. The Great Wall was only a start . . . © Reuters/Landov.

AMONG THE REALM'S GREAT CITIES . . . Shanghai

SAIL INTO THE mouth of the great Yangzi River, and you see little to prepare you for your encounter with China's largest city (population: 15.6 million). For that, you turn left into the narrow Huangpu River, and for the next several hours you will be spellbound. To starboard lies a fleet of Chinese warships. On the port side you pass oil refineries, factories, and crowded neighborhoods. Next you see an ultramodern container facility, white buildings, and rust-free cranes, built by Singapore. Soon rusty tankers and freighters line both sides of the stream. High-rise tenements tower behind the cluttered, chaotic waterfront where cranes, sheds, boatyards, piles of rusting scrap iron, mounds of coal, and stacks of cargo vie for space. In the river, your boat competes with ferries, barge trains, and cargo ships. Large vessels, anchored in midstream, are being offloaded by dozens of lighters tied up to them in rows. The air is acrid with pollution. The noise—bells, horns, whistles—is deafening.

What strikes you is the vastness and sameness of Shanghai's cityscape, until you pass beneath the first of two gigantic suspension bridges. Suddenly, everything changes. To the left, or east, lies *Pudong*, an ultramodern district with the space-age Oriental Pearl Television Tower rising like a rocket on its launchpad (see photo, p. 499) above a forest of modern, glass-and-chrome skyscrapers that make the Huangpu look like Hong Kong's Victoria Harbor. To the right, along the Bund, stand the remnants of Victorian buildings, monuments to the British colonialists who made Shanghai a treaty port and started the city on its way to greatness. Everywhere, construction cranes rise above the cityscape. Shanghai looks like one vast construction zone, complete with a 460-meter (1510-ft) office tower—called the World Financial Center—that is one of the world's tallest buildings.

The Chinese spent heavily to improve infrastructure in the Shanghai area, including brand-new Pudong International Airport, connected to the city by the world's first maglev (magnetic levitation) train—the fastest anywhere at 430 kph (267 mph). Pudong has become a magnet for

© H. J. de Blij, P. O. Muller, and John Wiley & Sons, Inc.

foreign investment, attracting as much as 10 percent of the annual total for the entire country. Shanghai's income is rising much faster than China's, and the Yangzi River Delta is becoming a counterweight to the massive Pearl River Delta development in China's South. Among other things, Shanghai is becoming China's "motor city," complete with a Formula One race track seating 200,000 at the center of an automobile complex where all levels of the industry, from manufacturing to sales, are being concentrated.

In 2010 Shanghai hosted World Expo on the banks of the Huangpu River (the maglev track was extended to the site), and the city's planners, who have already transformed the place from just another Open City to a symbol of the New China, enjoyed the opportunity to show the world what has been accomplished here in just a single generation.

design for China: the expansion of its coastal economic success into the deep interior. We return to this topic later, but consider Chongqing's relative location on the western margin of the Yangzi's great economic corridor and on the eastern edge of the vast and productive Sichuan Basin, now upstream from the country's greatest engineering project and about midway between Kunming to the south and Xian to the north. This city is likely to become the most important *growth pole* in China's interior.

Sichuan, its basin crossed by the Chang (called the Yangzi downstream) after the river emerges from the mountains of the west, frequently was a problem province for both China's dynastic rulers and the communist regime. Its thriving capital, Chengdu, was one of China's most active centers of dissent during the 1989 pro-democracy movement. With about 125 million inhabitants (including Chongqing) today, Sichuan Province would be one of the world's ten largest countries, but it does not even contain 10 percent of China's population. Sichuan's fertile, well-watered soils yield a huge variety of crops: grains (rice, wheat, corn), soybeans, tea, sugarcane, and many fruits and vegetables.

In many ways Sichuan is a country unto itself, isolated by topography and distance from the government in Beijing. The people here are Han Chinese, but cultural landscapes reveal the area's proximity to Southeast Asia. Along the great riverine axis of the Chang/Yangzi, China changes character time and again.

Basins of the Xi (West) and Pearl River

Finally, we get another lesson in Chinese geographic names: the southernmost river on which we focus is the Xi Jiang—which means *West* River! Upstream, it has other names, but downstream, where it reaches the coast,

it splits into a channel and an estuary. This estuary has the river's most famous name of all: the Pearl River Estuary. But although it really *is* an estuary, you will see it called the Pearl River *Delta* in some discussions. The reason is simple: here, a huge triangular area has become China's greatest manufacturing complex, flanking the "delta" from Hong Kong to Guangzhou to Zhuhai.

More on this later. Here we should know that the Xi River, by whatever name, is not much longer than the Liao, not as navigable as the Yangzi, and not as important agriculturally as the Huang, its tropical environs notwithstanding. With higher relief and less alluvial soil, farms here are smaller; rice can be double-cropped, but output is just a fraction of the Yangzi Basin's. Water supply is a problem in the interior, and this is not an area rich in natural resources, although there is some oil inland as well as offshore (Fig. 9-15). As economic geographers have often pointed out, this area does not even

FIGURE 9-15

© H. J. de Blij, P. O. Muller, and John Wiley & Sons, Inc.

possess many locational advantages; it is about as far from the country's core area as you can get in eastern China. But what has happened here, as we shall see later, has rightly been called a Pacific Rim miracle.

The four river-based subregions we have just discussed in many ways define what has long been called China Proper, the "real" China. But as Figure 9-5 reminds us, China Proper is only one part of China. We will return to the dramatic changes that continue to transform the geography of China's Pacific front. But before we do, we should take note of another China: the interior regions where China holds sway over huge territories sparsely populated by non-Han minorities that at times cause Beijing trouble (Fig. 9-16). One of these territories, mountainous Tibet (the Chinese call it *Xizang*), has come to world attention largely through the efforts of its exiled Buddhist leader, the Dalai Lama. In the second, mainly desert *Xinjiang*, China meets the Islam of Turkestan, an encounter that has not always gone smoothly. We now turn to consider each of these components of China's periphery.

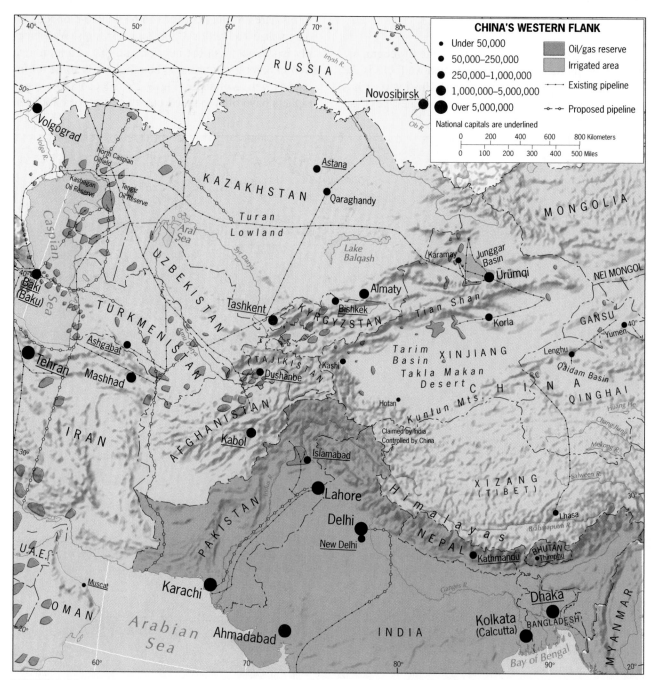

FIGURE 9-16

© H. J. de Blij, P. O. Muller, and John Wiley & Sons, Inc.

XIZANG (TIBET)

Take a close look at the photo on this page, and you are witness to a recent event that signified the penetration of China's most remote corner: the historic Buddhist outpost of Tibet (Xizang). The 2006 opening of the railroad that links Beijing with Xizang's capital, Lhasa, across some of the Earth's most forbidding terrain, symbolizes China's determination to bring Tibet firmly into Beijing's orbit.

Remote, isolated, encircled by snowcapped mountains, and cut by deep valleys, Tibet for centuries was the site of fortress-like Buddhist monasteries whose monks paid allegiance to their supreme leader, the Dalai Lama. Tibetan society was tradition-bound and changeless, but came under nominal Chinese control in 1720, during the Qing (Manchu) Dynasty. When the Qing order collapsed in the late nineteenth century, Tibet found itself free of Chinese rule and remained autonomous (though powerless) during the chaos that enveloped China during the first half of the twentieth century.

A new Chinese invasion of Tibet followed shortly after the communist victory: in 1950 Chinese forces took control of Xizang, and in 1959 they crushed a Tibetan uprising, ousted the Dalai Lama, and emptied the monasteries, in the process destroying much of Tibet's cultural heritage. Formal annexation followed in 1965, and Xizang was given the status of Autonomous Region. After Mao's death in 1976, the Chinese made some amends, returning looted religious treasures, permitting reconstruction of monasteries, and allowing Buddhist religious life to resume. But villagers continued to rebel, the Dalai Lama campaigned in exile for his people and faith, and for all intents and purposes Xizang remained what it had been since annexation—a colony of the Chinese Empire, an occupied society. (It is also worth noting that present-day Tibet is only part of the Buddhist culture area's earlier domain, which included what is now western China's Qinghai Province [see Fig. 9-9].)

As in other remote parts of China, Beijing's objective has been to integrate its minority peoples into the Chinese state, to establish a strong Chinese presence, and to display the virtues and assets of Chinese culture. In the case of Xizang, this was especially difficult because of the region's remoteness. Less than three million Tibetan people live scattered across a vast, frigid, mountainous and high-plateau area consisting not only of Tibet itself but also the neighboring Qaidam Basin in Qinghai Province (refer again to Fig. 9-11, and also note Fig. 9-16). Difficult roads through mountain passes and steep-sided valleys, seasonally impassable, slowed sinicization here more than anywhere else in the empire. But a mountain-defying railway would facilitate Chinese penetration, display China's technological prowess, and confirm the permanence of China's presence in Xizang.

In 2008, during the period leading up to the Beijing Olympic Games, an open revolt among Tibetans, involving attacks on Chinese residents and property, caused death and destruction in Lhasa and elsewhere. Organizers of this uprising, which triggered severe Chinese suppression, undoubtedly saw the coming Beijing Olympics as a way to draw worldwide attention to their cause. But China's heavy-handed rule, and a major involuntary relocation project steering more than 250,000 Tibetans (10 percent of the population) from their rural hamlets into "socialist villages," created an atmosphere of resentment easily exploited by the rebels.

China has good reasons to establish itself firmly in Xizang and the Qaidam Basin. Hydroelectric power potential, mineral deposits only partially known, and, in the Qaidam Basin, oil reserves and coal deposits make this more than just a cultural mission.

A symbolic event took place in China during July 2006: the inaugural journey of a train linking China's capital, Beijing, with Lhasa, the capital of Xizang (Tibet). Laid by 100,000 construction workers across icy mountains, the final 1150 kilometers (700 mi) of the track constitutes an engineering marvel, climbing as high as 5072 meters (16,640 ft) across permafrost-ridden terrain. No other railway in the world reaches such altitudes; the carriages of the "Sky Train," with a total capacity of 900 passengers, are sealed and pressurized like the fuselage of an airplane. But the train has its detractors. This $4.2 billion project, they say, will facilitate Chinese penetration of Xizang and enhance Beijing's economic and military control over the remote territory while Tibet's infrastructure needs, such as roadbuilding and electrification, are neglected. It is evidence that China will spend whatever it takes to bring in larger numbers of Han Chinese to overwhelm the local Buddhist population, suppress Tibetan separatism, and promote "modernization." As such, the train to Xizang symbolizes the future as seen in Beijing, not Lhasa. © AP/World Wide Photo/Color China Photo.

● XINJIANG

A vast, sprawling state with a history of imperial expansion is bound to encroach on neighboring peoples and cultures, and so it is with China. In the far southwest, Chinese secularism meets Tibetan Buddhism; in the distant (from Beijing) northwest, China meets Islamic Turkestan. Have another look at Figure 9-5 to see how deeply China penetrates Central Asia as a result of its acquisition of what is officially called the Xinjiang-Uyghur Autonomous Region. Huge, sparsely peopled, resource-rich, and strategically situated, there is nothing "autonomous" about this territory. China rules, and sometimes it seems as though China's rule is calculated to aggravate the locals. Arrive at the capital, Ürümqi—as far west of Beijing as Denver is from Washington, D.C.—and you don't have to change your watch. By order of the government, all official clocks in Xinjiang are set to Beijing time, so that the winter day ends around 2:00 p.m. That's what your hotel clock may tell you, but locals keep their own time, setting their clocks and watches two hours behind "Chinese" time.

With about one-sixth of China's territory but less than 2 percent of China's population, Xinjiang has better connections to China Proper than mountain-rimmed Tibet. Road and rail links along the Hexi Corridor through neighboring Gansu Province create access that has served to strengthen the Chinese presence in Xinjiang to the point that nearly half the population of over 21 million is now Chinese. Figure 9-17 shows where most of the Chinese are concentrated: Xinjiang's emerging core area lies centered on the district around and between Ürümqi, the army-built city of Shihezi, and the oil center of Karamay. Administration, military and police service, oil production, and irrigated farming of cotton and fruits employ the Han Chinese here. Ürümqi has its minority neighborhoods, but has the look and atmosphere of a Chinese city. Shihezi is so completely Chinese that it seems to be transplanted lock, stock, and—importantly—barrel. Meanwhile, the Uyghurs and other locals complain that all the best jobs seem to go to the Chinese.

But there is another core area in Xinjiang: that of the (mostly Muslim) minorities. In the far west lies the city of Kashi (locals call it Kashgar), where Uyghurs, Kazakhs, Kyrgyz, and other Central Asian and Mongolian peoples still form the majority and create a cultural landscape very different from that of the A.R.'s capital. Mosques and traditional markets make you think you are in Turkestan here—and among Uyghurs there are those who dream of real autonomy in their "Autonomous Region"and even independence. During the period prior to the 2008 Beijing Olympics, when Tibetans also saw their opportunity for world attention, Muslim militants exploded a truck amid a group of jogging Chinese policemen in Kashi, killing 16. In the aftermath of the Olympics, with less attention focused on China's human-rights issues, resentment simmered until in 2009 a minor workplace incident in Ürümqi precipitated a disastrous outbreak of violence between Uyghurs and Chinese in which, by official accounts, nearly 200 people died and many more were injured. Chinese retribution was swift and severe, but Uyghur grievances, ranging from religious strictures to work discrimination, were brought to world attention and damaged China's international reputation.

Valuable Asset

Xinjiang came under continuous Chinese control in the eighteenth century during the Qing (Manchu) Dynasty, and became a province of China in 1884 after the Qing armies gained a measure of control over Muslim warlords who were, however, allowed to continue governing their domains. The communist regime established direct rule in 1949, and Xinjiang gained its official name and status of Autonomous Region in 1955. As Figure 9-17 shows, two huge desert basins rimmed by three snowcapped mountain ranges create spectacular scenery dominated by vast plains and steep slopes, with mountain streams disappearing beneath sediments to reappear lower down to feed wells and *qanats*. Traditionally, the north (the Junggar Basin) has been more agricultural and the south (the Tarim Basin) more pastoral, but what makes Xinjiang far more important to China is the discovery of the major oil reserves shown on the map. And Xinjiang's remoteness allowed China to experiment with space technology, rocketry, and nuclear weaponry far from densely populated areas and shielded from foreign spies. When China makes—as it will—its move for the moon, its space vehicles will take off from Xinjiang.

But Xinjiang's relative location may be its most valuable asset of all. As a part of China, Xinjiang's westernmost borders lie separated by only one country—Kazakhstan—from the major energy reserves of the Caspian Sea Basin. As Figures 9-16 and 9-17 show, pipelines are already being laid across Kazakhstan into Xinjiang. As economically burgeoning China's thirst for energy grows, the security of such land links will be of incalculable value.

CHINA GLOBAL AND LOCAL

Before leaving China to study the other regions of East Asia, we focus on a phenomenon of global importance: the implications of the economic transformation that is

FIGURE 9-17   © H. J. de Blij, P. O. Muller, and John Wiley & Sons, Inc.

in the process of creating a new regional geography in the old China—and may be cutting China in half.

No one working or studying in China, or seeing it as a visitor, can fail to be impressed, even overwhelmed, by what is happening here. Even the Chinese communist leaders who in the 1970s decided to embark on their open-market policies could not have imagined that opening China to the world would create, by the turn of the millennium, a cadre of bur-

geoning cities with skylines to challenge Manhattan, industrial complexes to supply the world, superhighways crowded with cars, state-of-the-art bridges and tunnels, ultramodern airports, and every conceivable icon of globalization from stock markets to maglev and bullet trains. The geographic term *Pacific Rim* has become a synonym for unprecedented modernization from Dalian on the doorstep of the Northeast to Zhuhai in China's tropical south.

Consider this: the main terminal building of Beijing's airport is 3 kilometers (1.8 mi) long and has floor space larger than all the terminals of London's Heathrow Airport combined. The Hangzhou Bay Bridge near Shanghai is the world's longest sea-crossing bridge, 36 kilometers (22 mi) in length with six lanes of traffic. The world's highest-altitude railroad links China and Tibet. One of the world's largest hydroelectric projects harnesses the waters of the great Chang River. In 1975, China had no expressways; by 2020, China will have 70,000 kilometers (44,000 mi) of limited-access, four-lane superhighways. Between 2005 and 2020, China plans to build and open an additional 100 airports.

Economic development combining this pace and these dimensions is unprecedented in the world. In China's case, it generated what may be the largest population redistribution from the countryside to the cities in human history. But behind this great success story lie troubling byproducts. At the local level, the construction boom leveled villages, pushed farmers off their land, and destroyed traditional urban neighborhoods. At the national level, it created severe regional disparities. In terms of income, China's Gini coefficient (see box, page 27) rose markedly as the gap between per-capita annual incomes between rich coastal cities and poor interior villages grew. Before the economic downturn of 2008, China's national GNI approached U.S. $7000, but city dwellers on average were earning more than ten times what villagers were.

Two consequences for China are immediately obvious. First, in recent years the people in the inland villages, and even those in still-rural areas of the Pacific Rim region, have begun to give voice (and sometimes action) to their grievances. These range from corrupt communist mayors extorting money for favors to government-sanctioned land expropriation enabling well-connected businesspeople to build factories on villagers' fields (see the photo on page 501). Stream and air pollution by such factories, road building, and mining without local consent, plus restrictions on personal liberties, including the right to relocate, generate tens of thousands of protests annually, and some of these turn violent. According to police officials, there were over 80,000 such incidents in China in 2008, significantly more than in the previous year.

Second, the core-periphery contrast that is emerging in China between its globalizing Pacific Rim coreland and the lagging interior is taking on dangerous proportions clearly reflected in Figure 9-18. The "haves" in the coastal provinces and the "have-nots" in the interior, mapped here in terms of the contributions by their provinces to the national economy, seem to be walled off from each other. In the absence of representative government and with insufficient protection for the dis-

advantaged in China's forward rush, this spells serious peril for the future of the state.

Geography of Development

So where does China fit in the "development" indices of transnational banks and agencies? Is China a developed country by World Bank standards? Does it rank among the richer states, given its huge manufacturing output and available funds for building all this infrastructure?

When geographers approach the development process, they use many yardsticks to measure a country's achievements, including GNI per capita; percentage of workers in farming, manufacturing, or other kinds of employment; amount of energy consumed; efficiency of transport and communications; use of manufactured metals like aluminum and steel in the economy; productivity of the labor force; and social measures such as literacy, nutrition, medical services, and savings rates. All these data are calculated per person, yielding an overall measure that ranks the country among all those providing statistics. These data do not, however, tell us *why* some countries, or parts of countries, exhibit the level of development they do.

The **geography of development** leads us into the study of raw-material distributions, environmental conditions, cultural traditions, historic factors (such as the lingering effects of colonialism), and forces of location. In the emergence of China's Pacific Rim, relative location played a major role, as we will see.

Before the current wave of globalization began to affect development processes virtually everywhere on Earth, economists and economic geographers tried to identify the factors that persuaded some societies (but not others) to follow a particular path from traditional subsistence to mass production and consumption. One of these scholars, Walt Rostow, traced this sequence in a model that focused on cultural flexibility, progressive leadership, modernization, "takeoff," maturity (reflected by product specialization and foreign trade), and "mass consumption" (when most workers would be employed in service and related industries).

This model was especially relevant in the immediate postcolonial period, but it took little account of core-periphery contrasts and relationships within individual countries or regions, and it could not anticipate the impact of globalization processes that propelled certain parts of countries rapidly ahead while leaving other parts barely affected. China's case is especially relevant here, because its leadership was reformist, but not progressive (representative government is not seen as essential to development; some Chinese like to say that "democracy constrains dynamism"). Chinese leaders deliberately favored the coastal provinces and cities when they opened their country to market forces,

CHINA: GDP BY PROVINCE, 2003, IN BILLION RMB

- Over 20,000
- 10,000 to 20,000
- 7,500 to 10,000
- 5,000 to 7,500
- Under 5,000

Source: *China Statistical Yearbook 2003*

FIGURE 9-18

because that is where China had a history of contact with the outside world, and from those coastal provinces millions of citizens had departed for Southeast Asian (and other) countries. Residual links of family and friendship caused many of these **Overseas Chinese** to invest in their ancestral homeland, thus playing a crucial role in the Pacific Rim "miracle." Many of the growing core area's earliest factories were built with repatriated money.

Development in the current era, therefore, has taken on a new meaning. Contrary to Rostow's multi-stage model of adaptations and consensus, development under contemporary globalization has become a winner-take-all form of opportunism, often with little restraint and with much exploitation of those who find themselves in the path of progress. Scholars studying this phenomenon use the adjective *neoliberal* to describe both its "new" and "unrestricted" features: new in the globalizing world,

liberal in the absence or paucity of regulations. When traditional neighborhoods (called *hutongs*) in Beijing are swept away by bulldozers to make way for high-rise apartment buildings, and when farmers in Shaanxi Province are summarily ousted from their lands because some corporation with connections to the regime wants to build a factory, this is neoliberalism at work.

China's Evolving Economic Geography

The story of China's ongoing economic transformation will be told for centuries to come, but we are far from certain as to how it will turn out. Already, we are witnessing the metamorphosis of the world's most populous nation. Despite the dislocations of collectivization, communization, the Cultural Revolution, power struggles, policy reversals, and civil conflicts, China as a communist state in the aftermath of Mao managed to stave off famine and hunger, control its population growth rate, strengthen its military capacity, improve its infrastructure, and enhance its world position. It took over Hong Kong from the British and Macau from the Portuguese, closing the chapter on colonialism. When economic turbulence struck its Pacific Rim and Southeast Asian neighbors, China withstood the onslaught. Its trade with neighbors near and far is growing. Foreign investment in China has mushroomed. And still its government espouses communist dogma.

How could the government manipulate the coexistence of communist politics and market economics? That was the key question confronting Deng Xiaoping and his comrades when they took power in 1979. In terms of ideology, this objective would seem unattainable; an economic "open-door" policy would surely lead to rising pressures for political democracy.

Economic Zones of the Pacific Rim

But Deng thought otherwise. If China's economic experiments could be spatially separated from the bulk of the country, the political impact would be kept at bay. At the outset, as we noted, the new economic policies would apply mainly to China's bridgehead on the Pacific Rim, leaving most of the vast country comparatively unaffected. Accordingly, the government introduced a complicated system of **11** **Special Economic Zones (SEZs)**, so-called *Open Cities*, and *Open Coastal Areas*, which would attract technologies and investments from abroad and transform the economic geography of eastern China (Fig. 9-19).

In these economic zones, investors are offered many incentives. Taxes are low. Import and export regulations are eased. Land leases are simplified. The hiring of labor under contract is allowed. Products made in the economic zones may be sold on foreign markets and, under some restrictions, in China as well. Even Taiwanese enterprises may operate here. And profits made may be sent back to the investors' home countries.

When Deng's government made the decisions that would reorient China's economic geography, location was a prime consideration. Beijing wanted China to participate in the global market economy, but it also wanted to cause as little impact on interior China as possible—at least in the first stages. The obvious answer was to position the Special Economic Zones along the coast. Initially, four SEZs were established in 1980, all with particular locational properties (Fig. 9-19):

1. *Shenzhen*, adjacent to then-booming British Hong Kong on the Pearl River Estuary in Guangdong Province
2. *Zhuhai*, across from then still-Portuguese Macau, also on the Pearl River Estuary in Guangdong Province
3. *Shantou*, opposite southern Taiwan, a colonial treaty port, also in Guangdong Province, source of many Chinese now living in Thailand
4. *Xiamen*, on the Taiwan Strait, also a colonial treaty port (then known as Amoy in the local dialect), in Fujian Province, source of many Chinese now based in Singapore, Indonesia, and Malaysia

In 1988 and 1990, respectively, two additional SEZs were proclaimed:

5. *Hainan Island*, declared an SEZ in its entirety, its potential success linked to its location near Southeast Asia
6. *Pudong*, across the river from Shanghai, China's largest city, different from other SEZs because it was a giant state-financed project designed to attract large multinational companies.

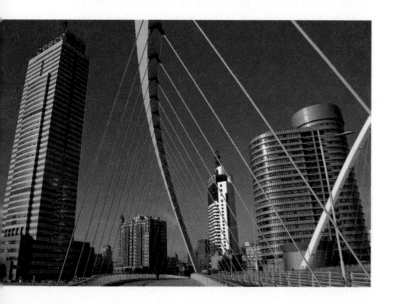

The success of SEZs along China's Pacific Rim is leading to still another gigantic project of this kind: the Binhai New Area (BNA) in Tianjin, the port city 120 kilometers (75 mi) southeast of the capital of Beijing. The government hopes to repeat the successes of Shenzhen and Pudong here, and this 10-square-kilometer (3.9-sq-mi) district will, according to official proclamations, become another international free trade zone that will focus on "international distribution, global procurement, and export processing." As usual, the Chinese are not wasting time: BNA started operating at the end of 2009, just three years after the first announcements describing the plan were issued. Its great advantage will be its proximity to Beijing and environs; and make no mistake, BNA will have a competitive impact on Pudong and Shenzhen as well as smaller SEZs along the coast. This photo shows some of the construction completed by mid-2009.
© Alamy.

FIGURE 9-19

© H. J. de Blij, P. O. Muller, and John Wiley & Sons, Inc.

And the process continues, with the newest SEZ authorized in 2006:

7. *Binhai New Area*, the coastal zone of the northern port city of Tianjin, a long-established Open City with considerable foreign investment, now elevated to SEZ status, projected to outperform even Shanghai–Pudong and Shenzhen itself (see photo at left).

The grand design of China's economic planners, therefore, was to stimulate economic growth in the coastal provinces and to capitalize (as it were) on the exchange opportunities created by location; the availability of funding; the proximity of foreign investors in Southeast Asia, Taiwan, Japan, and, importantly, still-British Hong Kong; the presence of cheap labor; and the promise of world markets eager for low-cost Chinese products.

No SEZ symbolizes the success of this plan more dramatically than a place called *Shenzhen* right next door to Hong Kong. In the late 1970s, Shenzhen was just a sleepy fishing and duck-farming village of about 20,000.

By 2010, it centered a booming metropolis of more than 8 million—the fastest growth of any urban settlement in the history of the world. Hundreds of skyscrapers, thousands of factories, and burgeoning infrastructure made Shenzhen the prototype of what an SEZ should be (see photo, p. 498).

Some of the other SEZs were not far behind, and non-SEZ cities and districts soon shared in the economic boom. Shenzhen, as Figure 9-20 shows, has its port on the Pearl River Estuary, locally called the *Pearl River Delta (PRD)* because virtually the entire estuary is flanked by a triangular juggernaut of urban-industrial complexes, the core area of Guangdong Province. At the north end of this *Pearl River Megalopolis* lies Guangzhou, the provincial capital, and across from Shenzhen and Hong Kong lies the Zhuhai SEZ adjacent to the Macau SAR. The dimensions of development here are almost unimaginable: this Megalopolis includes three cities of more than 5 million and a dozen cities of 1 million or more. Farther up China's east coast, across the

FIGURE 9-20   © H. J. de Blij, P. O. Muller, and John Wiley & Sons, Inc.

Huangpu River from old Shanghai, the *Pudong* District took off like an economic rocket; and Tianjin's *Binhai* New Area is only the latest in the Pacific Rim's chain of economic giants.

Hong Kong's Role

The process we have been discussing started with the economic success of Hong Kong while it was still a British dependency, empowered by a unique set of geographic circumstances. Here, on the left bank of the Pearl River Estuary, more than 7 million people (97 percent of Chinese descent) crowded onto 1000 square kilometers (400 sq mi) of fragmented, hilly territory under the tropical sun and created an economy larger than that of a hundred countries.

All of us who witnessed the incredible conversion of a fishing village with duck ponds and paddies into a metropolis of millions in a few years of frenzied economic and demographic growth wondered, inevitably, what kind of city Shenzhen would be. After all, shouldn't cities grow slowly, have histories etched in their architecture, develop traditional neighborhoods, and negotiate by consensus the pitfalls of planning and zoning? But look at Shenzhen only 30 years later, and it seems far more orderly, established, and accommodating than could have been imagined. Certainly Shenzhen has its richer and poorer neighborhoods—not everyone can afford to live in the CBD shown here. And when China was buffeted by the global economic downturn, Shenzhen was among SEZs where factories closed and workers had to leave. But this near-megacity is here to stay, an icon of the new China. © Photolibrary Group Limited.

Hong Kong (the name means "Fragrant Harbor") was ceded by the Chinese to Britain during the colonial period in three parts: the islands in 1841, the Kowloon Peninsula in 1860, and the New Territories (a lease) in 1898 (Fig. 9-21). Until the 1950s, Hong Kong was just another trading colony, but then the Korean War and the United Nations embargo on Communist China cut Hong Kong off from its hinterland. With a huge labor force and no domestic resources, the colony needed an alternative—and it came in the form of a textile industry based on imported raw materials. Soon Hong Kong textiles were selling on worldwide markets; capital was invested in light industries; and the industrial base diversified into electrical equipment, appliances, and countless other consumer goods. Meanwhile, Hong Kong quietly served as a back door to China, which rewarded the colony by providing fresh water and staple foods. The colony had become an **economic tiger**.

Hong Kong also became one of the world's leading financial centers. When China changed course and embarked on its "Open Door" policy in the 1980s, Hong Kong was ready. Hong Kong factory owners moved their plants to Shenzhen, where wages were even lower; Hong Kong banks financed Shenzhen industries; and Hong Kong's economy adjusted once again to a new era. Hong Kong's port, however, continued to transfer Shenzhen's products. During the 1990s, about 25 percent of all of China's trade passed through Hong Kong harbor.

When the British lease on Hong Kong expired in 1997 and the British agreed to yield authority to China, the Chinese promised to allow Hong Kong's freewheeling economy to continue, in effect creating "one country, two systems." At midnight on June 30, Hong Kong became the Xianggang Special Administrative Region (SAR). True to the letter (if not always the spirit) of this commitment, the name "Hong Kong" has continued in general use. And on December 20, 1999 China took control over the last remaining European colony—nearby Portuguese *Macau*—and made it a SAR as well, guaranteeing (as in Hong Kong) the continuation of its existing social and economic system for 50 years.

Coast and Interior

We noted earlier that the concentration of industrial and commercial expansion in the coastal provinces intensifies the regional disparities that have always marked this vast country. China's planners are now addressing this potentially dangerous problem by channeling economic development inland, thereby to reverse the growth of the gap between coast and interior. The key initiative involves the Yangzi-Chang corridor and includes Three Gorges Dam, the city of Chongqing, the Sichuan Basin, and Yunnan

FROM THE FIELD NOTES

"Shanghai in April 2009 showed no sign of any economic slowdown and gave much evidence of the environmental consequences. From my vantage point on the top floor of the Jin Mao Tower in Pudong District, buildings beyond the Huangpu River disappeared in a dense smog on a hot spring day that locals described to me as "pretty clear." China's prodigious economic growth has made the People's Republic the world's largest polluter of the atmosphere, surpassing the United States. Some projections suggest that, by 2020, China may be emitting twice as much pollution as America, but since China has more than four times the population, the U.S. would still lead in terms of pollutants per capita. That's just one reason the United States shouldn't invoke the 'China excuse' to weaken its own efforts to reduce pollution from motor vehicles, factories, and other sources. 'Enjoy this view of the Oriental Pearl Television Tower,' Li Sheng told me. 'We're building the Shanghai Tower right down there where you see the construction site, and it'll be in the way. And it'll be half as tall again as your Empire State Building.' Well, let's hope the view from the top will be clearer." © H. J. de Blij.

www.conceptcaching.com

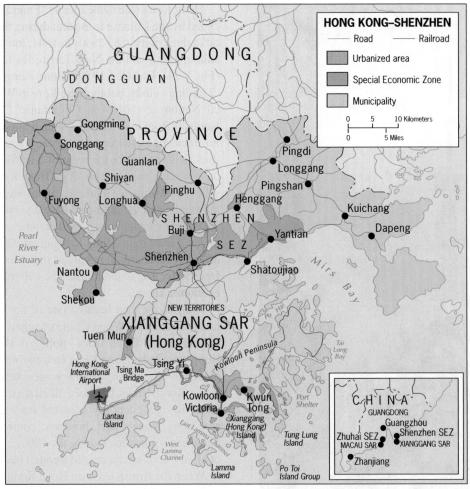

FIGURE 9-21   © H. J. de Blij, P. O. Muller, and John Wiley & Sons, Inc.

Province, where SEZ-type privileges are being extended to corporations as they are along the coast. As Figures 9-9 and 9-19 show, however, this project involves mainly central and southern China, but in many ways the greatest disparities lie in the interior of the north, in Shanxi, Shaanxi, and Gansu provinces. Channeling development westward in the north will prove to be far more difficult. Nonetheless, the need to improve the living conditions of farmers and others in the interior provinces is rising. China must also restrain the influx of migrants from countryside to city, but this makes things worse—and can only be a temporary remedy. Closing the prosperity gap will be the lasting solution.

Trouble Ahead?

Reducing the economic gap between China's Pacific Rim and the interior (starkly evident in Fig. 9-18) obviously is easier when the economy is booming than during a slowdown. China over the past 35 years has become a vital component of the global economy, the world's workshop, pouring out a stream of low-priced exports to markets across the planet. But the economic crisis that began in 2008 confronted China with the unthinkable: decelerating growth, factories closing, rising unemployment, protesting workers, plunging stock markets. How would China pursue its domestic goals, especially a reduction in its regional disparities, with a dragging economy?

To be sure, the Chinese (unlike their American trading partners) are ardent savers, and in 2009 China had foreign reserves amounting to nearly U.S. $2 trillion resulting from its huge trade surpluses and heavy foreign investment. Chinese planners can thus afford to pour money into additional infrastructure projects, but what China really needs is to create a domestic market for products no longer selling overseas. And that requires closing the still-widening gap between urban prosperity and rural poverty. A 2008 report showed that more than 500 million people in China subsist on U.S. $2 per day or less, while the government dithered over plans to give farmers more rights concerning their land. In a country

Here is a dramatic image of the "urban front," as geographers refer to the encroachment of cities and their suburbs on farmland nearby. This farm family rests outside their home near Lishui, about 130 kilometers (80 mi) from the coastal city of Wenzhou in Zhejiang Province, as truckloads of earth are dumped on their crops while builders convert their land into a construction site. Farmers are not always compensated fairly for the land they lose, but they have little power to slow the relentless advance of China's urbanization. Sometimes they remain on the village land until the day their homes are bulldozed away. © NG Image Collection.

with a U.S. $2 trillion reserve, 40 percent of all farm villages have no access to running water. Americans are used to Social Security and Medicare; Chinese peasants have nothing like it and save what they can for their needs in old age. That puts consumer goods way down on their list of priorities.

Grandiose infrastructure projects like dams, river diversions, bullet trains, and world-class bridges tend to be more satisfying to authoritarian regimes than piping water and laying electrical lines to rural settlements. But in China, property rights still mean the housing market in Beijing, not land in rural Shaanxi. In 2008, the communist regime with much fanfare unveiled a land reform plan that would address this dichotomy, but read the fine print and you will see that more flexible state-controlled leases, not outright ownership, is the key "reform." It seems to give farmers no better protection against land grabs by local officials (see photo above), who still translate "collective ownership" into "our prerogative," than was the case before. The successors to the authorities who were so pragmatic in their Pacific Rim projects in the 1970s are proving far more timid when it comes to the rural interior. In the end, China may pay a high price for their reluctance.

CHINA: GLOBAL SUPERPOWER?

China in the early twenty-first century appears likely to become more than an economic force of world proportions: China also appears on course to achieve global superpower stature. As long as it retains its autocratic form of government (which allowed it to impose dra-

conian population policies and comprehensive economic experiments without having to consult the electorate), China will be able to practice the kind of state capitalism that—in another guise—made South Korea an economic power (see page 524).

Unlike Japan or South Korea, however, China has no constraints on its military power. Its People's Liberation Army served as security forces during the pro-democracy turbulence of 1989. Today, China's military is the largest standing army in the world, with about 3 million soldiers and some 1.2 million reserves. The government has made a major investment in its military by raising its armed forces budget significantly since 2000. Already, China is a nuclear power and has a growing arsenal of medium-range and intercontinental ballistic missiles. It is also rapidly expanding its fleet of submarines, frigates, and destroyers, as well as its air force (currently numbering more than 2300 fighters and bombers). And in 2009, military officials were reportedly seeking to acquire China's first aircraft carrier.

During the twentieth century, the United States and the Soviet Union were locked in a 45-year Cold War that repeatedly risked nuclear conflict. That fatal exchange never happened, in part because it was a struggle between superpowers who understood each other comparatively well. While the politicians and military strategists were plotting, the cultural doors never closed: American audiences listened to Prokofiev and Shostakovich, watched Russian ballet, and read Tolstoy and Pasternak even as the Soviets applauded Van Cliburn, read Hemingway, and lionized American dissidents. In short, it was an *intracultural* Cold War, which reduced the threat of mutual destruction.

The twenty-first century may witness a far more dangerous geopolitical struggle in which the adversaries may well be the United States and China. U.S. power and influence still prevail in the western Pacific, but it is easy to discern areas where Chinese and American interests will diverge (Taiwan is only one example). American bases in Japan, thousands of American troops in South Korea, and American warships in the East and South China seas are potential grounds for dispute. All this might generate the world's first *intercultural* Cold War, in which the risk of fatal misunderstanding is incalculably greater than it was during the last.

How can such a Cold War be averted? Trade, scientific and educational links, and cultural exchanges are obvious remedies: the stronger our interconnections, the less likely is a deepening conflict. We Americans should learn as much about China as we can, to understand it better, to appreciate its cultural characteristics, and to recognize the historical-geographical factors underlying China's views of the West.

For more than 40 centuries, China has known authoritarianism of both the brutal and the benevolent kind, has been fractured by regionalism only to unify again, and has depended on communalism to survive environmental and despotic depredations. For nearly 25 centuries Confucianism has guided it. Time and again, China has been opened to, and ruled by, foreigners, and time and again it has retreated into isolationism when things went wrong. Such a reversal may no longer be possible, given what we have seen in this chapter.

But now a renewed force is rising in modernizing China: nationalism. A combination of factors drives this process, ranging from broad issues such as Taiwan to incidents like the mid-air collision in 2001 between what the Chinese regarded as a U.S. spy plane and one of China's jet fighters near the island of Hainan. One of the broad issues has geographic dimensions: China disputes Japanese ownership of islands in the East China Sea (and the surrounding waters beneath which oil reserves may be found). China's attitude toward Japan is further inflamed by Japanese refusals to acknowledge atrocities committed in China during World War II and by a recent Japanese prime minister's repeated visits to a war memorial containing the remains of Japanese war criminals. In 2005, angry anti-Japanese demonstrations throughout China followed the appearance of a Japanese history textbook containing a whitewashed version of Japanese wartime actions. Meanwhile, American scorn for Chinese human-rights failures inflames nationalist sentiments still more.

China today is on the world stage, globally engaged and internationally involved. Constraining the forces that tend to lead to competition and conflict between China and the United States must be a priority in the decades ahead, whether the issue is Taiwan, North Korea, or Japan. At stake are regional stability and world peace.

● MONGOLIA

We have treated China as a region, recognizing the existence of its internal subregions and emphasizing the emergence of its dual coast-interior regionalism as a challenge for the state's future. As noted, Xizang (Tibet) and Xinjiang already form part of China's East Asian sphere, but another look at Figure 9-5 reveals the presence of still another political entity in East Asia's deep interior: the state of Mongolia.

Many Chinese feel that Mongolia has no more right to its independence than either Tibet or Xinjiang, and that this vast land and its sparse population (only 2.8 million) should be ruled by China. From the late 1600s until the revolutionary days of 1911, Mongolia was indeed part of the Chinese Empire. The Soviets helped the Mongolians hold off Chinese attempts to regain control, and in the 1920s Mongolia became a communist People's Republic on the Soviet model. When the Soviet Union collapsed in 1991, free elections ushered in a new era of Mongolian independence.

Despite its historical associations, ethnic affinities, and cultural involvement with China, Mongolia's core area, including the capital of Ulaanbaatar, adjoins Russia's Eastern Frontier. The country's 800,000 herders and their millions of sheep follow nomadic tracks along the margins of the vast Gobi Desert. Periodic environmental fluctuations cause terrible hardships in this barren country.

China respects Mongolia's independence, but in recent years China's relationship with its landlocked neighbor has strengthened. Copper, iron, zinc, and coal reserves attract Chinese companies; road, bridge, and railroad building are funded by China. Nearly 200 Chinese businesses are active in Mongolia today, and public opinion polls reflect Mongolians' concern over China's growing influence. Mongolia is a weak country in a vulnerable part of inner Asia, a **buffer state** wedged between populous former **13** and future superpowers. Will it escape the fate of other Asian buffers?

● TAIWAN

Mongolia is not the only political entity on China's periphery to feel the pressures of history. Mongolia was part of China's sphere of influence but fell to Soviet influence; the island of Taiwan was part of China but fell to Japan's colonial expansion. Indeed, Taiwan's fate riles many Chinese far more than Mongolia's independence.

Taiwan's inhabitants were able to forge an economically thriving, politically democratic society well before China's momentous transformation, in which Taiwanese capital investment and technological know-how played a major role. Taiwan is regarded by Beijing's communist regime as a Chinese province temporarily "wayward"—that is, disobedient. After the late-1990s reunification with China of British Hong Kong and of Portuguese Macau as SARs, Taiwan remained the only major unresolved territorial issue in China's postcolonial, postwar sphere.

As noted earlier in the chapter, Taiwan is part of a three-unit East Asian region—that also includes Japan, by far the largest and most influential component, and South Korea—which we designate as the *Jakota Triangle*. It is a region in flux: Japan, South Korea, and Taiwan, as we will see, are each confronting their own specific challenges. Before we return to this regional designation on page 525, let us examine the geographies of each component. What makes them similar? How are they different?

When we see where Taiwan is located, it would seem to be inextricably bound up with China—it lies a mere 200 kilometers (125 mi) from the mainland. Taiwan has a history of immigration from coastal Chinese provinces and lies more than 1000 kilometers (625 mi) from the main islands of Japan (Fig. 9-1). But take note of something else this map shows: an arc of islands called the Ryukyus that comes as close to Taiwan on the east as China does on the west—and they are owned by Japan. Beginning in the late nineteenth century, Japan was the formidable colonial power of the Pacific Rim and conquered not only Taiwan but also Korea. The Japanese put a lasting imprint on their Pacific colonies, and after World War II, while China was in its weakened condition, invested heavily in Taiwan's economic development. But the People's Republic never softened its assertion of authority over Taiwan. It was (and remains) high-stakes competition.

Embattled Outpost

Beijing's claim to Taiwan has strengths as well as weaknesses. Han people settled on Taiwan, but the island had been settled thousands of years earlier by Malay-Polynesian groups who had split into mountain (east) and plains (west) inhabitants. When the Chinese emigrated from famine-stricken Fujian Province to Taiwan in the seventeenth century, they soon displaced and assimilated the plains people, but the indigenous highlanders held out until the Japanese colonial invasion. Formal Chinese control began in 1683 when the expansionist Qing (Manchu) Dynasty made Taiwan part of Fujian Province.

Rice and sugar became important exports to the mainland, and by the 1840s Taiwan's population was an estimated 2.5 million. In 1886 the island was declared a province of China, but by then the British had established two treaty ports there, the French had blockaded it, and the Japanese had sent expeditions to it. Nine years later, in 1895, China ceded Taiwan to Japan in the treaty settlement of the Sino-Japanese War, and it became a Japanese colony.

When China's aging Qing Dynasty was overthrown following the rebellion of 1911, a Nationalist government in 1912 proclaimed the Republic of China (ROC), led by Sun Yat-sen. Naturally, this government wanted to oust all colonialists, not only Europeans but also Japanese, and not just from the mainland but from Taiwan as well. On Taiwan, however, the Japanese held on, using the island as a source of food and raw materials as well as a market for Japanese products. To ensure all this, the Japanese launched a prodigious development program involving road and railroad construction, irrigation projects, hydroelectric schemes, mines (mainly for coal), and factories. Farmlands were expanded and farming methods improved. While mainland China was engulfed in conflict, Taiwan remained under Japanese control. Not until 1945, when Japan was defeated in the Second World War, did Taiwan become an administrative part of China again.

Republic of China?

But that post–World War II reattachment would be a brief one. As noted earlier, the Nationalists, under the leadership of Chiang Kai-shek, now faced the rising tide of communism on the mainland, and the ROC government fought a losing war. In 1949, the flag of the Republic of China was taken to a final stronghold: offshore Taiwan. There, based in the city of Taipei, Chiang and his followers, with their military might, weapons, and wealth taken from the mainland, established what they (and the world, led by the United States) proclaimed as the legitimate government of all China. On the island, the Nationalists, helped by the United States, began a massive reconstruction of the war-damaged infrastructure. In the international arena, the ROC represented the country; in the still-young United Nations, the Nationalists of Taiwan occupied China's seat. On the huge mainland, however, it was the communists in Beijing who ruled and called their country the People's Republic of China (PRC). Thus was born the two-Chinas dilemma, which still confronts the international community. When the ROC was ousted from the United Nations in 1971 to make way for the PRC, legitimacy was also transferred.

Globalization Impacts

All this might have remained an internal dispute but for two major developments: Taiwan's rise as a globally-connected economic tiger and domestically vibrant democracy on the Pacific Rim, and China's emergence on the world stage as an economic power and political force after decades of isolation. In Beijing, the communist rulers of the PRC sought ways to limit the ROC's international linkages. In Taipei, Taiwan's leaders debated the course of action they should take with regard to the PRC. Some wanted to declare independence. Others favored a slow but progressive reconciliation.

As we have seen repeatedly in other chapters, economic realities have a way of intruding into political schemes. As long as Taiwan's economy flourished, its leaders had the capacity to reward their supporters worldwide, keeping markets open, ensuring diplomatic ties, and allowing the "province" to operate much like a full-fledged state. But when economic times turn sour, as happened in 2008, politicians have to look for other options. Taiwanese voters showed much less appetite for confrontation with China on the issue of independence. President Ma Ying-jeou, who as a presidential candidate had infuriated Beijing by proposing a referendum on whether Taiwan should seek to reenter the United Nations under the ROC name, mended fences and began a campaign toward closer cooperation instead. By mid-2009, direct flights and shipping between island and mainland were operating, economic ties were strengthening, cultural exchanges were in progress, and two-way tourism was rapidly expanding. There seemed to be hope that one of the world's potential flashpoints was on its way toward resolution.

Taiwan SAR?

Certainly Taiwanese leaders and voters in 2009 were taking another look at the status and fortunes of the Hong Kong SAR. While the worldwide economic downturn had struck China as well as Taiwan (although more specialized Taiwan had the bigger problem), Hong Kong—still with its own currency, discrete economic plan, and "international" borders—was doing comparatively well. In 2009, in Hong Kong, you could attend the kind of public commemoration of the 1989 Tiananmen Square tragedy prohibited in the PRC. Hong Kong's democratic institutions might not match Taiwan's, but they certainly outperformed the PRC's sclerotic system. As a result, Taiwanese were openly debating the merits of reunification with the mainland under SAR-like terms, anticipating economic advantages inhibited by the island's uncertain

political future. In the spring of 2009, polls were suggesting that the electorate was about evenly split on the issue, with a substantial minority still undecided.

Thriving Economic Tiger

Taiwan, as Figure 9-22 shows, is not a large island. It is smaller than Switzerland but has a population much larger (23.1 million), most of it concentrated in a crescent lining the western and northern coasts. The Chungyang Mountains, an area of high elevations (some over 3000 meters [10,000 ft]), steep slopes, and dense forests, dominate the eastern half of the island. Westward, these mountains yield to a zone of hilly topography and, facing the Taiwan Strait, a substantial coastal plain. Streams from the mountains irrigate the paddyfields, and farm production has more than doubled since 1950 even though hundreds of thousands of farmers left the fields for work in Taiwan's expanding industries.

Today, the lowland urban-industrial corridor of western Taiwan is anchored by the capital, Taipei (Taibei), at the island's northern end and rapidly growing Kaohsiung (Gaoxiong) in the far south.* The Japanese developed Chilung (Jilong), Taipei's outport, to export nearby coal, but now the raw materials flow the other way. Taiwan imports raw cotton for its textile industry, bauxite (for aluminum) from Indonesia, oil from Brunei, and iron ore from Africa. Taiwan has a developing iron and steel industry, nuclear power plants, shipyards, a large chemical industry, and modern transport networks. Increasingly, however, it is exporting products of its high-technology industries: personal computers, telecommunications equipment, and precision electronic instruments. Taiwan has enormous brainpower, and many foreign firms join in the research and development carried on in such places as Hsinchu (Xinzhu) in the north, where the government has helped establish a technopole centered on the microelectronics and personal computer industries. In the south, the "science city" of Tainan specializes in microsystems and information technology.

The establishment of China's Special Economic Zones went a long way toward mitigating longstanding problems between Beijing and Taipei. China needed capital and Taiwan had it, and in the almost-anything-goes environment of the SEZs, Taiwanese entrepreneurs could buy or build factories just as "real" foreigners did. Importantly, as we noted earlier, the Xiamen SEZ was laid out directly across from Taiwan for this purpose (people

---

* The Taiwanese have retained the old Wade-Giles spelling of place names. China now uses the pinyin system (see box on page 459). Names in parentheses are written according to the pinyin system.

© H. J. de Blij, P. O. Muller, and John Wiley & Sons, Inc.

FIGURE 9-22

there will tell you that 80 percent of locals have relatives across the Taiwan Strait). As a result, thousands of Taiwanese-owned factories began operating in China, their owners and investors having a strong interest in avoiding violent conflict over political issues.

As one of the original East Asian economic tigers, Taiwan was the star of the region, recording higher per-capita annual incomes than many European countries, getting out of its postwar debt faster than anyone expected, and amassing trade surpluses that enabled investments far and wide (for instance, Taiwan became the largest foreign investor in the economy of Vietnam). But Taiwan proved to be as susceptible as other "tiger" economies to regional and global downturns. The U.S. and Japanese markets for high-tech exports, particularly computer chips, declined so rapidly in the late 1990s that thousands of firms long profitable in Taiwan had to

move their operations to lower-labor-cost mainland China. A decade later came the worldwide recession, which caused further shrinkage of an economy now facing the unfamiliar specters of rising unemployment and declining GDP.

One China?

If the prospects for a settlement of the long-term PRC-ROC quarrel have brightened in recent years, the obstacles to it remain daunting. The Nationalists who made the island their last stand in 1949 established a Republic of China that was authoritarian, but the ROC evolved into a democratic state; many Taiwanese proudly proclaim that the direct and popular election of 1996 was the first such vote not just in Taiwan or the ROC, but in the 4000-year history of the Chinese nation. So long as the PRC is not making progress in this arena and meddles in the legislature of Hong Kong, say these voters, Taiwan should resist reunification. But, if it wanted to, the PRC could overwhelm Taiwan militarily in short order; hundreds of missiles are poised on the mainland shore, ready to strike, in stark contrast to the factories Taiwan has built nearby. Obviously such an outbreak of hostilities would be catastrophic for both sides.

And so Taiwan's future hangs in the balance, a wayward province with higher aspirations, a quasi-state with global connections, a test of Chinese capacity for compromise and accommodation (see the Issue Box titled "Chinese Rule or Independence for Taiwan?"). In this era of devolutionary forces, the fate of Taiwan will be a harbinger of the world order beyond the first decade of the twenty-first century.

● JAPAN

More than a century before China embarked on its amazing quest to reverse its long-term isolation and economic stagnation, something even more spectacular happened in Japan. With a far smaller population, much less in the way of domestic resources, and a sequestered culture, Japan jumped onto the world stage with a bang. In 1868, a group of reform-minded modernizers seized power from an old guard, and by the end of the century Japan was a military and economic force. From the factories in and around Tokyo and from urban-industrial complexes elsewhere poured forth a stream of weapons and equipment the Japanese used to embark on colonial expansion. By the mid-1930s, Japan lay at

Chinese Rule or Independence for Taiwan?

SHOULD CHINA RULE TAIWAN?

"I am a graduate student in political geography here at East China Normal University, and I've been listening to your lecture about Chinese-American relations and the disagreement we have over Taiwan. You talk about this as though you have a right to stand between two parts of China, and with the power you have right now, I guess you can. You said that you hoped that mainland Chinese would become more 'reasonable' in their attitude toward the island province, but let me (as my professor would say) erect a hypothetical.

"Imagine that America wakes up one morning and discovers that a band of thugs have flown several 747s, loaded with weapons of all kinds, including those of mass destruction ready for use, into the San Juan, Puerto Rico airport. After taking the airport amid many casualties, they move into the city and enter administration buildings, threatening mass annihilation unless the local government resigns. By evening, Puerto Rico is under their control, American and Puerto Rican flags have been lowered and replaced by the banner of an 'independent' Puerto Rico.

"In Washington, your Congress meets and, realizing that the entire population of the island is at risk, makes the only move it can under the circumstances: it orders your armed forces to assemble a fleet to head for Puerto Rico, to blockade the island so as to create a basis for negotiation and a platform for possible military action. The fleet assembles near Miami and heads eastward through the channel between the Bahamas and Cuba.

"A day before reaching Puerto Rican waters, the path of the hastily assembled American fleet is blocked by a larger fleet from China, which has sailed from its base at Colón near the northern end of the Panama Canal. The commander of the American fleet communicates with his Chinese counterpart. 'We cannot let you pass,' states the Chinese admiral. 'We have not yet determined whether the invaders of Puerto Rico were invited by the people there to liberate them. We may have common ground with them. Please avoid an unnecessary conflict by going back to your bases.'

"How, sir, do you suppose the American people would feel toward China if a version of this scenario took place? They would be outraged at China's interference in a domestic matter. They would order your navy to attack. Well, we in China have had something like this happen with Taiwan. Thugs took it over. American aircraft carriers appeared in the Taiwan Strait. We've 'avoided conflict' for many decades, but that won't last forever. One way or another, Taiwan will submit to Beijing's rule. The Chinese people demand it."

TAIWAN SHOULD BE INDEPENDENT

"Instead of harping constantly on the 'reunification' issue, why don't Beijing's rulers get their own house in order first? Their political system is straight out of the discarded past, they're apparently allergic to democracy, and they can't seem to accept or accommodate any disobedience or dissent. I represent a Taiwanese high-tech corporation and I fly frequently between Taipei and Hong Kong and visit the so-called People's Republic via that route—and I can tell you, the Chinese could learn a lot about democracy, openness, freedom, transparency, and honesty from the Taiwanese. Politically, that 'One Country, Two Systems' promise in Hong Kong is eroding as we speak. Economically, the Chinese ought to acknowledge that a lot of the 'Chinese miracle' on the Pacific Rim is actually a 'Taiwanese miracle.' Without our money, skills, discipline, and rigor there wouldn't be much of a miracle, especially if you also subtract what Southeast Asia's Chinese have contributed.

"Of course, Taiwan is a great rallying point for Beijing's autocrats. Gives them a chance to rail against the Western world, against America, against social excesses (we even have freedom of religion in Taiwan, imagine that!), against international conspiracies. But the truth is that the voters of Taiwan long ago ousted the 'thugs' who took us over, forged a thriving economy and a vibrant democracy, and made this island an example of what might have been across the Taiwan Strait.

"Could I say a few words about the 'historic' justification for China's claim to Taiwan? First of all, this island was peopled by Southeast Asian groups long before any Chinese arrived here, so if it's a matter of who got here first, we're last on the list. Beijing didn't integrate Taiwan politically until the 1680s, after groups of hungry citizens started to cross the Strait. That's pretty late in China's vaunted 5000-year history. Taiwan didn't become a province until 1886, which lasted exactly nine years. From 1895 to 1945 Taiwan belonged to Japan as part of the war settlement, and then Beijing took it back only to lose it again four years later to the 'thugs' under Chiang Kai-shek. That makes 13 years as part of the mainland system and 103 years of something else.

"For a country that has expanded and contracted for millennia, has gained and lost territory from Tibet to Taiwan, you would think that China could live with an ethnically compatible, economically successful, politically stable, democratic and independent neighbor. Those stiff bureaucrats in Beijing should have better things to do than threaten force to 'reincorporate' their exemplary neighbor."

Regional ISSUE

Vote your opinion at www.wiley.com/college/deblij

the center of an empire that included all of the Korean Peninsula, the whole of China's Northeast (which the Japanese called Manchukuo), the Ryukyu Islands, and Taiwan, as well as the southern half of Russia's Sakhalin Island (called Karafuto). Not even a disastrous earthquake, which destroyed much of Tokyo in 1923 and killed 143,000 people, could slow the Japanese drive.

Colonial Wars and Recovery

World War II saw Japan expand its domain farther than the architects of the 1868 "modernization" could have anticipated. By early December 1941, Japan had conquered large parts of China Proper, all of French Indochina to the south, and most of the small islands in the western Pacific. Then, on December 7, 1941, Japanese-built aircraft carriers moved Tokyo's warplanes within striking range of Hawai'i, and the surprise attack on Pearl Harbor underscored Japan's confidence in its war machine. Soon the Japanese overran the Philippines, the (then) Netherlands East Indies, Thailand, and British Burma and Malaya, and drove a wide corridor through the heart of China to the border with Vietnam.

A few years later, Japan's expansionist era was over. Its armies had been driven from virtually all its possessions, and when American nuclear bombs devastated two Japanese cities in 1945, the country lay in ruins. But once again, Japan, aided this time by an enlightened U.S. postwar administration, surmounted disaster.

Japan's economic recovery and its rise to the status of world economic superpower was the success story of the second half of the twentieth century. Japan lost the war and its empire, but it scored many economic victories in a new global arena. Japan became an industrial giant, a technological pacesetter, a fully urbanized society, a political power, and an affluent nation. No city in the world today is without Japanese cars in its streets; few photography stores lack Japanese cameras and film; laboratories the world over use Japanese optical equipment. From microwave ovens to DVD players, from oceangoing ships to camcorders, Japanese goods flood the world's markets. Even today, while experiencing a severe downturn, the Japanese economy remains the second largest national economy in the world. As we will see later in this chapter, Japan faces several challenges, economic, political, and demographic. But even while in recession, the Japanese economy remains formidable.

Japan's brief colonial adventure helped lay the groundwork for other economic successes along the western Pacific Rim. The Japanese ruthlessly exploited Korean and Formosan (Taiwanese) natural and human resources, but they also installed a new economic order there. After World War II, this infrastructure facilitated an economic transition—and soon made both Taiwan and South Korea competitors on world markets.

FROM THE FIELD NOTES

"Visiting the Peace Memorial Park in Hiroshima is a difficult experience. Over this site on August 6, 1945 began the era of nuclear weapons use, and the horror arising from that moment in history, displayed searingly in the museum, is an object lesson in this time of nuclear proliferation and enhanced risk. In the museum is a model of the city immediately after the explosion (the red ball marks where the detonation occurred), showing the total annihilation of the entire area at an immediate loss of more than 80,000 people and the death from radiation of many more subsequently. In the park outside, the Atomic Bomb Memorial Dome, the only building to partially survive the blast, has become the symbol of Hiroshima's devastation and of the dread of nuclear war."
© H. J. de Blij.

Concept Caching

British Tutelage

In discussing the economic geography of Europe, we noted how the comparatively small island of Britain became the crucible of the Industrial Revolution, which gave it a huge head start and advantage over the continent, across which the Industrial Revolution spread decades later. Britain was not large, but its insular location, its local raw materials, the skills of its engineers, its large labor force, and its social organization combined to endow it not only with industrial strength but also with a vast overseas empire.

After the modernizers took control of Japan in 1868—an event known as the *Meiji Restoration* (the return of "enlightened rule" centered on the Emperor Meiji)—they turned to Britain for guidance in reforming their nation and its economy. In the decades that followed, the British advised the Japanese on the layout of cities and the construction of a railroad network, on the location of industrial plants, and on the organization of education. The British influence is still visible in the Japanese cultural landscape today: the Japanese, like the British, drive on the left side of the road. Consider how this affects the effort to open the Japanese market to U.S. automobiles!

The Japanese reformers of the late nineteenth century undoubtedly saw many geographic similarities between Britain and Japan. At that time, most of what mattered in Japan was concentrated on the country's largest island, Honshu (literally, "mainland"). The ancient capital, Kyoto, lay in the interior, but the modernizers wanted a coastal, outward-looking headquarters. So they chose the city of Edo, on a large bay where Honshu's eastern coastline turns sharply (Fig. 9-23). They renamed the place *Tokyo* ("eastern capital"), and little more than a century later it was the largest urban agglomeration in the world. Honshu's coasts were near mainland Asia, where raw materials and potential markets for Japanese products lay. The notion of a greater Japanese empire followed naturally from the British example.

Spatial Contrasts

But in other ways, the British and Japanese archipelagoes, at opposite ends of the Eurasian landmass, differed considerably. In total area, Japan is larger than Britain. In addition to Honshu, Japan has three other large islands—Hokkaido to the north and Shikoku and Kyushu to the south (Fig. 9-23)—as well as numerous small islands and islets, for a total land area of about 377,000 square kilometers (146,000 sq mi). Much of this area is mountainous and steep-sloped, geologically young, earthquake-prone, and studded with volcanoes. Britain has lower relief, is older geologically, does not suffer from severe earthquakes, and has no active volcanoes. And in terms of the raw materials for industry, Britain was much better endowed than Japan. Self-sufficiency in iron ore and high-quality coal gave Britain the head start that lasted a century.

Japan's Limitations

Japan's high-relief topography has been an ever-present challenge. All of Japan's major cities, except the ancient capital of Kyoto, are perched along the coast, and virtually all lie partly on artificial land claimed from the sea.

FROM THE FIELD NOTES

"The city of Kyoto, chronologically Japan's second capital (after Nara; before Tokyo), is the country's principal center of culture and religion, education and the arts. Tree-lined streets lead past hundreds of Buddhist temples; tranquil gardens provide solace from the bustle of the city. I rode the bullet train from Tokyo and spent my first day following a walking route recommended by a colleague, but got only part of the way because I felt compelled to enter so many of the temple grounds and gardens. And not only Buddhism, but also Shinto makes its mark on the cultural landscape. I passed under a *torii*—a gateway usually formed by two wooden posts topped by two horizontal beams turned up at their ends—which signals that you have left the secular and entered the sacred, and found this beautiful Shinto shrine with its orange trim and olive-green glazed tiles." © H. J. de Blij.

www.conceptcaching.com

Concept Caching

FIGURE 9-23

© H. J. de Blij, P. O. Muller, and John Wiley & Sons, Inc.

Sail into Kobe harbor, and you will pass artificial islands designed for high-volume shipping and connected to the mainland by automatic space-age trains. Enter Tokyo Bay, and the refineries and factories to your east and west stand on huge expanses of landfill that have pushed the bay's shoreline outward. With 127.7 million people, the vast majority (79 percent) living in towns and cities, Japan uses its habitable living space intensively—and expands it wherever possible.

As Figure 9-24 shows, farmland in Japan is both limited and regionally fragmented. Urban sprawl has invaded much cultivable land. In the hinterland of Tokyo lies the Kanto Plain; around Nagoya, the Nobi Plain; and surrounding Osaka, the Kansai District—each a major agricultural zone under relentless urban pressure. All three of these plains lie within Japan's fragmented but well-defined core area (delimited by the red line on the map), the heart of Japan's prodigious manufacturing complex.

FIGURE 9-24

© H. J. de Blij, P. O. Muller, and John Wiley & Sons, Inc.

From Realm to Region

For nearly a half-century, Japan's seemingly insurmountable economic lead, its almost unique (for so large a population) ethnic homogeneity, its insular situation, and its cultural uniformity justified its designation as a discrete geographic realm, not a region of a greater East Asia. But rapid change on the rim of the western Pacific—both within Japan and beyond it—has eroded the bases for this distinction. In the mid-1990s, Japan's economic engine faltered while those of its Pacific Rim neighbors accelerated; its social fabric was torn by unprecedented acts of violence; and its foreign relations were soured by disputes over wartime atrocities in China and Korea. Japan stood alone in the vanguard of Pacific Rim developments, but it is no longer discrete.

Early Directions

On the map, the Korean Peninsula seems to extend like an unfinished bridge from the Asian mainland toward southern Japan. Two Japanese-owned islands, Tsushima and Iki, in the Korea Strait almost complete the connection (Fig. 9-23). In fact, Korea was linked to Japan in prehistoric times, when sea levels were lower. Peoples and cultural infusions reached Japan from Asia time and again, even after rising waters inundated the land bridge between them.

The Japanese islands were inhabited even before the ancestors of the modern Japanese arrived from Asia. As we noted earlier, the Ainu, a people of Caucasian ancestry, had established themselves on all four of the major islands thousands of years before. But the Ainu could not hold their ground against the better organized and armed invaders, and they were steadily driven northward until they retained a mere foothold on Hokkaido. Today, the last vestiges of Ainu ancestry and culture are disappearing.

By the sixteenth century, Japan had evolved a highly individualistic culture, characterized by social practices and personal motivations based on the Shinto belief system. People worshiped many gods, venerated ancestors, and glorified their emperor as a divine figure, infallible and omnipotent. The emperor's military ruler, or *shogun*, led the campaign against the Ainu and governed the nation. The capital, Kyoto, became an assemblage of hundreds of magnificent temples, shrines, and gardens (see photo, p. 508). In their local architecture, modes of dress, festivals, theater, music, and other cultural expressions, the Japanese forged a unique society. Handicraft industries, based on local raw materials, abounded.

Nonetheless, in the mid-nineteenth century, Japan hardly seemed destined to become Asia's leading power. For 300 years the country had been closed to outside influences, and Japanese society was stagnant and tradition-bound. When the European colonizers first appeared on Asia's shores, Japan tolerated, even welcomed, their merchants and missionaries. But as the European presence in East Asia grew stronger, the Japanese began to shut their doors. Toward the end of the sixteenth century, the emperor decreed that all foreign traders and missionaries should be expelled. Christianity (especially Roman Catholicism) had gained a considerable foothold in Japan, but now this was feared as being a prelude to colonial conquest. Early in the seventeenth century, the shogun launched a massive and bloody campaign that practically stamped out Christianity.

Determined not to suffer the same fate as the Philippines, which by then had fallen to Spain, the Japanese henceforth allowed only minimal contact with Europeans. A few Dutch traders, confined to a little island near the city of Nagasaki, were for many decades the sole representatives of Europe in Japan. Thus Japan entered a long period of isolation, lasting past the middle of the nineteenth century.

Japan could maintain its aloofness, while other areas were being enveloped by the colonial tide, because of its strong central government and well-organized military, the natural protection its islands provided, and its remoteness along East Asia's difficult northern coast. Also, Japan's isolation was far less splendid than that of China, whose exquisite silks, prized teas, and skillfully made wares attracted traders and usurpers alike.

When Japan finally came face to face with the new weaponry of its old adversaries, it had no answers. In the 1850s, the steel-hulled "black ships" of the American fleet sailed into Japanese harbors, and the Americans extracted one-sided trade agreements. Soon the British, French, and Dutch were also on the scene, seeking similar treaties. When there was local resistance to their show of strength, the Americans quickly demonstrated their superiority by shelling parts of the Japanese coast. By the late 1860s, even as Japan's modernizers were about to overturn the old order, no doubt remained that Japan's protracted isolation had come to an end.

A Japanese Colonial Empire

When the architects of the Meiji Restoration confronted the challenge to build a Japan capable of competing against powerful adversaries in a changing world, they took stock of the country's assets and liabilities. Material assets, they found, were limited. To achieve industrialization, coal and iron ore were needed. In terms of coal, there was enough to support initial industrialization. Coalfields in Hokkaido and Kyushu were located near the coast; since the new industries also were on the coast, cheap water transportation was possible. As a result, the shores of the (Seto) Inland Sea (Fig. 9-24) became the sites of many factories. But there was little iron ore, certainly nothing like what was available domestically in Britain, and not enough to sustain the massive industrialization the reformers had in mind. This commodity would have to be purchased overseas and imported.

On the positive side, manufacturing—light manufacturing of the handicraft type—already was widespread in Japan. In cottage industries and in community workshops, the Japanese produced textiles, porcelain, wood products, and metal goods. The small ore deposits in the country were enough to supply these local industries. Power came from human arms and legs and from wheels driven by water; the chief source of fuel was charcoal. Importantly, Japan did have an industrial tradition and an experienced labor force that possessed appropriate manufacturing skills. All this was not enough to lead directly to industrial modernization, but it did hold promise for capital formation. The community and home workshops were integrated into larger units, hydroelectric plants were built, and some thermal (coal-fired) power stations were constructed in critical areas.

For the first time, Japanese goods (albeit of the light manufactured variety) began to compete with Western products on international markets. The Japanese planners resisted any infusion of Western capital, however. Although Western support would have accelerated industrialization, it would have cost Japan its economic autonomy. Instead, farmers were more heavily taxed, and the money thus earned was poured into the industrialization effort.

Another advantage for Japan's planners lay in the country's military tradition. Although the shogun's forces had been unable to repel the invasions of the 1850s, this was the result of outdated equipment, not lack of manpower or discipline. So while its economic transformation gathered momentum, the military forces, too, were modernized. Barely more than a decade after the Meiji Restoration, Japan laid claim to its first Pacific prize: the Ryukyu Islands (1879). A Japanese colonial empire was in the making.

When Japan gained control over its first major East Asian colonies, Taiwan and Korea, its domestic raw-material problems were essentially solved, and a huge labor pool fell under its sway as well. High-grade coal, high-quality iron ore, and other resources were shipped to the factories of Japan; and in Taiwan as well as in Korea, the Japanese built additional manufacturing plants to augment production. This, in turn, provided the equipment to sustain the subsequent drive into China and Southeast Asia.

Modernization

The reformers who set Japan on a new course in 1868 probably did not anticipate that three generations later their country would lie at the heart of a major empire sustained by massive military might. They set into motion a process of **modernization**, but they managed to build on, not replace, Japanese cultural traditions. We in the Western world tend to equate modernization with Westernization: urbanization, the spread of transport and communications facilities, the establishment of a market (money) economy, the breakdown of local traditional communities, the proliferation of formal schooling, the acceptance and adoption of foreign innovations. In the non-Western world, the process often is viewed differently. There, "modernization" is seen as an outgrowth of colonialism, the perpetuation of a system of wealth accumulation introduced by foreigners driven by greed. In this view, the local elites who replaced the colonizers in the newly independent states only continue the disrup-

tion of traditional societies, not their true modernization. Traditional societies, they argue, can be modernized without being Westernized.

In this context, Japan's modernization is unique. Having long resisted foreign intrusion, the Japanese did not achieve the transformation of their society by importing a Trojan horse; it was done by Japanese planners, building on the existing Japanese infrastructure, to fulfill Japanese objectives. Certainly Japan imported foreign technologies and adopted innovations from the British and others, but the Japan that was built, a unique combination of modern and traditional elements, was basically an indigenous achievement.

Relative Location

Japan's changing fortunes over the past century reveal the influence of **relative location** in the country's development. When the Meiji Restoration took place, Britain, on the other side of the Eurasian landmass, lay at the center of a global empire. The colonization and Europeanization of the world were in full swing. The United States was still a developing country, and the Pacific Ocean was an avenue for European imperial competition. Japan, even while it was conquering and consolidating its first East Asian colonies (the Ryukyus, Taiwan, Korea), lay remote from the mainstream of global change.

Then Japan became embroiled in World War II and dealt severe blows to the European colonial armies in Asia. The Europeans never recovered: the French lost Indochina, and the Dutch were forced to abandon their East Indies (now Indonesia). When the war ended, Japan was defeated and devastated, but at the same time the Japanese had done much to diminish the European presence in the Pacific Basin. Moreover, the global situation had changed dramatically. The United States, Japan's trans-Pacific neighbor, had become the world's most powerful and wealthiest country, whereas Britain and its global empire were fading. Suddenly Japan was no longer remote from the mainstream of global action: now the Pacific was becoming the avenue to the world's richest markets. Japan's relative location—its situation relative to the economic and political foci of the world—had changed. Therein lay much of the opportunity the Japanese seized after the postwar rebuilding of their country.

Japan's Spatial Organization

Imagine this: 128 million people crowded into a territory the size of Montana (population: 970,000), most of it mountainous, subject to frequent earthquakes and volcanism, with no domestic oilfields, little coal, few raw materials for industry, and not much level land for farming. If Japan today were an underdeveloped country in need of food relief and foreign aid, explanations would abound: overpopulation, inefficient farming, energy shortages.

True, only an estimated 18 percent of Japan's national territory is designated as habitable. And Japan's large population is crowded into some very big cities. Moreover, Japan's agriculture is not especially efficient. But Japan defeated the odds by calling on old Japanese virtues: organizational efficacy, massive productivity, dedication to quality, and adherence to common goals. Even before the Meiji Restoration, Japan was a tightly organized country of some 30 million citizens. Historians suggest that Edo, even before it became the capital and was renamed Tokyo, may have been the world's largest urban center. The Japanese were no strangers to urban life, they knew manufacturing, and they prized social and economic order.

Areal Functional Organization

All this proved invaluable to the modernizers when they set Japan on its new course. The new industrial growth of the country could be based on the urban and manufacturing development that was already taking place. As we noted, Japan does not possess major domestic raw-material sources, so no substantial internal reorganization was necessary. However, some cities were better sited and enjoyed better situations relative to those limited local resources and, more importantly, to external sources of raw materials than others. As Japan's regional organization took shape, a hierarchy of cities developed; Tokyo took and kept the lead, but other cities grew rapidly into industrial centers.

16 This process was governed by a geographic principle that Allen Philbrick called **areal functional organization**, a set of five interrelated tenets that help explain the evolution of regional organization, not only in Japan but throughout the world. Human activity, Philbrick reasoned, has spatial focus. It is concentrated in some locale, whether a farm or factory or store. Every one of these establishments occupies a particular location; no two of them can occupy exactly the same spot on the Earth's surface (even in high-rises, there is a vertical form of absolute location). Nor can any human activity proceed in total isolation, so that interconnections develop among these various establishments. This system of interconnections grows more complex as human capacities and demands expand. Each system (for example, farmers sending crops to market and buying equipment at service centers) forms a unit of areal functional organization. In the Introduction, we referred to functional regions as systems of spatial organization; we can map Philbrick's

units of areal functional organization as regions. These regions evolve because of what he called "creative imagination" as people apply their total cultural experience and their technological know-how when they organize and rearrange their living space. Finally, Philbrick suggests, we can recognize levels of development in areal functional organization, a ranking of places and regions based on the type, extent, and intensity of exchange.

In the broadest sense, we can divide regions of human organization into three categories: subsistence, transitional, and exchange. Japan's areal organization reflects the exchange category. Within each of these categories, we can also rank individual places on the basis of the number and kinds of activities they generate. Even in regions where subsistence activities dominate, some villages have more interconnections than others. The map of Japan (Fig. 9-24) showing its resources, urban settlements, and surface communications can tell us much about Japan's economy. It looks just like maps of other parts of the world where an exchange type of areal organization has developed—a hierarchy of urban centers ranging from the largest cities to the tiniest hamlets, a dense network of railways and roads connecting these places, and productive agricultural areas near and between the urban centers.

For Japan, however, the map shows us something else: Japan's external orientation, its dependence on foreign trade. All primary and secondary regions lie on the coast. Of all the cities in the million-size class, only Kyoto lies in the interior (Fig. 9-23). If we deduced that Kyoto does not match Tokyo–Yokohama, Osaka–Kobe, or Nagoya in terms of industrial development, that would be correct—the old capital remains a center of small-scale light manufacturing. Actually, Kyoto's ancient character has been deliberately preserved, and large-scale industries have been discouraged. With its old temples and shrines, its magnificent gardens, and its many workshop and cottage industries, Kyoto remains a link with Japan's premodern past.

Leading Economic Regions

As Figure 9-24 shows, Japan's dominant region of urbanization and industry (along with productive agriculture) is the *Kanto Plain*, which contains about one-third of the Japanese population and is focused on the Tokyo–Yokohama–Kawasaki metropolitan area. This gigantic cluster of cities and suburbs (the world's second largest urban agglomeration), interspersed with intensively cultivated farmlands, forms the eastern anchor of the country's elongated and fragmented core area. Besides its flatness, the Kanto Plain possesses other advantages: its fine natural harbor at Yokohama, its

Tokyo, at the center of one of the largest metropolises in the world, continues to change. Land-filling and bridge-building in the bay continue; skyscrapers sprout amid low-rise neighborhoods in this earthquake-prone area; traffic congestion worsens. The red-painted Tokyo Tower, a beacon in this part of the city, was modeled on the Eiffel Tower in Paris but, as a billboard at its base announces, is an improvement over the original: lighter steel, greater strength, less weight. Tokyo Bay, part of which can be seen from this vantage point, was the scene of one of history's most costly environmental disasters during the earthquake of 1923. A giant tsunami (seismic sea wave) swept up the bay from the epicenter even as landfills liquefied and buildings sank into the mud. The death toll exceeded 140,000; a much larger population than lived in Tokyo in 1923 is now at risk of a repeat. © Yann Arthus-Bertrand/Photo Researchers.

relatively mild and moist climate, and its central location with respect to the country as a whole. (The region's only disadvantage is its vulnerability to earthquakes—see box titled "When the Big One Strikes.") It has also benefited enormously from Tokyo's designation as the modern capital, which coincided with Japan's embarkation on its planned course of economic development. Many industries and businesses chose Tokyo as their headquarters in view of the advantages of proximity to the government's decision makers.

The Tokyo–Yokohama–Kawasaki conurbation has become Japan's leading manufacturing complex, producing more than 20 percent of its annual output. The raw materials for all this industry, however, come from far away. For example, the Tokyo area is among the chief steel producers in Japan, using iron ores from the Philippines, Malaysia, Australia, India, and even Africa; most of the coal is imported from Australia and North America, and the petroleum comes from Southwest Asia and Indonesia. The Kanto Plain cannot produce nearly enough food for its massive resident population. Imports must come from Canada, the United States, and Australia as well as other areas in Japan. Thus Tokyo depends completely on its external trading ties for food, raw materials, and markets for its products, which run the gamut from children's toys to high-precision optical equipment to the world's largest oceangoing ships.

The second-ranking economic region in Japan's core area is the Osaka–Kobe–Kyoto triangle—also known as the *Kansai District*—located at the eastern end of the

Inland Sea (Fig. 9-24). The Osaka–Kobe conurbation had an advantageous position with respect to Manchuria (Manchukuo) during the height of Japan's colonial empire. Situated at the head of the Inland Sea, Osaka was the major Japanese base for the China trade and for the exploitation of Manchuria, but it suffered when the empire was destroyed and it lost its trade connections with China after World War II. Kobe (like Yokohama, which is Japan's chief shipbuilding center) has remained one of the country's busiest ports, handling both Inland Sea traffic and extensive overseas linkages. Kyoto, as we noted, remains much as it was before Japan's great leap forward, a city of small workshop industries. The Kansai District is also an important farming area. Rice, of course, is the most intensively grown crop in the warm, moist lowlands, but this agricultural zone is smaller than the one on the Kanto Plain. Here is another huge concentration of people that must import food.

The Kanto Plain and Kansai District, as Figure 9-24 indicates, are the two leading primary regions within Japan's core. Between them lies the *Nobi Plain* (also called the *Chubu District*), focused on the industrial metropolis of Nagoya, Japan's leading textile producer. The map indicates some of the Nagoya area's advantages and liabilities. The Nobi Plain is larger than the lowlands of the Kansai District; thus its agricultural productivity is greater, though not as great as that of the Kanto Plain. But Nagoya has neither Tokyo's centrality nor Osaka's position on the Inland Sea: its connections to Tokyo are via the tenuous Sen-en Coastal Strip. Its westward connec-

When the Big One Strikes

THE TOKYO–YOKOHAMA–KAWASAKI urban area is the second biggest metropolis on Earth (only Mexico City is larger). But Tokyo is more than a gigantic city: it comprises one of the most densely concentrated financial and industrial complexes in the world. Two-thirds of Japan's businesses worth more than U.S. $50 million are clustered here, many with extensive overseas holdings. More than half of Japan's huge industrial profits are generated in the factories of this enormous metropolitan agglomeration.

But Tokyo has a worrisome environmental history because three active tectonic plates are converging here (Fig. 9-25). All Japanese know about the "70-year" rule: over the past three-and-a-half centuries, the Tokyo area has been struck by major earthquakes roughly every 70 years—in 1633, 1703, 1782, 1853, and 1923. The Great Kanto Earthquake of 1923 set off a firestorm that swept over the city and killed an estimated 143,000 people. Tokyo Bay virtually emptied of water; then a tsunami (seismic sea wave) roared back in, sweeping everything before it. That Japan could overcome this disaster was evidence of the strength of its economy.

Today, Tokyo is more than a national capital. It is a global financial and manufacturing hub in which so much of the world's wealth and productive capacity are concentrated that an earthquake comparable to the one of 1923 would have a calamitous effect worldwide. Ominously, Tokyo at the opening of the second decade of our new century is a much more vulnerable place than the Tokyo of the 1920s. True, building regulations are stricter, and civilian preparedness is better. But entire expanses of industries have been built on landfill that will liquefy; the city is honey-combed by underground gas lines that will rupture and stoke countless fires; congestion in the area's maze of narrow streets will hamper rescue operations; and many older high-rise buildings do not have the structural integrity that has lately emboldened builders to erect skyscrapers of 50 stories and more. Add to this the burgeoning population of the Kanto Plain—which surpasses 40 million on what may well be the most dangerous 4 percent of Japan's territory—and we realize that the next big earthquake in the Tokyo area will not be a remote, local news story.

© H. J. de Blij, P. O. Muller, and John Wiley & Sons, Inc.

FIGURE 9-25

tions are better, and the Nagoya area may be coalescing with the Osaka–Kobe conurbation. Still, the quality of Nagoya's port is not nearly as good as Tokyo's Yokohama or Osaka's Kobe, and it has had silting problems.

Westward from the three regions just discussed—which together constitute what is often called the *Tokaido Megalopolis*—extends the Seto Inland Sea, along whose shores the remainder of Japan's core area is continuing to develop. The most impressive growth has occurred around the western entry to the Inland Sea (the Strait of Shimonoseki), where *Kitakyushu*—a conurbation of five cities on northern Kyushu—constitutes the fourth Japanese manufacturing complex and primary economic region. Highway and railroad tunnels connect Honshu and Kyushu, but the northern Kyushu area does not have an urban-industrial equivalent on the Honshu side of the strait. The Kitakyushu conurbation includes

Yawata, site of northwest Kyushu's (rapidly declining) coal mines. The first steel plant in Japan was built there on the basis of this coal; for many years it was the country's largest. The advantages of transportation here at the western end of the Inland Sea are obvious: no place in Japan is better situated to do business with Korea and China. As relations with mainland Asia expand, this area will reap many benefits. Elsewhere on the Inland Sea coast, the Hiroshima–Kure urban area has a manufacturing base that includes heavy industry. And on the coast of the Korea Strait, Fukuoka and Nagasaki are the principal centers—respectively, an industrial city and a center of large shipyards.

Only one major Japanese manufacturing complex lies outside the belt extending from Tokyo in the east to Kitakyushu in the west—the secondary region centered on Toyama on the East Sea (Sea of Japan). The advantage

AMONG THE REALM'S GREAT CITIES . . . Tokyo

MANY URBAN AGGLOMERATIONS are named after the city that lies at their heart, and so it is with the second largest of all: Tokyo (26.7 million). Even its longer name—Tokyo-Yokohama-Kawasaki—does not begin to describe the congregation of cities and towns that form this massive, crowded metropolis that encircles the head of Tokyo Bay and continues to grow, outward and upward.

Near the waterfront, some of Tokyo's neighborhoods are laid out in a grid pattern. But the urban area has sprawled over hills and valleys, and much of it is a maze of narrow winding streets and alleys. Circulation is slow, and traffic jams are legendary. The train and subway systems, however, are models of efficiency—although during rush hours you must get used to the *shirioshi* pushing you into the cars to get the doors closed.

At the heart of Tokyo lies the Imperial Palace with its moats and private parks. Across the street, buildings retain a respectful low profile, but farther away Tokyo's skyscrapers seem to ignore the peril of earthquakes. Nearby lies one of the world's most famous avenues, the Ginza, lined by department stores and luxury shops. In the distance you can see an edifice that looks like the Eiffel Tower, only taller: this is the Tokyo Tower (see photo, p. 514), a multipurpose structure designed to test lighter Japanese steel, transmit television and radio signals, detect Earth tremors, monitor air pollution, and attract tourists.

© H. J. de Blij, P. O. Muller, and John Wiley & Sons, Inc.

Tokyo is the epitome of modernization, but Buddhist temples, Shinto shrines, historic bridges, and serene gardens still grace this burgeoning urban complex, a cultural landscape that reflects Japan's successful marriage of the modern and the traditional.

---

here is cheap power from nearby hydroelectric stations, and the cluster of industries reflects it: paper manufacturing, chemical factories, and textile plants. Figure 9-24 also gives an inadequate picture of the variety and range of industries that exist throughout Japan, many of them oriented to local (and not insignificant) markets. Thousands of manufacturing plants operate in cities and towns other than those shown on this map, even on the cold northern island of Hokkaido, which is connected to Honshu by the Seikan rail tunnel, the world's longest, beneath the treacherous Tsugaru Strait.

The map of Japan's areal organization shows that four primary regions dominate its core area. Each of them is primary because each duplicates to some degree the contents of the others. Each contains iron and steel plants, is served by one major port, and lies in or near a large productive farming area. What the map does not show is that each also has its own external connections for the overseas acquisition of raw materials and sale of finished

products. These linkages may even be stronger than those among the four internal regions of the core area. Only for Kyushu and its coal have domestic raw materials affected the nature and location of heavy manufacturing, and these resources are almost depleted. In the structuring of the country's areal functional organization, therefore, more than just the contents of Japan itself is involved. In this respect Japan is not unique: all countries that have exchange-type organization must adjust their spatial forms and functions to the external interconnections required for progress. But for few countries is this truer than Japan.

East and West, North and South

The previous discussion shows that Japan has its own Pacific Rim: almost all of Honshu's economic expansion is occurring on the east coast. Japanese geographers often call this eastern zone *Omote Nippon* ("Front Japan"), the Pacific-facing, megalopolitan, high-tech, dynamic sector that embodies the modern economic superpower—as opposed to *Ura Nippon* ("Back Japan"), the comparatively underdeveloped periphery of unkempt farms and poor villages bordering the East Sea (Sea of Japan) to the west. In a country that needs food as much as Japan does, the derelict state of farmlands in *Ura Nippon* is of much concern to the country's planners.

The congestion on Honshu is another concern. Efforts to link Hokkaido and Inland-Sea-facing Shikoku to the "mainland" have produced world-record tunnels and bridges such as the Seikan Tunnel and the Akashi Kaikyo Bridge (Fig. 9-24), designed to bring remote areas into the national sphere and to diffuse Honshu's economic primacy beyond the main island. This effort has had limited success, especially in Hokkaido, which constitutes nearly one-quarter of Japan's total area, but contains only about 5 percent of the national population. Farming remains the main pursuit on Hokkaido, producing wheat, dairy products, and meat; fishing and tourism also do well. But Hokkaido has not translated its improved linkages into competitive industries; locals will tell you that there is a "subsidy mentality" here, that Tokyo's investments have led to complacency rather than initiative. Only the Sapporo area, with its automobile parts and electronics factories, shares in the Honshu-led export boom.

Food and Population

Japan's economic modernization so occupies center stage that we can easily forget the country's considerable achievements in agriculture. Japan's planners, who are as interested in closing the food gap as in expanding industries, have created extensive networks of experiment stations to promote mechanization, optimal seed selection and fertilizer use, and information services to distribute knowledge about enhancing crop yields to farmers as rapidly as possible. Although this program has succeeded, Japan faces the unalterable reality of its stubborn topography: it simply lacks sufficient land to farm. Populous Japan has one of the highest physiologic population densities—2600 per square kilometer (6735 per sq mi)—on Earth. (We discuss this concept on pp. 414 and 446.)

The Japanese go to extraordinary lengths to maximize their limited farming opportunities, making huge investments in research on higher-yielding rice varieties, mechanization, irrigation, terracing, fertilization, and other practices. More than 90 percent of farmland is assigned to food crops. Hilly terrain is given over to cash crops,

FROM THE FIELD NOTES

"No matter how you arrive at Nagasaki—by sea, by air, by road—you are reminded of the site factor in the city's annihilation by atomic bomb on August 9, 1945. Unlike Hiroshima, Nagasaki sprawled over hills and into valleys, and destruction was almost total in this valley containing the CBD and opening onto the port, encircled by higher terrain that concentrated the devastation. The road up to Mount Inasa, from where this photo was taken in mid-2009, goes through narrow streets and hillside lanes, and a media tower on the mountaintop can be climbed for an even better perspective. Today Nagasaki exemplifies modern Japan, its challenging topography overcome by a network of tunnels, bridges, and viaducts, its economy propelled by its port functions (including a state-of-the-art liquefied natural gas terminal), shipyards, and tourism."
© H. J. de Blij

www.conceptcaching.com

such as tea and grapes (Japan has a thriving wine industry). Vegetable gardens ring all the cities.

But all this costs money, and Japan's food prices are high. Millions of Japanese have moved from the farms to the cities, depopulating the countryside, which needs farmers. Japan's political system gives underpopulated rural areas disproportionate influence, and produce prices are kept artificially high, all to induce farmers to stay on the land. It has not worked. Between 1920 and 2010, full-time farmers declined from 50 percent to less than 4 percent of the labor force. Yet the policies persist. In recent years, American farmers proved that they could provide rice to Japanese consumers at one-sixth the price the Japanese were paying for home-grown rice. But the government resisted pressure to open its markets to imports.

With their national diet so rich in starchy foods like rice, wheat, barley, and potatoes, the Japanese need protein to balance it. Fortunately, they can secure enough of it, not by buying it abroad but by harvesting it from rich fishing grounds near the Japanese islands. With their customary thoroughness, the Japanese have developed a fishing industry that is larger than that of the United States or any of the long-time fishing nations of northwestern Europe, and it now supplies the domestic market with a second staple after rice. Although mention of the Japanese fishing industry brings to mind a fleet of ships scouring the oceans and seas far from Japan, most of this huge catch (about one-seventh of the world's annual total) comes from waters within a few dozen kilometers of Japan itself. Where the warm Kuroshio and Tsushima currents meet colder water off Japan's coasts, a rich fishing ground yields sardines, herring, tuna, and mackerel in the warmer waters and cod, halibut, and salmon in the seas to the north. Japan's coasts have about 4000 fishing villages, and tens of thousands of small boats ply the waters offshore to bring home catches that are distributed to local and city markets. The Japanese also practice *aquaculture*—the "farming" of freshwater fish in artificial ponds and flooded paddy (rice) fields, of seaweeds in aquariums, and of oysters, prawns, and shrimp in shallow bays. They are even experimenting with cultivating algae for their food potential.

When you walk the streets of Japan's cities, you will notice something that Japanese vital statistics are reporting: the Japanese are getting taller and heavier. Coupled with this change is evidence that heart disease and cancer are rising as causes of mortality. The reason seems to lie in the changing diets of many Japanese, especially the younger people. When rice and fish were the staples for virtually everyone, and the calories available were limited but adequate, the Japanese as a nation were among the world's healthiest and longest-lived. But as fast-food establishments diffused throughout Japan, and Western (especially American) tastes for red meat and fried food spread among children and young adults, their combined impact on the population was soon evident. Coupled with the heavy cigarette smoking that prevails in all the Asian countries of the Pacific Rim, this is changing the region's medical geography.

Japan's Pacific Rim Prospects

As the twenty-first century unfolds, Japan remains the economic giant on the western Pacific Rim—but a giant with an uncertain future. Japanese products still dominate world markets, Japanese investments span the globe, but in the late 1990s Japan's economy faltered. Its growth rate declined. Many companies scaled back their operations; some went bankrupt. The banking system proved to be vulnerable. The first Pacific Rim economic tiger was ailing.

Japan's problems stemmed from a combination of circumstances at home and abroad. Overseas, the collapse of Southeast Asian currencies and economies hurt Japanese investments there. From automobile plants to toy factories, long-profitable operations turned into liabilities. In Japan itself, financial mismanagement, long tolerable because of the booming economy, now took its toll. In addition, competition from still-healthy economies, notably from Taiwan but also from struggling South Korea, undercut Japanese goods on world markets.

In the long run, all this may prove to have been a temporary setback, but Japan is unlikely to regain its former economic dominance on the Pacific Rim. Japan's dependence on foreign oil is a potential weakness that could further depress its economy should international supply lines be disrupted. Western competition for Japan's markets is growing. The countless investments Japan made around the world during its time of plenty, from skyscrapers and movie studios to golf courses and hotels, lost much of their value and became huge losses. There was a time when thriving European and American businesses (along with Middle Eastern and Russian investors) bought such properties, but the global recession beginning in 2008 caused many experts to look again at Japan's experience—to study just how long a recovery might take.

Were you to make a field trip to Japan to study its economic geography, little of this might be immediately evident. The cities still bustle with activity, raw materials continue to arrive at the ports, container traffic still stirs the docks. Even in a downturn, this mighty economy outproduces all others on the western Pacific Rim. And, indeed, that observation is relevant to the future. Japan's infrastructure has the capacity to propel an economic rebound. But it needs new opportunities.

Politico-Geographical Dilemma

One of these opportunities lies in the Russian Far East and beyond, in Siberia. As we noted in Chapter 2, the Russian Far East lies directly across from northern Honshu and Hokkaido and is a storehouse of raw materials awaiting exploitation. For this, the Japanese are ideally located—but political relations with the former Soviet Union, and now with Russia, have stymied all initiatives (see box titled "The Wages of War"). Japan should be one of the major beneficiaries of any substantial oil reserves near Sakhalin (see map, p. 139), but again the necessary connections may not be possible. Japan also has a major potential opportunity in Korea, especially if a unification process takes hold there. But relations with Korea continue to be eroded by Korean memories of Japanese behavior during World War II and by Japan's refusal to acknowledge its misdeeds.

The Wages of War

JAPAN AND THE former Soviet Union never signed a peace treaty to end their World War II conflict. Why? There are four reasons, and all of them are on the map just to the northeast of Japan's northernmost large island, Hokkaido (Fig. 9-23). Their names are Habomai, Shikotan, Kunashiri, and Etorofu. The Japanese call these rocky specks of the Kurile Island chain their "Northern Territories." The Soviets occupied them late in the war and never gave them back to Japan. Now they are part of Russia, and the Russians have not given them back either.

The islands themselves are no great prize. During World War II, the Japanese brought 40,000 forced laborers, most of them Koreans, to mine the minerals there. When the Red Army overran them in 1945, the Japanese were ordered out, and most of the Koreans fled. Today the population of about 50,000 is mostly Russian, many of them members of the military based on the islands and their families. At their closest point the islands are only 5 kilometers (3 mi) from Japanese soil, a constant and visible reminder of Japan's defeat and loss of land. Moreover, territorial waters bring Russia even closer, so that the islands' geostrategic importance far exceeds their economic potential.

All attempts to resolve this issue have failed. In 1956, Moscow offered to return the tiniest two, Shikotan and Habomai, but the Japanese declined, demanding all four islands back. In 1989, then-Soviet President Mikhail Gorbachev visited Tokyo in the hope of securing an agreement. The Japanese, it was widely reported, offered an aid-and-development package worth U.S. $26 billion to develop Russia's eastern zone—its Pacific Rim and the vast resources of the eastern Siberian interior. This would have begun the transformation of Russia's Far East, stimulated the ports of Nakhodka and Vladivostok, and made Russia a participant in the spectacular growth of the western Pacific Rim.

But it was not to be. Subsequently, Russian presidents Yeltsin, Putin, and Medvedev also were unable to come to terms with Japan on this issue, facing opposition from the islands' inhabitants and from their own governments in Moscow. And so the Second World War, 65 years-and-counting after its conclusion, continues to cast a shadow over this northernmost segment of the Asian Pacific Rim.

Population Change

All this must be seen against the backdrop of Japan's changing society. The country's fast-aging population peaked at just over 128 million in the late 2000s and has now begun a downward course toward a total of less than 120 million by 2025. Demographers project this decline to increase quite rapidly in the second quarter of the century, reaching barely 95 million in 2050; and by century's end, that total is expected to drop an additional 30 percent, yielding a 2100 population of only about 67 million. Already, for some time, Japan has faced a labor shortage (although the recent economic downturn produced rising unemployment, a novelty in the postwar period). Over the long term, Japan will need millions of immigrants to keep its economy going, but for culturally homogeneous, ethnically conscious Nippon, which has historically resisted immigration, this will be a wrenching reality. Still, Japanese leaders will have no choice. The country's accelerated "graying" will strain welfare systems and tax burdens alike.

Modernization, meanwhile, is further stressing Japan's society. The Japanese have a strong tradition of family cohesion and veneration of their oldest relatives, but this tradition is breaking down. Younger people are less willing than their parents and grandparents to accept overcrowded housing, limited comforts, and minimal privacy. In Japan, for all its material wealth, family homes tend to be small, cramped, often flimsily built, and sometimes lacking the basic amenities that other high-income countries consider indispensable. With their disposable income, the Japanese during the boom economy came to rank among the world's most-traveled tourists, and, having seen how Americans and Europeans of similar income levels live, they returned home dissatisfied.

As we just noted, Japan remains one of the most uniform, culturally homogeneous societies in today's interactive world. The Chinese closed their doors to foreigners intermittently; the Japanese have done so far more effectively. Other than the dwindling Ainu, there are no indigenous minorities. Koreans were brought to Japan during World War II to serve as forced laborers, but otherwise Japanese residency and citizenship imply an almost total ethnic sameness. During periods of labor shortage, the notion to invite foreign workers would not be entertained. When the United States, after the Vietnam War, sought new homes for hundreds of thousands of Vietnamese refugees, Japan accepted just a few hundred after much negotiation. Such jealously guarded ethnic-cultural uniformity is unlikely to be viable in the globalizing world of the twenty-first century.

Crossroads

Japan's future in the geopolitical framework of the western Pacific is also uncertain. More than six decades after the end of World War II, tens of thousands of American forces still are based on Japanese soil. China and South Korea accept this arrangement, which constrains Japan's rearming. But in Japan, where nationalism is rising, domestic military preparedness is a growing political issue. Over the past ten years, however, Japan's sense of security was shaken by developments in communist North Korea, which in 1998 (and again in 2009) fired a missile across Honshu into the Pacific Ocean and is suspected of developing nuclear weapons capacity; by admissions from North Korea that it had kidnapped Japanese citizens from streets and beaches, and forced them to teach Japanese to North Korean spies; and by the appearance of armed North Korean vessels in Japanese territorial waters. In 2003, the Japanese Parliament approved legislation providing for the participation of Japanese troops in the U.S.-led intervention in Iraq, a move that contradicted public opinion as reflected by surveys but temporarily strengthened American-Japanese government ties. Japan also joined the United States and three other countries (China, Russia, and South Korea) to negotiate an end to North Korea's nuclear threat.

Japan's economic success and political stability may have been facilitated by enlightened U.S. post–World War II policy, but the future will be different. In 2002, China overtook the United States as the leading exporter to the Japanese market; today, Japanese companies compete vigorously against American corporations in China's Pacific Rim. Japan continues to have image problems among Chinese and Koreans, but at the highest levels cooperation is growing. On the other hand, Japan and China are becoming fierce competitors over critical energy supplies. In 2005, the Japanese tried to outbid the Chinese over the routing of oil and natural gas pipelines from Russia: the Chinese wanted to position these pipelines to link to terminals in its Northeast, whereas Japan wanted to route them to terminals on the East Sea (Sea of Japan) at Nakhodka, directly across from northern Honshu. Japan today is the world's second-largest oil consumer, but China's consumption is rising rapidly and heightened competition for Russian exports is inevitable. This will obviously have an effect on Japanese-Chinese relations, political as well as economic, in the decades ahead.

Japan, therefore, is a pivotal nation on the western Pacific Rim, an economic superpower that stands at social and political crossroads whose future will profoundly influence not only East Asia and the Pacific, but the world as a whole.

● KOREA

On the Asian mainland, directly across the East Sea (Sea of Japan), lies the peninsula of Korea (Fig. 9-26), a territory about the size of the State of Idaho, much of it mountainous and rugged, and containing a population of 73 million. Unlike Japan, however, Korea has long been a divided country, and the Koreans a divided nation. For uncounted centuries Korea has been a pawn in the struggles of more powerful neighbors. It has been a dependency of China and a colony of Japan. When it was freed from Japan's oppressive rule at the end of World War II (1945), the victorious Allied powers divided Korea for administrative purposes. That division gave North Korea (north of the 38th parallel) to the forces of the Soviet Union and South Korea to those of the United States. In effect, Korea traded one master for two new ones. The country still awaits reunification because North Korea immediately fell under the communist ideological sphere and became a dictatorship in the familiar (but in this case extreme) pattern. South Korea, with massive American aid, became part of East Asia's capitalist perimeter. Once again, it was the will of external powers that prevailed over the desires of the Korean people.

This unusual photograph of the capital of North Korea, one of the world's most isolated countries, is reminiscent of images from an earlier communist era elsewhere. Pyongyang, on the Taedong River, displays the bleak, gray architecture of Chinese cities during the Mao period and Soviet 'socialist' cities of a half-century ago, its wide avenues nearly devoid of traffic. North Korea's dictatorial regime has impoverished the country even as it pursues nuclear weaponry and ballistic-missile technology, its elitism sustained through China's assistance. Over more than 60 years of autocratic mismanagement, uncounted millions have suffered famine and starvation even as South Korea prospered.   © On Asia.

FIGURE 9-26   © H. J. de Blij, P. O. Muller, and John Wiley & Sons, Inc.

War and Aftermath

In 1950, North Korea sought to reunite the country by force and invaded South Korea across the 38th parallel. This attack drew a United Nations military response led by the United States. Thus began the devastating Korean War (1950–1953) in which North Korea's forces pushed far to the south only to be driven back across its own half of Korea almost to the Chinese border. Then China's Red Army entered the war and drove the UN troops southward again. A cease-fire was arranged in 1953, but not before the people and the land had been ravaged in a way that was unprecedented even in Korea's violent past. The cease-fire line shown in Figure 9-26

High-Risk North Korea

THE KOREAN PENINSULA is physiographically well-defined, and its people—the Korean nation—have occupied it for thousands of years. But internal political division, foreign cultural intrusion, and colonial subjugation have created a turbulent history. Now Koreans are divided again, this time by ideology. Even as the capitalist South became one of the Pacific Rim's economic tigers, North Korea remained an isolated, stagnant state on a Stalinist-communist model abandoned long ago even in the former Soviet Union itself. For six and a half decades, North Korea has been ruled by two men: Kim Il Sung, the "Great Leader," and his son Kim Jong Il, the "Dear Leader." Total allegiance to the dictators is ensured by an omnipotent police and military. Self-imposed isolation is North Korea's hallmark. The state controls all aspects of life in North Korea.

Economic policies under the communist regime have been disastrous. Per-capita incomes in the North (population: 23.9 million) are less than one-tenth of those in the South; poverty is rampant and ubiquitous. The infant mortality rate is more that 5 times higher than in South Korea. Industrial equipment is at least a generation out of date. Exports are negligible. When a combination of failed agricultural policies and environmental extremes created famine in the late 1990s and early 2000s, the regime resisted relief efforts, even from South Korea. As during the Maoist period in China, the world will never know the number of casualties.

North Korea did progress on one front: nuclear capability and associated weaponry. In 1999, it fired a test missile across northern Japan, heightening international concern (an act that was repeated in 2009). The North Koreans have used their nuclear threat to negotiate a multibillion-dollar technical aid program to build reactors for peaceful purposes, with Japan and South Korea among the major contributors. But in 2002, a North Korean source admitted that his country had indeed pursued a nuclear weapons program and was now in possession of "at least one" nuclear bomb. The North Koreans refused to allow visitors to inspect its nuclear facilities at Yongbyon, 100 kilometers (60 mi) due north of the capital, Pyongyang. Other known nuclear installations are mapped in Figure 9-26.

Former President George W. Bush, in his 2002 State of the Union address, labeled North Korea part of the "axis of evil" to be dealt with as part of the War on Terror. But while Iraq was forced to admit UN weapons inspectors, North Korea, with 37,000 U.S. troops along its border with the South, remained exempt. Meanwhile, North Korea acknowledged having kidnapped Japanese youngsters from public places in Japan and having forced them to teach Japanese language and culture to North Korean spies. In 2002 this admission led to authorized visits to Japan by some of the survivors and a furor among the Japanese people.

Even as the "Dear Leader" was able to reject U.S. and Japanese demands that nuclear operations at Yongbyon be reviewed, he approved some small gestures toward his southern neighbor. Beginning in 2000, a few hundred members of families separated by the 1953 ceasefire line were allowed a few hours to meet each other. He also allowed the preliminary design of a railroad that would link North and South across the narrow demilitarized zone (DMZ). After authorizing the creation of a barbed-wire-enclosed "free-trade zone" in the far northeast at Rajin-Sonbong, where South Korean cruise ships could dock, Kim Jong Il approved another one in the northwest, at Sinuiju, in 2002. But when the Chinese billionaire who was to take charge of this free-trade zone was arrested for corruption in China, that plan fell apart.

Desperate North Koreans flee their country across the Chinese border and on the seas; in foreign countries they invade embassy compounds and plead for asylum. In 2010 the reunification of the two Koreas seemed far off; but so did German reunification in 1985, and then it happened a few years later. But the reunification of the two Koreas will involve much greater difficulties than that of the Germanies. North Korea is far worse off even than former communist East Germany; the cost will be staggering. South Koreans will face far greater sacrifices than West Germans did, with far fewer resources. Still, the boundary superimposed on the Korean nation is a leftover of the Cold War, and its elimination and the reintegration of North Korea into the international community is a laudable goal.

became a heavily fortified *de facto* boundary. For five decades, until quite recently, virtually no contact of any kind occurred across this border, which continues to divide families and economies alike.

As a result, the affluent Jakota Triangle region of East Asia, must include South Korea but exclude North Korea (see box titled "High-Risk North Korea"). Times may change, as they did after 1990 in formerly divided Germany, but by the turn of this century South Korea, with 45 percent of Korea's land area but with two-thirds of the Korean population, had emerged as one of the economic tigers on the Pacific Rim. If it were an undivided country, Korea might have achieved even more because North and South Korea exhibit what geographers call **regional complementarity**. This condition arises when two adjacent regions complement each other in economic-geographic terms. Here, North Korea has raw materials that the industries of South Korea need; South Korea produces food the North needs; North Korea produces chemical fertilizers farms in the South need.

After the mid-twentieth century, however, the two Koreas were cut off from each other, and they developed in opposite directions. North Korea, whose large coal and iron ore deposits attracted the Japanese and whose hydroelectric plants produce electricity that the South could use, carried on what limited trade it generated with the Chinese and (through 1991) the Soviets. South Korea's external trade links were with the United States, Japan, and Western Europe.

In recent years, South Korean leaders have tried a so-called sunshine policy that would foster cooperation between the North and the South through economic ventures. The most impressive result of this initiative is the Kaesong Industrial Complex, 60 kilometers (38 mi) from Seoul but located in North Korea, where 90 South Korean companies pay nearly 34,000 North Korean workers to produce goods for local and export markets. This goodwill gesture did not have the results hoped for by later South Korean governments, whose concern over North Korea's nuclear weapons program led to a reduction of the initiative. That North Korea should emerge as a nuclear-capable state amid the grinding poverty and desperation of its citizens, adding still another threat to regional stability in this part of the world, surely is one of the tragic ironies of our time.

South Korea

In the early postwar years, there were few indications that South Korea would emerge as a major economic force on the Pacific Rim and, indeed, on the world stage. Over 70 percent of all workers were farmers; agriculture was inefficient; the country was stagnant. But huge infusions of

AMONG THE REALM'S GREAT CITIES . . . Seoul

SEOUL (POPULATION: 9.9 MILLION), located on the Han River, is ideally situated to be the capital of all of Korea, North and South (its name means capital in the Korean language). Indeed, it served as such from the late fourteenth century until the early twentieth, but events in that century changed its role. Today, the city lies in the northwest corner of South Korea, for which it serves as capital; not far to the north lies the tense, narrow demilitarized zone (DMZ) that contains the cease-fire line with North Korea. That line cuts across the mouth of the Han, depriving Seoul of its river traffic. Its bustling ocean port, Incheon, has emerged as a result (Fig. 9-26).

Seoul's undisciplined growth, attended by a series of major accidents including the failure of a key bridge over the Han and the collapse of a six-story department store, reflects the unbridled expansion of the South Korean economy as a whole, as well as the political struggles that carried the country from autocracy to democracy. Central Seoul lies in a basin surrounded by hills to an elevation of about 300 meters (1000 ft), and this megacity has sprawled outward in all directions, even toward the DMZ (see photo, p. 4). An urban plan designed in the 1960s was overwhelmed by the immigrant inflow.

During the period of Japanese colonial control, Seoul's surface links to other parts of the Korean Peninsula were improved, and this infrastructure played a role

© H. J. de Blij, P. O. Muller, and John Wiley & Sons, Inc.

in the city's later success. Seoul is not only the capital but also the leading industrial center of South Korea, exporting sizeable quantities of textiles, clothing, footwear, and (increasingly) electronic goods. South Korea thrives as an economic tiger on the Pacific Rim, and Seoul is its heart.

aid, first from the United States and then Japan, coupled with the reorganization of farming and stimulation of industries (even if they lost money), produced a dramatic turnaround. Large feudal estates were parceled out to farm families in 3-hectare (7.5-acre) plots, and a program of massive fertilizer importation was begun. Production rose to meet domestic needs and in some years yielded surpluses.

The industrialization program was based on modest local raw materials, plentiful and capable labor, ready overseas (especially American) markets, and continuing foreign assistance. Despite corrupt dictatorial rule, political instability, and social unrest, South Korean regimes managed to sustain a rapid rate of economic growth that placed the country, by the late 1980s, among the world's top ten trading powers. To get the job done, the government borrowed heavily from overseas, and it controlled banks and large industries. South Korea's growth resulted from **state capitalism** rather than free-enterprise capitalism, but it had impressive results. South Korea became the world's leading shipbuilding nation; its automobile industry grew rapidly; and its iron and steel and chemical industries began to thrive. It has placed less emphasis, however, on smaller, high-technology industries, which are the strengths of other Pacific Rim economic tigers. So the future remains uncertain.

18

Three Industrial Areas

South Korea (population: 49.1 million) prospers today, and we can see its economic prowess on the map (Fig. 9-26). The capital, Seoul, with 10 million residents, now ranks among the world's megacities and anchors a huge industrial complex facing the Yellow Sea at the waist of the Korean Peninsula. Hundreds of thousands of farm families migrated to the Seoul area after the end of the Korean War (today only 18 percent of South Koreans remain on the land). They also moved to Busan, the nucleus of the country's second largest manufacturing zone, located on the Korea Strait opposite the western tip of Honshu. And the government-supported, urban-industrial drive here continues. Just 30 years ago, Ulsan City, 65 kilometers (40 mi) north of Busan along the coast, was a fishing center with perhaps 50,000 inhabitants; today its population exceeds a million, nearly half of them the families of workers in the Hyundai automobile factories and the local shipyards and docks.

The third industrial area shown in Figure 9-26, anchored by the city of Kwangju, has advantages of relative location that will spur its development, but it faces domestic problems. Kwangju lies at the center of the *Cholla* region, the southwest corner of South Korea (the southeast is called *Kyongsang*). For more than three

FROM THE FIELD NOTES

"South Korea was one of the four 'economic tigers' marking the upsurge of the western Pacific Rim in the 1980s and 1990s, its products ranging from computers to cars selling on world markets and its GNI rising rapidly. Walking through the prosperous central business district of globalizing Seoul you see every luxury name brand on display amid plentiful evidence of wealth and well-being. But you also see evidence of the survival of traditions long lost elsewhere, in the huge number of bookstores (South Koreans still read voraciously), in the stores selling traditional as well as modern musical instruments, in the school uniforms. And in other ways: when I got on a bus with all seats taken, a boy got up and offered me his place." © H. J. de Blij

www.conceptcaching.com

Concept Caching

decades beginning in the 1960s, the army generals who ruled South Korea all came from Kyongsang, and through state control of banks and other monopolies they bestowed favors on Kyongsang companies while denying access to potential competitors in Cholla. The regime in Seoul also thwarted infrastructure development in Cholla, even to the point of impeding plans for improved railroad links between Kwangju and the capital. This regional friction led to protests in Cholla and violent repression by the military rulers, the aftermath of which is fading only slowly in this part of the now-democratic country. These historic discriminations seem to be ending as South Korea's democracy matures and regional differences subside. One reason for this has to do with infrastructural improvements: a new high-speed train system, the Korea Train Express (KTX), now links Seoul with both Mokpo in the Cholla region and Busan in the Kyongsang region (Fig. 9-26), and there are plans to link Kwangju with Daegu, thereby establishing an interregional link in the south. South Koreans know what the bullet-train network did for Japan, and they expect the KTX to accelerate the integration of their country as well.

Pacific Rim Opportunities

Another sign of South Korea's political maturation is the government's proposal to move the capital from congested Seoul to an area about 150 kilometers (90 mi) to the south, near Taejon in Chungchong Province (Fig. 9-26). Leaders argue that moving the government to a more centralized location will be costly but will enhance efficiency, yet as the plan gains momentum look for politicians and others to raise obstacles.

Meanwhile, South Korea has moved far beyond the days when giant corporate conglomerates (known as *chaebol*) conspired to suppress competition, controlled the banks, and dictated economic policy to the government. It was state capitalism at its most extreme, and it proved to have weaknesses similar to planned communist economies. After the system collapsed and South Korea proved its capacity to thrive in the competitive arena of the Pacific Rim, the country became one of the region's economic tigers—not without economic problems, but with boundless potential. Seoul's metropolitan economy has become a high-tech hub; the adjacent port city of Incheon is now spawning New Songdo City, a special economic zone (see photo at right) designed to rival Shanghai's Pudong. China has surpassed the United States as South Korea's leading trade partner, and Songdo lies only about 300 kilometers (200 mi) across the Yellow Sea from the Chinese mainland—symbolizing the new century as well as the new Pacific Rim era that is transforming all countries in this realm.

A JAKOTA TRIANGLE?

In the introductory chapter we suggested that our world can be constructed as possessing a wealthy, generally privileged core and a less fortunate periphery, our criteria for this construct ranging from the income of countries to the standards of living in cities and from the prevalence of representative government to the effects of globalization (Fig. G-12). If you look carefully at that rather small-scale map, you will note that South Korea forms part of the global core but North Korea does not, that Japan does, but Russia does not, and that Taiwan is in, but China is out. As the map indicates, we used state (i.e., national) boundaries to delimit the core, and here in East Asia this produces some unusual challenges. First, Taiwan is not formally a state, although its political system is democratic and its economy is discrete from that of China. Close (and closer) as Taiwan's ties

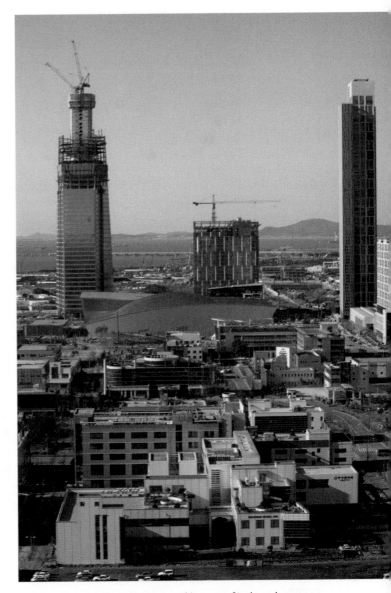

Seoul, the capital of South Korea, and its port of Incheon have different names on the map but are part of one huge conurbation. South Korea, taking a page from China's economic book, is creating here the Incheon Free Economic Zone of which the area shown in this photo, the New Songdo International City, will, in the words of government publicity, "take Korea into the next century." Given the current calendar, that is an ambitious objective indeed. © Reuters/Landov.

with China may be, the PRC is way behind Taiwan in terms of per-capita income and representative government. Second, the core boundary crosses the Korean Peninsula and coincides with the DMZ there, so that the South is an epitome of the global core and the North remains one of the global periphery's most disconnected societies. Third, there is the inescapable factor of dominant, historically expansionist Japan, its economy still the world's second largest (with a population less

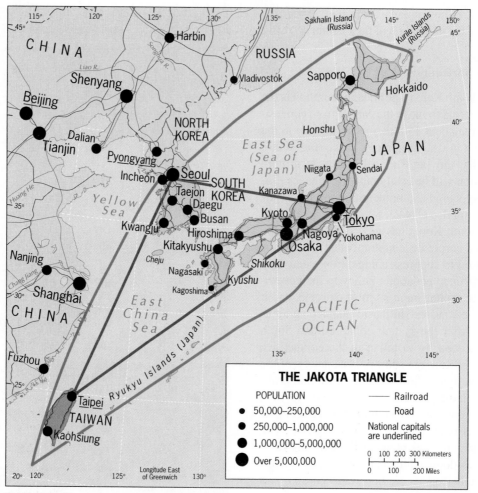

FIGURE 9-27

© H. J. de Blij, P. O. Muller, and John Wiley & Sons, Inc.

than 10 percent of China's) and its imperial role in South Korea as well as Taiwan still influential in the course of events.

Although there is much that divides these three components of the global core on its western margin, their economic successes, administrative achievements, and worldwide linkages—reflected by remarkable similarities of cultural landscapes and ways of life—suggest that they constitute parts of the informal region we call the *Jakota Triangle* (Fig. 9-27). Today, relations among the three capitals of the Triangle are not always cooperative: the Koreans and the Japanese argue over islands in the East Sea (Sea of Japan), and sometimes the Taiwanese accuse the Japanese of meddling in their party politics; neither South Korea nor Taiwan has forgotten Japan's colonial rule and associated misbehavior. But on the bigger issues, these three overachievers concur, as seen in their systems of government, huge consumption of raw

materials from all over the world, voluminous exports to foreign markets, and global financial linkages.

As we have noted repeatedly in other chapters, regions change over time. When China lay isolated from the world under Mao Zedong's rule, Japan formed a discrete world realm and an economically mighty outpost, and Taiwan's links with the PRC were minimal. When China's new era opened the PRC to the world, Japan's involvement and Taiwanese investment altered regional frameworks. But only when China moves toward representative government, Taiwan reformulates its relationships with Beijing, and the barriers between South and North Korea break down, will the Jakota Triangle concept become obsolete. At the start of the second decade of this century such changes may seem remote, but consider what has happened in just one generation in this fast-evolving geographic realm. In East Asia, anything seems possible.

What You Can Do

RECOMMENDATION: Join a Research Team! Many of the faculty members of your college or university, the professors you meet in class, have another life: they perform research and publish their findings in their discipline's scholarly literature. Some faculty may be working every summer at a base in some remote locale in a distant land; others will be using the library and the Internet. Often, when you read their publications, you will note that these are joint ventures (some articles in academic journals may have three, four, or even more authors). Such long-term research is likely to require lots of labor. If a professor's course or field is of special interest to you, ask her or him whether she or he is doing research in this particular area, and if so whether you can be of any assistance. Such help may start out in the library, but soon you may be working on interview teams or, who knows, joining a team of scientists overseas. Make your interest known, and an opportunity is likely to arise. You will learn things you would never learn in a classroom!

GEOGRAPHIC CONNECTIONS

1 East Asia is a realm of great cities, including many of the largest in the world. Names like Tokyo, Beijing, and Shanghai represent the megacity phenomenon like few others. But look at East Asia in the Data Table inside the back cover, and you will find that the countries of this realm display a range of urbanization levels from high (Japan and South Korea are in the lead) to low (China, for all its burgeoning cities, remains the least urbanized state in the realm). What geographic factors help explain these variable levels of urbanization among East Asia's countries? Why does China have more cities in the one-million-plus population category than any other country in the realm and yet the lowest percentage of urbanization? Do you expect the urban geography of this realm to change, and if so, where and how?

2 Consider the core-periphery and globalization concepts (discussed in the Introduction) in the context of the East Asian realm. If you were asked to define what region or regions of East Asia presently form part of the global "core," what criteria would you use and where would this core component appear on the map? In those areas that form part of the periphery, what has been the impact of globalization? Has globalization drawn these areas closer to the core, benefiting them, or has the economic gap widened between core and periphery in East Asia?

Geographic Literature on East Asia: The key introductory works on this realm were authored by Burks, Chow, Hodder, Hsieh & Lu, Karan, Pannell & Ma, Pannell & Veeck, Smith, Veeck et al., and Weightman. These works, together with a comprehensive listing of noteworthy books and recent articles on the geography of China, Japan, and other components of East Asia, can be found in the *References and Further Readings* section of this book's website at **www.wiley.com/college/deblij**.

SOUTHEAST ASIA

10

Roads
Railroads

Elevation (m)

	3000
	1500
	600
	300
	150
	0

below sea level
0
-150
-1500
-3000
-6000

CONCEPTS, IDEAS, AND TERMS

1	Tsunami
2	Buffer zone
3	Shatter belt
4	Overseas Chinese
5	Organic theory
6	State boundaries
7	Antecedent boundary
8	Subsequent boundary
9	Superimposed boundary
10	Relict boundary
11	State territorial morphology
12	Compact state
13	Protruded state
14	Elongated state
15	Fragmented state
16	Perforated state
17	Domino theory
18	*Entrepôt*
19	Growth triangle
20	Archipelago
21	Transmigration

REGIONS

● **MAINLAND SOUTHEAST ASIA**
● **INSULAR SOUTHEAST ASIA**

FIGURE 10-1   © H. J. de Blij, P. O. Muller, and John Wiley & Sons, Inc.

Where was this photo taken? Find out at www.wiley.com/college/deblij

In This Chapter

- Tsunamis and terror: A dangerous part of the world
- Chinese diaspora and China's comeback
- Why Myanmar suffers
- The world's most perfect king
- Democratic Indonesia takes a leading role

Photos: © H. J. de Blij

SOUTHEAST ASIA . . . the very name roils American emotions. Here the United States owned its only major colony. Here American forces triumphed over Japanese enemies. Here the United States fought the first war it ever lost. Here Islamic terrorists killed hundreds, drawing American troops and intelligence agents into the fray.

Here, too, American companies invested heavily when the Pacific Rim's economic opportunities beckoned and dormant economies turned into potential economic tigers. Southeast Asia, once remote and stagnant, has taken center stage in our globalizing world.

Southeast Asia's human drama plays itself out on a volatile, unsteady physical stage. Repeatedly in human history, gigantic volcanic eruptions here have had global consequences; one of them, the Toba eruption about 72,000 years ago, plunged the planet into darkness and, scientists conclude, came close to extinguishing humankind. In 1815, people around the world woke up one April morning to skies that darkened as the day went on, the result of ash and soot propelled into the stratosphere by the eruption of the Indonesian volcano named Tambora. A year later, that continuing eruption had shrouded the Earth in a blanket so thick that there literally was no summer. Crops would not ripen, and hunger and starvation afflicted millions. The human population at the time was just 1 billion; consider what would happen today under similar circumstances, with the human population about seven times as large.

Those globe-girdling Indonesian eruptions killed many hundreds of thousands of Indonesians living in the shadows of the realm's great volcanoes; at least another 30,000 perished during Krakatau's relatively minor eruption in 1883. But volcanic activity is only one of the hazards of this realm's physical geography. Here churns some of the

MAJOR GEOGRAPHIC QUALITIES OF

Southeast Asia

1. Southeast Asia extends from the peninsular mainland to the archipelagos offshore. Because Indonesia controls part of New Guinea, its functional region reaches into the neighboring Pacific geographic realm.

2. Southeast Asia, like Eastern Europe, has been a shatter belt between powerful adversaries and has a fractured cultural and political geography shaped by foreign intervention.

3. Southeast Asia's physiography is dominated by high relief, crustal instability marked by volcanic activity and earthquakes, and tropical climates.

4. A majority of Southeast Asia's more than 600 million people live on the islands of just two countries: Indonesia, with the world's fourth-largest population, and the Philippines. The rate of population increase in the Insular region of Southeast Asia exceeds that of the Mainland region.

5. Although the overwhelming majority of Southeast Asians have the same ancestry, cultural divisions and local traditions abound, which the realm's divisive physiography sustains.

6. The legacies of powerful foreign influences, Asian as well as non-Asian, continue to affect the cultural landscapes of Southeast Asia.

7. Southeast Asia's political geography exhibits a variety of boundary types and several categories of state territorial morphology.

8. The Mekong River, Southeast Asia's Danube, has its source in China and borders or crosses five Southeast Asian countries, sustaining tens of millions of farmers, fishing people, and boat owners.

9. The realm's giant in terms of territory as well as population, Indonesia, has not asserted itself as the dominant state because of mismanagement and corruption; but Indonesia has enormous potential.

Earth's most mobile crust, its colliding and subducting tectonic plates creating a steady roar of quakes and tremors (refer back to Figs. G-3 and G-4). Earthquakes with epicenters on land claim many victims, but even those occurring beneath the sea pose terrible risks. On December 26, 2004 one of those submarine earthquakes, in a subduction zone west of the Indonesian island of Sumatera,* generated a **tsunami** that thrust walls of water onto Indian [1] Ocean shores near and far, sweeping away entire towns and villages as well as their populations and killing as many as 300,000 people from nearby Indonesia to distant Somalia in Africa's Horn. Southeast Asia's fragmented map is the product of geologic forces and human risk-taking.

Defining the Realm

Southeast Asia is a realm of peninsulas and islands, a corner of Asia bounded by India on the northwest and China on the northeast (Fig. 10-1). Its western coasts are washed by the Indian Ocean, and to the east stretches the vast Pacific. From all these directions, Southeast Asia has been penetrated by outside forces. From India came traders; from China, settlers; from across the Indian Ocean, Arabs to engage in commerce and Europeans to build empires; and from across the Pacific, the Americans. Southeast Asia has been the scene of countless contests for power and primacy—the competitors have come from near and far.

Southeast Asia's geography in some ways resembles that of Eastern Europe. It is a mosaic of smaller countries on the periphery of one of the world's largest states. It has been a **buffer zone** between powerful adversaries. It is a [2] **shatter belt** in which stresses and pressures from without [3] and within have fractured the political geography. Like Eastern Europe, Southeast Asia exhibits great cultural diversity. It is a realm of hundreds of cultures, numerous languages and dialects, and several major religions.

---

*As in Africa, names and spellings have changed with independence. In this chapter we will use the contemporary spellings, except when we refer to the colonial period. Thus Indonesia's four major islands are Jawa, Sumatera, Kalimantan (the Indonesian part of Borneo), and Sulawesi. The Dutch called them Java, Sumatra, Dutch Borneo, and Celebes, respectively.

REGIONAL FRAMEWORK

The Southeast Asian realm borders South Asia and East Asia on land and the Pacific and Austral realms at sea (Fig. 10-2). Note that one political entity that forms part of it, Indonesia, extends beyond the realm's borders. The easternmost province of Indonesia is the western half of the Pacific island of New Guinea, where indigenous cultures are Papuan, not Indonesian. Today Indonesia rules what is in effect a Pacific island colony called Papua. We discuss this territory as part of our assessment of Indonesia, but we will also consider all of New Guinea as a component of the Pacific Realm in Chapter 12.

Because the politico-geographical map is so complicated, Figure 10-2 should be studied attentively. One good way to strengthen your mental map of this realm is to follow the mainland coastline from west to east. The westernmost state in the realm is Myanmar (called Burma before 1989 and still referred to by that name), the only country in Southeast Asia that borders both India and China. Myanmar shares the "neck" of the Malay Peninsula with Thailand, heart of the *Mainland* region. The south of the peninsula is part of Malaysia—except for Singapore, at the very tip of it. Facing the Gulf of Thailand is Cambodia. Still moving generally eastward, we reach Vietnam, a strip of land that extends all the way to the Chinese bor-

FIGURE 10-2

© H. J. de Blij, P. O. Muller, and John Wiley & Sons, Inc.

der. And surrounded by its neighbors is landlocked Laos, remote and isolated. This leaves the islands that constitute *Insular* Southeast Asia: the Philippines in the north and Indonesia in the south, and between them the offshore portion of Malaysia, situated on the largely Indonesian island of Borneo. Also on Borneo lies the ministate of Brunei, small but, as we will see, important in the regional picture. And finally, a new state has just appeared on the map in the realm's southeastern corner: East Timor, a former colony of Portugal annexed by Indonesia in 1976 and released to United Nations supervision in 1999. After a transition period, East Timor became an independent state in 2002 (its government wants the country to be known as *Timor-Leste*, the Portuguese way of writing it).

No Dominant State—Yet

No dominant state has arisen among the 11 countries of Southeast Asia. The largest country in terms of population as well as territory, Indonesia, has as-yet unrealized potential to overshadow the rest but not in the foreseeable future. Neither did any single, dominant core of indigenous culture develop here as it did in East Asia. In the river basins and on the plains of the mainland, as well as on the islands offshore, a flowering of cultures produced a diversity of societies whose languages, religions, arts, music, foods, and other achievements formed an almost infinitely varied mosaic—but none of those cultures rose to imperial power. The European colonizers forged empires here, often by playing one state off against another; the Europeans divided and ruled. Out of this foreign intervention came the modern map of Southeast Asia; only Thailand

(formerly Siam) survived the colonial era as an independent entity. Thailand was useful to two competing powers, the French to the east and the British to the west. It was a convenient buffer, and while the colonists carved pieces off Thailand's domain, the kingdom endured.

Indeed, the Europeans accomplished what local powers could not: the formation of comparatively large, multicultural states that encompassed diverse peoples and societies and welded them together. Were it not for the colonial intervention, it is unlikely that the 17,000 islands of far-flung Indonesia would today comprise the world's fourth-largest country in terms of population. Nor would the nine sultanates of Malaysia have been united, let alone with the peoples of northern Borneo across the South China Sea. For good or ill, the colonial intrusion consolidated a realm of few culture cores and numerous ministates into less than a dozen countries.

PHYSICAL GEOGRAPHY

As Figure 10-3 shows, Southeast Asia is a realm in which high relief dominates the physiography. From the Arakan Mountains in western Myanmar (Burma) to the glaciers (yes, glaciers!) of the Indonesian part of New Guinea, elevations rise above 3300 meters (10,000 ft) in numerous places. In Myanmar, Mount Victoria is such a peak; in northern Laos, Phou Bia comes close; on the Indonesian island of Sumatera (Sumatra under the old spelling) the tallest volcano is Kerinci; on Jawa (Java) Mounts Slamet and Semeru tower over the countryside; and on Borneo Mount Kinabalu, not a volcano but an erosional remnant, is higher than any other mountain in

When you take the road from Semarang on central Jawa's north coast toward Yogyakarta near the south coast, you have a constant companion to your left: towering Mount Merapi, its near-perfect volcanic cone rising above the densely-peopled, fertile rural area below. Merapi is an active volcano, and the people in its shadow are at continuing risk—not just from volcanic eruptions, but also from associated earthquakes in this especially unstable part of the Pacific Ring of Fire. In the spring of 2006, Mount Merapi came to life and thousands were evacuated from its most hazardous slopes. Then, in May, a severe earthquake, probably related to the volcanic activity, killed thousands. Merapi's recent spate of activity is continuing, but few people in the area are still alive who witnessed its most massive eruption in recent times, that of 1930 which resulted in a death toll of nearly 1400. © Bay Ismoyo/AFP/Getty Images, Inc.

FIGURE 10-3

© H. J. de Blij, P. O. Muller, and John Wiley & Sons, Inc.

this realm. In the Philippines, Mount Pinatubo at 1477 meters (4874 ft) is not especially high, but this volcano's eruptions have had a major impact on the country and, briefly in 1991, on global climate.

The relief map is a further reminder of the overlap here between the Pacific Rim and the Pacific Ring—the Pacific Rim of economic and political development and the Pacific Ring of Fire with its earthquakes and volcanic eruptions (see photo, p. 533). If you examine Figure G-4 closely, you will note that all the islands of this realm are affected by the latter except one, the largely Indonesian island of Borneo. Borneo is a slab of ancient crust, a minicontinent pushed high above sea level by tectonic forces and eroded into its present mountainous topography.

Four Mainland Rivers

As Figures 10-1 and 10-3 underscore, rivers rise in the highland backbones of the islands and peninsulas, and deposit their sediments as they wind their way toward the coast; the physiography of Sumatera demonstrates this

unmistakably. The volcanic hills, plateaus, and better-drained lowlands are fertile and, in the warmth of tropical climates, can yield multiple crops of rice.

On the peninsular mainland we see a pattern that is already familiar: rivers rising in the Asian interior that create alluvial plains and deltas. The Mekong River is the Danube of Southeast Asia: you can trace it all the way from China via Laos, Thailand, and Cambodia into southern Vietnam, where it forms a massive and populous delta. In the west, Myanmar's key river is the Irrawaddy; Thailand's is the Chao Phraya. In the north, the Red River Basin is the breadbasket of northern Vietnam.

No survey of the physical geography of Southeast Asia would be complete without reference to the realm's seas, gulfs, straits, and bays. Irregular and indented coastlines such as these, with thousands of islands near and far, create difficult problems when it comes to drawing maritime boundaries (the islands also form havens for rebels and criminals). As we note later, Southeast Asia has one of the most complex maritime boundary frameworks in the world, and it is rife with potential for conflict.

FIGURE 10-4   © H. J. de Blij, P. O. Muller, and John Wiley & Sons, Inc.

POPULATION GEOGRAPHY

Compared to the huge population numbers and densities in the habitable regions of South Asia and China, demographic totals for the countries of Southeast Asia, with the exception of Indonesia's 247 million, seem modest. Again, comparisons with Europe come to mind. Three countries—Thailand, the Philippines, and Vietnam—have populations between 65 and 95 million. Laos, quite a large country territorially (comparable to the United Kingdom), had barely more than 6 million inhabitants in 2010. Cambodia, substantially larger than Greece, had just over 15 million. It is noteworthy that of Southeast Asia's 602 million inhabitants, almost 57 percent live on the islands of Indonesia and the Philippines, leaving the realm's Mainland countries with less than 40 percent of the population. This spatial pattern is reflected in Figure 10-4: Jawa (Indonesia) and Luzon (the Philippines) dominate the island population clusters while the deltas of the four major rivers noted above constitute the leading mainland concentrations.

The Ethnic Mosaic

Southeast Asia's peoples come from a common stock just as most (Caucasian) Europeans do, but this has not prevented the emergence of regionally or locally discrete ethnic or cultural groups. Figure 10-5 displays the broad distribution of ethnolinguistic groups in the realm, but be aware that this is a generalization. At the scale of this map, numerous small groups cannot be depicted.

This map shows the rough spatial coincidence, on the mainland, between major ethnic group and modern political state. The Burman (Burmese) dominate in the country once known as Burma (today Myanmar); the Thai occupy the state once known as Siam (now Thailand); the Khmer form the nation of Cambodia and extend northward into Laos; and the Vietnamese inhabit the long strip of territory facing the South China Sea.

Territorially, by far the largest population is classified in Figure 10-5 as Indonesian, the inhabitants of the great archipelago that extends from Sumatera west of the Malay Peninsula to the Malukus (Moluccas) in the east and from the Lesser Sunda Islands in the south to the Philippines in the north. Collectively, all these peoples—the Filipinos, Malays, and Indonesians—shown in Figure 10-5 are known as Indonesians, but they have been divided by history and politics. Note, on the map, that the Indonesians in Indonesia itself include Javanese, Madurese, Sundanese, Balinese, and other large groups; hundreds of smaller ones are not displayed. In the Philippines, too, island isolation and contrasting ways of life are reflected in the cultural mosaic. Also part of this Indonesian ethnic-cultural complex are the

FIGURE 10-5

© H. J. de Blij, P. O. Muller, and John Wiley & Sons, Inc.

Malays, whose heartland lies on the Malay Peninsula but who form minorities in other areas as well. Like most Indonesians, the Malays are Muslims, but Islam is a more powerful force in Malay society than, in general, in Indonesian culture.

In the northern part of the Mainland region, numerous minorities inhabit remote parts of the countries in which the Burman, Thai, and Vietnamese dominate. Those minorities, as a comparison of Figures 10-2 and 10-5 indicates, tend to occupy areas on the peripheries of their countries, away from the core areas, where the terrain is mountainous and the forest is dense, and where the governments of the national states do not have complete control. This remoteness and sense of detachment give rise to notions of secession, or at least resistance to governmental efforts to establish authority, often resulting in bitter ethnic conflict.

Immigrants

Figure 10-5 also reminds us that, again like Eastern Europe, Southeast Asia is home to major ethnic minorities from outside the realm. On the Malay Peninsula, note the South Asian (Hindustani) cluster. Hindu communities with Indian ancestries exist in many parts of the peninsula, but in the southwest they form the majority in a small area. In Singapore, too, South Asians form a significant minority. These communities arose during the colonial period, but South Asians had arrived in this realm many centuries earlier, propagating Buddhism and leaving architectural and other cultural imprints on places as far away as Jawa and Bali.

The Chinese

By far the largest immigrant minority in Southeast Asia, however, is Chinese. The Chinese began arriving here during the Ming and early Qing (Manchu) dynasties, but the largest exodus occurred during the late colonial period (1870–1940), when as many as 20 million immigrated. The European powers at first encouraged this influx, using the Chinese in administration and trade. But soon these **Overseas Chinese** began to move into the major cities, where they established "Chinatowns" and gained control over much of the commerce. By the time the Europeans tried to reduce Chinese immigration, World War II was about to start and the colonial era would soon end.

Today, Southeast Asia is home to almost 30 million Overseas Chinese, three-quarters the world total. Their lives have often been difficult. The Japanese relentlessly persecuted those Chinese who lived in Malaya during World War II; during the 1960s, Chinese in Indonesia were accused of communist sympathies, and hundreds of thousands were killed. In the late 1990s, Indonesian mobs again attacked Chinese and their property, this time because of their relative wealth and because many Chinese became Christians during the colonial period, now targeted by Islamic throngs. Figure 10-6 shows the migration routes and current concentrations of Chinese in Southeast Asia. Most originated in China's Fujian and Guangdong provinces, and many invested much of their wealth in China when it opened up to foreign businesses. The Overseas Chinese of Southeast Asia played a major role in the economic miracle of the Pacific Rim.

FROM THE FIELD NOTES

"It was a relatively cool day in Hanoi, which was just as well because the drive from Haiphong had taken nearly four hours, much of the time waiting behind oxcarts, and the countryside was hot and humid. On my way to an office along Hang Bai Street I ran into this group of youngsters and their teacher on their way to the Ho Chi Minh Memorial. He right away realized the level of my broken French, and explained patiently that he taught history, and took his class on a field trip several times a year. 'They get all dressed up and it is a big moment for them,' he said, 'and there is a lot of history to be visited in Hanoi.' The kids had seen me take this picture and now they crowded around. I told them that I noticed that rouge seemed to be a favorite color in their outfits. 'But that should not surprise you, monsieur,' said the teacher, smiling. 'This is Vietnam, communist Vietnam, and red is the color that rallies us!' In the city at the opposite end of the country named after Ho Chi Minh (founder of his country's Communist Party, president from 1945 to 1969, victor over the French colonizers, and leader in the war against the United States), my experiences with teachers and students had been quite different, to the point that most called their city Saigon. But here in Hanoi the ideological flame still burned brightly." © H. J. de Blij (1992).

www.conceptcaching.com

Concept Caching

HOW THE POLITICAL MAP EVOLVED

The leading colonial competitors in Southeast Asia were the Dutch, French, British, and Spanish (with the last replaced by the Americans in their stronghold, the

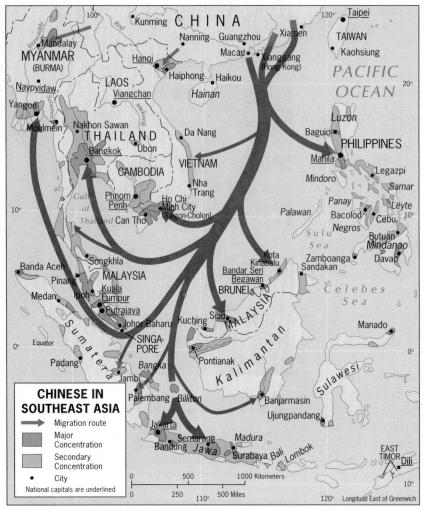

FIGURE 10-6

© H. J. de Blij, P. O. Muller, and John Wiley & Sons, Inc.

Laos. In recent years, Thailand and Cambodia have been skirmishing over a stretch of their common border.

The Colonial Imprint

French Indochina

The colonial powers divided their possessions into administrative units as they did in Africa and elsewhere. Some of these political entities became independent states when the colonial powers withdrew. France, one of the mainland's leading colonial powers, divided its Southeast Asian empire into five units. Three of these units lay along the east coast: Tonkin in the north next to China, centered on the basin of the Red River; Cochin China in the south, with the Mekong Delta as its focus; and between these two, Annam. The other two French territories were Cambodia, facing the Gulf of Thailand, and Laos, landlocked in the interior. Out of these five French dependencies there emerged the three states of *Indochina*. The three east-coast territories ultimately became one state, Vietnam; the other two (Cambodia and Laos) each achieved separate independence.

Philippines). The Japanese had colonial objectives here as well, but these came and went during the course of World War II.

The Dutch acquired the greatest prize: control over the vast archipelago now called Indonesia (formerly the Netherlands East Indies). France established itself on the eastern flank of the mainland, controlling all territory east of Thailand and south of China. The British conquered the Malay Peninsula, gained power over the northern part of the island of Borneo, and established themselves in Burma as well. Other colonial powers also gained footholds but not for long. The exception was Portugal, which held on to its eastern half of the island of Timor (Indonesia) until well after the Dutch had been ousted from their East Indies.

Figure 10-7 shows the colonial framework in the late nineteenth century, before the United States assumed control over the Philippines. Note that while Thailand survived as an independent state, it lost territory to the British in Malaya and Burma and to the French in Cambodia and

British Imperialism

The British ruled two major entities in Southeast Asia (Burma and Malaya) in addition to a large part of northern Borneo and many small islands in the South China Sea. Burma was attached to Britain's Indian Empire; from 1886 until 1937 it was governed from distant New Delhi. But when British India became independent in 1947 and split into several countries, Burma was not part of the grand design that created West and East Pakistan (the latter now Bangladesh), Ceylon (now Sri Lanka), and India. Instead, in 1948 Burma (now Myanmar) was given the status of a sovereign republic.

In Malaya, the British developed a complicated system of colonies and protectorates that eventually gave rise to the equally complex, far-flung Malaysian Federation. Included were the former Straits Settlements (Singapore was one of these colonies), the nine protectorates on the Malay Peninsula (former sultanates of the Muslim era), the

FIGURE 10-7

© H. J. de Blij, P. O. Muller, and John Wiley & Sons, Inc.

British dependencies of Sarawak and Sabah on the island of Borneo, and numerous islands in the Strait of Malacca and the South China Sea. The original Federation of Malaysia was created in 1963 by the political unification of recently independent mainland Malaya, Singapore, and the former British dependencies on the largely Indonesian island of Borneo. Singapore, however, left the Federation in 1965 to become a sovereign city-state, and the remaining units were later restructured into peninsular Malaysia and, on Borneo, Sarawak and Sabah. Thus the term *Malaya* properly refers to the geographic area of the Malay Peninsula, including Singapore and other nearby islands; the term *Malaysia* identifies the politico-geographical entity of which Kuala Lumpur is the capital city.

Netherlands "East Indies"

The Hollanders consolidated control of the "spice islands" through their Dutch East India Company, and the wealth that they extracted from what is today Indonesia brought the Netherlands its Golden Age. From the mid-seventeenth to the late-eighteenth century, the Dutch could develop their East Indies sphere of influence almost without challenge, for the British and French were preoccupied with the Indian subcontinent. By playing the princes of Indonesia's states against one another in the search for economic concessions and political influence, by placing the Chinese in positions of responsibility, and by imposing systems of forced labor in areas directly under its control, the Company had a ruinous effect on the Indonesian societies it subjugated. Java (Jawa), the most populous and productive island, became the focus of Dutch administration; from its capital at Batavia (now Jakarta), the Company extended its sphere of influence into Sumatra (Sumatera), Dutch Borneo (Kalimantan), Celebes (Sulawesi), and the smaller islands of the East Indies. This was not accomplished overnight, and the struggle for territorial control was carried on long after the Dutch East India Company had yielded its administration to the Netherlands government. Dutch colonialism thereby threw a girdle around Indonesia's more than 17,000 islands, paving the way for the creation of the realm's largest and most populous state (whose population approaches 250 million today).

From Spain to the United States

In the colonial tutelage of Southeast Asia, the Philippines, long under Spanish domination, had a unique experience. As early as 1571, the islands north of Indonesia were under Spain's control (they were named for Spain's King Philip II). Spanish rule began when Islam was reaching the southern Philippines via northern Borneo. The Spaniards spread their Roman Catholic faith with great zeal, and between them the soldiers and priests consolidated Hispanic dominance over the mostly Malay population. Manila, founded in 1571, became a profitable waystation on the route between southern China and western Mexico (Acapulco usually was the trans-Pacific destination for the galleons leaving Manila's port). There was much profit to be made, but the indigenous people shared little in it. Great landholdings were awarded to loyal Spanish civil servants and to men of the church. Oppression eventually yielded revolution, and Spain was confronted with a major uprising when the Spanish-American War broke out elsewhere in 1898.

As part of the settlement of that war, the United States replaced Spain in Manila. That was not the end of the revolution, however. The Filipinos now took up arms against their new foreign ruler, and not until 1905, after terrible losses of life, did American forces manage to "pacify" their new dominion. Subsequently, U.S. administration in the Philippines was more progressive than Spain's had been. In 1934, Congress passed the Philippine Independence Law, providing for a ten-year transition to sovereignty. But before independence could be arranged, World War II intervened. In 1941, Japan conquered the islands, temporarily ousting the Americans; U.S. forces returned in 1944 and, with strong Filipino support, defeated the Japanese in 1945. The agenda for independence was resumed, and in 1946 the sovereign Republic of the Philippines was proclaimed.

Today, all of Southeast Asia's states are independent, but centuries of colonial rule have left strong cultural imprints. In their urban landscapes, their education systems, their civil service, and countless other ways, this realm still carries the marks of its colonial past.

Cultural-Geographic Legacies

The French, who ruled and exploited a crucial quadrant of Southeast Asia, had a name for their empire: *Indochina*. That name would be appropriate for the rest of the realm as well because it suggests its two leading Asian influences for the past 2000 years. Periodic expansions of the Chinese Empire, notably along the eastern periphery, as well as the arrival of large numbers of Overseas Chinese settlers (discussed on p. 537 and mapped on p. 538), infused the region with cultural norms from the north. The Indians came from the west by way of the sea, as their trading ships plied the coasts, and settlers from India founded colonies on Southeast Asian shores in the Malay Peninsula, on the lower Mekong Plain, on Jawa and Bali, and on Borneo.

Religious Infusions from South Asia

With the migrants from the Indian subcontinent came their faiths: first Hinduism and Buddhism, later Islam. The Muslim religion, promoted by the growing number of Arab traders who appeared on the scene, became the dominant religion in Indonesia (where nearly 90 percent of the population adheres to Islam today). But in Myanmar, Thailand, and Cambodia, Buddhism remained supreme, and in all three countries the overwhelming majority of the people are now adherents. In culturally diverse Malaysia, the Malays are Muslims (to be a Malay *is* to be a Muslim), and almost all Chinese are Buddhists; but most Malaysians of Indian ancestry remain Hindus. Although Southeast Asia has generated its own local cultural expressions, most of what remains in tangible form has resulted from the infusion of foreign elements. For instance, the main temple at Angkor Wat, constructed in Cambodia during the twelfth century, remains a monument to the Indian architecture of that time.

The *Indo* part of Indochina, then, refers to the cultural imprints from South Asia: the Hindu presence, the importance of Buddhism (which came to Southeast Asia via Sri Lanka [Ceylon] and its seafaring merchants), the influences of Indian architecture and art (especially sculpture), writing and literature, and social structures and patterns.

Chinese Imprints and Influence

The *China* in the name Indochina signifies the role of the Chinese here. Chinese emperors coveted Southeast Asian lands, and China's power reached deep into the realm. Social and political upheavals in China sent millions of sinicized people southward. Chinese traders, pilgrims, seafarers, fishermen, and others sailed from southeastern China to the coasts of Southeast Asia and established settlements there. Over time, those settlements attracted more Chinese emigrants, and Chinese influence in the realm grew (see map, p. 538). Not surprisingly, relations between the Chinese settlers and the earlier inhabitants of Southeast Asia have at times been strained, even violent. The Chinese presence in Southeast Asia is longterm, but the invasion has continued into modern times. The economic power of Chinese minorities and their role in the political life of the realm have led to conflicts.

The Chinese initially profited from the arrival of the Europeans, who stimulated the growth of agriculture, trade, and industries; here the Chinese found opportunities they lacked at home. They harvested rubber, found

jobs on the docks and in the mines, cleared the bush, and transported goods in their sampans. They brought useful skills with them, and as tailors, shoemakers, blacksmiths, and fishermen, they prospered. The Chinese were also astute in business; soon, they not only dominated the region's retail trade but also held prominent positions in banking, industry, and shipping. Thus they have always been far more important than their modest numbers in Southeast Asia would suggest. The Europeans used them for their own designs but at times found the Chinese to be stubborn competitors—so much so that eventually they tried to impose restrictions on Chinese immigration. When the United States took control of the Philippines, it also sought to stop the influx of Chinese into those islands.

When the European colonial powers withdrew and Southeast Asia's independent states emerged, Chinese population sectors ranged from nearly 50 percent of the total in Malaysia (in 1963) to barely over 1 percent in Myanmar. In Singapore today, Chinese constitute 77 percent of the population of 4.9 million; when Singapore seceded from Malaysia in 1965, the Chinese component in Malaysia was substantially reduced (it is about 25 percent today). In Indonesia, the percentage of Chinese in the total population is not high (no more than 3 percent), but the Indonesian population is so large that even this small percentage indicates a Chinese sector of almost 7.5 million. As recently as 1998, this tiny minority still controlled 70 percent of Indonesia's economy. In Thailand, on the other hand, many Chinese have married Thais, and the Chinese minority of about 14 percent has become a cornerstone of Thai society, dominant in trade and commerce (see photo above).

The Chinese Presence in Southeast Asia

THE CHINESE ARE TOO INFLUENTIAL!

"It's hard to imagine that there was a time when we didn't have Chinese minorities in our midst. I think I understand how Latin Americans [sic] feel about their 'giant to the north.' We've got one too, but the difference is that there are a lot more Chinese in Southeast Asia than there are Americans in Latin America. The Chinese even run one country because they are the majority there, Singapore. And if you want to go to a place where you can see what the Chinese would have in mind for this whole region, go there. They're knocking down all the old Malay and Hindu quarters, and they've got more rules and laws than we here in Indonesia could even think of. I'm a doctor here in Bandung and I admire their modernity, but I don't like their philosophy.

"We've had our problems with the Chinese here. The Dutch colonists brought them in to work for them, they got privileges that made them rich, they joined the Christian churches the Europeans built. I'm not sure that a single one of them ever joined a mosque. And then shortly after Sukarno led us to independence they even tried to collaborate with Mao's communists to take over the country. They failed and many of them were killed, but look at our towns and villages today. The richest merchants and the money lenders tend to be Chinese. And they stay aloof from the rest of us.

"We're not the only ones who had trouble with the Chinese. Ask them about it in Malaysia. There the Chinese started a full-scale revolution in the 1950s that took the combined efforts of the British and the Malays to put down. Or Vietnam, where the Chinese weren't any help against the enemy when the war happened there. Meanwhile they get richer and richer, but you'll see that they never forget where they came from. The Chinese in China boast about their coastal economy, but it's coastal because that's where our Chinese came from and where they sent their money when the opportunity arose.

"As a matter of fact, I don't think that China itself cares much for or about Southeast Asia. Have you heard what's happening in Yunnan Province? They're building a series of dams on the Mekong River, our major river, just across the border from Laos and Burma. They talk about the benefits and they offer to destroy the gorges downstream to facilitate navigation, but they won't join the Mekong River Commission. When the annual floods cease, what will happen to Tonle Sap, to fish migration, to seasonal mudflat farming? The Chinese do what they want—they're the ones with the power and the money."

THE CHINESE ARE INDISPENSABLE!

"Minorities, especially successful minorities, have a hard time these days. The media portray them as exploiters who take advantage of the less fortunate in society, and blame them when things go wrong, even when it's clear that the government representing the majority is at fault. We Chinese arrived here long before the Europeans did and well before some other 'indigenous' groups showed up. True, our ancestors seized the opportunity when the European colonizers introduced their commercial economy, but we weren't the only ones to whom that opportunity was available. We banded together, helped each other, saved and shared our money, and established stable and productive communities. Which, by the way, employed millions of locals. I know: my family has owned this shop in Bangkok for five generations. It started as a shed. Now it's a six-floor department store. My family came from Fujian, and if you added it all up we've probably sent more than a million U.S. dollars back home. We're still in touch with our extended family and we've invested in the Xiamen SEZ.

"Here in Thailand we Chinese have done very well. It is sometimes said that we Chinese remain aloof from local society, but that depends on the nature of that society. Thailand has a distinguished history and a rich culture, and we see the Thais as equals. Forgive me, but you can't expect the same relationship in East Malaysia. Or, for that matter, in Indonesia, where we are resented because the 3 percent Chinese run about 60 percent of commerce and trade. But we're always prepared to accommodate and adjust. Look at Malaysia, where some of our misguided ancestors started a rebellion but where we're now appreciated as developers of high-tech industries, professionals, and ordinary workers. This in a Muslim country where Chinese are Christians or followers of their traditional religions, and where Chinese regularly vote for the Malay-dominated majority party. So when it comes to aloofness, it's not us.

"The truth is that the Chinese have made great contributions here, and that without the Chinese this place would resemble South or Southwest Asia. Certainly there would be no Singapore, the richest and most stable of all Southeast Asian countries, where minorities live in peace and security and where incomes are higher than anywhere else. In fact, mainland Chinese officials come to Singapore to learn how so much was achieved there. Imagine a Southeast Asia without Singapore! It's no coincidence that the most Chinese country in Southeast Asia also has the highest standard of living."

Vote your opinion at www.wiley.com/college/deblij

In general, Southeast Asia's Chinese communities remained aloof and formed their own separate societies in the cities and towns. They kept their culture and language alive through social clubs, schools, and even residential suburbs that, in practice if not by law, were Chinese in character. At one time they were caught in the middle between the Europeans and Southeast Asians when the local people were hostile toward both white people and the Chinese. Since the withdrawal of the Europeans, however, the Chinese have become the main target of this antagonism, which remained strong because of the Chinese involvement in money lending, banking, and trade monopolies (see the Issue Box entitled "The Chinese Presence in Southeast Asia"). Moreover, there is the specter of an imagined or real Chinese political imperialism along Southeast Asia's northern flanks.

The *China* in Indochina, therefore, represents a diversity of penetrations. Southern China was the source of most of the old invasions, and Chinese territorial consolidation provided the impetus for successive immigrants. Although Indian cultural influences remained strong, people throughout Southeast Asia adopted Chinese modes of dress, plastic arts, types of houses and boats, and other cultural attributes. During the past century, and especially during the last half-century, renewed Chinese immigration brought skills and energies that propelled these minorities to comparative wealth and influence.

SOUTHEAST ASIA'S POLITICAL GEOGRAPHY

Southeast Asia's diverse and fractured natural landscapes, as well as varied cultural influences, have combined to create a complex political geography. As we noted earlier, *political geography* focuses on the spatial expressions of political behavior; we were introduced to several aspects of it when we examined Western Europe's supranationalism, Eastern Europe's balkanization, Pakistan's irredentism toward Kashmiri Muslims, and the geopolitics of eastern Eurasia. Many political geographers have studied the fundamental causes of the cyclical rise and decline of states; their answers have ranged from environmental changes to biological forces. Friedrich Ratzel (1844–1904) conceptualized the state as a biological organism whose life, from birth through expansion and maturation to eventual senility and collapse, mirrors that of any living thing. **5** Ratzel's **organic theory** of state development held that nations, being aggregates of human beings, would over the long term live and die as their citizens did.

Other political geographers sought to measure the strength of the forces that bind states (centripetal forces) and that divide them (centrifugal forces), and thus to assess a state's chances to survive separatism of the kind Serbia and Somalia are experiencing today, to name just two examples. Even in today's era of modern warfare, a stretch of water still affects the course of events. Taiwan would not be what it is without the 200 kilometers (125 mi) of water that separates it from mainland China. Singapore's secession from (then) Malaya was facilitated because Singapore is an island; had it been attached to the Malay Peninsula, the centrifugal forces of separatism might not have prevailed. The role of physical geography in political events remains powerful.

In this section we pay particular attention to the boundaries and territories of the component states of Southeast Asia because this realm's political geography displays virtually every possible attribute of both. We focus first on the borders and then on the states' territorial shapes.

The Boundaries

Boundaries are sensitive parts of a state's anatomy: just as people are territorial about their individual properties, so nations and states are sensitive about their territories and borders. The saying that "good fences make good neighbors" certainly applies to states, but, as we know, the boundaries between states are not always "good fences."

Boundaries, in effect, are contracts between states. **6** That contract takes the form of a treaty that contains the *definition* of the boundary in the form of elaborate description. Next, cartographers perform the *delimitation* of the treaty language, drawing the boundary on official, large-scale maps. Throughout human history, states have used those maps to build fences, walls, or other barriers in a process called *demarcation*.

Classifying Boundaries

Once established, we can classify boundaries geographically. Some are sinuous, conforming to rivers (see photo, p. 544) or mountain crests (*physiographic*) or coinciding with breaks or transitions in the cultural landscape (*anthropogeographic*). As any world political map shows, many boundaries are simply straight lines, delimited without reference to physical or cultural features. These *geometric* boundaries can produce problems when the cultural landscape changes where they exist.

In general, the boundaries of Southeast Asia were better defined than those of several other postcolonial areas of the world, notably Africa, the Arabian Peninsula, and Turkestan. The colonial powers that established the original treaties tried to define boundaries to lie in remote and/or sparsely peopled areas, for example, across interior Borneo. Nevertheless, certain Southeast Asian boundaries have produced problems, among them the geometric boundary between Papua, the portion of New Guinea ruled by Indonesia, and Papua New Guinea, the eastern component of the island.

Even on a small-scale map of the kind we use in this chapter, we can categorize the boundaries of this realm. A comparison between Figures 10-2 and 10-5 reveals that the boundary between Thailand and Myanmar over long segments is anthropogeographic, notably where the name *Kayin* (*Karen*), the Myanmar minority, appears in Figure 10-5. Figure 10-1 shows that a large segment of the Vietnam–Laos boundary is physiographic-political, coinciding with the Annamite Cordillera (Highlands).

FROM THE FIELD NOTES

"I stood on the Laotian side of the great Mekong River which, during the dry season, did not look so great! On the opposite side was Thailand, and it was rather easy for people to cross here at this time of the year. But, the locals told me, it is quite another story in the wet season. Then the river inundates the rocks and banks you see here, it rushes past, and makes crossing difficult and even dangerous. The buildings where the canoes are docked are built on floats, and rise and fall with the seasons. The physiographic-political boundary between Thailand and Laos lies in the middle of the valley we see here."

© Barbara A. Weightman.

Concept Caching

www.conceptcaching.com

Genetic Classification System

Boundaries also can be classified genetically, that is, as their evolution relates to the cultural landscapes they traverse. A leading political geographer, Richard Hartshorne (1899–1992), proposed a four-level *genetic boundary classification*. All four of these boundary types can be observed in Southeast Asia.

Certain boundaries, Hartshorne reasoned, were defined and delimited before the present-day human landscape developed. In Figure 10-8 (upper-left map), the boundary between Malaysia and Indonesia on the island of Borneo is an example of the first boundary type, **[7]** the **antecedent boundary**. Most of this border passes through sparsely inhabited tropical rainforest, and the break in settlement can clearly be detected on the realm's population map (Fig. 10-4).

A second category of boundaries evolved as the cultural landscape of an area took shape, part of the ongoing process of accommodation. These **subsequent bound-** **[8]** **aries** are represented in Southeast Asia by the map in the upper right of Figure 10-8, which shows in some detail the border between Vietnam and China. This border is the result of a long process of adjustment and modification, the end of which may not yet have come.

The third category involves boundaries drawn forcibly across a unified or at least homogeneous cultural landscape. The colonial powers did this when they divided the island of New Guinea by delimiting a boundary in a nearly straight line (curved in only one place to accommodate a bend in the Fly River), as shown in the lower-left map of Figure 10-8. The **superimposed boundary** **[9]** they delimited gave the Netherlands the western half of New Guinea. When Indonesia became independent in 1949, the Dutch did not yield their part of New Guinea, which is peopled mostly by ethnic Papuans, not Indonesians. In 1962 the Indonesians invaded the territory by force of arms, and in 1969 the United Nations recognized its authority there. This made the colonial, superimposed boundary the eastern border of Indonesia and had the effect of extending Indonesia from Southeast Asia into the Pacific Realm. Geographically, all of New Guinea forms part of the Pacific Realm.

The fourth genetic boundary type is the so-called **relict boundary**—a border that has ceased to function **[10]** but whose imprints (and sometimes influence) are still evident in the cultural landscape. The boundary between the former North and South Vietnam (Fig. 10-8, lower-right map) is a classic example: once demarcated militarily, it has had relict status since 1976 following the reunification of Vietnam in the aftermath of the Indochina War (1964–1975).

Southeast Asia's boundaries have colonial origins, but they have continued to influence the course of events in postcolonial times. Take one instance: the physiographic boundary that separates the main island of Singapore from the rest of the Malay Peninsula, the Johor Strait (see Fig. 10-14). That physiographic-political boundary facilitated, perhaps crucially, Singapore's secession from the state of Malaysia. Without it, Malaysia might have been persuaded to stop the separation process; at the very least, territorial issues would have arisen to slow the

GENETIC POLITICAL BOUNDARY TYPES

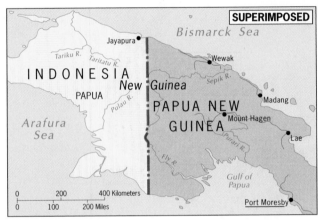

FIGURE 10-8

© H. J. de Blij, P. O. Muller, and John Wiley & Sons, Inc.

sequence of events. As it was, no land boundary needed to be defined. The Johor Strait demarcated Singapore and left no question as to its limits.

State Territorial Morphology

Boundaries define and delimit states; they also create the mosaic of often interlocking territories that give individual countries their shape, also known as their *morphology*. The **territorial morphology** of a state affects its condition, even its survival. Vietnam's extreme elongation has influenced its existence since time immemorial. As we will see, Indonesia has tried to redress its fragmentation into thousands of islands by promoting unity through the "transmigration" of Jawanese from the most populous island to many of the others.

Political geographers identify five dominant state territorial configurations, all of which we have encountered in our world regional survey but which we have not categorized until now. All but one of these shapes are represented in Southeast Asia, and Figure 10-9 provides the terminology and examples:

- **Compact states** have territories shaped somewhere **12** between round and rectangular, without major indentations. This encloses a maximum amount of territory within a minimum length of boundary. Southeast Asian example: Cambodia.

- **Protruded states** (sometimes called *extended*) have **13** a substantial, usually compact territory from which extends a peninsular corridor that may be landlocked or coastal. Southeast Asian examples: Thailand, Myanmar.

- **Elongated states** (also called *attenuated*) have terri- **14** torial dimensions in which the length is at least six times the average width, creating a state that lies astride environmental and/or cultural transitions. Southeast Asian example: Vietnam.

- **Fragmented states** consist of two or more territori- **15** al units separated by foreign territory or by water. Subtypes are mainland-mainland, mainland-island, and island-island. Southeast Asian examples: Malaysia, Indonesia, Philippines.

- **Perforated states** completely surround the territory **16** of other states, so that they have a "hole" in them. No

FIGURE 10-9   © H. J. de Blij, P. O. Muller, and John Wiley & Sons, Inc.

Southeast Asian example; the most illustrative current case is South Africa, perforated by Belgium-sized Lesotho (also see map on p. 309).

In the discussion that follows, we will have frequent occasion to refer to this geographic property of Southeast Asia's states. For so comparatively small a realm with so few states, Southeast Asia displays a considerable variety of state morphologies. When we link these features to other geographic aspects (such as relative location), we obtain useful insights into the regional framework.

One point of caution: states' territorial morphologies do not determine their viability, cohesion, unity, or lack thereof; they can, however, influence these qualities. Cambodia's compactness has not ameliorated its divisive political geography, for example. But as we will find in the pages that follow, shape plays a key role in the still-unfolding political and economic geography of Southeast Asia.

POINTS TO PONDER

● Myanmar is becoming increasingly important to China. An oil pipeline across Myanmar from a west-coast terminal will carry Middle East oil to Yunnan Province. China's only military base on the Indian Ocean lies in Myanmar.

● While democracy appears to be strengthening in Indonesia, it is faltering in Malaysia; in both countries, Islam is the dominant faith.

● The nation-building project in East Timor has essentially failed. The country is independent, but its people are mired in poverty and demographic disaster, and its institutions are failing.

● Thriving Singapore wants to lease one or two of Indonesia's more than 17,000 islands in order to expand its commercial, research, financial, and other operations.

Regions of the Realm

Southeast Asia's first-order regionalization must be based on its mainland-island fragmentation. But as we have noted there are physiographic, historical, and cultural reasons to include the Malaysian (southern) part of the Malay Peninsula in the Insular region, as shown in Figure 10-2. Using the political framework as our grid, we see that the regions of Southeast Asia are constituted as follows:

Mainland region   Vietnam, Cambodia, Laos, Thailand, Myanmar (Burma)

Insular region   Malaysia, Singapore, Indonesia, Brunei, East Timor, Philippines

● MAINLAND SOUTHEAST ASIA

Five countries form the Mainland region of Southeast Asia (Fig. 10-2): two of them protruded, sharing a narrow peninsula; one elongated along the Pacific coast; one compact and facing the Gulf of Thailand; and one landlocked. One religion, Buddhism, dominates cultural landscapes, but this is a multicultural, multiethnic region. Although still one of the less urbanized regions of the world, the Mainland region contains several major cities and the pace of urbanization is increasing. And as Figure 10-2 shows, two countries—Vietnam and Myanmar—possess more than one core area each. We approach the region from the east.

VIETNAM

Vietnam's economic transformation from postwar stagnation to present advancement is one of the realm's—and the world's—great geographic stories. In the aftermath of the war that ended in American (and South Vietnamese) defeat in 1975, devastated Vietnam could only watch as China's Pacific Rim burgeoned, its products flooding world markets and foreign investment rushing in. The United States placed an embargo on the country, isolating it and stifling its economy. The Vietnamese government experimented with communism even as an estimated two million citizens departed by any means they could (hundreds of thousands of "boat people" perished in the process), and for a few years this onetime breadbasket of the region faced famine. Then, in the mid-1980s, Vietnamese leaders decided to try what they called *doi-moi* economic reforms—a more gradual opening of their country to globalization's forces than China had adopted. By 1995, Vietnam's exports, from coffee to shrimps, from rice to footwear, and including electronic equipment, boats, and even some crude oil, were growing faster than China's. In that significant year, Washington during the Clinton Administration moved to normalize relations with Hanoi, opening an embassy and encouraging economic ties. In 2008, Vietnam's economy grew by more than 6 percent (population growth had leveled off at 1.2 percent—below today's global average), and Vietnamese economists calculate that the average annual growth rate for the first decade of this century was 7.5 percent, despite the global economic downturn.

Vietnam, as the Data Table inside the back cover indicates, has a long way to go, but the Vietnamese government is determined not to allow uncontrolled capitalism to create the kinds of wealth disparities among its 88.3 million citizens that are now seen in China. And indeed, the severest form of poverty—people living on the equivalent of one U.S. dollar a day or less—has dropped dramatically in Vietnam from more than 50 percent in 1990 to under 10 percent today while high-school attendance more than doubled and now exceeds 75 percent.

MAJOR CITIES OF THE REALM

City	Population* (in millions)
Bangkok, Thailand	7.0
Hanoi, Vietnam	4.8
Ho Chi Minh City (Saigon), Vietnam	5.9
Jakarta, Indonesia	10.1
Kuala Lumpur, Malaysia	1.5
Manila, Philippines	11.7
Phnom Penh, Cambodia	1.8
Singapore, Singapore	4.9
Viangchan, Laos	0.7
Yangon, Myanmar (Burma)	4.4

*Based on 2010 estimates.

Elongation and Integration

Travel up the crowded road from the northern port of Haiphong to the capital, Hanoi, or sail up the river to the southern metropolis officially known as Ho Chi Minh City—but called *Saigon* by almost everyone there—and you are quickly reminded of the cultural effects of Vietnam's elongation (Fig. 10-10). The French colonizers delimited Vietnam as a 2000-kilometer (1200-mi) strip of land extending from the Chinese border in the north to the tip of the Mekong Delta in the south. Substantially smaller than California, this coastal belt was the

FIGURE 10-10   © H. J. de Blij, P. O. Muller, and John Wiley & Sons, Inc.

domain of the Vietnamese (Fig. 10-5). The French recognized that Vietnam, whose average width is under 250 kilometers (150 mi), was not a homogeneous colony, so they divided it into three units: (1) *Tonkin*, land of the Red River Delta and centered on Hanoi in the north; (2) *Cochin China*, the Mekong Delta region and centered on Saigon in the south; and (3) *Annam*, focused on the

ancient city of Hué, in the middle (Fig. 10-7). Today, the Vietnamese prefer to use *Bac Bo, Nam Bo,* and *Trung Bo,* respectively, to designate these areas.

The Vietnamese (or Annamese, also Annamites, after their cultural heartland) speak the same language, although the northerners can easily be distinguished from southerners by their accent. As elsewhere in their colonial empire, the French made their language the *lingua franca* of Indochina, but their tenure was cut short by the Japanese, who invaded Vietnam in 1940. During the Japanese occupation, Vietnamese nationalism became a powerful force, and after the Japanese defeat in 1945, the French could not regain control. In 1954, the French suffered a disastrous final trouncing on the battlefield at Dien Bien Phu in the far northwest and were ousted from the country.

Even after its forces routed the colonizers, however, Vietnam did not become a unified state. Separate regimes took control: a communist one in Hanoi and a noncommunist counterpart in Saigon. Vietnam's pronounced elongation had made things difficult for the French; now it played its role during the postcolonial period. Note, in Figure 10-10, that Vietnam is widest in the north and south, with a narrow "waist" in its middle zone. North and South Vietnam were worlds apart.

The Indochina War

Many Americans still remember how the United States became involved in the inevitable conflict between the communists and noncommunists in Vietnam during the 1960s and 1970s. At first, American military advisors were sent to Saigon to help the shaky regime there cope with communist insurgents. When the tide turned against the South, the United States committed military forces and equipment to halt the further spread of communism (see box titled "Domino Theory"); at one time, more than 500,000 U.S. soldiers were in Vietnam.

The conduct of the war, its mounting casualties, and its apparent futility created severe social tensions in the United States. If you had been a student during that period, your college experience would have been radically different from what it is today. Protest rallies, "teach-ins," antidraft demonstrations, "sit-ins," marches, strikes, and even hostage-taking disrupted campus life. The Indochina War threatened the stability of American society. It drove an American president (Lyndon Johnson) from office in 1968 and destroyed the electoral chances of his vice president (Hubert Humphrey), who would not disavow the U.S. role in the war.

In 1975, the Saigon government fell, and the United States was ousted—just a little over two decades after the French defeat at Dien Bien Phu. When the last helicopter left from the roof of the U.S. embassy in Saigon amid scenes of desperation and desertion, it marked the end of a sequence of events that closely conformed to the *insurgent state model* we introduced in the context of both contemporary Colombia (pp. 248–250) and Sri Lanka (p. 452).

Vietnam in Transition

During the difficult years following the end of the war, Vietnam's prospects looked bleak not only because of the huge death toll (an estimated 3 million Vietnamese died during the conflict), devastated infrastructure, and chaotic emigration, but also because Hanoi found itself in a territorial conflict with China and faced still more strife.

Domino Theory

DURING THE INDOCHINA War (1964–1975), it was U.S. policy to contain communist expansion by supporting the efforts of the government of South Vietnam to defeat communist insurgents. Soon, the war engulfed North Vietnam as U.S. bombers attacked targets north of the border between North and South. In the later phases of the war, conflict spilled over into Laos and Cambodia. U.S. warplanes also used bases in Thailand. Like dominoes, one country after another fell to the ravages of the war or was threatened by it.

Some scholars warned that this domino effect could eventually affect not only Thailand but also Malaysia, Indonesia, and Burma (today Myanmar): the whole Southeast Asian realm, they predicted, could be destabilized. But, as we know, that did not happen. The war remained confined to Indochina. And the domino "theory" seemed invalid.

But is the theory totally without merit? Unfortunately, some political geographers to this day mistakenly define this idea in terms of communist activity. Communist insurgency, however, is only one way a country may be destabilized (Peru came close to collapse from it). But right-wing rebellion (Nicaragua's Contras), ethnic conflict (Bosnia), religious extremism (Algeria), and even economic and environmental causes can create havoc in a country. Properly defined, the **domino theory** holds that destabilization from any cause in one country can result in the collapse of order in a neighboring country, triggering a chain of events that can affect a series of contiguous states in turn.

A recent instance of the domino effect comes from Equatorial Africa, where the collapse of order in Rwanda spilled over into The Congo (then Zaïre) and also affected Burundi, Congo, and Angola.

17

"Although modernization has changed the skylines of Vietnam's major cities (Saigon more so than Hanoi), most of the country's urban areas look and function much as they have for generations. Main arteries throb with commerce and traffic (still mostly bicycles and mopeds); side streets are quieter and more residential. No high-rises here, but the problems of dense urban populations are nevertheless evident. Every time I turned into a side street, the need for street cleaning and refuse collection was obvious, and drains were clogged more often than not. 'We haven't allowed our older neighborhoods to be destroyed like they have in China,' said my colleague from Hanoi University, 'but we also don't have an economy growing so fast that we can afford to provide the services these people need.' Certainly the Pacific Rim contrast between burgeoning coastal China and slower-growing Vietnam—a matter of communist-government policy—is evident in their cities. 'As you see, nothing much has changed here,' she said. 'Not much evidence of globalization. But we're basically self-sufficient, nobody goes hungry, and the gap between the richest and the poorest here in Vietnam is a fraction of what it is now in China. We decided to keep control.' You would find more agreement on these points here in Hanoi than in Saigon, but the Vietnamese state is stable and progressing."   © H. J. de Blij.

Concept Caching

Not only Vietnamese, but tens of thousands of Chinese citizens of Vietnam fled the country. It was difficult for distant Hanoi to establish effective control over Saigon (renamed Ho Chi Minh City in honor of the North's wartime leader), and soon the two cities and their respective core areas began to grow apart. Indeed, most residents of Ho Chi Minh City continued to call their city Saigon; with about 8 percent of the country's population and a far larger share of its best-educated people, metropolitan Saigon recovered faster than Hanoi, and its old entrepreneurial spirit quickly revived. In the late 1980s, studying Saigon and Hanoi in turn was like visiting two different Southeast Asian countries. Saigon bustled with bicycle and scooter traffic, businesses of all kinds lined streets and alleys, modern buildings were going up, its river port a beehive of activity. A Special Economic Zone on the Chinese model, and a New Saigon business and residential district, signified the new era. Hanoi, by comparison, was quiet and staid (see photos above), its port a four-hour crawl away along a congested, two-lane road.

Today, more than two decades later, those differences and disparities are fading. Hanoi's answer to New Saigon is *Tu Liem*, a modern business-and-residential complex on the city's southwestern periphery complete with four-lane roads and modern amenities. Major hotels and a world-class convention center are changing the capital's skyline. The road to the port of Haiphong now has four lanes, and travel time is down to 90 minutes on a good day. But Saigon and environs still lead, in 2009 contributing more than 25 percent of Vietnam's industrial production and paying about one-third of its taxes. A new, vast development named *Saigon South* is under construction, and a tunnel is being dug beneath the Saigon River (see map, p. 551). The planners in Saigon do not have just Vietnam in mind when they talk about expansion and growth: they want nothing less than the transformation of Ho Chi Minh City into a regional hub for all of Southeast Asia.

During colonial times, central Saigon grew into a gracious and elegant city on the right bank of the Saigon River. Because a bridge was never built, the opposite side of the river remained essentially undeveloped. With the need for a new downtown, and in anticipation of a one-third jump in total population over the next decade, Ho Chi Minh City's leaders aim to build a world-class city center across the river, taking the pressure off what remains of

AMONG THE REALM'S GREAT CITIES . . . *Saigon*

OFFICIALLY, IT IS known as Ho Chi Minh City, renamed in 1975 after the ouster of American forces, but only bureaucrats and atlases call it that. To the people here, it is still *Saigon*, city on the Saigon River, served by, it seems, a thousand companies using the name. Saigon Taxi, Saigon Hotel, Saigon Pharmacy—there is hardly a city block without the name Saigon on some billboard.

The phenomenon is significant. Hanoi's northerners ordained the change, but this still is South Vietnam, and many southerners see the North as peopled by slow, dogmatic communists obstructing the hopes and plans for their vibrant city. According to some estimates, Saigon may be nearly twice as large as Hanoi (and possibly Mainland Southeast Asia's largest city), but like the capital it suffers from decades of deterioration, a broken-down infrastructure, and undependable public services. When, in the early 1990s, businesspeople could find no modern accommodations or working telephones, an Australian firm built an eight-story floating hotel with its own power plant, towed it up the Saigon River, and parked it on the CBD waterfront.

Therein lies Saigon's great advantage: large oceangoing vessels can reach the city in a three-hour sail up the meandering river. The Chinese developed its potential: in the 1960s, Chinese merchants controlled more than half of the South's exports and nearly all of its textile factories and foreign exchange. Saigon-Cholon (Cholon means "great market") was a Chinese success story, but after the war the Hanoi regime nationalized all Chinese-owned

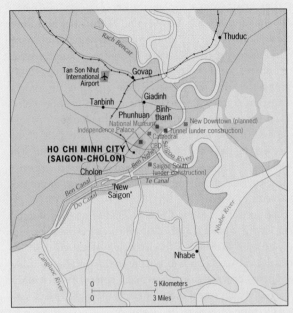

© H. J. de Blij, P. O. Muller, and John Wiley & Sons, Inc.

firms. Tens of thousands of middle-class Chinese left the city, its economy devastated.

Today, Saigon (official population: 5.9 million) is steadily recovering. A modern skyline is rising above the colonial one. A large Special Economic Zone has been created south of the city. Factories are being built. The Pacific Rim era is under way. Will the South rise again?

the historic city—much as Shanghai's Bund was saved by the development of Pudong (see box on p. 488). Three and a half decades after the end of its disastrous war, Vietnam is reaching its potential on its own terms.

CAMBODIA

Former French Indochina contained two additional entities—Cambodia and Laos. In Cambodia, the French possessed one of the greatest treasures of Hinduism: the city of Angkor, capital of the ancient Khmer Empire, and the temple complex known as Angkor Wat. The Khmer Empire prevailed here from the ninth to the fifteenth centuries, and King Suryavarman II built Angkor Wat to symbolize the universe in accordance with the precepts of Hindu cosmology. When the French took control of this area during the 1860s, Angkor lay in ruins, sacked long before by Vietnamese and Thai armies. The French

created a protectorate, began the restoration of the shrines, established Cambodia's permanent boundaries, and restored the monarchy under their supervision.

Geographically, Cambodia enjoys several advantages, most notably its compact shape. Compact states enclose a maximum of territory within a minimum of boundary and are without peninsulas, islands, or other remote extensions of the national spatial framework. Cambodia had the further advantage of strong ethnic and cultural homogeneity: 90 percent of its 15.2 million inhabitants are Khmers, with the remainder divided between Vietnamese and Chinese. As Figure 10-10 shows, Southeast Asia's greatest river, the Mekong, enters Cambodia from Laos and crosses it from north to south, creating a great bend before flowing into southern Vietnam (see box titled "The Mighty Mekong"). Phnom Penh, the country's capital, lies on the river. The ancient capital of Angkor lies in the northwest, not far from the *Tonle Sap* ("Great Lake"), a major lake linked

The Mighty Mekong

FROM ITS SOURCE among the snowy peaks of China's Qinghai and Xizang, the Mekong River rushes and flows some 4200 kilometers (2600 mi) to its delta in southernmost Vietnam. This "Danube of Southeast Asia" crosses or borders five of the realm's countries (see photo, p. 544), supporting rice farmers and fishing people, forming a transport route where roads are few, and providing electricity from dams upstream. Tens of millions of people depend on the waters of the Mekong, from subsistence farmers in Laos to apartment dwellers in China. The Mekong Delta in southern Vietnam is one of the realm's most densely populated areas and produces enormous harvests of rice.

But problems loom. China is building a series of dams across the Lancang (as the Mekong is called there) to supply Yunnan Province with electricity. Although such hydroelectric dams should not interfere with water flow, countries downstream worry that a severe dry spell in the interior would impel the Chinese to slow the river's flow to keep the reservoirs full. Cambodia is concerned over the future of the Tonle Sap, a large natural lake filled by the Mekong (Fig. 10-10). In Vietnam, farmers worry about salt water invading the paddies should the Mekong's level

drop. And the Chinese may not be the only dam builders in the future: Thailand has expressed an interest in building a dam on the Thai-Laos border where it is defined by the Mekong.

In such situations, the upstream states have an advantage over those downstream. Several international organizations have been formed to coordinate development in the Mekong Basin, including the Mekong River Commission (MRC) founded in the 1950s. China has offered to sell electricity from its dams to Thailand, Laos, and Myanmar. Coordinated efforts to reduce logging in the Mekong's drainage basin have had some effect. After consultations with the MRC, Australia built a bridge linking Laos and Thailand. There is even a plan to make the Mekong navigable from Yunnan to the coast, creating an alternative outlet for interior China.

Sail the Mekong today, however, and you are struck by the slowness of development along this artery. Wooden boats, thatch-roofed villages, and teeming paddies mark a river still crossed by antiquated ferries and flanked by few towns. Of modern infrastructure, one sees little. And yet the Mekong and its basin form the lifeline of mainland Southeast Asia's dominantly rural societies.

to, and filled by, the waters of the Mekong. Cambodia has a coastline on the Gulf of Thailand, but its core area lies in the interior; the Mekong is more important than the Gulf. Kompong Som, the port at the end of the railroad from Phnom Penh, has only limited facilities.

Postwar Problems

Cambodia's present social geography continues to suffer from the aftereffects of the Indochina War. Neither its compactness nor its isolation could overcome the disadvantage of relative location near a conflict that was bound to spill across its borders (Fig. 10-10). Internal disharmony fueled by external interference led in 1970 to the ouster of the last king by the military; in 1975, that regime was itself overthrown by communist revolutionaries, the so-called Khmer Rouge. These new rulers embarked on a course of terror and destruction in order to reconstruct Kampuchea (as they called Cambodia) as a rural society. They drove townspeople into the countryside where they had no place to live or work, emptied hospitals and sent the sick and dying into the streets, outlawed religion and family, and in the process killed as many as 2 million Cambodians (out of a popula-

tion of 8 million). In the late 1970s, Vietnam, having won its own war, invaded Cambodia to drive the Khmer Rouge away. But this action led to new terror, and a stream of refugees crossed the border into Thailand. Eventually, remnants of the Khmer Rouge managed to establish a base in the northwest of Cambodia, where their murderous leader, Pol Pot, who was never brought to account for his actions, committed suicide in 1998.

Cambodia's postwar trauma continues today. Once self-sufficient and able to feed others, it now must import food. Rice and beans are the subsistence crops, but the dislocation in the farmlands set production back severely, and continuing strife in the countryside has disrupted supply routes for years. If there is hope for the national budget of the future, it may lie in recently discovered offshore oil reserves near Cambodia's relatively short coastline. Income from oil exports is set to begin in 2011 and could approach U.S. $2 billion annually by the end of the decade, a significant boost to an otherwise weak economy.

Cambodians hope that tourism, focused chiefly on the great temple complex at Angkor, will also become a mainstay of the economy. But political stability is a prerequisite for this industry, and it has not yet returned to this ancient kingdom.

LAOS

North of Cambodia lies Southeast Asia's only landlocked country, Laos (Fig. 10-10). Interior and isolated, Laos changed little during 60 years of French colonial administration (1893–1953). Then, along with other French-ruled areas, it became an independent state. Soon this well-entrenched, traditional kingdom fell victim to rivalries between traditionalists and the victorious communists, and the old order collapsed.

Laos is typical of the poorest and most vulnerable countries in the global periphery, and is one of the lowest-income states in Southeast Asia. With 6.2 million people (more than half of them ethnic Lao), squeezed between Thailand and Vietnam and dependent on the sale of electricity from a growing number of dams, minerals from mines run by Australian and other foreign companies, and lumber from its dwindling forests, its natural environments are fast deteriorating, worsened further by the development of export-driven commercial agriculture (coffee, sugarcane) by Chinese, Thai, and Vietnamese leaseholders.

As Figure 10-10 shows, Laos's capital, Viangchan (Vientiane), lies on the Mekong River that forms much of the boundary with Thailand. An oil pipeline from Vietnam's coast provides its energy needs, but other than some textile factories there is as yet very little industry. Laos, about the size of Michigan, in 2010 was only 27 percent urbanized, had no railways, and contained fewer than 100 kilometers (60 mi) of paved road. Neither its farmers nor its leaders are prepared to cope with the intrusion of the forces of globalization.

THAILAND

In virtually every way, Thailand is the leading state of the Mainland region. In contrast to its neighbors, Thailand has been a strong participant in the Pacific Rim's economic development. Its capital, Bangkok, is the largest urban center in the Mainland region and one of the world's most prominent primate cities. The country's population, 66.8 million in 2010, is now growing at the slowest rate in this geographic realm. Over the past few decades, only political instability and uncertainty have inhibited economic progress. Although Thailand is a constitutional monarchy with an elected parliament, its progress toward stable democracy has a history of setbacks. The latest of these began in 2006, when the armed forces ousted a controversial prime minister and took control amid the usual promises of a return to representative government. By mid-2009, the country had witnessed serious civil disorder involving violence in the streets of Bangkok and the closure, by thousands of protesters, of both its domestic and international airports for several days as supporters of the supposedly infallible king clashed with the exiled prime minister's loyalists. Enormous damage was done to Thailand's stature and its tourist-dependent economy.

Thailand is a classic example of a protruded state. From a relatively compact heartland, in which lie the core area, capital, and major areas of productive capacity, a 1000-kilometer (600-mi) corridor of land, in places less than 30 kilometers (20 mi) wide, extends southward to the border with Malaysia (Fig. 10-11). The boundary that defines this protrusion runs down the length of the upper Malay Peninsula to the Kra Isthmus, where neighboring Myanmar peters out and Thailand fronts the Andaman Sea (an arm of the Indian Ocean) as well as the Gulf of Thailand. In the entire country, no place lies farther from the capital than the southern end of this tenuous protrusion.

Consider this spatial situation in the context of Figure 10-5 (p. 536). Note that the Malay ethnic population group extends from Malaysia more than 300 kilometers (200 mi) northward into Thai territory. In Thailand's six southernmost provinces, 85 percent of the inhabitants are Muslims (the figure for Thailand as a whole is less than 4 percent). The political border between Thailand and Malaysia is porous; in fact, you can cross it at will almost anywhere along the Kolok River by canoe, and inland it is a matter of walking a forest trail. For more than a century, the southern provinces have had closer ties with Malaysia across the border than with Bangkok 1000 kilometers (600 mi) away, and the Thai government has not tried to restrict movement or impose onerous rules on the Muslim population here.

The Restive Peninsular South

But in the new era of Islamic militancy and in view of rising violence in this southern frontier zone, Thailand's south is attracting national attention. The Pattani United Liberation Organization, named after the functional capital of the Muslim south, was active in the 1960s and 1970s but had been dormant since. Then a series of bombings at a Pattani hotel, several schools, a Chinese shrine, and a Buddhist temple in 2001 and 2002 raised fears that Muslim-inspired violence was on the rise again. In 2003, the notorious al-Qaeda terrorist Hambali, who worked to link al-Qaeda with its Southeast Asian counterpart, Jemaah Islamiyah, was captured in Thailand.

The middle years of the 2000s, however, brought the most ominous developments yet. A terrorist attack on an army base in 2004 killed four Buddhist soldiers and resulted in the theft of hundreds of rifles. And several

AMONG THE REALM'S GREAT CITIES . . . Bangkok

BANGKOK (POPULATION: 7.0 MILLION), mainland Southeast Asia's largest city, is a sprawling metropolis on the banks of the Chao Phraya River, an urban agglomeration without a center, an aggregation of neighborhoods ranging from the immaculate to the impoverished. In this city of great distances and high tropical temperatures, getting around is often difficult because roadways are choked with traffic. A (diminishing) network of waterways affords the easiest way to travel, and life focuses on the busiest waterway of all, the Chao Phraya. Ferries and water taxis carry tens of thousands of commuters and shoppers across and along this bustling artery, flanked by a growing number of high-rise office buildings, luxury hotels, and ultramodern condominiums. Many of these modern structures reflect the Thais' fondness for domes, columns, and small-paned windows, creating a skyline unique in Asia.

Gold is Buddhist Thailand's symbol, adorning religious and nonreligious architecture alike. From a high vantage point in the city, you can see hundreds of golden spires, pagodas, and façades rising above the townscape. The Grand Palace, where royal, religious, and public buildings are crowded inside a white, crenellated wall 2 kilometers (more than 1 mi) long (see photo, p. 556), is a gold-laminated city within a city embellished by ornate gateways, dragons, and statuary. Across the mall in front of the Grand Palace are government buildings sometimes tar-

© H. J. de Blij, P. O. Muller, and John Wiley & Sons, Inc.

geted by rioters in Bangkok's volatile political atmosphere. Not far away are Chinatown and myriad markets; Bangkok has a throbbing commercial life.

Decentralized, dispersed Bangkok has major environmental problems, ranging from sinking ground (resulting in part from excessive water pumping) to some of the world's worst air pollution. Yet industries, port facilities, and residential areas continue to expand as the city's hinterland grows and prospers with Pacific Rim investment. Bangkok is the pulse of Southeast Asia's Mainland region.

---

dozen people, including Buddhist monks and police officers, were killed in subsequent skirmishes when counterterrorism forces in the town of Tak Bai arrested six Muslims suspected of stealing weapons from the Thai armed forces. A riot outside the police station where the suspects were being held then led to a mass arrest and the deaths of nearly 80 Muslims during transfer to a military base. The government responded by declaring martial law in Thailand's three southernmost provinces (Fig. 10-11, inset map). By 2005, a series of renewed attacks by southern Muslims had provoked then-Prime Minister Thaksin to issue an emergency decree enabling him to more effectively combat the militants; but its effect was to worsen what was becoming a full-scale regional insurgency. When Thaksin, who had become the focal point of Muslim anger, was deposed by the military in 2006, the new regime immediately sought to negotiate a peace with the insurgents.

The border area nevertheless remained tense. The Thai government has claimed that some Muslim insurgents it arrested had received training in Malaysia's

northern State of Kelantan. When several hundred Muslims crossed the border from Thailand into Malaysia to escape the fighting, the Malaysian government refused to return them despite Thailand's claim that some of the refugees were in fact militant separatists. In 2007, Muslim insurgents set fire to the largest rubber-storage facility in Yala Province on the Malaysian border, causing major damage to the rubber-dominated economy in this part of the peninsula and proving that their campaign had not been suppressed.

Religious tensions in this increasingly troubled Islamic outpost should be seen in the context of its geography. Thailand's pronounced protrusion, its porous border with Muslim Malaysia (the central government now proposes the building of a "security fence" there), its dependence on foreign tourism, and the vulnerable location of its high-profile Andaman Sea coastal tourist facilities (which have rebounded from the 2004 tsunami disaster that locally claimed 5400 lives), combine to raise concern over the stability and security of this distant part of the national territory.

FIGURE 10-11

© H. J. de Blij, P. O. Muller, and John Wiley & Sons, Inc.

Thailand's Core and Pacific Rim Prospects

As Figure 10-2 shows, Thailand occupies the heart of the Mainland region of Southeast Asia. While Thailand has no Red, Mekong, or Irrawaddy Delta, its central lowland is watered by a set of streams that flow off the northern highlands and the Khorat Plateau in the east. One of these streams, the Chao Phraya, is the Rhine of Thailand; from the head of the Gulf of Thailand to Nakhon Sawan, this river is a busy highway filled with traffic (Fig. 10-11). Barge trains loaded with rice head for the coast, ferry boats head upstream, and freighters transport tin and tungsten (of which Thailand is among the world's largest producers). Bangkok sprawls on both sides of the lower Chao Phraya, here flanked by skyscrapers, pagodas, factories, boatsheds, ferry landings, luxury hotels, and modest dwellings in crowded confusion. On the right bank of the Chao Phraya, Bangkok's west side, lie the city's remaining klong (canal) neighborhoods, where waterways and boats form the transport system. Bangkok is still known as the Venice of Asia, although many klongs have been filled in and paved over to serve as roadways.

During the Pacific Rim boom of the 1990s, Thailand's economy grew rapidly. Relative location, natural environment, and social conditions all contributed to this growth. Situated at the head of the Gulf of Thailand, the country's core area opens to the South China Sea and hence the Pacific Ocean. The Gulf itself has long been a rich fishing ground and more recently a source of oil (the major refinery lies in the port area of Laem Chabang, an industrial zone just to the north of the tourist resort of Pattaya). But the key to Thailand's economic explosion was the Thai workforce, laboring under often dreadful conditions for low wages to produce goods sold on foreign markets at cheap prices. This labor force attracted major foreign investment that, in turn, stimulated service industries.

For a time, Thailand became a Pacific Rim economic tiger, producing goods ranging from plastic toys to "Japanese" cars. But Thailand's success story turned out to have its downs as well as ups. While the workers toiled, government mismanagement of the economy produced a recession in the late 1990s that resembled the one we have experienced recently in the United States and the rest of the global core, except that Thailand's currency, the *baht*, began the kind of slide in value that economists fear will affect the U.S. dollar but by the end of 2009 had not. As the *baht* lost value, banks collapsed, real estate values plunged, jobs were lost, exports declined, and Thailand's economy looked nothing like that of a Pacific Rim tiger.

Eventually Thailand achieved a recovery, helped by the improvements in its infrastructure made during the pre-recession years. Investments in highways, ports, power supply, airports, and housing all served to expedite this comeback, but once again Thailand's very success sowed seeds of trouble. Infrastructural betterment focused on Bangkok and Thailand's core area, but tended to leave the countryside behind. That created an opportunity for political advantage, and even as Thailand's per-capita GNI recovered to nearly U.S. $8,000 in 2008, a populist leader had become prime minister, and the country was disastrously divided between his mainly rural followers and those of the king, the army, and the wealthy urban elite. By the end of the century's first decade, Thailand's balance of trade was again favorable, but its political life was dangerously out of balance.

Tourism Bright and Dark

Under normal circumstances, Thailand's natural environments and cultural landscapes continue to attract millions of visitors annually. The magnificent architecture of Bud-

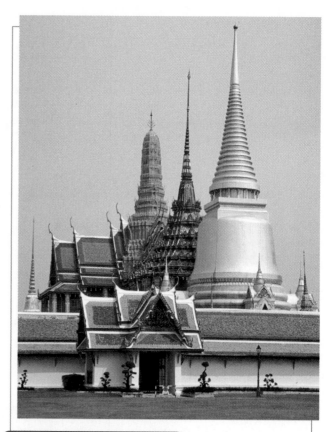

FROM THE FIELD NOTES

"Thailand exhibits one of the world's most distinctive cultural landscapes, both in its urban and its rural areas. Graceful pagodas, stupas, and spirit houses of Buddhism adorn city and countryside alike. Gold-layered spires rise above, and beautify cities and towns that would otherwise be drab and featureless. The architecture of Buddhism has diffused into public architecture, so that many secular buildings, from businesses to skyscrapers, are embellished by something approaching a national style. The grandest expression of the form undoubtedly is a magnificent assemblage of structures within the walls of the Grand Palace in the capital, Bangkok, of which a small sample is shown here. Climb onto the roof of any tall building in the city's center, however, and the urban scene will display hundreds of such graceful structures, interspersed with the modern and the traditional townscape."   © H. J. de Blij.

www.conceptcaching.com

Concept Caching

dhism (the country's old name, *Siam*, was a Chinese word meaning golden-yellow), swathed in gold and shining in the tropical sun, is without equal, as the photo on this page underscores. The warm coasts and beaches have long made this country a favorite destination for tourists from Germany to Japan. One of the world's most famous resorts is Phuket, on the distant Andaman coast near the southwestern end of Thailand's Malay Peninsula protrusion.

Tourism is Thailand's single leading source of foreign revenues, but there also is a dark side to the industry. The

Thai people's relaxed attitude toward sex has made Thai tourism in large part a sex industry, candidly advertised in such foreign markets as Japan where "sex tours" attract planeloads of male participants. When the AIDS pandemic reached Thailand in the 1980s, it changed no habits, and by the mid-1990s the country was suffering one of the world's worst outbreaks of the disease. After peaking at more than 1 million in the early 1990s, the latest number of HIV cases reported at press time was about 610,000 in mid-2008.

Thailand's territorial morphology creates both problems and opportunities. Problems include the effective integration of the national territory through surface communications, which must cover long distances to reach remote locales; the influx of refugees from neighboring countries dislocated by internal conflict; and control over contraband (opium) drug production and trade in the interior "Golden Triangle" where Myanmar, Thailand, and Laos meet. The government's efforts to speed the development of rural areas, notably the northeast, through megaprojects such as large dams have disrupted local life, damaged ecologies, and deepened the split between city and countryside. On the other hand, Thailand's lengthy Kra Isthmus creates economic opportunities where southernmost Thailand lies close to Malaysia and Indonesia, raising hopes for a three-country development scheme in that area. But these hopes—as well as those anticipating Thailand's elevation from lower- to upper-middle income status (see Fig. G-11)—depend on good governance and political stability. In recent years, these have not been attainable.

MYANMAR

Thailand's also-protruded neighbor, Myanmar (still referred to as Burma in some anti-regime quarters), is one of the world's poorest countries where, it seems, time has stood still for centuries. Long languishing under one of the world's most corrupt and brutal military dictatorships, Myanmar shares with Thailand the neck of the Malay Peninsula—but look again at Figure 10-11 and note the contrast in surface communications.

Morphology and Structure

Myanmar's territorial morphology is complicated by a shift in the Burmese core area that took place during colonial times. Prior to the colonial period, the focus of embryonic Burma lay in the so-called dry zone between the Arakan Mountains and the Shan Plateau, which covers the country's triangular eastern extension toward the Laotian border (Fig. 10-3). The urban focus of the state

was Mandalay, which had a central situation and relative proximity to the non-Burmese highlands all around. Then the British developed the agricultural potential of the Irrawaddy Delta, and Rangoon (now called Yangon) became the capital and hub of the colony. The Irrawaddy waterway links the old and the new core areas, but the center of gravity has shifted to the south.

The political geography of Myanmar constitutes a particularly good example of the role and effect of territorial morphology on internal state structure (Fig. 10-12). Not only is Myanmar a protruded state: its Irrawaddy core area is also surrounded on the west, north, and east by a horseshoe of great mountains—where many of the country's 11 minority peoples had their homelands before the British occupation. The colonial boundaries had the effect of incorporating these peoples into the domain of the Burman people, who today constitute just over two-thirds of the population of 50.1 million. When the British departed and Burma became independent in 1948, the minorities traded one master for another. In 1976, nine of these indigenous peoples formed a union to demand the right of self-determination in their homelands.

As Figure 10-5 indicates, the peripheral peoples of Myanmar occupy a significant part of the state. A closer look at Figure 10-12 shows that their domains have the status of internal State, of which there are seven; the Burman-dominated areas are designated as Divisions. The Shan of the northeast and far north, who are related to the neighboring Thai, account for about 9 percent of the population, or 4.5 million. The Kayin (Karen), who constitute just over 7 percent (3.5 million), live in the neck of Myanmar's protrusion and have proclaimed their desire to create an autonomous territory within a federal Myanmar. The Mon (2.3 percent; 1.2 million) were in what is today Myanmar long before the Burmans, and introduced Buddhism to the area; they want the return of ancestral lands from which they were ousted.

But the military junta refuses even to consider such aspirations. Repression, corruption, and incompetence have produced a poverty-stricken, malnourished, fearful populace in a country once able to export large cargoes of staples. Rampant diseases rage in a state where the rulers spend 2 percent of the health budget on the people and 40 percent on the army. Opposition has come from courageous political leaders (Aung San Suu Kyi of the National League for Democracy, whose resistance won her the 1991 Nobel Peace Prize, has been under intermittent house arrest for decades) and from many of Myanmar's 400,000 Buddhist monks, who have repeatedly challenged the junta and did so again in 2007. Hundreds were killed by the "security forces"; when the senior author of this book gained entry to Myanmar in April 2008, Yangon's barbed-wire barricades and sandbagged street-corner foxholes were still in place, and

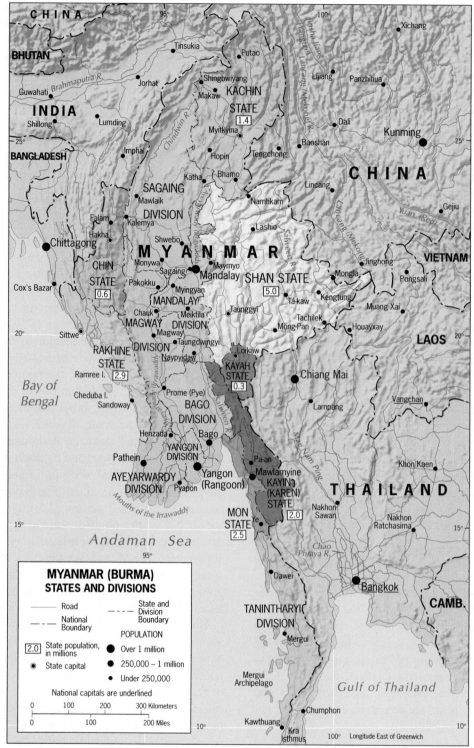

FIGURE 10-12

© H. J. de Blij, P. O. Muller, and John Wiley & Sons, Inc.

bullet holes pocked the walls of buildings along the monks' protest-march route.

Contrasts between the luxurious villas of families with links to the regime and the squalid, overcrowded tenements housing most of Yangon's population mark the urban landscape of a city of 4.4 million losing its archi-

tectural heritage through decay and dereliction (see the cover of this book). In 2005, news reports began to suggest that the junta was in the process of abandoning Yangon as Myanmar's capital in favor of a site about 300 kilometers (190 mi) due north, about halfway to Mandalay, named Naypyidaw. According to visitors to this new

Yangon, Myanmar. "Little did I realize, as I walked the central business district of Yangon (Myanmar) in mid-April 2008, that these streets would be turned into chaos by Cyclone Nargis just days later. This is Sule Pagoda Road, a few blocks north of the famed Octagonal Pagoda, the golden landmark in the center of town. Many signs in English remind you that this was once a British colony."   © H. J. de Blij

Concept Caching

www.conceptcaching.com

capital in 2007, much of the construction of government facilities was taking place underground. In 2008, Naypyidaw seemed to consist of modest ministry buildings, pastel-colored residential compounds, and eight-lane avenues, but no access to any underground structures was provided and no questions on the matter were answered. The new zoo, however, was a visitor "must"— yet the animals outnumbered the visitors!

The regime is reported to proclaim that the new capital's more central location would be more efficient administratively than Yangon and that its subsurface configuration would offer better protection against (unspecified) enemies. More immediately visible, however, is the junta's reliable friend: China. Chinese construction, from the bridge linking the city of Yangon with its port to the pipeline that will send Myanmar's natural gas to China's Yunnan Province, is crucial to the junta; China is also the junta's biggest arms supplier by far. Another pipeline, for Middle East oil, will link the port of Sittwe on the Bay of Bengal to Yunnan as well. Relative location is one of Myanmar's obvious assets.

In May 2008 the junta had one more opportunity to display its disregard for its subjects. A massive tropical cyclone struck the low-lying Irrawaddy Delta and curved inland, killing an estimated 100,000 people (no casual-

ty count was made available, or perhaps even compiled, by the regime). Myanmar's rulers refused the transfer of assistance from French and American warships soon on the scene in the Andaman Sea; it was slow to grant visas to foreign aid workers; and it delayed giving clearance to relief-supplying aircraft.

Myanmar today ranks among the world's poorest, least-developed, and most corrupt (with Somalia) countries in the world, not because of its intrinsic indigence but because a rapacious military regime, tolerated and even commercially engaged by the "international community," represses its peoples, crushes their aspirations, and demonstrates the liabilities of life in the global periphery.

INSULAR SOUTHEAST ASIA

On the peninsulas and islands of Southeast Asia's southern and eastern periphery lie six of the realm's 11 states (Fig. 10-2). Few regions in the world contain so diverse a set of countries. Malaysia, the former British colony, consists of two major areas separated by hundreds of kilometers of South China Sea. The realm's southernmost state, Indonesia, sprawls across thousands of islands from Sumatera in the west to New Guinea in the east. North of the Indonesian archipelago lies the Philippines, a country that once was a U.S. colony. These are three of the most severely fragmented states on Earth, and each has faced the challenges that such politico-spatial division brings. This Insular region of Southeast Asia also contains two small but important sovereign entities: a city-state and a sultanate. The city-state is Singapore, once a part of Malaysia (and one instance in which internal centrifugal forces were too great to be overcome). The sultanate is Brunei, an oil-rich Muslim territory on the island of Borneo that seems transplanted from the Persian Gulf. In addition, a third small entity, East Timor, achieved independence in 2002. Few parts of the world are more varied or more interesting geographically.

MAINLAND-ISLAND MALAYSIA

The state of Malaysia represents one of the three subtypes of fragmented states listed earlier: the mainland-island type, in which one part of the national territory lies on a continent and the other on an island. Malaysia is a colonial political artifice that combines two quite disparate components into a single state: the southern end of the Malay Peninsula and the northern part of the island of Borneo. These are known, respectively, as West Malaysia and East Malaysia (Fig. 10-2). The name *Malaysia* came into use in 1963, when the original Federation of Malaya, on the Malay Peninsula, was expanded to incorporate the areas of Sarawak and Sabah in Borneo. When the name *Malaya* is used, it refers to the peninsular part of the Federation, whereas Malaysia refers to the total entity.

Ethnic Components

The Malays of the peninsula, traditionally a rural people, displaced older indigenous communities there and today make up about 58 percent of the country's population of 28.6 million. They possess a strong cultural identity expressed in adherence to the Muslim faith, a common language, and a sense of territoriality that arises from their perceived Malayan origins and their collective view of Chinese, Indian, European, and other foreign intruders.

The Chinese came to the Malay Peninsula and to northern Borneo in substantial numbers during the colonial period, and today they constitute about one-fourth of Malaysia's population (they are the largest single group in Sarawak).

Hindu South Asians were in this area long before the Europeans, and for that matter before the Arabs and Islam arrived on these shores. Today they still form a substantial minority of 8 percent of the population, clustered, like the Chinese, on the western side of the peninsula (Fig. 10-5).

Upon independence in 1957, the majority Malays, fearful of being marginalized by the successful Chinese and Indians in their country, adopted a constitution that,

in effect, discriminated against the latter two in favor of the Malays. The constitution enshrined privileges for Malays, including government contracts, state jobs, university admissions, and other forms of "empowerment" even as it guaranteed religious freedom in a secular state. A political party, the United Malays National Organisation (UNMO), has formed the government ever since, and even had the support of some Chinese and Indians whose businesses depended on government connections and social stability.

Serious ethnic violence in 1969 resulted in an expansion of Malay entitlements, which over time yielded charges of official racism, corruption, and the use of "security laws" to silence critics of the government, Malay as well as non-Malay. As to official secularism, moderate Malay Islam evolved into more fundamentalist strains, and this revivalism, as we will see, has geographic dimensions. Many citizens of Malaysia today are concerned that the country's three main cultural sectors are leading increasingly separate lives even as the government's policies are widening the income gap among Malays themselves.

The Dominant Peninsula

Always in Malaysia's modern history, the Malay Peninsula (West Malaysia) has been the country's dominant component with its core area, 11 of its 13 States, and fully 80 percent of its population. This is where the Malay-dominated government has most strictly controlled economic and social policies while pushing the country's modern-

---

FROM THE FIELD NOTES

"Walking along Tuanku Abdul Rahman Street in Kuala Lumpur, I had just passed the ultramodern Sultan Abdul Samad skyscraper when this remarkable view appeared: the old and the new in a country that seems to have few postcolonial hang-ups and in which Islam and democracy coexist. The British colonists designed and effected the construction of the Moorish-Victorian buildings in the foreground (now the City Hall and Supreme Court); behind them rises the Bank of Commerce, one of many banks in the capital. Look left, and you see the Bank of Islam, not a contradiction here in economically diversified Malaysia. And just a few hundred yards away stands St. Mary's Cathedral, across the street from still another bank. Several members of the congregation told me that the church was thriving and that there was no sense of insecurity here. 'This is Malaysia, sir,' I was told. 'We're Muslims, Buddhists, Christians. We're Malays, Chinese, Indians. We have to live together. By the way, don't miss the action at the Hard Rock Café on Sultan Ismail Street.' Now there, I thought, was a contradiction as remarkable as this scene." © H. J. de Blij.

www.conceptcaching.com

ization. During the Asian economic boom of the 1990s, Malaysia's planners embraced the notion of symbols: the capital, Kuala Lumpur, was endowed with what was then the world's tallest building; a space-age airport outpaced Malaysia's needs; a high-tech administrative capital was built at Putrajaya; and a nearby development was called Cyberjaya—all part of a so-called *Multimedia Supercorridor* to anchor Malaysia's core area (Fig. 10-13).

The chief architect of this program was Malaysia's long-term and autocratic head of state, Mahathir bin Mohamad, leader of the UNMO. Mahathir had the support not only of the great majority of the country's Malays, but also that of many in the important and influential ethnic Chinese minority, which saw him as the only acceptable alternative to the more fundamentalist Islamic party challenging his rule. But Malaysia's headlong rush to modernize caused a backlash among more conservative Muslims, which, in 2001, led to Islamist election victories in two States, tin-producing Kelantan and energy-rich but socially poor Terengganu. As the fundamentalist governments in those two States imposed strict religious laws, Malaysians talked of two "corridors" marking their country: the Multimedia Supercorridor in the west and the Mecca Corridor in the east (Fig. 10-13). But after Mahathir's resignation and the appointment of Abdullah Badawi as his more moderate successor, Islamist fervor in the Mecca Corridor, which had featured calls for a *jihad* in Malaysia, began to wane. In the 2004 elections, the Islamist party lost the two State legislatures it had gained in 2001, and its leader even lost his seat in the federal parliament.

Even as revivalist Islam suffered electoral setbacks, however, conservative Muslims continued their infringement of the secular state. In 2007, Terengganu State established official links between a network of informants and the State's Islamic Morality Police to report instances of illegal contact between members of the opposite sex in public places like parks and beaches. Also in 2007, Malaysia's Supreme Court revealed the extent to which the constitution's guarantee of religious freedom had been eroded by refusing to endorse an Islamic woman's request to change her religion (specified on her official identity card) from Muslim to Christian. The Mecca Corridor still merits its name.

The Malay Peninsula's primacy began long ago, during colonial times, when the British created a substantial economy based on rubber plantations,

STATES OF WEST MALAYSIA

	'Multimedia Supercorridor'	
	States with greatest Islamic strength	
	States with largest Chinese and Hindu presence	

— Highway
— Main road
— Railroad
✈ Airport

POPULATION
● 50,000–250,000
● 250,000–1,000,000
● 1,000,000–5,000,000
National capital is underlined

FIGURE 10-13   © H. J. de Blij, P. O. Muller, and John Wiley & Sons, Inc.

palm-oil extraction, and mining (tin, bauxite, copper, iron). The Strait of Malacca (Melaka) became one of the world's busiest and most strategic waterways, and Singapore, at the southern end of it, a cornerstone of the young state. The formal incorporation of Singapore occurred in 1963, an important moment in the historical geography of Malaysia because this was also the year when two British colonial territories on the Indonesian island of Borneo, Sabah and Sarawak, were made part of it and the country's name was changed from Malaya to Malaysia.

The marriage between Malaya and Singapore, however, did not work well. Singapore, like Malaysia, was (and still is) a multicultural society, but unlike Malaysia the great majority of its citizens are of Chinese ancestry. After two difficult years, Singapore in 1965 seceded from Malaysia and subsequently became one of the world's richest states.

Malaysian Borneo

The decision to combine the 11 Sultanates of Malaya with the States of Sabah and Sarawak on Borneo, creating the country now called Malaysia, had far-reaching consequences for the Federation. These two States make up 60 percent of Malaysia's territory (although they represent only 20 percent of the population). They endowed Malaysia with major energy resources and huge stands of timber. They also complicated Malaysia's ethnic makeup because each State is home to more than two dozen indigenous groups (in fact, the immigrant Chinese form the largest single group in Sarawak). These locals complain that the federal government in Kuala Lumpur treats East Malaysia as a colony, and politics here are contentious and fractious. It is likely that Malaysia will eventually confront devolutionary forces here in East Malaysia.

SINGAPORE

In 1965, a fateful event occurred in Southeast Asia. Singapore, crown jewel of British colonialism in this realm, seceded from the newly independent Malaysian Federation and became a sovereign state, albeit a ministate (Fig. 10-14). With its magnificent relative location, its human resources, and its firm government, Singapore

George Town, the capital of Malaysia's Pinang State, is sometimes referred to as Malaysia's new Singapore—but the city does not yet have the skyline to match. This photo, taken from the top of the Komtar Tower in 2007, nonetheless reveals George Town's modern, high-rise profile, second only to Kuala Lumpur in the Malaysian Federation.   © Alamy

FIGURE 10-14

© H. J. de Blij, P. O. Muller, and John Wiley & Sons, Inc.

then overcame the limitations of space and the absence of raw materials to become one of the economic tigers on the Pacific Rim.

With a mere 619 square kilometers (239 sq mi) of territory and a population of 4.9 million, space is at a premium in Singapore, and this is a constant worry for the government. The Jurong Reclamation Project has recently joined several islands and enlarged Singapore's territory (Fig. 10-14), but Singapore's leaders are always trying to find additional ways to add space. One plan is to ask Indonesia to, in effect, lease one of its many islands to Singapore, allowing Singapore to convert it into a gigantic industrial complex administered from the city-state. Already, Singapore has built factories on the nearby Indonesian islands of Batam and Bintan (Fig. 10-14), part of a chain of industrial parks, ports, and resorts it has

constructed from China to Thailand. The proximity of Batam and Bintan makes them especially attractive for large-scale development of this kind, and Indonesia may find Singapore's offer too lucrative to decline. Exploratory discussions (relating only to Bintan) have been under way since 2007.

In any case, Singapore's expanding economy requires space-conserving high-technology and service industries in this new era. At first, benefiting from its relative location on the key corner of Asia, Singapore became one of the world's busiest ports (by number of ships served) even before independence. It thrived as an *entrepôt* between the Malay Peninsula, Southeast Asia, Japan, and other emerging economic powers on the Pacific Rim and beyond. Crude oil from Southeast Asia still is unloaded and refined at Singapore, then shipped to Asian destinations. Raw rubber from the adjacent peninsula and from Indonesia's island of Sumatera is shipped to Japan, the United States, China, and elsewhere. Timber from Malaysia, rice, spices, and other foodstuffs are processed and forwarded via Singapore. In return, automobiles, machinery, and equipment are imported into Southeast Asia through Singapore.

But that is the old pattern. The Singaporean economy today is propelled by international banking and finance, communications, and tourism. Accordingly, the city-state's planners are redirecting the domestic economy toward a high-technology industrial future. In Singapore the government tightly controls business as well as other aspects of life. (Some newspapers and magazines have been banned for criticizing the regime, and there are even fines for such things as eating on the subway and failing to flush a public toilet.) Its overall success after secession has tended to keep the critics quiet: while GNI per capita from 1965 to 2007 multiplied by a factor of more than 17 to reach U.S. $32,470, that of neighboring Malaysia reached only $13,570. Among other things, Singapore became (and for many years remained) the world's largest producer of disk drives for small computers.

To accomplish its revival, Singapore has moved in several directions. First, it is concentrating on three growth areas: information technology, automation, and biotechnology. [19] Second, there are notions of a **growth triangle** involving Singapore's developing neighbors, Malaysia and Indonesia; those two countries would supply the raw materials and cheap labor, and Singapore the capital and technical know-how. Third, Singapore opened its doors to capitalists of Chinese ancestry who left Hong Kong when China took it over and who wanted to relocate their enterprises here. Singapore's population is 77 percent Chinese, 14 percent Malay, and 8 percent South Asian. The government is Chinese-dominated, and its policies have served to sustain Chinese control. Indeed, Singapore's combination of authoritarianism and economic success often is cited in China itself as proving that communism and market economies can coexist.

INDONESIA'S ARCHIPELAGO

The world's fourth-largest country in terms of human numbers is also the globe's most expansive **archipelago**. [20] Spread across a chain of more than 17,000 islands, Indonesia's 247.2 million people live both separated and clustered—separated by water and clustered on islands large and small.

The complicated map of Indonesia (Fig. 10-15) requires close attention. Five large islands dominate the archipelago territorially, but one of these, *New Guinea* in the far east, is not part of the Indonesian culture sphere, although its western half is under Indonesian control. The other four major islands are collectively known as the Greater Sunda islands: *Jawa* (Java), smallest but by far the most populous and important; *Sumatera* (Sumatra) in the west, directly across the Strait of Malacca from Malaysia; *Kalimantan*, the Indonesian sector of large, compact, minicontinent Borneo; and wishbone-shaped, distended *Sulawesi* (formerly Celebes) to the east. Extending eastward from Jawa are the Lesser Sunda Islands, including Bali and, near the eastern end, Timor. Another important island chain within Indonesia is the Maluku (Molucca) Islands, between Sulawesi and New Guinea. The central water body of Indonesia is the Java Sea.

Rich and Broken Brunei

BRUNEI IS AN anomaly in Southeast Asia—an oil-exporting Islamic sultanate far from the Persian Gulf. Located on the north coast of Borneo, sandwiched between Malaysian Sarawak and Sabah (Fig. 10-8, upper-left map), the Brunei sultanate is a former British-protected remnant of a much larger Islamic kingdom that once controlled all of Borneo and areas beyond. Brunei achieved full independence in 1984. With a mere 5770 square kilometers (2228 sq mi)—slightly larger than Delaware—and only 425,000 people, Brunei is dwarfed by the other political entities of Southeast Asia (Fig. 10-1). But the discovery of oil here in 1929 (and natural gas in 1965) heralded a new age for this remote territory.

Today, Brunei is one of the largest oil producers in the British Commonwealth, and recent offshore discoveries suggest that production will increase. As a result, the population is growing rapidly through immigration (67 percent of Brunei's residents are Malay, 15 percent Chinese), and the sultanate enjoys one of the highest standards of living in Southeast Asia. In 1998, however, a financial scandal of gigantic proportions was revealed, involving the disappearance of more than U.S. $15 billion and the collapse of Brunei's largest private company, which had been under the direction of the sultan's youngest brother. This financial calamity continued to buffet the country's economy over the past decade and was reflected by a steep drop in its per-capita GNI.

Most of Brunei's inhabitants live and work near the oilfields in the western corner of the country and in the capital, Bandar Seri Begawan. Evidence of profligate spending is everywhere, ranging from sumptuous palaces to magnificent mosques to luxurious hotels. In this respect Brunei offers a stark contrast to neighboring East Malaysia, but even within Brunei there are spatial differences. Brunei's interior remains an area of subsistence agriculture (a small minority of indigenous groups survive here) and rural isolation, its villages a world apart from the modern splendor of the coast.

Brunei's territory is not only small: it is fragmented. A sliver of Malaysian land separates the larger west from the east (where the capital lies), and Brunei Bay provides a water link between the two parts. It has been proposed that Brunei purchase this separating corridor from the Malaysians, but no action has yet been taken.

As the map shows, Indonesia does not control all of the archipelago. In addition to East Malaysia's territory on northern Borneo, there also is Brunei on that island's northwest-facing coast (see box titled "Rich and Broken Brunei"). And on the eastern part of the island of Timor, a long struggle against Indonesian rule led, in 1999, to a referendum on independence. The overwhelming majority of East Timorese voted in favor of sovereignty, and after guidance from the United Nations and military intervention led by Australia, East Timor achieved sovereignty in 2002 (this country is discussed separately on pp. 572–573).

The Major Islands

Indonesia is a Dutch colonial creation, and the Hollanders chose Jawa as their colonial headquarters, making Batavia (now Jakarta) their capital. Today, *Jawa* remains the country's core area (Fig. 10-2). With about 140 million inhabitants—56 percent of Indonesia's population—Jawa is one of the world's most densely settled places (see Fig. 10-4). It also is the most highly urbanized part of a country in which 52 percent of the people still live on the land, and the Pacific Rim boom of the 1990s had strong impact here. The megacity of Jakarta (10.1 million), on the northwestern coast, became the heart of a larger conurbation now known as *Jabotabek*, consisting of *Ja*karta as well as *Bo*gor, *Ta*ngerang, and *Bek*asi. During the 1990s the population of this megalopolis grew from 15 to 20 million, and it is predicted to exceed 30 million by the time official numbers for 2010 are reported. Already, Jabotabek is home to one-eighth of Indonesia's entire population and 25 percent of its urban population. Thousands of factories, their owners taking advantage of low prevailing wages, were built in Jabotabek, straining its infrastructure and overburdening the port of Jakarta. On an average day, hundreds of ships lie at anchor, awaiting docking space to offload raw materials and take on finished products.

Always in Indonesia, Jawa is where the power lies. As an ethnic-cultural group (though itself heterogeneous), the Jawanese constitute about 45 percent of the country's population. In the center of the island, a politically powerful sultanate centers on Yogyakarta. Jawa also is the main base for the country's two national Islamic movements (Indonesia is home to more Muslims than any other nation on Earth).

Major active volcanoes in or near densely populated areas always raise concern, and on the island of Jawa such worries focus on Mount Merapi (photo p. 533), located in the middle of the Semarang–Yogyakarta–Surakarta

FIGURE 10-15

triangle (Fig. 10-15). In 2006, Mount Merapi showed signs of an impending eruption, and thousands of people were moved from the slopes of the nearly 3000-meter (10,000-ft) high volcano that, in 1867, caused death and destruction over a wide area. Smaller, but also deadly, outbursts occurred in 1930 and 1994. In the end, the 2006 threat did not materialize, but the towering mountain may someday join Tambora and Krakatau in the annals of Indonesian catastrophes.

Sumatera, Indonesia's westernmost island, forms the western shore of the busy Strait of Malacca; Singapore lies across the Strait from approximately the middle of the island. Although much larger than Jawa, Sumatera contains only about one-third as many people (49 million). In colonial times the island became a base for rubber and palm-oil plantations; its high relief makes possible the cultivation of a wide range of crops, and neighboring Bangka and Belitung yield petroleum and natural gas. Palembang is the key urban center in the south, but current attention focuses on the north. There, the Batak people accommodated colonialism and Westernization and made Medan one of Indonesia's Pacific Rim boom cities.

Farther north the Aceh fought the Dutch into the twentieth century and, after Indonesia became a sovereign country, demanded autonomy and even outright independence for their State. Rebels fought the Indonesian army to a costly stalemate, and thousands died in the conflict—which would probably continue today but for a dramatic turn of events. The seafloor epicenter of the December 26, 2004 Indian Ocean tsunami lay near the far northern coast of Sumatera, and Aceh was directly in the path of the most powerful ocean waves. Entire towns

© H. J. de Blij, P. O. Muller, and John Wiley & Sons, Inc.

were swept away and tens of thousands died; Banda Aceh, the capital city, was devastated (Fig. 10-15). The international relief effort opened Aceh to foreigners in ways the Indonesians had long prevented, and the Indonesian army as well as the rebels were engaged in rescue missions rather than warfare. This combination of circumstances facilitated a truce and negotiations resulting in an agreement under which the rebels agreed to drop their demand for independence and the Indonesian army withdrew.

The settlement in Aceh Province had positive effects throughout Indonesia and was viewed by some cultural geographers as a template for the troubled provinces on the island of Papua (see p. 570). But in Aceh itself the blessing was mixed. In 2003, the government of Aceh became the first of Indonesia's 33 provinces to put Sharia

Islamic law onto the books, but it was Aceh's post-tsunami self-government that made implementation possible. So an area once known for its open society and bustling economy introduced a new force of "Sharia officers" to police social behavior, arresting unmarried couples, drunks, and gamblers and hauling them to public canings at mosques. In 2006, the caning of a man arrested for drinking at the beach was televised nationally. Indonesia's moderate Muslims, used to the country's historic open-minded interpretation of Islam, are worried that Indonesia's secular 1945 Constitution will not be enough to protect the country from the kind of Muslim revivalism that is spreading elsewhere.

Kalimantan is the Indonesian part of the minicontinent of Borneo, a slab of the Earth's crystalline crust whose backbone of mountains is of erosional, not

JAKARTA, CAPITAL OF Indonesia and the realm's second-largest city, sometimes is called the Kolkata (Calcutta) of Southeast Asia. Stand on the elevated highway linking the port to the city center and see the villages built on top of garbage dumps by scavengers using what they can find in the refuse, and the metaphor seems to fit. There is poverty here unlike that in any other Southeast Asian metropolis.

But there are other sides to Jakarta. Indonesia's economic progress has made its mark here, and the evidence is everywhere. Television antennas and satellite dishes rise like a forest from rusted, corrugated-iron rooftops. Cars (almost all, it seems, late-model), mopeds, and bicycles clog the streets, day and night. A meticulously manicured part of the city center contains a cluster of high-rise hotels, office buildings, and apartments. Billboards advertise planned communities on well-located, freshly cleared land.

Jakarta's population is a cross-section of Indonesia's, and the silver domes of Islam rise above the cityscape alongside Christian churches and Hindu temples. The city always was cosmopolitan, beginning as a conglomerate of villages at the mouth of the Ciliwung River under Islamic rule, becoming a Portuguese stronghold and later the capital of the Dutch East Indies (under the name of Batavia). Advantageously situated on the northwestern coast of Jawa, Indonesia's most populous island, Jakarta bursts at the seams with growth. Sail into the port, and hundreds of vessels, carrying flags from Russia to Argentina, await berths. Travel to the outskirts, and huge shan-

© H. J. de Blij, P. O. Muller, and John Wiley & Sons, Inc.

tytowns are being expanded by a constant stream of new arrivals. So vast is the human agglomeration—nobody really knows how many millions have descended on this megacity (the official figure is 10.1 million)—that the majority live without adequate (or any) amenities.

A seemingly permanent dome of metallic-gray air, heavy with pollutants, rests over the city. The social price of economic progress is high, but Jakartans are willing to pay it.

volcanic, origin. Compact and massive—at over 750,000 square kilometers (300,000 sq mi) it is larger than Texas—Borneo has a deep, densely forested interior that is a last refuge of some 35,000 orangutans. Elephants, rhinoceroses, and tigers still survive even as the loggers constrict their habitat (photo p. 571). Numerous other species of fauna—insects, reptiles, mammals, birds—inhabit the rich mountain and lowland forests. So much of Borneo's Pleistocene heritage has survived principally because of its comparatively small human population. The Indonesian part, Kalimantan, constitutes 28 percent of the national territory, but, with 13 million people, it contains just over 5 percent of Indonesia's population.

Figure 10-15 shows that the main areas of human activity lie in the west, southeast, and east. All towns of any size, including Pontianak in the west and Balikpapan in the east, are on or near the coast; the rivers still form important routes into the interior. Parks and reserves to protect flora and fauna cover but a small portion of the vast territory. Not only the flora and fauna, but also the human population has ancient roots. Indigenous peoples, including some of the Dayak clans, continue their slash-and-burn subsistence in the interior; others have become sedentary farmers. But Indonesia's transmigration policy has brought about half a million Jawanese and Madurese to Kalimantan, and they are laying out farms and digging drainage canals, bringing drastic change to rural areas. Of Kalimantan's four provinces, only the southernmost exports significant raw materials, including oil from offshore reserves, coal, iron ore, and some gold and diamonds; but in the national picture, Indonesian Borneo remains little developed.

Sulawesi is an island that looks like a set of intersecting mountain ranges rising from the sea. Its north-

FROM THE FIELD NOTES

"I drove from Manado on the Minahasa Peninsula in northeastern Sulawesi to see the ecological crisis at Lake Tondano, where a fast-growing water hyacinth is clogging the water and endangering the local fishing industry. On the way, in the town of Tomolon, I noticed this side street lined with prefabricated stilt houses in various stages of completion. These, I was told, were not primarily for local sale. They were assembled from wood taken from the forests of Sulawesi's northern peninsula, then taken apart again and shipped from Manado to Japan. 'It's a very profitable business for us,' the foreman told me. 'The wood is nearby, the labor is cheap, and the market in Japan is insatiable. We sell as many as we can build, and we haven't even begun to try marketing these houses in Taiwan or China.' At least, I thought, this wood was being converted into a finished product, unlike the mounds of logs and planks I had seen piled up in the ports of Borneo awaiting shipment to East Asia."   © H. J. de Blij.

Concept Caching

www.conceptcaching.com

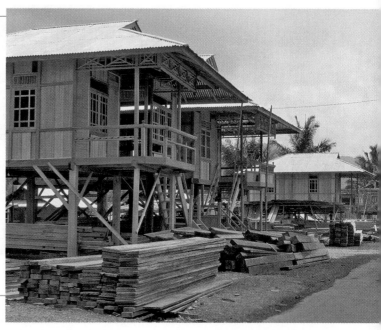

ern end is a 800-kilometer (500-mi) chain of extinct, dormant, and active volcanoes known as the Minahasa Peninsula, its future northern extension toward the Philippine Sea already marked by a ribbon of seafloor volcanoes just rising above the surface. So rugged is the relief on this island that you cannot take surface transportation from Palu to Ujungpandang. The ethnic mosaic is quite complex, with seven major groups inhabiting more or less isolated parts of the island as well as a sizeable cluster of Jawanese transmigrants who are concentrated in Sulawesi Tengah (Fig. 10-15). Sulawesi is unusual in Indonesia in that its Muslim and Christian numbers are just about equal. Since 1999, the level of strife between local Christians and immigrant Jawanese Muslims has risen and has destabilized this remote province of Sulawesi—forcing Christians who once lived peacefully among Muslim villagers to flee and culturally segregating settlements.

The two most populous and developed parts of Sulawesi are the southwestern peninsula centered on Ujungpandang and the eastern end of the Minahasa Peninsula, where Manado is the urban focus. Subsistence farming occupies most of the population of 19 million, but logging and wood products, some mining, and fishing augment the cash economy. Cultural landscapes are varied; in Minahasa, Christian churches remain numerous and form reminders of the colonial period, when relations between this province and the Dutch were favorable. During the liberation period, Minahasans sided with the Dutch (as many of their neighbors in the Malukus did) and called their peninsula Holland's "twelfth province."

It was an unfulfilled wish, souring relations between Jakarta and Manado for years afterward. The religious strife in Sulawesi Selatan (Fig. 10-15), however, has been more enduring: in 2004 the Indonesian military succeeded in pacifying the area, but animosities between Christians and Muslims, migrants and locals run deep, and only time and improving livelihoods are likely to soften them.

Papua, like East Timor, fell to Indonesia well after the Dutch colonial era had ended, but unlike East Timor, the United Nations approved Indonesia's takeover in 1969. Until 2002, this province was known by its Bahasa Indonesian name of *Irian Jaya* (Guinea West), but in that year the government of President Megawati Sukarnoputri bowed to the demands of indigenous leaders of the Papuan communities on the island and altered its name. As the map shows, the eastern half of the island of New Guinea comprises the independent state of Papua New Guinea. The Papuans are divided not only into numerous ethnolinguistic groups but also into two political entities.

Papua is an Indonesian province, but it lies in the Pacific geographic realm, not in Southeast Asia. As such, it has about 22 percent of the country's land area but barely 1 percent of its population, including more than 200,000 ethnic Indonesians, most from Jawa and many of them concentrated in the provincial capital with the non-Papuan name of Jayapura. Forested, high-relief, glacier-peaked Papua, lying as it does in a different geographic realm, is a world apart from teeming Jawa. It has what is reputedly the world's richest gold mine and the

FROM THE FIELD NOTES

"Getting to Ambon was no easy matter, and my boat trip was enlivened by an undersea earthquake that churned up the waters and shook up all aboard. The next morning our first view of the town of Ambon, provincial capital of Maluku, showed a center dominated by a large mosque with a modern minaret and a large white dome to the right of the main street leading to the port. My host from the local university told me that, as on many of the islands of the Malukus, about half of the people were Christians, not Muslims, and that some large churches were situated in the outskirts. 'But the relationships between Muslims and Christians are worsening,' he said. It had to do with the arrival of Jawanese, some of whom were 'agitators' and Islamic fundamentalists, stirring up religious passions. 'Have you heard of the Taliban?' he asked. 'Well, we have similar so-called religious students here, educated in Islamic schools that teach extremism.' . . . We drove from the town into the countryside to the west, toward Wakisihu and past the old Dutch Fort Rotterdam. 'Let me show you something,' he said as we turned up a dirt road. At its end, in the middle of a field, sat a large single structure called University of Islam, distinguished by a pair of enormous stairways. 'Can you call a single building like this where all they teach is Islam a university?' he asked. 'What they teach here is Islamic fundamentalism and intolerance.' He was prophetic. Weeks later, religious conflict broke out, the mosque in town was burned along with Christian churches; his own university lay in ruins."   © H. J. de Blij.

www.conceptcaching.com

second-largest open-pit copper mine; its forests not only harbor the indigenous peoples but also sustain a major logging industry.

Sanctioned as Indonesia's rule in Papua may be, local opposition has been much in evidence for decades. The *Organisasi Papua Merdeka* (Free Papua Movement, or FPM) and occasional rebel attacks remind the Jakarta government of its role here. In 1999, during the turmoil in the other provinces, the FPM held rallies in Jayapura to promote its cause, and in 2000 a congress was held where independence was demanded and a Papuan flag, the Morning Star, displayed. Later that year, Indonesian troops killed demonstrators in Merauke on the province's south coast. In 2003, the Indonesian government divided Papua into three administrative provinces. Shortly thereafter, several executives of a metals corporation accused of polluting Papuan waters and poisoning local residents were arrested, only to be exonerated months later, but not before the world was alerted to the allegations of corporate misdeeds in remote Papua. All this

deepened Papuan opposition to Jakarta's rule, and the time may come when Indonesia will rue its acquisition of Papua, a potential source of the kinds of centrifugal forces an archipelagic state must constantly confront.

Diversity in Unity

Indonesia's survival as a unified state is as remarkable as India's and Nigeria's. With more than 300 discrete ethnic clusters, over 250 languages, and just about every religion practiced on Earth (although Islam is predominant), actual and potential centrifugal forces are powerful here. Wide waters and high mountains perpetuate cultural distinctions and differences. Not surprisingly, Indonesia's national motto is *bhinneka tunggal ika*: diversity in unity.

What Indonesia has achieved is etched against the country's continuing cultural complexity. There are dozens of distinct aboriginal cultures; virtually every coastal community has its own roots and traditions. And the majority, the rice-growing Indonesians, include not only the numerous Jawanese—who have their own cultural identity—but also the Sundanese (who constitute 14 percent of Indonesia's population), the Madurese (8 percent), and others. Perhaps the best impression of the cultural mosaic comes from the string of islands that extends eastward from Jawa to Timor (Fig. 10-15). The rice-growers of Bali adhere to a modified version of Hinduism, giving the island a unique cultural atmosphere; the population of Lombok is mainly Muslim, with some Balinese Hinduism; Sumbawa is a Muslim community; Flores is mostly Roman Catholic. In western Timor, Protestant groups dominate; in the now-independent east, where the Portuguese ruled, Roman Catholicism prevails. Nevertheless, Indonesia nominally is the world's largest Muslim country: overall, just under 90 percent of the people adhere to Islam, and in the cities the silver domes of neighborhood mosques rise above the townscape. But Islam is not (perhaps not yet) the issue it is in Malaysia, where observance generally is stricter and where minorities fear Islamization as Malay power grows.

Transmigration and the Outer Islands

As noted earlier, Indonesia's population of 247.2 million makes it the world's fourth most populous country, but Jawa, we also noted, contains 56 percent of it. With almost 140 million people on an island the size of Louisiana, population pressure is enormous here. Moreover, Indonesia's annual rate of population growth remains a bit above the realm average of 1.3 percent. To deal with this problem, and at the same time to strengthen the core's power over outlying areas, the Indonesian government long pursued a policy known as **transmigration**, inducing Jawanese (and Madurese from the adjacent island of Madura) to relocate to other islands. Several million Jawanese have moved to locales as distant as the Malukus, Sulawesi, and Sumatera; many Madurese were resettled in Kalimantan.

In fact, the Dutch colonialists began the Transmigration Program, but then President Sukarno abandoned it and it was not revived until 1974, when President Suharto saw economic and political opportunities in it. Over the next 25 years, nearly 8 million people were relocated; in fact, the World Bank gave Indonesia almost U.S. $1 billion to facilitate the program before it pulled out amid reports of forced migration.

The results were mixed. World Bank project reviews suggested that the vast majority of the transmigrants were happy with the land, schools, medical services, and other amenities with which they were provided, but independent surveys painted a different picture. About half of the migrants never managed to rise above a life of sub-

Indonesia still has vast areas of rainforest from Sumatera to Papua, but the forest is under assault in many parts of this populous country, with disastrous impact on biodiversity, soils, and rivers. High-quality wood fetches high prices in markets such as Japan and China, and for decades Indonesia's corrupt authoritarian regime actively participated in the plunder. More recently, democratically-elected governments have sought to curb the rate of destruction, but corruption continues to impede these efforts. Here in Kalimantan, on the Indonesian side of Borneo, about 10,000 square kilometers (4,000 sq mi) of rainforest are lost every year. This depressing photograph shows the frontier of clear-cutting in Kalimantan Barat (West Kalimantan). © Wayne G. Lawler/Photo Researchers, Inc.

sistence; many never received even the most basic needs for survival in their new environment; land was simply taken away from indigenous inhabitants, notably in Kalimantan; the newcomers did much damage to the traditional cultures they invaded; and the program led to massive forest and wetland destruction. And thousands of migrants died in clashes with local inhabitants.

In 2001, the Transmigration Program was terminated as one of the final acts of the short-lived government of President Abdurrahman Wahid, but the damage it did will trouble Indonesia for generations. From northern Sumatera to western Kalimantan to central Sulawesi to the islands of the Malukus to Papua, cultural strife and its polarizing effect will challenge Indonesian governments of the future.

Indonesia accounts for 41 percent of the population (and about the same proportion of the territory) of the Southeast Asian realm, and is more than two and a half times larger than the second most populated country. Its political system is maturing, there are signs that its brand of Islam is more moderate than most, its economy is strengthening, poverty is declining albeit slowly, and corruption is being countered. Thus Indonesia may be approaching a time when it assumes the regional role commensurate with its dimensions and capacities.

EAST TIMOR

As Figure 10-15 shows, Timor is the easternmost of the major Lesser Sunda Islands, and during the Dutch colonial period the Portuguese maintained a colony on the eastern half of it. When Indonesia shook off Dutch rule in the late 1940s, the Portuguese hung on in Timor, but in 1975 Indonesian forces overran the colony and annexed it formally in 1976. Indonesian rule, however, was even less benign than the Portuguese had been, and soon a bitter struggle for independence was under way. In 1999, after much death and destruction, East Timor's 800,000 inhabitants were allowed, under UN supervision, to vote on independence. They voted overwhelmingly in favor, provoking Indonesia's armed forces to redouble their ruthless repression. East Timor's already modest infrastructure was devastated, the death toll

mounted daily, and only foreign intervention, led by Australia (which, among all the world's states, had been Indonesia's sole supporter in the matter), established a degree of order. East Timor became an independent state in 2002, the 198th member of the United Nations.

Subsequently, East Timor's leaders proclaimed the official name of their country to be *Timor-Leste*, the Portuguese version of "Timor East." As in the case of Côte d'Ivoire in West Africa, the French name for Ivory Coast, we use the two options interchangeably.

Nation-Building Nightmare

Nation-building in this Connecticut-sized country continues to be a difficult proposition. Figure 10-16 helps explain why Australia supported Indonesia in its effort

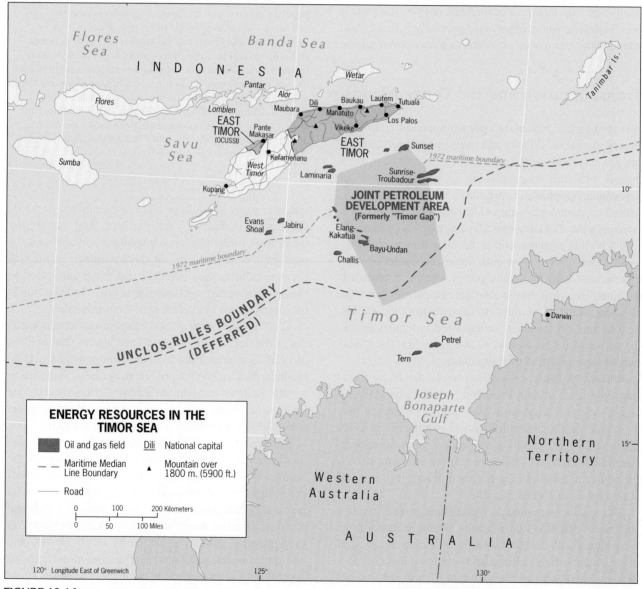

FIGURE 10-16

© H. J. de Blij, P. O. Muller, and John Wiley & Sons, Inc.

to absorb East Timor: by doing so, Australia would benefit from Indonesia's compliance in the delimitation of maritime boundaries in the Timor Sea, giving Australia a larger share of the known oil and gas reserves than an independent East Timor would be likely to relinquish. In the end, difficult negotiations between powerful and technologically capable Australia and weak and incapable East Timor produced a settlement that gave East Timor an acceptable share of future revenues—but not before Australia's international reputation for fairness was further damaged. On the map, the red maritime boundary is the one delimited under United Nations regulations; the blue boundary represents the bilateral agreement between Indonesia and Australia. At issue is the "gap" between them, most of it on East Timor's side but now a "joint petroleum development area".

But the billions of dollars in prospect cannot overcome the devastation and abject poverty that bog down the nation-building project here. As Figure 10-16 shows, East Timor is a fragmented country, with a dominant east (where the coastal capital, Dili, is located) and a small exclave on the north coast of Indonesian West Timor called Ocussi (sometimes mapped as "Ocussi-Ambeno Province"). Although there is a road from this exclave to the main territory, relations between East Timor and Indonesia make its use impractical. Worse, the three-year United Nations buildup to independence in 2002 launched a society totally unprepared to go it alone. Good government has been in short supply. The freedom fighters who were demobilized upon independence were organized into police and army units that soon fell apart; widely available weapons empowered gangs and criminals, ethnic hostilities erupted, returned exiles and locals were at odds, and East Timor's still-fragile institutions collapsed. East Timor is a predominantly agricultural country, but while United Nations experts were trying to erect political and social institutions (a judiciary, banking system, medical establishment, public education) farming was neglected and poverty worsened. Moreover, East Timor now has about 100,000 people concentrated in squalid refugee camps. And to make matters worse, its population of 1.2 million continues to grow explosively: according to the latest (2008) data, the rate of natural increase was a disastrous 3.1 percent annually, and the country also exhibited one of the highest fertility rates in the world with 6.7 children per woman.

Given the limited dimensions of East Timor, its still relatively small population, its prospective oil income, and foreign assistance, the failure of nation-building here is a warning to the world that independence and freedom do not guarantee well-being and progress.

THE PHILIPPINES

North of Indonesia, across the South China Sea from Vietnam, and south of Taiwan lies an archipelago of more than 7000 islands (only a few hundred of them larger than one square kilometer [0.4 sq mi] in area) inhabited by 94.3 million people. The inhabited islands of the Philippines can be viewed as three groups: (1) Luzon, largest of all, and Mindoro in the north, (2) the Visayan group in the center, and (3) Mindanao, second largest, in the south (Fig. 10-17). Southwest of Mindanao lies a small group of islands, the Sulu Archipelago, nearest to Indonesia, where a Muslim-based insurgency has kept the area in turmoil.

Few of the generalizations we have been able to make for Southeast Asia could apply in the Philippines without qualification. The country's location relative to the mainstream of change in this part of the world has had much to do with this situation. The islands, inhabited by peoples of Malay ancestry with Indonesian strains, shared with much of the rest of Southeast Asia an early period of Hindu cultural influence, which was strongest in the south and southwest and diminished northward. Next came a Chinese invasion, felt more strongly on the largest island of Luzon in the northern part of the Philippine archipelago. Islam's arrival was delayed somewhat by the position of the Philippines well to the east of the mainland and to the north of the Indonesian islands. The few southern Muslim beachheads were soon overwhelmed by the Spanish invasion during the sixteenth century. Today the Philippines, adjacent to the world's largest Muslim state (Indonesia), is 83 percent Roman Catholic, 9 percent Protestant, and only 5 percent Muslim.

Muslim Insurgency

The Muslim population, concentrated on the southeastern flank of the archipelago (Fig. 10-17), has long protested its marginalization in this predominantly Christian country. Over the past 30 years, a half-dozen Muslim organizations have promoted the Muslim cause with tactics ranging from political pressure to violent insurgency. The campaigns of violence have affected all of the Muslim areas, from southern Mindanao to Palawan, and the strategies of the rebels have ranged from bombings to kidnappings. Densely forested Basilan Island is the headquarters of a group known as Abu Sayyaf, which demanded a separate Muslim state and is believed to have received support through the al-Qaeda network. The counterterror operations extended into this part of the Philippines when American troops joined Filipino forces to pursue Abu Sayyaf terrorists. Also active on behalf of Muslim causes in this area are the Moro National Liberation Front and the Moro Islamic Liberation Front; the Philippine government has

FIGURE 10-17   © H. J. de Blij, P. O. Muller, and John Wiley & Sons, Inc.

the "pure" Chinese minority numbers less than 2 percent of the population (far lower than in most Southeast Asian countries), a much larger portion of the Philippine population carries a decidedly Chinese ethnic imprint. What has happened is that the Chinese have intermarried, producing a sort of Chinese-mestizo element that constitutes more than 10 percent of the total population. In another cultural sphere, the country's ethnic mixture and variety are paralleled by its great linguistic diversity. Nearly 90 Malay languages, major and minor, are spoken by the 94 million people of the Philippines; only about 1 percent still use Spanish. At independence in 1946, the largest of the Malay languages, Tagalog, or Pilipino, became the country's official language, and the educational system promotes its general use. English is learned as a subsidiary language and remains the chief *lingua franca*; an English-Tagalog hybrid ("Taglish") is increasingly heard today, cutting across all levels of society.

The widespread use of English in the Philippines, of course, results from a half-century of American rule and influence, beginning in 1898 when the islands were ceded to the United States by Spain under the terms of the treaty that followed the Spanish-American War. The United States took over a country in open revolt against its former colonial master and proceeded to destroy the Filipino independence struggle, now directed against the new foreign rulers. It is a measure of the subsequent success of U.S. administration in the Philippines that this was the only dependency in Southeast Asia that sided against the Japanese during World War II in favor of the colonial power. The U.S. rule had its good and bad features, but the Americans did initiate reforms that were long overdue, and they were already in the process of negotiating a future independence for the Philippines when the war intervened in 1941.

The Philippines' population, concentrated where the good farmlands lie in the plains, is clustered in three general areas (Fig. 10-4): (1) the northern and south-central

entered into negotiations with both. Cease-fire agreements, however, have not held. The Muslim challenge is pirating a disproportionate share of Manila's operating budget.

People and Culture

Out of the Philippines melting pot, where Malay, Arab, Chinese, Japanese, Spanish, and American elements have met and mixed, has emerged the distinctive Filipino culture. It is not a homogeneous or a unified culture, but in Southeast Asia it is in many ways unique. One example of its absorptive qualities is demonstrated by the way the Chinese infusion has been accommodated: although

AMONG THE REALM'S GREAT CITIES . . . Manila

MANILA, CAPITAL OF the Philippines and the realm's largest city, was founded by the Spanish invaders of Luzon more than four centuries ago. The colonists made a good choice in terms of site and situation. Manila sprawls at the mouth of the Pasig River where it enters one of Asia's finest natural harbors. To the north, east, and south a crescent of mountains encircles the city, which lies just 1000 kilometers (600 mi) southeast of China's Hong Kong.

Manila, named after a flowering shrub in the local marshlands, is bisected by the Pasig, which is bridged in numerous places. The old walled city, Intramuros, lies to the south. Despite heavy bombardment during World War II, some of the colonial heritage survives in the form of churches, monasteries, and convents. St. Augustine Church, completed in 1599, is one of the city's landmarks.

The CBD of Manila lies on the north side of the Pasig River. Although Manila has a well-defined commercial center with several avenues of luxury shops and modern buildings, the skyline does not reflect the high level of energy and activity common in Pacific Rim cities on the opposite side of the South China Sea. Neither is Manila a city of notable architectural achievements. Wide, long, and straight avenues flanked by palm, banyan, and acacia trees give it a look similar to San Juan, Puerto Rico.

In 1948, a newly built city immediately to the northeast of Manila was inaugurated as the de jure capital of the Philippines and called Quezon City. The new facilities were eventually to house all government offices, but many

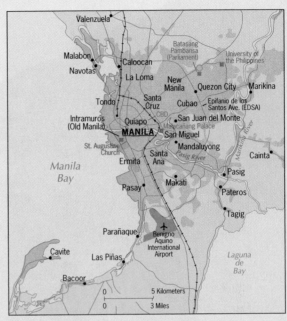

© H. J. de Blij, P. O. Muller, and John Wiley & Sons, Inc.

functions of the national government never made the move. In the meantime, Manila's growth overtook Quezon City's, so that it became part of the Greater Manila metropolis (which today is home to 11.7 million). Although the proclamation of Quezon City as the Philippines' official capital was never rescinded, Manila remains the de facto capital of the country today.

part of Luzon, (2) the southeastern extension of Luzon, and (3) the islands of the Visayan Sea between Luzon and Mindanao. Luzon is the site of the capital, Manila-Quezon City (11.7 million, one-eighth of the entire national population), a major metropolis facing the South China Sea. Alluvial as well as volcanic soils, together with ample moisture in this tropical environment, produce self-sufficiency in rice and other staples and make the Philippines a net exporter of farm products despite a fairly high population growth rate of 2.1 percent.

Perhaps more than any other people, Filipinos take jobs in foreign countries in massive numbers, proving their capacity to succeed in jobs they cannot find at home. More than 8 million people—nearly 10 percent of the country's entire population—are thus employed. Not only do hundreds of thousands of Filipinos work in the shipping industry around the world, but Filipino domestic workers find jobs from Dubai to Dubuque. They have a major impact on the Philippines' economy in the form of remittances,

regularly ranking the country among the world's three or four leading recipients of such monetary inflows.

Prospects

The Philippines seems to get little mention in discussions of developments on the Pacific Rim, and yet it would appear to be well positioned to share in the Pacific Rim's economic growth. Governmental mismanagement and political instability have slowed the country's participation, but during the 1990s the situation improved. Despite a series of jarring events—the ouster of U.S. military bases, the damaging eruption of a volcano near the capital, the violence of Muslim insurgents, and a dispute over the nearby Spratly Islands in the South China Sea—the Philippines made substantial economic progress during the decade. Its electronics and textile industries (mostly in the Manila hinterland) expanded continuously, and

In the Philippines, rice is second only (by volume) to sugarcane in terms of total production of agricultural commodities, but rice does not figure prominently among export revenues. The export economy today is dominated by high-tech products, with textiles in distant second place and farm products yielding less than 4 percent by value. Rice being the domestic staple, much of the countryside is draped by paddies, many of them meticulously terraced as seen here on the island of Luzon—but productivity compared to Vietnam and Thailand is low. © Photolibrary/Corbis

more foreign investment is arriving. The centrally positioned Visayan island of Cebu experienced particularly rapid growth, based on its central location, good port, expanded airport, large and literate labor force, and cadre of managers experienced in the service industries. A free-trade zone attracted foreign companies from the United States and Japan, and now products from toys to semiconductors flow from Cebu's manufacturers to world markets. South Korean investment is expanding the shipbuilding industry. And with their ability in English, Philippine workers are competing with their Indian counterparts in the outsourcing business in a growing number of call centers run by American companies.

But agriculture continues to dominate the Philippines' economy, unemployment remains high, further land reform is badly needed, and social restructuring (reducing the controlling influence over national affairs by a comparatively small group of families) must occur. Still, progress is being made. The country now is a lower-middle-income economy, and given a longer period of stability and success in reducing the population growth rate, it will rise to the next level and finally take its place among Pacific Rim growth poles.

In recent years, the population issue has divided this dominantly Roman Catholic society, with the government promoting family planning and the clergy opposing it. But behind this debate lies another of the Philippines' assets: in a realm of mostly undemocratic regimes, the Philippines has come out of its period of authoritarian rule a rejuvenated, if not yet robust, democracy.

What You Can Do

RECOMMENDATION: Participate in National Geography Awareness Week! By an Act of Congress, a week in November is designated National Geography Awareness Week, and all over the country organizations ranging from colleges to corporations mark the occasion by offering programs highlighting geography's importance and usefulness. Activities include competitions, exhibits, lectures, GIS demonstrations, and field trips, and faculty and students alike make a special effort to reach out to the general public to show what professional geographers do. This always requires assistance, from posting announcements to meeting speakers and from supervising exhibits to planning field trips. How about asking your local Geography Department what is being planned for next November, and lending a hand?

GEOGRAPHIC CONNECTIONS

1 In this chapter we divide the Southeast Asian geographic realm into a Mainland region and an Insular region. A major factor in this regionalization is religion: Buddhism dominates in Mainland Southeast Asia, and Islam in Insular Southeast Asia. As Figure 7-3 reveals, however, the religious geography of Southeast Asia is more complicated than that, and important vestiges of one major religion cannot even be shown at this small scale. Look very carefully at the political boundaries of Southeast Asia and the religious regions in Figure 7-3, and identify those countries that experience significant internal religious division of a regional nature. How is such division finding expression?

Is religious regionalism absent in countries dominated by a single faith?

2 When we studied colonialism and its legacies in Subsaharan Africa, it was noted that, although most African states have been independent for around half a century, the imprint of colonial power still remains—and continues to divide the realm. Southeast Asia, too, was occupied by colonial powers (with one significant exception), but the cultural impress of colonialism appears to be less durable here. To what geographic factors do you attribute this contrast between two formerly colonized realms?

Geographic Literature on Southeast Asia: The key introductory works on this realm were authored by Dixon, Fisher, Fryer, Leinbach & Ulack, McGregor, Rigg, Spencer & Thomas, and Weightman. These works, together with a comprehensive listing of noteworthy books and recent articles on the geography of Mainland and Insular Southeast Asia, can be found in the *References and Further Readings* section of this book's website at **www.wiley.com/college/deblij**.

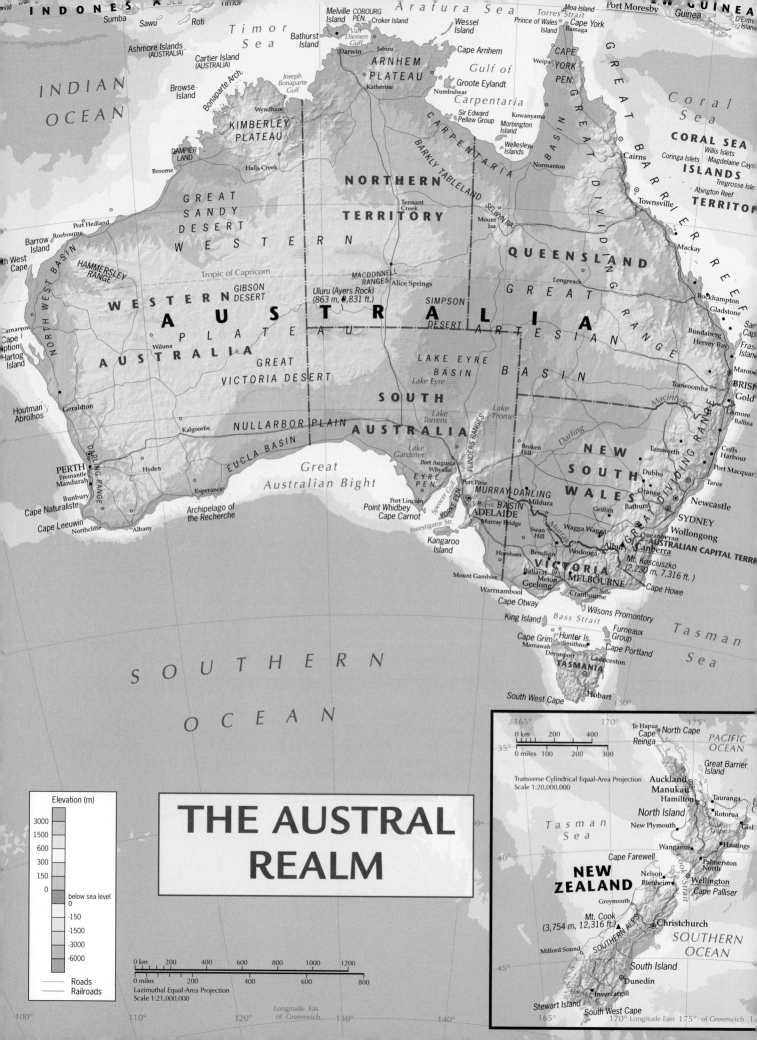

THE AUSTRAL REALM

INDONES A

Sumba Sawu Roti Timor

Arafura Sea

GUINEA

Guinea D'Entre Island

Timor Sea

Melville Island COBOURG PEN. Croker Island Wessel Island Torres Strait Moa Island Port Moresby

Prince of Wales Cape York Bamaga Island

Ashmore Islands (AUSTRALIA) Cartier Island (AUSTRALIA) Bathurst Island Van Diemen Gulf Darwin Jabiru Cape Arnhem Weipa CAPE YORK PEN.

INDIAN OCEAN

Browse Island Bonaparte Arch. Joseph Bonaparte Gulf Katherine **ARNHEM PLATEAU** Groote Eylandt Numbulwar

Wyndham Sir Edward Pellew Group Mornington Island Kowanyama *Gulf of Carpentaria* **GREAT**

KIMBERLEY PLATEAU Wellesley Islands Normanton **BASIN**

DAMPIER LAND Halls Creek **NORTHERN**

Coral Sea

CORAL SEA ISLANDS TERRITOR

Willis Islets Coringa Islets Magdelaine Cays Tregrosse Isle Abington Reef

Broome **GREAT SANDY DESERT** **TERRITORY** BARKLY TABLELAND Mount Isa SELWYN RA. Cairns

Port Hedland Barrow Island Roebourne **WESTERN** Tennant Creek Townsville Mackay

th West Cape HAMMERSLEY RANGE Tropic of Capricorn MACDONNELL RANGES Alice Springs **QUEENSLAND** Longreach Rockhampton Gladstone

Carnarvon Cape ption Hartog Island **WESTERN** GIBSON DESERT Uluru (Ayers Rock) (863 m, 2,831 ft.) SIMPSON DESERT **GREAT DIVIDING RANGE** Bundaberg Hervey Bay Fras Islan

Houtman Abrolhos Geraldton Wiluna **AUSTRALIA** PLATEAU **ARTESIAN** Toowoomba Maroo BRIS Gold

AUSTRALIA LAKE EYRE BASIN Lake Eyre **BASIN** Macintyre Lismore Ballina

GREAT VICTORIA DESERT **SOUTH** Lake Torrens Lake Frome **NEW** Toowoomba

Kalgoorlie Lake Gairdner **AUSTRALIA** Broken Hill Darling **SOUTH** Tamworth Coffs Harbour

NULLARBOR PLAIN FLINDERS RANGES **WALES** Dubbo Port Macquar

PERTH Hyden EUCLA BASIN EYRE PEN. Port Augusta Whyalla Mildura Orange Taree

Fremantle Mandurah Esperance Port Pirie Griffith Bathurst Newcastle

Bunbury Archipelago of the Recherche Point Whidbey Cape Carnot SPENCER GULF YORKE PEN. Vincent Gulf MURRAY-DARLING BASIN ADELAIDE Wagga Wagga Queanbeyan SYDNEY Wollongong

Cape Naturaliste *Great Australian Bight* Port Lincoln Investigator Str. Murray Bridge Swan Hill Murray Albury Canberra AUSTRALIAN CAPITAL TERR

Cape Leeuwin Northcliffe Albany Kangaroo Island Horsham Wodonga Mt. Kosciuszko (2,230 m, 7,316 ft.)

DARLING RANGE Bendigo **VICTORIA** Cape Howe

SOUTHERN Ballarat Melton MELBOURNE Sale Cranbourne

Mount Gambier Geelong Warrnambool Cape Otway Wilsons Promontory

King Island Bass Strait Furneaux Group Cape Portland *Tasman Sea*

OCEAN Cape Grim Marrawah Hunter Is. Smithton Devonport Launceston

Stanley **TASMANIA** South West Cape Hobart

Elevation (m)

3000
1500
600
300
150
0
below sea level 0
-150
-1500
-3000
-6000

— Roads
— Railroads

0 km 200 400 600 800 1000 1200
0 miles 200 400 600 800

Lazimuthal Equal-Area Projection
Scale 1:21,000,000

Longitude Eas of Greenwich

New Zealand inset

0 km 200 400
0 miles 100 200 300

PACIFIC OCEAN

Te Hapua Cape Reinga North Cape Great Barrier Island

Transverse Cylindrical Equal-Area Projection
Scale 1:20,000,000

Auckland Manukau Hamilton Tauranga Rotorua

North Island New Plymouth Lake Taupo Gist

Tasman Sea Wanganui Hastings

Cape Farewell Palmerston North

NEW ZEALAND Nelson Blenheim Wellington Cape Palliser

Greymouth Cook Strait

Mt. Cook (3,754 m, 12,316 ft.) SOUTHERN ALPS Christchurch

Milford Sound **South Island** **SOUTHERN OCEAN**

South West Cape Dunedin

Stewart Island Invercargill

100° 110° 120° 130° 140° 150° 165° 170° 170° Longitude East 175° of Greenwich

35° 40° 45°

11

CONCEPTS, IDEAS, AND TERMS

1. Austral
2. Southern Ocean
3. Subtropical Convergence
4. West Wind Drift
5. Biogeography
6. Wallace's Line
7. Aboriginal population
8. Outback
9. Federation
10. Unitary state
11. Import-substitution industries
12. Aboriginal land issue
13. Immigration policies
14. Environmental degradation
15. Peripheral development

REGIONS

● **AUSTRALIA**
    CORE AREA
    OUTBACK
● **NEW ZEALAND**

Where was this photo taken? Find out at **www.wiley.com/college/deblij**

In This Chapter

- Australia's amazing biogeography
- Australia's changing population
- Aboriginal claims to land and resources
- China covets Australian commodities
- Foreign policy dilemmas downunder
- New Zealand's matchless physiography

Photos: © H. J. de Blij

FIGURE 11-1 © H. J. de Blij, P. O. Muller, and John Wiley & Sons, Inc.

THE AUSTRAL REALM is geographically unique (Fig. 11-1). It is the only geographic realm that lies entirely in the Southern Hemisphere. It is also the only realm that has no land link of any kind to a neighboring realm and is thus completely surrounded by ocean and sea. It is second only to the Pacific as the world's least populous realm. Appropriately, its name refers to its location (**Austral** means south)—a location far from the sources of its dominant cultural heritage but close to its newfound economic partners on the western Pacific Rim.

Defining the Realm

Two countries constitute the Austral Realm: Australia, in every way the dominant one, and New Zealand, physiographically more varied—but demographically much smaller (Fig. 11-2)—than its giant partner to the north-west. Between them lies the Tasman Sea. To the west lies the Indian Ocean, to the east the Pacific, and to the south the frigid Southern Ocean.

This southern realm is at a crossroads. On the doorstep of populous Asia, its Anglo-European legacies are now infused by other cultural strains. Polynesian Maori in New Zealand and Aboriginal communities in Australia are demanding better terms of life. Pacific Rim markets are buying huge quantities of raw materials. Japanese and other Asian tourists fill hotels and resorts. Queensland's tropical Gold Coast resembles Honolulu's Waikiki. The streets of Sydney and Melbourne display a multicultural panorama unimagined just two generations ago. All these changes have stirred political debate. Issues ranging from immigration quotas to indigenous land rights dominate, exposing social fault lines (city versus Outback in Australia, North and South in New Zealand). Aborigines and Maori were here first, and the Europeans came next. Now Asia looms in Australia's doorway.

MAJOR GEOGRAPHIC QUALITIES OF

The Austral Realm

1. Australia and New Zealand constitute a geographic realm by virtue of territorial dimensions, relative location, and dominant cultural landscape.

2. Despite their inclusion in a single geographic realm, Australia and New Zealand differ physiographically. Australia has a vast, dry, low-relief interior; New Zealand is mountainous.

3. Australia and New Zealand are marked by peripheral development—Australia because of its aridity, New Zealand because of its topography.

4. The populations of Australia and New Zealand are not only peripherally distributed but also highly clustered in urban centers.

5. The realm's human geography is changing—in Australia because of Aboriginal activism and Asian immigration, and in New Zealand because of Maori activism and Pacific-Islander immigration.

6. The export of livestock products dominates the economic geography of Australia and New Zealand (and in Australia, wheat production and mining also dominate).

7. Australia and New Zealand are being integrated into the economic framework of the western Pacific Rim, principally as suppliers of raw materials.

LAND AND ENVIRONMENT

Physiographic contrasts between massive, compact Australia and elongated, fragmented New Zealand are related to their locations with respect to the Earth's tectonic plates (consult Fig. G-3). Australia, with some of the geologically most ancient rocks on the planet, lies at the center of its own plate, the Australian Plate. New Zealand, younger and less stable, lies at the convulsive convergence of the Australian and Pacific plates. Earthquakes are rare in Australia, and volcanic eruptions are unknown; New Zealand has plenty of both. This locational contrast is also reflected by differences in relief (Fig. 11-1). Australia's highest relief occurs in what Australians call the Great Dividing Range, the mountains that line the east coast from the Cape York Peninsula to southern Victoria, with an outlier in Tasmania. The high-

FIGURE 11-2   © H. J. de Blij, P. O. Muller, and John Wiley & Sons, Inc.

est point along these old, now eroding mountains is Mount Kosciusko, 2228 meters (7316 ft) tall. In New Zealand, entire ranges are higher than this, and Mount Cook reaches 3764 meters (12,349 ft).

West of Australia's Great Dividing Range, the physical landscape generally has low relief, with some local exceptions such as the Macdonnell Ranges near the center; plateaus and plains dominate (Fig. 11-3). The Great Artesian Basin is a key physiographic region, providing underground water sources in what would otherwise be desert country; to the south lies the continent's predominant river system, the drought-stressed Murray-Darling. The area mapped in tan as *Western Plateau and Margins* in the inset map of Figure 11-3 contains much of Australia's mineral wealth.

Climates

Figure G-7 reveals the effects of latitudinal location and interior isolation on Australia's climatology. In this respect, Australia is far more varied than New Zealand, its climates ranging from tropical in the far north, where rainforests flourish, to Mediterranean in parts of the south. The interior is dominated by desert and steppe conditions, the steppes providing the grasslands that sustain tens of millions of livestock. Only in the east does Australia have an area of humid temperate climate, and here lies most of the country's economic core area.

New Zealand, by contrast, is totally under the influence of the Southern and Pacific oceans, creating moderate, moist conditions, temperate in the north and colder in the south.

The Southern Ocean

Twice now we have referred to the **Southern Ocean**, but try to find this ocean on maps and globes published by famous cartographic organizations such as the National Geographic Society and Rand McNally. From their maps you would conclude that the Atlantic, Pacific, and Indian oceans reach all the way to the shores of Antarctica. Australians and New Zealanders know better. They experience the frigid waters and persistent winds of this great weather-maker on a daily basis.

For us geographers, it is a good exercise to turn the globe upside down now and then. After all, the usual orientation is quite arbitrary. Modern mapmaking started in the Northern Hemisphere, and the cartographers put their hemisphere on top and the other at the bottom. That is now the norm, and it can distort our view of the world. In bookstores in the Southern Hemisphere, you sometimes see tongue-in-cheek maps showing Australia and Argentina at the top, and Europe and Canada at the bottom. But this matter has a serious side. A reverse view of the globe shows us how vast the ocean encircling Antarctica is. The Southern Ocean may be remote, but its existence is real.

Where do the northward limits of the Southern Ocean lie? This ocean is bounded not by land but by a marine transition called the **Subtropical Convergence**. Here the cold, extremely dense waters of the Southern Ocean meet the warmer waters of the Atlantic, Pacific, and Indian oceans. It is quite sharply defined by changes in temperature, chemistry, salinity, and marine fauna. Flying over it, you can actually observe it in the changing colors of the water: the Antarctic side is a deep gray, the northern side a greenish blue.

Although the Subtropical Convergence moves seasonally, its position does not vary far from latitude 40° South, which also is the approximate northern limit of Antarctic icebergs. Defined this way, the great Southern Ocean is a huge body of water that moves clockwise (from west to east) around Antarctica, which is why we also call it the **West Wind Drift**.

AUSTRALIA: PHYSIOGRAPHY

| 0 | 200 | 400 | 600 | 800 Kilometers |
| 0 | 100 | 200 | 300 | 400 Miles |

INDONESIA

New Guinea

PAPUA NEW GUINEA

Arafura Sea

Torres Strait

Timor

EAST TIMOR

Sumba

Timor Sea

INDIAN OCEAN

15°

Cape York Penin.

Gulf of Carpentaria

Coral Sea

15°

Arnhem Land

Atherton Plateau

15°

Kimberley Plateau

NORTHERN TERRITORY

Carpentaria Basin

Great Barrier Reef

King Leopold Ranges

Barkly Tableland

Gregory Range

Great Dividing Range

Dampier Land

Canning Basin

Macdonnell Ranges

Selwyn Range

QUEENSLAND

Great Sandy Desert

James Range

Great Artesian Basin

PACIFIC OCEAN

Gibson Desert

Simpson Desert

Grey Range

Hamersley Range

Musgrave Ranges

Lake Eyre

WESTERN AUSTRALIA

Stuart Range

North West Basin

Great Victoria Desert

Flinders Range

Darling

NEW SOUTH WALES

30°

Yilgarn Plateau

Nullarbor Plain

SOUTH AUSTRALIA

Murray-Darling River Basin

Lachlan

Blue Mts.

AUSTRALIAN CAPITAL TERRITORY

Darling Range

Eucla Basin

Eyre Penin.

Murray

Mt. Kosciusko 2,223 m. (7,316 ft.)

30°

Great Australian Bight

Kangaroo Island

VICTORIA

Snowy Mts.

Bass Strait

Tasman Sea

SOUTHERN OCEAN

TASMANIA

120°

135°

45°

120°

130°

140°

150°

10°

105°

45°

150°

Longitude East of Greenwich

Darwin

INDIAN OCEAN

PACIFIC OCEAN

Kimberley Plateau

Carpentaria Basin

20°

Great Sandy Desert

Macdonnell Ranges

Great Dividing Range

20°

Great Artesian Basin

Brisbane

Great Victoria Desert

Lake Eyre

Darling R.

30°

Perth

Nullarbor Plain

Sydney

Eucla Basin

Murray-Darling River Basin

Canberra

Adelaide

Murray

Mt. Kosciusko 2,223 m. (7,316 ft.)

SOUTHERN OCEAN

Melbourne

40°

110°

Longitude East of Greenwich

TASMANIA

Bass Strait

Tasman Sea

40°

120°

130°

140°

150°

Hobart

PHYSICAL LANDSCAPES

Eastern Uplands   Central Lowlands   Western Plateau & Margins

| 0 | 400 | 800 | 1200 | 1600 Kilometers |
| 0 | | 500 | | 1000 Miles |

FIGURE 11-3

© H. J. de Blij, P. O. Muller, and John Wiley & Sons, Inc.

582

Biogeography

One of this realm's defining characteristics is its wildlife. Australia is the land of kangaroos and koalas, wallabies and wombats, possums and platypuses. These and numerous other *marsupials* (animals whose young are born very early in their development and then carried in an abdominal pouch) owe their survival to Australia's early isolation during the breakup of Gondwana (see Fig. 6-3). Before more advanced mammals could enter Australia and replace the marsupials, as happened in other parts of the world, the landmass was separated from Antarctica and India, and today it contains the world's largest assemblage of marsupial fauna.

Australia's vegetation also has distinctive qualities, notably the hundreds of species of eucalyptus trees native to this geographic realm. Many other plants form part of Australia's unique flora, some with unusual adaptation to the high temperatures and low humidity that characterize much of the continent.

The study of fauna and flora in spatial perspective combines the disciplines of biology and geography in a field
5 known as **biogeography**, and Australia is a giant laboratory for biogeographers. In the Introduction (pp. 12–17), we noted that several of the world's climatic zones are named after the vegetation that marks them: tropical savanna, steppe, tundra. When climate, soil, vegetation, and animal life reach a long-term, stable adjustment, vegetation forms the most visible element of this ecosystem.

Biogeographers are especially interested in the distribution of plant and animal species, and in the relationships between plant and animal communities and their natural environments. (The study of plant life is called *phytogeography*; the study of animal life *zoogeography*.) These scholars seek to explain the distributions the map reveals. In 1876 one of the founders of biogeography, Alfred Russel Wallace, published a book entitled *The Geographical Distribution of Animals* in which he fired the first shot in a long debate: where does the zoogeographic boundary of Australia's fauna lie? Wallace's fieldwork in the area revealed that Australian forms exist not only in Australia itself but also in New Guinea and in some islands to the northwest. So Wallace proposed that the faunal boundary should lie between Borneo and Sulawesi, and just east of Bali (Fig. 11-4).

6 **Wallace's Line** soon was challenged by other researchers, who found species Wallace had missed and who visited islands Wallace had not. There was no question that Australia's zoogeographic realm ended somewhere in the Indonesian archipelago, but where? Western Indonesia was

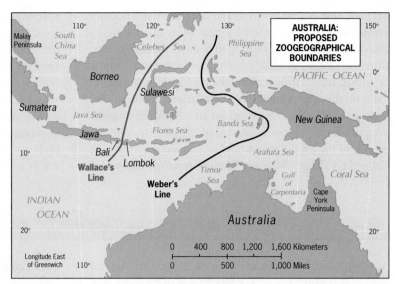

FIGURE 11-4   © H. J. de Blij, P. O. Muller, and John Wiley & Sons, Inc.

the habitat of nonmarsupial animals such as tigers, rhinoceroses, and elephants, as well as primates; New Guinea clearly was part of the realm of the marsupials. How far had the more advanced mammals progressed eastward along the island stepping stones toward New Guinea? The zoogeographer Max Weber found evidence that led him to postulate his own *Weber's Line*, which, as Figure 11-4 shows, lay very close to New Guinea.

The Human Impact

Not all research in zoogeography or phytogeography deals with such large questions. Much of it focuses on the relationships between particular species and their

POINTS TO PONDER

● Climate change seems to be worsening the droughts that transform ground fires into high-intensity firestorms, exacerbating an already-serious environmental hazard in Australia.

● Some geographers suggest that Australia by the middle of this century may more than double its current population and have as many as 50 million inhabitants, as well as one of the most multicultural societies in the world—its water-supply problems notwithstanding.

● Australia's embrace of China entails costs. In 2008 the Chinese embassy in Canberra organized gangs of Chinese students throughout the country to obstruct local pro-Tibet demonstrations.

habitats, that is, the environment they normally occupy and of which they constitute a part. Such environments change, and the changes can spell disaster for the species. But it is not always clear whether environmental change or human intervention caused the problem. In Australia, **7** the arrival of the **Aboriginal population** (about 50,000 years ago) appears to have caused an ecosystem collapse because there is no evidence of significant climate change at the time. In an article in *Science* not long ago, Gifford H. Miller and his colleagues reported that wide-

spread burning of the existing forest, shrub, and grass-land vegetation across Australia appears to have led to the spread of desert scrub and to the rapid extinction of most of the continent's large mammals soon after the human invasion occurred. The species that survived faced a second crisis when European colonizers intro-duced their livestock, leading to the further destruction of remaining wildlife habitats. Survivors include marsu-pials such as the koala "bear" and the wombat, but the list of extinctions is far longer.

Regions of the Realm

Australia is the dominant component of the Austral Realm, a continent-scale country in a size category that also includes China, Canada, the United States, and Brazil. For two reasons, however, Australia has fewer regional divisions than do the aforementioned countries: Australia's relatively uncomplicated physiography and its diminutive human numbers. Our discussion, therefore, uses the core-periphery concept as a basis for investigating Australia and focuses on New Zealand as a region by itself.

● AUSTRALIA

On January 1, 2001, Australia celebrated its 100th birthday as a state, the Commonwealth of Australia, recognizing (still) the British monarch as the head of state and enter-ing its second century with a strong economy, stable political framework, high standard of living for most of its people, and favorable prospects ahead. Positioned on the Pacific Rim, nine-tenths as large as the 48 contigu-ous U.S. States, well endowed with farmlands and vast pastures, major rivers, ample underground water, mi-nerals, and energy resources, served by good natural harbors, and populated by 21.6 million mostly well-

educated people, Australia is one of the most fortunate countries on Earth. Indeed, as Figure G-12 reminds us, Australia today is part of the global core of states with the highest standards of living.

Sharing the Bounty

Not everyone in Australia shares adequately in all this good fortune, however, and the less advantaged made their voices heard during the celebrations. The country's indigenous (*Aboriginal*) population, though today a small minority (2.5 percent) of about 550,000, remains dispro-portionately disadvantaged in almost every way, from lower life expectancies to higher unemployment than av-erage, from lower high school graduation rates to higher imprisonment ratios. But, as we shall see, the nation is now embarked on a campaign to address these ills, with actions ranging from public demonstrations supporting reconciliation to official expressions of regret for past mistreatment and from enhanced social services to favor-able court decisions involving Aboriginal land claims.

When Australia was born as a federal state, its per-capita GNI, as reckoned by economic geographers, was the highest in the world. Australia's bounty fueled a huge flow of exports to Europe, and the Australians prospered. That golden age could not last forever, and eventually the country's share of world trade declined. Still, Australia today ranks among the top 20 countries in the world in terms of GNI, and for the vast majority of Australians life is comfortable.

The statistics bear witness to Australia's current good fortune. In terms of the indicators of development dis-cussed in Chapter 9, Australia is far ahead of all its west-ern Pacific Rim competitors except Japan. As Australians celebrated their first century, they were, on average, earn-ing far more than Thais, Malaysians, or Koreans. In terms of consumption of energy per person, the number of automobiles and miles of roads, levels of health, and

MAJOR CITIES OF THE REALM

City	Population* (in millions)
Adelaide, Australia	1.2
Auckland, New Zealand	1.2
Brisbane, Australia	1.9
Canberra, Australia	0.4
Melbourne, Australia	3.8
Perth, Australia	1.5
Sydney, Australia	4.4
Wellington, New Zealand	0.4

*Based on 2010 estimates.

literacy, Australia had all the properties of a developed country. Australian cities and towns, where 87 percent of all Australians live, are not encircled by crowded shantytowns. Nor is the Australian countryside inhabited by a poverty-stricken peasantry.

Distance

Australians often talk about distance. One of their leading historians, Geoffrey Blainey, labeled it a "tyranny"—an imposed remoteness from without and a divisive part of life within. Even today, Australia is far from nearly everywhere on Earth. A jet flight from Los Angeles to Sydney takes 14 hours nonstop and is correspondingly expensive. Seaborne freighters carrying products to European markets take ten days to two weeks to get there. Inside Australia, distances also are of continental proportions, and Australians pay the price—literally. Until some upstart private airlines started a price war, Australians paid more per mile for their domestic flights than air passengers anywhere else in the world.

But distance also was an ally, permitting Australians to ignore the obvious. Australia was a British progeny, a European outpost. Once you had arrived as an immigrant from Britain or Ireland, there were a wide range of environments, magnificent scenery, vast open spaces, and seemingly limitless opportunities. When the Japanese Empire expanded, Australia's remoteness saved the day. When immigration became an issue, Australia in its comfortable isolation could adopt an all-white admission policy that was not officially terminated until 1976. When boat people by the hundreds of thousands fled Vietnam in the aftermath of the Indochina War, almost none reached Australian shores.

Immigrants

Today Australia is changing and rapidly so. Immigration policy now focuses on the would-be immigrants' qualifications, skills, financial status, age, and facility with the English language. With regard to skills, high-technology specialists, financial experts, and medical personnel are especially welcome. Relatives of earlier immigrants, as well as a quota of genuine asylum-seekers, also are readily admitted. In recent years, total immigration has been between about 120,000 and 180,000 annually, which keeps the country's population growing. According to the latest data, Australia's declining natural rate of increase today stands at 0.7 percent.

Already, Australia's changed immigration policies have dramatically altered cultural landscapes, especially in the urban areas (see photo, p. 593). The country is fast becoming a truly multicultural society. In Sydney, for

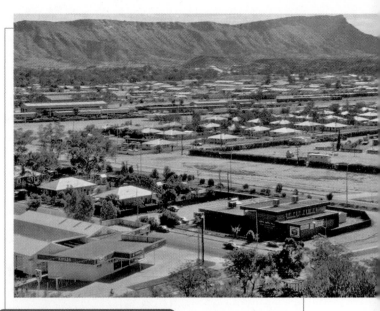

FROM THE FIELD NOTES

"My most vivid memory from my first visit to Alice Springs is spotting vineyards and a winery in this parched, desert environment as the plane approached the airport. I asked a taxi driver to take me there, and got a lesson in economic geography. Drip irrigation from an underground water supply made viticulture possible; the tourist industry made it profitable. None of this, however, is evident from the view seen here: a spur of the Macdonnell Ranges overlooks a town of bare essentials under the hot sun of the Australian desert. What Alice Springs has is centrality: it is the largest settlement in a vast area Australians often call 'the centre.' Not far from the midpoint on the nearly 3200-kilometer (2000-mi) Stuart Highway from Darwin on the Northern Territory's north coast to Adelaide on the Southern Ocean, Alice Springs also was the northern terminus of the Central Australian Railway (before the line was extended north to Darwin in 2003), seen in the middle distance. The shipping of cattle and minerals is a major industry here. You need a sense of humor to live here, and the locals have it: the town actually lies on a river, the intermittent Todd River. An annual boat race is held, and in the absence of water the racers carry their boats along the dry river bed. No exploration of Alice Springs would be complete without a visit to the base of the Royal Flying Doctor Service, which brings medical help to outlying villages and homesteads." © H. J. de Blij.

www.conceptcaching.com

instance, one in five residents is now of Asian ancestry. Overall, more than a quarter of Australia's 2010 population is foreign-born, and another quarter consists of first-generation Australians.

Core and Periphery

Australia is a large landmass, but its population is heavily concentrated in a core area that lies in the east and southeast, most of which faces the Pacific Ocean (here named the Tasman Sea between Australia and New Zealand). As Figure 11-5 shows, this crescent-like Australian heartland extends from north of the city of Brisbane to the vicinity of Adelaide and includes the largest city, Sydney, the capital, Canberra, and the second-largest city, Melbourne. A secondary core area has developed in the far southwest, centered on Perth and its outport, Fremantle. Beyond lies the vast periphery, which the Australians call the **Outback** (see photo, p. 585).

To better understand the evolution of this spatial arrangement, it helps to refer again to the map of world climates (Fig. G-7). Environmentally, Australia's most favored strips face the Pacific and Southern oceans, and they are not large. We can describe the country as a

AUSTRALIA:
POLITICAL DIVISIONS, CAPITALS,
AND COMMUNICATIONS

POPULATION

● Under 50,000

● 50,000–250,000

● 250,000–1,000,000

● 1,000,000–5,000,000

Canberra National capital

Sydney State capital

—— Railroad

—— Main Road

Australian core area

| 0 | 200 | 400 | 600 | 800 | 1000 Kilometers |
| 0 | | 250 | | 500 Miles | |

FIGURE 11-5

© H. J. de Blij, P. O. Muller, and John Wiley & Sons, Inc.

coastal rimland with cities, towns, farms, and forested slopes giving way to the vast, arid, interior Outback. On the western flanks of the Great Dividing Range lie the extensive grassland pastures that catapulted Australia into its first commercial age—and on which still graze one of the largest sheep herds on Earth (over 160 million sheep, producing more than one-fifth of all the wool sold in the world). Where it is moister, to the north and east, cattle by the millions graze on ranchlands. This is frontier Australia, over which livestock have ranged for nearly two centuries.

Aborigines and the British

Aboriginal Australians reached this landmass as long as 50,000 years ago, crossed the Bass Strait into Tasmania, and had developed a patchwork of indigenous cultures when Captain Arthur Phillip sailed into what is today Sydney Harbor (1788) to establish the beginnings of modern Australia. The Europeanization of Australia doomed the continent's Aboriginal societies. The first to suffer were those situated in the path of British settlement on the coasts, where penal colonies and free towns were founded. Distance protected the Aboriginal communities of the northern interior longer than elsewhere; in Tasmania, the indigenous Australians were exterminated in just decades after having lived there for perhaps 45,000 years.

The Seven Colonies

Eventually, the major coastal settlements became the centers of seven different colonies, each with its own hinterland; by 1861, Australia was delimited by its now-familiar pattern of straight-line boundaries (Fig. 11-5). Sydney was the focus for New South Wales; Melbourne,

Sydney's rival, anchored Victoria. Adelaide was the heart of South Australia, and Perth lay at the core of Western Australia. Brisbane was the nucleus of Queensland, and Hobart was the seat of government in Tasmania. The largest clusters of surviving Aboriginal people were in the so-called Northern Territory, with Darwin, on Australia's tropical north coast, its colonial city. Notwithstanding their shared cultural heritage, the Australian colonies were at odds not only with London over colonial policies but also with each other over economic and political issues. The building of an Australian nation during the late nineteenth century was a slow and difficult process.

A Federal State

On New Year's Day of 1901, following years of difficult negotiations, the Australia we know today finally emerged: the Commonwealth of Australia, consisting of six States and two Federal Territories (Table 11-1; Fig. 11-5). The two Federal Territories are the Northern Territory, assigned to protect the interests of the large Aboriginal population concentrated there, and the Australian Capital Territory, carved from southern New South Wales to accommodate the federal capital of Canberra, inaugurated in 1927.

Australia's six States are New South Wales (capital Sydney), at 7.0 million the most populous and politically powerful; Queensland (Brisbane), with the Great Barrier Reef offshore and tropical rainforests in its north; Victoria (Melbourne), small but populous by Australian standards with 5.3 million inhabitants; South Australia (Adelaide), where the Murray–Darling river system reaches the sea; Western Australia (Perth) with just over

TABLE 11-1

States and Territories of Federal Australia, 2010

State	Area (1000 sq km [1000 sq mi])	Population (millions)	Capital	Population (millions)
New South Wales	801.6 (309.5)	7.0	Sydney	4.4
Queensland	1727.3 (666.9)	4.4	Brisbane	1.9
South Australia	983.9 (379.9)	1.6	Adelaide	1.2
Tasmania	67.9 (26.2)	0.5	Hobart	0.2
Victoria	227.7 (87.9)	5.3	Melbourne	3.8
Western Australia	2525.5 (975.1)	2.2	Perth	1.5
Territory				
Australian Capital Territory	2.3 (0.9)	0.4	Canberra	0.4
Northern Territory	1346.3 (519.8)	0.2	Darwin	0.1

2 million people in an area of more than 2.5 million square kilometers (nearly 1 million sq mi); and Tasmania (Hobart), the island across the Bass Strait from the mainland and in the path of the storms of the Southern Ocean.

Successful Federation

In earlier chapters, we have referred to the concept of federalism, an idea with ancient Greek and Roman roots familiar to Americans, Canadians, South Asians, and, more recently, Russians. It is a notion of communal association whose name comes from the Latin *foederis*, implying alliance and coexistence, a union of consensus and common interest—a **federation**. It stands in contrast to the idea that states should be centralized, or *unitary*. For this, too, the ancient Romans had a term: *unitas*, meaning unity. Most European countries are **unitary states**, including the United Kingdom of Great Britain and Northern Ireland. Although the majority of Australians came from that tradition (a kingdom, no less), they managed to overcome their differences and establish a Commonwealth that was, in effect, a federation of States with different viewpoints, economies, and objectives, separated by vast distances along the rim of an island continent. And yet, the experiment succeeded.

An Urban Culture

During this century of federal association, the Australians developed an urban culture. Despite those vast open spaces and romantic notions of frontier and Outback, more than 85 percent of all Australians live in cities

and towns. On the map, Australia's areal functional organization is similar to Japan's: large cities lie along the coast, the centers of manufacturing complexes as well as the foci of agricultural areas. Contributing to this situation in Japan was mountainous topography; in Australia, it was an arid interior. There, however, the similarity ends. Australia's territory is 20 times larger than Japan's, and Japan's population is almost six times that of Australia's. Japan's port cities are built to receive raw materials and to export finished products. Australia's cities forward minerals and farm products from the Outback to foreign markets and import manufactures from overseas. Distances in Australia are much greater, and spatial interaction (which tends to decrease with increasing distance) is less. In comparatively small, tightly organized Japan, you can travel from one end of the country to the other along highways, through tunnels, and over bridges with utmost speed and efficiency. In Australia, the overland trip from Sydney to Perth, or from Darwin to Adelaide (using the newly completed railway link via Alice Springs), is time-consuming and slow. Nothing in Australia compares to Japan's high-speed bullet trains.

The Cities

For all its vastness and youth, Australia nonetheless developed a remarkable cultural identity, a sameness of urban and rural landscapes that persists from one end of the continent to the other. Sydney, often called the New York of Australia, lies on a spectacular estuarine site, its compact, high-rise central business district overlooking a port bustling with ferry and freighter traffic. Sydney is a vast, sprawling metropolis of 4.4 million, with multiple outlying centers studding its far-flung suburbs; brash

MORE THAN TWO centuries ago, Sydney was founded by Captain Arthur Phillip as a British outpost on one of the world's most magnificent natural harbors. The free town and penal colony that struggled to survive evolved into Australia's largest city. Today metropolitan Sydney (4.4 million) is home to more than one-fifth of the country's entire population. An early start, the safe harbor, fertile nearby farmlands, and productive pastures in its hinterland combined to propel Sydney's growth. Later, as road and railroad links made Sydney the focus of Australia's growing core area, industrial development and political power augmented its primacy.

With its incomparable setting and mild, sunny climate, its many urban beaches, and its easy reach to the cool Blue Mountains of the Great Dividing Range, Sydney is one of the world's most liveable cities. Good public transportation (including an extensive cross-harbor ferry system from the doorstep of the waterfront CBD), fine cultural facilities headed by the multi-theatre Opera House complex, and many public parks and other recreational facilities make Sydney attractive to visitors as well. A healthy tourist trade, much of it from Japan and other Asian countries, bolsters the city's economy. Sydney's hosting of the 2000 Olympic Games was further testimony to its rising visibility.

Increasingly, Sydney also is a multicultural city. Its small Aboriginal sector is being overwhelmed by the arrival of large numbers of Asian immigrants. The Sydney suburb of Cabramatta symbolizes the impact: more than half of its nearly 100,000 residents were born elsewhere, mostly in Vietnam. Unemployment is high, drug use is a

© H. J. de Blij, P. O. Muller, and John Wiley & Sons, Inc.

problem, and crime and gang violence persist. Yet, despite the deviant behavior of a small minority, tens of thousands of Asian immigrants have established themselves in some profession. As the photo and caption on p. 593 reveal, today things are definitely improving in this dynamic community.

These developments underscore Sydney's coming of age. The end of Australia's isolation has brought Asia across the country's threshold, and again the leading metropolis will show the way.

modernity and reserved British ways blend here. Melbourne (3.8 million), sometimes regarded as the Boston of Australia, prides itself on its more interesting architecture and more civil ways. Brisbane, the capital of Queensland, which also anchors Australia's Gold Coast and adjoins the Great Barrier Reef, is the Miami of Australia; unlike Miami, however, its residents can find nearby relief from the summer heat in its immediate hinterland mountains (as well as at its beaches). Perth, Australia's San Diego, is one of the world's most isolated cities, separated from its nearest Australian neighbor by two-thirds of a continent and from Southeast Asia and Africa by thousands of kilometers of ocean.

And yet, each of these cities—as well as the capitals of South Australia (Adelaide), Tasmania (Hobart), and, to a lesser extent, the Northern Territory (Darwin)—exhibits an Australian character of unmistakable quali-

ty. Life is orderly and unhurried. Streets are clean, slums are few, graffiti rarely seen. By American and even European standards, violent crime (though rising) is uncommon. Standards of public transport, city schools, and health-care provision are high. Spacious parks, pleasing waterfronts, and plentiful sunshine make Australia's urban life more acceptable than that almost anywhere else in the world. Critics of Australia's way of life say that this very pleasant state of affairs has persuaded Australians that hard work is not really necessary. But a few days' experience in the commercial centers of the major cities contradicts that assertion: the pace of life is quickening. The country's cultural geography evolved as that of a European outpost, prosperous and secure in its isolation. Now Australia must reinvent itself as a major link in an Australo-Asian chain, a Pacific partner in a transformed regional economic geography.

Economic Geography

Because of its early history as a treasure trove of raw materials, Australia to this day is seen worldwide as a country whose economy depends primarily on its exportable natural resources. But in fact, Australia's economy depends overwhelmingly on services, not commodity exports. Tourism alone contributes around 5 percent—about the same as the value of mineral exports. The action is in those bustling coastal cities far more than in the Outback.

When Australia became established as a state, however, it needed goods from overseas, and here the tyranny of distance played a role. Imports from Britain (and later the United States) were expensive mainly because of transport costs. This encouraged local entrepreneurs to set up their own industries to produce such goods more cheaply. Economic geographers call these activities **import-substitution industries**, and this is how local industrialization got its start.

When prices of foreign goods became lower because transportation was more efficient and therefore cheaper, local businesses demanded protection from the colonial governments, and high tariffs were erected against imported goods. Local products now could continue to be made inefficiently because their market was guaranteed. If Japan could not afford this, how could Australia? We can see the answer on the map (Fig. 11-6). Even before federation in 1901, all the colonies could export valuable minerals whose earnings shored up those inefficient, uncompetitive local industries. By the time the colonies unified, pastoral industries were also contributing income. So the miners and the farmers paid for those imports Australians could not produce themselves, plus the products made in the cities. No wonder the cities grew: here were secure manufacturing jobs, jobs in state-run service enterprises, and jobs in the growing government bureaucracy. When we noted earlier that Australians once had the highest per-capita GNI in the world, this was initially achieved in the mines and on the farms, not in the cities.

But the good times had to come to an end. In the 1980s, the prices of farm products began to fluctuate, and international market competition increased. The cost of mining, transporting, and shipping ores and minerals also rose. Australians like to drive (there are more road kilometers per person in Australia than in the United States), and expensive petroleum imports were needed. Meanwhile, the government-protected industries had been further fortified by strong labor unions. Not surprisingly, the economy slumped, the national debt rose, inflation grew, and unemployment crept upward. It was not a good time for immigration reform, and Australian industry lagged.

More recently, Australia has been booming again, and this time too minerals fueled the economic surge. Over the past nearly 20 years, Australia first had a raw-material-hungry Japan at the top of its customer list; when the Japanese economy slowed, there to pick up the slack was China with its even deeper pockets. In early 2009, unemployment was at a 30-year low, union membership and power were in decline, immigration numbers were rising, and inflation was Australia's biggest concern—before it started to worry about the local impact of the world economic downturn that began in 2008. That global recession reminded Australians that, in their services- and commodities-dominated economy, domestic industries were not gaining much traction—a potential problem for the future.

Agricultural Mainstay

Australia has material assets of which other countries can only dream, and minerals and fuels are only part of the inventory. In agriculture, sheep-raising was the earliest commercial venture, but it was the technology of refrigeration that brought world markets within reach of Australian beef producers. Wool, meat, and wheat have long been the country's big three income earners; Figure 11-6 displays the vast pastures in the east, north, and west that constitute the ranges of Australia's huge herds. The zone of commercial grain farming forms a broad crescent extending from northeastern New South Wales through Victoria into South Australia, and covers much of the hinterland of Perth. Keep in mind the scale of this map: Australia is only slightly smaller than the 48 contiguous States of the United States! Commercial grain farming in Australia is big business. As the climatic map would suggest, sugarcane grows along most of the warm, humid coastal strip of Queensland, and Mediterranean crops (including grapes for Australia's highly successful wines) cluster in the hinterlands of Adelaide and Perth. Mixed horticulture concentrates in the basin of the Murray River system, including rice, grapes, and citrus fruits, all under irrigation. And, as elsewhere in the world, dairying has developed near the large urban areas. With its considerable range of environments, Australia yields a diversity of crops.

Mineral Wealth

Australia's mineral resources, as Figure 11-6 shows, also are diverse. Major gold discoveries in Victoria and New South Wales produced a ten-year gold rush starting in 1851 and ushered in a new economic era. By the middle of that decade, Australia was producing 40 percent of the world's gold. Subsequently, the search for more gold led to the discoveries of other minerals. New finds are still being made today, and even oil and natural gas have been found both inland and offshore (see the symbols

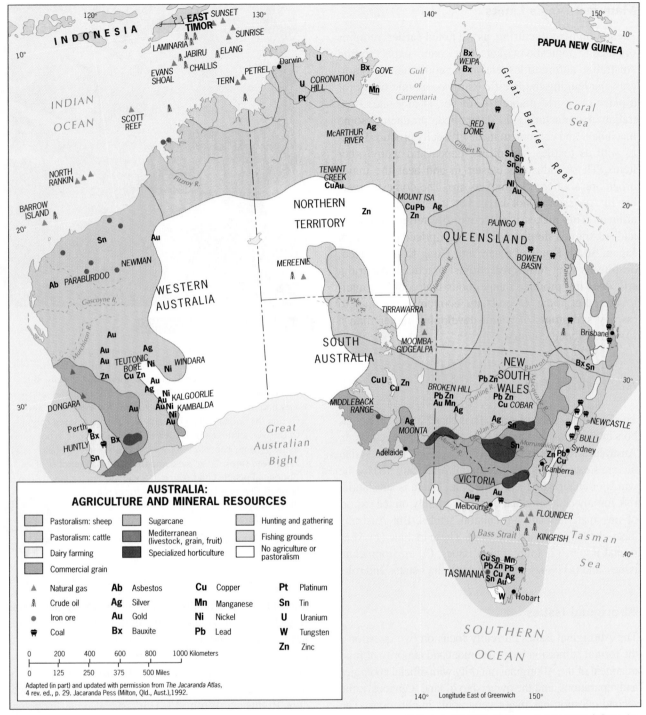

AUSTRALIA:
AGRICULTURE AND MINERAL RESOURCES

▨ Pastoralism: sheep	▨ Sugarcane	▨ Hunting and gathering
▨ Pastoralism: cattle	▨ Mediterranean (livestock, grain, fruit)	▨ Fishing grounds
▨ Dairy farming	▨ Specialized horticulture	▨ No agriculture or pastoralism
▨ Commercial grain		

▲ Natural gas	**Ab** Asbestos	**Cu** Copper	**Pt** Platinum
⚒ Crude oil	**Ag** Silver	**Mn** Manganese	**Sn** Tin
● Iron ore	**Au** Gold	**Ni** Nickel	**U** Uranium
⛏ Coal	**Bx** Bauxite	**Pb** Lead	**W** Tungsten
			Zn Zinc

```
0      200    400    600    800    1000 Kilometers
0      125    250    375    500 Miles
```

Adapted (in part) and updated with permission from *The Jacaranda Atlas*, 4 rev. ed., p. 29. Jacaranda Pess (Milton, Qld., Aust.),1992.

FIGURE 11-6

© H. J. de Blij, P. O. Muller, and John Wiley & Sons, Inc.

in Fig. 11-6 in the Bass Strait between Tasmania and the mainland, and off the northwestern coast of Western Australia). Coal is mined at many locations, notably in the east near Sydney and Brisbane but also in Western Australia and even in Tasmania; before coal prices fell, this was a valuable export. Major deposits of metallic and nonmetallic minerals abound—from the complex at Broken Hill and the mix of minerals at Mount Isa to the huge nickel deposits at Kalgoorlie and Kambalda, the copper of Tasmania, the tungsten and bauxite of far northern Queensland, and the asbestos of Western Australia. A closer look at the map reveals the wide distribution of iron ore (the red dots), and for this raw material as for many others, Japan was Australia's best customer for many years; today, however, China is Australia's leading customer.

Manufacturing's Limits

Australian manufacturing, as we noted earlier, remains oriented to domestic markets. One cannot expect to find Australian automobiles, electronic equipment, or cameras challenging the Pacific Rim's economic tigers for a place on world markets—not yet, at any rate. Australian manufacturing is diversified, producing some machinery and equipment made of locally produced steel as well as textiles, chemicals, paper, and many other items. These industries cluster in and near the major urban areas where the markets are. The domestic market in Australia is not large, but it remains relatively affluent. This makes it attractive to foreign producers, and Australia's shops are full of high-priced goods from Japan, South Korea, Taiwan, and Hong Kong. Indeed, despite its long-term protectionist practices, Australia still does not produce many goods that could be manufactured at home. Overall, the economy continues to display symptoms of a still-developing rather than a fully developed country.

Australia's Future

The Commonwealth of Australia is changing, but its neighbors in Southeast and South Asia are changing even faster. Australia's European bonds are weakening and its Asian ties are strengthening, but Australia will be buffeted by the winds of change on the Pacific Rim. There is substantial risk for any economy that depends strongly on the export of raw materials and agricultural staples. But Australia faces challenges at home as well. These include (1) Aboriginal issues, (2) immigration issues, (3) environmental issues, and (4) issues involving Australia's status and role.

Aboriginal Issues

The Aboriginal issues currently focus on two questions: the formal admission by government and majority of mistreatment of the Aboriginal minority with official apologies and reparations, and the ownership of land. Obviously, the second question has major geographic implications. The Aboriginal population of 550,000 (including many of mixed ancestry) has been gaining influence in national affairs, and during the 1980s Aboriginal leaders began a campaign to obstruct exploration on what they designated as ancestral and sacred lands. Until 1992, Australians had taken it for granted that Aborigines had no right to land ownership, but in that year the Australian High Court made the first of a series of rulings in favor of Aboriginal claimants. A subsequent court decision implied that vast areas (probably as much as 78 percent of the whole con-

FROM THE FIELD NOTES

"While in Darwin I had to visit a government office, and waiting in line with me was this Aboriginal girl, her brother, and her father. They lived about two hours from the city, she said (which I gathered was the length of their bus ride), and they had to come here from time to time for filing forms and visiting the doctor. When I asked whether they enjoyed coming to Darwin, all three shook their heads vigorously. 'It is not a friendly place,' she said. 'But we must come here.' In fact, I had been surprised at the comparatively small number of Aboriginal people I had seen in Darwin, far fewer than in Alice Springs. Certainly laid-back Alice Springs is a very different place from larger, busier, and more modern Darwin." © H. J. de Blij.

www.conceptcaching.com

tinent) could potentially be subject to Aboriginal claims (Fig. 11-7). Although the **Aboriginal land issue** remains largely (though not exclusively) an Outback issue, it has had a polarizing effect on politics and society as views on the matter in the large cities tend to differ sharply from those in the interior (see the Issue Box entitled "Indigenous Rights and Wrongs"). This issue also has the potential to overwhelm Australia's court system and to inhibit economic growth.

Lately, Australians have begun debating ways to bring so-called market-driven incentives to Aboriginal areas. At present, much Aboriginal land is administered communally by Aboriginal Land Councils, which makes the people land-rich but keeps them dirt-poor because the system in effect prevents private enterprise, including even the construction of family-owned housing. Some Aboriginal leaders are arguing that the system perpetuates an unde-

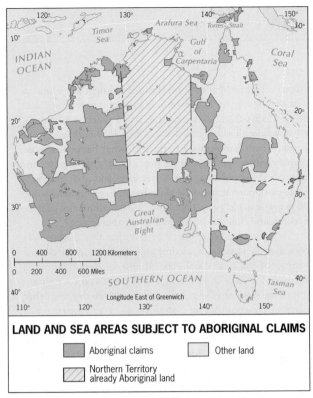

LAND AND SEA AREAS SUBJECT TO ABORIGINAL CLAIMS

Aboriginal claims   Other land

Northern Territory already Aboriginal land

© H. J. de Blij, P. O. Muller, and John Wiley & Sons, Inc.

FIGURE 11-7

tory that chronicled rampant alcoholism, domestic violence, unemployment, school truancy, and a general breakdown of Aboriginal culture. The then-prime minister, calling this a national emergency, ordered the federal government to take control of 60 Aboriginal communities in the Northern Territory through police and army patrols, health checks, and other measures. While there were accusations of political grandstanding and questions about sincerity, there could be no doubt that Australia's conscience was once again troubled by the fate of the country's first inhabitants left behind in the era of modernization.

Immigration Issues

The immigration issue is older than Australia itself. Fifty years ago, when Australia had less than half the population it has today, 95 percent of the people were of European ancestry, and more than three-quarters of them came from the British Isles. Eugenic (race-specific) **immigration** 13 **policies** maintained this situation until the 1970s. Today, the picture is dramatically different: of 21.6 million Australians, only about one-third have British-Irish origins, and Asian immigrants outnumber both European immigrants and the natural increase each year. During the early 1990s, about 150,000 legal immigrants arrived in Australia annually, most from Hong Kong, Vietnam, China, the Philippines, India, and Sri Lanka. This engendered a national debate over Australia's ability to absorb such numbers, and immigration quotas were subsequently reduced to about 80,000 annually. Additional measures, such as restrictions on the foreign ownership of real estate, reflected fears of

sirable dependence on federal handouts, keeps jobs from becoming available, and restrains economic incentive. But not all Aboriginal leaders agree, worrying that privatization of any kind will undermine the Land Councils and create a small minority of better-off businesspeople, leaving the great majority of tradition-bound Aborigines even worse off. Still, the fact that the idea has some Aboriginal support is a sign of changing times in the relationships between Australian communities.

Also, in 2007 Australia was shaken by a report on conditions in Aboriginal desert camps in the Northern Territory.

The name Cabramatta conjures up varied reactions among Australians. During the 1950s and 1960s, many immigrants from southern Europe settled in this western suburb of Sydney, attracted by affordable housing. During the 1970s and 1980s, Southeast Asians arrived in large numbers, and during this period Cabramatta, in the eyes of many, became synonymous with gang violence and drug dealing. More recently, however, Cabramatta's ethnic diversity has come to be viewed in a more favorable light, and is now seen as the "multicultural capital of Australia," a tourist attraction and proof of Australia's capacity to accommodate non-Europeans. Meanwhile, Cabramatta has been spruced up with Oriental motifs of various kinds. This 'Freedom Gate' in the Vietnamese community is flanked by a Ming horse and a replica of a Forbidden City lion—all reflecting better times for an old transit point for immigrants and refugees. © Trip/Eric Smith.

Indigenous Rights and Wrongs

THE LEAST WE SHOULD DO IS APOLOGIZE

"I can't believe this. Here we are more than a dozen years beyond our two-hundredth-year anniversary, the evidence of our mistreatment of the Aborigines is everywhere, and we can't, as a nation, say we're sorry? How could we get ourselves into a public debate over this? What does it take to offer sincere apologies to the original Australians and to tell them (and not only to tell them) that we'll do better?

"We should remind ourselves more often than we do of what happened here. When our ancestors arrived on these shores there were more than a million Aboriginal people in Australia, with numerous clans and cultures. They had many virtues, but one in particular: they didn't appropriate land. Land was assigned to them by the creator (what they call the Dreaming), and their relationship to it was spiritual, not commercial. They didn't build fences or walls. Neither had they adopted some bureaucratic religion. Just think, a world where land was open and free and religion was local and personal. Wouldn't that do something for the Middle East and Northern Ireland! Well, we showed up and started claiming and fencing off the land that, under British rules, was there for the taking. If the Aborigines got in the way, they were pushed out, and if they resisted, they got killed. You don't even want to think about what happened in Tasmania. A campaign of calculated extermination. Between 1800 and 1900 the Aboriginal population dropped from 1 million to around 50,000. And when we became a nation, they weren't even accorded citizenship.

"None of this stopped some white Australian men from getting Aboriginal women pregnant. But from 1910 on church and state managed to make things even worse. They took the young children away from their mothers and put them in institutions, where they would be "Europeanized" and then married off to white partners, so that they would lose their Aboriginal inheritance. This, if you'll believe it, went on into the 1960s! Think of the scenes, these kids being kidnapped by armed officials from their mothers. And we have to argue about saying we're sorry?

"Well, at least we're discussing it in the open, and the misdeeds of the past are coming out. Aboriginal Australians finally got the vote in 1962, and they're able to get elected to parliament. I work in a State government office that assists Asian immigrants here in Sydney, and I only wish that we'd done for the Aborigines what we've done for the immigrants we admit."

ENOUGH IS ENOUGH

"[Former] Prime Minister John Howard was right when he said that apologizing for history should be a private matter. I was born on this Outback sheep station 25 years ago, and we employ a dozen Aboriginal workers, most of whom were born in this area. I had nothing to do with what happened more than a century ago, or even the first 75 years of the past century. What I would have done is irrelevant, and what my great-great-grandparents may have done has nothing to do with me. All over the world people are born into situations not of their making, and they should all go around apologizing?

"And as a matter of fact I believe that this country has bent over backwards, in my time at least, to undo the alleged wrongs of the past. Look, the Aboriginal minority counts about 550,000 or 2.5 percent of the population. Compare this to the map: they've got the Northern Territory and other parts of the country too, and that's close to 15 percent of Australia. And now we're going to give them more? I could see disaster in the making when that guy Eddie Mabo got the High Court to agree to his so-called 'native title' claim to land with which he and his people hadn't done a thing. Since when do Aboriginal customs take precedence over the law of the land? But even worse was that decision supporting the Wiks up there in Cape York. That was unbelievable: native title on land already held in pastoral lease? That affects all of us. Pretty soon the whole of Australia will be targeted by these Aboriginal claimants and their lawyers. And by the way, in the old days those people moved around all the time. Who's to say what clans owned what land when it comes to claiming native title? These court cases are going to tie us up in legal knots for generations to come.

"And now I gather that some Aboriginal people are calling for a treaty between themselves and the government, for what purpose I'm not sure—I guess it'll be like the one the Kiwis made with the Maori there. Is this really what we need if we're going to put the past behind us? Look, I like the fellows working here, but you've got to realize that no laws or treaties are going to change all of the problems they have. They're getting preference for many kinds of government employment, remedial help in all kinds of areas, but still they wind up in jail, leaving school, giving up their jobs. Yes, I'm sure wrongs were done in the past. But apologizing isn't going to make any difference. They have their opportunity now. In a lot of countries they would have never gotten it: Aussies are pretty decent people. It's up to them to make the best of it."

Regional ISSUE

Vote your opinion at www.wiley.com/college/deblij

what extremist groups called the "Asianization" of the country and the "corruption" of its Anglo-Saxon heritage.

But by 2008, the situation was quite different. Economically booming Australia faced a shortage of skilled workers—and not just doctors and dentists but also plumbers and electricians. The expanding economy called for hiring mining personnel ranging from surveyors to truck drivers; an Australian airline recruited foreign pilots and mechanics; a building contractor advertised overseas for air-conditioning specialists and electricians. Prime Minister Kevin Rudd had already announced that Australia would issue residency permits to more than 190,000 immigrants in 2009, about 70 percent of them selected for their skills. According to media reports, another 100,000 persons would be allowed to enter Australia on short-term visas issued on the basis of "urgent professional need."

Australians realize that such a proportionately huge influx creates tensions that must be accommodated. The success of many previous Asian settlers demonstrates the opportunities available in this free and open society with its dependable social institutions; Sydney, the main urban recipient of that Asian influx, has become a mosaic of ethnic districts some of which have gone through stages of gang violence, drug dealing, and vandalism but have stabilized and even prospered over time. Yet economic booms do not last forever, and multiculturalism will remain a long-term challenge for the fast-changing Australia of the twenty-first century.

Environmental Issues

14 **Environmental degradation** is almost synonymous with Australia. To both Aborigines and European settlers, its ecology paid a heavy price. Great stands of magnificent forest were destroyed. In Western Australia, centuries-old trees were simply "ringed" and left to die, so that the sun could penetrate through their leafless crowns to nurture the grass below. Then the sheep could be driven into these new pastures. In Tasmania, where Australia's native eucalyptus tree reaches its greatest dimensions (comparable to North American redwood stands), tens of thousands of acres of this irreplaceable treasure have been lost to chain saws and pulp mills. Many of Australia's unique marsupial species have been destroyed, and many more are endangered or threatened. "Never have so few people wreaked so much havoc on the ecology of so large an area in so short a time," observed a geographer in Australia recently. But awareness of this environmental degradation is growing. In Tasmania, the "Green" environmentalist political party has become a force in State affairs, and its activism has slowed deforestation, dam-building, and other "development" projects. Still, many Australians fear the environmentalist movement as an obstacle to economic growth at a time when the economy needs stimulation. This, too, is an issue for the future.

Another environmental problem involves Australia's wide and long-term climatic variability. In a dominantly arid continent, droughts in the moister fringes are the worst enemy, and Australia's history is replete with devastating dry spells. Australia is vulnerable to El Niño events (see p. 273), but recent global warming may also be playing a role in the process. One such serious drought, regarded as the worst in living memory, endured throughout the 2000s

During the first decade of the twenty-first century Australia was desiccated by its worst drought on record, with calamitous consequences ranging from deadly forest fires to parched farmlands. In the southeast, water levels in the Murray-Darling river system (70 percent of whose water is used for irrigated agriculture) fell to their lowest levels ever recorded. Lake Hume on the Victoria-New South Wales border, whose dam is the largest on the Murray River, fell below one-third of capacity by 2009. Although federal and State governments cooperated in seeking ways to ration what little water was left for farming, the drought's impact on production spelled disaster. In this photo, once-submerged trees at the bottom of Lake Hume are exposed as the water level continues to fall.   © AFP photo/Newscom.

and imperils the entire region watered by the Murray–Darling river system, Australia's breadbasket (Fig. 11-6). By some estimates, an area that accounts for 40 percent of Australia's agriculture and contains 85 percent of its irrigated land is on the verge of ruin. Part of this is caused by nature, but other factors have to do with the increasing demand for water in Australia's burgeoning urban areas and by excessive damming, well-drilling, and water diversion in the Murray–Darling system's upstream tributaries. The Australian government wants to institute a coordinated drainage-basin-control program for the Murray–Darling region, but this requires cooperation from above (the State governments involved) and below (local farmers who survive by diverting water to raise their crops). So far, the plan has not been adopted, but if the drought continues to deepen, Australia's environmental crisis will leave no alternative—and even then it may be too little too late.

Status and Role

Several issues involving Australia's status at home, relations with neighbors, and position in the world are stirring up national debate as well. A persistent domestic question is whether Australia should become a republic, ending the role of the British monarch as the head of state, or continue its status quo in the British Commonwealth.

Relations with neighboring *Indonesia* and *East Timor* have been complicated. For many years, Australia had what may be called a special relationship with Indonesia, whose help it needs in curbing illegal seaborne immigration. It was also profitable for Australia to counter international (UN) opinion and recognize Indonesia's 1976 annexation of Portuguese East Timor, for in doing so Australia could deal directly with Jakarta for the oil reserves under the Timor Sea (see Fig. 10-16). Thus Australia gave neither recognition nor support to the rebel movement that fought for independence in East Timor. But this story had a relatively happy ending. When the East Timorese campaign for independence succeeded in 1999 and Indonesian troops began an orgy of murder and destruction, Australia sent an effective peacekeeping force and spearheaded the United Nations effort to stabilize the situation. Today, a new chapter has opened in Australia's relations with these northern neighbors.

Australia also has a long-term relationship with *Papua New Guinea (PNG)*, as we will note in Chapter 12. This association, too, has gone through difficult times. In recent years, the inhabitants of Papua New Guinea have strongly resisted privatization, World Bank involvement, and globalization generally. Australia assists PNG in a number of spheres, but its motives are sometimes questioned. In 2001, the construction of a projected gas pipeline from PNG to the Australian State of Queensland precipitated fighting among tribespeople over land rights, resulting in dozens of casualties, and Australian public opinion reflected doubts over the appropriateness of this venture.

When political violence and disorder overtook the *Solomon Islands* east of Papua New Guinea in 2003, Australian forces intervened: a failing state in Australia's neighborhood could become a base for terrorist activity.

Immediately after the 9/11 attack on U.S. targets in New York and Washington, Australian leaders expressed strong support for the American campaign against terrorism, but they made it a point also to assure the Indonesian government that the War on Terror was not a war on Islam. Although the great majority of Australians supported this stance, their will was severely tested in 2002 when a terrorist attack on a nightclub on the Indonesian island of Bali killed 88 vacationing Australians—and again in 2005 when another attack in Bali was followed by the apprehension of a terrorist cell preparing for an assault in Australia itself.

Territorial dimensions, raw-material wealth, and relative location have helped determine Australia's place in the world and, more specifically, on the Pacific Rim. Australia's population, still barely 20 million in the second decade of the twenty-first century, is smaller than Malaysia's and only slightly larger than that of the Caribbean island of Hispaniola. But Australia's importance in the international community far exceeds its human numbers.

● NEW ZEALAND

Two thousand four hundred kilometers (1500 mi) east-southeast of Australia, in the Pacific Ocean across the Tasman Sea, lies New Zealand. In an earlier age, New Zealand would have been part of the Pacific geographic realm because its population was Maori, a people with Polynesian roots. But New Zealand, like Australia, was invaded and occupied by Europeans. Today, its population of 4.4 million is about 75 percent European, and the Maori form a minority of about 660,000, with many of mixed Euro-Polynesian ancestry.

New Zealand consists of two large mountainous islands and many scattered smaller islands (Fig. 11-8). The two large islands, with the South Island somewhat larger than the North Island, look diminutive in the great Pacific Ocean, but together they are larger than Britain. In contrast to Australia, the two main islands are mostly mountainous or hilly, with several peaks rising far higher than any on the Australian landmass. The South Island has a spectacular snowcapped range appropriately called the Southern Alps, with numerous peaks reaching beyond 3300 meters (10,000 ft). The smaller North Island has proportionately more land under low relief, but it also has an area of central highlands along whose lower slopes lie the pastures of New Zealand's chief dairying district. Hence, while Australia's land lies relatively low in elevation and exhibits much low relief, New Zealand's is on the average high and is dominated by rugged terrain.

Human Spatial Organization

Thus the most promising areas for habitation are the lower-lying slopes and lowland fringes on both islands. On the North Island, the largest urban area, Auckland, occupies a comparatively low-lying peninsula. On the South Island, the largest lowland is the agricultural Canterbury Plain, centered on Christchurch. What makes these lower areas so attractive, apart from their availability as cropland, is their magnificent pastures. The range of soils and pasture plants allows both summer and winter grazing. Moreover, the Canterbury Plain, the chief farming region, also produces a wide variety of vegetables, cereals, and fruits. About half of all New Zealand is pasture land, and much of the farming provides fodder for the pastoral industry. Sixty million sheep and eight million cattle dominate these livestock-raising activities, with wool, meat, and dairy products providing about two-thirds of the islands' export revenues.

FIGURE 11-8   © H. J. de Blij, P. O. Muller, and John Wiley & Sons, Inc.

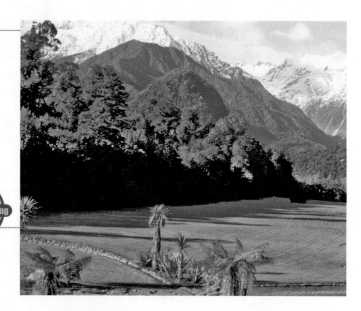

Despite their contrasts in size, shape, physiography, and history, New Zealand and Australia have much in common. Apart from their joint British heritage, they share a sizeable pastoral economy, a small local market, the problem of great distances to world markets, and a desire to stimulate (through protection) domestic manufacturing. The high degree of urbanization in New Zealand (86 percent of the total population) again resembles Australia: substantial employment in city-based industries, mostly the processing and packing of livestock and farm products, and government jobs.

In terms of population geography, New Zealand 15 shares with Australia its pattern of **peripheral development** (Fig. 11-2), imposed not by desert but by high rugged mountains. The country's major cities—Auckland and the capital of Wellington (together with its satellite of Hutt) on the North Island, and Christchurch and Dunedin on the South Island—are all located on the coast, and the entire rail and road system is peripheral in its configuration (Fig. 11-8). This is more pronounced on the South Island than in the north because the Southern Alps are New Zealand's most formidable barrier to surface communications.

The Maori Factor and New Zealand's Future

Like Australia, New Zealand has a history of difficult relations with its indigenous population. The Maori, who account for 15 percent of the country's population today, appear to have reached these islands during the tenth century AD. By the time the European colonists arrived, they had had a tremendous impact on the islands' ecosystems, especially on the North Island where most Maori lived. In 1840, the Maori and the British signed a treaty at Waitangi that granted the colonists sovereignty over New Zealand but guaranteed Maori rights over established tribal lands. Although the British abrogated parts of the treaty in 1862, the Maori had reason to believe that vast reaches of New Zealand, as well as offshore waters, were theirs.

As in Australia, judicial rulings since 1990 have supported the Maori position, which led to expanded claims and growing demands. One of the Maori's persistent complaints is the slow pace of integration of the minority into modern New Zealand society. Although Maori claims encompass much of the South Island, they also cover prominent sites in the major cities. Today, the Maori question is the leading national issue in the country.

Dominant cultural heritage and prevailing cultural landscape form two criteria on which the delimitation of the Austral Realm is based. But in both Australia and New Zealand, the cultural mosaic is changing, and the convergence with neighboring realms is proceeding.

What You Can Do

RECOMMENDATION: Become a member! Just by taking this course, you are joining a still-small minority of geographically literate Americans (Canadians are far ahead), and you may wish to consider joining some organizations that support this cause. The National Geographic Society (1145 Seventeenth Street N.W., Washington, DC 20036) invites members to join and sends them its famous Magazine with its remarkable foldout maps. Gamma Theta Upsilon is a students' geographical society (c/o James D. Lowry, Department of Geography, University of New Orleans, New Orleans, LA 70148) that publishes *The Geographical Bulletin*, a journal that invites students to submit their work for publication. The American Geographical Society (120 Wall Street, Suite 100, New York, NY 10005-3904) publishes two journals that can be appreciated by the general public, *The Geographical Review* and *Focus*. And if you want to get really technical, there is the Association of American Geographers (1710 Sixteenth Street N.W., Washington, DC 20009-3198), which publishes the quarterly *Annals of the Association of American Geographers*, *The Professional Geographer*, and the monthly *AAG Newsletter* that tells you what's going on in the discipline and where the jobs are. Student members, at appropriately reduced annual dues, are especially welcome!

GEOGRAPHIC CONNECTIONS

1 Both Australia and New Zealand are plural societies, and there were indigenous peoples in both countries when the European settlers arrived. In virtually every respect, however, the historical geographies of contact and interaction in the two countries have progressed in very different ways. The differences relate to contrasts in indigenous as well as immigrant cultures, divergent population numbers, contrasting territorial dimensions and physiographies, and dissimilar governmental policies and practices. Comment on the differences between Australia and New Zealand in this context. Are there ways in which indigenous demands are similar? How are they expressed on the map?

2 Australia may have a modest population, but its territorial dimensions, resource base, economic wealth, military capacity, and relative location make it a regional power. Its relationships with its neighbors, however, have at times been problematic. To what extent does relative location affect Australia's interaction with Indonesia? How has Australia's relationship with East Timor, before and after the latter's independence in 2002, been affected by its links with Indonesia, and what was the key issue in this context? What interests does Australia maintain in Papua New Guinea? Why would Australia be involved in problems in the Solomon Islands?

Geographic Literature on the Austral Realm: The key introductory works on this realm were authored by Barrett & Ford, Heathcote, McKnight, Morton & Johnston, O'Connor et al., and Robinson et al. These works, together with a comprehensive listing of noteworthy books and recent articles on the geography of Australia and New Zealand, can be found in the *References and Further Readings* section of this book's website at **www.wiley.com/college/deblij**.

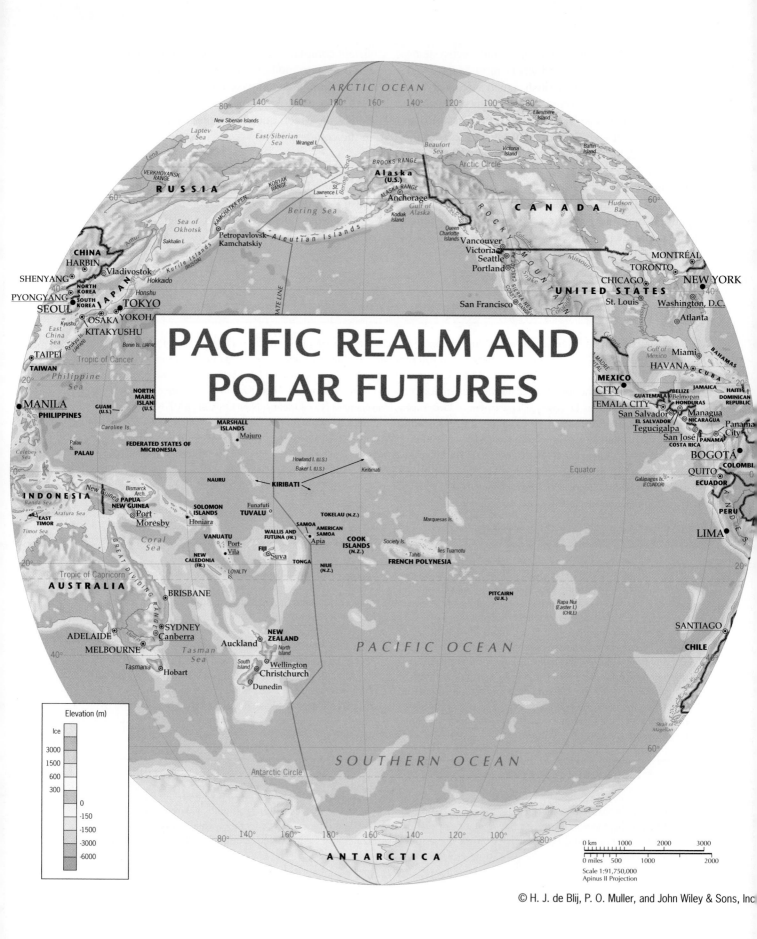

PACIFIC REALM AND POLAR FUTURES

12

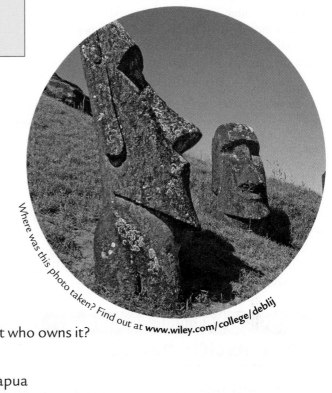

Where was this photo taken? Find out at **www.wiley.com/college/deblij**

In This Chapter

- Water, water everywhere—but who owns it?
- Islands high and low
- The conundrum of divided Papua
- Colonists and foreigners: Still there, still dominant
- Native Hawaiians hope for a sliver of sovereignty
- Polar regions south and north

Photos: © H. J. de Blij.

FIGURE 12-1   *Map:* © H. J. de Blij, P. O. Muller, and John Wiley & Sons, Inc.

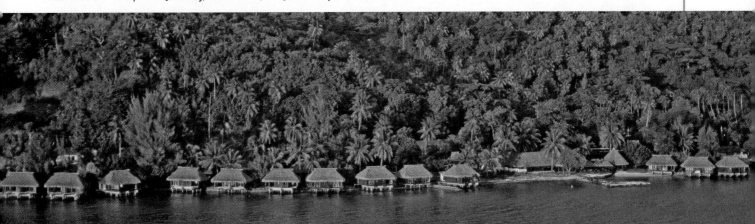

N THIS FINAL chapter, we focus on three of the world's vast expanses so different from those discussed in previous chapters that we need an entirely new perspective. Water and ice, not land and soil, dominate the physiography. Environmental changes going on here affect the entire planet.

The largest of these expanses, the *Pacific Realm*, covers almost half the globe from the Bering Sea to the Southern Ocean and from Ecuador to Indonesia. The Pacific Ocean, larger than all the world's land areas combined, is studded with tens of thousands of islands of which New Guinea is by far the largest. The two polar areas to which it is linked could hardly be more different: the *Arctic* is water, frozen at the surface, so that you can reach the North Pole in a submarine; the *Antarctic* is land, weighed down by the planet's largest accumulation of permanent ice. The surface journey to the South Pole has cost many an explorer his life.

Climate change is altering the geography everywhere, as is enhanced technology of mineral (especially oil and gas) exploration and exploitation. Rising sea levels would imperil thousands of populated low-elevation islands. Areas of ocean floor once beyond the scope of human intervention are now coming within reach. Melting ice is clearing new maritime shipping routes. But, as we will see, international rules to govern who owns what in these remote frontiers are subject to dispute. When land boundaries jut outward into the sea, it depends who draws the maps. We should be aware of the potential for competition, even conflict.

Defining the Realm

Our survey begins with the Pacific geographic realm that covers nearly an entire hemisphere of this world, the one commonly called the Sea Hemisphere (Fig. 12-1). This Sea Hemisphere meets the Russian and North American realms in the far north and merges into the Southern Ocean in the south. Despite the preponderance of water, this fragmented, culturally complex realm does possess regional identities. It includes the Hawaiian Islands, Tahiti, Tonga, and Samoa—fabled names in a world apart.

In terms of modern cultural and political geography, Indonesia and the Philippines are not part of the Pacific Realm, although Indonesia's political system reaches into it; nor are Australia and New Zealand part of it. Before the European invasion and colonization, Australia would have been included because of its Aboriginal population and New Zealand because of its Maori population's Polynesian affinities. But the Europeanization of their countries has engulfed black Australians and Maori

MAJOR GEOGRAPHIC QUALITIES OF

The Pacific Realm

1. The Pacific Realm's total area is the largest of all geographic realms. Its land area, however, is the smallest, as is its population.

2. The island of New Guinea, with 9.3 million people, alone contains over three-quarters of the Pacific Realm's population.

3. The Pacific Realm, with its wide expanses of water and numerous islands, has been strongly affected by United Nations Law of the Sea provisions regarding states' rights over economic assets in their adjacent waters.

4. The highly fragmented Pacific Realm consists of three regions: Melanesia (including New Guinea), Micronesia, and Polynesia.

5. Melanesia forms the link between Papuan and Melanesian cultures in the Pacific.

6. The Pacific Realm's islands and cultures may be divided into volcanic high-island cultures and coral-based low-island cultures.

7. In Micronesia, U.S. influence has been particularly strong and continues to affect local societies.

8. In Polynesia, local cultures nearly everywhere are severely strained by external influences. In Hawai'i, as in New Zealand, indigenous culture has been largely submerged by Westernization.

9. Indigenous Polynesian culture continues to exhibit a remarkable consistency and uniformity throughout the Polynesian region, its enormous dimensions and dispersal notwithstanding.

New Zealanders, and the regional geography of Australia and New Zealand today is decidedly not Pacific. In New Guinea, on the other hand, Pacific peoples remain numerically and culturally the dominant element.

In examining the Pacific Realm, it is important to keep in mind the dimensional contrasts this part of the world presents. The realm is vast, but its total land area is a mere 975,000 square kilometers (377,000 sq mi), about the size of Texas and New Mexico, and over 90 percent of this lies in New Guinea.* The population is so widely scattered that a distribution map would be impractical, and it only totals a bit more than 12 million (about as many people as there are in Beijing, China).

COLONIZATION AND INDEPENDENCE

The Pacific islands were colonized by the French, British, and Americans; an indigenous Polynesian kingdom in the Hawaiian Islands was annexed by the United States and is now the fiftieth State. Still, today, the map is an assemblage of independent and colonial territories (Fig. 12-1). Paris controls New Caledonia and French Polynesia. The United States administers Guam and American Samoa, the Line Islands, Wake Island, Midway Islands, and several smaller islands; the United States also has special relationships with other territories, former dependencies that are now nominally independent. The British, through New Zealand, have responsibility for the Pitcairn group of islands, and New Zealand administers and supports the Cook, Tokelau, and Niue Islands. Easter Island, that storied speck of land in the southeastern Pacific, is part of Chile. Indonesia rules Papua, the western half of the island of New Guinea.

Other island groups have become independent states. The largest are Fiji, once a British dependency, the Solomon Islands (also formerly British), and Vanuatu (until 1980 ruled jointly by France and Britain). Also on the current map, however, are such microstates as Tuvalu, Kiribati, Nauru, and Palau. Foreign aid is crucial to the survival of most of these countries. Tuvalu, for example, has a total area of around 25 square kilometers (10 sq mi), a population of some 10,000, and a per-capita GNI of about U.S. $1500, derived from fishing, copra sales, and some tourism. But what really keeps Tuvalu going is an international trust fund set up by Australia, New Zealand, the United Kingdom, Japan, and South Korea.

Annual grants from that fund, as well as money sent back to families by workers who have left for New Zealand and elsewhere, allow Tuvalu to survive.

THE PACIFIC REALM AND ITS MARINE GEOGRAPHY

Certain land areas may not be part of the Pacific Realm (we cited the Philippines and New Zealand), but the Pacific Ocean extends from the shores of North and South America to mainland East and Southeast Asia and from the Bering Sea to the Subtropical Convergence. This means that several seas, including the East Sea (Sea of Japan), the East China Sea, and the South China Sea, are part of the Pacific Ocean. As we will see below, this relationship matters. Pacific coastal countries, large and small, mainland and island, compete for jurisdiction over the waters that bound them.

The Pacific Realm and its ocean, therefore, form an ideal place to focus on **marine geography**. This field encompasses a variety of approaches to the study of oceans and seas; some marine geographers focus on the biogeography of coral reefs, others on the geomorphology of beaches, and still others on the movement of currents and drifts. A particularly interesting branch of marine geography has to do with the definition and delimitation of political boundaries at sea. Here geography meets political science and maritime law.

The State at Sea

Littoral (coastal) states do not end where atlas maps suggest they do. States have claimed various forms of jurisdiction over coastal waters for centuries, closing off bays and estuaries and ordering foreign fishing fleets to stay away from nearby fishing grounds. Thus arose the notion of the **territorial sea**, where all the rights of a coastal state would prevail. Beyond lay the **high seas**, free, open, and unfettered by national interests.

It was in the interest of colonizing, mercantile states to keep territorial seas narrow and high seas wide, thus interfering as little as possible with their commercial fleets. In the seventeenth and eighteenth centuries the territorial sea was 3, 4, or at most 6 nautical miles wide, and the colonizing powers claimed the same widths for their colonies (1 nautical mile = 1.85 kilometers [1.15 statute miles]).*

---

*The figures in the Data Table inside the back cover do not match these totals because only the political entity of Papua New Guinea is listed, not the Indonesian Province of Papua that occupies the western part of the island. Here, as Figure 10-2 shows, the political and the realm boundaries do not coincide.

*Here, and in the rest of this section, we will give all distances involving maritime boundaries in nautical miles only. Throughout the modern history of maritime law, the latter unit has been the only one used for this type of boundary-making. Any distance can be converted to its kilometric equivalent by multiplying the number of nautical miles by 1.85.

In the twentieth century these constraints weakened. States without trading fleets saw no reason to limit their territorial seas. States with nearby fishing grounds traditionally exploited by their own fleets wanted to keep the increasing number of foreign trawlers away. States with shallow **continental shelves**, offshore continuations of coastal plains, wished to control the resources on and below the seafloor, made more accessible by improving technology. States disagreed on the methods by which offshore boundaries, whatever their width, should be defined. Early efforts by the League of Nations in the 1920s to resolve these issues had only partial success, mainly in the technical area of boundary delimitation.

Scramble for the Oceans

In 1945, the United States helped precipitate what has become known as the "scramble for the oceans." President Harry S Truman issued a proclamation that claimed U.S. jurisdiction and control over all the resources "in and on" the continental shelf down to its margin, around 100 fathoms (600 ft or 183 m) deep. In some areas, the shallow continental shelf of the United States extends more than 200 miles (300 km) offshore, and Washington did not want foreign countries drilling for oil just beyond the 3-mile territorial sea.

Few observers foresaw the impact the Truman Proclamation would have, not only on U.S. waters but on the

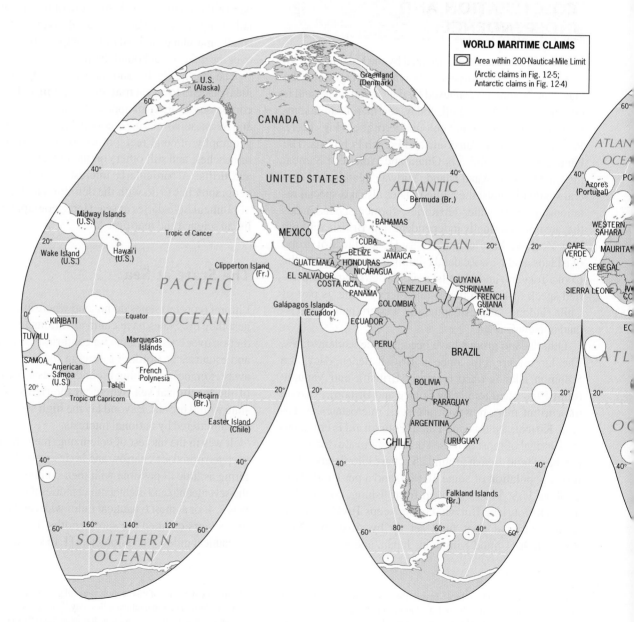

FIGURE 12-2

oceans everywhere, including the Pacific. It set off a rush of other claims. In 1952, a group of South American countries, some with little continental shelf to claim, issued the Declaration of Santiago, claiming exclusive fishing rights up to a distance of 200 nautical miles off their coasts. Meanwhile, as part of the Cold War competition, the Soviet Union urged its allies to claim a 12-mile territorial sea.

UNCLOS Intervention

Now the United Nations intervened, and a series of meetings of UNCLOS (United Nations Conference on the Law of the Sea) began. These meetings addressed issues ranging from the closure of bays to the width and delimitation of the territorial sea, and after three decades of negotiations they achieved a convention that changed the political and economic geography of the oceans forever. Among its key provisions were the authorization of a 12-mile territorial sea for all countries and the establishment of a 200-mile **Exclusive Economic Zone (EEZ)** [5] over which a coastal state would have total economic rights. Resources in and under this EEZ (fish, oil, minerals) belong to the coastal state, which could either exploit them or lease, sell, or share them as it saw fit.

These provisions had a far-reaching impact on the oceans and seas of the world (Fig. 12-2), and especially so on the Pacific. Unlike the Atlantic Ocean, the Pacific

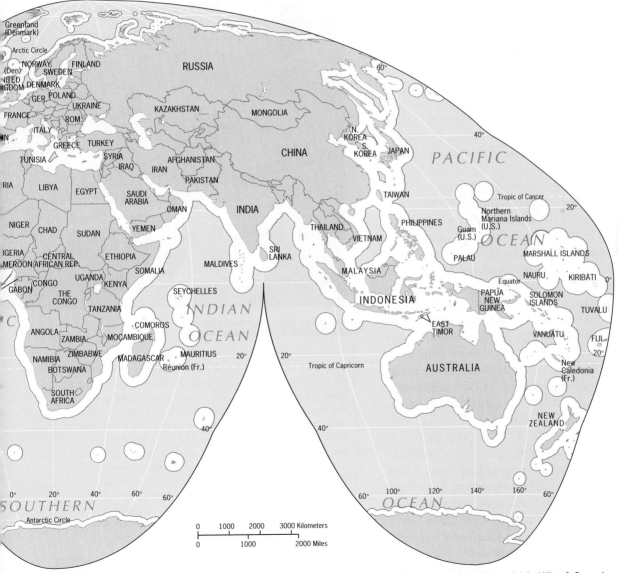

© H. J. de Blij, P. O. Muller, and John Wiley & Sons, Inc.

is studded with islands large and small, and a microstate consisting of one small island suddenly acquired an EEZ covering 160,000 square nautical miles. European colonial powers still holding minor Pacific possessions (notably France) saw their maritime jurisdictions vastly expanded. Small low-income archipelagos could now bargain with large, rich fishing nations over fishing rights in their EEZs. And for all the UNCLOS Convention's provisions for the "right of innocent passage" of shipping through EEZs and via narrow straits, the world's high seas have obviously been diminished.

Maritime Boundaries

The extension of the territorial sea to 12 nautical miles and the EEZ to an additional 188 nautical miles created **6** new **maritime boundary** problems. Waters less than 24 nautical miles wide separate many countries all over the **7** world, so that **median lines**, equidistant from opposite shores, have been delimited to establish their territorial seas. And even more countries lie closer than 400 nautical miles apart, requiring further maritime-boundary delimitation to determine their EEZs. In such maritime regions as the North Sea, the Caribbean Sea, and the East China and South China seas, a maze of maritime boundaries emerged, some of them subject to dispute. Political changes on land can lead to significant modifications at sea.

A case in point involves newly independent East Timor and its neighbor across the Timor Sea, Australia. As we noted in Chapter 10, Australia had divided the waters and seafloor of the Timor Sea with Indonesia while recognizing Indonesia's 1976 annexation of East Timor. When East Timor achieved independence in 2002, the so-called *Timor Gap* (mapped in Fig. 10-16) became an issue: where was the median line that would divide Timorese and Australian claims to the oil and gas reserves in this zone? The Australians argued that since the line defined with Indonesia predated East Timor's independence, it should continue in effect, giving Australia the bulk of the energy resources. But UNCLOS regulations, to which Australia subscribed, required a redelimitation, giving East Timor a much larger share. This led Australia to withdraw from UNCLOS in 2002 so that it would not be bound by its rules. After difficult negotiations, the Australians in 2005 offered East Timor a half share of the energy revenues from the natural gas reserve that was in dispute, in addition to income already flowing from another major gasfield in the so-called Joint Petroleum Development Area. In return, East Timor's government agreed to defer the maritime-boundary issue for 50 years, leaving a future generation to solve this question. It was estimated that, over the next 30 years, East Timor might receive U.S. $13 billion from these sources and,

depending on the life and capacity of the reserves, perhaps even more. This will go a long way toward meeting the young state's most urgent needs.

EEZ Implications

The UNCLOS provisions created opportunities for some states to expand their spheres of influence. Wider territorial-sea and EEZ allocations raised the stakes: claiming an island now entailed potential control over a huge maritime area. In Chapters 9 and 10, we referred to several island disputes off mainland East Asia, which involve Japan and Russia, Japan and South Korea, Japan and China, China and Vietnam, and China and the Philippines. Ownership of many islands there is uncertain, and small specks of is-

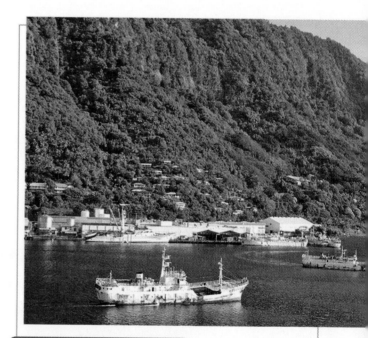

FROM THE FIELD NOTES

"Having arrived late at night, I did not have a perspective yet of the setting of Pago Pago—but even before I got up I knew what the leading industry here had to be. An intense, overpowering smell of fish filled the air. Sure enough, daylight revealed a huge tuna processing plant right across the water, with fishing boats arriving and departing continuously. American Samoa has a huge Pacific Ocean EEZ, one of the richest fishing grounds in the whole realm, and seafood exports rank ahead of tourism as the leading source of external income for the territory, an industry that is mostly American-owned. As the photograph shows, American Samoa is a group of 'high-islands' of volcanic origin, with considerable relief and environmental variety. Tuvalu, the country I had just left, was the complete opposite—a 'low-island' territory whose EEZ has been leased to Japanese fishing fleets, which do their processing on board and employ no locals."
© H. J. de Blij.

land territory have become large stakes in the scramble for the oceans. For example, in the case of the Spratly Islands (Fig. 10-2), six countries claim ownership, including both Taiwan and China. China's island claims in the South China Sea support Beijing's contention that this body of water is part and parcel of the Chinese state—a position that worries other states with coasts facing it.

Figure 12-2 reveals what EEZ regulations have meant to Pacific Realm countries such as Tuvalu, Kiribati, and Fiji, with nearly circular EEZs now surrounding the clusters of islands in this vast ocean space. Japan, Taiwan, and other fishing nations have purchased fishing rights in these EEZs from the island governments. Nonetheless, violations of EEZ rights do occur; recently, Vanuatu and the Philippines were at odds over unauthorized Filipino fishing in Vanuatu's EEZ.

The process of boundary delimitation continues. In an earlier edition of this book, we included a map of the South China Sea and nearby waters off East and Southeast Asia, showing the median-line boundaries delimited according to UNCLOS specifications. But that map changed when coastal states engaged in bilateral and multilateral negotiations—and argued over island ownership with major boundary implications. The Pacific—and world—map of maritime boundaries remains a work in progress (see the Issue Box entitled "Who Should Own the Oceans?").

This point is illustrated by a development that occurred late in 1999, when the implications of provisions in Article 76 of the 1982 UNCLOS Convention were re-examined. Where a continental shelf extends beyond the 200-mile limit of the EEZ, the Convention apparently allows a coastal state to claim that extension as a "natural prolongation" of its landmass. Although there is as yet no

regulation that would also allow the state to extend its EEZ to the edge of the continental shelf, it is not difficult to foresee such an amendment to the Convention. As it stands, states are delimiting their proposed "natural prolongations" under a current deadline, which puts poorer states at a disadvantage since the required marine surveys are very expensive. In any case, a recent article in *Science* reported that the big winners are likely to include the United States, Canada, Australia, New Zealand, Russia, and India—although a total of 60 coastal countries could ultimately benefit to some extent from the provision.

Thus the "scramble for the oceans" continues, and with it the constriction of the world's high seas and open waters. And the scramble may be entering a new stage and geographic arena: the Arctic Ocean, where oil, natural gas, fishing grounds, and maritime routes all figure in the coming competition.

Regions of the Realm

Sail across the Pacific Ocean, and one spectacular vista follows another. Dormant and extinct volcanoes, sculpted by erosion into basalt spires draped by luxuriant tropical vegetation and encircled by reefs and lagoons, tower over azure waters. Low atolls with nearly snow-white beaches, crowned by stands of palm trees, seem to float in the water. Pacific islanders, where foreign influences have not overtaken them, appear to take life with enviable ease.

More intensive investigations reveal that such Pacific sameness is more apparent than real. Even the Pacific Realm, with its long sailing traditions, its still-diffusing populations, and its historic migrations, has a durable regional framework. Figure 12-3 outlines the three regions that constitute the Pacific Realm:

Melanesia: Papua (Indonesia), Papua New Guinea, Solomon Islands, Vanuatu, New Caledonia (France), Fiji

Micronesia: Palau, Federated States of Micronesia, Northern Mariana Islands, Republic of the Marshall Islands, Nauru, western Kiribati, Guam (U.S.)

Polynesia: Hawaiian Islands (U.S.), Samoa, American Samoa, Tuvalu, Tonga, eastern Kiribati, Cook and other New Zealand-administered islands, French Polynesia

Ethnic, linguistic, and physiographic criteria are among the bases for this regionalization of the Pacific Realm, but we should not lose sight of the dimensions. Not only is the land area small, but as we also noted earlier the 2010 population of this entire realm (including Indonesia's Papua Province) was only slightly more than 12 million—9.6 million without Papua—about the same as one very large metropolitan area. Even fewer people live in this realm than in another vast area of far-flung settlements—the oases of North Africa's Sahara.

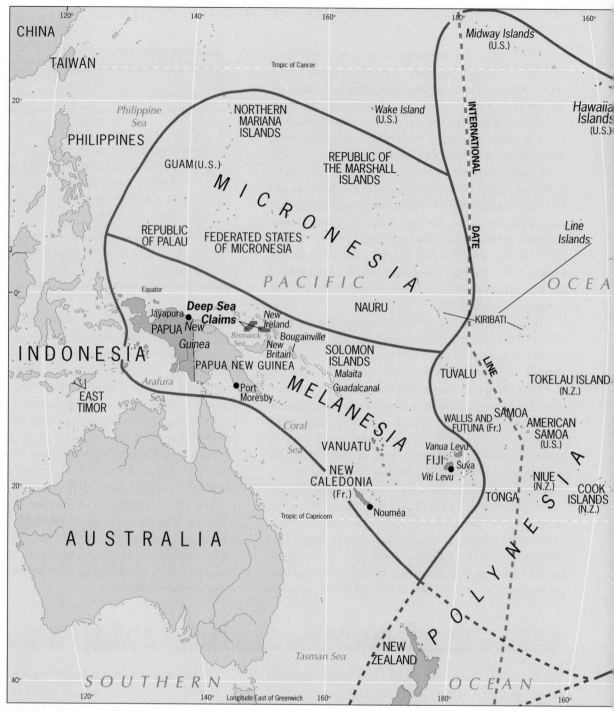

FIGURE 12-3

● MELANESIA

The large island of New Guinea lies at the western end of a Pacific region that extends eastward to Fiji and includes the Solomon Islands, Vanuatu, and New Caledonia (Fig. 12-3). The human mosaic here is complex, both ethnically and culturally. Most of the 9.3 million people of New Guinea (including the Indonesian part, the Province of Papua, and the independent state of Papua New Guinea) are Papuans, and a large minority is Melanesian. Altogether

there are as many as 700 communities speaking different languages; the Papuans are most numerous in the densely forested highland interior and in the south, while the Melanesians inhabit the north and east. The region as a whole has more than 8 million inhabitants, making this the most populous Pacific region by far.

With 6.8 million people today, Papua New Guinea (PNG) became a sovereign state in 1975 after nearly a century of British and Australian administration. Almost all of PNG's limited development is taking place along the coasts,

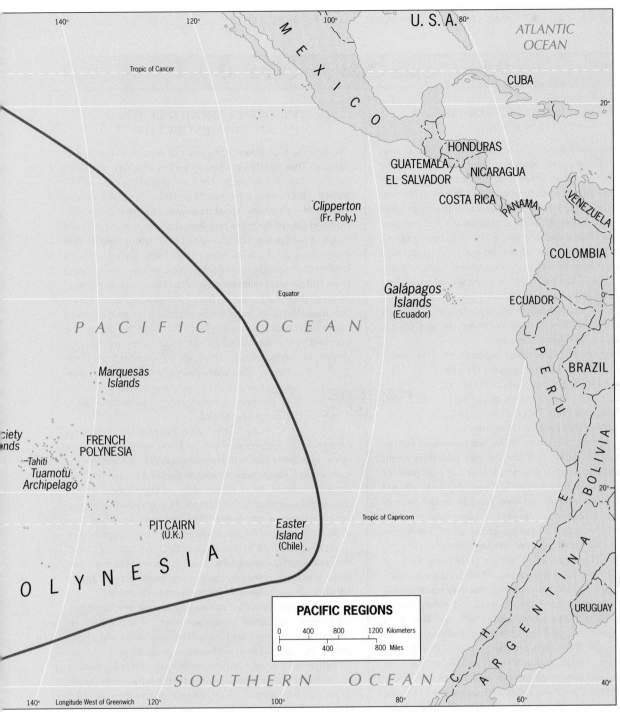

Tropic of Cancer

U.S.A.

ATLANTIC
OCEAN

MEXICO

CUBA

20°

HONDURAS

GUATEMALA

EL SALVADOR   NICARAGUA

COSTA RICA

PANAMA

VENEZUELA

Clipperton
(Fr. Poly.)

COLOMBIA

Equator

Galápagos
Islands
(Ecuador)

ECUADOR

0°

PACIFIC   OCEAN

P
E
R
U

BRAZIL

Marquesas
Islands

ciety
nds

FRENCH
POLYNESIA

B
O
L
I
V
I
A

Tahiti

Tuamotu
Archipelago

20°

PITCAIRN
(U.K.)

Easter
Island
(Chile)

Tropic of Capricorn

OLYNESIA

C
H
I
L
E

A
R
G
E
N
T
I
N
A

URUGUAY

PACIFIC REGIONS

| 0 | 400 | 800 | 1200 Kilometers |

| 0 | 400 | 800 Miles |

SOUTHERN   OCEAN

40°

140°   Longitude West of Greenwich   120°   100°   80°   60°

© H. J. de Blij, P. O. Muller, and John Wiley & Sons, Inc.

while most of the interior remains hardly touched by the changes that transformed neighboring Australia. Perhaps four-fifths of the population lives in what we may describe as a self-sufficient subsistence economy, growing root crops and hunting wildlife, raising pigs, and gathering forest products. Old traditions of the kind lost in Australia persist here, protected by remoteness and the rugged terrain.

Welding this disparate population into a nation is a task hardly begun, and PNG faces numerous obstacles in addition to its cultural complexity. Not only are

MAJOR CITIES OF THE REALM

City	Population* (in millions)
Honolulu, Hawai'i (U.S.)	0.9
Nouméa, New Caledonia	0.2
Port Moresby, Papua New Guinea	0.3
Suva, Fiji	0.2

*Based on 2010 estimates.

Who Should Own the Oceans?

WE WHO LIVE HERE SHOULD OWN THE WATERS

"Without exclusive fishing rights in our waters up to 200 nautical miles offshore, our economy would be dead in the water, no pun intended, sir. As it is, it's bad enough. I'm an official in the Fisheries Department of the Ministry of Natural Resources Development here in Tuvalu, and this modest building you're visiting is the most modern in all of the capital, Funafuti. That's because we're the only ones making some money for the state other than our Internet domain (*tv*), which has been the big story around here since we commercialized it. We don't have much in the way of natural resources and we export some products from our coconut and breadfruit trees, but otherwise we need donors to help us and major fishing nations to pay us for the use of our 750,000 square kilometers of ocean.

"Which brings me to one of our big concerns. Three-quarters of a million square kilometers sound like a lot, but look at the map. It's a mere speck in the vast Pacific Ocean. Enormous stretches of our ocean seem to belong to nobody. That is, they're open to fishing by nations that have the technology to exploit them, but we don't have those fast boats and factory ships they use nowadays. Still, the fish they take there might have come to our exclusive grounds, but we don't get paid for them. It seems to us that the Pacific Ocean should be divided among the countries that exist here, not just on the basis of nearby Exclusive Economic Zones but on some other grounds. For example, islands that are surrounded by others, like we are, ought to get additional maritime territory someplace else. Islands that aren't surrounded should get much more than just 200 nautical miles. To us the waters of the Pacific mean much more than they do to other countries, not only economically but historically too. I don't like Japanese and American and Norwegian fishing fleets on these nearby 'high seas.' They don't obey international regulations on fishing, and they overfish wherever they go. When there's nothing left, they'll simply end their agreement with us, and we'll be left without income from a natural resource over which we never had control.

"I don't expect anything to change, but we feel that the era of the 'high seas' should end and all waters and subsoil should be assigned to the nearest states. The world doesn't pay much attention to our small countries, I realize. But we've put up with colonialism, world wars, nuclear weapons testing, pollution. It's time we got a break."

THE OCEAN SHOULD BE OPEN AND UNRESTRICTED

"Ours is the Blue Planet, the planet of oceans and seas, and the world has fared best when its waters were free and open for the use of all. Encroachment from land in the form of 'territorial' seas (what a misnomer!) and other jurisdictions only interfered with international trade and commercial exploitation and caused far more problems than they solved. Already much of the open ocean is assigned to coastal countries that have no maritime tradition and don't know what to do with it other than to lease it. And when coastal countries lie closer than 400 nautical miles to each other, they divide the waters between them for EEZs and don't even leave a channel for international passage. Yes, I realize that there are guarantees of uninhibited passage through EEZs. But mark my words: the time will come when small coastal states will start charging vessels, for example cruise ships, for passage through 'their' waters. That's why, as the first officer of a passenger ship, I am against any further allocation of maritime territory to coastal states. We've gone too far already.

"In the 1960s, when those UNCLOS meetings were going on, a representative from Malta named Arvid Pardo came up with the notion that the whole ocean and seafloor beyond existing jurisdictions should be declared a 'common heritage of mankind' to be governed by an international agency that would control all activities there. The idea was that all economic activity on and beneath the high seas would be controlled by the UN, which would ensure that a portion of the proceeds would be given to 'Geographically Disadvantaged States' like landlocked countries, which can't claim any seas at all. Fortunately nothing came of this—can you imagine a UN agency administering half the world for the benefit of humanity?

"So you can expect us mariners to oppose any further extensions of land-based jurisdiction over the seas. The maritime world is complicated enough already, what with choke points, straits, narrows, and various kinds of prohibitions. To assign additional maritime territory to states on historic or economic grounds is to complicate the future unnecessarily and to invite further claims. From time to time I see references to something called the 'World Lake Concept,' the preliminary assignment of all the waters of the world to the coastal states through the delimitation of median lines. Such an idea may be the natural progression from the EEZ concept, but it gives me nightmares."

Regional ISSUE

Vote your opinion at www.wiley.com/college/deblij

hundreds of languages in use, but close to half the population is illiterate. English, the official language, is used by the educated minority but is of little use beyond the coastal zone and its towns. The capital, Port Moresby, has less than 350,000 inhabitants, reflecting the very low level of urbanization (13 percent) in this developing economy.

Yet Papua New Guinea is not without economic opportunities. Oil was discovered in the 1980s, and by the late 1990s crude oil was PNG's largest export by value. Gold now ranks second, followed by copper, silver, timber, and several agricultural products including coffee and cocoa, reflecting the country's resource and environmental diversity. Pacific Rim developments have affected even PNG: most exports go to the nearest neighbor, Australia, but Japan ranks close behind.

Turning eastward, it is a measure of Melanesia's cultural fragmentation that as many as 120 languages are spoken in the approximately 1000 islands that make up the Solomon Islands (about 80 of these islands support almost all the people, numbering about 525,000). Inter-island, historic animosities among the islanders were worsened by the events of World War II, when thousands of Malaitans were moved by U.S. forces to Guadalcanal. This started a postwar cycle of violence that led Australia to intervene in 2003.

New Caledonia, still under French rule, is in a very different situation. Only 43 percent of the population of about 250,000 are Melanesian; 37 percent are of French ancestry, many of them descended from the inhabitants of the penal colony France established here in the nineteenth century. Nickel mines, based on reserves that rank among the world's largest, dominate New Caledonia's export economy (see photo below). The mining industry attracted additional French settlers, and social problems arose. Most of the French population lives in or near the capital city of Nouméa, steeped in French cultural landscapes, in the southeastern quadrant of the island. Melanesian demands for an end to colonial rule have led to violence, and the two communities are still in the process of coming to terms.

On its eastern margins, Melanesia includes one of the Pacific Realm's most interesting countries, Fiji. On two larger and over 100 smaller islands live nearly 900,000 Fijians, of whom 51 percent are Melanesians and 44 percent South Asians, the latter brought to Fiji from India during the British colonial occupation to work on the sugar plantations. When Fiji achieved independence in 1970, the native Fijians owned most of the land and held political control, while the Indians were concentrated in the towns (chiefly Suva, the capital; see photos, p. 612) and dominated commercial life. It was a recipe for trouble, and it was not long in coming when, in a later election, the politically active Indians outvoted the Fijians for seats in the parliament. A coup by the Fijian military was followed by a revision of the constitution, which

FROM THE FIELD NOTES

"Arriving in the capital of New Caledonia, Nouméa, in 1996, was an experience reminiscent of French Africa 40 years earlier. The French tricolor was much in evidence, as were uniformed French soldiers. European French residents occupied hillside villas overlooking palm-lined beaches, giving the place a Mediterranean cultural landscape. And New Caledonia, like Africa, is a source of valuable minerals. It is one of the world's largest nickel producers, and from this vantage point you could see the huge treatment plants, complete with concentrate ready to be shipped (left, under conveyor). What you cannot see here is how southern New Caledonia has been ravaged by the mining operations, which have denuded whole mountainsides. Working in the mines and in this facility are the local Kanaks, Melanesians who make up about 43 percent of the population of about 240,000. Violent clashes between Kanaks and French have obstructed government efforts to change New Caledonia's political status in such a way as to accommodate pressures for independence as well as continued French administration." © H. J. de Blij.

www.conceptcaching.com

Concept Caching

awarded a majority of seats to ethnic Fijians. Before long, however, the Fijian majority splintered, but a coalition government for some time proved the constitution workable—until 2000. In 1999, Fiji's first prime minister of Indian ancestry had taken office, angering some ethnic Fijians to the point of staging another coup. The following year the prime minister and members of the government were taken hostage at the parliament building in Suva, and the perpetrators demanded that Fijians would henceforth govern the country.

This action, and Fiji's inability to counter it, had a devastating impact on the country. Foreign trading partners stopped buying Fijian products. The tourist industry suffered severely. And in the end, although the coup leaders were ousted and arrested, they secured a deal that gave ethnic Fijians the control they had sought. The coup continues to cast a shadow across this country. The sugar industry was damaged by the nonrenewal of Indian-held

leases by Fijian landowners, which also resulted in a substantial movement of Indians from the countryside to the towns, where unemployment is already high. Fiji's future remains clouded.

Melanesia, the most populous region in the Pacific Realm, is bedeviled by centrifugal forces of many kinds. No two countries present the same form of multiculturalism; each has its own challenges to confront, and some of these challenges spill over into neighboring (or more distant foreign) islands.

● MICRONESIA

North of Melanesia and east of the Philippines lie the islands that constitute the region known as Micronesia (Fig. 12-3). The name (*micro* means small) refers to the size of the islands: the 2000-plus islands of Micronesia

FROM THE FIELD NOTES

"Back in Suva, Fiji more than 20 years after I had done a study of its CBD, I noted that comparatively little had changed, although somehow the city seemed more orderly and prosperous than it was in 1978. At the Central Market I watched a Fijian woman bargain with the seller over a batch of taro, the staple starch-provider in local diets. I asked her how she would serve it. 'Well, it's like your potato,' she said. 'I can make a porridge, I can cut it up and put it in a stew, and I can even fry pieces of it and make them look like the French fries they give you at the McDonald's down the street.' Next I asked the seller where his taro came from. 'It grows over all the islands,' he said. 'Sometimes, when there's too much rain, it may rot—but this year the harvest is very good.' . . . Next I walked into the crowded Indian part of the CBD. On a side street I got a reminder of the geographic concept of agglomeration. Here was a colonial-period building, once a hotel, that had been converted into a business center. I counted 15 enterprises, ranging from a shoestore to a photographer and including a shop where rubber stamps were made and another selling diverse tobacco products. Of course (this being the Indian sector of downtown Suva) tailors outnumbered all other establishments." © H. J. de Blij.

are not only tiny (many of them no larger than 1 square kilometer [half a square mile]), but they are also much lower-lying, on an average, than those of Melanesia. Some are volcanic islands (high islands, as the people call them), but they are outnumbered by islands composed of coral, the low islands that barely lie above sea level. Guam, with 550 square kilometers (210 sq mi), is Micronesia's largest island, and no island elevation anywhere in Micronesia reaches 1000 meters (3300 ft).

The high-island/low-island dichotomy is useful not only in Micronesia, but also throughout the realm. Both the physiographies of these islands and the economies they support differ in crucial ways. **High islands** wrest substantial moisture from the ocean air; they tend to be well watered and have good volcanic soils. As a result, agricultural products show some diversity, and life is reasonably secure. Populations tend to be larger on these high islands than on the **low islands**, where drought is the rule and fishing and the coconut palm are the mainstays of life. Small communities cluster on the low islands, and over time many of these have died out. The major migrations, which sent fleets to populate islands from Hawai'i to New Zealand, tended to originate in the high islands.

Until the mid-1980s, Micronesia was largely a United States Trust Territory (the last of the post–World War II trusteeships supervised by the United Nations), but that status has now changed. As Figure 12-3 shows, today Micronesia is divided into countries bearing the names of independent states. The Marshall Islands, where the United States tested nuclear weapons (giving prominence to the name *Bikini*), now is a republic in "free association" with the United States, having the same status as the Federated States of Micronesia and (since 1994) Palau. This arrangement was tested in 2008, when Marshall Islands voters ousted its pro-American administrators and installed a government that intends to reopen the issue of compensation for victims of the nuclear tests and also plans to initiate diplomatic relations with China, ending those with Taiwan. This is no minor matter for Washington, because the United States continues to operate missile-defense tests at Kwajalein, the Marshall Islands' largest atoll.

The Northern Mariana Islands are a commonwealth "in political union" with the United States. In effect, the United States provides billions of dollars in assistance to these countries, in return for which they commit themselves to avoid foreign policy actions that are contrary to U.S. interests. There are other conditions: Palau, for example, granted the United States rights to existing military bases for 50 years following independence.

Also part of Micronesia is the U.S. territory of Guam, where independence is not in sight and where U.S. military installations and tourism provide the bulk of income, and the remarkable Republic of Nauru With a population of barely 15,000 and 20 square kilometers (8 sq mi) of land, this microstate got rich by selling its phosphate deposits to Australia and New Zealand, where they are used as fertilizer. Per-capita incomes rose to U.S. $11,500, and made Nauru one of the Pacific's high-income societies. But the phosphate deposits ran out, and an island scraped bare faced an economic crisis.

In this region of tiny islands, most people subsist on farming or fishing, and virtually all the countries need infusions of foreign aid to survive. The natural economic complementarity between the high-island farming cultures and the low-island fishing communities all too often is negated by distance, spatial as well as cultural. Life here may seem idyllic to the casual visitor, but for the Micronesians it often is a daily challenge.

● POLYNESIA

To the east of Micronesia and Melanesia lies the heart of the Pacific, enclosed by a great triangle stretching from the Hawaiian Islands to Chile's Easter Island to New Zealand. This is Polynesia (Fig. 12-3), a region of numerous islands (*poly* means many), ranging from volcanic mountains rising above the Pacific's waters (Mauna Kea on Hawai'i reaches beyond 4200 meters [nearly 13,800 ft]), clothed by luxuriant tropical forests and drenched by well over 250 centimeters (100 in) of rainfall each year, to low coral atolls where a few palm trees form the only vegetation and where drought is a persistent problem. The Polynesians have somewhat lighter-colored skin and wavier hair than do the other peoples of the Pacific Realm; they are often also described as having an excellent physique. Anthropologists differentiate between these original Polynesians and a second group, the Neo-Hawaiians, who are a blend of Polynesian, European, and Asian ancestries. In the U.S. State of Hawai'i—actually an archipelago of more than 130 islands—Polynesian culture has been both Europeanized and Orientalized.

Its vastness and the diversity of its natural environments notwithstanding, Polynesia clearly constitutes a geographic region within the Pacific Realm. Polynesian culture, though spatially fragmented, exhibits a remarkable consistency and uniformity from one island to the next, from one end of this widely dispersed region to the other. This consistency is particularly expressed in vocabularies, technologies, housing, and art forms. The Polynesians are uniquely adapted to their maritime environment, and long before European sailing ships began to arrive in their waters, Polynesian seafarers had learned to navigate their wide expanses of ocean in huge double canoes as long as 45

Tourists visiting Pacific islands see pristine beaches and meticulously maintained hotel facilities, but Pacific islanders live with other realities, including the pollution of their environment. From the distant deck of a cruise ship, the lagoons of Kiribati's Tarawa Atoll look unspoiled, but grab a bike and ride to some villages, and here is what you are likely to see. Already there is not enough room to dispose of local trash, and daily the tide contributes more. Look closely at piles like this, and you will find the spoils of "civilization" ranging from garbage-filled plastic bags thrown overboard to fish traps, netting, styrofoam, shards of glass, and pieces of wood. Pollution is one of this realm's most serious problems. © Caroline Penn/Corbis Images.

meters (150 ft). They traveled hundreds of kilometers to favorite fishing zones and engaged in inter-island barter trade, using maps constructed from bamboo sticks and cowrie shells and navigating by the stars. However, modern descriptions of a Pacific Polynesian paradise of emerald seas, lush landscapes, and gentle people distort harsh realities. Polynesian society was forced to get used to much loss of life at sea when storms claimed their boats; families were ripped apart by accident as well as migration; hunger and starvation afflicted the inhabitants of smaller islands; and the island communities were often embroiled in violent conflicts and cruel retributions.

The political geography of Polynesia is complex. In 1959, the Hawaiian Islands became the fiftieth State to join the United States. The State's population is now 1.4 million, with over 80 percent living on the island of Oahu. There, the superimposition of cultures is symbolized by the panorama of Honolulu's skyscrapers against the famous extinct volcano at nearby Diamond Head. The Kingdom of Tonga became an independent country in 1970 after seven decades as a British protectorate; the British-administered Ellice Islands were renamed Tuvalu, and along with the Gilbert Islands to the north (now renamed Kiribati), they received independence from Britain in 1978. Other islands continued under French control (including the Marquesas Islands and Tahiti), under New Zealand's administration (Rarotonga), and under British, U.S., and Chilean flags.

In the process of politico-geographical fragmentation, Polynesian culture has suffered severe blows. Land developers, hotel builders, and tourist dollars have set Tahiti on a course along which Hawai'i has already traveled far. The Americanization of eastern Samoa has created a new society different from the old (see photo, p. 606). Polynesia has lost much of its ancient cultural consistency;

today, the region is a patchwork of new and old—the new often bleak and barren, with the old under intensifying pressure.

The countries and cultures of the Pacific Realm lie in an ocean on whose rim a great drama of economic and political transformation will play itself out during the twenty-first century. Already, the realm's own former margins—in Hawai'i in the north and in New Zealand in the south—have been so recast by foreign intervention that little remains of the kingdoms and cultures that once prevailed. Now the Pacific world faces changes far greater even than those brought here by European colonizers. Once upon a time the shores and waters of the Mediterranean Sea formed an arena of regional transformation that changed the world. Then it was the Atlantic, avenue of the Industrial Revolution and stage of fateful war. Now it seems to be the turn of the Pacific as the world's largest country and next superpower (China) faces the richest and most powerful (the United States). Giants will jostle for advantage in the Pacific; how will the weak societies of the Pacific Realm fare?

ANTARCTIC PARTITION

South of the Pacific Realm lies Antarctica and its encircling Southern Ocean. The combined area of these two geographic expanses constitutes 40 percent of the entire planet—two-fifths of the Earth's surface containing one one-thousandth of the world's population.

Is Antarctica a geographic realm? In physiographic terms, yes, but not on the basis of the criteria we use in this book. Antarctica is a continent, nearly twice as large as Australia, but virtually all of it is covered by a dome-shaped ice sheet 3 kilometers (nearly 2 mi) thick

FIGURE 12-4

© H. J. de Blij, P. O. Muller, and John Wiley & Sons, Inc.

near its center. The continent often is referred to as the "white desert" because, despite all of its ice and snow, annual precipitation is low—less than 15 centimeters (6 in) per year. Temperatures are frigid, with winds so strong that Antarctica also is called the "home of the blizzard." For all its size, no functional regions have developed here, no towns, no transport networks except the supply lines of research stations. And Antarctica remains a frontier, even a scientific frontier still slowly giving up its secrets. Underneath all that ice lie some 70 lakes of which Lake Vostok is the largest at 14,000 square kilometers (more than 5400 sq mi). It may be as much as 600 meters (2000 ft) deep. No one has yet seen a sample of its water.

Like virtually all frontiers, Antarctica has always attracted pioneers and explorers. Whale and seal hunters destroyed huge populations of Southern Ocean fauna during the eighteenth and nineteenth centuries, and explorers planted the flags of their countries on Antarctic shores. Between 1895 and 1914, the quest for the South Pole became an international obsession; Roald Amundsen, the Norwegian, reached it first in 1911. All this led to national claims in Antarctica during the interwar period (1918–1939).

The geographic effect was the partitioning of Antarctica into pie-shaped sectors centered on the South Pole (Fig. 12-4). In the least frigid area of the continent, the Antarctic Peninsula (photo p. 616), British, Argentinean, and Chilean claims overlapped—and still do. One sector, Marie Byrd Land (shown in neutral beige on the main map), was never formally claimed by any country.

Why should states be interested in territorial claims in so remote and difficult an area? Both land and sea contain raw materials that may some day become crucial: proteins in the waters, and fuels and minerals beneath the ice. Antarctica (14.2 million square kilometers [5.5 million sq mi]) is, as already noted, almost twice as large as Australia, and the Southern Ocean (see p. 581) is nearly as large as the North and South Atlantic. However distant actual exploitation may be, countries want to keep their stakes here.

But the claimant states (those with territorial claims) recognize the need for cooperation. During the late 1950s, they joined in the International Geophysical Year (IGY) that launched major research programs and established a number of permanent research stations throughout the continent. This spirit of cooperation led to the 1961 signing of the **Antarctic Treaty**, which ensures **10** continued scientific collaboration, prohibits military activities, protects the environment, and holds national claims in abeyance. In 1991, when the treaty was extended under the terms of the Wellington Agreement, concerns were raised that it does not do enough to control future resource exploitation.

Notice, in Figure 12-4, that the map shows no maritime claims off the pie-shaped sectors that blanket most of Antarctica. In fact, some of the claimant states did draw maps that extended their claims into the Southern Ocean, but the 1991 Wellington Agreement terminated that initiative and restricted the existing claims to the landmass. It is of course possible that those claims could be reinstated should the Wellington Agreement collapse; national self-interest has abrogated international treaties before. But delimiting maritime claims off Antarctica is especially difficult for practical reasons. As Figure 12-4 shows, Antarctica in many places is flanked by *ice shelves*, permanent slabs of floating ice (such as the Ronne and the Ross). These ice shelves are replenished on the landward side and break off (*calve*) on the seaward side, where huge icebergs float away into the Southern Ocean to become part of the wide ring of *pack ice* that encircles the continent. From where would any territorial sea or EEZ be measured? From the inner edge of the ice shelf? That would put the territorial "sea" boundary on the ice shelf! From the outer edge of the ice shelf? Measuring a territorial sea or EEZ from an unstable perimeter would not work either. No UNCLOS regulations would be applicable in situations like this, so no matter what the maps showed, such claims are legally unsupportable. That is just as well. The last thing the world needs is a scramble for Antarctica.

In an age of growing national self-interest and increasing raw-material consumption, the possibility exists that Antarctica and its offshore waters may yet become an arena for international rivalry. Until now, its remoteness and its forbidding environments have saved it from that fate. The entire world benefits from this because evidence is mounting that Antarctica plays a critical role in the global environmental system, so that human modifications may have worldwide (and unpredictable) consequences.

GEOPOLITICS IN THE ARCTIC

To observe how different the physiographic as well as the political situation is in the Arctic, consider the implications of Figure 12-5. Not only does the North Pole lie on the floor of a relatively small body of water grandiosely called the Arctic Ocean, but the entire Arctic is ringed by countries whose EEZs, delimited under UNCLOS rules, would allocate much of the ocean floor (the *subsoil*, to use its legal designation) to those states. And, again under UNCLOS rules, states can expand their rights to the seafloor even farther than the 200-mile EEZ if they can prove their continental shelves continue beyond that limit, up to 350 nautical miles offshore.

As the map shows, the Siberian Continental Shelf is by far the largest in the Arctic; indeed, it is the largest in the

COMPETITION IN THE ARCTIC
- ▬·▬· EEZ Boundary
- ▬·▬· Treaty Boundaries
- ●●●● Disputed Boundaries
- ☐ Continental Shelf

R U S S I A

140°
160° E
180°
160° W
140°
120°
100°
80°
70°
60°
40°
20° W
0
20° E
40°
60°
80°
100°

East Siberian Sea

Siberian Continental Shelf

North Land (Severnaya Zemlya)

Franz Josef Land

Barents Sea

Lomonosov Ridge

Svalbard I. (Norway)

NORWAY

FINLAND

SWEDEN

North Pole

ARCTIC OCEAN

Wrangel I.

Chukchi Sea

Chukchi Shelf

Greenland Sea

U.S.A.

GREENLAND (Denmark)

Beaufort Sea

Baffin Bay

C A N A D A

MELTING ICE ON ARCTIC WATERS, 1980–2008

Lomonosov Ridge

North Pole

ARCTIC OCEAN

SHIPPING ROUTES

Lomonosov Ridge

North Pole

ARCTIC OCEAN

- ☐ 1980–2000 average extent
- ☐ 2008 extent
- ▬ North West (Canadian) Passages
- ▬ Northern (Russian) Maritime Passage

FIGURE 12-5

© H. J. de Blij, P. O. Muller, and John Wiley & Sons, Inc.

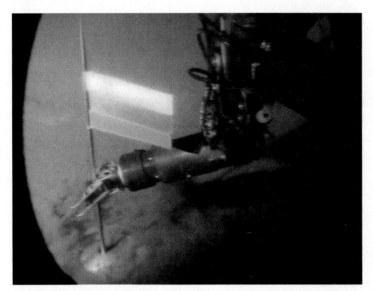

On August 3, 2007, tens of millions of Russian television viewers witnessed live the scene pictured here: the operator of a Russian mini-submarine planting a metal Russian flag at the North Pole, on the seafloor of the Arctic Ocean beneath the permanent polar ice. As Figure 12-5 shows, the North Pole lies far beyond the Russian continental shelf, but in the upcoming international negotiations relating to national claims in the Arctic region, the Russians will claim that the Lomonosov "Ridge" actually constitutes an extension of their continental shelf, thus entitling Russia to draw international boundaries up to, and even beyond, the North Pole. Other countries with actual and pending Arctic claims make less dramatic but equally assertive moves in the run-up to negotiations that have the potential to transform the map of the Arctic region. With new sea routes and newly-exploited oil and gas reserves in prospect, diplomacy will be difficult but crucial.
© Visarkryeziu/AP/Wide World Photos

world. It extends from Russia's north coast beneath the waters of the Arctic Ocean and under the permanent ice at the ocean's center. And because several island groups off the Siberian mainland (such as Franz Josef Land, North Land, and the New Siberian Islands) belong to Russia, the Russians can claim virtually the entire Siberian Continental Shelf under existing regulations. But that is not enough: the Russians want to extend their claim all the way to the North Pole itself, where in August 2007 they sent a submarine to plant a Russian flag nearly 4000 meters (13,000 ft) below the sea (see photo above).

To bolster their claim, the Russians assert that the Lomonosov Ridge, a submerged mountain range, is a "natural extension" of their Siberian Continental Shelf. Few if any neutral observers would agree that this ridge, or several others rising from the deep floor of the Arctic Ocean, is a continental-shelf landform. But this is the stuff of international disputes over maritime boundaries. The legal wrangling will go on for decades.

Another look at Figure 12-5 reveals the other states with frontage on the Arctic Ocean: Norway, through its Svalbard Islands (including Spitsbergen); Denmark, because it still has ultimate authority over autonomous Greenland; Canada, with tens of thousands of kilometers of Arctic coastline but proportionately less continental shelf; and the United States, with Alaska's limited but important continental shelves to its north and west. The United States stands alone among these candidates by not having ratified the UNCLOS treaties, so that the Americans do not feel bound by the regulations to which everyone else adheres. In any case, you can quickly see that an Antarctic-type, pie-shaped sector partitioning would not

work here, but it is also clear that the evolving contest among claimant states is fraught with risk. And the stakes are high: estimates of the quantity of oil and natural gas reserves to be found below Arctic water and ice run as high as 25 percent of the world's remaining total. No matter what method is used to calculate which country gets how much, the big winner will be Russia, no matter what the resolution of its Lomonosov claim will be.

Disputation and Navigation

Even as a dispute over Arctic maritime boundaries looms, another issue is emerging, this one related to the effects of recently sustained global warming. The Greenland Ice Cap, in some ways a smaller version of the Antarctic one, is in a melting phase, and the zone of seasonally expanding and contracting Arctic sea ice has also shown significant losses in both surface extent and thickness. Researchers of the Greenland Ice Core Project reported in a 2007 issue of *Science* that during a previous warming, more than 450,000 years ago, the southernmost part of Greenland was covered by boreal forest—suggesting that the ice cap as well as the sea ice may disappear altogether if current global warming continues.

The ecological consequences would be far-reaching, threatening the habitats of polar bears, whales, walruses, seals, and other species. But the geographic implications would be extensive as well. When the sea ice melts in summer and does not recover in winter, once-blocked waterways open up and change accessibilities (Fig. 12-5, subsidiary maps). The Northwest Passage, that fabled northerly route between the Atlantic and Pacific oceans, has been the object of hope and despair for centuries, causing much loss of human life. But in 2007, for the first time in recorded history, it was briefly but completely ice-free. Consider what it would mean if vessels bound for East Asia from Western Europe or the U.S. eastern seaboard could sail year-round from the Labrador Sea to the Bering Sea via this waterway!

But who owns the waterways vacated by the ice? Canada calls the Northwest Passage a domestic waterway, giving it the right to control shipping. The United States holds that the Northwest Passage is an international waterway, and American and Russian ships have sailed through it without Canada's permission. The Canadians have confirmed their position by starting work on a

deepwater port at Nanisivik on Nunavut's Baffin Island at the eastern entrance to the Northwest Passage, and by building an army base at Resolute Bay to the west; they also announced the construction of eight Arctic patrol vessels to police the route. The United States, meanwhile, is opening a new Coast Guard station to patrol the narrow Bering Strait. The lines, if not yet the battle lines, are being drawn in the warming Arctic.

Humanity's growing numbers, escalating demands, environmental impacts, and technological capacities are transforming even the most remote recesses of our resilient planet. No international resolution is more critical than the avoidance of destructive conflict in fragile polar environments.

What You Can Do

RECOMMENDATION: Attend an AAG Meeting! The Association of American Geographers is the largest organization for professional geographers and students in North America, and it holds an Annual Meeting every spring (its 105th meeting took place in Las Vegas in March 2009). The meeting sites rotate around North America: Washington, D.C. in 2010 (April 14–18) and Seattle in 2011 (April 12–16), so one of them is likely to convene relatively close to your college or home in the near future. Several thousand geographers and students participate, and while these meetings have their professional side, they are also a lot of fun, and there are jobs available as well, as the AAG can tell you. If you're within a reasonable distance, why not join the crowd and enjoy the exhibits, plenary lectures, and parties, and meet some of the people who keep geography in the limelight! Details are available on the AAG website at **www.aag.org**.

GEOGRAPHIC CONNECTIONS

1 One of the most important achievements of the United Nations has been a series of conferences out of which grew the UN Convention on the Law of the Sea (UNCLOS). This Convention redrew the maritime map of the world, stabilizing what had been a scramble for the oceans, limiting the width of the Territorial Sea, and introducing the concept of the Exclusive Economic Zone (EEZ). How has the EEZ changed the economic geography of the Pacific geographic realm? To what extent has the EEZ functioned to expand as well as limit states' offshore activities? Considering the map of the Pacific Realm, what countries are favored over others in terms of maritime holdings?

2 The Pacific Realm has the least land and the smallest population of all the geographic realms, but its cultural distinctiveness is second to none. Its largest territorial component and largest population cluster constitute the island of New Guinea, which is divided by a geometric, superimposed boundary between Indonesia and independent Papua New Guinea (PNG) (Fig. 10-8, lower-left map). Although Papuans on both sides of this border dream of a united homeland, the two entities are going in different directions. How did this division come about? Why was it allowed to continue during the period of decolonization? What demographic and cultural-geographic contrasts between the eastern (PNG) and western (Indonesian) sides of the boundary can you identify?

Geographic Literature on the Pacific Realm: The key introductory works on this realm were authored by Bier, Glassner, McKnight, Peake, and Woodcock. These works, together with a comprehensive listing of noteworthy books and recent articles on the geography of Melanesia, Micronesia, and Polynesia, can be found in the *References and Further Readings* section of this book's website at **www.wiley.com/college/deblij**.

APPENDIX A

METRIC (STANDARD INTERNATIONAL [SI]) AND CUSTOMARY UNITS AND THEIR CONVERSIONS

Appendix A provides a table of units and their conversions from older (British) units to metric (SI) units for the measures of length, area, and temperature that are used in this book.

NOTE: Appendices B, C, and D may be found on the book's website at www.wiley.com/college/deblij

Length

Metric Measure

1 kilometer (km) = 1000 meters (m)
1 meter (m) = 100 centimeters (cm)
1 centimeter (cm) = 10 millimeters (mm)

Nonmetric Measure

1 mile (mi) = 5280 feet (ft)
= 1760 yards (yd)
1 yard (yd) = 3 feet (ft)
1 foot (ft) = 12 inches (in)
1 fathom (fath) = 6 feet (ft)

Conversions

1 kilometer (km) = 0.6214 mile (mi)
1 meter (m) = 3.281 feet (ft)
= 1.094 yards (yd)
1 centimeter (cm) = 0.3937 inch (in)
1 millimeter (mm) = 0.0394 inch (in)
1 mile (mi) = 1.609 kilometers (km)
1 foot (ft) = 0.3048 meter (m)
1 inch (in) = 2.54 centimeters (cm)
= 25.4 millimeters (mm)

Area

Metric Measure

1 square kilometer (km^2) = 1,000,000 square meters (m^2)
= 100 hectares (ha)
1 square meter (m^2) = 10,000 square centimeters (cm^2)
1 hectare (ha) = 10,000 square meters (m^2)

Nonmetric Measure

1 square mile (mi^2) = 640 acres (ac)
1 acre (ac) = 4840 square yards (yd^2)
1 square foot (ft^2) = 144 square inches (in^2)

Conversions

1 square kilometer (km^2) = 0.386 square mile (mi^2)
1 hectare (ha) = 2.471 acres (ac)
1 square meter (m^2) = 10.764 square feet (ft^2)
= 1.196 square yards (yd^2)
1 square centimeter (cm^2) = 0.155 square inch (in^2)
1 square mile (mi^2) = 2.59 square kilometers (km^2)
1 acre (ac) = 0.4047 hectare (ha)
1 square foot (ft^2) = 0.0929 square meter (m^2)
1 square inch (in^2) = 6.4516 square centimeters (cm^2)

Temperature

To change from Fahrenheit (F) to Celsius (C)
$$°C = \frac{°F - 32}{1.8}$$
To change from Celsius (C) to Fahrenheit (F)
$$°F = °C \times 1.8 + 32$$

GLOSSARY

Aboriginal land issue The legal campaign in which Australia's **indigenous peoples** have claimed title to traditional land in several parts of that country. The courts have upheld certain claims, fueling Aboriginal activism that has raised broader issues of indigenous rights.

Aboriginal population See **indigenous peoples**.

Absolute location The position or place of a certain item on the surface of the Earth as expressed in degrees, minutes, and seconds of **latitude**, 0° to 90° north or south of the equator, and **longitude**, 0° to 180° east or west of the *prime meridian* passing through Greenwich, England (a suburb of London).

Accessibility The degree of ease with which it is possible to reach a certain location from other locations. *Inaccessibility* is the opposite of this concept.

Acculturation Cultural modification resulting from intercultural borrowing. In **cultural geography**, the term refers to the change that occurs in the **culture** of **indigenous peoples** when contact is made with a society that is technologically superior.

Advantage The most meaningful distinction that can now be made to classify a country's level of economic **development**. Takes into account geographic **location, natural resources**, government, political stability, productive skills, and much more.

Agglomeration **Process** involving the clustering or concentrating of people or activities.

Agrarian Relating to the use of land in rural communities or to agricultural societies in general.

Agriculture The purposeful tending of crops and livestock in order to produce food and fiber.

Alluvial Referring to the mud, silt, and sand (collectively *alluvium*) deposited by rivers and streams. *Alluvial plains* adjoin many larger rivers; they consist of the renewable deposits that are laid down during floods, creating fertile and productive soils. Alluvial **deltas** mark the mouths of rivers such as the Nile (Egypt) and the Ganges (Bangladesh).

Altiplano High-elevation plateau, basin, or valley between even higher mountain ranges, especially in the Andes of South America.

Altitudinal zonation Vertical regions defined by physical-environmental zones at various elevations (see Fig. 4-13), particularly in the highlands of South and Middle America. See *tierra caliente, tierra templada, tierra fría, tierra helada*, and *tierra nevada*.

American Manufacturing Belt North America's near-rectangular Core Region, whose corners are Boston, Milwaukee, St. Louis, and Baltimore. This region dominated the industrial geography of the United States and Canada during the industrial age; still a formidable economic powerhouse that remains the realm's geographic heart.

Animistic religion The belief that inanimate objects, such as hills, trees, rocks, rivers, and other elements of the natural landscape, possess souls and can help or hinder human efforts on Earth.

Antarctic Treaty International cooperative agreement on the use of Antarctic territory (see pp. 615–616).

Antecedent boundary A political boundary that existed before the **cultural landscape** emerged and stayed in place while people moved in to occupy the surrounding area.

Anthracite coal Hardest and highest carbon-content coal, and therefore of the highest quality.

Apartheid Literally, *apartness*. The Afrikaans term for South Africa's pre-1994 policies of racial separation, a system that produced highly segregated socio-geographical patterns.

Aquaculture The use of a river segment or an artificial pond for the raising and harvesting of food products, including fish, shellfish, and even seaweed. The Japanese pioneered the practice, which is now spreading globally, particularly along heavily populated coastlines.

Aquifer An underground reservoir of water contained within a porous, water-bearing rock layer.

Arable Land fit for cultivation by one farming method or another. See also **physiologic density**.

Archipelago A set of islands grouped closely together, usually elongated into a *chain*.

Area A term that refers to a part of the Earth's surface with less specificity than **region**. For example, *urban area* alludes generally to a place where urban development has occurred, whereas *urban region* requires certain specific criteria on which such a designation is based (e.g., the spatial extent of commuting or the built townscape).

Areal functional organization A geographic principle for understanding the evolution of regional organization, whose five interrelated tenets are applied to the spatial development of Japan on p. 513.

Areal interdependence A term related to **functional specialization**. When one area produces certain goods or has certain raw materials and another area has a different set of raw materials and produces different goods, their needs may be *complementary*; by exchanging raw materials and products, they can satisfy each other's requirements.

Arithmetic density A country's population, expressed as an average per unit area, without regard for its **distribution** or the limits of **arable** land. See also **physiologic density**.

Aryan From the Sanskrit *Arya* (meaning "noble"), a name applied to an ancient people who spoke an **Indo-European language** and who moved into northern India from the northwest.

Atmosphere The Earth's envelope of gases that rests on the oceans and land surface and penetrates open spaces within soils. This layer of nitrogen (78 percent), oxygen (21 percent), and traces of other gases is densest at the Earth's surface and thins with altitude.

Austral South.

Autocratic A government that holds absolute power, often ruled by one person or a small group of persons who control the country by despotic means.

Balkanization The fragmentation of a **region** into smaller, often hostile political units. Named after the historically contentious Balkan Peninsula of southeastern Europe.

Barrio Term meaning "neighborhood" in Spanish. Usually refers to an urban community in a Middle or South American city; also applied to low-income, inner-city concentrations of Hispanics in such western U.S. cities as Los Angeles.

(*Note:* Words in **boldface type** within an entry are defined elsewhere in this Glossary)

Bauxite   Aluminum ore; usually deposited at shallow depths in the wet tropics.

Biogeography   The study of *flora* (plant life) and *fauna* (animal life) in spatial perspective.

Birth rate   The *crude birth rate* is expressed as the annual number of births per 1000 individuals within a given population.

Bituminous coal   Softer coal of lesser quality than **anthracite**, but of higher grade than **lignite**. When heated and converted to coking coal or *coke*, it is used to make steel.

Break-of-bulk point   A location along a transport route where goods must be transferred from one carrier to another. In a port, the cargoes of oceangoing ships are unloaded and put on trains, trucks, or perhaps smaller river boats for inland distribution. An **entrepôt**.

Buffer state   See **buffer zone**.

Buffer zone   A country or set of countries separating ideological or political adversaries. In southern Asia, Afghanistan, Nepal, and Bhutan were parts of a buffer zone between British and Russian-Chinese imperial spheres. Thailand was a *buffer state* between British and French colonial domains in mainland Southeast Asia.

Caliente   See *tierra caliente*.

Cartogram   A specially transformed map not based on traditional representations of **scale** or area. Examples are found in Figures G-9, 5-7, and 7-15.

Cartography   The art and science of making maps, including data compilation, layout, and design. Also concerned with the interpretation of mapped patterns.

Caste system   The strict **social stratification** and segregation of people—specifically in India's Hindu society—on the basis of ancestry and occupation.

Cay   A low-lying small island usually composed of coral and sand. Pronounced *kee* and often spelled "key."

Central business district (CBD)   The downtown heart of a central city; marked by high land values, a concentration of business and commerce, and the clustering of the tallest buildings.

Centrality   The strength of an urban center in its capacity to attract producers and consumers to its facilities; a city's "reach" into the surrounding region.

Centrifugal forces   A term employed to designate forces that tend to divide a country—such as internal religious, linguistic, ethnic, or ideological differences.

Centripetal forces   Forces that unite and bind a country together—such as a strong national culture, shared ideological objectives, and a common faith.

Cerrado   Regional term referring to the fertile savannas of Brazil's interior Central-West that make it one of the world's most promising agricultural frontiers (mapped in Fig. 5-5). Soybeans are the leading crop, and other grains and cotton are expanding. Inadequate transport links to the outside world remain a problem.

Chaebol   Giant corporation controlling numerous companies and benefiting from government connections and favors, dominant in South Korea's **economic geography**; key to that country's **development** into an **economic tiger**, but more recently a barrier to free-market growth.

Charismatic   Personal qualities of certain leaders that enable them to capture and hold the popular imagination, securing the allegiance and even the devotion of the masses. Mahatma Gandhi, Mao Zedong, and Franklin D. Roosevelt were good examples in the twentieth century.

China Proper   The eastern and northeastern portions of China that contain most of the country's huge population; mapped in Figure 9-5.

Choke point   A narrowing of an international waterway causing marine traffic congestion, requiring reduced speeds and/or sharp turns, and increasing the risk of collision as well as vulnerability to attack. When the waterway narrows to a distance of less than 38 kilometers (24 mi), this necessitates the drawing of a **median line (maritime) boundary**. Examples

are the Hormuz Strait between Oman and Iran at the entrance to the Persian Gulf, and the Strait of Malacca between Malaysia and Indonesia.

City-state   An independent political entity consisting of a single city with (and sometimes without) an immediate **hinterland**. The ancient city-states of Greece have their modern equivalent in Southeast Asia's Singapore.

Climate   The long-term conditions (over at least 30 years) of aggregate **weather** over a region, summarized by averages and measures of variability; a synthesis of the succession of weather events we have learned to expect at any given location.

Climate change theory   An alternative to the **hydraulic civilization theory**; holds that changing **climate** (rather than a monopoly over **irrigation** methods) could have provided certain cities in the ancient Fertile Crescent with advantages over other cities.

Climate region   A **formal region** characterized by the uniformity of the **climate** type within it. Figure G-7 maps the global distribution of such regions.

Climatology   The geographic study of **climates**. Includes not only the classification of climates and the analysis of their regional distribution, but also broader environmental questions that concern climate change, interrelationships with soil and vegetation, and human–climate interaction.

Coal   See **anthracite coal, bituminous coal, fossil fuels**, and **lignite**.

Collectivization   The reorganization of a country's **agriculture** under communism that involves the expropriation of private holdings and their incorporation into relatively large-scale units, which are farmed and administered cooperatively by those who live there.

Colonialism   Rule by an autonomous power over a subordinate and an alien people and place. Though often established and maintained through political structures, colonialism also creates unequal cultural and economic relations. Because of the magnitude and impact of the European colonial thrust of the last few centuries, the term is generally understood to refer to that particular colonial endeavor. Also see **imperialism**.

Command economy   The tightly controlled economic system of the former Soviet Union, whereby central planners in Moscow assigned the production of particular goods to particular places, often guided more by socialist ideology than the principles of **economic geography**.

Commercial agriculture   For-profit **agriculture**.

Common market   A **free-trade area** that not only has created a **customs union** (a set of common tariffs on all imports from outside the area) but also has eliminated restrictions on the movement of capital, labor, and enterprise among its member countries.

Compact state   A politico-geographical term to describe a state that possesses a roughly circular, oval, or rectangular territory in which the distance from the geometric center to any point on the boundary exhibits little variance. Poland and Cambodia are examples of this shape category.

Complementarity   Exists when two regions, through an exchange of raw materials and/or finished products, can specifically satisfy each other's demands. See also **areal interdependence**.

Confucianism   A philosophy of ethics, education, and public service based on the writings of Confucius (*Kongfuzi*); traditionally regarded as one of the cornerstones of Chinese **culture**.

Congo   See the footnote on p. 289, which makes the distinction between the two Equatorial African countries, Congo and The Congo.

Coniferous forest   A forest of cone-bearing, needleleaf evergreen trees with straight trunks and short branches, including spruce, fir, and pine. See also **taiga**.

Contagious diffusion   The distance-controlled spreading of an idea, innovation, or some other item through a local population by contact from person to person—analogous to the communication of a contagious illness.

Conterminous United States   The 48 **contiguous** or adjacent States that occupy the southern half of the North American realm. Alaska is not contiguous to these States because western Canada lies in between; neither is Hawai'i, separated from the mainland by over 3000 kilometers (2000 mi) of ocean.

Contiguous Adjoining; adjacent.

Continental drift The slow movement of continents controlled by the processes associated with **plate tectonics**.

Continental shelf Beyond the coastlines of many landmasses, the ocean floor declines very gently until the depth of about 660 feet (200 m). Beyond the 660-foot line the sea bottom usually drops off sharply, along the *continental slope*, toward the much deeper mid-oceanic basin. The submerged continental margin is called the continental shelf, and it extends from the shoreline to the upper edge of the continental slope.

Continentality The variation of the continental effect on air temperatures in the interior portions of the world's landmasses. The greater the distance from the moderating influence of an ocean, the greater the extreme in summer and winter temperatures. Continental interiors also tend to be dry when the distance from oceanic moisture sources becomes considerable.

Conurbation General term used to identify a large multimetropolitan complex formed by the coalescence of two or more major **urban areas**. The Atlantic Seaboard **Megalopolis**, extending along the northeastern U.S. coast from southern Maine to Virginia, is a classic example.

Copra The dried-out, fleshy interior of a coconut that is used to produce coconut oil.

Cordillera Mountain chain consisting of sets of parallel ranges, especially the Andes in northwestern South America.

Core See **core area**; **core-periphery relationships**.

Core area In geography, a term with several connotations. *Core* refers to the center, heart, or focus. The core area of a **nation-state** is constituted by the national heartland, the largest population cluster, the most productive region, and the part of the country with the greatest **centrality** and **accessibility**—probably containing the capital city as well.

Core-periphery relationships The contrasting spatial characteristics of, and linkages between, the *have* (core) and *have-not* (periphery) components of a national or regional **system**.

Corridor In general, refers to a spatial entity in which human activity is organized in a linear manner, as along a major transport route or in a valley confined by highlands. More specifically, the politico-geographical term for a land extension that connects an otherwise **landlocked state** to the sea.

Cultural diffusion The **process** of spreading and adopting a cultural element, from its place of origin across a wider area.

Cultural ecology The multiple interactions and relationships between a **culture** and its **natural environment**.

Cultural environment See **cultural ecology**.

Cultural geography The wide-ranging and comprehensive field of geography that studies spatial aspects of human **cultures**.

Cultural landscape The forms and artifacts sequentially placed on the **natural landscape** by the activities of various human occupants. By this progressive imprinting of the human presence, the physical (natural) landscape is modified into the cultural landscape, forming an interacting unity between the two.

Cultural pluralism See **plural(istic) society**.

Cultural revival The regeneration of a long-dormant **culture** through internal renewal and external infusion.

Culture The sum total of the knowledge, attitudes, and habitual behavior patterns shared and transmitted by the members of a society. This is anthropologist Ralph Linton's definition; hundreds of others exist.

Culture area See **culture region**.

Culture hearth Heartland, source area, innovation center; place of origin of a major **culture**.

Culture region A distinct, culturally discrete spatial unit; a **region** within which certain cultural norms prevail.

Customs union A **free-trade area** in which member countries set common tariff rates on imports from outside the area.

Death rate The *crude death rate* is expressed as the annual number of deaths per 1000 individuals within a given population.

Deciduous A deciduous tree loses its leaves at the beginning of winter or the onset of the dry season.

Definition In **political geography**, the written legal description (in a treaty-like document) of a boundary between two countries or territories. See also **delimitation**.

Deforestation The clearing and destruction of forests (especially tropical rainforests) to make way for expanding settlement frontiers and the exploitation of new economic opportunities.

Deglomeration Deconcentration.

Delimitation In **political geography**, the translation of the written terms of a boundary treaty (the **definition**) into an official cartographic representation (map).

Delta Alluvial lowland at the mouth of a river, formed when the river deposits its alluvial load on reaching the sea. Often triangular in shape, hence the use of the Greek letter whose symbol is Δ.

Demarcation In **political geography**, the actual placing of a political boundary on the **cultural landscape** by means of barriers, fences, walls, or other markers.

Demographic transition model Multi-stage **model**, based on Western Europe's experience, of changes in population growth exhibited by countries undergoing industrialization. High **birth rates** and **death rates** are followed by plunging death rates, producing a huge net population gain; this is followed by the convergence of birth and death rates at a low overall level. See Figure 8-8.

Demography The interdisciplinary study of population—especially **birth rates** and **death rates**, growth patterns, longevity, **migration**, and related characteristics.

Desert An arid area supporting sparse vegetation, receiving less than 25 centimeters (10 in) of precipitation per year. Usually exhibits extremes of heat and cold because the moderating influence of moisture is absent.

Desertification **Process** of **desert** expansion into neighboring **steppelands** as a result of human degradation of fragile semiarid environments.

Development The economic, social, and institutional growth of national **states**.

Devolution The **process** whereby regions within a **state** demand and gain political strength and growing autonomy at the expense of the central government.

Dhows Wooden boats with characteristic triangular sails, plying the seas between Arabian and East African coasts.

Dialect Regional or local variation in the use of a major language, such as the distinctive accents of many residents of the U.S. South.

Diffusion The spatial spreading or dissemination of a **culture** element (such as a technological innovation) or some other phenomenon (e.g., a disease outbreak). For the various channels of outward geographic spread from a source area, see **contagious, expansion, hierarchical**, and **relocation diffusion**.

Distance decay The various degenerative effects of distance on human spatial structures and interactions.

Diurnal Daily.

Divided capital In **political geography**, a country whose central administrative functions are carried on in more than one city is said to have divided capitals. The Netherlands and South Africa are examples.

Domestication The transformation of a wild animal or wild plant into a domesticated animal or a cultivated crop to gain control over food production. A necessary evolutionary step in the development of humankind: the invention of **agriculture**.

Domino theory The belief that political destabilization in one **state** can result in the collapse of order in a neighboring state, triggering a chain of events that, in turn, can affect a series of **contiguous** states.

Double cropping The planting, cultivation, and harvesting of two crops successively within a single year on the same plot of farmland.

Dry canal An overland rail and/or road **corridor** across an **isthmus** dedicated to performing the transit functions of a canalized waterway. Best adapted to the movement of containerized cargo, there must be a port at each end to handle the necessary **break-of-bulk** unloading and reloading.

Ecology The study of the many interrelationships between all forms of life and the natural environments in which they have evolved and continue to develop. The study of *ecosystems* focuses on the interactions between specific organisms and their environments. See also **cultural ecology**.

Economic geography The field of geography that focuses on the diverse ways in which people earn a living and on how the goods and services they produce are expressed and organized spatially.

Economic restructuring The transformation of China into a market-driven economy in the post-Mao era, beginning in the late 1970s.

Economic tiger One of the burgeoning beehive countries of the western **Pacific Rim**. Following Japan's route since 1945, these countries have experienced significant modernization, industrialization, and Western-style economic growth since 1980. Three leading economic tigers are South Korea, Taiwan, and Singapore. The term is increasingly used more generally to describe any fast-developing economy.

Economies of scale The savings that accrue from large-scale production wherein the unit cost of manufacturing decreases as the level of operation enlarges. Supermarkets operate on this principle and are able to charge lower prices than small grocery stores.

Ecosystem See **ecology**.

Ecumene The habitable portions of the Earth's surface where permanent human settlements have arisen.

Elite A small but influential upper-echelon social class whose power and privilege give it control over a country's political, economic, and cultural life.

El Niño-Southern Oscillation (ENSO) A periodic, large-scale, abnormal warming of the sea surface in the low latitudes of the eastern Pacific Ocean that has global implications, disturbing normal **weather** patterns in many parts of the world, especially South America.

Elongated state A **state** whose territory is decidedly long and narrow in that its length is at least six times greater than its average width. Chile and Vietnam are two classic examples.

Emigrant A person **migrating** away from a country or area; an outmigrant.

Empirical Relating to the real world, as opposed to theoretical abstraction.

Enclave A piece of territory that is surrounded by another political unit of which it is not a part.

Endemism Referring to a disease in a host population that affects many people in a kind of equilibrium without causing rapid and widespread deaths.

Entrepôt A place, usually a port city, where goods are imported, stored, and transshipped; a **break-of-bulk point**.

Environmental degradation The accumulated human abuse of a region's **natural landscape** that, among other things, can involve air and water pollution, threats to plant and animal ecosystems, misuse of **natural resources**, and generally upsetting the balance between people and their habitat.

Epidemic A local or regional outbreak of a disease.

Escarpment A cliff or very steep slope; frequently marks the edge of a plateau.

Estuary The widening mouth of a river as it reaches the sea; land subsidence or a rise in sea level has overcome the tendency to form a **delta**.

Ethanol The leading U.S. biofuel that is essentially alcohol distilled from corn mash. Most of it is produced in the historic Corn Belt centered on Iowa. Many risks and problems accompany this energy source, however.

Ethnic cleansing The slaughter and/or forced removal of one **ethnic** group from its homes and lands by another, more powerful ethnic group bent on taking that territory.

Ethnicity The combination of a people's **culture** (traditions, customs, language, and religion) and racial ancestry.

European state model A **state** consisting of a legally defined territory inhabited by a population governed from a capital city by a representative government.

European Union (EU) **Supranational** organization constituted by 27 European countries to further their common economic interests. In alphabetical order, these countries are: Austria, Belgium, Bulgaria, Cyprus, the Czech Republic, Denmark, Estonia, Finland, France, Germany, Greece, Hungary, Ireland, Italy, Latvia, Lithuania, Luxembourg, Malta, the Netherlands, Poland, Portugal, Romania, Slovakia, Slovenia, Spain, Sweden, and the United Kingdom.

Exclave A bounded (non-island) piece of territory that is part of a particular **state** but lies separated from it by the territory of another state. Alaska is an exclave of the United States.

Exclusive Economic Zone (EEZ) An oceanic zone extending up to 200 **nautical miles** from a shoreline, within which the coastal **state** can control fishing, mineral exploitation, and additional activities by all other countries.

Expansion diffusion The spreading of an innovation or an idea through a fixed population in such a way that the number of those adopting grows continuously larger, resulting in an expanding area of dissemination.

Extraterritoriality The politico-geographical concept suggesting that the property of one **state** lying within the boundaries of another actually forms an extension of the first state.

Failed state A country whose institutions have collapsed and in which anarchy prevails. Somalia is a current example.

Fatwa Literally, a legal opinion or proclamation issued by an Islamic cleric, based on the holy texts of Islam, long applicable only in the *Umma*, the realm ruled by the laws of Islam. In 1989, the Iranian ayatollah Khomeini extended the reach of the *fatwa* by condemning to death a British citizen and author living in the United Kingdom.

Favela Shantytown on the outskirts or even well within an urban area in Brazil.

Fazenda Coffee plantation in Brazil.

Federal state A political framework wherein a central government represents the various subnational entities within a **nation-state** where they have common interests—defense, foreign affairs, and the like—yet allows these various entities to retain their own identities and to have their own laws, policies, and customs in certain spheres.

Federation See **federal state**.

Fertile Crescent Crescent-shaped zone of productive lands extending from near the southeastern Mediterranean coast through Lebanon and Syria to the **alluvial** lowlands of Mesopotamia (in Iraq). Once more fertile than today, this is one of the world's great source areas of **agricultural** and other innovations.

First Nations Name given Canada's **indigenous peoples** of American descent, whose U.S. counterparts are called Native Americans.

Fjord Narrow, steep-sided, elongated, and inundated coastal valley deepened by glacier ice that has since melted away, leaving the sea to penetrate.

Floodplain Low-lying area adjacent to a mature river, often covered by **alluvial** deposits and subject to the river's floods.

Forced migration Human **migration** flows in which the movers have no choice but to relocate.

Formal region A type of **region** marked by a certain degree of homogeneity in one or more phenomena; also called *uniform region* or *homogeneous region*.

Forward capital Capital city positioned in actually or potentially contested territory, usually near an international border; it confirms the **state's** determination to maintain its presence in the region in contention.

Fossil fuels The energy resources of **coal**, natural gas, and petroleum (oil), so named collectively because they were formed by the geologic compression and transformation of tiny plant and animal organisms.

Four Motors of Europe Rhône-Alpes (France), Baden-Württemberg (Germany), Catalonia (Spain), and Lombardy (Italy). Each is a high-technology-driven region marked by exceptional industrial vitality and economic success not only within Europe but on the global scene as well.

Fragmented state A **state** whose territory consists of several separated parts, not a **contiguous** whole. The individual parts may be isolated from each other by the land area of other states or by international waters. The United States and Indonesia are examples.

Francophone French-speaking. Quebec constitutes the heart of Francophone Canada.

Free-trade area A form of economic integration, usually consisting of two or more **states**, in which members agree to remove tariffs on trade among themselves. Usually accompanied by a **customs union** that establishes common tariffs on imports from outside the trade area, and sometimes by a **common market** that also removes internal restrictions on the movement of capital, labor, and enterprise.

Free Trade Area of the Americas (FTAA) The ultimate goal of **supranational** economic integration in North, Middle, and South America: the creation of a single-market trading bloc that would involve every country in the Western Hemisphere between the Arctic shore of Canada and Cape Horn at the southern tip of Chile.

Fría See *tierra fría*.

Frontier Zone of advance penetration, usually of contention; an area not yet fully integrated into a national **state**.

FTAA See **Free Trade Area of the Americas**.

Functional region A **region** marked less by its sameness than by its dynamic internal structure; because it usually focuses on a central node, also called *nodal region* or *focal region*.

Functional specialization The production of particular goods or services as a dominant activity in a particular location. See also **local functional specialization**.

Fundamentalism See **revivalism (religious)**.

Gentrification The upgrading of an older residential area through private reinvestment, usually in the downtown area of a central city. Frequently, this involves the displacement of established lower-income residents, who cannot afford the heightened costs of living, and conflicts are not uncommon as such neighborhood change takes place.

Geographic change Evolution of **spatial** patterns over time.

Geographic realm The basic spatial unit in our world regionalization scheme. Each realm is defined in terms of a synthesis of its total human geography—a composite of its leading cultural, economic, historical, political, and appropriate environmental features.

Geography of development The subfield of economic geography concerned with spatial aspects and regional expressions of **development**.

Geometric boundaries Political boundaries **defined** and **delimited** (and occasionally **demarcated**) as straight lines or arcs.

Geomorphology The geographic study of the configuration of the Earth's solid surface—the world's landscapes and their constituent landforms.

Ghetto An intraurban region marked by a particular **ethnic** character. Often an inner-city poverty zone, such as the black ghetto in U.S. central cities. Ghetto residents are involuntarily segregated from other income and racial groups.

Glaciation See **Pleistocene Epoch**.

Globalization The gradual reduction of regional contrasts at the world **scale**, resulting from increasing international cultural, economic, and political exchanges.

Green Revolution The successful recent development of higher-yield, fast-growing varieties of rice and other cereals in certain developing countries.

Gross domestic product (GDP) The total value of all goods and services produced in a country by that state's economy during a given year.

Gross national product (GNP) The total value of all goods and services produced in a country by that state's economy during a given year, plus all citizens' income from foreign investment and other external sources.

Growth pole An urban center with certain attributes that, if augmented by a measure of investment support, will stimulate regional economic **development** in its **hinterland**.

Growth triangle An increasingly popular economic **development** concept along the western **Pacific Rim**, especially in Southeast Asia. It involves the linking of production in growth centers of three countries to achieve benefits for all. For example, Singapore would supply capital and technical know-how to lead high-technology manufacturing projects based on raw materials and inexpensive labor supplied by adjacent areas of Malaysia (Johor State) and Indonesia (Riau Islands).

Hacienda Literally, a large estate in a Spanish-speaking country. Sometimes equated with the **plantation**, but there are important differences between these two types of agricultural enterprise (see p. 203).

Heartland theory The hypothesis, proposed by British geographer Halford Mackinder during the early twentieth century, that any political power based in the heart of Eurasia could gain sufficient strength to eventually dominate the world. Furthermore, since Eastern Europe controlled access to the Eurasian interior, its ruler would command the vast "heartland" to the east.

Hegemony The political dominance of a country (or even a region) by another country. The former Soviet Union's postwar grip on Eastern Europe, which lasted from 1945 to 1990, was a classic example.

Helada See *tierra helada*.

Hierarchical diffusion A form of **diffusion** in which an idea or innovation spreads by trickling down from larger to smaller adoption units. An urban **hierarchy** is usually involved, encouraging the leapfrogging of innovations over wide areas, with geographic distance a less important influence.

Hierarchy An order or gradation of phenomena, with each level or rank subordinate to the one above it and superior to the one below. The levels in a national urban hierarchy are constituted by hamlets, villages, towns, cities, and (frequently) the **primate city**.

High island Volcanic islands of the Pacific Realm that are high enough in elevation to wrest substantial moisture from the tropical ocean air (see **orographic precipitation**). They tend to be well watered, their volcanic soils enable productive agriculture, and they support larger populations than **low islands**—which possess none of these advantages and must rely on fishing and the coconut palm for survival.

High seas Areas of the oceans away from land, beyond national jurisdiction, open and free for all to use.

Highveld A term used in South Africa to identify the high, grass-covered plateau that dominates much of the country. The lowest-lying areas (mainly along the narrow coastlands) in South Africa are called *lowveld*; areas that lie at intermediate elevations are the *middleveld*.

Hinterland Literally, "country behind," a term that applies to a surrounding area served by an urban center. That center is the focus of goods and services produced for its hinterland and is its dominant urban influence as well. In the case of a port city, the hinterland also includes the inland area whose trade flows through that port.

Historical inertia   A term from manufacturing geography that refers to the need to continue using the factories, machinery, and equipment of heavy industries for their full, multiple-decade lifetimes to cover major initial investments—even though these facilities may be increasingly obsolete.

Holocene   The current *interglacial* epoch (the warm period of glacial contraction between the glacial expansions of an **ice age**); extends from 10,000 years ago to the present. Also known as the *Recent Epoch*.

Human evolution   Long-term biological maturation of the human species. Geographically, all evidence points toward East Africa as the source of humankind. Our species, *Homo sapiens*, emigrated from this hearth to eventually populate the rest of the **ecumene**.

Humus   Dark-colored upper layer of a soil that consists of decomposed and decaying organic matter such as leaves and branches, nutrient-rich and giving the soil a high fertility.

Hydraulic civilization theory   The theory that cities able to control **irrigated** farming over large **hinterlands** held political power over other cities. Particularly applies to early Asian civilizations based in such river valleys as the Chang (Yangzi), the Indus, and those of Mesopotamia.

Hydrologic cycle   The **system** of exchange involving water in its various forms as it continually circulates between the **atmosphere**, the oceans, and above and below the land surface.

Ice age   A stretch of geologic time during which the Earth's average atmospheric temperature is lowered; causes the equatorward expansion of continental ice sheets in the higher latitudes and the growth of mountain glaciers in and around the highlands of the lower latitudes.

Immigrant   A person **migrating** into a particular country or area; an in-migrant.

Immigration policies   Australia's policies to regulate immigration. The issue continues to roil Australian society and is discussed on pp. 593–595.

Imperialism   The drive toward the creation and expansion of a **colonial** empire and, once established, its perpetuation.

Import-substitution industries   The industries local entrepreneurs establish to serve populations of remote areas when transport costs from distant sources make these goods too expensive to import.

Inaccessibility   See **accessibility**.

Indentured workers   Contract laborers who sell their services for a stipulated period of time.

Indigenous peoples   Native or *aboriginal* peoples; often used to designate the inhabitants of areas that were conquered and subsequently colonized by the **imperial** powers of Europe.

Indo-European languages   The major world language family that dominates the European **geographic realm** (Fig. 1-8). This language family is also the most widely dispersed globally (Fig. G-10), and about half of humankind speaks one of its languages.

Industrial Revolution   The term applied to the social and economic changes in agriculture, commerce, and especially manufacturing and urbanization that resulted from technological innovations and specialization in late-eighteenth-century Europe.

Informal sector   Dominated by unlicensed sellers of homemade goods and services, the primitive form of capitalism found in many developing countries that takes place beyond the control of government.

Infrastructure   The foundations of a society: urban centers, transport networks, communications, energy distribution systems, farms, factories, mines, and such facilities as schools, hospitals, postal services, and police and armed forces.

Insular   Having the qualities and properties of an island. Real islands are not alone in possessing such properties of **isolation**: an **oasis** in the middle of a **desert** also has qualities of insularity.

Insurgent state   Territorial embodiment of a successful guerrilla movement. The establishment by antigovernment insurgents of a territorial base in which they exercise full control; thus a **state** within a state.

Intercropping   The planting of several types of crops in the same field; commonly used by **shifting cultivators**.

Interglacial   See **Pleistocene Epoch**.

Intermontane   Literally, between the mountains. Such a location can bestow certain qualities of natural protection or **isolation** to a community.

Internal migration   **Migration** flow within a country, such as ongoing westward and southward movements toward the **Sunbelt** in the United States.

International migration   **Migration** flow involving movement across an international boundary.

Intervening opportunity   In trade or **migration** flows, the presence of a nearer opportunity that greatly diminishes the attractiveness of sites farther away.

Inuit   **Indigenous** peoples of North America's Arctic zone, formerly known as Eskimos.

Irredentism   A policy of cultural extension and potential political expansion by a **state** aimed at a community of its nationals living in a neighboring state.

Irrigation   The artificial watering of croplands.

Islamic Front   The southern border of the African Transition Zone that marks the religious **frontier** of the **Muslim** faith in its southward penetration of Subsaharan Africa (see Fig. 6-19).

Islamization   Introduction and establishment of the **Muslim** religion (see Fig. 7-5). A **process** still under way, most notably along the **Islamic Front**, that marks the southern border of the African Transition Zone.

Isohyet   A line connecting points of equal rainfall total.

Isolated state   See **von Thünen's Isolated State model**.

Isolation   The condition of being geographically cut off or far removed from mainstreams of thought and action. It also denotes a lack of receptivity to outside influences, caused at least partially by poor **accessibility**.

Isotherm   A line connecting points of equal temperature.

Isthmus   A **land bridge**; a comparatively narrow link between larger bodies of land. Central America forms such a link between Mexico and South America.

Jakota Triangle   The easternmost region of the East Asian realm, consisting of *Ja*pan, (South) *Ko*rea, and *Tai*wan.

Juxtaposition   Contrasting places in close proximity to one another.

Karst   The distinctive natural landscape associated with the chemical erosion of soluble limestone rock.

Land alienation   One society or culture group taking land from another. In Subsaharan Africa, for example, European **colonialists** took land from **indigenous** Africans and put it to new uses.

Land bridge   A narrow **isthmian** link between two large landmasses. They are temporary features—at least in terms of geologic time—subject to appearance and disappearance as the land or sea level rises and falls.

Land Hemisphere   The half of the globe containing the greatest amount of land surface, centered on Western Europe (Fig. 1-3). In **geomorphology**, this can also refer to the position of the African continent, which lies central to the world's landmasses (Fig. 6-3).

Land reform   The spatial reorganization of **agriculture** through the allocation of farmland (often expropriated from landlords) to **peasants** and tenants who never owned land.

Land tenure   The way people own, occupy, and use land.

Landlocked   An interior **state** surrounded by land. Without coasts, such a country is disadvantaged in terms of **accessibility** to international trade routes, and in the scramble for possession of areas of the **continental shelf** and control of the **exclusive economic zone** beyond.

Language family   Group of languages with a shared but usually distant origin.

"Latin" American city model   The Griffin-Ford model of intraurban spatial structure in the Middle American and South American realms, diagrammed in Figure 5-8.

Latitude Lines of latitude are **parallels** that are aligned east-west across the globe, from 0° latitude at the equator to 90° North and South latitude at the poles.

Leached soil Infertile, reddish-appearing, tropical soil whose surface consists of oxides of iron and aluminum; all other soil nutrients have been dissolved and transported downward into the subsoil by percolating water associated with the heavy rainfall of moist, low-latitude climates.

Leeward The protected or downwind side of a **topographic** barrier with respect to the winds that flow across it.

Lignite Low-grade, brown-colored variety of **coal**.

Lingua franca A "common language" prevalent in a given area; a second language that can be spoken and understood by many peoples, although they speak other languages at home.

Littoral Coastal or coastland.

Llanos The interspersed **savanna** grasslands and scrub woodlands of the Orinoco River's wide basin that covers much of interior Colombia and especially Venezuela.

Local functional specialization A hallmark of Europe's **economic geography** that later spread to many other parts of the world, whereby particular people in particular places concentrate on the production of particular goods and services.

Location Position on the Earth's surface; see also **absolute location** and **relative location**.

Location theory A logical attempt to explain the locational pattern of an economic activity and the manner in which its producing areas are interrelated. The agricultural location theory that underlies the **von Thünen model** is a leading example.

Loess Deposit of very fine silt or dust that is laid down after having been windborne for a considerable distance. Notable for its fertility under **irrigation** and its ability to stand in steep vertical walls.

Longitude Angular distance (0° to 180°) east or west as measured from the *prime meridian* (0°) that passes through the Greenwich Observatory in suburban London, England. For much of its length across the mid-Pacific Ocean, the 180th meridian functions as the *international date line*.

Low island Low-lying coral islands of the Pacific Realm that—unlike **high islands**—cannot wrest sufficient moisture from the tropical ocean air to avoid chronic drought. Thus productive agriculture is impossible, and their modest populations must rely on fishing and the coconut palm for survival.

Lusitanian The Portuguese sphere, which by extension includes Brazil.

Madrassa **Revivalist (fundamentalist)** religious school in which the curriculum focuses on Islamic religion and law and requires rote memorization of the Qu'ran (Koran), Islam's holy book. Founded in former British India, these schools were most numerous in present-day Pakistan but have **diffused** as far as Turkey in the west and Indonesia in the east.

Maghreb The region occupying the northwestern corner of Africa, consisting of Morocco, Algeria, and Tunisia.

Main Street Canada's dominant **conurbation** that is home to nearly two-thirds of the country's inhabitants; extends southwestward from Quebec City in the mid-St. Lawrence Valley to Windsor on the Detroit River.

Mainland-Rimland framework The twofold regionalization of the Middle American realm based on its modern cultural history. The Euro-Amerindian *Mainland*, stretching from Mexico to Panama (minus the Caribbean coastal strip), was a self-sufficient zone dominated by **hacienda land tenure**. The Euro-African *Rimland*, consisting of that Caribbean coastal zone plus all of the Caribbean islands to the east, was the zone of the **plantation** that relied heavily on trade with Europe. See Figure 4-7.

Maquiladora The term given to modern industrial plants in Mexico's U.S. border zone. These foreign-owned factories assemble imported components and/or raw materials, and then export finished manufactures, mainly to the United States. Import duties are disappearing under **NAFTA**, bringing jobs to Mexico and the advantages of low wage rates to the foreign entrepreneurs.

Marchland An area or **frontier** of uncertain boundaries that is subject to various national claims and an unstable political history. Refers specifically to the movement of various armies, refugees, and migrants across such zones.

Marine geography The geographic study of oceans and seas. Its practitioners investigate both the physical (e.g., coral-reef **biogeography**, ocean–**atmosphere** interactions, coastal **geomorphology**) as well as human (e.g., **maritime boundary-making**, fisheries, beachside development) aspects of oceanic environments.

Maritime boundary An international boundary that lies in the ocean. Like all boundaries, it is a vertical plane, extending from the seafloor to the upper limit of the air space in the atmosphere above the water.

Median-line boundary An international **maritime boundary** drawn where the width of a sea is less than 400 **nautical miles**. Because the **states** on either side of that sea claim **exclusive economic zones** of 200 nautical miles, it is necessary to reduce those claims to a (median) distance equidistant from each shoreline. **Delimitation** on the map almost always appears as a set of straight-line segments that reflect the configurations of the coastlines involved.

Medical geography The study of health and disease within a geographic context and from a spatial perspective. Among other things, this geographic field examines the sources, **diffusion** routes, and distributions of diseases.

Megacity Informal term referring to the world's most heavily populated cities; in this book, the term refers to a **metropolis** containing a population of greater than 10 million.

Megalopolis When spelled with a lower-case *m*, a synonym for **conurbation**, one of the large coalescing supercities forming in diverse parts of the world. When capitalized, refers specifically to the multimetropolitan (*Bosnywash*) corridor that extends along the northeastern U.S. seaboard from north of Boston to south of Washington, D.C. (Fig. 3-9).

Mercantilism Protectionist policy of European **states** during the sixteenth to the eighteenth centuries that promoted a **state**'s economic position in the contest with rival powers. Acquiring gold and silver and maintaining a favorable trade balance (more exports than imports) were central to the policy.

Meridian Line of **longitude**, aligned north-south across the globe, that together with **parallels** of **latitude** forms the global grid system. All meridians converge at both poles and are at their maximum distances from each other at the equator.

Mestizo Derived from the Latin word for *mixed*, refers to a person of mixed European (white) and Amerindian ancestry.

Métis **Indigenous** Canadian people of mixed native and European ancestry.

Metropolis Urban **agglomeration** consisting of a (central) city and its suburban ring. See also **urban (metropolitan) area**.

Metropolitan area See **urban (metropolitan) area**.

Migration A change in residence intended to be permanent. See also **forced, internal, international,** and **voluntary migration**.

Migratory movement Human relocation movement from a source to a destination without a return journey, as opposed to cyclical movement (see also **nomadism**).

Model An idealized representation of reality built to demonstrate its most important properties. A **spatial** model focuses on a geographical dimension of the real world, such as the **von Thünen model** that explains agricultural location patterns in a commercial economy.

Modernization In the eyes of the Western world, the Westernization **process** that involves the establishment of **urbanization**, a market (money) economy, improved circulation, formal schooling, adoption of foreign innovations, and the breakdown of traditional society. Non-Westerners mostly see "modernization" as an outgrowth of **colonialism** and often argue that traditional societies can be modernized without being Westernized.

Monsoon Refers to the seasonal reversal of wind and moisture flows in certain parts of the subtropics and lower-middle latitudes. The *dry monsoon* occurs during the cool season when dry offshore winds prevail. The *wet monsoon* occurs in the hot summer months, which produce onshore winds that bring large amounts of rainfall. The air-pressure differential over land and sea is the triggering mechanism, with windflows always moving from areas of relatively higher pressure toward areas of relatively lower pressure. Monsoons make their greatest regional impact in the coastal and near-coastal zones of South Asia (Fig. 8-4), Southeast Asia, and East Asia.

Mosaic culture   Emerging cultural-geographic framework of the United States, dominated by the fragmentation of specialized social groups into homogeneous communities of interest marked not only by income, race, and **ethnicity** but also by age, occupational status, and lifestyle. The result is an increasingly heterogeneous socio-spatial complex, which resembles an intricate mosaic composed of myriad uniform—but separate—tiles.

Mulatto   A person of mixed African (black) and European (white) ancestry.

Multilingualism   A society marked by a mosaic of local languages. Constitutes a **centrifugal force** because it impedes communication within the larger population. Often a *lingua franca* is used as a "common language," as in many countries of Subsaharan Africa.

Multinationals   Internationally active corporations capable of strongly influencing the economic and political affairs of many countries in which they operate.

Muslim   An adherent of the Islamic faith.

Muslim Front   See **Islamic Front**.

NAFTA (North American Free Trade Agreement)   The **free-trade area** launched in 1994 involving the United States, Canada, and Mexico.

Nation   Legally a term encompassing all the citizens of a **state**, it also has other connotations. Most definitions now tend to refer to a group of tightly knit people possessing bonds of language, **ethnicity**, religion, and other shared **cultural** attributes. Such homogeneity actually prevails within very few states.

Nation-state   A country whose population possesses a substantial degree of **cultural** homogeneity and unity. The ideal form to which most **nations** and **states** aspire—a political unit wherein the territorial state coincides with the area settled by a certain national group or people.

NATO (North Atlantic Treaty Organization)   Established in 1950 at the height of the Cold War as a U.S.-led **supranational** defense pact to shield postwar Europe against the Soviet military threat. NATO is now in transition, expanding its membership while modifying its objectives in the post-Soviet era. Its 28 member-states (as of mid-2009) are: Albania, Belgium, Bulgaria, Canada, Croatia, Czech Republic, Denmark, Estonia, France, Germany, Greece, Hungary, Iceland, Italy, Latvia, Lithuania, Luxembourg, Netherlands, Norway, Poland, Portugal, Romania, Slovakia, Slovenia, Spain, Turkey, United Kingdom, and the United States.

Natural hazard   A natural event that endangers human life and/or the contents of a **cultural landscape**.

Natural increase rate   Population growth measured as the excess of live births over deaths per 1000 individuals per year. Natural increase of a population does not reflect either **emigrant** or **immigrant** movements.

Natural landscape   The array of landforms that constitutes the Earth's surface (mountains, hills, plains, and plateaus) and the physical features that mark them (such as water bodies, soils, and vegetation). Each **geographic realm** has its distinctive combination of natural landscapes.

Natural resource   Any valued element of (or means to an end using) the environment; includes minerals, water, vegetation, and soil.

Nautical mile   By international agreement, the nautical mile—the standard measure at sea—is 6076.12 feet in length, equivalent to approximately 1.15 statute miles (1.85 km).

Near Abroad   The 14 former Soviet republics (listed in Table 2-1) that, together with the dominant Russian Republic, constituted the USSR. Since the 1991 breakup of the Soviet Union, Russia has asserted a sphere of influence in these now-independent countries, based on its proclaimed right to protect the interests of ethnic Russians who were settled there in substantial numbers during Soviet times.

Neocolonialism   The term used by developing countries to underscore that the entrenched **colonial** system of international exchange and capital flow has not changed in the postcolonial era—thereby perpetuating the huge economic advantages of the developed world.

Network (transport)   The entire regional **system** of transportation connections and nodes through which movement can occur.

Nevada   See *tierra nevada*.

New World Order   A description of the international system resulting from the collapse of the Soviet Union in which the balance of nuclear power theoretically no longer determines the destinies of **states**.

Nomadism   Cyclical movement among a definite set of places. Nomadic peoples mostly are **pastoralists**.

North American Free Trade Agreement   See **NAFTA**.

Nucleation   Cluster; **agglomeration**.

Oasis   An area, small or large, where the supply of water (from an **aquifer** or a major river such as the Nile) permits the transformation of the immediately surrounding **desert** into productive cropland.

Occidental   Western. Also see *Oriental*.

Offshore banking   Term referring to financial havens for foreign companies and individuals, who channel their earnings to accounts in such a country (usually an "offshore" island-state) to avoid paying taxes in their home countries.

Oligarchs   Opportunists in post-Soviet Russia who used their ties to government to enrich themselves.

OPEC (Organization of Petroleum Exporting Countries)   The international oil *cartel* or syndicate formed by a number of producing countries to promote their common economic interests through the formulation of joint pricing policies and the limitation of market options for consumers. The 12 member-states (as of late 2009) are: Algeria, Angola, Ecuador, Iran, Iraq, Kuwait, Libya, Nigeria, Qatar, Saudi Arabia, United Arab Emirates (UAE), and Venezuela.

Organic theory   Friedrich Ratzel's theory of **state** development that conceptualized the state as a biological organism whose life—from birth through maturation to eventual senility and collapse—mirrors that of any living thing.

Oriental   The root of the word "oriental" is from the Latin for *rise*. Thus it has to do with the direction in which one sees the sun "rise"—the east; *oriental* therefore means Eastern. *Occidental* originates from the Latin for fall, or the "setting" of the sun in the west; occidental therefore means Western.

Orographic precipitation   Mountain-induced precipitation, especially when air masses are forced to cross **topographic** barriers. Downwind areas beyond such a mountain range experience the relative dryness known as the **rain shadow effect**.

Outback   The name given by Australians to the vast, peripheral, sparsely settled interior of their country.

Outer city   The non-central-city portion of the American **metropolis**; no longer "sub" to the "urb," this outer ring was transformed into a full-fledged city during the late twentieth century.

Overseas Chinese   The more than 50 million ethnic Chinese who live outside China. Over half live in Southeast Asia (see Fig. 10-6), and many have become quite successful. A large number maintain links to China and as investors played a major economic role in stimulating the growth of **SEZs** and Open Cities in China's **Pacific Rim**.

Pacific Rim   A far-flung group of countries and parts of countries (extending clockwise on the map from New Zealand to Chile) sharing the following criteria: they face the Pacific Ocean; they evince relatively high levels of economic development, industrialization, and urbanization; their imports and exports mainly move across Pacific waters.

Pacific Ring of Fire   Zone of crustal instability along **tectonic plate** boundaries, marked by earthquakes and volcanic activity, that ring the Pacific Ocean Basin (see Fig. G-4).

Paddies (paddyfields)   Ricefields.

Pandemic   An outbreak of a disease that spreads worldwide.

Pangaea   A vast, singular landmass consisting of most of the areas of the present-day continents. This supercontinent began to break up more than 200 million years ago when still-ongoing **plate** divergence and **continental drift** became dominant **processes** (see Fig. 6-3).

Parallel   An east-west line of **latitude** that is intersected at right angles by **meridians** of **longitude**.

Pastoralism   A form of **agricultural** activity that involves the raising of livestock.

Peasants   In a **stratified** society, peasants are the lowest class of people who depend on **agriculture** for a living. But they often own no land at all and must survive as tenants or day workers.

Peninsula   A comparatively narrow, finger-like stretch of land extending from the main landmass into the sea. Florida and Korea are examples.

Peon (*peone*)   Term used in Middle and South America to identify people who often live in serfdom to a wealthy landowner; landless **peasants** in continuous indebtedness.

Per capita   Capita means *individual*. Income, production, or some other measure is often given per individual.

Perforated state   A **state** whose territory completely surrounds that of another state. South Africa, which encloses Lesotho and is perforated by it, is a classic example.

Periodic market   Village market that opens every third day or at some other regular interval. Part of a regional network of similar markets in a preindustrial, rural setting where goods are brought to market on foot and barter remains a major mode of exchange.

Peripheral development   Spatial pattern in which a country's or region's development (and population) is most heavily concentrated along its outer edges rather than in its interior. Australia, with its peripheral population distribution, is a classic example (note its **core area** in Fig. 11-5), and nearby New Zealand exhibits a similar configuration of people and activities.

Periphery   See **core-periphery relationships**.

Permafrost   Permanently frozen water in the near-surface soil and bedrock of cold environments, producing the effect of completely frozen ground. Surface can thaw during brief warm season.

Physical geography   The study of the geography of the physical (natural) world. Its subfields include **climatology, geomorphology, biogeography, soil geography, marine geography**, and water **resources**.

Physical landscape   Synonym for **natural landscape**.

Physiographic political boundaries   Political boundaries that coincide with prominent physical features in the **natural landscape**—such as rivers or the crest ridges of mountain ranges.

Physiographic region (province)   A **region** within which there prevails substantial **natural-landscape** homogeneity, expressed by a certain degree of uniformity in surface **relief, climate**, vegetation, and soils.

Physiography   Literally means *landscape description*, but commonly refers to the total **physical geography** of a place; includes all of the natural features on the Earth's surface, including landforms, **climate**, soils, vegetation, and water bodies.

Physiologic density   The number of people per unit area of **arable** land.

Pilgrimage   A journey to a place of great religious significance by an individual or by a group of people (such as a pilgrimage [*hajj*] to Mecca for **Muslims**).

Plantation   A large estate owned by an individual, family, or corporation and organized to produce a cash crop. Almost all plantations were established within the tropics; in recent decades, many have been divided into smaller holdings or reorganized as cooperatives.

Plate tectonics   Plates are bonded portions of the Earth's mantle and crust, averaging 100 kilometers (60 mi) in thickness. More than a dozen such plates exist (see Fig. G-3), most of continental proportions, and they are in motion. Where they meet one slides under the other, crumpling the surface crust and producing significant volcanic and earthquake activity; a major mountain-building force.

Pleistocene Epoch   Recent period of geologic time that spans the rise of humankind, beginning about 2 million years ago. Marked by *glaciations* (repeated advances of continental ice sheets) and milder *interglacials* (ice sheet contractions). Although the last 10,000 years are known as the **Holocene** Epoch, Pleistocene-like conditions seem to be continuing and we are most probably now living through another Pleistocene interglacial; thus the glaciers likely will return.

Plural(istic) society   A society in which two or more population groups, each practicing its own **culture**, live adjacent to one another without mixing inside a single **state**.

Polder   Land reclaimed from the sea adjacent to the shore of the Netherlands by constructing dikes and then pumping out the water trapped behind them.

Political geography   The study of the interaction of geographic space and political **process**; the spatial analysis of political phenomena and processes.

Pollution   The release of a substance, through human activity, which chemically, physically, or biologically alters the air or water it is discharged into. Such a discharge negatively impacts the environment, with possible harmful effects on living organisms—including humans.

Population decline   A decreasing national population. Russia, which now loses about half a million people per year, is the best example. Also see **population implosion**.

Population density   The number of people per unit area. Also see **arithmetic density** and **physiologic density** measures.

Population distribution   The way people have arranged themselves in geographic space. One of human geography's most essential expressions because it represents the sum total of the adjustments that a population has made to its natural, cultural, and economic environments. A population distribution map is included in every chapter in this book.

Population explosion   The rapid growth of the world's human population during the past century, attended by accelerating *rates* of increase.

Population geography   The field of geography that focuses on the spatial aspects of **demography** and the influences of demographic change on particular places.

Population implosion   The opposite of **population explosion**; refers to the declining populations of many European countries and Russia in which the **death rate** exceeds the **birth rate** and **immigration** rate.

Population movement   See **migration** and **migratory movement**.

Population projection   The future population total that demographers forecast for a particular country. For example, in the table at the end of this book such projections are given for all the world's countries for 2025.

Population (age-sex) structure   Graphic representation (*profile*) of a population according to age and gender.

Postindustrial economy   Emerging economy, in the United States and a number of other highly advanced countries, as traditional industry is increasingly eclipsed by a higher-technology productive complex dominated by services, information-related, and managerial activities.

Primary economic activity   Activities engaged in the direct extraction of **natural resources** from the environment such as mining, fishing, lumbering, and especially **agriculture**.

Primate city   A country's largest city—ranking atop the urban **hierarchy**—most expressive of the national culture and usually (but not always) the capital city as well.

Process   Causal force that shapes a spatial pattern as it unfolds over time.

Productive activities   The major components of the spatial economy. For individual components see: **primary economic activity, secondary economic activity, tertiary economic activity, quaternary economic activity**, and **quinary economic activity**.

Protruded state Territorial shape of a **state** that exhibits a narrow, elongated land extension (or *protrusion*) leading away from the main body of territory. Thailand is a leading example.

Push-pull concept The idea that **migration** flows are simultaneously stimulated by conditions in the source area, which tend to drive people away, and by the perceived attractiveness of the destination.

Qanat In **desert** zones, particularly in Iran and western China, an underground tunnel built to carry **irrigation** water by gravity flow from nearby mountains (where **orographic precipitation** occurs) to the arid flatlands below.

Quaternary economic activity Activities engaged in the collection, processing, and manipulation of *information*.

Quinary economic activity Managerial or control-function activity associated with decision-making in large organizations.

Rain shadow effect The relative dryness in areas downwind of mountain ranges caused by **orographic precipitation**, wherein moist air masses are forced to deposit most of their water content as they cross the highlands.

Rate of natural population increase See **natural increase rate**.

Realm See **geographic realm**.

Refugees People who have been dislocated involuntarily from their original place of settlement.

Region A commonly used term and a geographic concept of central importance. An **area** on the Earth's surface marked by specific criteria, which are discussed on pp. 6–10.

Regional boundary In theory, the line that circumscribes a **region**. But razor-sharp lines are seldom encountered, even in nature (e.g., a coastline constantly changes depending on the tide). In the **cultural landscape**, not only are regional boundaries rarely self-evident, but when they are ascertained by geographers they most often turn out to be **transitional** borderlands.

Regional character The personality or "atmosphere" of a **region** that makes it distinct from all other regions.

Regional complementarity See **complementarity**.

Regional concept The geographic study of **regions** and regional distinctions, as discussed on pp. 9–10.

Regional disparity The spatial unevenness in standard of living that occurs within a country, whose "average," overall income statistics invariably mask the differences that exist between the extremes of the wealthy **core** and the poorer **periphery**.

Regional geography Approach to geographic study based on the spatial unit of the **region**. Allows for an all-encompassing view of the world, because it utilizes and integrates information from geography's topical (**systematic**) fields; diagrammed in Figure G-14.

Regional state A "natural economic zone" that defies political boundaries and is shaped by the global economy of which it is a part; its leaders deal directly with foreign partners and negotiate the best terms they can with the national governments under which they operate.

Regionalism The consciousness of and loyalty to a **region** considered distinct and different from the **state** as a whole by those who occupy it.

Relative location The regional position or **situation** of a place relative to the position of other places. Distance, **accessibility**, and connectivity affect relative location.

Relict boundary A political boundary that has ceased to function, but the imprint of which can still be detected on the **cultural landscape**.

Relief Vertical difference between the highest and lowest elevations within a particular area.

Religious revivalism See **revivalism (religious)**.

Relocation diffusion Sequential **diffusion process** in which the items being diffused are transmitted by their carrier agents as they relocate to new areas. The most common form of relocation diffusion involves the spreading of innovations by a **migrating** population.

Restrictive population policies Government policy designed to reduce the **rate of natural population increase**. China's one-child policy, instituted in 1979 after Mao's death, is a classic example.

Revivalism (religious) Religious movement whose objectives are to return to the foundations of that faith and to influence state policy. Often called *religious fundamentalism*; but in the case of Islam, **Muslims** prefer the term *revivalism*.

Rift valley The trough or trench that forms when a thinning strip of the Earth's crust sinks between two parallel faults (surface fractures).

Rural-to-urban migration The dominant **migration** flow from countryside to city that continues to transform the world's population, most notably in the less advantaged geographic realms.

Russification Demographic resettlement policies pursued by the central planners of the Soviet Empire, whereby ethnic Russians were encouraged to emigrate from the Russian Republic to the 14 non-Russian republics of the USSR.

Sahel Semiarid **steppeland** zone extending across most of Africa between the southern margins of the arid Sahara and the moister tropical **savanna** and forest zone to the south. Chronic drought, **desertification**, and overgrazing have contributed to severe famines in this area since 1970.

Savanna Tropical grassland containing widely spaced trees; also the name given to the tropical wet-and-dry climate type (*Aw*).

Scale Representation of a real-world phenomenon at a certain level of reduction or generalization. In **cartography**, the ratio of map distance to ground distance; indicated on a map as a bar graph, representative fraction, and/or verbal statement. *Macroscale* refers to a large area of national proportions; *microscale* refers to a local area no bigger than a county.

Scale economies See **economies of scale**.

Secondary economic activity Activities that process raw materials and transform them into finished industrial products; the *manufacturing* sector.

Sedentary Permanently attached to a particular area; a population fixed in its location; the opposite of **nomadic**.

Separate development The spatial expression of South Africa's "grand" **apartheid** scheme, whereby nonwhite groups were required to settle in segregated "homelands." The policy was dismantled when white-minority rule collapsed in the early 1990s.

Sequent occupance The notion that successive societies leave their cultural imprints on a place, each contributing to the cumulative **cultural landscape**.

Shantytown Unplanned slum development on the margins of cities in disadvantaged countries, dominated by crude dwellings and shelters mostly made of scrap wood and iron, and even pieces of cardboard.

Sharecropping Relationship between a large landowner and farmers on the land whereby the farmers pay rent for the land they farm by giving the landlord a share of the annual harvest.

Sharia The criminal code based in Islamic law that prescribes corporal punishment, amputations, stonings, and lashing for both major and minor offenses. Its occurrence today is associated with the spread of **religious revivalism** in **Muslim** societies.

Shatter belt **Region** caught between stronger, colliding external cultural-political forces, under persistent stress, and often fragmented by aggressive rivals. Eastern Europe and Southeast Asia are classic examples.

Shifting agriculture Cultivation of crops in recently cut and burned tropical-forest clearings, soon to be abandoned in favor of newly cleared nearby forest land. Also known as *slash-and-burn agriculture*.

Sinicization Giving a Chinese cultural imprint; Chinese **acculturation**.

Site The internal locational attributes of an urban center, including its local spatial organization and physical setting.

Situation The external locational attributes of an urban center; its **relative location** or regional position with reference to other nonlocal places.

Social stratification See **stratification (social)**.

Southern Ocean   The ocean that surrounds Antarctica (see discussion on p. 581).

Spatial   Pertaining to space on the Earth's surface. Synonym for *geographic(al)*.

Spatial diffusion   See **diffusion**.

Spatial interaction   See **complementarity, intervening opportunity,** and **transferability**.

Spatial model   See **model**.

Spatial process   See **process**.

Spatial system   The components and interactions of a **functional region**, which is defined by the areal extent of those interactions. See also **system**.

Special Administrative Region (SAR)   Status accorded the former dependencies of Hong Kong and Macau that were taken over by China, respectively, from the United Kingdom in 1997 and Portugal in 1999. Both SARs received guarantees that their existing social and economic systems could continue unchanged for 50 years following their return to China.

Special Economic Zone (SEZ)   Manufacturing and export center within China, created since 1980 to attract foreign investment and technology transfers. Seven SEZs—all located on China's Pacific coast—currently operate: Shenzhen, adjacent to Hong Kong; Zhuhai; Shantou; Xiamen; Hainan Island, in the far south; Pudong, across the river from Shanghai; and Binhai New Area, next to the port of Tianjin.

Squatter settlement   See **shantytown**.

State   A politically organized territory that is administered by a sovereign government and is recognized by a significant portion of the international community. A state must also contain a permanent resident population, an organized economy, and a functioning internal circulation system.

State boundaries   The borders that surround **states** which, in effect, are derived through contracts with neighboring states negotiated by treaty. See **definition, delimitation,** and **demarcation**.

State capitalism   Government-controlled corporations competing under free-market conditions, usually in a tightly regimented society. South Korea is a leading example. Also see *chaebol*.

State formation   The creation of a **state** based on traditions of human **territoriality** that go back thousands of years.

State planning   Involves highly centralized control of the national planning process, a hallmark of communist economic systems. Soviet central planners mainly pursued a grand political design in assigning production to particular places; their frequent disregard of the principles of economic geography contributed to the eventual collapse of the USSR.

State territorial morphology   A **state's** geographical shape, which can have a decisive impact on its spatial cohesion and political viability. A **compact** shape is most desirable; among the less efficient shapes are those exhibited by **elongated, fragmented, perforated,** and **protruded** states.

Stateless nation   A national group that aspires to become a **nation-state** but lacks the territorial means to do so; the Palestinians and Kurds of Southwest Asia are classic examples.

Steppe   Semiarid grassland; short-grass prairie. Also the name given to the semiarid climate type (**BS**).

Stratification (social)   In a layered or stratified society, the population is divided into a **hierarchy** of social classes. In an industrialized society, the working class is at the lower end; **elites** that possess capital and control the means of production are at the upper level. In the traditional **caste system** of Hindu India, the "untouchables" form the lowest class or caste, whereas the still-wealthy remnants of the princely class are at the top.

Subduction   In **plate tectonics**, the **process** that occurs when an oceanic plate converges head-on with a plate carrying a continental landmass at its leading edge. The lighter continental plate overrides the denser oceanic plate and pushes it downward.

Subsequent boundary   A political boundary that developed contemporaneously with the evolution of the major elements of the **cultural landscape** through which it passes.

Subsistence   Existing on the minimum necessities to sustain life; spending most of one's time in pursuit of survival.

Subtropical Convergence   A narrow marine **transition zone**, girdling the globe at approximately latitude 40°S, that marks the equatorward limit of the frigid **Southern Ocean** and the poleward limits of the warmer Atlantic, Pacific, and Indian oceans to the north.

Suburban downtown   In the United States (and increasingly in other advantaged countries), a significant concentration of major urban activities around a highly accessible suburban location, including retailing, light industry, and a variety of leading corporate and commercial operations. The largest are now coequal to the American central city's **central business district (CBD)**.

Sunbelt   The popular name given to the southern tier of the United States, which is anchored by the mega-States of California, Texas, and Florida. Its warmer climate, superior recreational opportunities, and other amenities have been attracting large numbers of relocating people and activities since the 1960s; broader definitions of the Sunbelt also include much of the western United States, particularly Colorado and the coastal Pacific Northwest.

Superimposed boundary   A political boundary emplaced by powerful outsiders on a developed human landscape. Usually ignores preexisting cultural-spatial patterns, such as the border that still divides North and South Korea.

Supranational   A venture involving three or more **states**—political, economic, and/or cultural cooperation to promote shared objectives. The **European Union** is one such organization.

System   Any group of objects or institutions and their mutual interactions. Geography treats systems that are expressed spatially, such as in **functional regions**.

Systematic geography   Topical geography: **cultural, political, economic geography,** and the like.

Taiga   The subarctic, mostly **coniferous** snowforest that blankets northern Russia and Canada south of the **tundra** that lines the Arctic shore.

Takeoff   Economic concept to identify a stage in a country's **development** when conditions are set for a domestic Industrial Revolution.

Taxonomy   A **system** of scientific classification.

Technopole   A planned techno-industrial complex (such as California's Silicon Valley) that innovates, promotes, and manufactures the products of the **postindustrial** informational economy.

Tectonics   See **plate tectonics**.

Templada   See *tierra templada*.

Terracing   The transformation of a hillside or mountain slope into a step-like sequence of horizontal fields for intensive cultivation.

Territoriality   A country's or more local community's sense of property and attachment toward its territory, as expressed by its determination to keep it inviolable and strongly defended.

Territorial sea   Zone of seawater adjacent to a country's coast, held to be part of the national territory and treated as a component of the sovereign **state**.

Tertiary economic activity   Activities that engage in *services*—such as transportation, banking, retailing, education, and routine office-based jobs.

Tierra caliente   The lowest of the **altitudinal zones** into which the human settlement of Middle and South America is classified according to elevation. The *caliente* is the hot humid coastal plain and adjacent slopes up to 750 meters (2500 ft) above sea level. The natural vegetation is the dense and luxuriant tropical rainforest; the crops include sugar and bananas in the lower areas, and coffee, tobacco, and corn along the higher slopes.

Tierra fría   Cold, high-lying **altitudinal zone** of settlement in Andean South America, extending from about 1800 meters (6000 ft) in elevation up to nearly 3600 meters (12,000 ft). **Coniferous** trees stand here; upward they change into scrub and grassland. There are also important pastures within the *fría*, and wheat can be cultivated.

Tierra helada In Andean South America, the highest-lying habitable **altitudinal zone**—ca. 3600 to 4500 meters (12,000 to 15,000 ft)—between the tree line (upper limit of the *tierra fría*) and the snow line (lower limit of the *tierra nevada*). Too cold and barren to support anything but the grazing of sheep and other hardy livestock.

Tierra nevada The highest and coldest **altitudinal zone** in Andean South America (lying above 4500 meters [15,000 ft]), an uninhabitable environment of permanent snow and ice that extends upward to the Andes' highest peaks of more than 6000 meters (20,000 ft).

Tierra templada The intermediate **altitudinal zone** of settlement in Middle and South America, lying between 750 meters (2500 ft) and 1800 meters (6000 ft) in elevation. This is the "temperate" zone, with moderate temperatures compared to the *tierra caliente* below. Crops include coffee, tobacco, corn, and some wheat.

Topography The surface configuration of any segment of **natural landscape**.

Toponym Place name.

Transculturation Cultural borrowing and two-way exchanges that occur when different **cultures** of approximately equal complexity and technological level come into close contact.

Transferability The capacity to move a good from one place to another at a bearable cost; the ease with which a commodity may be transported.

Transition zone An area of spatial change where the **peripheries** of two adjacent **realms** or **regions** join; marked by a gradual shift (rather than a sharp break) in the characteristics that distinguish these neighboring geographic entities from one another.

Transmigration The now-ended policy of the Indonesian government to induce residents of the overcrowded, **core-area** island of Jawa to move to the country's other islands.

Treaty ports **Extraterritorial enclaves** in China's coastal cities, established by European colonial invaders under unequal treaties enforced by gunboat diplomacy.

Triple Frontier The turbulent and chaotic area in southern South America that surrounds the convergence of Brazil, Argentina, and Paraguay. Lawlessness pervades this haven for criminal elements, which is notorious for money laundering, arms and other smuggling, drug trafficking, and links to terrorist organizations, including money flows to the Middle East.

Tropical deforestation See **deforestation**.

Tropical savanna See **savanna**.

Tsunami A seismic (earthquake-generated) sea wave that can attain gigantic proportions and cause coastal devastation. The tsunami of December 26, 2004, centered in the Indian Ocean near the Indonesian island of Sumatera (Sumatra), produced the first great natural disaster of the twenty-first century.

Tundra The treeless plain that lies along the Arctic shore in northernmost Russia and Canada, whose vegetation consists of mosses, lichens, and certain hardy grasses.

Turkestan Northeasternmost region of the North Africa/Southwest Asia realm. Known as Soviet Central Asia before 1992, its five (dominantly Islamic) former Soviet Socialist Republics have become the independent countries of Kazakhstan, Uzbekistan, Turkmenistan, Kyrgyzstan, and Tajikistan. Today Turkestan has expanded to include a sixth state, Afghanistan.

Unitary state A **nation-state** that has a centralized government and administration that exercises power equally over all parts of the **state**.

Urbanization A term with several connotations. The proportion of a country's population living in urban places is its level of urbanization. The **process** of urbanization involves the movement to, and the clustering of, people in towns and cities—a major force in every geographic realm today. Another kind of urbanization occurs when an expanding city absorbs rural countryside and transforms it into suburbs; in the case of cities in disadvantaged countries, this also generates peripheral **shantytowns**.

Urban (metropolitan) area The entire built-up, nonrural area and its population, including the most recently constructed suburban appendages. Provides a better picture of the dimensions and population of such an area than the delimited municipality (central city) that forms its heart.

Urban realms model A spatial generalization of the contemporary large American city. It is shown to be a widely dispersed, multinodal metropolis consisting of increasingly independent zones or *urban realms*, each focused on its own **suburban downtown**; the only exception is the shrunken central realm, which is focused on the central city's **central business district** (see Fig. 3-10).

Veld See **highveld**.

Voluntary migration Population movement in which people relocate in response to perceived opportunity, not because they are forced to **migrate**.

Von Thünen's Isolated State model Explains the location of **agricultural** activities in a commercial economy. A **process** of spatial competition allocates various farming activities into concentric rings around a central market city, with profit-earning capability the determining force in how far a crop locates from the market. The original (1826) Isolated State model now applies to the continental scale (see Fig. 1-5).

Wahhabism A particularly virulent form of (Sunni) **Muslim revivalism** that was made the official faith when the modern **state** of Saudi Arabia was founded in 1932. Adherents call themselves "Unitarians" to signify the strict fundamentalist nature of their beliefs.

Wallace's Line As shown in Figure 11-4, the zoogeographical boundary proposed by Alfred Russel Wallace that separates the marsupial fauna of Australia and New Guinea from the nonmarsupial fauna of Indonesia.

Weather The immediate and short-term conditions of the **atmosphere** that impinge on daily human activities.

West Wind Drift The clockwise movement of water as a current that circles around Antarctica in the **Southern Ocean**.

Wet monsoon See **monsoon**.

Windward The exposed, upwind side of a **topographic** barrier that faces the winds that flow across it.

World geographic realm See **geographic realm**.

This Data Table is a valuable resource, and should be consulted throughout your reading. Like all else in this book, this table is subject to continuous revision and modification. Compared to previous editions, we have deleted some indices, elaborated others, and introduced new ones. For example, in a world with ever-slower population growth, the so-called Doubling Time index—the number of years it will take for a population to double in size based on its current rate of natural increase—has lost most of its utility. On the other hand, when it comes to **Life Expectancy** and **Literacy**, general averages conceal significant differences between male and female rates, which in turn reflect conditions in individual countries, so we report these by gender. Also in this edition, we continue to use the **Corruption Index**, not available for all countries but an important reflection of a global problem. The **Big Mac Price** index, a measure introduced by the journal *The Economist*, tells you much more than what a hamburger with all the trimmings would cost in real dollars in various countries of the world—it also reflects whether those countries' currencies are overvalued or undervalued. And the final column, in which we formerly used to reported the per-capita GNP (Gross National Product), now reveals the **GNI**, that is, the Gross National Income per per-

son and what this would buy in each country. In the language of economic geographers, this is called the GNI-PPP, the per-capita Gross National Income in terms of its Purchasing Power Parity.

Indexes that may not be immediately obvious to you include **Arithmetic Population Density**, the number of people per square kilometer in each country; **Physiologic Population Density**, the number of people per square kilometer of agriculturally productive land; **Birth** and **Death Rates** per thousand in the population, resulting in the national population's rate of **Natural Increase**; a population's **Infant Mortality**, the number of deaths per thousand in the first year of life, thus reflecting largely the number of deaths at birth; **Child Mortality**, the deaths per thousand of children in their first five years; the **Corruption Index**, based on Transparency International data in which 10.0 is perfect and 0.1 is the worst; the **Big Mac Price** index, which tells you why Malaysia in 2009 was the best place to buy a hamburger in U.S. dollars; and the **Per Capita GNI ($US)**, the GNI-PPP index referred to above, which tells you how spendable income varies around the globe. For additional details on sources and data, please consult the Data Sources section of the Preface.

Continued

	Land Area Sq km	Land Area Sq mi	Population 2010 (Millions)	Population 2025 (Millions)	Pop. Density Arithmetic	Pop. Density Physiologic	Birth Rate	Death Rate	Natural Increase %	Infant Mortality per 1,000	Child Mortality per 1,000	Life Exp. Male (years)	Life Exp. Female (years)	Percent Urban Pop.	Literacy Male %	Literacy Female %	Corruption Index	Big Mac Price ($US)	Per Capita GNI ($US)
EAST ASIA																			
REALM	**11,774,215**	**4,546,050**	**1573.4**	**1696.6**	**134**	**1083**	**12**	**7**	**0.5**	**21.0**	**24**	**72**	**76**	**50**	**94.9**	**84.7**			**$8,380**
China	9,572,855	3,696,100	1338.0	1476.0	140	998	12	7	0.5	23.0	27	71	75	45	92.3	77.4	3.6	1.83	$5,370
Japan	377,799	145,869	127.7	119.3	338	2600	9	9	0.0	2.8	4	79	86	79	100.0	100.0	7.3	3.23	$34,600
Korea, North	120,541	46,541	23.9	25.8	198	1240	16	7	0.9	21.0	28	68	73	60	99.0	99.0			
Korea, South	99,259	38,324	49.1	49.1	495	2603	10	5	0.5	4.0	4	76	82	82	99.2	96.4	5.6	2.39	$24,750
Mongolia	1,566,492	604,826	2.8	3.3	2	178	21	6	1.5	41.0	50	61	67	59	99.2	99.3	3.0		$3,160
Taiwan	36,180	13,969	23.1	23.1	640	2558	9	6	0.3	4.6		75	81	78	97.6	90.2	5.7	2.32	
SOUTHEAST ASIA																			
REALM	**4,494,790**	**1,735,448**	**601.8**	**709.2**	**134**	**642**	**20**	**7**	**1.3**	**31.0**	**37**	**68**	**72**	**45**	**93.0**	**86.1**			**$4,440**
Brunei	5,770	2,228	0.4	0.5	72	3578	19	3	1.6	7.0	8	72	77	72	94.7	88.2			$49,900
Cambodia	181,040	69,900	15.2	20.6	84	382	26	8	1.8	67.0	87	59	66	15	79.7	53.4	1.8		$1,690
East Timor	14,869	5,741	1.2	1.7	79	655	42	11	3.1	88.0	110	59	61	22			2.2		$3,190
Indonesia	1,904,561	735,355	247.2	291.9	130	763	21	6	1.5	34.0	43	69	72	48	91.9	82.1	2.6	1.74	$3,580
Laos	236,800	91,429	6.2	8.7	26	871	34	10	2.4	70.0	88	59	63	27	73.6	50.5	2.0		$1,940
Malaysia	329,750	127,317	28.6	34.6	87	361	21	5	1.6	9.0	11	72	76	68	91.5	83.6	5.1	1.52	$13,570
Myanmar/Burma	676,577	261,228	50.1	55.4	74	463	19	10	0.9	70.0	98	58	64	31	89.0	80.6	1.3		$3,730
Philippines	299,998	115,830	94.3	120.2	314	953	26	5	2.1	25.0	31	66	72	63	95.5	95.2	2.3	2.07	$3,730
Singapore	619	239	4.9	5.3	7848	392384	11	5	0.6	2.4	3	78	83	100	96.4	88.5	9.2	2.61	$32,470
Thailand	513,118	198,116	66.8	70.2	130	325	13	8	0.5	16.0	17	68	75	36	97.2	94.0	3.5	1.77	$7,880
Vietnam	331,689	128,066	88.3	100.1	266	1210	17	5	1.2	16.0	19	71	75	27	95.7	91.0	2.7		$2,550
AUSTRAL REALM																			
REALM	**8,012,942**	**3,093,814**	**26.0**	**29.6**	**3**	**45**	**14**	**7**	**0.7**	**4.8**	**5.1**	**79**	**84**	**87**	**100.0**	**100.0**			**$32,164**
Australia	7,741,184	2,988,888	21.6	24.7	3	40	14	7	0.7	4.7	5	79	84	87	100.0	100.0	8.7	2.19	$33,340
New Zealand	270,529	104,452	4.4	4.9	16	135	15	7	0.8	5.0	6	78	82	86	100.0	100.0	9.3	2.48	$26,340
PACIFIC REALM																			
REALM	**975,341**	**376,804**	**9.6**	**11.8**	**17**	**806**	**29**	**9**	**2.0**	**50.0**	**67**	**58**	**63**	**22**	**65.9**	**52.1**			**$1,526**
Federated States of Micronesia	699	270	0.1	0.1	149	286	26	6	2.0	40.0	49	67	67	22	67.0	87.2			$3,710
Fiji	18,270	7,054	0.9	0.9	51	317	21	6	1.5	17.0	19	66	71	51	95.0	90.9			$4,370
French Polynesia	3,999	1,544	0.3	0.3	77	964	18	4	1.4	6.8		73	77	53	94.9	95.0			
Guam	549	212	0.2	0.2	375	1706	19	4	1.5	10.7		75	82	93	99.0	99.0			
Marshall Islands	179	69	0.1	0.1	596	3506	38	6	3.2	23.0	25	64	67	68	92.4	90.0			
New Caledonia	18,581	7,174	0.2	0.3	11	1105	18	5	1.3	7.0		73	80	58	57.4	58.3			
Papua New Guinea	462,839	178,703	6.8	8.6	15	1464	31	10	2.1	62.0	81	54	60	13			2.0		$1,500
Samoa	2,841	1,097	0.2	0.2	74	171	29	6	2.3	20.0	24	72	74	22	100.0	100.0	4.4		$3,570
Solomon Islands	28,899	11,158	0.5	0.7	18	607	34	8	2.6	48.0	63	62	63	17	62.4	44.9	2.9		$1,400
Vanuatu	12,191	4,707	0.2	0.4	17	172	31	6	2.5	27.0	33	66	69	21	57.3	47.8	2.9		$2,890

| | Land Area | | Population | | Population Density | | Birth Rate | Death Rate | Natural Increase % | Infant Mortality per 1,000 | Child Mortality per 1,000 | Life Expectancy | | Percent Urban Pop. | Literacy | | Corruption Index | Big Mac Price ($US) | Per Capita GNI ($US) |
	Sq km	Sq mi	2010 (Millions)	2025 (Millions)	Arithmetic	Physiologic						Male (years)	Female (years)		Male %	Female %			
Bahrain	689	266	0.8	1.0	1201	13345	20	3	1.7	8.0	9	73	77	100	91.0	82.7	5.4		$34,310
Egypt	1,001,445	386,660	78.1	95.9	78	2599	27	6	2.1	33.0	40	70	74	43	66.6	43.7	2.8	2.34	$5,400
Iran	1,633,182	630,575	74.4	88.0	46	414	20	5	1.5	32.0	36	69	72	67			2.3		$10,800
Iraq	438,319	169,236	30.9	43.3	71	543	34	10	2.4	94.0	116	56	60	67	70.7	45.0	1.3		
Israel	21,059	8,131	7.7	9.3	368	1751	21	5	1.6	3.5	5	79	82	92	97.9	94.3	6.0	3.69	$25,930
Jordan	89,210	34,444	6.1	7.7	68	1363	28	4	2.4	24.0	28	71	73	83	94.9	84.4	5.1		$5,160
Kazakhstan	2,717,289	1,049,151	16.0	17.1	6	49	21	10	1.1	29.0	33	61	72	53	99.1	96.1	2.2		$9,700
Kuwait	17,819	6,880	2.8	3.6	157	15734	21	2	1.9	8.0	9	77	79	98	84.3	79.9	4.3		$49,970
Kyrgyzstan	198,499	76,641	5.4	6.5	27	387	24	7	1.7	50.0	57	62	70	35	98.6	95.5	1.8		$1,950
Lebanon	10,399	4,015	4.1	4.6	396	1276	19	5	1.4	19.0	22	69	74	87	92.3	80.4	3.0		$10,050
Libya	1,759,532	679,359	6.6	8.1	4	373	24	4	2.0	21.0	23	71	76	77	90.9	67.6	2.6		$11,500
Morocco	446,548	172,413	32.1	36.6	72	327	21	6	1.5	43.0	46	68	72	56	61.9	36.0	3.5		$3,990
Oman	212,459	82,031	2.8	3.1	13	1325	24	3	2.1	10.0	11	73	75	71	80.4	61.7	5.5		$19,740
Palestinian Territories	6,260	2,417	4.5	6.2	716	35797	37	4	3.3	25.0	29	72	73	72					
Qatar	11,000	4,247	0.9	1.1	84	8429	17	2	1.5	7.0	8	74	76	100	80.5	83.2	6.5		
Saudi Arabia	2,149,680	829,996	29.6	35.7	14	688	29	3	2.6	16.0	20	74	78	81	84.1	67.2	3.5	2.66	$22,910
Sudan	2,505,798	967,494	41.1	54.3	16	234	33	12	2.1	81.0	127	56	59	38	36.0	14.0	1.6		$1,880
Syria	185,179	71,498	20.9	26.8	113	376	28	4	2.4	19.0	22	71	75	50	88.3	60.4	2.1		$4,370
Tajikistan	143,099	55,251	7.6	9.5	53	761	27	5	2.2	65.0	77	64	69	26	99.6	98.9	2.0		$1,710
Tunisia	163,610	63,170	10.5	12.1	64	201	17	6	1.1	19.0	22	72	76	65	81.4	60.1	4.4		$7,130
Turkey	774,816	299,158	76.8	87.8	99	261	19	6	1.3	23.0	25	69	74	62	93.6	76.7	4.6	3.13	$12,090
Turkmenistan	488,099	188,456	5.4	6.5	11	276	24	6	1.8	74.0	82	58	67	47	98.8	96.6	1.8		$6,640
United Arab Emirates	83,600	32,278	4.6	6.2	55	5524	15	2	1.3	7.0	8	77	81	83	75.5	79.5	5.9		
Uzbekistan	447,397	172,741	28.1	33.3	63	524	24	7	1.7	48.0	55	63	70	36	98.5	96.0	1.8		$1,680
Western Sahara	252,120	97,344	0.5	0.8	2	5	28	8	2.0	53.0		62	66	81					
Yemen	527,966	203,849	23.6	35.2	45	1493	41	9	3.2	77.0	102	60	62	30	67.4	25.0	2.3		$2,200
SOUTH ASIA																			
REALM	**4,487,762**	**1,732,734**	**1569.5**	**1877.7**	**146**	**291**	**25**	**8**	**1.7**	**58.0**	**73**	**64**	**66**	**28**	**66.1**	**39.9**			**$2,576**
Bangladesh	143,998	55,598	152.4	180.1	1058	1679	24	7	1.7	52.0	67	62	64	24	51.7	29.5	2.1		$1,340
Bhutan	47,001	18,147	0.7	0.9	16	520	30	7	2.3	40.0	60	66	67	31	61.1	33.6	5.2		$4,980
India	3,287,576	1,269,340	1186.4	1407.7	361	633	24	8	1.6	57.0	75	65	66	28	68.6	42.1	3.4		$2,740
Maldives	300	116	0.3	0.4	1029	10287	19	4	1.5	16.0	19	72	73	27	96.3	96.4	2.8		$5,040
Nepal	147,179	56,826	28.1	36.5	191	909	29	9	2.0	48.0	61	63	64	17	59.1	21.8	2.7		$1,040
Pakistan	796,098	307,375	180.8	228.9	227	783	31	8	2.3	75.0	93	62	64	35	57.6	27.8	2.5		$2,570
Sri Lanka	65,610	25,332	20.8	23.2	317	1093	19	7	1.2	15.0	19	67	75	15	94.5	88.9	3.2		$4,210

Continued

| | Land Area | | Population | | Population Density | | Birth Rate | Death Rate | Natural Increase % | Infant Mortality per 1,000 | Child Mortality per 1,000 | Life Expectancy | | Percent Urban Pop. | Literacy | | Corruption Index | Big Mac Price ($US) | Per Capita GNI ($US) |
	Sq km	Sq mi	2010 (Millions)	2025 (Millions)	Arithmetic	Physiologic						Male (years)	Female (years)		Male %	Female %			
Gabon	267,668	103,347	1.4	1.7	5	269	27	12	1.5	58.0	88	56	58	84	79.8	62.2	3.1		$13,080
Gambia	11,300	4,363	1.7	2.3	149	711	38	11	2.7	93.0	124	57	59	54	43.8	29.6	1.9		$1,140
Ghana	238,538	92,100	25.0	33.7	105	455	32	10	2.2	71.0	112	58	59	48	79.5	61.2	3.9		$1,330
Guinea	245,860	94,927	10.9	15.7	44	738	42	14	2.8	113.0	183	52	55	30	55.1	27.0	1.6		$1,120
Guinea–Bissau	36,120	13,946	1.8	2.9	50	385	50	19	3.1	117.0	196	43	47	30	53.0	21.4	1.9		$470
Ivory Coast	322,459	124,502	21.7	26.2	67	293	38	14	2.4	100.0	142	50	53	48	54.6	38.5	2.0		$1,590
Kenya	580,367	224,081	40.2	51.3	69	865	40	12	2.8	77.0	117	53	53	19	89.0	76.0	2.1		$1,540
Lesotho	30,349	11,718	1.8	1.7	60	541	27	25	0.2	91.0	112	35	36	24	73.6	93.6	3.2		$1,890
Liberia	111,369	43,000	4.2	6.8	37	932	50	18	3.2	133.0	190	45	47	58	69.9	36.8	2.4		$290
Madagascar	587,036	226,656	20.0	28.0	34	680	38	10	2.8	75.0	120	57	60	30	87.7	72.9	3.4		$920
Malawi	118,479	45,745	14.5	20.4	122	582	48	16	3.2	80.0	125	45	59	17	74.5	46.7	2.8		$750
Mali	1,240,185	478,838	13.6	20.6	11	273	48	15	3.3	96.0	161	54	59	31	47.9	33.2	3.1		$1,040
Mauritania	1,025,516	395,954	3.4	4.5	3	328	35	9	2.6	77.0	121	59	62	40	50.6	29.5	2.8		$2,010
Mauritius	2,041	788	1.3	1.4	646	1242	14	7	0.7	15.4	18	69	76	42	87.7	81.0	5.5		$11,390
Moçambique	801,586	309,494	21.3	27.5	27	663	41	20	2.1	108.0	158	42	44	29	59.9	28.4	2.6		$690
Namibia	824,287	318,259	2.1	2.3	3	260	25	15	1.0	47.0	67	48	47	35	82.9	81.2	4.5		$5,120
Niger	1,266,994	489,189	15.6	26.3	12	308	46	15	3.1	81.0	171	58	56	17	23.5	8.3	2.8		$630
Nigeria	923,766	356,668	155.6	205.4	168	495	43	18	2.5	100.0	194	46	47	47	72.3	56.2	2.7		$1,770
Réunion	2,510	969	0.8	1.0	328	2185	19	5	1.4	8.0		72	80	92	84.8	89.2			
Rwanda	26,340	10,170	10.1	14.6	384	915	43	16	2.7	86.0	143	47	48	18	73.7	60.6	3.0		$860
São Tomé and Príncipe	961	371	0.2	0.2	220	511	35	8	2.7	77.0	119	63	66	58	70.2	39.1	2.7		$1,630
Senegal	196,720	75,954	13.4	18.0	68	570	39	10	2.9	61.0	118	60	64	41	47.2	27.6	3.4		$1,640
Seychelles	451	174	0.1	0.1	227	1512	18	7	1.1	11.0	12	67	77	53	82.9	85.7	4.8		$8,670
Sierra Leone	71,740	27,699	5.8	7.6	81	1007	48	23	2.5	158.0	267	48	49	37	50.7	22.6	1.9		$660
Somalia	637,658	246,201	9.5	14.3	15	744	46	19	2.7	117.0	189	47	49	37	85.8	84.5	1.0		
South Africa	1,221,034	471,444	49.1	51.5	40	309	23	15	0.8	45.0	58	48	52	59	68.3	46.0	4.9	1.66	$9,560
Swaziland	17,361	6,703	1.1	1.0	63	576	31	31	0.0	85.0	116	33	34	24	80.9	78.7	3.6		$4,930
Tanzania	945,087	364,900	42.1	58.2	45	890	38	15	2.3	75.0	118	50	52	25	84.1	66.6	3.0		$1,200
Togo	56,791	21,927	7.2	9.9	127	294	38	10	2.8	91.0	140	56	60	40	72.2	42.6	2.7		$800
Uganda	241,040	93,066	31.1	56.4	129	379	48	16	3.2	76.0	121	47	48	13	77.7	57.1	2.6		$920
Zambia	752,607	290,583	12.7	15.5	17	241	43	22	2.1	100.0	164	38	37	37	85.2	71.2	2.8		$1,220
Zimbabwe	390,759	150,873	13.8	16.0	35	392	31	21	1.0	60.0	92	40	40	37	95.5	89.9	1.8		
REALM	**19,318,887**	**7,459,064**	**592.4**	**733.9**	**63**	**687**	**26**	**7**	**1.9**	**46.0**	**51**	**66**	**70**	**53**	**83.1**	**64.5**			**$8,309**
NORTH AFRICA/SOUTHWEST ASIA																			
Afghanistan	652,086	251,772	34.4	50.3	53	440	47	21	2.6	163.0	254	43	43	20	51.0	20.8	1.5		
Algeria	2,381,730	919,591	36.0	43.3	15	503	22	4	1.8	27.0	30	71	74	63	75.1	51.3	3.2		$5,490

Continued

	Land Area Sq km	Land Area Sq mi	Population 2010 (Millions)	Population 2025 (Millions)	Population Density Arithmetic	Population Density Physiologic	Birth Rate	Death Rate	Natural Increase %	Infant Mortality per 1,000	Child Mortality per 1,000	Life Expectancy Male (years)	Life Expectancy Female (years)	Percent Urban Pop.	Literacy Male %	Literacy Female %	Corruption Index	Big Mac Price ($US)	Per Capita GNI ($US)
Panama	75,519	29,158	3.5	4.2	46	516	20	4	1.6	15.0	19	73	78	64	92.6	91.3	3.4		$8,340
Puerto Rico	8,951	3,456	4.0	4.1	450	5005	12	8	0.4	9.2		74	82	94	93.7	94.0	5.8		
Saint Lucia	619	239	0.2	0.2	328	1172	15	7	0.8	19.4	26	71	76	28			7.1		$7,090
St. Vincent & the Grenadines	391	151	0.1	0.1	260	930	17	8	0.9	17.6		70	74	40			6.5		$5,720
Trinidad and Tobago	5,131	1,981	1.3	1.4	256	1068	14	8	0.6	24.0	27	67	71	12	99.0	97.5	3.6		$14,580
REALM	**17,867,238**	**6,898,579**	**397.5**	**460.9**	**22**	**356**	**20**	**6**	**1.4**	**23.0**	**25**	**69**	**76**	**81**	**90.1**	**89.0**			**$9,290**

SOUTH AMERICA

	Land Area Sq km	Land Area Sq mi	Population 2010 (Millions)	Population 2025 (Millions)	Population Density Arithmetic	Population Density Physiologic	Birth Rate	Death Rate	Natural Increase %	Infant Mortality per 1,000	Child Mortality per 1,000	Life Expectancy Male (years)	Life Expectancy Female (years)	Percent Urban Pop.	Literacy Male %	Literacy Female %	Corruption Index	Big Mac Price ($US)	Per Capita GNI ($US)
Argentina	2,780,388	1,073,514	40.6	46.3	15	146	19	8	1.1	13.3	15	71	79	91	96.9	96.9	2.9	3.30	$12,990
Bolivia	1,098,575	424,162	10.4	13.3	9	474	29	8	2.1	51.0	62	63	67	64	92.1	79.4	3.3		$4,140
Brazil	8,547,360	3,300,154	200.6	228.9	23	335	20	6	1.4	24.0	26	69	75	83	85.5	85.4	3.5	3.45	$9,370
Chile	756,626	292,135	17.1	19.1	23	754	14	5	0.9	8.8	10	75	81	87	95.9	95.5	6.9	2.51	$12,590
Colombia	1,138,906	439,734	45.7	53.8	40	1002	20	6	1.4	19.0	23	69	76	72	91.8	91.8	3.8		$6,640
Ecuador	283,560	109,483	14.4	17.5	51	460	26	6	2.0	25.0	28	72	78	62	93.6	90.2	2.0		$7,040
French Guiana	89,999	34,749	0.2	0.3	2	235	32	4	2.8	10.4		72	79	76	83.6	82.3			
Guyana	214,969	83,000	0.8	0.8	4	191	21	9	1.2	48.0	64	63	68	28	99.0	98.1	2.6		$2,600
Paraguay	406,747	157,046	6.5	8.0	16	265	27	6	2.1	36.0	43	69	73	57	94.4	92.2	2.4		$4,380
Peru	1,285,214	496,224	28.7	34.0	22	745	21	6	1.5	24.0	28	68	73	76	94.7	85.4	3.6	2.54	$7,240
Suriname	163,270	63,039	0.5	0.5	3	312	17	7	1.0	16.0	17	66	73	74	95.9	92.6	3.6		$6,000
Uruguay	177,409	68,498	3.3	3.5	19	268	14	9	0.5	10.5	12	72	79	94	97.4	98.2	6.9		$11,040
Venezuela	912,046	352,143	29.1	34.9	32	797	25	4	2.1	16.5	19	70	76	88	93.3	92.7	1.9		$11,920

SUBSAHARAN AFRICA

	Land Area Sq km	Land Area Sq mi	Population 2010 (Millions)	Population 2025 (Millions)	Population Density Arithmetic	Population Density Physiologic	Birth Rate	Death Rate	Natural Increase %	Infant Mortality per 1,000	Child Mortality per 1,000	Life Expectancy Male (years)	Life Expectancy Female (years)	Percent Urban Pop.	Literacy Male %	Literacy Female %	Corruption Index	Big Mac Price ($US)	Per Capita GNI ($US)
REALM	**21,786,509**	**8,411,818**	**769.5**	**1106.8**	**32**	**421**	**40**	**15**	**2.5**	**87.0**	**121**	**49**	**51**	**35**	**73.1**	**59.9**			**$1,828**
Angola	1,246,693	481,351	17.7	26.2	14	473	47	21	2.6	132.0	180	41	44	57	55.6	28.5	1.9		$4,400
Benin	112,620	43,483	9.9	14.5	88	548	42	12	3.0	98.0	155	54	57	41	47.8	23.6	3.1		$1,310
Botswana	581,727	224,606	1.8	2.2	3	316	24	14	1.0	44.0	54	50	49	57	74.4	79.8	5.8		$12,420
Burkina Faso	274,000	105,792	16.1	23.7	59	490	45	15	3.0	89.0	163	49	52	16	31.2	13.1	3.5		$1,120
Burundi	27,829	10,745	9.4	15.0	339	789	46	16	3.0	107.0	177	47	50	10	56.3	40.5	1.9		$330
Cameroon	475,439	183,568	19.4	25.5	41	255	36	13	2.3	74.0	127	51	52	57	81.8	69.2	2.3		$2,120
Cape Verde Islands	4,030	1,556	0.5	0.7	130	1185	30	5	2.5	28.0	38	68	74	59	84.3	65.3	5.1		$2,940
Central African Republic	622,978	240,533	4.6	5.5	7	244	38	19	1.9	102.0	155	43	44	38	59.6	34.5	2.0		$740
Chad	1,283,994	495,753	10.7	13.9	8	277	44	17	2.7	106.0	179	46	48	27	66.9	40.8	1.6		$1,280
Comoros	2,230	861	0.7	1.1	332	626	36	8	2.8	69.0	92	62	66	28	63.5	49.1	2.5		$1,150
Congo	341,998	132,046	4.0	5.6	12	1165	37	13	2.4	75.0	119	52	54	60	87.5	74.4	1.9		$2,750
Congo, The	2,344,848	905,351	70.7	109.7	30	754	44	13	3.1	92.0	138	49	55	33	86.6	67.7	1.7		$290
Djibouti	23,201	8,958	0.8	1.1	36	3573	30	12	1.8	67.0	101	53	55	87	65.0	38.4	3.0		$2,260
Equatorial Guinea	28,050	10,830	0.6	0.9	23	252	39	10	2.9	91.0	152	59	60	39	92.5	74.5	1.7		$21,230
Eritrea	117,598	45,405	5.3	7.7	45	1128	40	10	3.0	59.0	91	54	59	21	43.9	33.4	2.6		$400
Ethiopia	1,104,296	426,371	83.1	110.5	75	684	40	15	2.5	77.0	121	48	51	16	83.7	70.0	2.6		$780

Continued

	Land Area Sq km	Land Area Sq mi	Population 2010 (Millions)	Population 2025 (Millions)	Pop. Density Arithmetic	Pop. Density Physiologic	Birth Rate	Death Rate	Natural Increase %	Infant Mortality per 1,000	Child Mortality per 1,000	Life Expectancy Male (years)	Life Expectancy Female (years)	Percent Urban Pop.	Literacy Male %	Literacy Female %	Corruption Index	Big Mac Price ($US)	Per Capita GNI ($US)
Slovakia	49,010	18,923	5.4	5.2	110	324	10	10	0.0	6.1	7	70	78	56	100.0	100.0	5.0		$19,330
Slovenia	20,251	7,819	2.0	2.1	99	707	10	9	0.1	3.1	4	74	81	48	98.6	96.8	6.7		$26,640
Spain	505,988	195,363	46.7	46.2	92	237	11	9	0.2	3.7	4	77	83	77	100.0	100.0	6.5		$30,110
Sweden	449,959	173,730	9.2	9.9	21	293	12	10	0.2	2.5	3	79	83	84	100.0	100.0	9.3	4.58	$35,840
Switzerland	41,290	15,942	7.6	8.1	185	1540	10	8	0.2	4.0	5	79	84	68	99.5	97.4	9.0	5.60	$43,080
Ukraine	603,698	233,089	45.6	41.7	76	128	10	16	-0.6	11.0	14	62	74	68	100.0	100.0	2.5		$6,810
United Kingdom	244,878	94,548	61.8	68.8	252	971	13	9	0.4	4.9	6	77	81	80	97.6	89.2	7.7	3.30	$34,370
RUSSIAN REALM																			
REALM	**17,261,421**	**6,664,672**	**157.7**	**146.5**	**9**	**123**	**12**	**14**	**-0.2**	**10.0**	**11**	**61**	**73**	**71**	**99.7**	**99.1**			**$13,512**
Armenia	29,800	11,506	3.1	3.3	105	526	15	9	0.6	26.0	29	68	75	64	99.4	98.1	2.9		$5,900
Azerbaijan	86,599	33,436	8.9	9.7	103	468	18	6	1.2	12.0	14	70	75	52	98.9	95.9	1.9		$6,370
Georgia	69,699	26,911	4.6	4.2	66	441	11	10	0.1	16.0	18	70	79	53	99.7	99.4	3.9		$4,770
Russia	17,075,323	6,592,819	141.0	129.3	8	138	12	15	-0.3	9.0	10	60	73	73	99.8	99.2	2.1	1.73	$14,400
NORTH AMERICA																			
REALM	**19,599,647**	**7,567,466**	**341.9**	**393.3**	**17**	**139**	**14**	**8**	**0.6**	**6.5**	**7.9**	**76**	**81**	**79**	**95.7**	**95.4**			**$44,790**
Canada	9,970,600	3,849,670	33.6	37.6	3	67	11	7	0.4	5.4	6	78	83	81	95.7	95.3	8.7	3.36	$35,310
United States	9,629,047	3,717,796	308.2	355.7	32	160	14	8	0.6	6.6	8	75	81	79			7.3	3.54	$45,850
MIDDLE AMERICA																			
REALM	**2,714,579**	**1,048,105**	**196.3**	**225.6**	**72**	**447**	**21**	**6**	**1.6**	**24.0**	**28**	**71**	**76**	**69**	**90.3**	**87.1**			**$9,489**
Antigua and Barbuda	440	170	0.1	0.1	232	1287	17	7	1.0	20.0	23	71	75	31					$12,610
Bahamas	13,880	5,359	0.3	0.4	22	2209	17	6	1.1	14.0	15	69	75	83	95.4	96.8			
Barbados	430	166	0.3	0.3	706	1811	14	8	0.6	14.0	15	73	79	38	98.0	96.8	7.0		$10,880
Belize	22,960	8,865	0.3	0.4	14	342	27	4	2.3	18.0	21	71	74	50	95.5	95.7	2.9		$5,100
Costa Rica	51,100	19,730	4.6	5.6	90	1002	16	4	1.2	9.7	11	76	81	59	96.5	96.4	5.1		$8,340
Cuba	110,859	42,803	11.3	11.2	102	248	10	7	0.3	5.3	7	75	79	76	84.0	83.7	4.3		
Dominica	751	290	0.1	0.1	135	675	16	9	0.7	16.0	20	72	77	73			6.0		$5,650
Dominican Republic	48,731	18,815	10.3	12.1	211	679	24	6	1.8	32.0	38	69	75	67	81.6	76.1	3.0		$5,050
El Salvador	21,041	8,124	7.5	9.1	355	909	24	6	1.8	24.0	28	68	74	60			3.9		$4,840
Grenada	339	131	0.1	0.1	302	943	19	7	1.2	17.0	21	66	69	31	89.7	90.5			$6,010
Guadeloupe	1,709	660	0.4	0.5	238	1585	15	7	0.8	8.0	11	75	82	100	76.2	61.1			
Guatemala	108,888	42,042	14.5	20.0	133	739	34	6	2.8	34.0		66	73	47	51.0	46.5	3.1		$4,120
Haiti	27,749	10,714	9.4	11.7	340	1030	29	11	1.8	57.0	76	56	60	43	72.5	72.0	1.4		$1,050
Honduras	112,090	43,278	7.6	9.8	68	378	27	5	2.2	23.0	27	69	74	46	82.5	90.7	2.6		$3,160
Jamaica	10,989	4,243	2.8	3.0	251	1005	17	6	1.1	21.0	25	70	75	52	96.0	97.1	3.1		$5,050
Martinique	1,101	425	0.4	0.4	368	1839	13	7	0.6	6.0		76	83	98	93.1	89.1			
Mexico	1,958,192	756,062	111.0	123.8	57	405	20	5	1.5	19.0	23	73	78	76	96.6	96.6	3.6	2.30	$12,580
Netherlands Antilles	800	309	0.2	0.2	253	2534	14	7	0.7	5.0		71	79	92	64.2	64.4			
Nicaragua	129,999	50,193	5.9	6.8	46	208	26	5	2.1	29.0	35	68	74	59			2.5		$2,080